5th UTP-UMP-UAF Symposium on Energy Systems (SES2019)

IOP Conference Series: Materials Science and Engineering
Volume 863

Kuantan, Malaysia
1 - 2 October 2019

ISBN: 978-1-7138-3617-9
ISSN: 1757-8981

Printed from e-media with permission by:

Curran Associates, Inc.
57 Morehouse Lane
Red Hook, NY 12571

proceedings.com

Some format issues inherent in the e-media version may also appear in this print version.

Additional copies of this publication are available from:

Curran Associates, Inc.
57 Morehouse Lane
Red Hook, NY 12571 USA
Phone: 845-758-0400
Fax: 845-758-2633
Email: curran@proceedings.com
Web: www.proceedings.com

TABLE OF CONTENTS

IOP Conf. Series: Materials Science and Engineering **863** (2020) 011001 doi:10.1088/1757-899X/863/1/011001

Preface the 5th SES2019

The 5th UTP-UMP-UAF Symposium on Energy Systems 2019 (SES2019) is a two-day symposium broadly targeted to gather experts, professionals and researchers primarily from three institutions in related areas of Energy Systems to share recent advances and development accomplished. Held via tele-conference at Universiti Malaysia Pahang (UMP), Universiti Teknologi PETRONAS (UTP), and University of Agricultural Faisalabad (UAF), the symposium is aimed to provide a unique platform for interactive exchange of latest state-of-the-art knowledge among the professionals in the Energy Systems domain. Besides that, the symposium also accepts other field of knowledge such as Mechanical, Chemical, Electrical and other field of engineering. In addition, the symposium becomes a networking event to spur future collaborations between researchers among the universities.

We would like to thank the Organizing Committee and the Universities for supporting the symposium. We also would like to thanks to reviewers and everyone who contributed along the symposium and publication of these proceeding.

The Chairman of SES2019
Associate Professor Dr Puteri Sri Melor Megat Yusoff, UTP Malaysia.
Professor Dr. Ir. Hassan Ibrahim, UMP Malaysia.
Associate Professor Dr. Abdul Ghaffar, UAF Pakistan.

IOP Conf. Series: Materials Science and Engineering **863** (2020) 011002 doi:10.1088/1757-899X/863/1/011002

Peer review statement

All papers published in this volume of *IOP Conference Series: Materials Science and Engineering* have been peer reviewed through processes administered by the proceedings Editors. Reviews were conducted by expert referees to the professional and scientific standards expected of a proceedings journal published by IOP Publishing.

IOP Conf. Series: Materials Science and Engineering **863** (2020) 012001 doi:10.1088/1757-899X/863/1/012001

Simulation study on the effect of fuel injection and air intake boundary setup on the brake torque response by using comprehensive vehicle model for natural gas vehicle (NGV)

M F A Rahim[1,2,*], A A Jaafar[2], Z Taha[2] and R Mamat[3]

[1] Automotive Engineering Research Group, Faculty of Mechanical and Automotive Engineering Technology, Universiti Malaysia Pahang, 26600, Pekan, Pahang, Malaysia.
[2] Innovative Manufacturing, Mechatronics and Sport Laboratory, Faculty of Manufacturing and Mechatronic Engineering Technology, Universiti Malaysia Pahang, 26600, Pekan, Pahang, Malaysia.
[3] Advance Fluid Research Group, Faculty of Mechanical and Automotive Engineering Technology, Universiti Malaysia Pahang, 26600, Pekan, Pahang, Malaysia.

*E-mail: mfadzil@ump.edu.my

Abstract. An analytical, dynamic and comprehensive vehicle model, replicated a high-pressure direct injection compressed natural gas (HPDI-CNG) engine in a vehicle is proposed with the objective to study the effect of fuel injection and air intake boundary condition on the brake torque response. The model simulated the output torque in transient simulations of the natural gas vehicle (NGV) in speed-sweep mode test. The vehicle model has coupled an analytical engine model developed in Simulink with a Simscape driveline model which consists of a clutch model, a simple transmission gear and a simplified vehicle model. In all problem, independent input to the model is the throttle opening ramp. The adjusted parameter in the first study is the methods to calculate the mass of fuel inlet which is based on i) fully experimental data, ii) measured air to fuel ratio (AFR) and (iii) constant injector mass flow rate. In the second problem, the pressure limit of the manifold absolute pressure (MAP) is adjusted as (i) fully experimental data (ii) 1.2 bar (iii) 1.5 bar and (iv) unlimited pressure limit. The dependent parameter in all problem is engine output torque. The results have been compared with the actual data of torque from chassis dynamometer measurement. For the effect of fuel inlet boundary set up, in the fully predictive mode, the model predicted an almost constant maximum torque at a value about 70 Nm, whereas the measured data only produced the same peak value at a very limited instant. If the model used measured AFR as the fuel input boundary, the model overpredicted the maximum peak torque of 70 Nm. In the study on the effect of MAP limit, the maximum torque for 1.2 bar, 1.5 bar and unlimited pressure set up has produced a maximum torque of 45, 62 and 70 Nm, respectively. Results of the first study showed that the use of constant injector mass flow rate has a tendency to simulate an ideal engine acceleration process. The prediction is closed to the measured data if the fuel mass is calculated based on the measured AFR. However, the use of measured AFR in our opinion have reduced the model predictability. In the second study, the increased MAP limit significantly increased the maximum brake torque response. However, the model cannot predict the abnormalities found in the experimental data. The use of MAP limit demonstrated the sensitivity of the output torque on the maximum value of the engine MAP. The results indicated that the injection fuel inlet boundary and the MAP limit have a strong significance on the model prediction and need to be improved for future use of the model.

1. Introduction

A new configuration of high-pressure direct injection (HPDI) compressed natural gas (CNG) engine is developed. Gasoline direct injection (GDI) injector was converted to operate with CNG fuel [1]. A model is required to predict the theoretical engine torque as a fundamental performance parameter. Moreover, the use of model-based analysis is a state of art in engineering application nowadays such as in product development and process optimization. Based on previous studies, engine modelling can be categorised based on the modelling technique. There are three general categories which are analytical modelling, the mean-value modelling and finally the system identification modelling [2]. The earliest method is to model the engine processes based on the physical equation for each of the processes. This method is called analytical modelling [3]. The more complex or the more detail process is modelled, the number of equations or the complexities of the equation will increase since each process is considered to produce a realistic prediction [4]. The mean value modelling utilised a simpler approach where certain complex phenomena such as intake flow dynamics or torque generation process are modelled by an empirical equation derived based on experimental data [5]. The generalized empirical model can be coupled with other analytical models for other processes. This model is simpler but the use of the empirical equation required model calibration to be fully generalized for another type of engine. The calibration might be expensive and tedious to perform in practical usage. The system identification engine model is the simplest form of modelling technique for the internal combustion engine. In addition, this type of model has been proven to be the most accurate type of model among others [6], [7]. However, the identification required reliable data for the learning process and lose insight information as the technique required just the input-output data pairs [2]. Hence, an analytical engine model based on thermodynamic single-zone modelling is preferable to analyse detail interaction between parameters. Analytical engine models are mostly consisted the solutions of in-cylinder thermodynamic processes such as progressive combustion, heat transfer, intake and exhaust flow dynamics [8], throttle dynamics [2], and dynamic work transfer from the cylinder piston to the crankshaft [9]. The engine modelled processes may be also coupled with a vehicle model and look-up tables which represent the element of the electronic control unit (ECU) to accurately calculate the engine input parameters. Recently, the model is also expected to be able to simulate the transient condition of engine operation [5,10] to gain a more realistic insight of actual driving condition. Based on the discussed findings, this study is carried out with the purpose to evaluate the effect of fuel injection inlet and air intake boundary condition on the brake torque response by using the new comprehensive model of HPDI-CNG engine. Special case studies have been executed to demonstrate the significance of these critical parameters which affect the prediction accuracies.

2. Modelling method

The model developed in the study is proposed to simulate the function of a converted HPDI-CNG prototype vehicle, brand Proton, model Gen.2 which has the specification in table 1. The original gasoline port injection of the engine had been converted to run on CNG by using high-pressure direct injection approach. Since the study was conducted solely based on this vehicle prototype, the proposed model in the current study is developed as similar as possible to the vehicle prototype. The developed analytical engine model in the current study is made of two types of model i) Simulink model ii) Simscape model. Simulink blocks symbolize basic mathematical operations. And linked Simulink blocks are equivalent to a mathematical model or representation of a system under study. Whereas, Simscape blocks dedicated to the modelling of physical components in their actual physical properties.

2.1. The simulink model

The Simulink model consisted of the following; the throttle dynamics model to predict the mass flow at the throttle [2], the intake manifold model to predict the manifold absolute pressure which is vital to the ECU [2], and the intake and exhaust valve dynamics which predict the valve curtain flow area [8]. In all section, the one-dimensional compressible flow equation is used and is given by equation (1) [8].

$$\dot{m}_a = \frac{C_d A_{cur} p_o}{(RT)^{0.5}} \left(\frac{p_{st}}{p_o}\right)^{\frac{1}{k}} \left\{ \frac{2k}{k-1} \left[1 - \left(\frac{p_{st}}{p_o}\right)^{\frac{k-1}{k}} \right]^{\frac{1}{2}} \right\} \tag{1}$$

The equation (1) is built by the discharge flow coefficient C_d upstream stagnation pressure, p_o, static downstream pressure, p_{st}, the pressure ratio, (p_{st}/p_o) and the valve curtain area, A_{cur}.

Table 1. Original baseline vehicle and engine specification.

Parameters	Value
Dimension	
Overall length (mm)	4310
Overall width (mm)	1725
Overall height (mm)	1435
Wheelbase (mm)	2600
Tracks (mm)	1475 mm/1470
Weight (kg)	1175 (Auto transmission : 1195)
Engine	
Code	Proton S4PH CAMPRO
Type	In-line 4 cylinders, DOHC, 4 valves per cylinder
Bore/stroke (mm)	76/78
Displacement (cc)	1597
Compression ratio	10:1
Max power (kW)	82 kW @ 6000 rpm
Max. Torque (Nm)	148 Nm @ 4000 rpm
Fuel Delivery	Multipoint EFI
Fuel Cut-Off Point (rpm)	7000 r
ECU	Proton EMS700

The combustion model based on Wiebe method is used to predict the heat release curve [11,12]. The heat transfer model based on Newton heat convection is used to predict the heat loss to the coolant [13] in combination with the empirical model of Hohenberg to calculate the convective heat transfer coefficient [14]. The variation of cylinder volume and cylinder surface area are solved by the solution of kinematics and dynamics of the crank slider mechanism which consider the piston-pin and crankshaft offset [15]. All the sub-models are incorporated within the framework of the modified first law of thermodynamics to solve the in-cylinder pressure and temperature. The cylinder pressure equation is given by equation (2).

$$\frac{dp}{d\theta} = \left(\frac{k-1}{V}\right)\left[Q_{in}\frac{df}{d\theta} - \frac{hA}{6N}\left(T_{b,g} - T_w\right)\right] - \left(k\frac{p}{V}\right)\frac{dV}{d\theta} \tag{2}$$

Major parameters in equation (2) are specific heat ratio, k, the heat supplied to the cylinder as a function of crank angle, $Q_{in}\frac{df}{d\theta}$, the heat loss from the cylinder to the coolant, $\frac{hA}{6N}\left(T_{b,g} - T_w\right)$, and the cylinder pressure variation due to the boundary work, $\left(k\frac{p}{V}\right)\frac{dV}{d\theta}$. The cylinder pressure is then used to determine the dynamic force acting to the piston and transferred to the crankshaft to produce torque pulse. The crankshaft dynamics are modelled in details based on the approach taken by Zweiri et. al [16]. The model considered detail fluctuation of friction components such as piston ring assembly friction torque, the pumping torque, the skirt friction torque. The outcome of the above model is the torque pulse which is then averaged to produce the net mean brake torque and the acceleration of the crankshaft which is used to calculate the engine speed by integration. These outputs are then converted to physical signal and coupled to the SimScape model. The equation for the crankshaft dynamics is given by equation (3).

$$\ddot{\theta}_1 = \frac{1}{\tau\dot{\theta}_1}\left\{T_{ind} - \left[M_{pass}rG(\theta_1)G_1(\theta_1) + \frac{1}{2}\frac{\partial J(\theta_1)}{\partial \theta_1}\right]\dot{\theta}_1^{\ 2} - \sum_{k=1}^{5} T_{fk} - T_S - T_D - T_L\right\} \quad (3)$$

This equation can be coupled to any type of coupling mechanism as founded in the original work of Zweiri et. al [16]. The terms which construct the above equation are the moment of inertia of the engine and crankshaft assembly, τ. This value is given by the equation (4).

$$\tau = J_{cran}(\theta_1) + M_{pass}rG(\theta_1)G_2(\theta_1) \quad (4)$$

Other parameters are crankshaft rotational acceleration and rotational speed, $\ddot{\theta}_1$ and $\dot{\theta}_2$, the portion of the reciprocating torque of the engine, $M_{pass}rG(\theta_1)G_1(\theta_1)$, the rate of variation of engine inertia due to inner components, $\frac{\partial J(\theta_1)}{\partial \theta_1}$, the sum of friction torque due to internal engine assemblies, $\sum T_{fk}$, the sum of stiffness torque at the coupling, $\sum T_S$, the sum of damping torque at the coupling, $\sum T_D$, and The sum of external load implied on the engine, $\sum T_L$. The Simulink model inputs are listed in Table 2. The wall surface temperature is assumed to have constant values through-out the engine processes. The total modelled equation in this Simulink model is about 50 equations per each engine cylinder.

Table 2. Important input to the Simulink model.

Input parameters	Values
Cylinder wall surface temperature (°C)	420
Bore size (mm)	76
Cylinder stroke (mm)	78
Number of cylinders	4
Compression ratio	10:1
connecting rod length (mm)	132
maximum valve lift (mm)	10
valve diameter (int./ exh.) (mm)	38/32
valve opening angle (int./ exh.) (° BTDC/ °BTDC)	12/ 45
valve closing angle (int./ exh.) (° ABDC/ °ATDC)	48/ 10
engine rotational speed	Feedback from the model output
Instantaneous crank angle	Feedback from the model output

2.2. The Simscape model

The SimScape model consists of the clutch model, the simple manual transmission model (considering the test gear only), the final drive model and finally attached to a vehicle model. The input to these models is the acceleration and torque from the crankshaft whereas the model output is the lateral vehicle speed and acceleration as well as the torque at the wheel. The SimScape model directly represents the external load on the engine which needed to be overcome by the engine output. This mimics closely the actual operation of a real vehicle. However, this study will only discuss the engine output torque as the actual focus is on the engine performance parameter. Table 3 presents the inputs to the vehicle model by Simscape.

Table 3. Input parameters for the vehicle body used in the simulation.

Parameters	Values
Mass (kg)	1050-1245
Number of the wheel per axle	2
Horizontal distance CG-front axle (m)	1.3
Horizontal distance CG rear axle (m)	1.3
CG height above ground (m)	0.5
Frontal area (m^2)	3
Drag coefficient	0.4
Initial velocity (mph)	0-20

IOP Conf. Series: Materials Science and Engineering **863** (2020) 012001 doi:10.1088/1757-899X/863/1/012001

2.3. Multi-cylinder engine synchronization
The actual baseline engine design is a four-cylinder inline engine configuration. Therefore, the formulations of thermo-fluid dynamics presented in Equation (1) to (4) must be accounted for the four cylinders which acting on the same crankshaft. In order to simulate the phasing of the cylinder, the crank angle input to each cylinder must be phased in relative to the crankshaft angle of cylinder 1, θ_1 following the engine firing order, 1-3-4-2. Hence, the input crankshaft angle for cylinder 3, 4 and 2 is added with the following phase values $\theta_1 + \pi$, $\theta_1 + 2\pi$ and $\theta_1 + 3\pi$ respectively.

2.4. ECU maps for ignition and injection setup
The developed engine model in the study is governed by the throttle input. The action of the throttle determines the manifold absolute pressure, the amount of air intake, and consequently the cylinder mixture pressure. In addition to that, the throttle action also controlled the amount of fuel injected to the cylinder. This is governed by the use of maps or lookup table which are a common approach in the feedforward control strategy. As in actual ECU of the prototype vehicle, the selection of lookup table values is based on the measured manifold absolute pressure and engine rotational speed. In fact, the table values are similar to the actual table in the actual ECU of the prototype vehicle.

2.5. Mean brake torque solution
The actual form of torque transferred from the piston-connecting rod assembly to the crankshaft is in the form of torque pulse. In practical, the damping of the torsional vibration is accomplished by the cylinder balancing procedure. A well-balanced engine cylinder will create a smooth torque transfer which is closed to a mean value of the torque pulse. The balancing calculation is tedious and the computing time will become expensive. Therefore, throughout the study, a cycle-averaging approach has been utilized to obtain the mean brake torque. A variable frequency mean-value quantifier from Simscape toolbox has been used. The instantaneous frequency of the engine is determined from engine speed and used as the quantifier input.

2.6. Simulation procedure
The simulation is performed based on speed-sweep test method by providing the ramp input to the throttle pedal of the vehicle. Once throttled, the engine will start to accelerate the vehicle from idle speed to the maximum achievable speed by the engine. As the vehicle speed is changed, all other parameters including the engine input parameters and the output parameters are changes altogether. This test is considered as a transient test as the engine input is changing with time. Two studies have been performed to evaluate the vehicle performances and the setup is presented in table 4 and table 5.

Table 4. The simulated cases to study the effect of fuel inlet boundary.

Case	Fuel inlet calculation method	Model's mode
Case 1	Fully measured data	Experiment (baseline data)
Case 2	A measured AFR as the fuel input calculation source and theoretically calculated airflow.	The model with measured input data
Case 3	A constant injector mass flow rate of CNG as the fuel input calculation source and theoretically calculated airflow.	The fully predictive model

Table 5. The simulated cases to study the effect of intake air inlet boundary.

Case	Manifold Absolute Pressure (MAP) for Upper Limit
Case 1	1.0 bar (actual ambient pressure)
Case 1	1.2 bar
Case 2	1.5 bar
Case 3	Unlimited

These two cases are crucial since the actual baseline vehicle prototype has been modified to accommodate the CNG direct injection system. The only system which has been added is the fuel injection system for CNG, Furthermore, the ECU has been replaced with a new programmable ECU. The base maps of the ECU have been prepared by professional motorsport tuner in the previous study. Hence, the parameter of the fuel injection system has changed and the performance of the vehicle needs to be reevaluated. Figure 1 presents the layout of the HPDI-CNG engine model.

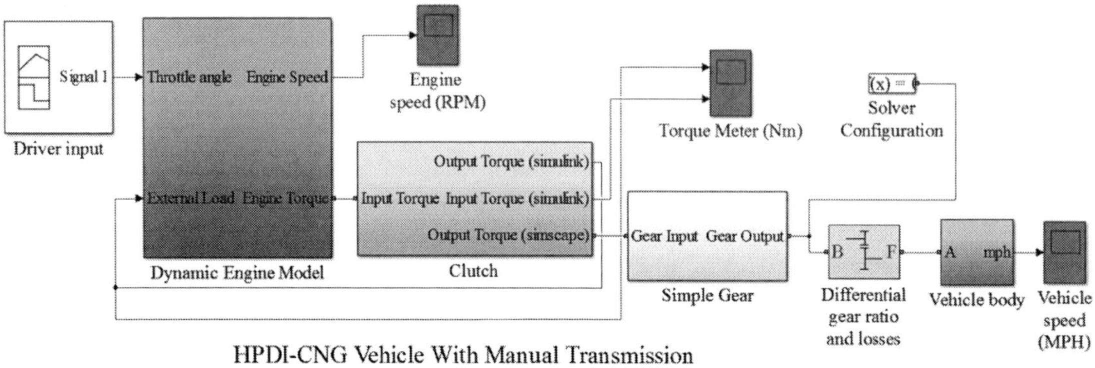

Figure 1. Overview of the HPDI-CNG engine model layout.

3. Results and discussion

Figure 2 presents the effect of different type of fuel inlet boundary input. Based on the results, the experimental data of Case 1 shown that the torque fluctuated along with the data points. These data points represent the recorded data at an interval of 0.019 seconds where the test was completed in about 18 seconds. From the beginning, it was expected that the fluctuations of the torque were due to the fluctuations of the measured AFR since the AFR affect the combustion pressure and consequently, affect the output torque. Measured AFR from the engine testing produced a randomly fluctuated trends which occurred because of the inefficient driver of the injector. This can be considered as part of the abnormal combustion trend. In order to examine this postulation, the measured AFR data has been imposed as the model input for Case 2. It was proved that the fluctuated, measured AFR data has resulted in a fluctuated engine torque. Even though the model's prediction cannot predict accurately the maximum and minimum value of measured torque, the prediction still able to capture the increment and decrement trends of the brake torque response with a certain amount of delay which is estimated to be about 2.5 seconds in the measured data. The delay between the simulated and measured response exists because the torque response is also dependent on other extraneous parameters such as the engine speed and manifold absolute pressure (MAPS). Different speed and MAP were used by the ECU in the simulations and experiments. In case 3, the response of the brake torque shown an increasing trend at the early of the test time but then produced an almost constant torque until the end of simulation time. When a constant fuel mass flow rate is used as the source for the fuel input boundary calculation, the model produced an almost constant AFR, hence producing an almost constant torque response. When the torque is constant, the speed became constant. And after the speed is feedback into the ECU, it influenced a constant injection duration. Case 3 represents an ideal combustion process of the engine.

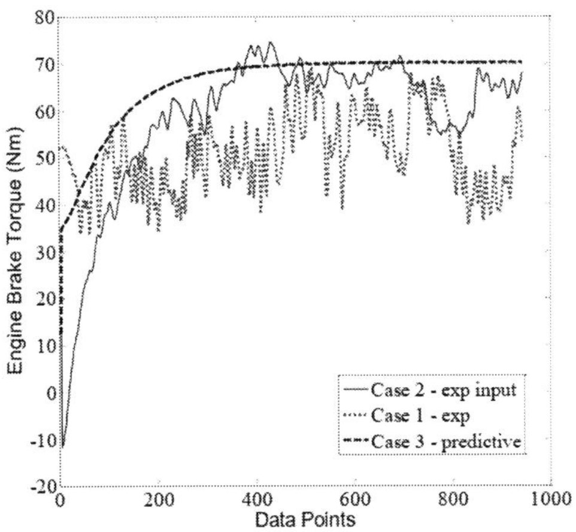

Figure 2. Effect of a different method of fuel inlet boundary input on the engine brake torque response in the speed-sweep test procedure.

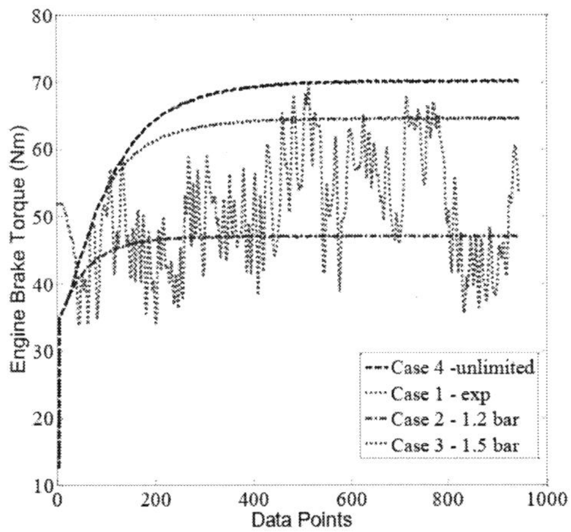

Figure 3. Effect of manifold absolute pressure limit on the engine brake torque response in the speed-sweep test procedure.

Figure 3 presents the plot of manifold absolute pressure limit effect on the engine brake torque response for the speed-sweep test procedure. For all simulated cases, the results trend shown a brake torque response which increased from zero at the start of the simulation until the peak torque is achieved and then constant until the end of the simulated time. Basically, the model took a minor time to adjust the initial torque response based on the model's initial conditions and then the torque prediction sharply increased to a value of about 35 Nm. The simulations are performed based on the full predictive model with a constant mass flow rate of fuel. The discussion of figure 2 has explained how the torque has become constant along the simulated time. Again, the model predicted ideal cases hence it cannot predict the abnormal combustion phenomena existed in the experimental result. Results in figure 3 also demonstrate that the engine torque response is highly affected by the MAP limit setting in the model. As the MAP limit is increased, the maximum engine brake torque response increased significantly. The maximum brake torque response is predicted when the MAP limit is set as unlimited at a value of 70 Nm. This is easily understood since the limit of MAP allows the use of greater manifold pressure to calculate the intake of fresh air. The higher the intake manifold pressure, then the easier for the engine to induce fresh air into the cylinder. When the amount of fresh air is increased, more fuel is allowed to be oxidized and produced greater heat release to be transformed as useful brake torque. As the MAP is restricted to 1.5 bar and 1.2 bar, the maximum torque produced is limited to a value of about 62 Nm and 45 Nm respectively. In actual condition, the MAP sensor only detected the manifold pressure value up to 1.04 bar only. The maximum value is slightly higher than the ambient pressure of 1.01 bar. This is true for normal aspirated (NA) internal combustion engine. The simulated cases aren't out of practice but have demonstrated how sensitive the output torque to the MAP limit setup. Review on the other MAP calculation parameters such as the estimated manifold volume and intake air temperature is suggested to further improve the calculation accuracy.

4. Conclusion

The comprehensive model developed in the study unable to predict the abnormal combustion existed in the actual engine. The model predicted the ideal combustion process through the use of a fully predictive setting. However, if the model is provided with actual input data from the experiment, the prediction is improved in which the results are getting closer to the measured engine torque trends. For the effect of fuel inlet boundary set up, in the fully predictive mode, the model predicted an almost constant maximum torque at a value about 70 bar, whereas the measured data only produced the same peak value at a very limited instant. If the model used measured AFR as the fuel input boundary, the model overpredicted the maximum peak pressure of 70 bar. In the study on the effect of MAP limit, the maximum torque for 1.2 bar, 1.5 bar and unlimited pressure set up has produced a maximum torque of 45 bar, 62 bar and 70 bar respectively. The simulated case studies for both problems proved that the fuel boundary setup and the limit of manifold absolute pressure in the intake air calculation are vital since the changes in the boundary types and values affected the results significantly in term of magnitudes and trends. However, the use of measured AFR as the input to calculate the amount of fuel supplied to the cylinder has reduced the predictability of the model. And a suitable limit of MAP is needed for future use of the model. Finally, the model needed to be improved in both aspects of set up to improve the simulation accuracy.

Acknowledgements

We would like to express our sincere gratitude to The Ministry of Education and Universiti Malaysia Pahang who has funded the study under research grant numbered FRGS/1/2017/TK03/UMP/03/2 and RDU1703145, members of Automotive Laboratory, and Innovative Manufacturing, Mechatronics and Sports laboratory which facilitate the study from the beginning.

References

[1] Z. Taha, M. Abdul Rahim, and R. Mamat, "Injection characteristics study of high-pressure direct injector for compressed natural gas (CNG) using an experimental and analytical method."

[2] A. Pezouvanis, "ENGINE MODELLING FOR VIRTUAL MAPPING," University of Bradford, 2010.

[3] S. R. Krishnan and K. K. Srinivasan, "Proceedings of the Institution of Mechanical Engineers, Part D : Journal of Automobile Engineering," 2010.

[4] A. Chow and M. L. Wyszynski, "Thermodynamic modelling of complete engine systems - a review," *Proc. Inst. Mech. Eng. Part D-Journal Automob. Eng.*, vol. 213, no. D4, pp. 403–415, 1999.

[5] M. Grahn, J. O. Olsson, and T. McKelvey, "A diesel engine model for dynamic drive cycle simulations," *IFAC Proc. Vol.*, vol. 18, no. PART 1, pp. 11833–11838, 2011.

[6] S. J. Rutherford and D. J. Cole, "Modelling nonlinear vehicle dynamics with neural networks," *Int. J. Veh. Des.*, vol. 53, no. 4, pp. 260–287, 2010.

[7] K. Shimojo, Y. Kitamura, M. Sato, M. S. Vogels, M. Reumueller, and L. Lackner, "Global dynamic modelling for gasoline engine," *IFAC Proc. Vol.*, vol. 7, no. PART 1, pp. 695–699, 2013.

[8] S. Sitthiracha, "An analytical model of spark ignition engine for performance prediction," in *The 20th Conference of Mechanical Engineering Network of Thailand*, 2006, no. October, pp. 1–60.

[9] Y. H. Zweiri, J. F. Whidborne, and L. D. Seneviratne, "Complete Analytical Model of a Single-Cylinder Diesel Engine for Non-linear Control and Estimation," 1999.

[10] M. Grahn and M. Grahn, *Model-Based Diesel Engine Management System Optimization A Strategy for Transient Engine Operation*. 2013.

[11] S. Rousseau, B. Lemoult, and M. Tazerout, "Combustion characterization of natural gas in a lean-burn spark-ignition engine," *Proc. Inst. Mech. Eng. Part D J. Automob. Eng.*, vol. 213, no. 5, pp. 481–489, Jan. 1999.

[12] J. I. Ghojel, "Review of the development and applications of the Wiebe function: A tribute to the contribution of Ivan Wiebe to engine research," *Int. J. Engine Res.*, vol. 11, no. 4, pp. 297–312, 2010.

[13] E. Neshat and R. K. Saray, "Effect of different heat transfer models on HCCI engine simulation," *Energy Convers. Manag.*, vol. 88, pp. 1–14, 2014.

[14] H. S. Soyhan, H. Yasar, H. Walmsley, B. Head, G. T. Kalghatgi, and C. Sorusbay, "Evaluation of heat transfer correlations for HCCI engine modelling," *Appl. Therm. Eng.*, vol. 29, no. 2–3, pp. 541–549, 2009.

IOP Conf. Series: Materials Science and Engineering **863** (2020) 012001 doi:10.1088/1757-899X/863/1/012001

[15] K. Tripathi and H. Ranjangaonkar, "Kinematic and Dynamic Modeling and Simulation of Four Stroke Petrol Engine," *Int. J. Eng. Res. Appl.*, vol. 4, no. 6, pp. 54–57, 2014.
[16] Y. H. Zweiri, J. F. Whidborne, and L. D. Seneviratne, "Detailed analytical model of a single-cylinder {D}iesel engine in the crank-angle domain," *Proc. IMechE, Part D J. Automob. Eng.*, vol. 215, no. 11, pp. 1197–1216, 2001.

IOP Conf. Series: Materials Science and Engineering **863** (2020) 012002 doi:10.1088/1757-899X/863/1/012002

Probabilistic finite element analysis of the deflection on a beam

B Meyghani[1], M Awang[2,*] and C S Wu[1,*]

[1] MOE Key Lab for Liquid-Solid Structure Evolution and Materials Processing, Institute of Materials Joining, Shandong University, 17923, Jingshi Road, Jinan, 250061, People's Republic of China.
[2] Department of Mechanical Engineering, Faculty of Engineering, Universiti Teknologi PETRONAS, 32610 Bandar Seri Iskandar, Perak, Malaysia.

*E-mail: mokhtar_awang@utp.edu.my and wucs@sdu.edu.cn

Abstract. The reliability of complex systems can be quantified using probabilistic finite element analysis. A widely accepted approach is interlacing of probabilistic methods with commercial finite element solvers for evaluating the fatigue life. In this paper, the deflection of a beam is calculated using finite element method (ABAQUS® commercial software), then the results of the simulated model are employed in MATLAB®, which is used to determine the probability, reliability and the reliability index. It was found that the reliability index is around 1.2428, while the reliability value is calculated to be around 0.8930. From the calculations the values of the probability of failure (POF) was found to be around 0.1070. Finally, the results of the calculated maximum deflection have been compared and a good agreement is found.

1. Introduction

Investigation of the probability of failure (POF) which is define as the probability of the failure at a given time in an equipment is significant, because it affects the risk analyses. POF analysis is safely executed using numerical simulation, which gives more accurate predictions in complex systems [1-3]. Globally, computational numerical simulations are now employed in the design and performance testing of engineering structures, because it reduces testing times and saves costs. However, when the research is related to the complex structures, it is important to develop the numerical model to suit the intended POF analysis, because of several uncertainties that may be present in the structural parameters or loadings definitions.

The Monte Carlo method is a probabilistic method and it can be used to process large works with high to low probabilities, however this method is not suitable for fitting of complex and extensive computations [4]. Other methods like the approximate fast probability integration (FPI) and the advanced mean value (AMV) have been proven to be more suitable for complex structures with fewer computational time and better accuracy [5]. The AMV also identifies the input parameters that affect the probabilistic sensitivity and reliability.

However, due to the above-mentioned problems there is still a lack in knowledge for finding a proper method for simulating uncertainties in geometry, loads, material behavior, random variables and other user-defined parameters that are used to predict the probabilistic response, sensitivity measures of the system and reliability. In this paper ABAQUS® commercial software is used to find the deflection of a beam. Then, MATLAB® software is employed in order to calculate the governing equations of the probabilistic behavior. Finally, the POF, the reliability as well as, the reliability index for the model is found.

2. Methodology

2.1. Mathematical model (First order reliability method)
The performance function of form is written as shown in equation (1).

$$g(X) = X_1 - X_2 \tag{1}$$

Where, the transformation of the normally distributed random variables X_1 and X_2, becomes

$$X = \mu + \sigma U \tag{2}$$

Therefore, the performance function becomes

$$g(X) = \mu_1 + \sigma_1 U_1 - (\mu_2 + \sigma_2 U_2) \tag{3}$$

The search process starts from the initial point $U_0 = (0, 0)$. At this starting point

$$\nabla g(U_0) = \left[\frac{\partial g}{\partial U_1}, \frac{\partial g}{\partial U_2}\right] \tag{4}$$

A cantilever beam is illustrated in figure 1. For a cantilevered beam the model fails when the tip displacement exceeds the allowable displacement (D_0). The performance function at the tip of the displacement is denoted as,

$$g = D_0 - \frac{4L^3}{Ewt}\sqrt{(\frac{P_y}{t^2})^2 + (\frac{P_x}{w^2})^2} \tag{5}$$

Where D_0=3" is the maximum deflection, E=30×106 psi is the modulus of elasticity, L=100" is the length, P=1000 (±100) is the external force, w=2" (±0.2) is the width, and t=4" (±1) is the height of the cross section.

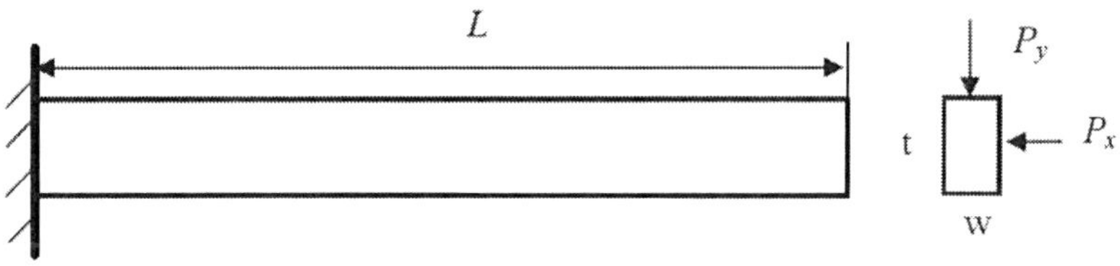

Figure 1. The cantilever beam.

The POF at the tip of the displacement becomes

$$P_f = P\left\{g = D_0 - \frac{4L^3}{Ewt}\sqrt{(\frac{P_y}{t^2})^2 + (\frac{P_x}{w^2})^2} \leq 0\right\} \tag{6}$$

The normally distributed variables are transformed into standard normal variables using

$$U = (U_x, U_y) = \left[\frac{P_x - \mu_{p_x}}{\sigma_{p_x}}, \frac{P_y - \mu_{p_y}}{\sigma_{p_x}}\right] \tag{7}$$

or

$$X = (P_x, P_y) = (\mu_{p_x} + U_x\mu_{p_x}, \mu_{p_y} + U_y\mu_{p_y}) \tag{8}$$

The transformed performance function in U-space becomes

$$g = D_0 - \frac{4L^3}{Ewt}\sqrt{(\frac{\mu_{p_y} + U_y\mu_{p_y}}{t^2})^2 + (\frac{\mu_{p_x} + U_x\mu_{p_x}}{w^2})^2} \quad (9)$$

Where, the gradient of g (U) can be obtained from equation (10).

$$\nabla g(U) = \frac{4L^3}{Ewt}\left(\frac{(\mu_x + U_x\sigma_x)\sigma_x}{w^4\sqrt{(\frac{\mu_y + U_y\sigma_y}{t^2})^2 + (\frac{\mu_x + U_x\sigma_x}{w^2})^2}}, \frac{(\mu_y + U_y\sigma_y)\sigma_y}{t^4\sqrt{(\frac{\mu_y + U_y\sigma_y}{t^2})^2 + (\frac{\mu_x + U_x\sigma_x}{w^2})^2}}\right) \quad (10)$$

2.2. Finite element descriptions

ABAQUS® has been chosen for modelling and analyzing the deflection of the beam due to its flexibility in creating geometry and material modelling [6-10]. The model of beam in 3D is shown in figure 2. The three-level design is used to find the number of runs for the simulation. The material property is assumed the same as the mathematical model. In addition, B21 (2-node linear beam in a plane) is assumed to be as the element type.

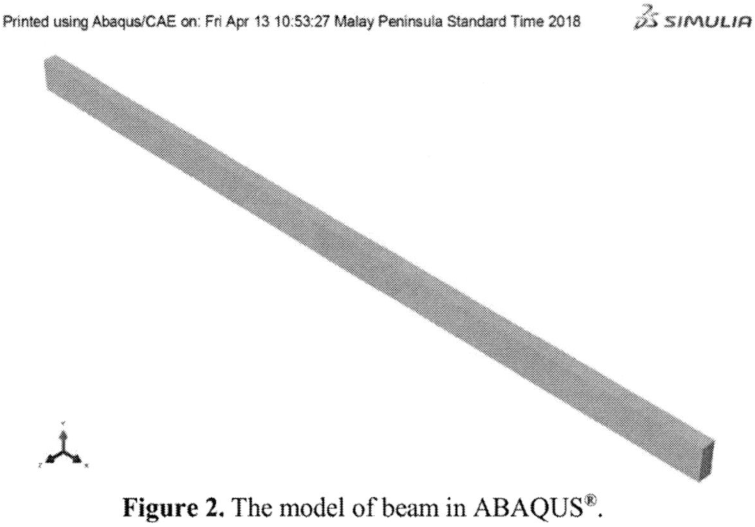

Figure 2. The model of beam in ABAQUS®.

3. Results and discussion

A total of 13 runs was needed for the three-level three-factorial Box–Behnken experimental design, as shown in Table 1. The model takes the following form as shown by equation (11).

$$y = \beta_0 + \beta_1 x_1 + \beta_2 x_2 + \beta_3 x_3 + \beta_{11} x_1^2 + \beta_{22} x_2^2 + \beta_{33} x_3^2 + \beta_{12} x_1 x_2 + \beta_{13} x_1 x_3 + \beta_{23} x_2 x_3 \quad (11)$$

Where the predicted response is y, the model constant is β_0; the independent variables are x1, x2 and x3; the linear coefficients are β_1, β_2 and β_3; the cross product coefficients are β_{12}, β_{13} and β_{23}; and the quadratic coefficients are β_{11}, β_{22} and β_{33}.

Table 1. The maximum deflection calculated by ABAQUS®.

Run number	Calculated deflection in ABAQUS®
1	1.042
2	0.852
3	1.273
4	1.042
5	2.222
6	0.48
7	2.716
8	0.587
9	2.743
10	0.593
11	2.245
12	0.485
13	1.042

A comparison between the predicted values for the maximum deflection by MATLAB® and the numerical displacement measured by ABAQUS® is shown in figure 3. Achieving the goodness of fit (R-square) of 0.98 shows a good agreement between the results of both software in different numerical displacements. The process is run continues until convergence of the solution which occurred in iteration 6. Table 2 shows the maximum probability point (MPP) and the complete convergence history. The deformed shape of the model can be seen in figure 4.

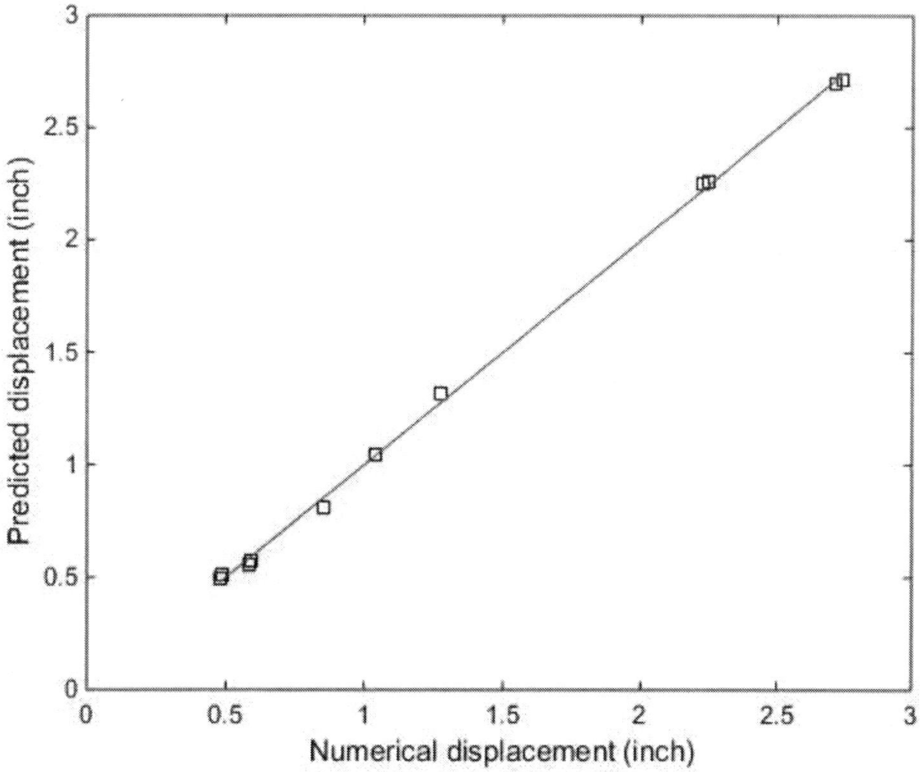

Figure 3. A comparison between the maximum deflection in ABAQUS® (straight line) and MATLAB® (squares).

IOP Conf. Series: Materials Science and Engineering **863** (2020) 012002 doi:10.1088/1757-899X/863/1/012002

Table 2. Maximum probability point (MPP) search history.

Iteration Number	Value for Pressure	Value for (w)	Value for (t)	∇g	|∇g|	U (Space)	β (Reliability index)
0	1050	2.1	4.5	(-0.0713, 0.0739, 0.5114)	0.5216	(0.7287, -0.7552, -5.2268)	5.3311
1	1.0729e+03	1.8490	-1.22	(-0.6430, 0.6570, 5.9455)	6.0161	(0.2731, -0.2791, -2.5252)	2.5552
2	1.0273e+03	1.9442	1.4748	(-0.3764, 0.3814, 3.3591)	3.4015	(0.1639, -0.1661, -1.4630)	1.4815
3	1.0164e+03	1.9668	2.5370	(-0.2718, 0.2754, 2.3562)	2.3877	(0.1427, -0.1446, -1.2370)	1.2535
4	1.0143e+03	1.9711	2.7630	(-0.2496, 0.2529, 2.1432)	2.1724	(0.1427, -0.1446, -1.2255)	1.2422
5	1.0143e+03	1.9711	2.7745	(-0.2484, 0.2517, 2.1326)	2.1617	(0.1428, -0.1447, -1.2254)	1.2422

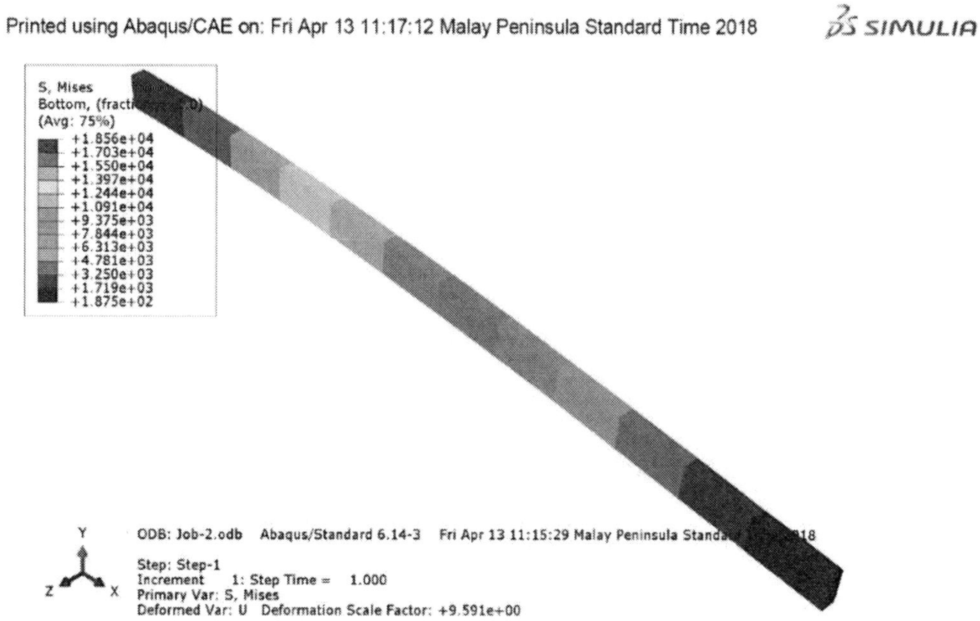

Figure 4. The deformed model created in ABAQUS® software.

The MPP is found at (0.1428, -0.1447, -1.2254) with the reliability index of β=1.2422. Then, the probability of failure (P_f) can be determined based on equation (12).

$$P_f = \Phi(\beta) \tag{12}$$

Therefore,

$$P_f = \Phi(1.2422) = 0.1070$$

It should be noted that, the reliability is calculated as,

$$R = 1 - P_f = 1 - 0.1070 = 0.8930$$

4. Conclusion

A computational tool was used to determine the rate of the deflection, reliability and POF of a beam. Three different variables including pressure width and the thickness were selected to change in different simulations in order to calculate the maximum values for the deflection of the beam in ABAQUS® commercial software. The original random variables of the probabilistic model were transformed from X-space to U-space. Also, the MPP in U-space, reliability index β and reliability R = F (β) were determined. The reliability index was found to be 1.2428, the reliability was equal to 0.8930 and the POF was measured to be 0.1070.

Acknowledgments

The authors would like to acknowledge the fellowship of the government of China, Shandong University from the International Postdoctoral Exchange Program and the Universiti Teknologi PETRONAS (UTP), Malaysia for the financial support from YUTP-FRG grant cost center 0153AA-H18. Moreover, the authors would like to thank Professor Wallace Kaufman for his endless support.

References

[1] X. Du, "First order and second reliability methods," *Probabilistic engineering design*, pp. 1-33, 2005.

[2] P. Shoaei and M. Mahsuli, "Reliability-based design of steel moment frame structures isolated by lead-rubber bearing systems," in *Structures*, 2019, pp. 765-778.

[3] F. Mirzaie, M. Mahsuli, and M. A. Ghannad, "Probabilistic analysis of soil - structure interaction effects on the seismic performance of structures," *Earthquake Engineering & Structural Dynamics,* vol. 46, pp. 641-660, 2017.

[4] D. Riha, M. Enright, H. Millwater, Y.-T. Wu, and B. Thacker, "Probabilistic engineering analysis using the NESSUS software," in *41st Structures, Structural Dynamics, and Materials Conference and Exhibit*, 2000, p. 1512.

[5] Y.-T. Wu and P. H. Wirsching, "New algorithm for structural reliability estimation," *Journal of Engineering Mechanics,* vol. 113, pp. 1319-1336, 1987.

[6] B. Meyghani, M. Awang, and S. Emamian, "A comparative study of finite element analysis for friction stir welding application," *ARPN Journal of Engineering and Applied Sciences,* vol. 11, pp. 12984-12989, 2016.

[7] B. Meyghani, M. B. Awang, M. Momeni, and M. Rynkovskaya, "Development of a Finite Element Model for Thermal Analysis of Friction Stir Welding (FSW)," in *IOP Conference Series: Materials Science and Engineering*, 2019, p. 012101.

[8] B. Meyghani and M. Awang, "A Comparison Between the Flat and the Curved Friction Stir Welding (FSW) Thermomechanical Behaviour," *Archives of Computational Methods in Engineering,* pp. 1-14, 2019.

[9] B. Meyghani and M. Awang, "Developing a Finite Element Model for Thermal Analysis of Friction Stir Welding (FSW) Using Hyperworks," in *Advances in Material Sciences and Engineering*, Singapore, 2020, pp. 619-628.

[10] B. Meyghani, M. Awang, P. Bokam, B. Plank, C. Heinzl, and K. Siow, "Finite element modeling of nano porous sintered silver material using computed tomography images," *Materialwissenschaft und Werkstofftechnik,* vol. 50, pp. 533-538, 2019.

IOP Conf. Series: Materials Science and Engineering **863** (2020) 012003 doi:10.1088/1757-899X/863/1/012003

A review of ultra-high temperature materials for thermal protection system

H K M Al-Jothery[1,2,*], T M B Albarody[1], P S M Yusoff[1], M A Abdullah[1] and A R Hussein[1]

[1] Mechanical Engineering Department. Universiti Teknologi PETRONAS, 32610 Seri Iskandar, Perak, Malaysia.
[2] Mechanical Engineering Dept. University of Al-Qadisiyah, 58001 Al-Diwaniyah, Qadisiyah, Iraq.

*E-mail: husam_16005123@utp.edu.my

Abstract. Ultra-High Temperature Materials (UHTMs) are at the base of entire aerospace industry; these high stable materials at temperatures exceeding 1600 °C are used to manage the heat shielding to protect vehicles and probes during the hypersonic flight through reentry trajectory against aerodynamic heating and reducing plasma surface interaction. Those materials are also recognized as Thermal Protection System Materials (TPSMs). The structural materials used during the high-temperature oxidizing environment are mainly limited to SiC, oxide ceramics, and composites. In addition to that, silicon-based ceramic has a maximum-use at 1700 °C approximately; as it is an active oxidation process over low temperature and water vapor environment condition. However, a great emphasis is required for developing structural materials in oxidation and rapid heating environment where the temperature is greater than 1700 °C. This review covers briefly all main types of Thermal Protection Systems (TPSs) and all the materials are used to fabricate them with the maximum operational temperatures. Also, it covers the promised UHTMs (SiC, ZrB_2, HfB_2, SiB_6 and B_4C) which are currently using for several aerospace applications, especially for TPS. Besides, it discusses the oxidation of SiC, B_4C, SiB_6, ZrB_2 and HfB_2. Therefore, the carbides and borides of the transition metals, Zr and Hf have a high-melting temperature and good stability in forming high-melting temperature oxides.

1. Introduction

Space flight, including reentry into Earth's atmosphere, entry into planetary atmospheres, and hypersonic flight all require the use at high temperatures and in severe environments for the protection of both vehicles and crews. Any vehicle system typically requires a complex array of materials to provide thermal protection. Materials' selection is governed strongly by environment and trajectory, as well as the location on the vehicle, so as to minimize weight and fuel consumption and maximize payload. A wide range of materials and concepts has been used in flight or is being considered for future vehicles and missions. However, this article focuses on an ultra-high temperature materials (UHTM), which are used for thermal protection system (TPS) [1].

Generally, the Thermal protection system (TPS) is rarely mentioned for space vehicles. This may be due to the fact that heat shields and shuttle tiles occur only during the manned spacecraft mission. In addition, to ensure a successful launch and operations of each spacecraft (manned or unmanned); thermal protection systems are highly essential.

A study highlighted key issues that provide successful re-entry of space vehicles a substantial improvements in terms of operational efficiency, reliability, and a reduction is cost [2]. Moreover, the study stated that during the re-entry, the temperature recorded at the Orbiter's external surface was 1,648 °C (3,000 °F). In order to protect the excessive temperature of the underlying metal of a structure, the thermal protection system design provides a smooth, and aerodynamic surface. Henceforth, this proceeded to more researches focusing on the materials that sustain higher melting point and after effects of aerospace environments [3].

Moreover, it is a tiny palette of materials drawn for an engineer or designer. Therefore, this paper defines UHTM with their usefulness in a real structure (load-bearing) application, where the generation of highly recorded temperature, occurs rapidly, by an aerodynamic heating and ionized gas (plasma) [4]. The author states that this allows a quick elimination of most materials that have higher melting point. Meanwhile, oxides are reasonable to be considered during an oxidizing environment only, as it has poor thermal shock resistance due to high thermal expansion and lower thermal conductivity. Thus, eliminating them from further reviewing and keeping SiC, B_4C, SiB_6, ZrB_2 and HfB_2. This is due to their usable UHTM in protective heat shield of re-entry aerospace vehicles is highly promising and reliable.

2. Types of thermal protection system

Until today, the TPS is used in aerospace vehicles to cover essentially the entire orbiter surface, and it consisted of different types of tiles in varying locations based on amount of required heat protection. Those types of TPS are briefly summarizing as shown in table 1 which are mainly dividing to Blanket Insulation, Tile Insulation, Ablator and Hot Structure. Each TPS has maximum operational temperature which depends on the materials are used to fabricate it.

2.1. Silicon carbide (SiC)

In nano-scale perspective, Silicon carbide (SiC) was astonishingly smart and adaptive in its characteristics. Not only that, it showed a great strength and modulus, dense equal to that of an aluminum; while ensuring a low cost in comparison. In terms of aerospace application, SiC can be used as a particulate filter in a massive ceramic composite which indicates as an Ultra High Temperature Ceramics (UHTCs), or it can act as a matric that is in reinforced composited (CMCs). Various studies stated that it is a mandatory for aerospace sector to include composites during the transfer of mechanical loads at a high temperatures occurs (up to 1,600 °C in air); with that, none of metallic material or intermetallic materials can be used [5-8].

SiC is usually mixed with carbon fibers to obtain CMCs with the non-oxide matric materials for greater temperature applications. A study suggested that there was no degradation reported when the carbon fibers were tested with the temperature over 2600 °C in non-oxidizing atmosphere [9]. This shows that carbon fibers are protected from oxidation reactions; becoming a useful material in space vehicle applications as temperature rises up to 1600 °C during the re-entry phase in atmosphere. The key role of SiC matrix in CMCs is protection of carbon fibers from oxidation processes; which activates at 500 °C temperature, leading to the formation of a protective silica-based glassy layer.

Other than that, SiC can be used to protect carbon fiber reinforces carbon composites (C/C); in the form of coating. Furthermore, the Reinforced Carbon-Carbon (RCC) has been used as TPS for applications such as wing leading edges and the nose cap of the Space Shuttle, due to the temperature ranges of 1500-1600 °C whilst the re-entry phase in atmosphere occurs. A few researches suggested that providing an oxidation resistance for reuse capabilities, the outer layers of the RCC should be converted to silicon carbide [10-14].

Table 1. The materials of different types of TPC and maximum working temperature.

No.	TPS	Type	Material	Max. Temp.	Ref.
1	Blanket Insulation	Advance Flexible Reusable Surface Insulation (AFRSI)	Al_2O_3, Borosilicate / C-9	649 °C	[3, 15, 16]
2	Blanket Insulation	Tailorable Advanced Blanket Insulation (AFRSI)	Al_2O_3, Borosilicate / PCC	704 °C	[3, 15, 16]
3	Tile Insulation	Low Temperature Reusable Surface Insulation (LRSI)	SiO_2 /white SiO_2	700 °C	[3, 15]
4	Tile Insulation	Low Temperature Reusable Surface Insulation (LRSI)	ZrO_2/ $TaSi_2$–SiO_2-borosilicate	857 °C	[17]
5	Tile Insulation	High Temperature Reusable Surface Insulation (HRSI)	SiO_2 / B_2O_3- SiB_4	1260 °C	[3, 15]
6	Tile Insulation	Fibrous Refractory Composite Insulation (FRCI)	SiO_2-Alumino-borosilicate/C9	1360 C°	[3, 18]
7	Tile Insulation	Alumina Enhanced Thermal Barrier (AET)	Al_2O_3-SiO2-Alumino-borosilicate/SiO_2-SiB_6-$MeSi_2$	1371 °C	[3, 18]
8	Ablator	Low Temperature Ablator	phenolic resin-Si-C-C	1800 °C	[19]
9	Ablator	High Temperature Ablator	SiC-C-C-ZrB_2-ZrC-SiC	2300 °C	[20]
10	Hot Structure	Reinforced Carbon-Carbon Composites (RCC)	C-C–ZrB_2–SiC/SiC	1500 °C	[21]
11	Hot Structure	Reinforced Carbon-Carbon Composites (RCC)	C-C/SiC–$MoSi_2$–ZrB_2	1500 °C	[22]
12	Hot Structure	Reinforced Carbon-Carbon Composites (RCC)	SiC-C-C/SiB_6–$MoSi_2$	1470 °C	[23]
13	Hot Structure	Ceramic Matrix Composites (CMC)	ZrB_2-SiC	2200 °C	[24]
14	Hot Structure	Ceramic Matrix Composites (CMC)	HfB_2-SiC	2500 °C	[25]

2.2. Boron carbide (B_4C)
Boron carbide B_4C, has been applied in engineering materials for various purposes such as wear parts, and abrasives; due to its strength in hardness and exceptional wear resistance. A few researchers reported that boron carbide has a higher temperature thermoelectric material and a first-wall material for a fusion reactor [26-28]. Another study stated the reason that due to its stability in compounds, with a melting point of 2350 °C, and a low density (2.52 g/cc) [29].

In addition to that, B_4C material promises its reliability with various distinctive applications that require elevated mechanical properties [30-34]. A few researchers reported that it has the potential to hinder a higher temperature which is required to sinter, on the covalent bonding that prevails in boron carbide [35-37]. Other than that, densified B_4C can slowly oxidizes at 600 °C, resulting in the formation of a thin B_2O_3 film; cracking after it reaches a cool temperature. Other researches stated that the oxidation process at 1200 °C, there is limited by diffusion of reagents through the oxide layer [38, 39]. On the other hand, another study stated that combining particles of ZrB_2–B_4C showed an effective inhibiting oxidation of C–C composites, during a 1500 °C temperature [40]. In addition, boron carbide powders are self-bonded, and they do not require additives by means of the Spark Plasma Sintering-processing at a faster, and cheaper way; retaining a fine grain structure.

2.3. Silicon boride (SiB_6)
There are few registered silicon boride phases in the X-ray cards of the International Center for Diffraction Data. These include silicon tetraboride, SiB_4 and silicon hexaboride, SiB_6 demonstrated potential due to their outstanding chemical resistance properties. In addition, SiB_6 has a greater stability reported at higher temperature than SiB_4 in the B-Si binary alloy phase diagram [38, 40].

Furthermore, SiB_6 has been an attraction to a few applications in material industry such as thermoelectric applications at a higher temperature. This is due to the fact that their favorable properties, including, higher melting points, chemical stability, higher electrical conductivity, lower thermal conductivity and also, high Seebeck coefficient at a higher temperature. A study reported about Si-B composite, indicating their good thermoelectric property at a higher temperature [41].

At high temperature, SiB_6 showed marvelous self-healing property because of the formation of borosilicate phase (silica and boron trioxide). Besides, a study stated that this is can remain as a flux after being oxidized during the preparation process to fusion a sintering temperature [42]. The authors add that it can also seal the micro-defects, also known as cracks or holes, and ensure an improvement on the oxidation resistance and thermal resistance [43].

2.4. Zirconium and hafnium diborides (ZrB_2, HfB_2)
It is known that the UHTC materials have become known worldwide. As the ZrB_2 and HfB_2-based UHTCs are the mostly used and researched upon; their characteristics of good oxidation resistance from room temperature to 2000 °C. In comparison, HfC and ZrC have higher melting points than HfB_2 and ZrB_2; the diborides are known to have a greater thermal conductivity than the carbides [44, 45]. Another study discussed the combination of high-temperature capabilities along with their thermal conductivity, which makes HfB_2 and ZrB_2 particularly attractive to use in sharp wing leading edges and nose tips [46].

The advantage of diboride-based UHTC materials in performance is obtained from their high-temperature capability and high thermal conductivity. Convective energy that enters the surface near the stagnation region is conducted away to cooler regions of the leading edge, where it can be radiated back to the environment figure 1. The higher the thermal conductivity of the leading-edge material, the more efficient this process becomes [48]. The UHTC leading edge then behaves much like a passive heat pipe, to move energy through, and eventually out of, the system [47].

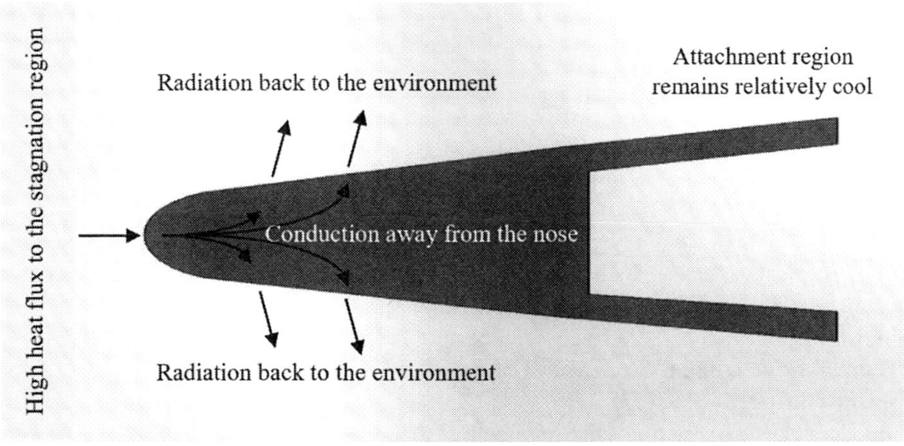

Figure 1. UHTC leading-edge thermal management concept [47].

Practically, high-temperature oxidation resistance of a pure diboride materials is not sufficient for aerothermal flight environment. The best oxidation performance is found for monolithic materials hot-pressed from mixtures of ceramic powders containing a silica former as a minor component. SiC matrix reinforced metal borides, such as ZrB_2, and HfB_2 have been commonly referred to as Ultra High Temperature Ceramics. UHTCs represent a class of promising materials that can be used in extreme applications due to their high melting point and relatively good oxidation resistance in re-entry conditions. ZrB_2, and HfB_2 are characterized by high melting temperatures (3250 and 3400 °C respectively), solid state stability, good thermo-chemical, and thermo-mechanical properties [48-51].

Paul, reported on five distinctive UHTC compositions, used for impregnation, viz. ZrB_2, ZrB_2–20 vol% SiC, ZrB_2–20 vol% SiC–10 vol% LaB_6, HfB_2 and HfC. Due to high temperature oxidation resistance occurrence, the purpose to build oxyacetylene torch to test facility at temperatures above 2500 °C and the results are compared with that of a C–C benchmark composite. Based on the high temperature oxidation testing, a conclusion can be drawn that impregnation of C_f preforming with UHTC powders, significantly improves the high temperature oxidation resistance of the composites compared to C–C composites. Hf-based UHTC powders provides superior oxidation protection as compared to Zr-based compositions, even though less could be impregnated into the pre-forms due to a larger mean particle size. Meanwhile, an addition of SiC and LaB_6 does not enhance the oxidation resistance at the very high temperatures, >2500 °C. A study of the two Hf-based compounds investigated and reported that HfB_2 composites were deemed to have better oxidation performance as the oxidation products were adherent to the base composite [25].

3. Conclusion

This paper covers briefly all main types of Thermal Protection Systems (TPSs) and all the materials are used to fabricate them with the maximum operational temperatures. Then, it focuses on the promised carbides and borides Ultra-High Temperature Materials (UHTMs) (SiC, ZrB_2, HfB_2, SiB_6 and B_4C) which are currently using for several aerospace applications, especially for TPS. Ultra-High Temperature Material (UHTM) compounds, which portray a unique set of properties, such as, extreme high temperature, high hardness, and stability against chemical, and also the strength against high temperatures. Besides, it discusses the oxidation of SiC, B_4C, SiB_6, ZrB_2 and HfB_2. The structural materials used during the high-temperature oxidizing environment are mainly limited to SiC, oxide ceramics, and composites. In addition to that, silicon-based ceramic has a maximum-use at 1700 °C approximately; as it is an active oxidation process over low temperature and water vapor environment condition. Therefore, the carbides and borides of the transition metals, Zr and Hf demonstrate the high potential for aerospace applications due to a high-melting temperature and good stability in forming high-melting temperature oxides.

Acknowledgments
The author acknowledges the support of Universiti Teknologi PETRONAS, Malaysia and University of Al-Qadisiyah, Iraq for this work.

References

[1] Blockley R and Shyy W, *Encyclopedia of Aerospace Engineering*: Wiley Online Library, 2010.
[2] Behrens B and Müller M, Technologies for Thermal Protection Systems Applied on Re-Usable Launcher, *Acta Astronautica* (2004) **55** 529-536.
[3] Natali M, Kenny J M, and Torre L, Science and Technology of Polymeric Ablative Materials for Thermal Protection Systems and Propulsion Devices: A Review, *Progress in Materials Science* (2016) **84** 192-275.
[4] Simonenko E, Sevast'yanov D, Simonenko N, Sevast'yanov V, and Kuznetsov N, Promising Ultra-High-Temperature Ceramic Materials for Aerospace Applications, *Russ. J. Inorg. Chem* (2013) **58** 1669-1693.
[5] Kovalev V and Kolesnikov A, Experimental and Theoretical Simulation of Heterogeneous Catalysis in Aerothermochemistry (a Review), *Fluid Dynamics* (2005) **40** 669-693.
[6] Monteverde F, Guicciardi S, and Bellosi A, Advances in Microstructure and Mechanical Properties of Zirconium Diboride Based Ceramics, *Materials Science and Engineering: A* (2003) **346** 310-319.
[7] Levine S R, Opila E J, Halbig M C, Kiser J D, Singh M, and Salem J A, Evaluation of Ultra-High Temperature Ceramics Foraeropropulsion Use, *Journal of the European Ceramic Society* (2002) **22** 2757-2767.
[8] Albarody T M B, Al-Jothery H K M, Hussein A R, Teng L Z, Muhsan A S, and Ahmad F, "Ultrafast-Contactless Plasma Arc Sintering," Conference on AIP Conference Proceedings, 2018, 080008.
[9] Morino Y, Yoshinaka T, Auweter-Kurtz M, Hilfer G, Speckmann H-D, and Sakai A, Erosion Characteristics of Sic Coated C/C Materials in Arc-Heated High Enthalpy Air Flow, *Acta Astronautica* (2002) **50** 149-158
[10] Pichon T, Barreteau R, Soyris P, Foucault A, Parenteau J, Prel Y, and Guedron S, Cmc Thermal Protection System for Future Reusable Launch Vehicles: Generic Shingle Technological Maturation and Tests, *Acta Astronautica* (2009) **65** 165-176.
[11] Reimer T and Laux T, "Thermal and Mechanical Design of the Expert C/C-Sic Nose," Conference on Thermal Protection Systems and Hot Structures, 2006.
[12] Pichon T, Soyris P, Foucault A, Parenteau J, Prel Y, and Guedron S, "C/Sic Based Rigid External Thermal Protection System for Future Reusable Launch Vehicles: Generic Shingle, Pre-X/Flpp Anticipated Development Test Studies," Conference on Thermal Protection Systems and Hot Structures, 2006.
[13] Paulmier T, Balat-Pichelin M, and Le Quéau D, Structural Modifications of Carbon–Carbon Composites under High Temperature and Ion Irradiation, *Applied surface science* (2005) **243** 376-393.
[14] Chamberlain A L, Fahrenholtz W G, Hilmas G E, and Ellerby D T, High-Strength Zirconium Diboride-Based Ceramics, *Journal of the American Ceramic Society* (2004) **87** 1170-1172
[15] Launius R D and Jenkins D R, *Coming Home: Reentry and Recovery from Space*: Government Printing Office, 2012.
[16] Kourtides D, Carroll C, Smith D, Guzinski M, Marschall J, Pallix J, Ridge J, and Tran D, Evaluation of Thermal Control Coatings for Flexible Ceramic Thermal Protection Systems, (1997).

[17] Shao G, Wu X, Kong Y, Shen X, Cui S, Guan X, Jiao C, and Jiao J, Microstructure, Radiative Property and Thermal Shock Behavior of Tasi 2–Sio 2-Borosilicate Glass Coating for Fibrous Zro 2 Ceramic Insulation, *Journal of Alloys and Compounds* (2016) **663** 360-370.
[18] Hinkle K A, Staszak P R, and Watts E, Advanced Ceramic Materials Development and Testing, *AIAA Paper* (1996) 96-1527.
[19] Yin R, Cheng H, Hong C, and Zhang X, Synthesis and Characterization of Novel Phenolic Resin/Silicone Hybrid Aerogel Composites with Enhanced Thermal, Mechanical and Ablative Properties, *Composites Part A: Applied Science and Manufacturing* (2017) **101** 500-510.
[20] Xin Y, Qizhong H, Zhean S, Xin C, Liang X, Ping Z, and Jun L, Ablative Property and Mechanism of C/C-Zrb 2-Zrc-Sic Composites Reinforced by Sic Networks under Plasma Flame, *Corrosion Science* (2016) **107** 9-20.
[21] Zhang Y-L, Fei T, Zeng W-Y, Yang B-X, Li H-J, and Li K-Z, Microstructure and Oxidation Behavior of C/C–Zrb 2–Sic Composites Coated with Sic Coating at High Temperature, *Corrosion Science* (2015) **100** 421-427.
[22] Fu Q-G, Jing J-Y, Tan B-Y, Yuan R-M, Zhuang L, and Li L, Nanowire-Toughened Transition Layer to Improve the Oxidation Resistance of Sic–Mosi 2–Zrb 2 Coating for C/C Composites, *Corrosion Science* (2016) **111** 259-266.
[23] Zhang Y-L, Huang J-F, Zhu K-J, Cao L-Y, Li C-Y, Zhou L, Zhang B-Y, Kong W-H, and Zhang B, Effects of Voltage on Microstructure and Oxidation Resistance of Sib 6–Mosi 2 Coating Deposited by Pulse Arc Discharge Deposition, *Applied Surface Science* (2015) **340** 43-48.
[24] Han J, Hu P, Zhang X, Meng S, and Han W, Oxidation-Resistant Zrb 2–Sic Composites at 2200 C, *Composites Science and Technology* (2008) **68** 799-806.
[25] Paul A, Venugopal S, Binner J, Vaidhyanathan B, Heaton A, and Brown P, Uhtc–Carbon Fibre Composites: Preparation, Oxyacetylene Torch Testing and Characterisation, *Journal of the European Ceramic Society* (2013) **33** 423-432.
[26] Narushima T, Goto T, Maruyama M, Arashi H, and Iguchi Y, Oxidation of Boron Carbide-Silicon Carbide Composite at 1073 to 1773 K, *Materials Transactions* (2003) **44** 401-406.
[27] Buzhinskij O and Semenets Y M, Thick Boron Carbide Coatings for Protection of Tokamak First Wall and Divertor, *Fusion Engineering and design* (1999) **45** 343-360.
[28] Thévenot F, Boron Carbide—a Comprehensive Review, *Journal of the European Ceramic society* (1990) **6** 205-225.
[29] Gasch M J, Ellerby D T, and Johnson S M, "Ultra High Temperature Ceramic Composites," in *Handbook of Ceramic Composites*, ed: Springer, 2005, pp. 197-224.
[30] Kim H W, Koh Y H, and Kim H E, Densification and Mechanical Properties of B4c with Al2o3 as a Sintering Aid, *Journal of the American Ceramic Society* (2000) **83** 2863-2865.
[31] Levin L, Frage N, and Dariel M, The Effect of Ti and Tio 2 Additions on the Pressureless Sintering of B 4 C, *Metallurgical and Materials Transactions A* (1999) **30** 3201-3210.
[32] Sigl L, Processing and Mechanical Properties of Boron Carbide Sintered with Tic, *Journal of the European Ceramic Society* (1998) **18** 1521-1529.
[33] Schwetz K A, Sigl L S, and Pfau L, Mechanical Properties of Injection Molded B4c–C Ceramics, *Journal of solid state chemistry* (1997) **133** 68-76.
[34] Skorokhod V, Vlajic M, and Krstic V, Mechanical Properties of Pressureless Sintered Boron Carbide Containing Tib 2 Phase, *Journal of materials science letters* (1996) **15** 1337-1339.
[35] Lee H and Speyer R F, Hardness and Fracture Toughness of Pressureless-Sintered Boron Carbide (B4c), *Journal of the American Ceramic Society* (2002) **85** 1291-1293.
[36] Goldstein A, Geffen Y, and Goldenberg A, Boron Carbide–Zirconium Boride in Situ Composites by the Reactive Pressureless Sintering of Boron Carbide–Zirconia Mixtures, *Journal of the American Ceramic Society* (2001) **84** 642-644.
[37] Brar N, Rosenberg Z, and Bless S, Applying Steinberg's Model to the Hugoniot Elastic Limit of Porous Boron Carbide Specimens, *Journal of applied physics* (1991) **69** 7890-7891.

[38] Lee H and Speyer R F, Pressureless Sintering of Boron Carbide, *Journal of the American Ceramic Society* (2003) **86** 1468-1473.
[39] Grady D, Shock-Wave Strength Properties of Boron Carbide and Silicon Carbide, *Le Journal de Physique IV* (1994) **4** C8-385-C8-391.
[40] Corral E L and Walker L S, Improved Ablation Resistance of C–C Composites Using Zirconium Diboride and Boron Carbide, *Journal of the European Ceramic Society* (2010) **30** 2357-2364
[41] Olesinski R and Abbaschian G, The B– Si (Boron-Silicon) System, *Journal of Phase Equilibria* (1984) **5** 478-484.
[42] Chen L, Goto T, Mukaida M, Niino M, and Hirai T, Phase Diagram and Thermoelectric Property of Si-B System Ceramics, *Journal of the Japan Society of Powder and Powder Metallurgy* (1994) **41** 1299-1303.
[43] Matsushita J and Komarneni S, High Temperature Oxidation of Silicon Hexaboride Ceramics, *Materials research bulletin* (2001) **36** 1083-1089.
[44] Opeka M M, Talmy I G, Wuchina E J, Zaykoski J A, and Causey S J, Mechanical, Thermal, and Oxidation Properties of Refractory Hafnium and Zirconium Compounds, *Journal of the European Ceramic Society* (1999) **19** 2405-2414.
[45] Fridlender B, Neshpor V S, Ordan'yan S S, and Unrod V, Thermal Conductivity and Thermal Diffusivity of Binary Alloys of the Zrc-Zrb 2 System at High Temperatures, *Teplofizika vysokikh temperatur* (1979) **17** 1210-1215.
[46] Kolodziej P, Bowles J V, and Roberts C, Optimizing Hypersonic Sharp Body Concepts from a Thermal Protection System Perspective, *8th, AIAA Paper* (1998).
[47] Squire T H and Marschall J, Material Property Requirements for Analysis and Design of Uhtc Components in Hypersonic Applications, *Journal of the European Ceramic Society* (2010) **30** 2239-2251.
[48] Fahrenholtz W G, Hilmas G E, Talmy I G, and Zaykoski J A, Refractory Diborides of Zirconium and Hafnium, *Journal of the American Ceramic Society* (2007) **90** 1347-1364.
[49] Fahrenholtz W G, Thermodynamic Analysis of Zrb2–Sic Oxidation: Formation of a Sic-Depleted Region, *Journal of the American Ceramic Society* (2007) **90** 143-148.
[50] Hilmas G, Fahrenholtz W, Chamberlain A L, and Ellerby D T, Oxidation of Zrb_2-Sic Ceramics under Atmospheric and Reentry Conditions, (2005).
[51] Fahrenholtz W G, The Zrb2 Volatility Diagram, *Journal of the American Ceramic Society* (2005) **88** 3509-3512.

IOP Conf. Series: Materials Science and Engineering **863** (2020) 012004 doi:10.1088/1757-899X/863/1/012004

Study on the effect of ultrasonic wave amplitude on de-emulsification of crude oil to enhance production process

C H Lim, S A Sulaiman*, A Japper-Jaafar and P T Bhaskoro

Department of Mechanical Engineering, Universiti Teknologi PETRONAS, 32610 Bandar Seri Iskandar, Perak, Malaysia.

*E-mail: shaharin@utp.edu.my

Abstract. One of the most challenging issue faced by the oil and gas industry is the formation of crude oil emulsion. The emulsion formation will lead to inconsistency in the production performance, which directly affects the economic growth of the industry. Current, conventional de-emulsification methods such as thermal, demulsifiers, and electrical methods are found ineffective in solving crude oil emulsion problem relativity to expected settling time. Therefore, this research aims to propose ultrasonic wave application method for better separation rate to solve crude oil emulsion problem. The effects of ultrasonic waves are investigated by designed experiment under different operating temperature. The optimum operating conditions are investigated and identified through optimization methods in Design Expert Software. The goals of this research are to achieve maximum water and oil separation and minimize the rag layer thickness. Based on the experiments and results, the best operating condition is at operating temperature of 60 °C at 40 µm of ultrasonic frequency of 20 kHz. These parameters were able to provide 73.3% of water separation rate and 20% volume fraction of oil layer after eight hours of bottle test.

1. Introduction

Crude oil emulsions can be technically defined as heterogeneous liquids system that consists of two immiscible liquids that came into contact and did not dissolve with each other [1]. An undisclosed oilfield in the South China Sea is facing emulsion problem as excess water is injected to the well for the enhanced oil recovery purpose [2, 3]. Besides, the neighbouring oilfields operated by a different company are also facing the same problem [4]. Emulsion is formed when oil and water are mixed vigorously in the process of entering perforation in the casing with the presence of huge different in pressure between the reservoir and well head [1]. Akpabio and Ekott [5] claimed that emulsion were naturally formed at the well head, chokes and valves section due to the action of the shear stress and the pressure drop in the system. Ultrasonic wave is a mechanical wave that require medium to propagated [6]. In the application of ultrasonic, amplitude are the contributing factor to the separation of crude oil emulsion [7-9]. A certain value of amplitude (P_A) pass through a medium at hydrostatic pressure (P_h) can caused oscillation to the molecules as shown in figure 1 [6]. During the oscillation cycle, the pressure is positive and the distance between the molecules decrease in compression cycle, whereas it is vice versa in rarefaction cycle. Thus, its yield can be represented by $P_c = P_h - P_A$ [6]. At certain amplitude, the average distance between molecules exceeds the critical distance and it will form a void and induce cavitation bubble in it as the droplets are not intact with each other [6]. Therefore, a higher amplitude of ultrasonic will result in a large formation of bubbles known as cavitation effect [7]. The objective of this research is to obtain the separation between water and oil at maximum value and minimize the rag layer thickness.

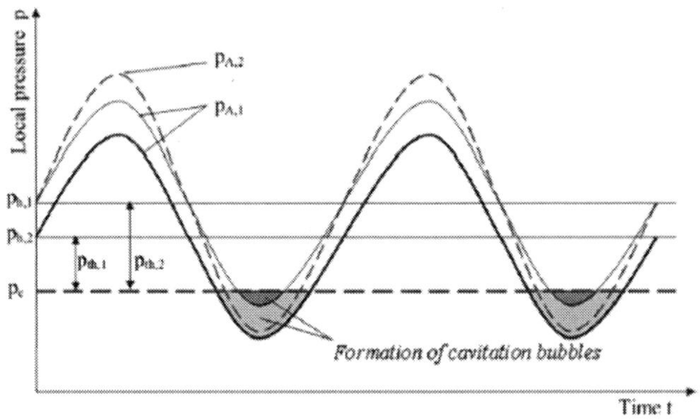

Figure 1. Relation of local pressure with cavitation [6].

2. Methodology and experimental design

The crude oil emulsion was prepared by mixing crude oil and formation water. The crude oil used in this study was obtain from crude oil terminal that served the oilfield. The crude oil was shaken well before mixed with formation water. One litre of formation water was prepared to represent the oil field environment. Firstly, seven chemical components were measured in g/L and pour into one litre beaker. Table 1 shows the chemical components used during the preparation of water formation. One litre of ultra-pure water, which was obtained from Elga PURELAB Flex 3 Water Purification System, was added gradually to the beaker. The mixture was agitated vigorously using IKA R 1389 three blade propeller at the speed of 400 rpm and ambient temperature of 23 °C for a duration of one hour.

Table 1. Basic properties of the crude oil sample.

Chemical Substance	Concentration (g/L)
Sodium Bicarbonate	5.1260
Potassium Chloride	0.2646
Sodium Chloride	6.0114
Barium Chloride Dihydrate	0.0067
Strontium Chloride Hexahydrate	0.0141
Magnesium Chloride Hexahydrate	0.0750
Calcium Chloride Dihydrate	0.2344

Water in oil (w/o) emulsion was prepared in 100 ml beaker in the ratio of 30:70 of water and oil. About 50 ml was prepared for each crude oil emulsion study. Firstly, the oil and water were mixed together and heated up to 60 °C. Once the temperature reached 60 °C, the mixture was stirred with IKA T25 DS2 stirrer at 12,000 rpm for duration of 15 minutes at mixing temperature of 60 °C. The stable emulsion is formed and underwent ultrasonic irradiation treatment instantly to prevent the break change of crude oil emulsion stated.

3. Results and discussion

The crude oil emulsion stability of base sample was determined through eight hours of bottle test. No ultrasonic wave application was conducted on the base sample. The volume fraction of oil and rag layer was recorded and is presented in figure 2. The volume fraction was calculated from the volume of each layer divided by the total volume. Rapid oil separation occurred after 60 minutes of bottle test based on gravity settling. The best separation achieved in base sample was after six hours with 20% volume fraction of crude oil layer and 80% volume fraction of rag layer. The base sample took six hours to reach stability. However, it was an ineffective separation as there was no water layer formation.

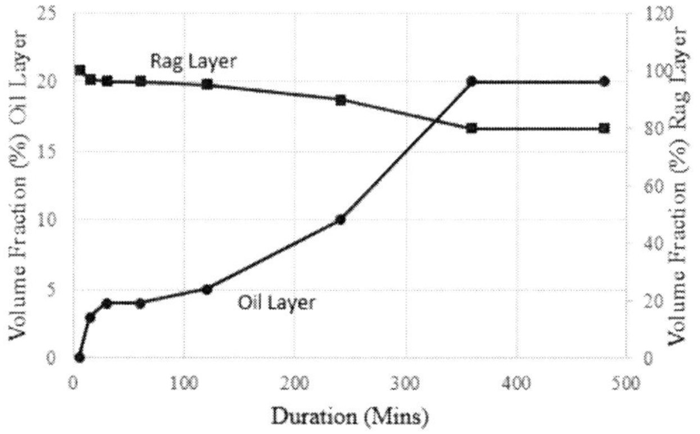

Figure 2. Oil and rag layer settling rate for base sample.

Figure 3(a) shows the lowest operating temperature of 30 °C. The expected increase in oil layer is observed on the volume fraction with respect to the time. At the first hour, the oil layer increases equally and it start to behave differently after first hour. Samples A1 and A3 crude oil layer reached it stability after 6 hours of bottle test. However, Sample A2 that underwent ultrasonic intensity of 60% did not achieve stability. Water-in-oil emulsion Sample A1 underwent 40% of ultrasonic irradiation showed the highest oil layer volume fraction of 16% compared to other ultrasonic intensity. Nevertheless, it is considered ineffective as the volume fraction of oil in base sample is 20%. Figure 4(b) shows the results at heating temperature of 45 °C. Based on the graph, the volume fraction of oil layer increases gradually with respect to the time. After 8 hours of bottle test, oil layer thickness did not achieve stability and increased exponentially. This indicates that coalescence and flocculation process were in progress. Sample B1, B2 and B3 clearly showed that oil layer thickness was inversely proportional to the ultrasonic intensity. Hence, B1 showed the highest oil layer thickness whereas B3 have the lowest oil layer thickness were expected due to the differential in amplitude. Nevertheless, volume fraction of oil layer in Sample B1 (16%) had not surpassed base sample (20%).

In figure 3(b), the oil layer formation for three samples are valid and in-line with the experiments of Gaikwad and Pandit [9] experimental result stating that the increase of the pressure amplitude of the ultrasonic prompted more cavitation effect. As a result, the formation of bubble and busting process increased aggressively and caused the w/o emulsion to breakup and enhance emulsification instead of de-emulsification process. Therefore, lower ultrasonic amplitude is desirable to provide sufficient energy for the new interface formation [9]. Figure 3(c) illustrates the change of oil layer and water layer volume fraction of Samples C1, C2 and C3 over period of 8 hours at operating temperature of 60 °C. The formation of oil layer was increased rapidly after five minutes of bottle test in Samples C1 to C3. Samples C3 have the highest oil layer volume fraction of 24% which higher than base sample by 4%. Besides, Samples C1 have the same oil layer volume fraction with base sample and it is expected to increase after eight hours of bottle test. The water layer formation was showed in Samples C1 and C2. Sample C3 does not exhibit any sign of water formation during 8 hours of bottle test. Water layer can be observed after 15 minutes of bottle test with a volume fraction of 22% in sample C1 as compared to sample C2 which formed after 30 minutes with a volume fraction of 12% at the end of bottle test. The results show significant improvement compared to no water separation in the base sample.

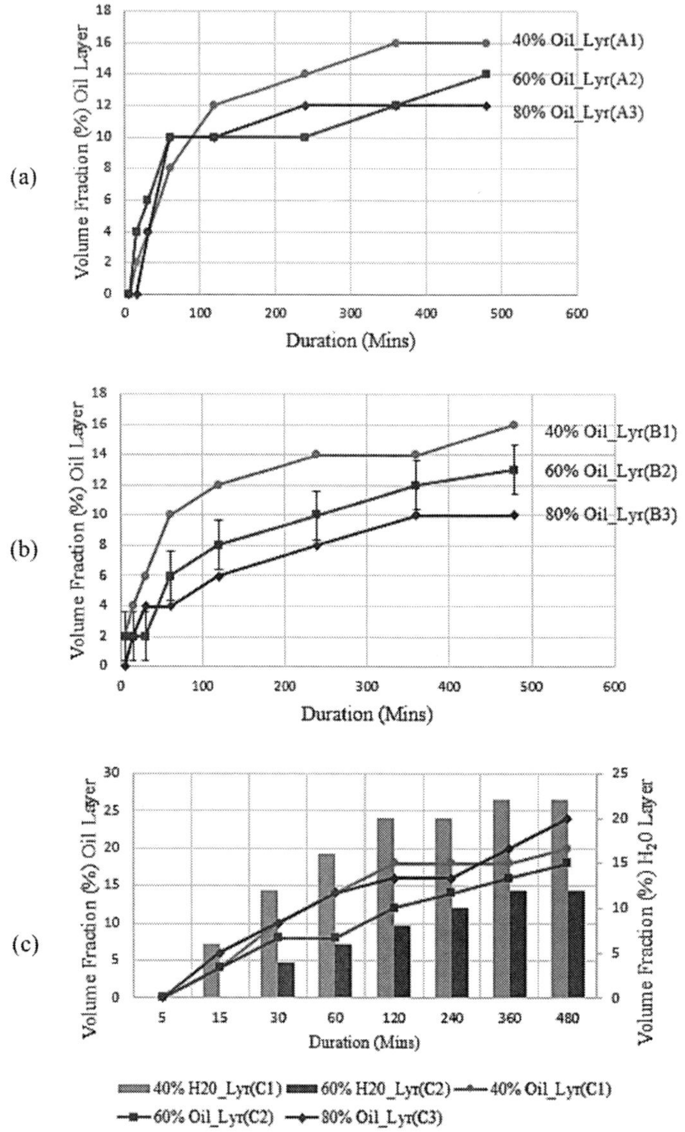

Figure 3. Volume fraction of oil layer separated (%) at temperature 30, 45 and 60 °C.

This phenomena was proven true by Kokal [10], Smith and Arnold [11]. They state that high temperature increases the collision rate between droplets by reducing the interfacial viscosity and enhanced coalescence rate. Sample C3 show no water layer formation due to high ultrasonic amplitude. These occurrence due to the increasing in the intensity of shock wave and caused the emulsion to breaking up and reduced coalescence rate [8, 9]. It can be deduced that ultrasonic is able to accelerate the separation rate of crude oil emulsion at heating temperature of 60 °C at 40% of ultrasonic amplitude.

Figure 4 shows the change of volume fraction of oil layer at different heating temperature at threshold amplitude of 40%. The threshold amplitude was focused because it gave the highest cavitation rate compared to other amplitude level. Based on figure 4, Sample C1 shows the highest volume fraction of oil layer. The trend was proven by Kokal [10] finding that increase in temperature enhanced the separation rate of crude oil. As the temperature increases, crude oil particles will absorb the energy and converts to kinetic energy. It allowed crude oil to vibrate and move due to differential in density of crude oil and water. In molecular structure of view, increased temperature resulting in destabilization of interfacial film and allowed coalescence process to occur [10]. Hence, coalescence rate increase at high thermal energy. In short, heat accelerated the de-emulsification process.

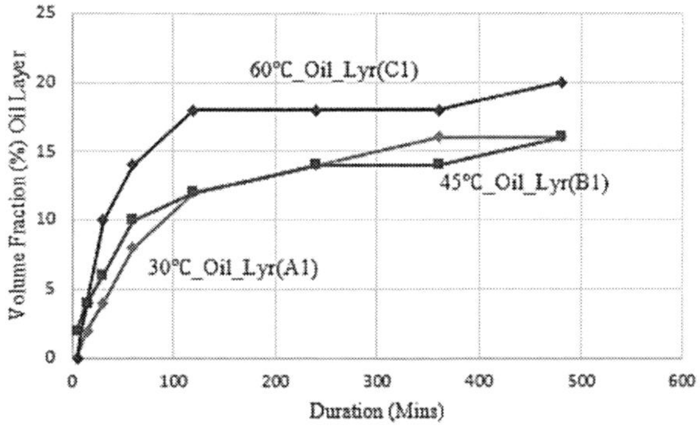

Figure 4. Volume fraction (%) of oil layer at threshold amplitude of 40% and at different heating temperature.

Based on the pattern, it clearly showed that higher temperature will increase the oil layer separation rate as it can be seen at high operating temperature of 60 °C, the volume fraction of oil layer is approximately 20%. As the temperature increase, the viscosity decreased and the hydrodynamic force big enough to enhance the separation rate. Figure 5 shows the distribution of water droplets in crude oil emulsion before and after bottle in Sample C1 which under operating temperature of 60 °C at 40% ultrasonic amplitude. After eight hours of bottle test, the water droplets size have increased from 5.22 μm to 18.32 μm and formed more uniform structure.

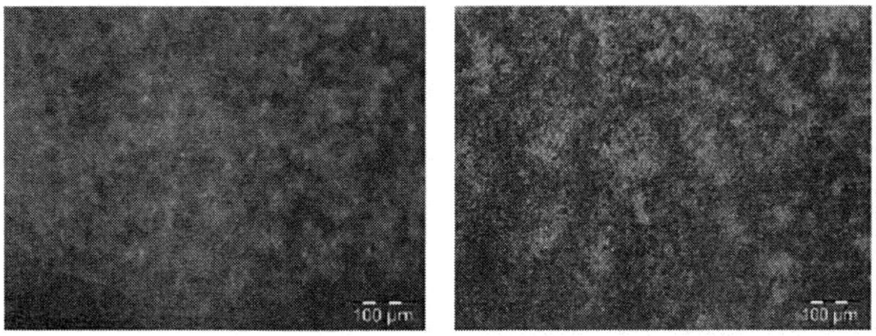

(a) (b)

Figure 5. Water droplets in rag layer for Sample C1 for (a) before bottle test and (b) after eight hours of bottle test.

The reason for this was due to the effect of temperature and the irradiation of ultrasonic. Under the heating effect, the water and oil droplets gain kinetic energy to vibrate and move vigorously in the system. Increase of operating temperature, caused interfacial tension between the particle and viscosity of oil to decrease, which increased the ease of dispersion of one phase into others. With all the experimental data, the optimum operating conditions for the ultrasonic wave amplitude de-emulsification process was analysed by using Design Expert software. Based on the optimization analysis, interaction effect between operating temperature and ultrasonic amplitude collated with crude oil, water and rag layers were identified. The models yield an empirical equation that valid in the range of operating temperature at 30 °C to 60 °C and ultrasonic wave amplitude of 40% to 80%. Where OL is the volume fraction of oil layer, WL is the volume fraction of water layer, RL is the volume fraction of rag layer, A is the operating temperature (°C) and B is the ultrasonic wave amplitude (%).

4. Conclusion

The optimum parameter for ultrasonic de-emulsification is at heating temperature of 60 °C at 40% of ultrasonic amplitude. At 40 μm of ultrasonic amplitude give the best sinusoidal wave to the crude oil emulsion. This statement is justified and proven by past researchers that the threshold amplitude and temperature are the key factors in affecting the quality of separation rate of crude oil emulsion. There was no water separation in crude oil emulsion after 8 hours of bottle test under ultrasonic irradiation treatment in Samples A and B which were conducted in heating temperature of 30 °C and 45 °C. Based on the experimental result and design expert optimization, Water formation in crude oil emulsion is strongly affected by the change of amplitude whereas oil layer formation is affected by the change of crude oil temperature.

References

[1] Udonne, J., *Chemical treatment of emulsion problem in crude oil production.* Journal of Petroleum and Gas Engineering Vol, 2012. **3**(7): p. 135-141.

[2] Vai Yee, H., et al., *Managing Chemical Flooding, and By-Products with Produced Water Re-Injection (PWRI) System in Offshore Environment*, in *SPE Enhanced Oil Recovery Conference.* 2013, Society of Petroleum Engineers: Kuala Lumpur, Malaysia.

[3] Najamudin, K.E., et al., *Chemical EOR Produced Water Management at Malay Basin Field*, in *Offshore Technology Conference-Asia.* 2014, Offshore Technology Conference: Kuala Lumpur, Malaysia.

[4] Mat, H., et al., *Study on demulsifier formulation for treating Malaysian crude oil emulsion*, in *Department of Chemical Engineering.* 2006, Universiti Teknologi Malaysia: Malaysia.

[5] Akpabio, E.J. and E.J. Ekott, *Application of Physico-Technological Principles in Demulsification of Water-In-Crude Oil System.* Indian Journal of Science and Technology, 2013. **6**(1): p. 3898-3902.

[6] Sauter, C., et al., *Influence of hydrostatic pressure and sound amplitude on the ultrasound induced dispersion and de-agglomeration of nanoparticles.* Ultrasonics Sonochemistry, 2008. **15**(4): p. 517-523.

[7] Hielscher, T., *Ultrasonic production of nano-size dispersions and emulsions.* Dans European Nano Systems Workshop - ENS 2005, Paris : France, 2007.

[8] Behrend, O. and H. Schubert, *Influence of hydrostatic pressure and gas content on continuous ultrasound emulsification.* Ultrasonics Sonochemistry, 2001. **8**(3): p. 271-276.

[9] Gaikwad, S.G. and A.B. Pandit, *Ultrasound emulsification: effect of ultrasonic and physicochemical properties on dispersed phase volume and droplet size.* Ultrasonics Sonochemistry, 2008. **15**(4): p. 554-563.

[10] Kokal, S.L., *Crude oil emulsions: A state-of-the-art review.* SPE Production & Facilities, 2005. **20**(01): p. 5-13.

[11] Smith, H.V. and K.E. Arnold, *Crude oil emulsions.* Petroleum Engineering Handbook. 3rd ed., Social of Petroleum Engineers, 1987: p. 19.1-19.34.

IOP Conf. Series: Materials Science and Engineering **863** (2020) 012005 doi:10.1088/1757-899X/863/1/012005

The effects of temperature, residence time and particle size on a charcoal produced from coconut shell

R K Ahmad[1,2,*], S A Sulaiman[1], M Inayat[1] and H A Umar[1]

[1] Department of Mechanical Engineering, Universiti Teknologi PETRONAS, 32610 Bandar Seri Iskandar, Perak, Malaysia.
[2] Department of Agricultural Engineering, Bayero University, Kano 3011, Gwarzo Road, Kano, Nigeria.

*E-mail: rabi_17000319@utp.edu.my, rkahmad.age@buk.edu.ng

Abstract. Coconut shell charcoal is an important product obtained from coconut shells, which are widely used as domestic and industrial fuel. Wastes of coconut residues (shells, husks and coir) from processing of coconut milk and oil industries are widely generated everywhere all over the world. They have the potential of use as a fuel in renewable energy for different applications. In this study, the production of charcoal from coconut shell through the process of pyrolysis was conducted to determine the influence of particle size, temperature and residence time on the charcoal weight loss. Pyrolysis is a thermochemical conversion method which converts biomass into more valuable products in form of solid (charcoal, bio-char and activated carbon), liquid (oil) and gas in the absence of oxygen. The process was carried out under various process conditions of temperature (300 °C, 400 °C, and 500 °C), residence time (15 minutes, 30 minutes, and 60 minutes), and particle size (5.0 mm and 25.0 mm). It was found that for both the temperature and the residence time, the weight loss from the coconut shell increase linearly with increase in temperature and residence time. The particle size has influence in percentage weight loss. The resulting charcoal obtained can be utilized as an alternative energy resource for different applications.

1. Introduction

Presently fossil fuels are sources of conventional energy and energy demand is increasing due to ever increasing population [1]. Current fossil fuels reserves are insufficient to overcome the increasing demands of energy. It is predicted that energy sources will be scared in near future if their use continues at the same pace. Therefore, renewable energy sources such as biomass, wind, solar, and hydro are appeared as alternative source to fulfill the increasing demands of energy and cut greenhouse gases discharge into environment [2, 3].

Biomass is a clean and renewable energy source that available in abundantly in nature from a variety of sources. It has the advantage of being nearly carbon neutral and the potential to replace conventional fossil fuels in the energy marketplace [4]. The use of biomass is increasing in developing countries because most of their economies largely depend on agriculture. Agricultural residues are the most potential and common biomass worldwide considering their abundant availability, but they are not fully utilized in harnessing energy. There are several barriers to be overcome for technological advancement and commercialization to progress. One of such is the conversion of high-quality fuel from biomass through thermal conversion like carbonization for energy utilization. In carbonization of biomass, charcoal is the main solid product generated from slow pyrolysis. It is a process where biomass is heated to obtain solid residues with increasing element content of carbon by pyrolysis in the absence or controlled presence of oxygen. The properties of charcoal are controlled by the type of raw material used and the manufacturing process [5, 6]. Pyrolysis is a thermochemical conversion process which converts biomass into more valuable products such as charcoal, bio-char, activated carbon (solid), oil (liquid), and gases in the absence of oxygen [7, 8].

Charcoal, a valuable product of biomass is an energy dense fuel which has a wide market acceptance and that can easily be transported. It has been employed in domestic roles for many years, in the manufacture of several industrial chemicals like carbon tetrachloride, sodium cyanide, and carbon disulfide, as an activated carbon and other high carbon products like carbon electrode [9]. Charcoal acts as a major player in the global carbon market as a result of rich, clean, and simple production technology which offers significant advantages over raw biomass fuels because the process of carbonization reduces the particulate emissions, and reduces the risk of developing health and environmental hazards [10-12]. The major charcoal production is the conventional method where the biomass is allowed to be heated for several days. There are many methods of charcoal production, but the charcoal piles, earth kilns/pit kilns, and pit methods are the most widely used methods of charcoal production. In the traditional kilns, there are large heat losses during long period of operation and it does not have good insulation. To sustain the pyrolysis process, a major fraction of biomass is used as fuel for burning. Depending on the type of raw material and the conditions produced, commercial charcoal contains almost 75% of carbon, 20% of volatile, and 5% of ash [12].

Again, the operating conditions in making charcoal are not well recognized in the production processes because charcoal making is a traditional industry across the world. The methods are inefficient due to long process time, poor process control, low char yield, energy loss and environmental pollution problems due to the release of gases into the atmosphere [12]. Whereas, in emerging technology of thermochemical conversion method of pyrolysis, the process parameters like residence time, heating rate, and temperature are controlled sophisticatedly.

Coconut (*Cocos nucifera L.)* is an important and useful perennial product from agriculture, which is grown throughout the year when compared with other fruits [13]. Among ninety-two countries worldwide produce almost ten million hectares of coconut product where the world's largest coconut producer (Indonesia) account for almost 75% of the production, followed by Philippines, and India [14]. Coconut biomass is generated as a waste from the processing of coconut in the coconut milk, and coconut oil industry everywhere in the world. These wastes are in the form of shells, husks, and coir which have the potential to utilize as a fuel in renewable energy for different applications. The shell of coconut is gotten from coconut milk, water, and oil processing industries as a by-product. It is produced in abundance everywhere as a biomass waste throughout the tropical countries worldwide [15]. As a result of coconut economic importance and potentials like in the health benefits, its demand increases, this leads to the growth of coconut oils and milk market. Coconut shells are attractive biomass fuel and are also a good source of charcoal production that can be utilized and serve as a potential source of carbon [14]. Thus, its utilization has always been a major challenge for scientific applications. The coconut shells generated from oil and the milk industry poses a risk to health as well as environmental issues if left discarded without proper utilization. Handling the waste of coconut shells is quite complicated. Incineration may lead to various environmental and health-related problems because it is flammable and has a strong odor. These coconut shells can be better utilized to convert into charcoal as a fuel for different applications. The aim of this study is to investigate the effect of different parameters on charcoal production using coconut shells as a potential feedstock for various applications prior it use for commercial application.

2. Materials and method

2.1. Materials and sample preparation

The feedstock used in this study was coconut shells and the tools include: Machete, oven, granulator, sieve, porcelain dish and pyrolysis reactor. The coconut shells were chosen for current study, mainly because of its abundance availability, suitability for thermochemical processes, and ease of operation, low initial moisture content, less feedstock preparation effort, high density, and energy content. Coconut shells were obtained from a coconut seller in a grocery store nearby the Universiti Teknologi PETRONAS, Seri Iskandar, Perak, Malaysia. The outer fibrous portion of the coconut shell was removed with a machete. The moisture content of the coconut shells, which is the key factor for thermal conversion, was determined according to the procedure given by ASTM E871-82 standard procedure [16]. The sample was dried in an oven at 105 °C for 24 hours in order to determine the initial moisture content of the feedstock. The sample was crushed using Granulator and sieved to a particle size of 25.0 mm and 5.0 mm by using a sieve. Figure 1 (a) and (b) show the sample of the coconut shells before the pyrolysis process.

Figure 1. Prepared coconut shells (a) 25.0 mm and (b) 5.0 mm particle size.

The study was conducted at biomass research laboratory, Universiti Teknologi PETRONAS, Seri Iskandar, Perak, Malaysia. The process was carried out under various process conditions as shown in table 1. These values were chosen as they were within the range of pyrolysis process (carbonization) as suggested by Prabir, [7]. The feedstock (coconut shells) was weigh to 100 g for each experiment and sample was heated gradually at different specific temperature and residence time in a pyrolysis reactor (figure 2).

Table 1. Pyrolysis parameters and process conditions.

Parameter		**Process conditions**	
Temperature (°C)	300	400	500
Residence time (minutes)	15	30	60
Particle size (mm)	5	25	

2.2. Experimental setup

The experiment was carried out using an electrical-powered downdraft pyrolysis reactor (figure 2). A nitrogen gas was connected with a rotameter to the reactor. The condensable and non-condensable gases pass through a 0 °C salty ice cooling system. The non-condensable gases further pass through gas cleaning system and the gas compositions were analysed via Emerson X-stream X2GP Online Gas Analyser, which records the concentrations of the major syngas components (CO, CO_2, CH_4 and H_2) in a computer. At the end of the process, the reactor was switched off and allowed to cool to collect the produced charcoal.

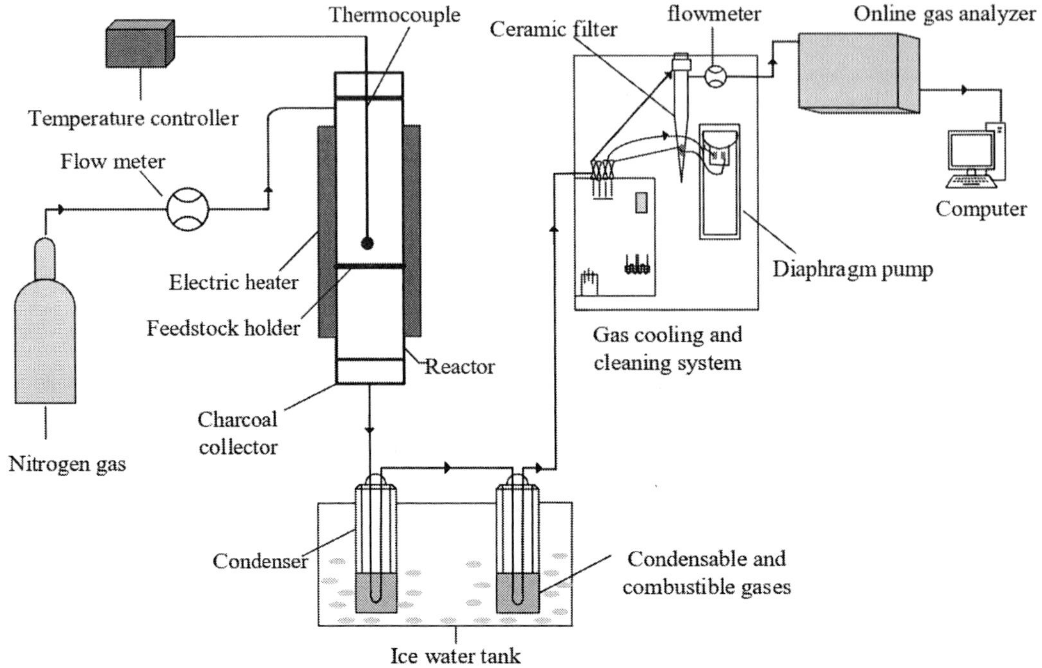

Figure 2. Experimental setup.

As the feedstock (100 g) was loaded to the pyrolysis reactor, the temperature was raised from 100 °C to desired temperature and residence time. Nitrogen gas (flow rate of 20 mL/minute) was used to obtain the inert condition and sweep away the volatile vapors. While the breakdown of the coconut shells structure continues at a specific temperature, the vapor gives off as condensable vapors and the non-condensable gases, which include carbon monoxide, hydrogen, methane, and carbon dioxide gas. The heating process was stopped at the desired temperature and residence time. The charcoal was collected at and placed in a desiccator to avoid moisture gain or loss. The gas was condensed by a condenser with an ice cooling system to collect the liquid that contain bio-oil and water bio-oil. The results were saved and the process was proceeds to the next step.

3. Results and discussion

This section reports the results and discussion of the analysis carried out. The pyrolysis of coconut shells were carried out as per standard test procedure. Temperature of 400°C was chosen to study the effect of pyrolysis parameter on the charcoal weight loss because it was the optimum and relatively highest temperature of pyrolysis for the production of charcoal as described by Basu [7].

3.1. Effect of residence time on the charcoal weight loss

Figure 3 shows the effect of residence time on percentage of weight loss of coconut shells charcoal at a constant temperature of 400 °C. It shows the higher the residence time of the pyrolysis process the higher the weight loss from the coconut shells. Figure 4(a), shows that feedstock partially pyrolyzed, and weight reduction was 48.88%. Some of the feedstock has not undergone complete process due to the short residence time, which was 15 minutes. Eventually increasing the residence time from 35 to 45 minutes as shown by figure 4(b) and figure 4(c), respectively, it slightly changes the weight loss of the coconut shells charcoal from 65.15% to 65.3%. This could be due to nearly completion of carbonization stage. Residence time of the product in the reactor is an important factor [7]. Duration of carbonization process has a significant effect on the development of the carbon's porous networks. The time should be sufficient to eliminate all the moisture and most of the volatile components in order to allow the formation of pores[17].

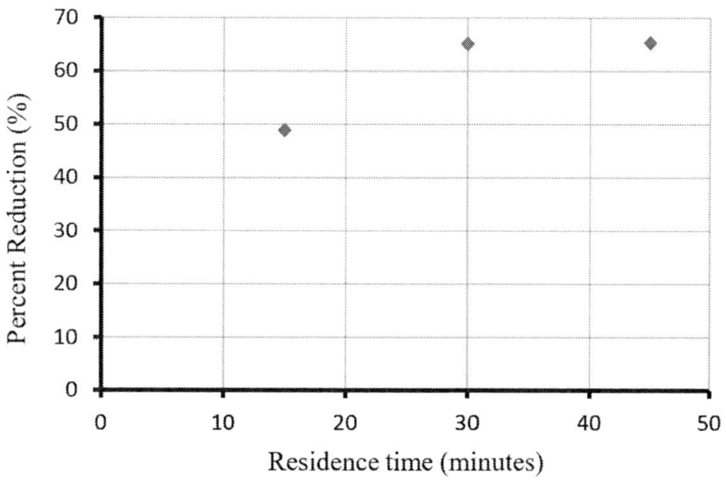

Figure 3. Effect of residence time on the weight loss at 400 °C pyrolysis temperature.

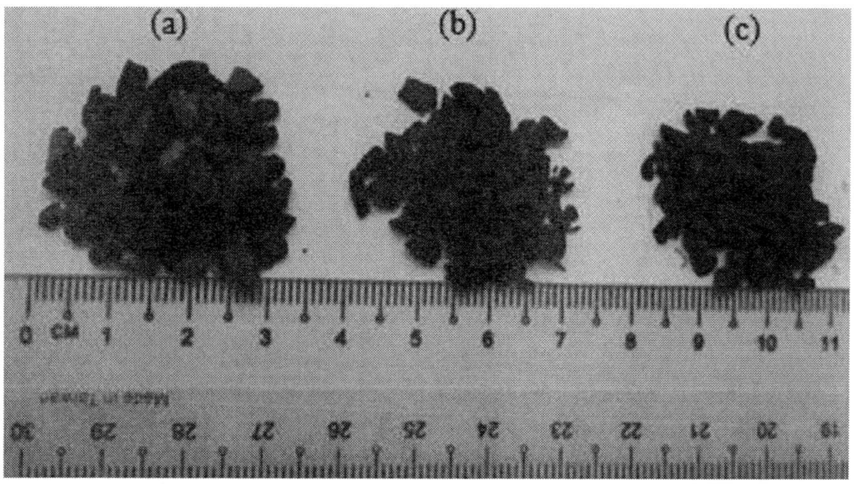

Figure 4. Coconut shells charcoal at 400 °C pyrolysis temperature at (a) 15 minutes residence time, (b) 30 minutes residence time, and (c) 45 minutes residence time.

3.2. Effect of temperature on the charcoal weight loss

Figure 5 shows the effect of temperature on percentage of weight loss of coconut shells charcoal at a constant residence time of 30 minutes. At 300 °C as shown in figure 6(a), the percentage of the weight loss is lower when compared to 400 °C in figure 6(b) and 500 °C in figure 6(c). This could be as a result of the temperature difference. This shows the process has not completed at low temperature. When the reactor temperature was 400 °C and 500 °C, the percentage weight loss indicates an increase of weight loss from 63.9% to 71.33%. This indicates that the carbonization has nearly reached the maximum pyrolysis temperature. Reaction temperature plays an important role in slow pyrolysis thermal decomposition process. It is necessary to bring the biomass particles to an optimum temperature and minimize their exposure to higher temperatures that favor other product. Feedstock particle is heated at a definite rate from ambient to a maximum pyrolysis temperature. The feedstock is held in the reactor until the operation is over. The pyrolysis temperature affects both composition and yield of the product [7, 18].

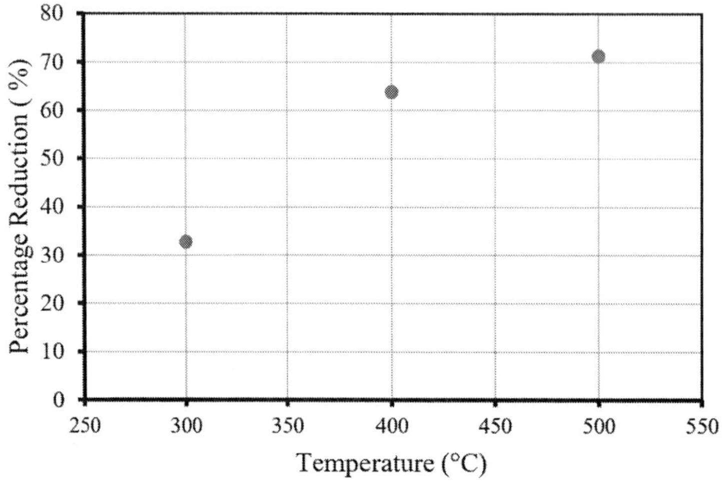

Figure 5. Effect of temperature on the weight loss at 30 minutes residence time and 25 mm particle size.

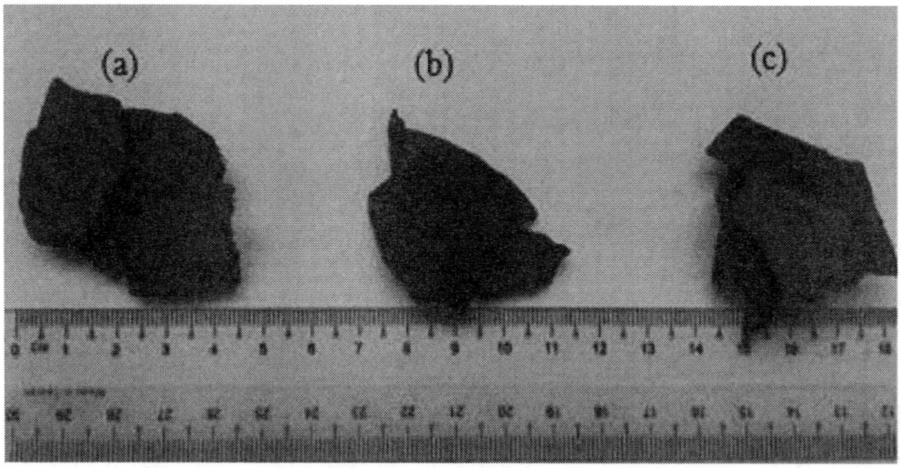

Figure 6. Coconut shells charcoal at (a) 300 °C, (b) 400 °C, and (c) 500 °C pyrolysis temperature with 30 minutes residence time and 25 mm particle size.

3.3. Effect of particle size on the charcoal weight loss

The effect of particle size on the charcoal weight loss can be seen clearly from figures 3 and 5. It shows that the weight loss from figure 3 is higher than that of figure 5 at the same temperature and residence time which are 65.15% and 63.9%, respectively. This shows that the heat transfer in smaller particles is uniform. Particle size greatly affects the heating rate of solid fuel. It is an important parameter which controls the rates of drying and primary pyrolysis process during biomass decomposition. The composition, size, shape, and physical structure of the biomass affect the heating rate of pyrolysis product. Fine particles have a uniform heating rate due to negligible extra particle and intra particle heat transfer resistance, this causes the drying and primary pyrolysis occur more or less uniformly throughout the particles. As a result, it enables the primary volatiles and moisture leave the particle with minor interaction between each other and limits the extent of the secondary reactions inside the particle [7, 19].

4. Conclusion

This study sets out to determine the effect of pyrolysis process conditions (temperature, residence time, and particle size) on the charcoal weight loss. It can be concluded that, higher the residence time and the temperature of the pyrolysis process, the higher the weight loss from the coconut shells. The particle size has influence also on the percentage weight loss based on this study. The results obtained can highlight the use of optimum conditions for charcoal production in terms of its production behaviour and the process efficiencies by knowing how it interacts at various operating conditions. This is important in designing a reactor system for its production. This process can greatly reduce the issue of waste disposal and emission of greenhouse gases, thereby reducing the environmental health issues as the coconut shells cannot be disposed rather it can be pyrolyzed.

References

[1] Sulaiman S A, Inayat M, Basri H, Guangul F M and Atnaw S M 2016 *Journal of Mechanical Engineering and Sciences (JMES)* **10**:2177-87.
[2] Jafri N, Wong W Y, Doshi V, Yoon L Y, Cheah K H 2018 *Process Safety and Environmental Protection* **118**:152-66.
[3] Bajwa D S, Peterson T, Sharma N, Shojaeiarani J, Bajwa S G 2018 *Renewable and Sustainable Energy Reviews* **96**:296-305.
[4] Meng N, Dennis Y C L, Michael K H L and Sumathy K 2006 *Fuel Processing Technology* **87**:461 – 72.
[5] Moore G R, Rlankenhorn P R, Beall F C and Kline D E 1974 *Wood and Fibre* **6**.
[6] Lavras M G. Mechanical and physical properties of Eucalyptus charcoal from pyrolysis under different conditions: Universidade Federal de Lavras; 2016.
[7] Basu P 2010 *Biomass Gasification and Pyrolysis: Practical Design and Theory*.(Elsevier, The Boulevard, Langford Lane Kidlington, Oxford, OX5 1GB, UK: Elsevier Inc. All rights reserved).
[8] Fardhyanti D S, and Damayanti A 2017 *International Journal of Chemical and Molecular Engineering* **11**:651-4.
[9] Oyedun O A, Lam K L and Hui C W 2012 *Chinese Journal of Chemical Engineering* **20**:455-60.
[10] Action P. *Fuel from the Fields: Charcal from Agricultural Waste*. 2009.
[11] Tippayawong N, Saengow N, Chaiya E, and Srisang N 2010 *International Journal of Energy and Environment (IJEE)* **1**:911-8.
[12] Clarke G C. Analysis of charcoal production systems in Guyana. 2010.
[13] Kongnine D M, Kpelou P, Baneto M and Napo K 2018 *International Journal of Advance Reasearch* **6**:234-8.
[14] Yerima I, and Grema M Z 2018 *The Journal of Middle East and North Africa Sciences* **4**:11-5.
[15] Zafar S 2016 *Bio Energy consult*. Available from: https://www.bioenergyconsult.com/coconut-biomass.
[16] ASTM. Standard Test Method for Moisture Analysis of Particulate Wood Fuels ASTM E871 1982(R2013). 2014.
[17] Gratuito M K B, Panyathanmaporn T, Chumnanclang R A, Sirunantawittaya N, Dutta A 2008 *Bioresource Technology* **99**:4887-95.
[18] Guangul F M, Sulaiman S A and Raghavan V R 2012 *AIP Conference*.
[19] Neves D, Thunman H, Matos A, Tarelho L, and Gómez-Barea A 2011 *Progress in Energy and Combustion Science* **37**:611-30.

Fluidized bed drying of stingless bee pot-pollen: Performance of swirling distributor

L A Halim, M F Basrawi*, A S M Yudin and S N Faizal

Meliponini Engineering Laboratory (MePEL), Faculty of Mechanical and Automotive Engineering Technology, Universiti Malaysia Pahang, 26600 Pekan, Pahang, Malaysia.

*E-mail: mfirdausb@ump.edu.my

Abstract. Pot-pollen is pollen collected by stingless bee, mixed with honey and bee secretion, and then stored in cerumen pots. Raw pot-pollen contains high moisture content which can lead to excessive fermentation and spoilage. Drying of pot-pollen is needed to preserve them. Typically, elevated temperature can increase the drying performance. However as pot-pollen is heat sensitive, heating above 40 °C is not preferable. Hence, the objective of this study is to explore drying performance of fluidized bed dryer with swirling distributor. The experiment was conducted using three different distributors, a perforated distributor, 45° and 67° swirling distributor, at 1.0 m/s and 1.5 m/s superficial air velocities. The weight of pot-pollen used in this experiment is 50 grams. The drying experiment was conducted at ambient conditions, with no heating being applied to the pot-pollen samples. It was found that 67° swirling distributor at 1.5 m/s superficial air velocity has the highest improvement in terms of drying performance, able to reduce the moisture content from 30.5 % to 18 % for 30 minutes drying time. Hence, using swirling distributor is a suitable enhancement to fluidized bed drying especially for heat sensitive food materials.

1. Introduction

Stingless bees are native to Malaysia and locally known as "kelulut". They are important pollinator especially in tropical region such as Malaysia [1]. As stingless bee is usually for its honey, meliponiculture or beekeeping tend to neglect other bee products. Similar to honey bees, stingless bee collect pollen from variety of plant species [2]. The pollen is mixed with honey and bee secretion before stored in cerumen pots, coining the term pot-pollen. Often used interchangeably with bee bread (which refer to honey bee stored pollen), the term pot-pollen exclusively refer to pollen stored in stingless bees nest [3]. Pot pollen has high nutritive value as it contain high level of proteins, carbohydrates, lipids, vitamins and phenolic compounds [4]. Besides, pot pollen has been shown to have therapeutic, and medicinal benefits [5–10].

Raw pot-pollen are not usually harvested by farmers due to difficulty in preserving and storing them. This is due to its high moisture content which can lead to excessive fermentation and spoilage [11]. Thus, drying of pot-pollen is needed to preserve and store them. Dried pot-pollen can then be consumed or processed further into various products. One of the method is using fluidized bed dryer. Fluidized Bed Dryer (FBD) employs the concept of fluidization to dry solid particles. FBD is currently considered as the most effective and practical method to dry solid particles [12]. In terms of technology, FBD is the fourth and latest generation of dehydration technology. FBD are widely used for drying of wet particles and grains owing to their advantages over conventional drying techniques. This is because of its advantages compared to conventional drying.

Meliponiculture or stingless bee farming as an industry is growing steadily in Malaysia. Hence, some technological development is needed in order to develop them further. At present, few research on technical and engineering application for meliponiculture especially in Malaysia. The research focuses more on hive temperature regulation [14, 15], honey processing [16, 17] as well as engineering economic analysis for stingless bee keeping [18]. However, there is little study done on pot-pollen drying. This is critical in order to increase the value of pot-pollen through suitable drying method.

In a fluidized bed, drying temperature is the most significant parameter towards drying rate [19–21]. However, pot-pollen is a heat sensitive food material. Drying temperature for pot-pollen is to be kept below 40 °C to retain its nutritional properties [22]. Swirling flow in a fluidized bed dryer can improve particulate mixing, increase mass and heat transfer [23, 24]. This can be done without applying heating to the bed material.

Thus, the objective of this paper is to study the performance of swirling distributor compared to perforated distributor for fluidized bed drying of stingless bee pot-pollen, without any heating, and at different levels of superficial air velocities.

2. Methodology

The drying experiments were conducted at the Meliponini Engineering Laboratory (MePEL), Faculty of Mechanical and Manufacturing Engineering, Universiti Malaysia Pahang (UMP). Stingless bee pot-pollen sample used is sourced from a local stingless bee farm. Prior to experiment, the pot-pollen sample is cleaned and stored at -10 °C. The schematic of the FBD used is shown in figure 1.

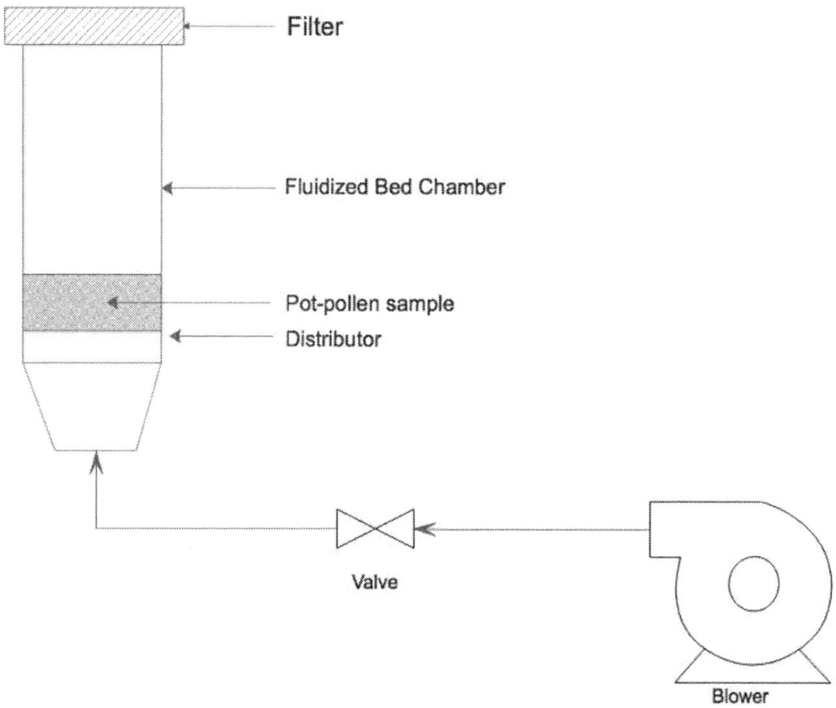

Figure 1. Schematic of fluidized bed dryer for pot pollen drying.

Three different distributor types were tested in this experiment. First, a perforated distributor, followed by a 45° swirling distributor, and 67° swirling distributor as shown in figure 2.

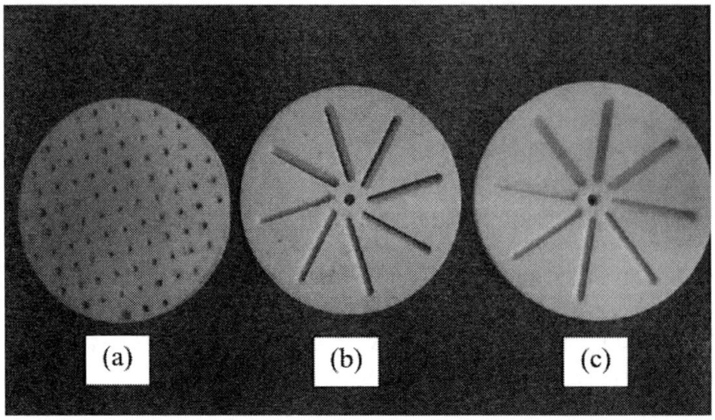

Figure 2. Type of distributor (a) perforated, (b) 45° swirling, and (c) 67° swirling.

For each experiment run, 50 grams of raw pot-pollen was weighed using a mass balance (EL-02H). The pot-pollen then was dried for 30 minutes at superficial air velocity of 1.0 m/s, measured using a hot air anemometer (YK-2400AH). The experiment was repeated for all three different types of distributor, and at two levels of superficial air velocities, 1.0 m/s and 1.5 m/s.

The moisture content of raw pot-pollen and dried pot-pollen samples were determined using hot air oven method. Each pot-pollen sample was weighed to 2 grams before put in hot air oven (DZF-6050) at 100 °C for 24 hours. The moisture content level will be the difference of the initial mass and the dried mass as shown in equation 1 [25].

$$Moisture\ content, MC\ [\%] = \frac{A - B}{B} \times 100 \qquad (1)$$

where, A is the initial mass in g of pot-pollen sample before oven drying and B is the final mass of pot-pollen after dried in the oven.

3. Results and discussion

Figure 3 and Figure 4 show the moisture content level of stingless bee pot-pollen dried at 1.0 m/s and 1.5 m/s, respectively. The reduction in moisture content represents the drying rate for each drying conditions.

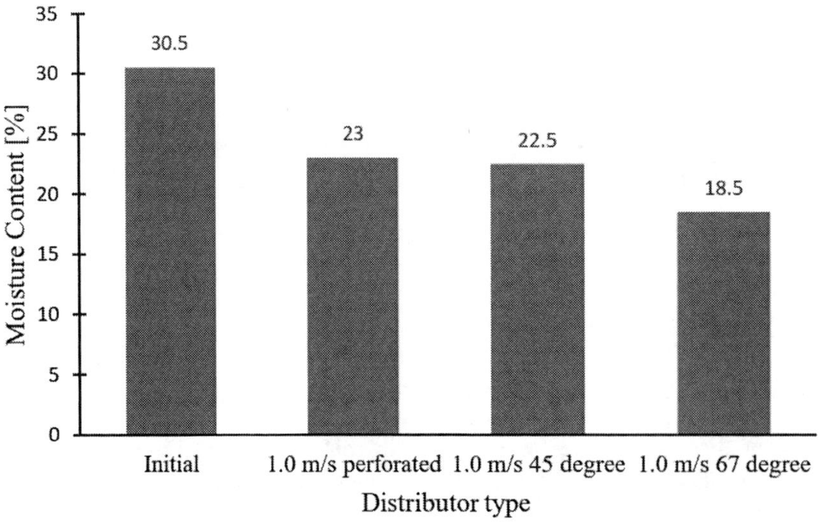

Figure 3. Moisture content for pot-pollen dried at 1.0 m/s.

The moisture content for the raw pot-pollen used for the drying experiment has been determined to be at 30.5%. This is common range of moisture content for stingless bee pot-pollen [3, 26–28].

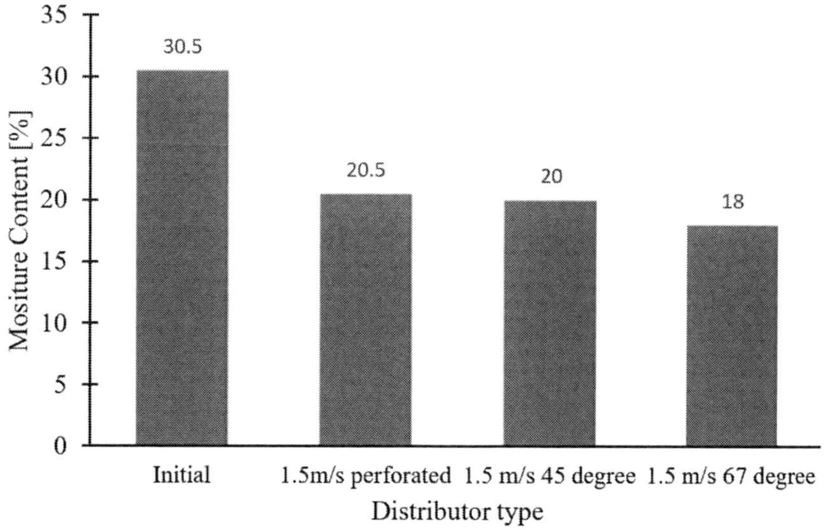

Figure 4. Moisture content for pot-pollen dried at 1.5 m/s.

The moisture content for the raw pot-pollen used for the drying experiment has been determined to be at 30.5%. This is common range of moisture content for stingless bee pot-pollen [3, 26–28]. The drying rate is represented by the final moisture content level of dried pot-pollen sample. All the conditions tested have managed to reduce the moisture content level. At both air velocities tested, 67° swirling distributor has the best drying performance, followed by 45° swirling distributor and perforated distributor. This can be explained by swirling flow enhance the mass and heat transfer during drying.

Increasing superficial air velocity also improves the drying rate for all distributor types. For perforated distributor, increasing air velocity result in additional 2.5% decrease in moisture content. In contrast, the effect is less significant with the 67° swirling distributor where only 0.5% difference when air velocity is increased. Such result can be attributed to the pot-pollen has reached the region of falling drying rate [29, 30].

The most significant reduction are from the 67° swirling distributor at 1.5 m/s, where the moisture content has been reduced by a staggering 12.5%. Thus, the drying process is done for only 30 minutes, this is considered as very quick drying process. In contrast, perforated distributor at 1.0 m/s has the least reduction, at 7.5% lower than initial moisture content. Hence, 67° swirling distributor has the best performance across the three distributor types tested. Thus, it is possible to enhance the drying performance of fluidized bed dryer without heating by using swirling distributor.

4. Conclusion

In this investigation, 50 grams of stingless bee pot pollen sample is dried using fluidized bed dryer for 30 minutes using three different distributor types and at two levels of superficial air velocities. From the three types of distributor tested, it was found that 67° swirling distributor has the best drying performance, where it managed to reduce the moisture content of raw pot-pollen sample from 30.5 % to 18.5 % and 18.0 % at 1.0 m/s and 1.5m/s, respectively. Hence, for heat sensitive food material such as stingless bee pot-pollen, drying rate can be improved by using swirling distributor in the fluidized bed dryer.

Acknowledgments
The authors would like to thank Universiti Malaysia Pahang for providing assistance for this research through the access of Post Graduate Research Scheme PGRS190342 and Ministry of Higher Education grant RDU190192

References
[1] Mohd N, Mohd F, Sajap A S, Rosliza J and Suri R 2010 *Int. Semin. Enhanc. Funct. Biodivers. Relev. to Sustain. Food Prod. ASPAC* 1–11
[2] Ghazi R, Zulqurnain N S and Azmi W A 2018 *Pot-Pollen in Stingless Bee Melittology* (Cham: Springer International Publishing) pp 77–88
[3] Vit P, D'Albore G R, Barth O M, Peña-Vera M and Pérez-Pérez E 2018 *Pot-Pollen in Stingless Bee Melittology* (Cham: Springer International Publishing) pp 361–75
[4] Urcan A C, Marghitas L Al, Dezmirean D S, Bobis O, Bonta V, Muresan C I and Margaoan R 2017 *Bull. Univ. Agric. Sci. Vet. Med. Cluj-Napoca. Anim. Sci. Biotechnol.* **74** 6
[5] Sulbarán-Mora M, Pérez-Pérez E and Vit P 2018 *Pot-Pollen in Stingless Bee Melittology* (Cham: Springer International Publishing) pp 391–9
[6] Salles J, Cardinault N, Patrac V, Berry A, Giraudet C, Collin M L, Chanet A, Tagliaferri C, Denis P, Pouyet C, Boirie Y and Walrand S 2014 B *Nutrients* **6** 5500–16
[7] Kieliszek M, Piwowarek K, Kot A M, Błażejak S, Chlebowska-Śmigiel A and Wolska I 2018 *Trends Food Sci. Technol.* **71** 170–80
[8] Abouda Z, Zerdani I, Kalalou I, Faid M and Ahami M T 2011 *Res. J. Microbiol.* **6** 376–84
[9] Akhir R A M, Bakar M F A and Sanusi S B 2017 *AIP Conference Proceedings* **1891** p 020090
[10] Feas X, Vazquez-Tato M P, Estevinho L, Seijas J A and Iglesias A 2012 *Molecules* **17** 8359–77
[11] Halim L A, Basrawi F, Chong J Z, Oumer A N, Yudin A S M, Yusof M H bin and Sulaiman S A 2018 *MATEC Web Conf.* **225** 04007
[12] Hall C W 1988 *Handbook of Industrial Drying* vol 6
[13] Vega-Mercado H, Marcela Góngora-Nieto M and Barbosa-Cánovas G V. 2001 *J. Food Eng.* **49** 271–89
[14] Ramli A S, Luqman A H, Basrawi F, Oumer A N, Aziz A A and Mustafa Z 2017 *MATEC Web Conf.* **131** 03013
[15] Basrawi F, A Hamid A R, Bahari R, Mohd Noordin M N and Yusof M H 2017 *MATEC Web Conf.* **131** 2–7
[16] Ramli A S, Basrawi F, Daing Idris D M N, bin Yusof M H, Khalil Ibrahim T, Mustafa Z and Sulaiman S A 2017 *MATEC Web Conf.* **131** 03014
[17] Ramli A S, Basrawi F, Ibrahim T K, Yusof M H Bin, Oumer A N, Johari N A, Muhamad A, Mamat M R and Habib K 2018 *Dry. Technol.* **0** 1–7
[18] Basrawi F, Ahmad A H, Daing Idris D M N, Maarof M R M, Chand M and Ramli A S 2017 *MATEC Web Conf.* **131** 04005
[19] Vizcarra-Mendoza M G, Ruiz-Martinez R S, Martinez-Vera C, Iruegas-Evaristo A and Carrillo-Guerreno J M 1998 *Dry. Technol.* **16** 1843–53
[20] Reyes A, Vega R and Garcia G 2008 *Dry. Technol.* **26** 476–86
[21] Li H W and Guo H 2017 *Powder Technol.* **308** 451–60
[22] Akhmetova R, Sibgatullin J, Garmonov S and Akhmetova L 2012 *Procedia Eng.* **42** 1822–5
[23] Yudin A S M, Anuar S and Oumer A N 2016 *Adv. Powder Technol.* **27** 2102–11
[24] Özbey M and Söylemez M S 2005 *Energy Convers. Manag.* **46** 1495–512
[25] Ismail W I W, Hussin N N, Mazlan S N F, Hussin N H and Radzi M N F M 2018 *IOP Conf. Ser. Mater. Sci. Eng.* **440**
[26] Souza B, Roubik D, Barth O, Heard T, EnrÍquez E, Carvalho C, Villas-Bôas J, Marchini L, Locatelli J, Persano-Oddo L, Almeida-Muradian L, Bogdanov S F and Vit P 2006 *Interciencia* **31** 867–75
[27] Chuttong B, Phongphisutthinant R, Sringarm K, Burgett M and Barth O M 2018 *Pot-Pollen in*

Stingless Bee Melittology (Cham: Springer International Publishing) pp 313–24
[28] De O A R M, Da S S G and Carvalho C A L 2018 *Pot-Pollen in Stingless Bee Melittology* (Cham: Springer International Publishing) pp 349–60
[29] Cubillos F and Reyes A 2003 *Dry. Technol.* **21** 1185–96
[30] Thomas P P and Varma Y B G 1992 *Powder Technol.* **69** 213–22

IOP Conf. Series: Materials Science and Engineering **863** (2020) 012007 doi:10.1088/1757-899X/863/1/012007

Effect of reaction time and microwave power on coil temperature during microwave-metal interaction pyrolysis of plastics

H Arshad[1,*], S A Sulaiman[1], Z Hussain[2] and M N Z Moni[1]

[1] Department of Mechanical Engineering, Universiti Teknologi PETRONAS, 32610 Seri Iskandar, Perak, Malaysia.
[2] Department of Chemistry, Abdul Wali Khan University, 23200 Mardan, Pakistan.

*E-mail: haroonarshad00@gmail.com

Abstract. Rapid rise in production of plastic waste has posed a threat to environment due to its nonbiodegradable nature. In recent years, microwave assisted pyrolysis of plastic waste has emerged as a promising solution toward reduction of waste and recovery of value-added products and fuels. Previous works on microwave-metal interaction pyrolysis estimated only the peak value of coil temperature at a constant microwave power without monitoring complete reaction. The current study was directed toward investigating the effect of reaction time and variable microwave power on the coil temperature during microwave-metal interaction pyrolysis of plastics. The experiment was performed on individual plastics PS (polystyrene), PP (polypropylene) and LDPE (low density polyethylene) for comparative analysis. Pyrolysis was carried out for a reaction time of 30 minutes at different values of microwave power in the range of 500-2500 W. Type K thermocouple was used to monitor temperature of metal coil. Fluctuation in temperature of coil was found to be a consequence of interaction of coil and thermocouple with the microwaves. Maximum heating rate was observed in the first 5 minutes of microwave exposure. The slopes of average temperature versus microwave power were found to be nearly equal estimated as: PS = 0.24 °C/W, PP = 0.20 °C /W, LDPE = 0.18 °C /W, which indicated consistency in heating process for each plastic, achieved by the current set-up. Further, it was inferred that nature of plastic pyrolyzed had insignificant influence on coil temperature due to absence of direct contact between plastic and the coil.

1. Introduction

Over the decades, plastic has dominated the market due to its extensive application as household items, packaging, automobiles and numerous other areas. According to statistic, demand for plastic has increased by 5% every year since 1990 [1]. This has resulted in generation of enormous amounts of plastic wastes difficult to dispose. Disposal through landfill is not much encouraged due to limited availability of land area and adverse effects on surrounding environment caused by nonbiodegradable nature of plastic waste. Plastics of widespread use mainly consists of low density polyethylene (LDPE), high density polyethylene (HDPE), polypropylene (PP), polystyrene (PS), polyethylene-terephthalate (PET) and polyvinyl chloride (PVC). Another option of plastic waste disposal is incineration. However, this method is not suitable with regard to environmental concerns. Based on previous report, incineration produce toxic emissions which are harmful to life [2]. Recycling of waste can be used to solve the disposal problem but this again requires expensive input for recovery and separation of waste components [3].

With development of modern methods of waste disposal, tertiary recycling through pyrolysis has found great application toward energy recovery from waste. Pyrolysis is a process that involves thermochemical breakdown of organic material at elevated temperatures 350 to 900 °C in oxygen free environment, resulting in oil and gas yields with char as the solid residue. The oil fraction may be separated into condensable hydrocarbons comprising paraffins, isoparaffins, olefins, naphthenes and aromatics, and a non-condensable high calorific value gas [4]. Recently, microwave assisted pyrolysis (MAP) has emerged as a promising option with added benefits of rapid and uniform heating over conventional heating. MAP of plastic waste has great potential to produce value added chemicals and fuels demonstrated in a range of studies [3–8]. A novel technique in the field of microwave assisted pyrolysis was successfully introduced by research group of Zahid et al. [9]. This method was based on microwave-metal interaction pyrolysis using metal coil as a reactor for breakdown of PS. The study proved this alternative technique to be effective and competent to previous works where dielectric material had to be used as an adsorbent of microwaves to generate essential heating.

Past studies on microwave-metal interaction pyrolysis estimated only the peak value of coil temperature at a constant microwave power without monitoring complete reaction. The current study is directed toward investigating the effect of reaction time and variable microwave power on the coil temperature during microwave-metal interaction pyrolysis of PS, PP and LDPE. This work was based on modification of experimental set-up used in previous studies on microwave-metal interaction pyrolysis [9, 11, 12]. Effort was made to estimate the temperature of iron coil through proper insertion of thermocouple aligned to touch the coil. The variation in coil temperature as a function of reaction time and microwave power was used to study the heating behavior inside the microwave field.

2. Experimental setup

The virgin plastic samples of PS, PP and LDPE were used in the experiments. The plastic was purchased from Lotte Chemical Titan (M) Sdn. Bhd. The samples were in the form of pellets of size in the range of 2 to 3 mm. The sample size for plastic pyrolysis was taken to be 20 grams in all the runs. During the preliminary study, calorific values of sample plastics were measured as: PS = 40.9, MJ/kg, PP = 46.7 MJ/kg and LDPE = 46.4 MJ/kg, respectively. The thermal degradation range of each plastic was estimated from Thermogravimetric Analysis (TGA): PS = (378-439 °C), PP = (407-483 °C) and LDPE = (420 to 499 °C), respectively. The metal coil of iron wire was fabricated with internal diameter 5 cm, height 5 cm and gauge 1.5 mm. The wire was purchased from a local market in Malaysia, commercially available as binding wire made of mild steel. Iron purity was up to 99.8% as provided by the manufacturer. The lab-scale multimode microwave oven was fabricated by Uni 10 Sdn. Bhd. that operated at a frequency of 2.45 GHz with two magnetrons rated 1.5 kW each. The microwave power was variable and continuous in the range of 250 to 2640 W. The microwave produced continuous heating based on pulse width modulation using invertor power supply. Figure 1 shows schematic of the reaction assembly. The dimensions of the heating chamber were width 40 cm, height 50 cm and depth 46 cm.

The microwave oven was modified for the microwave-metal interaction. A galvanized iron (GI) pipe was used as a reactor to place the sample. This GI pipe was installed from the top centre of the microwave chamber such that, the base of the pipe was surrounded by the iron coil placed inside a clay pot. The pot was raised to the centre of the chamber using a metallic cylindrical support. Type K thermocouple was inserted through the top of the chamber adjustable along the vertical direction and touching the coil to give temperature reading. The thermocouple probe was inbuilt in the customized design of the microwave system and the temperature reading was displayed at the control panel display screen. Clay pot provided heat insulation to the reactor and can withstand high temperatures. The function of the coil was to provide microwave flux for longer duration due to the repeated reflections of the microwaves trapped inside the coil. As a result, the GI pipe reactor was heated to high temperatures within a short time interval. The idea of using cylindrical metal support was to readily reflect the microwaves so that the essential microwave-metal interaction heating occurs only between coil and GI pipe.

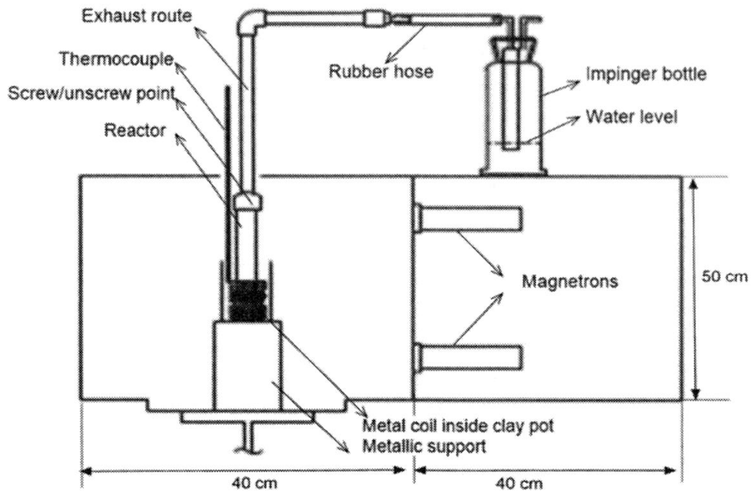

Figure 1. Microwave-metal interaction pyrolysis set-up.

Heat generated through microwave-metal interaction using iron coil was used to pyrolyze the plastic samples. Temperature of coil was monitored for a total reaction time of 30 minutes at different values of microwave power in the range of 500-2500 W and for each plastic PS, PP and LDPE, selected for study. To investigate the consistency of heating process, average temperature attained by the coil during the entire reaction at constant microwave power was determined and plotted against the variable microwave power for each plastic.

3. Results and discussion

Figure 2 shows variation of coil temperature with the reaction time and microwave power. Microwave power was increased from 500 to 2500 W to study the temperature distribution during pyrolysis of PS, PP and LDPE, respectively. It was evident from the plots that heating rate was maximum in the first 5 min of microwave exposure and later stabilized with small fluctuations until the end of reaction period. This trend was observed in all the plastic samples.

During the pyrolysis of PS, an increase in the value of microwave power, caused the slope or heating rate to shoot up rapidly. Similar behavior was observed in case of PP and LDPE. Results of variation in coil temperature at minimum and maximum microwave powers for each plastic are summarized in table 1.

Table 1. Variation in coil temperature at minimum and maximum microwave powers for each plastic.

Plastic sample	Rise in coil temperature (°C) after 5 min of microwave exposure	
	500 W (Min.)	2500 W (Max.)
PS	154	937
PP	365	748
LDPE	253	796

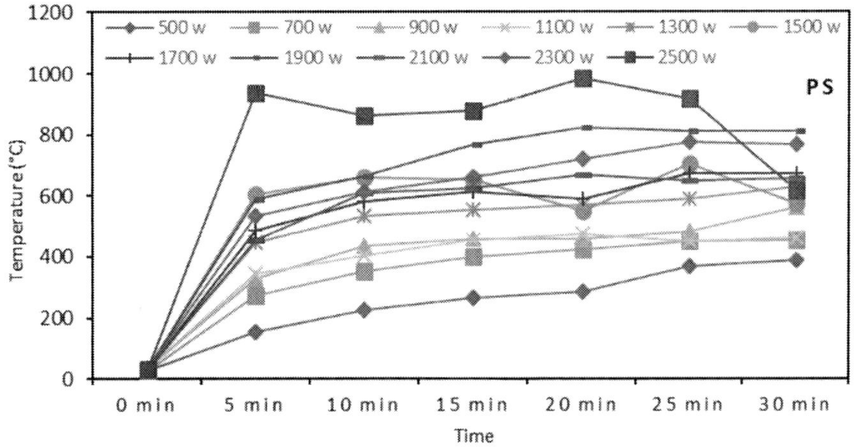

(a)

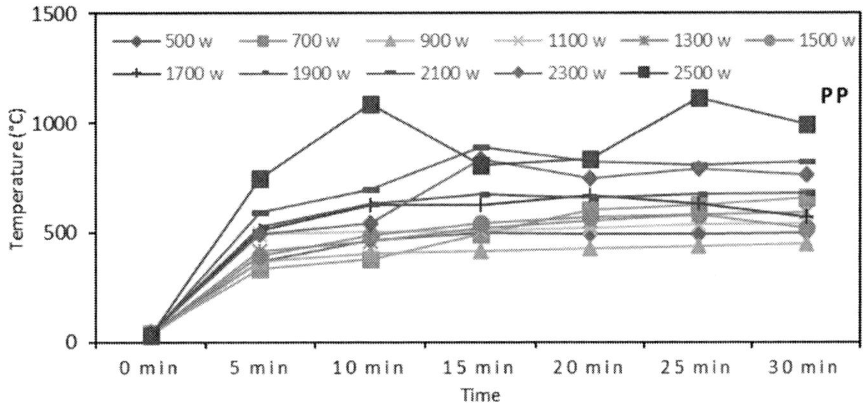

(b)

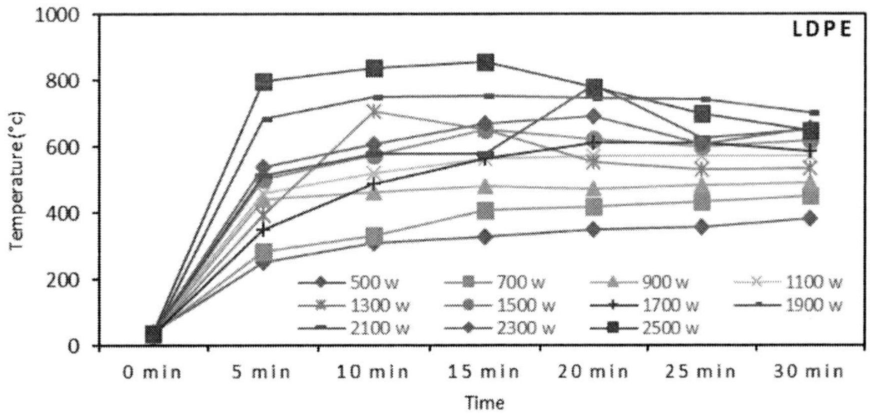

(c)

Figure 2. Variation of coil temperature with reaction time and microwave power for (a) PS, (b) PP and (c) LDPE.

The significant variation in coil temperature at same power of microwave heating in different plastic samples can be attributed to interaction of microwaves with metallic probe (Type K thermocouple), in addition to microwave interaction with iron coil. This is caused due to trips and sparks commonly encountered in microwave-metal interactions [10].

Repeated study on different plastics was carried out under the same conditions of microwave power and reaction time to reveal the distribution of fluctuations in the heating process. The production of oil during the process confirmed that temperature attained by coil was sufficient to breakdown the plastic. The process of thermal pyrolysis was carried out in the range of 300 to 700 °C.

Variation in average coil temperature with microwave power is presented in figure 3. Average temperature rise was calculated at different powers in the range of 500-2500 W for PS, PP and LDPE. It was observed that average temperature was directly proportional to microwave power in case of each plastic. Slopes of average temperature versus microwave power were estimated as: PS = 0.24 °C/W, PP = 0.20 °C /W, LDPE = 0.18 °C /W, which represented similar values with insignificant deviation showing consistency of heating process for each plastic, achieved by the current set-up. Further, it was inferred that nature of plastic pyrolyzed had insignificant influence on coil temperature. This is also obvious to the fact that plastic is not in direct contact with the coil but is heated inside the GI pipe used in the current set-up.

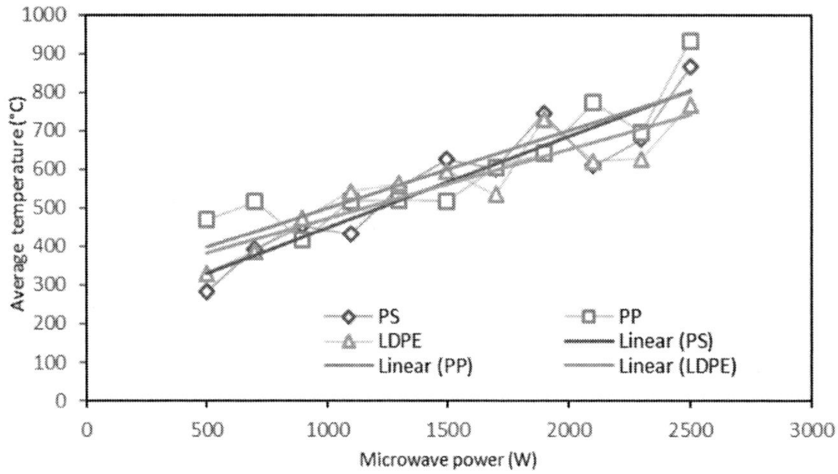

Figure 3. Variation of coil average temperature with different microwave powers and plastic samples.

4. Conclusions

The current investigation found that maximum heating occurred in the first 5 minutes of microwave exposure for entire range of microwave power between 500 to 2500 W for each plastic sample comprising PS, PP and LDPE. The average temperature directly proportional to microwave power for each plastic. Slopes of average temperature versus microwave power were estimated as: PS = 0.24 °C/W, PP = 0.20 °C /W, LDPE = 0.18 °C /W, which indicated consistency in heating process for each plastic achieved by the range of power rating 500 to 2500 W.

Acknowledgement

The work is supported by Yayasan Universiti Teknologi PETRONAS grant No. 0153AA-E06.

References

[1] Lettieri P and Al-Salem S M 2011 *Waste* 233–42.
[2] Aguado J, Serrano D P and Escola J M 2008 *Ind. Eng. Chem. Res.* **47** 7982–92.
[3] Sharuddin S. D. Anuar, Abnisa F, Daud W M A Wan and Aroua M K 2016 *Energy Convers. Manag.* **115** 308–26.

[4] Panda A K, Singh R K and Mishra D K 2010 *Renew. Sustain. Energy Rev.* **14** 233– 48.
[5] Kunwar B, Cheng H N, Chandrashekaran S R and Sharma B K 2016 *Renew. Sustain. Energy Rev.* **54** 421–28.
[6] Undri A, Rosi L, Frediani M and Frediani P 2011 *Microwave Heating* 207–32.
[7] Undri A, Rosi L, Frediani M and Frediani P 2014 *Fuel* **116** 662–71.
[8] Aishwarya K N and Sindhu N 2016 *Procedia Technol.* **25** 990–7.
[9] Hussain Z, Khan K M and Hussain K 2010 *J. Anal. Appl. Pyrolysis* **89** 39–43.
[10] Suriapparao D V and Vinu R 2015 *J. Anal. Appl. Pyrolysis* **113** 701–12.
[11] Hussain Z, Khan K M, Basheer N and Hussain K 2011 *J. Anal. Appl. Pyrolysis* **90** 53–5.
[12] Bashir N, Hussain K, Hussain Z, Naz M Y, Ibrahim K A and Abdel-salam N M 2018 *Chem.Eng. Process. Process Intensif.* **130** 140–7.

Gasification of raw and torrefied empty fruit bunch using bubbling fluidized bed

N H Abdul Halim, S Saleh and N A F Abdul Samad*

Faculty of Chemical & Process Engineering Technology, Universiti Malaysia Pahang, Lebuhraya Tun Razak, 26300 Gambang, Pahang, Malaysia.

*E-mail: asmafazli@ump.edu.my

Abstract. An experimental study of bubbling fluidized bed gasification process using empty fruit bunch (EFB) as a feedstock has been performed in order to study the effect of gasification temperature on gasification performance. Two types of EFB have been used, which are raw EFB and torrefied EFB at 300 °C. The composition of the synthesis gas, gas yield, low heating value (LHV), cold gas efficiency (CGE), and carbon conversion (CC) was determined and analyzed. In terms of synthesis gas, the composition of hydrogen and carbon monoxide gases were increased while methane and carbon monoxide gases were decreased for raw and torrefied EFB when the temperature is increased from 650 to 900 °C. Meanwhile, the synthesis gas yield was significantly enhanced and reached the maximum value of 1.21 and 1.45 MJ/Nm^3 for raw and torrefied EFB respectively, which indicates that the torrefaction process improves the synthesis gas yield. Besides, the LHV for the torrefied EFB at highest gasification temperature is 11.08 MJ/Nm^3 which is higher than raw EFB (10.81 MJ/Nm^3). However, the value of CGE and CC of torrefied EFB are lower than raw EFB with the different about 3% and 5% respectively.

1. Introduction

Currently, the world is depending on the fossil fuel as the energy resources which mostly utilized as the transportation fuel. Besides, the utilization of the fossil fuel itself will contributes to the climate changes and global warming. Thus, it is important to find another alternative source in order to reduce the dependency of fossil fuel. Biomass is one of the bioenergy resources which provide clean and green renewable energy for the consumers. Biomass is an organic matter which derived from living things and it can be converted into fuels or useful chemical by biological and thermochemical conversions. The oil palm waste is one of the biomass resources which consists of empty fruit bunch (EFB), palm kernel shell (PKS), palm mesocarp fibre (PMF) and oil palm fronds (OPF). The palm oil industries in the world or even in Malaysia generate over millions ton of wastes either in solid or liquid residues. From palm oil plantation, only 10% of them are being utilized while 90% of them are wastes [1]. However, approximately only 10% of the wastes are utilized commercially for new products like bio-fertilizers [2]. Currently, PKS and PMF are being utilized for the steam production as fuels while the rest are being left rotten. Therefore, a proper waste management and utilization is necessary in order to reduce the amount of waste and also to maintain the sustainability of the environment [3]. One of the methods is using thermochemical conversion technology such as biomass gasification which converting solid biomass into liquid and gases product through series of oxidation reaction.

Gasification is one of the processes used to produce synthesis gases which consist of hydrogen (H_2), carbon monoxide (CO), methane (CH_4) and carbon dioxide (CO_2). Gasification process usually is conducted at high temperature (>600 °C) with the present of the gasifying agent such as air and/or steam using the gasifier. There are several types of gasifier that can be used to conduct the gasification process. One of the widely used gasifier is bubbling fluidized bed since it can gives good temperature distribution and higher reaction rates compare to other gasifiers [4]. Even though it has complex operation, it also has large fuel inventory which provides safety, reliability and stability. In addition, the bubbling fluidized gasifier gives more advantages as it gives higher cold gas efficiency as well as specific capacity and lower residence time which suitable and easier to evaluate the gasification performances.

Although many of gasification experiments have been performed but it is important to note that most of previous studies use raw biomass as the gasification feedstock. However raw biomass usually content high moisture content and low values of fixed carbon and ash which reduce the efficiency of the gasification performances. In order to improve the quality of the biomass, the pre-treatment process is needed by using the torrefaction method [5]. This method removes the moisture content of the biomass by carrying out at torrefaction temperature in the range of 200 – 300 °C under the absence of the oxygen which resulting into increment of fixed carbon, volatile matter and ash content [6]. As consequence, torrefied biomass is considered more uniform compared to raw biomass and make it more easier to be utilized as feedstock in the gasification process [7]. This study focuses on utilization of two different types of EFB namely raw and torrefied EFB at 300 °C, as the biomass feedstock. The objective of this work is to study the production of synthesis gas and the gasification performances of raw and torrefied EFB by varying the gasification temperature. The synthesis gas composition is evaluated and the gas yields, lower heating value (LHV), cold gas efficiency (CGE) and carbon conversion (CC) are determined and analyzed by varying the reactor temperature from 650 to 900 °C.

2. Materials and methods

2.1. Sample preparation

The biomasses used in this study were raw Empty Fruit Bunch (EFB) and torrefied EFB. The raw EFB were obtained from Lepar Hilir Palm Oil Mill, Kuantan, Pahang. Initially, the biomass samples were dried using an oven for about 4 hours at 105 °C to reduce the moisture content and also to avoid biomass degradation that influence the quality of the sample. Then, the sample were grinded and sieved to the particle size of 0.5 to 1.0 mm. Accordingly, the torrefied EFB has been obtained based on torrefaction experiment of raw EFB at temperature of 300 °C with a residence time of 30 minutes. The torrefaction process was carried out using a vertical-stainless steel reactor with 39.7 cm long and 1.9 cm internal diameter as shown in figure 1. The grinded and dried sample was placed in the reactor. The reactor was flushed for 5 minutes with 10 mL/min nitrogen in order to create the inert atmosphere. After that, the biomass sample was heated by an electric furnace to the desired temperature of 300 °C for 30 minutes and then it was cooled to ambient temperature. Finally, the torrefied biomass was placed in air-tight container to keep the value of the moisture content from increasing [8].

The properties of proximate analysis, ultimate analysis and high heating value (HHV) for raw and torrefied EFB are shown in table 1. The moisture content is reduced after the torrefaction process from 15.77 wt% to 4.63 wt% due to the drying process which contributing to the loss of the water content in the biomass. Besides, the value of the volatile matter also decreased from 67.01 to 48.44 wt%. Meanwhile, the value of fixed carbon and ash are increased due to the breakdown of hydrocarbon bonds which contributing to the volatile loss and further increasing the fixed carbon and ash content. For the ultimate analysis, the result shows that the percentage of carbon and nitrogen were increasing and the other gases shows opposite trends. Moreover, the HHV value also increased from 15.49 to 19.60 MJ/kg due to the lower moisture content and loss of more volatiles. Both results of proximate and ultimate analysis as well as HHV show that the biomass properties are improved by using torrefaction process.

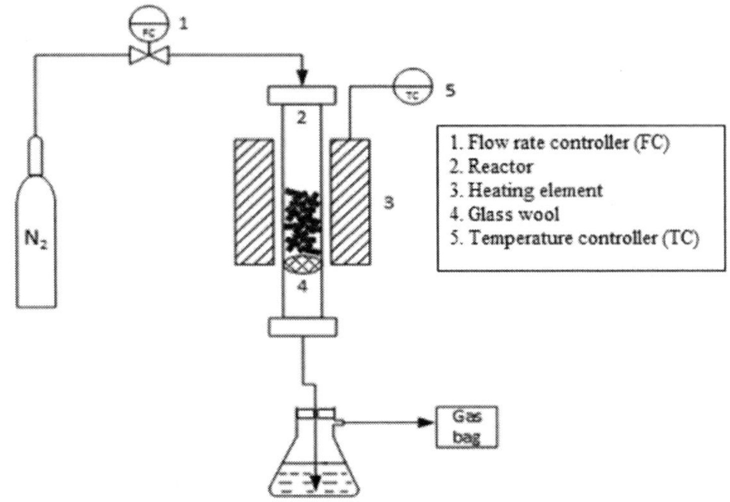

Figure 1. Schematic diagram of vertical tubular reactor.

Table 1. The proximate and ultimate analysis of the raw and torrefied EFB.

Proximate and Ultimate Analysis (wt%)	Raw EFB	Torrefied EFB at 300 °C
Moisture Content	15.77	4.63
Volatile Matter	67.01	48.44
Fixed Carbon	13.37	39.23
Ash	3.85	7.70
C	43.53	54.63
H	7.20	5.63
N	1.73	6.37
S	0.46	0.21
O	47.09	36.04
HHV (MJ/kg)	15.49	19.60

2.2. Gasification experimental setup

Experimental set-up used for the gasification process is shown in figure 2. The fluidized reactor was made from stainless steel with a bed diameter of 60 mm and height of 850 mm. Silica sand was used as a bed material in the fluidized bed reactor. The feedstock was fed into the reactor using a screw feeder conveyor at a rate of 0.25 to 0.4 kg/h. Steam was used as the gasifying agent and flushed into the bottom of the reactor. The fluidized bed reactor was heated to the desired gasification temperature using an electric furnace. Eight different gasification temperatures were used in this work (650, 700, 750, 800, 850 and 900 °C) to study the effects of gasification temperature on the production of synthesis gas. The produced gas exited at the top of the reactor and entered the cyclone separator for separating tar and ash. The gas exited the cyclone and underwent a cleaning and drying process using a dry ice trap and cotton filter. The dry and clean gas was then collected using a gas sampling bag and analysed using an Agilent 6890N gas chromatograph with a thermal conductivity detector (GC–TCD). The standard gas mixture was used as calibration for GC–TCD and nitrogen gas was used as the carrier gas for the analysis.

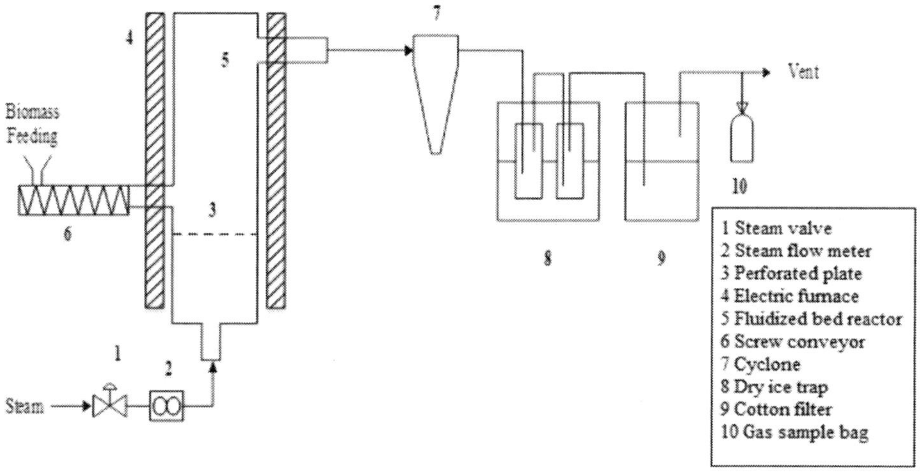

Figure 2. Schematic diagram of fluidized bed gasifier.

2.3. *Product gas analysis*

In this work, the gasification performance is evaluated by determining the synthesis gas yield, lower heating value (LHV), cold gas efficiency (CGE) and carbon conversion (CC) for both raw and torrefied EFB. The volume of the gas produced from the experimental work was used to obtain synthesis gas yield that is expressed as

$$Syn_{yield} = \frac{Vol_{gas}}{Mass_{feedstock}} \tag{1}$$

One of the energy contents in biomass is expressed as lower calorific value which includes all the sensible energy except for the heat of condensation of water. The LHV of the synthesis gas was calculated from the gas composition recorded at different temperature for both raw and torrefied EFB. The LHV of product gas is defined as

$$LHV_{yield} = (30x_{CO} + 25.7x_{H_2} + 85.4x_{CH_4}) \times 4.2 \tag{2}$$

where x is represents the mole fraction of the gas species. Furthermore, CGE is a crucial index to evaluate the performance of the biomass gasification [9]. The CGE is calculated by

$$CGE(\%) = \frac{LHV_{gas} \times Syn_{yield}}{HHV_{feedstock}} \times 100 \tag{3}$$

where HHV is the higher heating value of feedstock. In addition to CGE, carbon conversion (CC) is also analysed and expressed as

$$CC(\%) = \left[1 - \frac{\dot{m}_{product\ gas}\left(y_{CO_2}\frac{12}{44} + y_{CO}\frac{12}{28} + y_{CH_4}\frac{12}{16}\right)}{\dot{m}_{fuel}y_C}\right] \times 100 \tag{4}$$

where y is the mass fraction of the gas species. CC is the parameter that determines the solid feed converted into the permanent gases.

3. Results and discussion

3.1. *Effect of gasification temperature on synthesis gas composition*

The gas chromatography analysis was performed in order to determine each of the synthesis gas composition which are H_2, CO, CH_4 and CO_2. Figure 3 shows the effect of the gasification temperature on each of the synthesis gas produced. Based on figures 3 to 6, increment of gasification temperature from 650 to 900 °C contributes to the increase of hydrogen percentage from 15.28 mol% to 29.40 mol%

for raw EFB and 10.40 mol% to 21.50 mol% for torrefied EFB respectively. Moreover, the percentage of the carbon monoxide showed some increment for both raw and torrefied EFB which are from 36.42 mol% to 54.94 mol% and 36.74 mol% to 54.67 mol% respectively. The higher temperature is known to favour the reactants in exothermic and products in endothermic reactions [10]. Thus, the increasing of carbon monoxide can be clarified by taking into account the endothermic reduction reaction that will reach the equilibrium at higher temperature. The endothermic reactions involved in this process are water-gas shift reaction ($C + H_2O \leftrightarrow CO + H_2$ $\Delta H° = +131.4$ kJ/mol) and Boudouard reaction ($C + CO_2 \leftrightarrow 2CO$ $\Delta H° = +176.6$ kJ/mol) which produce the hydrogen and carbon monoxide that contributes to the increasing of the percentage of both gases. Meanwhile, the percentage of carbon dioxide is decreased when the reactor temperature is increased. This could be explained from the increasing of the hydrogen and carbon monoxide where the carbon dioxide would be consumed in Boudouard reaction [7]. Besides, the torrefied EFB shows higher percentage of carbon monoxide compare to raw EFB. This is due to the improvement in the carbon content of torrefied EFB from 43.5% to 54.6% as shown in table 1 which improving carbon conversion to produce more carbon monoxide. However, the torrefied EFB produced smaller amount of hydrogen, carbon dioxide and methane compared to the raw EFB with the different is lesser than 8% due to the reduction of the volatile matter from 67% to 48.4%. Generally, the biomass with lower volatile matter have lower reactivity in gasification applications. This shows that the torrefaction method is insignificant for the production of those gases.

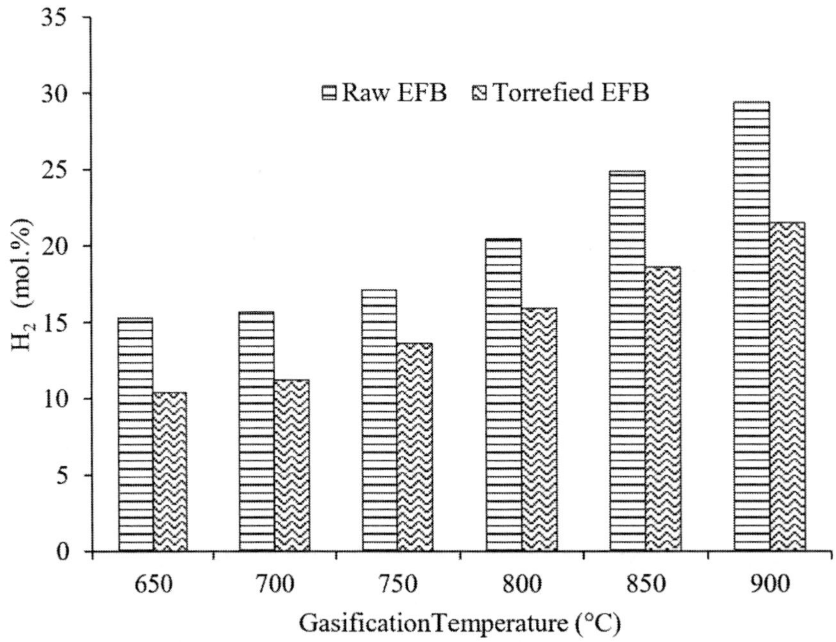

Figure 3. Composition of H_2 for raw and torrefied EFB at different temperature.

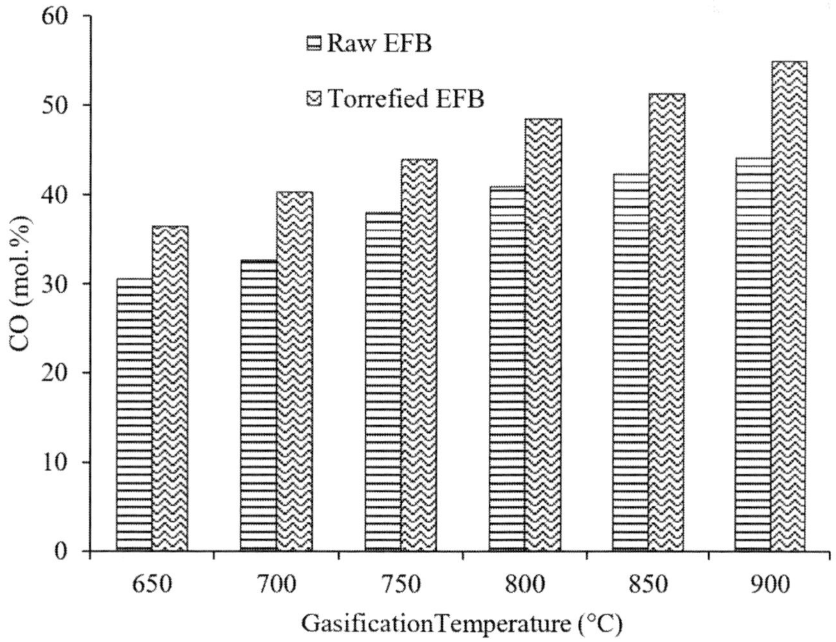

Figure 4. Composition of CO for raw and torrefied EFB at different temperature.

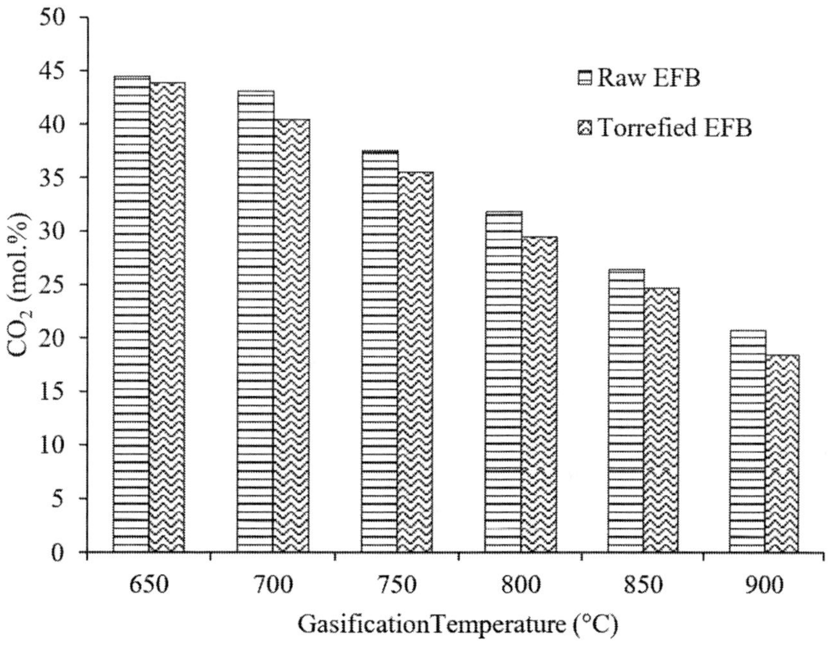

Figure 5. Composition of CO_2 for raw and torrefied EFB at different temperature.

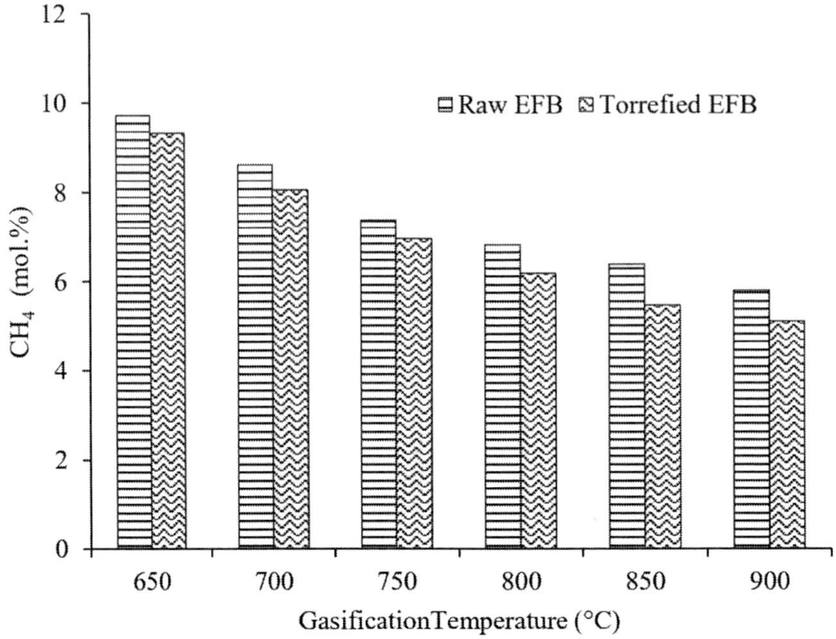

Figure 6. Composition of CH_4 for raw and torrefied EFB at different temperature.

3.2. Synthesis gas yields, lower heating value (LHV), cold gas efficiency (CGE) and carbon conversion (CC)

The synthesis gas yields, LHV, CGE and CC are essential criteria for measuring the efficiency of the biomass conversion for the gasification application. As shown in figure 7, the synthesis gas yield for both raw and torrefied EFB are increased when the temperature is increased. This is due to the further cracking of liquid and enhanced char reaction with gasifying agent which reduce the amount of liquid and char. However, the synthesis gas for torrefied EFB is higher than the raw EFB as the volume of gases produced for torrefied EFB is higher than raw EFB. In addition, the LHV for torrefied EFB is slightly higher than raw EFB which is not more than 0.3 MJ/Nm^3 as shown in figure 8. Referring to equation (4), the LHV is depending on the mole fractions of carbon monoxide, hydrogen and methane which are higher in torrefied EFB compare to raw EFB. Moreover, the carbon conversion of both raw and torrefied EFB as shown in figure 9 are increased due to the increment of the rate of devolatilization.

From the torrefaction method, the HHV for EFB has been increased from 15.49 to 19.60 MJ/kg which explains the increment of carbon conversion rate. Moreover, the CGE of torrefied EFB will be affected as well as shown on the equation (3). As shown in figure 10, the maximum CGE of raw EFB (84.26%) is slightly higher than torrefied EFB (81.88%) but both are increased with the increasing of temperature. The CGE is defined as ratio of the flow of energy in the gases and energy in biomass. Referring to equation (3), the CGE is increased when the syngas yield and LHV are increased but shown a decrement when the value HHV is increased.

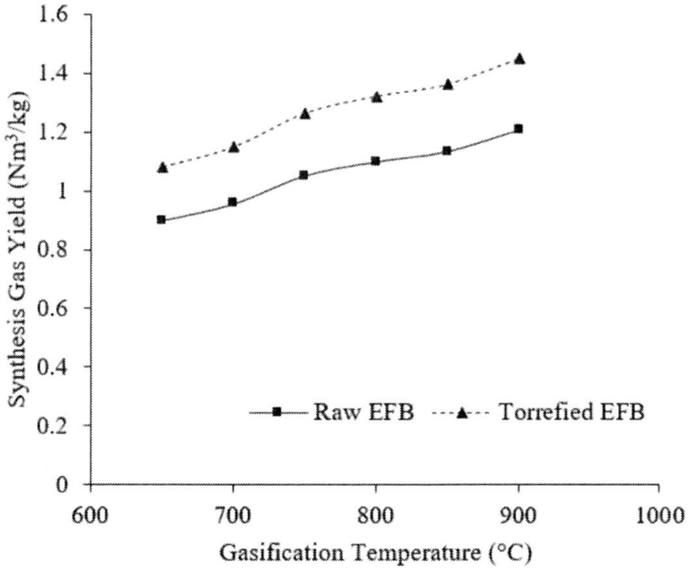

Figure 7. Comparison of synthesis gas yields between raw and torrefied EFB in increasing temperature.

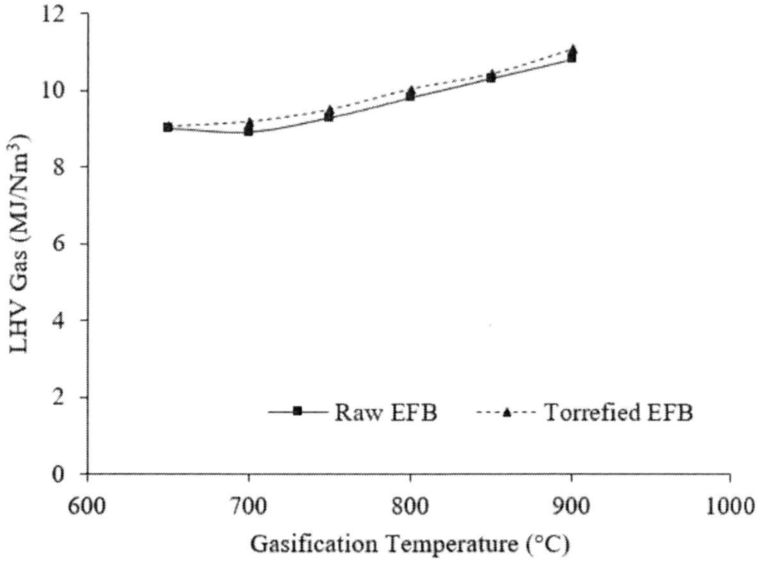

Figure 8. Comparison of lower heating value between raw and torrefied EFB in increasing temperature.

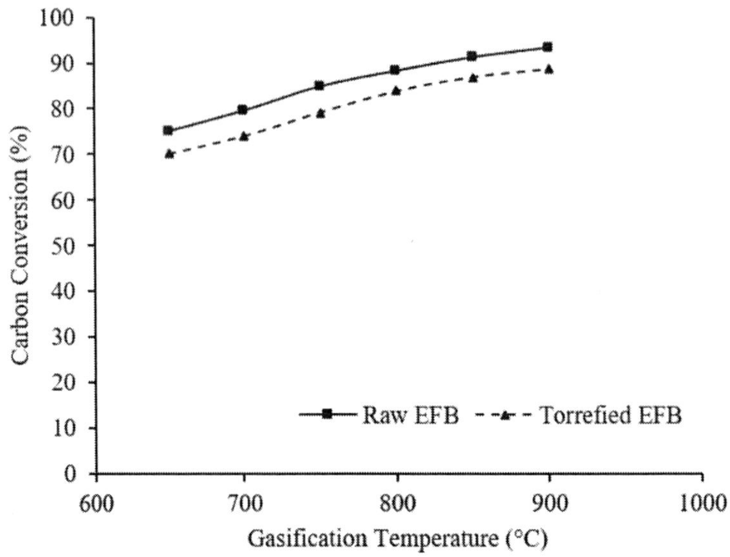

Figure 9. Comparison of carbon conversion between raw and torrefied EFB in increasing temperature.

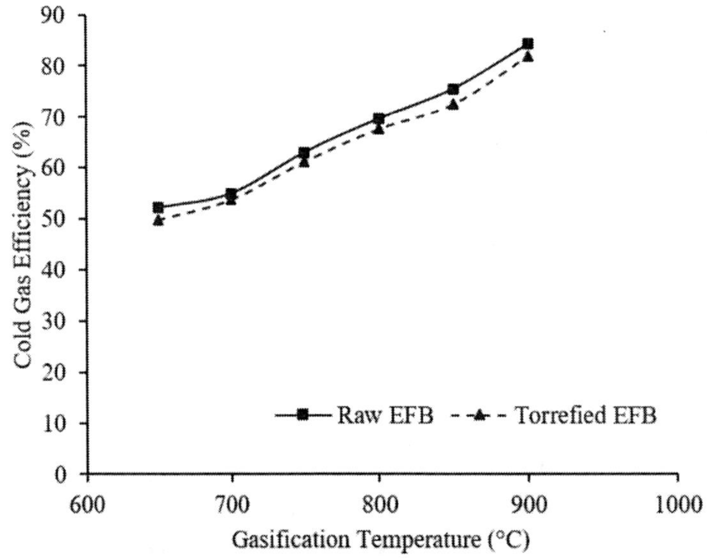

Figure 10. Comparison of cold gas efficiency between raw and torrefied EFB in increasing temperature.

4. Conclusion

Bubbling fluidized bed gasification experiments were conducted at different gasification temperature using raw and torrefied EFB as the feedstocks. Experimental results show that the increasing temperature from 650 to 900 °C would increase the performances of the gasification in terms of synthesis gas yield, lower heating value, cold gas efficiency and carbon conversion. Meanwhile, with the increasing of temperature, the compositions of hydrogen and carbon monoxide are increased while methane and carbon dioxide compositions are decreased. Torrefied EFB at 300 °C produces higher amount of carbon monoxide compare to other gases which are 54.9% at gasification temperature of 900 °C. In terms of feedstock comparison, it shows that the EFB undergo torrefaction process produces a higher synthesis gas yield and LHV compare to raw EFB. However, the value of CGE and CC for torrefied EFB are lower compared to raw EFB.

References

[1] Harun N H H M, Samad N A F A and Saleh S 2017 Development of Kinetics Model for Torrefaction of Empty Fruit Bunch from Palm Oil Waste *Energy Procedia* **105** 744–9
[2] Karl J and Pröll T 2018 Steam gasification of biomass in dual fluidized bed gasifiers: A review *Renew. Sustain. Energy Rev.* **98** 64–78
[3] Jamin N A, Saleh S and Samad N A F A 2019 Properties prediction of torrefied municipal solid waste using linear correlation model *Energy Procedia* **158** 836–41
[4] Mirmoshtaghi G 2016 *Modeling and simulation bed gasifiers*
[5] Li Y H and Chen H H 2018 Analysis of syngas production rate in empty fruit bunch steam gasification with varying control factors *Int. J. Hydrogen Energy* **43** 667–75
[6] Abdul Samad N A F, Jamin N A and Saleh S 2017 Torrefaction of Municipal Solid Waste in Malaysia *Energy Procedia* **138** 313–8
[7] Muslim M B, Saleh S and Abdul Samad N A F 2017 Torrefied Biomass Gasification: A Simulation Study by Using Empty Fruit Bunch *MATEC Web Conf.* **131** 03008
[8] Harun N H H M, Wahid F R A A, Saleh S and Samad N A F A 2017 Effect of torrefaction on palm oil waste chemical properties and kinetic parameter estimation *Chem. Eng. Trans.* **56** 1195–200
[9] Kuo P C, Wu W and Chen W H 2014 Gasification performances of raw and torrefied biomass in a downdraft fixed bed gasifier using thermodynamic analysis *Fuel* **117** 1231–41
[10] Rudolfsson M, Stelte W and Lestander T A 2015 Process optimization of combined biomass torrefaction and pelletization for fuel pellet production - A parametric study *Appl. Energy* **140** 378–84

IOP Conf. Series: Materials Science and Engineering **863** (2020) 012009 doi:10.1088/1757-899X/863/1/012009

Integrated reservoir study using well-test deconvolution analysis and well-logging data in a gas condensate carbonate reservoir

R Geravand[1,2], J Foroozesh[2,*], A Nakhaee[1,3,*] and M Abbasi[3]

[1] Department of Petroleum Engineering, Kish International Campus, University of Tehran, Iran.
[2] Department of Petroleum Engineering, Universiti Teknologi PETRONAS, 32610 Bandar Seri Iskandar, Perak, Malaysia.
[3] Institute of Petroleum Engineering, College of Engineering, University of Tehran, Iran.

*E-mail: jalal.foroozesh@gmail.com, anakhaee@ut.ac.ir

Abstract. This paper presents an integrated approach to characterize a gas-condensate carbonate reservoir using the combination of well-test deconvolution analysis, and well logging interpretation. Each set of data from the subject well, namely, petrophysical logs, production logs, surface well tests and drill-stem tests was analysed. Well test analyses was implemented using both pressure-derivative and variable-rate deconvolution techniques. An integrated analysis and interpretation of the results from different sources was performed. This approach was especially beneficial in reducing some uncertainties such as the layered nature of the reservoir, the heterogeneous behaviour, and boundary-dominated flow regimes. The application of deconvolved derivative increased the radius of investigation and duration of original constant-rate pressure from 24 hours to 140 hours, enabling us to detect the possible fault effect at late-times. This method was also used for performance evaluation of an acid treatment job in the wellbore. It was found that after acidizing, the total skin factor was severely reduced from 33 to 1.2, indicating a successful performance of the acidizing job.

1. Introduction

Identifying the rock and flow properties of a hydrocarbon reservoir is necessary in reservoir studies to manage the reservoir characteristic. Engineers have used several tools and methods in order to study the characteristic of the reservoir. The integration of various data from different resources including wireline formation tester, petrophysical logging, production logging, and drill stem test for gas condensate in carbonate reservoir is quite challenging. The integrated reservoir studies can be found in open literature source. In 1994, Domelen et al. provided integration of pressure transient analysis, core flood testing and dynamic simulation model to design and evaluate the fracture acidizing in a high-temperature carbonate formation [1]. Clarkson and Jensen conducted an integrated study to evaluate unconventional gas reservoirs [2]. Meanwhile, Noah and Shazly presented the integration of well logs and core analysis results to analyse the petrophysical properties of the rock in an oil sandstone reservoir [3]. The well-test data analysis (using deconvolution technique) together with production logging data to investigate the remedial water shut-off operation in a gas-condensate layer reservoir was implemented by Kgogo [4]. Deconvolution offers a powerful technique that has extensively improved the quality of the well-test data analysis. Gringarten presented the practical use of deconvolution, which had hardly ever been used before by engineers in well test analysis [5]. Ilk et al. and ILK applied a B-spline deconvolution algorithm to several field cases including wellbore storage dominated pressure data, multi-rate test, permanent downhole gauge data and production data [6, 7].

Experience with multi-disciplinary reservoir characterization has shown that the individual analysis of field data such as well test and logging data particularly in a complex gas condensate carbonate reservoirs is not always satisfactory and often contains some ambiguities, which needs to be confirmed with another complementary data. Furthermore, the complexity of well test analysis in such reservoirs and also the short duration of constant-pressure build-ups, have made us to consider the deconvolution approach for analysing the pressure transient data as well, that can increase the radius of investigation and help to identify the possible boundary effect dominated flow regimes at late times. In this paper, the main objective is to conduct an integrated reservoir study and characterization using a combination set of data including drill stem test (DST) as a well test tool, wireline formation tester, production logging and petrophysical logging from a gas condensate carbonate reservoir in Middle East. This study also intended to obtain a more reliable result that can contributes in the future practice with more accurate estimation of the well condition and reservoir specification for the well performance prediction and nearby wells/field development plan.

In the following sections, the methodology used in this work is discussed first. Then, the individual and integrated well-testing analysis and well logging interpretation in a vertical well in the gas condensate field will be elaborated. Finally, the conclusions will be presented.

2. Methodology

In this paper, the commercial softwares of Ecrin-Saphir, Emeraude and Geolog for well test analysis, production logging and petrophysical logging interpretation, respectively have been used. In this work, each single reservoir data from the subject well was individually analysed. Then, the integration of all the interpretation results performed in order to achieve more accurate findings. Well test analysis implemented using both pressure-derivative and variable-rate deconvolution methods. Deconvolution process can convert the observed variable-rate pressure response into a constant-rate pressure response, which makes more data available for analysis compared to the original data set [8-11]. The pressure/rate deconvolution model is given by the well-known convolution integral:

$$p(t) = p_0 - \int_0^t q(\tau)\frac{dp_u(t-\tau)}{dt}d\tau \tag{1}$$

In the equation above, $p(t)$ and $q(t)$ are the measured pressure and flow rate, and p_0 is the reservoir initial pressure. The deconvolved curve $p_u(t)$ corresponds to the constant/unit rate drawdown pressure response [12-14].

3. Results and discussion

A series of data collected from the reservoir layer of an appraisal/vertical well in a gas condensate carbonate reservoir, are analysed here. The reservoir interval is from depth 2909-3027 m (the reservoir thickness is 118 m), which was perforated uniformly from depth 2920-3015 m. In this section, firstly the interpretation results for the wireline formation tester and petrophysical logging are presented and then the well test analysis and production logging are discussed as well.

The formation pressure test has been run to record the sand-face pressure of the well. Figure 1 shows all 14 test points along with petrophysical logs including resistivity, lithology and porosity log across the open hole logging interval. This plot (from left to right) presents Gamma Ray (GR) in track 1, True Vertical Depth in track 2, formation quartz-pressure (QCP) in track 3, drawdown mobility with blue colour points and resistivity curves in track 4 and neutron, Photoelectric (PE)and density in track 5. The interpretation results identified that most of the test points have relatively high mobility while interval 2968-2972 m has the highest mobility in comparison to other zones. Gas gradient 0.108 has been fitted on test points with moderate quality points in zone 2945-2972 m as shown in figure 1, track 3. Below 2980 m, the test points are scattered and no gradient line can be fitted on them. From petrophysical log interpretation, it can be inferred that the reservoir layer mostly consisted of dolomite, limestone with some portion of anhydrite, which divided the reservoir zone into three sub-zone (upper, middle and lower). The middle zone (2945-2972 m) has the highest reservoir quality, which is mostly consisted of limestone-dolomite.

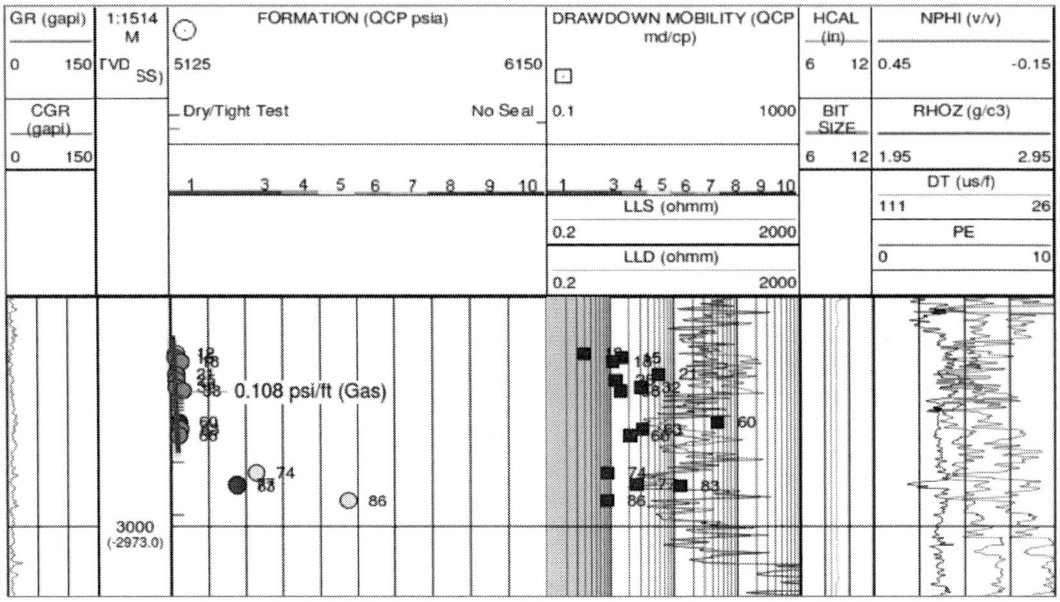

Figure 1. Test points from wireline formation tester along with open hole petrophysical logs.

DST has been conducted to determine the well and reservoir parameters including (kh product, skin factor, and reservoir pressure) and also to characterize reservoir behaviour such as layering, heterogeneities and flow regimes diagnosis. It is worth for mentioning that a stimulation acid job has been done during DST, to remove the formation damage. The acid treatment job has been followed by post-acid clean-up, intermediate build up, flow after flow test and the final build up respectively, as shown in the history plot in figure 2.

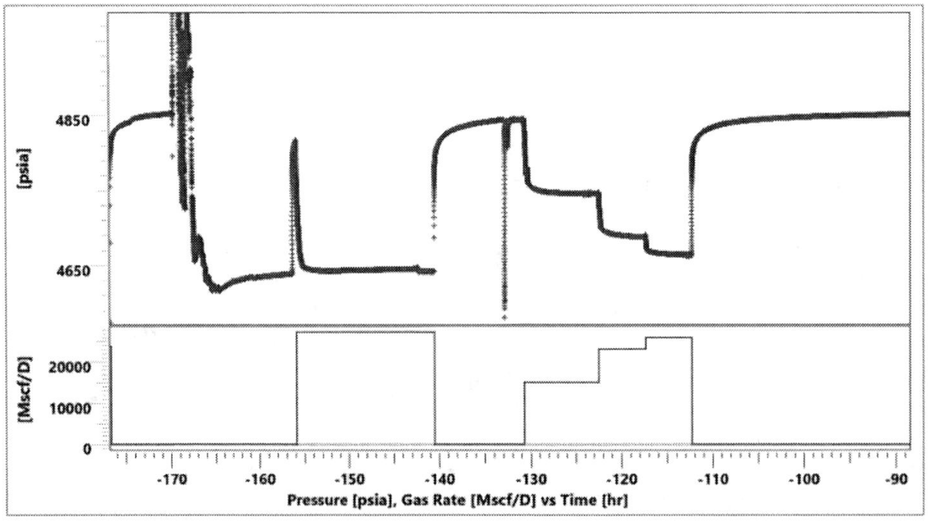

Figure 2. Drill stem test history plot.

There were three main build up (BU) which needed to be considered, initial BU before acidizing, intermediate BU and final BU after acid treatment job. At initial and final BU, the well had been shut-in at the surface which shows higher wellbore storage effects in comparison to intermediate BU in which well is shut at downhole as shown in figure 3(a). The wide separation between initial BU (before acid) and intermediate or final BU (after acid) in both pressure-derivative and semi-log type-curves signifies the high positive skin before stimulation in the well as shown in figure 3(a) and (b). The total skin factor is reduced from 33 to 1.2 after acidizing. Then, the infinite acting radial flow, IARF (#1) occurs at earlier time in intermediate BU compared to initial and final BU, as the graph shows the IARF (#2) at a later time because of the wellbore storage effect (Figure 3(a)). At final BU with duration 24 hr, two radial flow stabilization can be seen at late time, however, it needs to be verified with other reservoir data. Moreover, the initial and intermediate BU periods are not long enough to see the heterogeneous behaviour. The model interpretation of final BU is supposed to be two layers with two kh product in which the skin factor is around 0.9 and 9.8 for each zone indicating that one layer has a high skin. The total kh product for the reservoir system is estimated to be around 3500 md.ft.

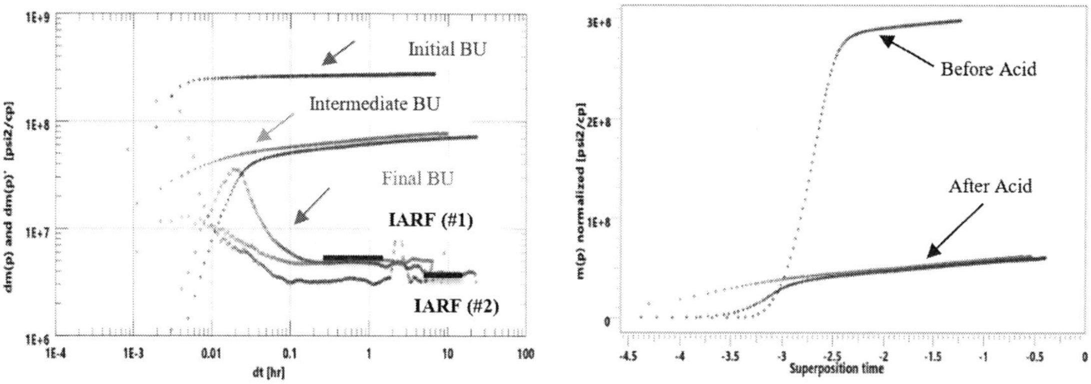

(a) (b)

Figure 3. (a) Log-log pressure derivative plot and (b) Semi-log plot.

Then, deconvolution analysis was run to verify the pressure-derivative results as shown in figure 4. The deconvolved derivative increased the radius of investigation (ROI) and the duration of original constant-rate pressure increased from 24 hr to 140 hr, which enabled us to detect the boundary dominated flow regime at late time showing probably the effect of fault/faults.

Production logging (PL) was also run while performing DST, to determine the zonal contribution of each zone and conduct the flow profile of the well, as shown in figure 5. Based on density, water hold-up, pressure and temperature sensors, it is understood to have a stagnant column in the bottom of well, below depth 2985 m which identifies with a sharp change in pressure gradient, density around 1 g/cc and hold up number around 1. The temperature proposes an upward crossflow from the bottom of the layer to the middle zone. Reservoir zone 2953-2979 with 85 percent downhole zonal contribution, is the most contributor zone in production, which is approximately consistent with formation pressure test and petrophysical log results, and also the heterogeneity effect identified on derivative plot of the well test analysis. The uppermost and the bottommost inflow zones have no contribution in production.

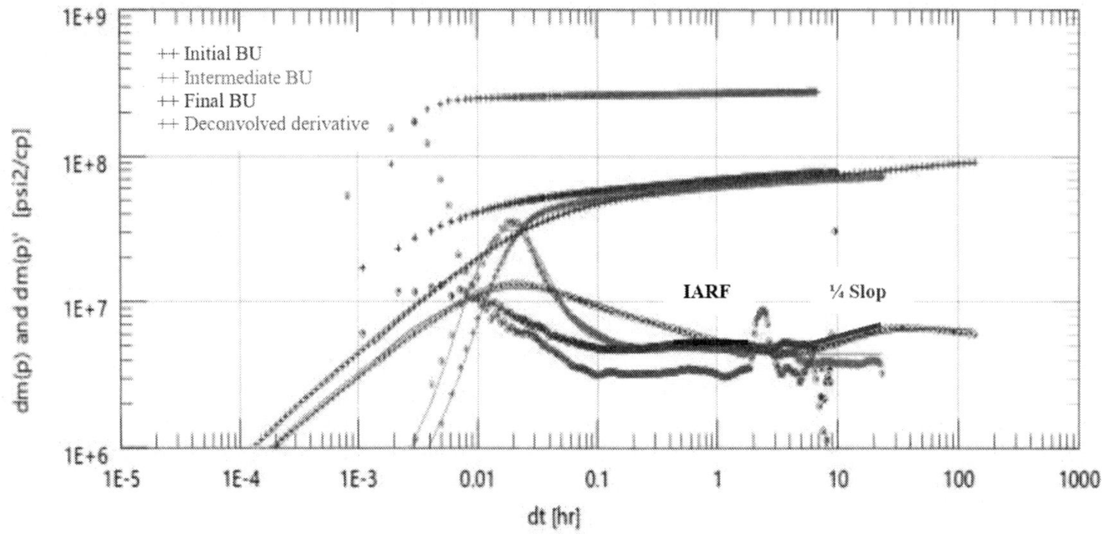

Figure 4. The boundary dominated flow regime or possible fault effect on deconvolved derivative.

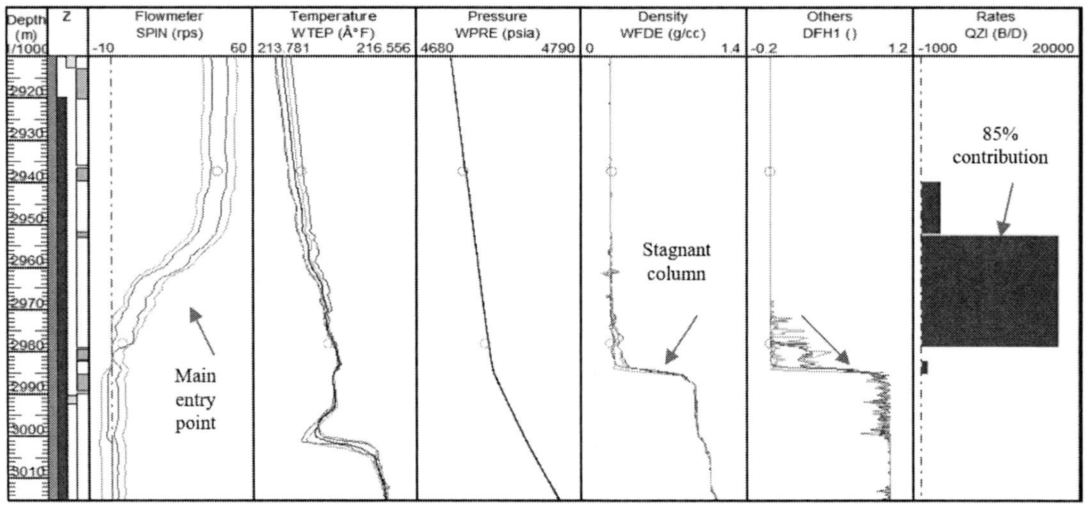

Figure 5. Production logging interpretation plot, the indication of layering behaviour and flowing survey.

4. Integrated study findings

As mentioned before, the final BU derivative showed two radial flow stabilization which indicated a multi-layered reservoir behaviour with different skin factor and kh, while the boundary dominated effects could not be seen. However, the deconvolved derivative enabled us to identify one IARF followed by ¼ slope and then slightly rolls over at late time which is possibly an indication of limited/leaky faults. Therefore, to overcome the issue/ uncertainty between well test results from both pressure-derivative and deconvolution in such complex reservoir, integration with the well logging interpretation results is taken into account as well.

The production logging interpretation results presented different reservoir behaviour in three distinct zones including the lower part of perforation below depth 2985 m, the middle zone around 2940-2985 (divided into three sub-zone with zonal contribution 11, 85 and 4 percent in production, respectively), and upper zone above 2940 m which suggests to have a multi-layered reservoir. In other words, PL was also presented the heterogeneous behaviour (the porous media with two or more than two storativity and mobility) while one reservoir layer (middle zone) has different sub-layers zonal contribution leading to have mobility contrast between sub-layers. After that, this mobility contrast verified with the results of formation pressure tester and petrophysical logging, which considered the middle zone as the highest mobility zone as was shown in figure 1. The uppermost and the bottommost inflow zones had no contribution in production, particularly the uppermost inflow zone had no reliable test points which could act as a barrier/fault so that this fault effect on deconvolved derivatives did not allow the second IARF in final BU pressure-derivative, to be appeared as well.

To summarize, the integrated well-test and well-log results verified the heterogeneous reservoir behaviour and the possible fault/boundary dominated effects, although it was a very complicated and complex case.

5. Conclusion

- The heterogeneous reservoir behaviour (three distinct zones including below depth 2985 m, the middle zone around 2940-2985, and upper zone above 2940 m) was verified by the integrated reservoir study of the well test analysis and well log interpretation results.
- Reservoir zone 2953-2979 with 85 percent production, was the most contributor. It involved the highest mobility zone, 2945-2972 m with limestone-dolomite lithology.
- The total flow capacity for the reservoir system estimated to be around 3500 md.ft.
- The results showed a high positive total skin factor before stimulation. It was reduced from 33 to 1.2 after acidizing, which indicated the good performance of the acid treatment job. Nevertheless, the two-layers reservoir model proposed that one layer had a high skin as it verified by the PL findings that showed a stagnant water column at the bottom of well.
- The deconvolved derivative increased the ROI and duration of original constant-rate pressure from 24 hours to 140 hours, which enabled us to detect the boundary dominated flow regime at late times, probably, to be a limited fault effect.

References

[1] Van Domelen M, Gdanski R, and Finley D 1994 The application of core and well testing to fracture acidizing treatment design: a case study, *Society of Petroleum Engineers, European Production Operations Conference and Exhibition*.

[2] Clarkson C R, Jensen J L, and Blasingame T 2011 Reservoir engineering for unconventional reservoirs: what do we have to consider? *North American Unconventional Gas Conference and Exhibition*.

[3] NOAH A and Shazly T 2014 Integration of Well Logging Analysis with Petrophysical Laboratory Measurements for Nukhul Formation at Lagia-8 Well, Sinai, Egypt, *American Journal of Research Communication,* **2**, 139-166.

[4] Kgogo T C 2011 Well test analysis of low permeability medium-rich to rich gas condensate homogeneous and layered reservoirs, PhD Thesis, Imperial College London.

[5] Gringarten A C 2010 Practical use of well-test deconvolution *SPE Annual Technical Conference and Exhibition*.

[6] Ilk D 2005 Deconvolution of variable rate reservoir performance data using B-splines, Texas A&M University.

[7] Ilk D 2010 Well performance analysis for low to ultra-low permeabilityreservoir systems, Texas A&M University.

[8] Von Schroeter T, Hollaender F, and Gringarten A C 2001 Deconvolution of well test data as a nonlinear total least squares problem *SPE Annual Technical Conference and Exhibition*.

[9] Von Schroeter T, Hollaender F, and Gringarten A C 2002 Analysis of well test data from permanent downhole gauges by deconvolution *SPE Annual Technical Conference and Exhibition*.

[10] Levitan M 2005 Practical Application of Pressure/Rate Deconvolution to Analysis of Real Well Tests *SPE 84290-PA,* **8**, 113–121.

[11] Gringarten A C 2008 From straight lines to deconvolution: The evolution of the state of the art in well test analysis, *SPE Reservoir Evaluation & Engineering,* **11**, 41-62.
[12] Onur M, Cinar M, Ilk D, Valko P P, Blasingame T A, and Hegeman P S 2008 An investigation of recent deconvolution methods for well-test data analysis, *SPE Journal,* **13**, 226-247.
[13] Liu W, Liu Y, Han G, Zhang J, and Wan Y 2017 An improved deconvolution algorithm using B-splines for well-test data analysis in petroleum engineering, *Journal of Petroleum Science and Engineering,* **149**, 306-314.
[14] Ecrin 4.10. Software Tutorial 2008 *Kappa Engineering, France.*

IOP Conf. Series: Materials Science and Engineering **863** (2020) 012010 doi:10.1088/1757-899X/863/1/012010

Investigation of gap losses in a modified compound parabolic concentrating collector with evacuated tube receiver

J Akhter*, S I Gilani, H H Al-Kayiem and I Al-Hassan

Department of Mechanical Engineering, Universiti Teknologi PETRONAS, 32610 Bandar Seri Iskandar, Perak, Malaysia.

*E-mail: javed.akhter_g03559@utp.edu.com

Abstract. Compound parabolic concentrating (CPC) collectors paired with evacuated tube receivers have great potential to supply solar thermal energy for numerous applications operating in the medium temperature range. Optical losses resulting from the gap between inner absorber tube and cusp of reflectors badly affect the performance of such collectors. In this paper, a modified design and optical performance of a CPC collector with evacuated tube receiver is presented. Effect of reflectors truncation on the collector's parameters including aperture width, concentration ratio and energy collection is also evaluated. Ray tracing simulation is performed to assess the instantaneous and daily energy collection. Optical performance of a cusp type CPC is compared with that of a modified CPC with w-shaped cavity at the bottom of reflectors to evaluate gap losses. Optical power density and daily average energy collection by cusp type CPC was 26% higher than CPC with w-shaped cavity due to relatively larger aperture size. CPC with w-shaped cavity was 11% more efficient than CPC with cusp type reflectors.

1. Introduction

Solar thermal energy at medium temperature can be obtained by converging solar radiation on smaller area using concentrating type collectors. Conventional concentrating collectors such as parabolic trough, parabolic dish or central tower receiver systems need accurate tracking mechanisms and large-scale installations for efficient heat collection [1]. Moreover, these collectors are imaging type and only intercept direct radiations. Compound parabolic concentrators (CPCs) are non-imaging devices which can deliver solar thermal energy in the medium temperature range (90-250 °C) without active tracking mechanisms. Ability of CPC collectors to collect diffuse radiations provide additional advantage especially in tropical regions where percentage of diffuse radiations is relatively high [2, 3]. Many researchers have designed and evaluated the performance of CPC collectors for various low-to-medium temperature applications such as industrial process heat [4], solar heating and cooling [5], water desalination, waste water treatment and purification [6], methanol reforming and hydrogen production [7], building integrated water heating systems [8] and photovoltaic/thermal hybrid systems [9].

A CPC is combination of two parabolas such that incident radiations falling on the entry aperture of the reflectors are directed to receiver placed at the exit aperture. Ideally, all solar radiations falling on the entry aperture at incidence angles between $\pm\theta_a$ (called acceptance half angle) are directed to the receiver placed at exit aperture. The maximum achievable optical concentration ratio (C_i) for an ideal CPC is determined by equation (1) [10].

$$C_i = \frac{1}{\sin \theta_a} \tag{1}$$

The geometric concentration (C_g) of a CPC is the ratio of aperture area to the receiver area. For a 2-D CPC with equal length of reflectors and receiver, it can be defined as ratio of aperture width to the receiver perimeter as given in equation (2).

$$C_g = \frac{A_{apr}}{A_{abs}} = \frac{W_{apr}}{2\pi r_o} \tag{2}$$

Potential application of CPC for solar energy collection was first presented by Roland Winston [11]. Extensive calculations carried out by Winston et al. [12] and Rabl et al. [13, 14] laid the foundation for future research related to CPCs. Many researchers have presented various designs of CPCs for improved optical and thermal performance. Derrick et al. [15] compared different designs of concentrators with tubular absorbers and calculated annual energy collection for east-west (E-W) and north-south (N-S) orientations. However, the research study did not consider thermal losses in the performance evaluation.

Integration of evacuated tube receiver (ETR) with CPC improved the performance of these collectors by reducing thermal losses from the receiver. However, large gap between inner absorber tube of ETR and reflector's cusp result in optical losses which badly affect the optical performance. In this paper, a modified design of CPC with a w-shaped cavity at the bottom is presented and compared with standard cusp type reflectors. The effect of design modification on collector parameters and optical performance is evaluated using Monte Carlos ray tracing simulations.

2. Methodology

A modified design of CPC with evacuated tube receiver is presented in which a w-shaped cavity is made at the bottom of reflectors to accommodate the envelope glass tube and minimize the optical gap losses resulting from large opening between inner absorber tube of the ETR and cusp of the concentrators. Reflectors height was truncated to save material and manufacturing cost with minor decrease in aperture width and concentration ratio. Different truncation levels were considered and their effect on the geometric parameters and energy collection was evaluated. Optical analysis of the newly designed CPC was carried out using Monte Carlo ray tracing simulations in TracePro 7.8.1 software. Instantaneous power density at the receiver of both designs is compared and daily average energy collection during the effective operating time is also analysed.

2.1. Geometric design of CPC

The size of a CPC collector depends on receiver size and acceptance angle which in turn defines the desired concentration ratio. The CPC profile with extended cusp type reflectors and evacuated tube receiver is shown in figure 1. Any point C on the reflector can be defined by considering angle φ (between lines from the origin O to E and O to D) and distance $\rho(\varphi)$ (line CE between point C and tangent at point E on the periphery of the absorber tube). The profile curve of a CPC with evacuated tube receiver is expressed by equation (3) as presented by [16].

$$\begin{cases} x = r * \sin\varphi - \rho(\varphi) * \cos\varphi \\ y = -r * \cos\varphi - \rho(\varphi) * \sin\varphi \end{cases} \tag{3}$$

For the extended cusp design incorporating size of envelope glass tube, the distance '$\rho(\varphi)$' can be determined by using string method as given by [17].

$$\frac{\rho}{r} = \begin{cases} \varphi + \gamma & \text{for } 0 \le \varphi \le \frac{1}{2}\pi + \theta_a \\ \dfrac{\frac{1}{2}\pi + \theta_a + \varphi + 2\gamma - \cos(\varphi - \theta_a)}{1 + \sin(\varphi - \theta_a)} & \text{for } \frac{1}{2}\pi + \theta_a < \varphi \le \frac{3}{2}\pi - \theta_a \end{cases} \quad (4)$$

where $\phi = \cos^{-1}(r/R)$ and $\gamma = \frac{\sqrt{R^2-r^2}}{r} - \phi$

The lower part in reflector profile ($\phi \le \varphi \le \frac{1}{2}\pi + \theta_a$) is involute and upper part ($\frac{1}{2}\pi + \theta_a < \varphi \le \frac{3}{2}\pi - \theta_a$) is parabolic.

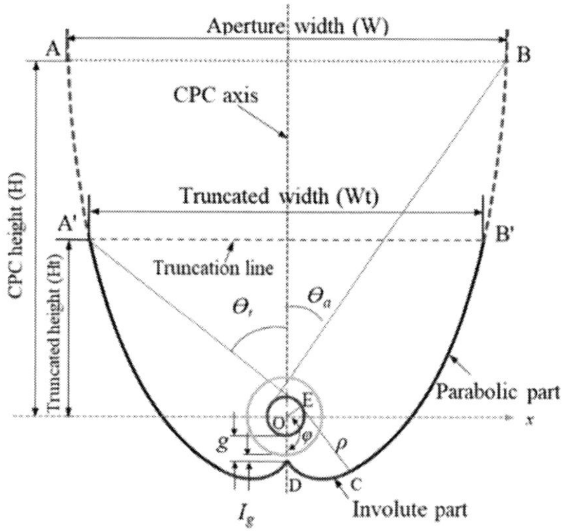

Figure 1. Profile of a 2-D CPC with evacuated tube receiver.

It is noteworthy that tubular receiver is in contact with the concentrator at cusp in an ideal CPC collector. However, considering the practical constraints small gap (g) between absorber and reflector was provided to accommodate glass tube. Some additional space between glass tube and cusp of reflector was also desired to avoid conduction heat loss. Therefore, profile of the concentrator was generated for the ETR with absorber radius (r) and glass tube radius (R).

2.2. *Reflectors modifications*

In order to minimise the effect of opening between inner absorber tube and reflectors in the form of gap losses, a w-shaped cavity was made at the bottom of the concentrators. For the modified design, following equation is used to determine the distance $\rho(\varphi)$.

$$\rho = r \begin{cases} \varphi - \gamma & \text{for } 2\phi \le \varphi \le \frac{1}{2}\pi + \theta_a \\ \dfrac{\frac{1}{2}\pi + \theta_a + \varphi - 2\gamma - \cos(\varphi - \theta_a)}{1 + \sin(\varphi - \theta_a)} & \text{for } \frac{1}{2}\pi + \theta_a < \varphi \le \frac{3}{2}\pi - \theta_a \end{cases} \quad (5)$$

where $\gamma = 2\phi - ((R+g)^2 - r^2)^{1/2}/r$ and $\phi = \cos^{-1}((r/(R+g))$. The design of w-shaped cavity made at the bottom of CPC reflectors is shown in figure 2.

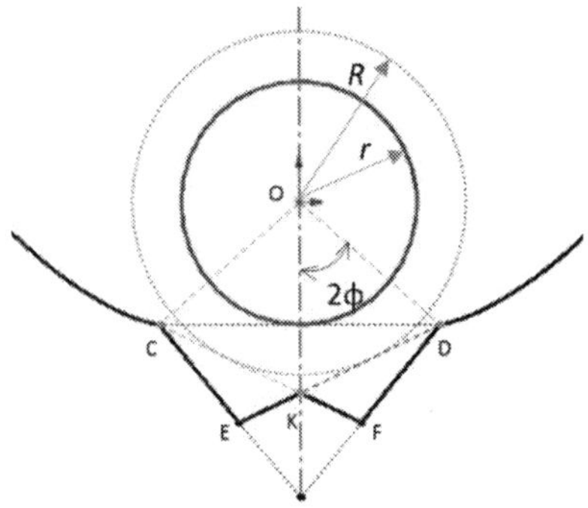

Figure 2. W-shaped cavity at the bottom of modified CPC with ETR.

The design of a CPC reflector depends on receiver size and acceptance angle which defines the desired concentration ratio. Figure 3 shows variation of reflector width and height as well as CR for different acceptance angles for the selected receiver. In the present design, 30° acceptance half angle is considered which corresponds to 4 hours of operating time of the full height CPC collector. The effective operating time can be increased by truncation of reflectors which enable direct radiation to reach the receiver beyond acceptance angle.

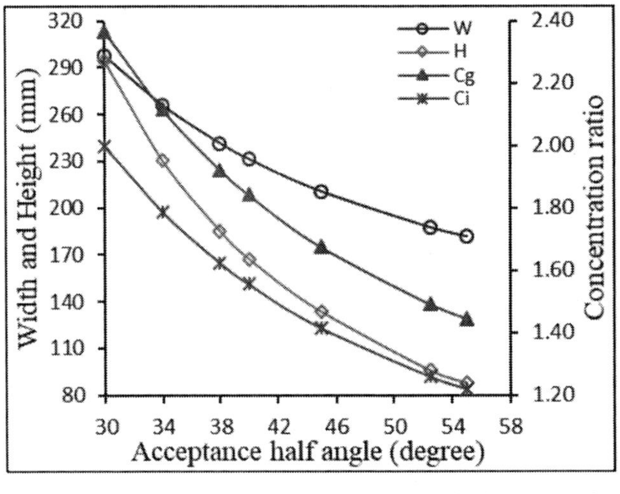

Figure 3. Effect of acceptance angle on width, height and CR.

The reflectors' height of an ideal CPC is very large which leads to high material and manufacturing cost. However, top portions of a CPC become almost parallel to CPC axis which contribute very less in the radiations concentration. Therefore, upper portions of reflectors are truncated to save cost of CPC. As a result, view field of the absorber is also increased and thus more direct and diffuse radiations, outside the acceptance half angle, could reach the receiver.

2.3. Effect of reflectors truncation

Since the upper portions of a CPC refelctors contribute very less in the concentration of incident radiations, therfore, upper sides of the reflectors were truncated to considerably reduce the material and manufacturing cost with minor decrease in CR. The effect of different truncation levels on width, height, width/height ratio, CR and effective acceptance angle for direct radiations in both types of the CPC troughs is shown in figure 4. Solid lines indicate the variation in width, height, width/height, CR and effective acceptance angle (for direct radiations only) in case of cusp type reflectors while dashed lines demonstrate the effect of truncation on corresponding parameters of cavity type CPC. The width/height ratio first decrease slowly and then steeply after about 50% truncation which shows that variation in aperture width is more than height at higher truncation levels. Effective acceptance angle for direct radiations also increases by truncating reflectors which in turn improves the overall effective operating time of the collector. Concentration ratio decreases slightly by increasing the truncation level as shown in figure 4(b).

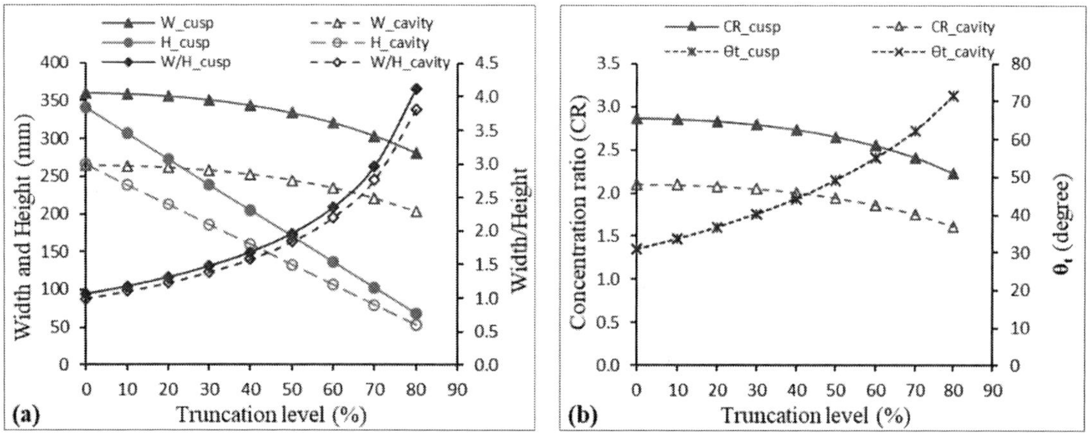

Figure 4. Effect of reflectors truncation on (a) width, height and width/height, (b) CR and effective acceptance angle for beam radiations.

The upper sides of the original height were truncated by 40% with only about 4.6% corresponding decrease in the aperture width and CR. The vertical height of the truncated reflectors was 0.205 m with 0.344 m corresponding aperture width for cusp type CPC. In case of cavity reflector, the truncated height and aperture width were 0.160 m and 0.252 m respectively. The length of reflectors was 2.5 m in both designs. Construction details of the designed CPC troughs with full and truncated heights are shown in table 1.

Table 1. Construction parameters of CPC.

Parameter	Unit	Design Value			
		Cusp reflector		Cavity reflector	
		Original	Modified	Original	Modified
Reflector size	M	H=0.341 W=0.360	H_T= 0.205 W_T= 0.344	H=0.266 W=0.264	H_T= 0.160 W_T= 0.252
Aperture area	(m^2)	0.90	0.86	0.66	0.63
Concentration ratio (CR)	-	2.86	2.74	2.10	2.01
Acceptance half angle (θ_a)	degree	30	30+20*	30	30+20*
Absorber Tube	M	L=2.5 d =0.04	L=2.5 d =0.04	L=2.5 d =0.04	L=2.5 d =0.04
Glass Tube	M	L=2.42 D =0.75	L=2.42 D =0.75	L=2.42 D =0.75	L=2.42 D =0.75

*increase in acceptance angle is only for direct radiations

2.4. Ray Tracing Simulation

Monte Carlo ray tracing method was used to plot ray paths and evaluate optical performance of the designed collectors. Geometric models of the CPC developed in Solidworks 2016 were imported into the ray tracing software TracePro 7.8.1. Material and surface properties for the collector components (given in table 2) were applied. Radiations source was defined, and parallel radiations were considered since the source was taken at long distance from the collector. Collector location (4.385N and 100.979E) was set and orientation was adjusted in north-south direction.

Table 2. Properties of CPC components for optical ray tracing.

Component	Material	Reflectivity	Absorptivity	Transmissivity
Reflector	Aluminium	0.92	0.08	-
Glass Tube	Borosilicate glass	-	0.002	0.98
Absorber	AlN coated Stainless Steel	-	0.98	-

3. Results and discussion

3.1. Ray path diagrams

Ray paths of solar radiations were obtained for different incidence angles (0°, 20°, 30°, 35°, 45° and 55°) at aperture planes of both cusp and cavity type reflectors as shown in figure 5 and figure 6 respectively. It is evident from figure 5(a) that the maximum radiations were concentrated at lower side of the absorber tube for normal incidence of solar radiations. When the radiations source (the sun) moved away from the zenith, the angle of incidence increased and concentration region shifted upward until the incidence angle approached acceptance angle of the CPC where the radiations were tangent to the absorber tube following edge ray principle as shown in figure 5(c). When the incidence angle further increased beyond the acceptance angle, only direct radiations could reach the absorber tube as shown in figure 5(d-f) which were made possible due to truncation of reflectors height. This additional advantage of increased view field partially compensates for the reduction in CR due to truncation.

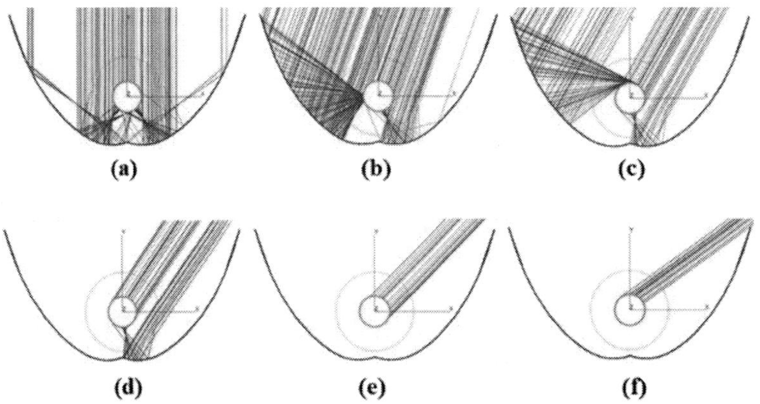

Figure 5. Ray paths of absorbed radiations in CPC with cusp reflector at different incident angles (a) 0°, (b) 20°, (c) 30°, (d) 35°, (e) 45° and (f) 55°.

The w-shaped cavity formed at the bottom of the reflectors made room for accommodating the envelope glass tube. In this way, the effect of gap between the reflectors and absorbers was reduced and some of the radiations which were previously passing through the gap and escaping to the environment were reflected back to the absorber surface. In this way optical gap losses were reduced and almost constant power was available at the absorber for incidence angles within the range of acceptance angle. Direct radiations were also intercepted beyond acceptance angle due to reflectors truncation as shown in figure 6(d-f).

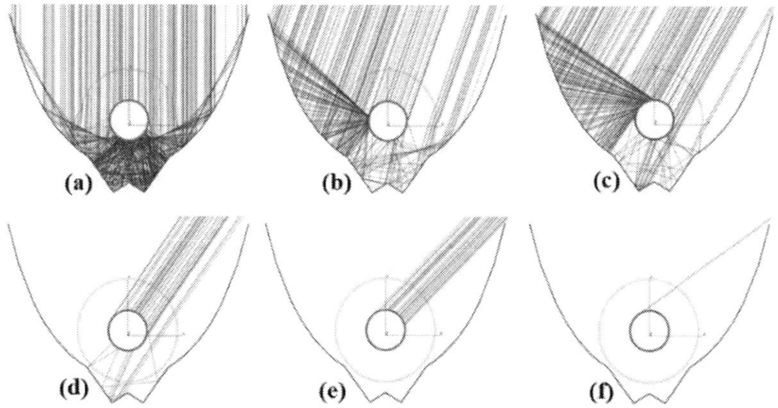

Figure 6. Ray paths of absorbed radiations in CPC with cavity at different incident angles (a) 0°, (b) 20°, (c) 30°, (d) 35°, (e) 45° and (f) 55°.

3.2. *Optical analysis*

The ray tracing simulations were carried out to evaluate the optical performance of both designs. The power density available at the outer surface of the absorbers of both cusp and cavity reflectors is shown in shown in figure 7. The maximum power density (about 500 W/m^2) was achieved at incidence angle close to acceptance angle which dropped to about 400 W/m^2 at noon time as shown in figure 7(a). Further increase in incidence angle beyond acceptance half angle resulted in sharp decrease of the available power due to limitation of radiations concentration beyond the acceptance half angle. However, direct radiations could reach the absorber until incidence angle reached the value of θ_t (angle between the optical axis and edge ray of truncated reflector). In case of cusp type reflectors, the optical gap losses were highest (about 100 W/m^2) at noon time. These gap losses were significantly abridged by using cavity type reflectors as shown in figure 7(b).

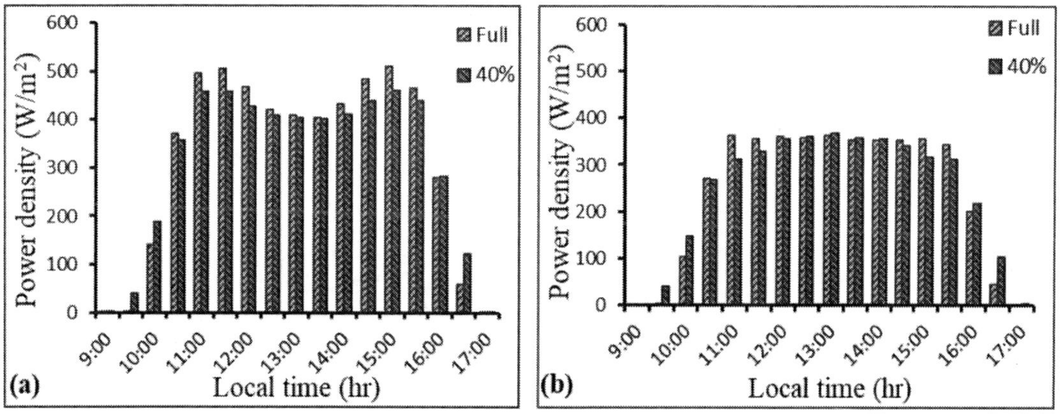

Figure 7. Instantaneous power density on receiver with (a) cusp reflectors, and (b) cavity reflectors.

The optical efficiency defined by ratio of power available at the absorber surface to incident at aperture plan was also determined for both designs. Similar trends were observed for optical efficiency in case of cusp as well as cavity type reflectors for different truncation levels as shown in figure 8. Even though, the total optical power was comparatively lower in case of cavity type CPC due to smaller aperture area, the available power density was almost uniform during the concentration time zone of the collector. Cavity type CPC was found about 5% more efficient than cusp type reflectors.

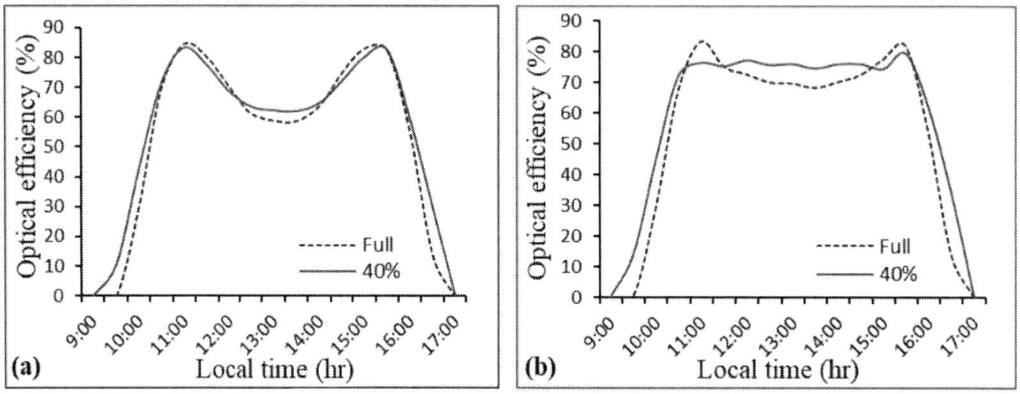

Figure 8. Optical efficiency of CPC with (a) cusp, and (b) cavity reflectors.

Comparison of daily average energy collection during operating time of the collectors showed that full height CPC with cusp type reflectors achieved 9.23 MJ which was 30% more than cavity type CPC. Similarly, 40% truncated CPC with cusp type reflectors intercepted about 9.00 MJ which was 26.8% higher than corresponding value for cavity type reflectors as shown in figure 9. It is noteworthy here that even though average daily energy capture by cusp type reflectors was higher due to larger aperture area, the optical efficiency of cavity type CPC was almost 7% higher than cusp type reflectors for different truncation levels. The reason can be attributed to reduced gap losses due to w-shaped cavity at the bottom which trapped large amount of the radiations passing through the space between absorber and envelop glass tube and re-direct to the absorber tube.

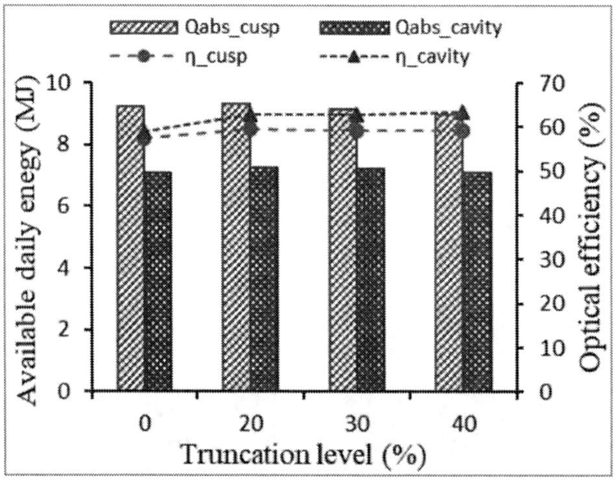

Figure 9. Comparison of daily average energy collection and optical efficiency of CPC with cusp and cavity reflectors.

4. Conclusion

In this paper, design and optical performance of a CPC paired with evacuated tube receiver is presented. Two variants of reflectors i.e. extended cusp type and w-shaped cavity type were considered. Acceptance half angle was selected to be 30° which corresponds to 4 hours of operating time of the full height CPC trough. Truncation level of the concentrators was selected to be 40% to save material and manufacturing cost with only about 4.6% decrease in CR. Ray tracing analysis of both collectors showed that about 26% gap losses occurred in case of cusp type reflectors which were abridged by using w-shaped cavity at the bottom of the reflectors. Comparative analysis of daily average energy capture revealed that full height CPC with cusp type reflectors achieved 9.23 MJ which was 30% more than cavity type reflectors. Similarly, 40% truncated CPC with cusp type reflectors intercepted about 9.00 MJ which was 26% higher than corresponding value of cavity type CPC. The optical efficiency of cavity type CPC was almost 7% higher than cusp type reflectors for different truncation levels.

Acknowledgements
The authors are grateful to the Universiti Teknologi PETRONAS, Malaysia to provide financial support through fundamental research grant scheme (cost centre 0153AA-H28) to execute this project.

References

[1] Xu X, Vignarooban K, Xu B, Hsu K, Kannan A M 2016 *Renewable and Sustainable Energy Reviews* **53**:1106-31.
[2] Rabl A, Gallagher J O, Winston R 1980 *Solar Energy* **25**:335-51.
[3] Ratismith W, Inthongkhum A 2012 *The 25th International Conference on Efficiency, Cost, Optimization and Simulation of Energy Conversion Systems and Processes* Perugia.
[4] Buttinger F, Beikircher T, Pröll M, Schölkopf W 2010 *Solar Energy* **84**:1166-74.
[5] Fischer S, Kovacs P, Lampe C, Serrats E M. White paper on concentrating collectors. Solar Rating and Certification Procedures2013.
[6] Ajona J I, Vidal A 2000 *Solar Energy* **68**:109–20.
[7] Gu X, Taylor R A, Morrison G, Rosengarten G 2014 *Applied Energy* **119**:467-75.
[8] Gang P, Guiqiang L, Xi Z, Jie J, Yuehong S 2012 *Solar Energy* **86**:1280-6.
[9] Hedayatizadeh M, Ajabshirchi Y, Sarhaddi F, Safavinejad A, Farahat S, Chaji H 2013 *International Journal of Green Energy* **10**:494-522.
[10] Tíba C, Fraidenraich N 2011 *Renewable Energy* **36**:2547-53.
[11] Winston R 1974 *Solar Energy* **16**:89-95.
[12] Winston R, Welford W T 1975 *SoJar Energy* **17**:255-8.
[13] Rabl A 1976 *Solar Energy* **18**:497-511.
[14] Rabl A, Goodman N B, Winston R 1979 *Solar Energy* **22**:373-81.
[15] Derrick G H, Bassett I M, Mills D R 1986 *Solar Energy* **37**:195-203.
[16] Kim Y S, Balkoski K, Jiang L, Winston R 2013 *Applied Energy* **111**:1071-9.
[17] Wang Q, Wang J, Tang R 2016 *Energies* **9**:795.

Characterisation of oil palm trunk and frond as fuel for biomass thermochemical

H A Umar[1,*], S A Sulaiman[1], R K Ahmad[1,2] and S N Tamili[1]

[1] Department of Mechanical Engineering, Universiti Teknologi PETRONAS, 32610 Bandar Seri Iskandar, Perak, Malaysia.
[2] Faculty of Engineering, Bayero University Kano, PMB 3011 Kano, Nigeria.

*E-mail: ummihadiza@gmail.com

Abstract. The rate of oil palm production in Malaysia increases annually and as a result, the oil palm wastes, especially oil palm trunk (OPT) and oil palm fronds (OPF) remain abundant. A suitable way of converting this abundant waste to renewable energy is through thermochemical conversion. Thus, this study investigates the characteristics of OPT and OPF biomass, for use as feedstock in thermochemical processes like gasification, pyrolysis, and combustion. The analysis carried out includes; ultimate (CHNSO) and proximate (thermogravimetric) analysis, calorific value, field emission scanning electron microscopy (FESEM) and x-ray fluorescence (XRF). Both feedstocks exhibited potential for use as fuel in biomass thermochemical conversion. The CHNSO analysis showed the presence of sufficient carbon, hydrogen and oxygen elements in both feedstocks, with carbon being the highest 45.42% in OPT and 43.35% in OPF. The percentages of nitrogen and sulphur which are required to be less for a good fuel were also obtained in low quantities for both fuel; 0.47% and 0.13% in OPT and 0.76% and 0.45% in OPF, respectively. The thermogravimetric analysis revealed both feedstocks to be having high volatile matter 62.28% in OPT and 66.10% in OPF. Meanwhile, sufficient fixed carbon content of 26.18% in OPT and 25.68% in OPF with low ash content of 9.82% in OPT and 6.32% in OPF were obtained in the analysis. FESEM and XRF were used to investigate the surface morphology, elemental and mineralogical nature of the samples. The findings were compared with those of other biomass and non-biomass materials. The EDX graph showed the presence of carbon and oxygen in a higher amount while in the XRF analysis CaO and K_2O were the major oxides present in both OPT and OPF, with a low amount of SiO making the feedstocks less prone to agglomeration during thermochemical conversion.

1. Introduction

Global concerns on the depletion of fossil fuels and the need for cleaner energy and safer environment are rekindling researcher's efforts on the lookout for more renewable energy resources that can be utilised. Among the renewable energy sources, biomass is considered an excellent option due to its abundance. Thermochemical processes like liquefaction, combustion, pyrolysis, and gasification have proven to be an effective means of biomass energy conversion [1]. It is desirable to partake characterisation tests to ascertain the suitability of using biomass feedstock as fuel for energy generation. The oil palm tree is considered as one of the major biomass sources in Malaysia, and the country happens to be amongst the world's largest producers of palm oil, with an increasing plantation area reported as 5.23 million hectares in 2013, 5.39 million hectares in 2014, 5.64 million hectares in 2015, and 5.74 million hectares in 2016. The oil palm planted area in 2017 reached 5.81 million

IOP Conf. Series: Materials Science and Engineering **863** (2020) 012011 doi:10.1088/1757-899X/863/1/012011

hectares, an increase of 1.3% as against that of the previous year. Sarawak overtook Sabah as the largest oil palm planted state, with 1.56 million hectares or 26.8% of the total Malaysian oil palm planted area, followed by Sabah with 1.55 million hectares or 26.6% and Peninsular Malaysia with 2.70 million hectares or 46.6% [2]. Recent concerns on accelerating replanting activity, improving oil extraction rate, expanding mill capacity, etc. are expected to further increase the total oil palm biomass availability in Malaysia. This situation has presented a huge opportunity for the utilization of the oil palm biomass in various applications including renewable energy [3].

The major parts of the oil palm tree utilised as biomass sources include the oil palm frond (OPF), empty fruit bunch (EFB), palm kernel shell (PKS) and oil palm trunk (OPT). This study focused on OPF and OPT, the OPF is usually available throughout the year during the pruning, or fruits harvesting period while OPT is only available in abundance during replanting periods. The Malaysian oil palm sector produces approximately 80 million dry tonnes of solid biomass, which contributes to 85.5% of the total dry biomass share in the country [4-6]. Generally, oil palm mills generate several oil palm wastes. The oil palm wastes contribute about RM6379 millions of energy annually [7]. However, more effort is needed to optimise the utilization of oil palm wastes for cogeneration in Malaysia. The knowledge of the characteristics of such biomass is of immense importance to determine their feasibility for use in thermochemical processes or as raw materials for use in industries. The elemental study of biomass helps to point out the levels of inorganic elements that may be released during the thermal processing of the biomass [8]. This information helps in the modelling, analysis, and designing of energy conversion processes along with the prediction of the economic feasibility of establishing such gasification systems. This study intends to characterise OPT and OPF for use as feedstock in thermochemical processes through processes that include proximate and ultimate analysis, calorific value, and elemental analysis.

2. Methodology

2.1. Sample preparation

The OPT and OPF samples were obtained from Felcra plantation, Bota Kanan Perak, Malaysia as shown in figure 1. The samples were collected as replanting exercise was going on in the plantation. The samples were collected raw from the site, and size reduction for proper drying was immediately carried out. The samples were oven-dried immediately at a temperature of 105 °C for 24 hours to remove excess moisture and were later sun-dried for 6 hrs, for ten days. The dried samples were later granulated, after which they were ground again and sieved to a particle size of 250 μm as depicted by figure 2.

Figure 1. Images of (a) oil palm frond and (b) oil palm trunk in drying process.

Figure 2. Ground and sieved samples of (a) OPT and (b) OPF.

2.2 Characterisation methods

The heating value of a material can be expressed in either lower (LHV) or higher heating value (HHV). The former is obtained when the heat of the combustion of the sample was determined relative to gaseous water while the latter is obtained when the product of the heat of combustion of the sample was determined relative to liquid water. The between the two values gives the latent heat of the product water [9]. In the given study, the Leco AC-350 bomb calorimeter was used for the determination of heating values.

The CHNSO elemental analysis determines the basic composition of a feedstock in terms of the percentage proportion of carbon (C), hydrogen (H), nitrogen (N), sulfur (S) oxygen (O) and ash in order to establish its suitability for energy production and for the calculation of stoichiometric air to fuel ratio [10]. The analysis was conducted according to the ASTM D3176-15 standard by using a Leco CHNS-932 model analyser. The weight percentage of carbon, hydrogen, nitrogen, and sulphur were reported directly from the analysis and oxygen was determined by difference [11]. The ash value was obtained from the thermogravimetric analysis. The thermogravimetric analysis is used to determine the amount of moisture content (M), volatile matter (VM), ash (A) and fixed carbon content (FC) present in the feedstock. In this study, the analysis was carried out using a TGA analyser (Pyris - 1 TGA) based on ASTM E 1131-98 standard test method for compositional analysis by thermogravimetry. Thermogravimetric analysis is an empirical technique in which the mass of a substance is heated at a controlled rate and the weight loss is recorded as a function of temperature and time [12].

FESEM microstructures, EDX analysis, and XRF analysis were used in this study to determine the elemental distribution and composition of OPT and OPF. The microstructure of each sample was determined using a field emission scanning electron microscope (FESEM) at different magnifications. The weight percentage of each element was determined using energy dispersive X-ray spectroscopy (EDX). The XRF analysis was used to determine the inorganic composition of the biomass samples, SUPRA 55VP model Field Emission Scanning Electron Microscopes (FESEM) analyser in combination with Energy Dispersive X-ray Spectroscopy (EDS) in non-conducting specimens mode was used.

3. Results and discussion

3.1. Calorific value

The amount of heat energy released from the combustion of the sample is proportional to the heating value of the material or calorific value. The higher a fuel heating value the higher the energy that will be dissipated during combustion and as such for thermal conversion purposes, fuels with high calorific value are more desirable. Table 1 tabulates the calorific value of OPT and OPF along with some selected biomass, coal and plastics. The range of calorific value of biomass feedstocks is given between 15-20 MJ/kg [8]. The oil palm trunk and frond have a calorific value of 17.41 MJ/kg and 17.48 MJ/kg which falls within that range and is also very similar to other biomass like the date palm frond, empty fruit bunch, rice husk and cotton stalk. The calorific value of the plastic feedstock is seen to be higher than that of both OPT and OPF which may be due to the difference in chemical composition. Nevertheless, the calorific values of OPT and OPF were found to be nearly similar to peat and lignite coal confirming to their potential for use as both domestic and industrial fuel.

Table 1. Calorific value of OPT and OPF and some selected coal, biomass and plastic types.

Type	Sample	Calorific value (MJ/kg)	Ref
Coal	Peat	14	[13]
	Lignite	16.3	[11]
	Bituminous	23.25	[11]
Biomass	Redwood waste wood	21.29	[6]
	OPF	17.28	[5]
	OPF	17.48	Current study
	Rice husk	14.79	[13]
	Wheat straw	16.59	[14]
	Municipal waste	19.83	[6]
	Animal waste	17.10	[6]
	Oil palm trunk	17.41	Current study
	Date palm frond	17.57	[12]
	Empty Fruit bunch	18.60	[11]
	Saw dust pellet	20.45	[11]
	Sugar cane bagasse	15.25	[13]
	Cotton stalk	18.26	[11]
Plastics	Polyethylene	43.00	[14]
	Mixed Plastic	35.00	[14]
	PVC	22.90	[15]
	Polystyrene	38.60	[15]

3.2. CHNSO elemental analysis

The CHNSO elemental analysis was conducted using a Leco-932 analyser and the values of C, H, N, S were determined directly from the result. The value of oxygen was determined by the difference method. Table 2 shows the analysis result for OPT and OPF, and other biomasses found from the literature for comparison. The CHNSO elemental analysis showed the presences of high carbon content in both OPT and OPF which was comparable to that of the low-rank coal. The carbon contents were also similar to most forestry biomass and agricultural biomass. The level of nitrogen and sulphur which may cause environmental effect is seen less in both OPT and OPF as compared to other biomass and is seen as negligible compared to that of coal.

Table 2. CHSNO elemental analysis of OPT, OPF and other feedstock.

Material	C	H	N	S	O	Reference
Peat	55	6	0.38	0.25	35	[13]
Lignite	50	4	2	0.90	43.1	[13, 16]
Bituminous	68	4	3	0.90	24.10	[13, 16]
Date palm frond	40.48	5.63	0.28	0.00	53.61	[17]
Corn stover	43.65	5.56	0.61	0.01	50.17	[16]
Rice husk	38.74	5.83	0.55	0.06	54.82	[18]
Wheat straw	42.11	6.53	0.58	0.32	40.51	[16]
Sesame stalk	41.34	6.57	0.81	0.29	45.16	[19]
OPF	44.58	4.53	0.71	0.07	48.80	[8]
OPF	45.42	6.35	0.47	0.13	47.63	Current study
OPT	43.35	6.26	0.76	0.45	49.18	Current study
Empty fruit bunch	40.73	5.75	1.40	0.22	51.90	[18]
Palm kernel shell	49.65	6.13	0.41	0.48	43.33	[18]
Sugarcane bagasse	42.93	5.82	0.68	0.06	54.82	[18]
Coconut shell	43.00	6.30	0.75	0.05	49.90	[18]

3.3. *Thermogravimetric analysis*

The proximate analysis result of OPF and OPT is given in table 3 with that of other biomass and coal types. The weight loss with a change in temperature of the two biomasses is shown in figure 3.

Table 3. Thermogravimetric analysis of OPT and OPF samples and some selected biomass and coal.

Sample	Volatile matter	Fixed Carbon	Ash	Reference
Peat	65	26.5	8.5	[13]
Lignite	34.41	48.39	17.2	[16]
Bituminous	18.69	61.86	19.63	[16]
OPT	62.28	26.18	9.82	Present study
OPF	66.10	25.68	6.32	Present study
Coconut shell	78	19.48	5.52	[18]
Corn stover	75.17	19.25	5.58	[16]
Sesame Stalk	72.2	17.0	6.6	[19]
Rice Husk	72.2	12.0	15.8	[18]
Sugar cane bagasse	80.19	15.91	3.9	[18]
Red wood	83.5	16.1	0.4	[12]

From the results, comparing OPF and OPT with the other biomass it is seen that the volatile and fixed carbon contents are within the range of the other biomass, though the amount of volatile matter and fixed carbon present in OPF is quiet higher than OPT. Conversely, OPT has more ash content than OPF which indicates the need for ash handling during gasification of OPT, noting that rice husk and the coal types have more ash content than the OPT.

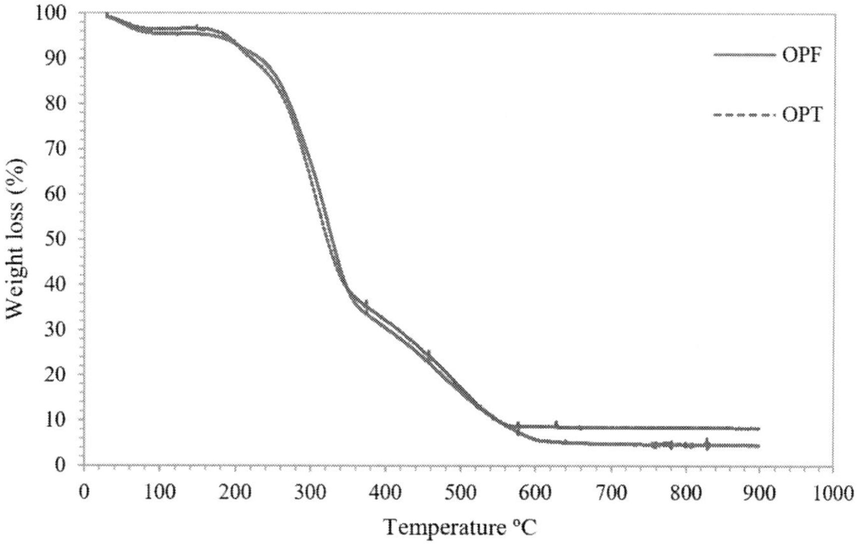

Figure 3. Percentage weight loss of OPT and OPF with temperature.

3.4. EDX analysis

The microstructure of each sample is shown in figures 4 to 6, as determined by field emission scanning electron microscope (FESEM). The weight percentage of each element was determined using energy dispersive X-ray spectroscopy (EDX) as shown in table 4.

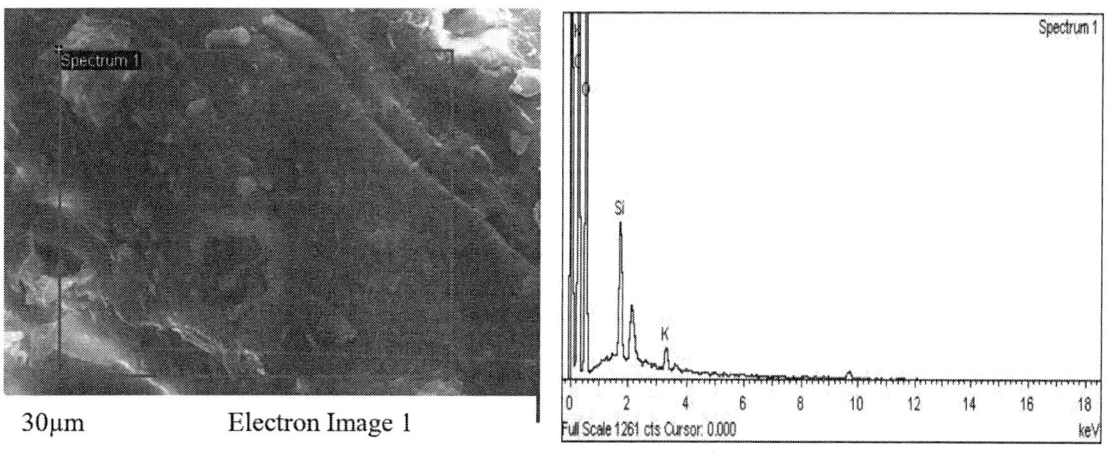

(a) (b)

Figure 4. (a) FESEM image and (b) EDX spectrum of raw OPT.

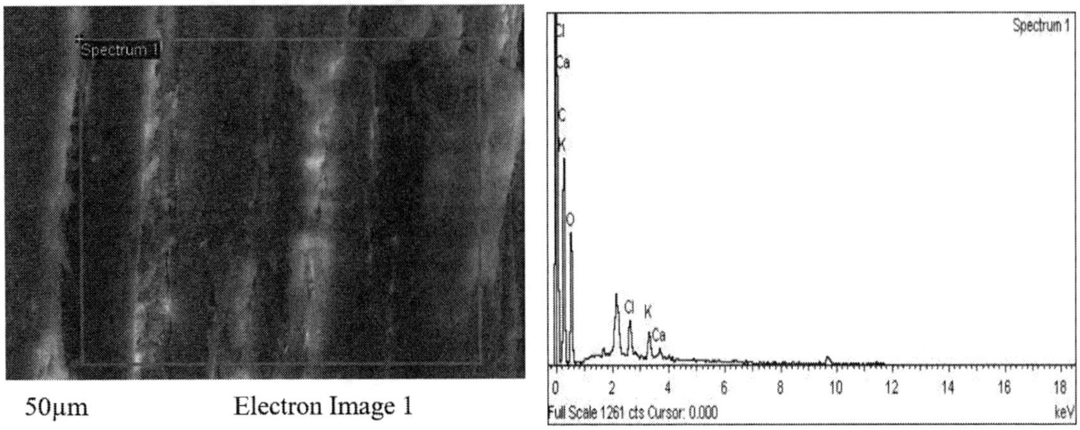

..... (a) (b)

Figure 5. (a) FESEM image (b) EDX spectrum of raw OPF.

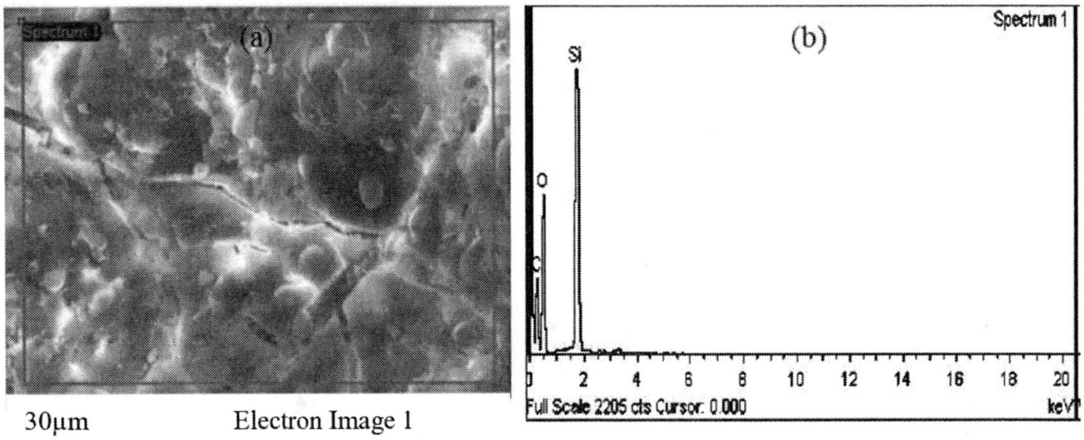

Figure 6. (a) FESEM image (b) EDX spectrum of Raw Rice Husk [8].

Table 4. EDX result for OPT and OPF raw samples.

Elements	OPT		OPF		Rice Husk [8]	
	Weight %	Atomic%	Weight%	Atomic%	Weight%	Atomic%
C	53.86	61.32	53.39	61.77	32.21	42.51
O	44.20	37.79	42.09	36.55	45.40	49.98
Cl	-	-	2.18	0.85	-	-
K	0.4	0.14	1.72	0.61	0.77	0.31
Ca	-	-	0.62	0.21	-	-
Na	-	-	-	-	-	-
Mg	-	-	-	-	-	-
Si	1.54	0.75	-	-	21.62	12.20

The carbon weight obtained from EDX for both OPT and OPF was higher than that obtained from the ultimate analysis. The EDX analysis predicted the carbon content of OPT to be 53.86% and that of OPF to be 53.39% while from the ultimate analysis the carbon content was 43.35% and 45.42% for OPT and OPF respectively. This may be because of the untraced hydrogen and nitrogen in EDX analysis. It is of immense importance to consider the presence of inorganic elements like Al, Na, K, Si, Mg and Ca present in the biomass samples as they have to be considered in the design of thermochemical processing systems. The high contents of such elements may lead to critical problems during the gasification process operation. From table 3, the OPT contains very little amount of K, 0.4% so also the rice husk with 0.77% while that of OPF was quiet higher being 1.72%. It is noteworthy that potassium form salts with different elements and may cause corrosion and erosion problems in the gasification process on the downstream equipment. The presence of Ca is not seen in OPT and rice husk, but in the case of OPF, it has about 0.62%. The Si content in OPT is relatively negligible (1.54%) compared to the one present in rice husk, while in the case of OPF there was no trace of it. However silicon content as low as 1% may lead to problems during the combustion of biomass due to the production of slagging in silicates form [20]. A high amount of Si present in the rice husk about 21.62% suggests that using the rice husk for the gasification process may require ash handling and syngas cleaning systems. Usually, non-woody biomass has a high amount of silicon compared to woody biomass [21].

From the XRF result in table 5, CaO and K_2O form part of the major oxides found in both OPT and OPF constituting 18.7% and 36.5% CaO, 44.5% and 32.1% K_2O respectively. The CaO level in rice husk (3.13%) and wheat straw (3.12%) is less compared to that present in OPT, OPF and wood ash (39.03%), while the K_2O for wood ash (13.34%) and wheat straw (18.85%) are quiet higher compared to that found in rice husk (8.78%). The amount of silicon oxide is highest in rice husk about 73.20%, then wheat straw 65.45% with lower levels found in OPT, OPF and, wood. The higher amount of K_2O and SiO present in a feedstock might the result into the formation of a compound K_2O-SiO which may accelerate agglomeration problem [22]. However, the presence of bed materials like Al_2O_3, MgO, CaO, Fe_2O_3 reduces such problems by forming a high-temperature compound. Therefore the high content of CaO in OPF and wood ash may help to suppress agglomeration problems than in rice husk and wheat straw making them less preferable for gasification process. As for OPT, the presence of Fe_2O_3 oxide in conjunction with the CaO will also hinder agglomeration during the gasification.

Table 5. XRF Analysis of OPT, OPF and Rice Husk.

Oxides	**Current Result**		**Literature Results [8]**		
	OPT	**OPF**	**Rice Husk**	**Wood Ash**	**Wheat Straw**
CaO	18.7	36.5	3.14	39.03	3.12
K_2O	44.5	32.1	8.78	13.34	18.85
Cl	6.65	15.2	2.38	0.06	3.58
P_2O_5	5.37	4.45	7.55	3.85	1.17
SiO_2	8.33	4.33	73.20	11.72	65.45
SO_3	8.30	1.95	3.35	-	-
MgO	1.15	1.93	0.91	6.03	2.10
Fe_2O_3	4.82	1.90	0.32	1.32	0.44
MnO	0.451	1.23	0.31	6.03	0.08
NiO	0.397	0.147	-	-	-
CuO	0.114	0.111	0.04	-	-
ZnO	0.215	0.110	0.05	-	-
Pd	0.07	0.0860	-	-	-
Rb_2O	0.169	0.0109	-	-	-

4. Conclusion

Oil palm trunk and frond have been characterised for use in thermochemical processes like gasification and combustion. The ultimate analysis showed that OPT and OPF had a carbon content of 43.35% and 45.42% respectively. The heating value analysis exhibited their potential for use as solid fuels as their calorific values 17.41 MJ/kg and 17.48 MJ/kg lies within the biomass calorific value range and was closely similar to that of low-rank coals. The proximate analysis indicated fewer ash contents in the samples with OPT having higher amount 9.82% than OPF 6.32%. From the elemental analysis, the

EDX result exhibited a higher amount of carbon for both OPT and OPF, in the XRF result, the presence of CaO and K_2O in high levels were detected for both samples. The presence of SiO in small traces in both feedstocks might lead to agglomeration. However, the presence of high CaO present in OPF may dampen the process to form a high-temperature compound. Also in the case of OPT, the CaO and Fe_2O3 present will counter the effect of agglomeration. Overall OPT and OPF have exhibited remarkable properties for use in thermochemical conversion as they have sufficient C, H, O content, high calorific value and volatile matter, and low ash content and less prone to sintering and agglomeration. This makes OPT and OPF a suitable feedstock for power generation in processes like gasification, combustion, and pyrolysis.

Acknowledgement
The authors wish to acknowledge and thank Universiti Teknologi PETRONAS, Malaysia, for the financial and technical assistance for this research work.

References

[1] P. Basu, *Biomass gasification and pyrolysis: practical design and theory*. Academic press, 2010.
[2] m. P. O. Board, "<Overview_of_Industry_2017.pdf>," http://bepi.mpob.gov.my/images/overview/Overview_of_Industry_2017.pdf2017.
[3] S. K. Loh, "The potential of the Malaysian oil palm biomass as a renewable energy source," *Energy Conversion and Management,* vol. 141, pp. 285-298, 2017.
[4] M. S. Umar, P. Jennings, and T. Urmee, "Strengthening the palm oil biomass Renewable Energy industry in Malaysia," *Renewable Energy,* vol. 60, pp. 107-115, 2013/12/01/ 2013.
[5] M. S. Umar, P. Jennings, and T. Urmee, "Generating renewable energy from oil palm biomass in Malaysia: The Feed-in Tariff policy framework," *Biomass and Bioenergy,* vol. 62, pp. 37-46, 2014/03/01/ 2014.
[6] D. Darshini, P. Dwivedi, and K. Glenk, "Capturing stakeholders′ views on oil palm-based biofuel and biomass utilisation in Malaysia," *Energy Policy,* vol. 62, pp. 1128-1137, 2013/11/01/ 2013.
[7] M. Z. Jaafar, W. H. Kheng, and N. J. E. p. Kamaruddin, "Greener energy solutions for a sustainable future: issues and challenges for Malaysia," vol. 31, no. 11, pp. 1061-1072, 2003.
[8] F. M. Guangul, S. A. Sulaiman, and V. R. Raghavan, "Elemental and thermo-chemical analysis of oil palm fronds for biomass energy conversion," in *AIP Conference Proceedings*, 2012, vol. 1440, no. 1, pp. 1197-1205: AIP.
[9] R. Stahl, E. Henrich, H. Gehrmann, S. Vodegel, and M. Koch, 2004 "Definition of a standard biomass," *Renewable Fuels for Advanced Power Trains*.
[10] D. J. Bushnell, C. Haluzok, and A. Dadkhah-Nikoo, 1990 "Biomass Fuel Characterization: Testing and Evaluating the Combustion Characteristics of Selected Biomass Fuels: Final Report May 1, 1988-July, 1989," USDOE Bonneville Power Administration, Portland, OR (USA); Oregon State Univ.
[11] F. M. Guangul, 2013 "Gasification of OPF with preheated Inlet Air " Ph.D, Mechanical Engineering, Universiti Teknologi Petronas.
[12] T. Reed and A. Das, 1988, *Handbook of biomass downdraft gasifier engine systems: Biomass Energy Foundation.*
[13] H. Sadig, S. A. Sulaiman, M. N. Z. Moni, and L. D. Anbealagan, 2017, "Characterization of date palm frond as a fuel for thermal conversion processes," in *MATEC Web of Conferences*, **131**, p. 01002: EDP Sciences.
[14] A. K. Panda, R. K. Singh, D. J. R. Mishra, and S. E. Reviews, 2010 "Thermolysis of waste plastics to liquid fuel: A suitable method for plastic waste management and manufacture of value added products—A world prospective," vol. **14**, no. 1, pp. 233-248.
[15] D. A. T. a. M. J. Castaldi, 2016. "Determining accurate heating values of non-recycled plastics (NRP) ", C. U. O. N. York, Ed.

[16] Y. Demirel, 2012, "Energy and energy types," in *Energy*: Springer, pp. 27-70.

[17] R. Nasser *et al.*, 2016 "Chemical analysis of different parts of date palm (Phoenix dactylifera L.) using ultimate, proximate and thermo-gravimetric techniques for energy production," **9**, no. 5, p. 374,.

[18] S. S. Abdullah and S. J. J. o. A. S. Yusup, "Method for screening of Malaysian biomass based on aggregated matrix for hydrogen production through gasification," vol. **10**, no. 24, pp. 3301-3306, 2010.

[19] L. Cuiping, W. Chuangzhi, and H. Haitao, 2004 "Chemical elemental characteristics of biomass fuels in China," *Biomass and bioenergy,* vol. **27,** no. 2, pp. 119-130.

[20] A. Akhtar, T. Ivanova, I. Jiříček, V. J. J. o. R. Krepl, and S. Energy, "Detailed characterization of waste from date palm (Phoenix dactylifera) branches for energy production: Comparative evaluation of heavy metals concentration," vol. **11**, no. 1, p. 013102, 2019.

[21] M. Öhman *et al.*, 2011 "Slag formation during combustion of biomass fuels with low phosphorus content," in *European Biomass Conference & Exhibition: 06/06/2011-10/06/2011*, pp. 1267-1270: ETA-Renewable Energies.

[22] Z. A. B. Z. Alauddin, P. Lahijani, M. Mohammadi, and A. R. Mohamed, 2010 "Gasification of lignocellulosic biomass in fluidized beds for renewable energy development: A review," *Renewable and Sustainable Energy Reviews,* vol. **14**, no. 9, pp. 2852-2862.

IOP Conf. Series: Materials Science and Engineering **863** (2020) 012012 doi:10.1088/1757-899X/863/1/012012

Investigation of stability, dispersion, and thermal conductivity of functionalized multi-walled carbon nanotube based nanofluid

O A Hussein[1,2], K Habib[1], M Nasif[1], R Saidur[3,4] and A S Muhsan[5,*]

[1] Mechanical Engineering Department, Universiti Teknologi PETRONAS, 32610 Bandar Seri Iskandar, Perak, Malaysia.
[2] Mechanical Engineering Department, Tikrit University, Tikrit, Iraq.
[3] Faculty of Science and Technology, Sunway University, 47500 Petaling Jaya, Malaysia.
[4] Department of Engineering, Lancaster University, LA1 4YW, United Kingdom.
[5] Petroleum Engineering Department, Universiti Teknologi PETRONAS, 32610 Bandar Seri Iskandar, Perak, Malaysia.

*E-mail: ali.samer@utp.edu.my

Abstract. In the attempt of preparing multi-walled carbon nanotube (MWCNTs), covalent functionalisation (CF-MWCNTs) were applied. The stable thermal conductivity was measured as a function of temperature. A number of techniques, such as FTIR, FESEM and UV-vis spectrophotometer were employed to characterise both dispersion stability and morphology of functionalised materials. By using ultrasonic test time, the highest stability of nanofluids was achieved at 60 minutes. As a result, the thermal conductivity displayed by CF-MWCNTs was higher than distilled water. In conclusion, improvement in thermal conductivity and stability displayed by CF-MWCNTs was higher, while the best thermal conductivity improvement was recorded at 31%.

1. Introduction

Nanofluid refers to the stable and homogenous suspension exhibited by nanoparticles (NPs) within conventional working fluids such as ethylene glycol, oil, and water [1, 2]. Prior studies reported that addition of small weight concentration of NPs can improve the efficiency of heat transfer and thermal conductivity in base fluid, hence the suitability for a wide range applications, such as solar collectors, cooling appliances, and heat exchangers [3, 4]. Nanoparticles (NPs) are made of either metal oxide (e.g. Al_2O_3, SiO_2, CuO, and TiO_2) [5 – 7], carbon-based particles such as carbon nanotube (CNT) [8, 9], graphene oxide (GO) [10], or graphene nanoplatelets (GNPs) [11, 12]. Numerous types of NPs have been proposed since past decades to generate nanofluids; wherein CuO, Al_2O_3, ZnO, and SiO_2 are more commonly employed than nanomaterials based on metal oxide [13 – 15]. Both graphene and carbon-based NPs (e.g. CNT) [16] have been assessed empirically [12, 17, 18]. Simply put, nanofluids have garnered much attention due to its chemical and physical properties, as well as its nanometre size. More importantly, nanofluids have improved thermal conductivity, hence its suitability to serve as heat-exchanging fluid. Upon inclusion of NP base fluid, a number of empirical studies have also improved convective heat transfer, in comparison to pure base fluid.

However, due to its large surface area with strong attraction force of van der Waals, the preparation of these nanofluids suffered from lack of colloidal stability and agglomeration followed by sedimentation of the nano additives [19].Therefore, this work aims to improve the suspension stability, dispersibility and thermal conductivity of nanofluid based MWCNTs using chemical functionalization approach by introducing carboxyl (-COOH) and hydroxyl (-OH) functional groups to MWCNTs surfaces.

2. Experimental methodology

The pristine MWCNTs used in this study was purchased from, XG Sciences, Lansing, MI, USA with a diameter of 15 nm, length of 5 µm and purity of 95%. Sulfuric acid 95-97% (H_2SO_4) and nitric acid 65% (HNO_3) were purchased from Sigma-Aldrich Co., Selangor, Malaysia, and were used as the functionalization media. Following description focuses on the nanofluid preparation and method used and also the measurement for analysis of the nanofluid.

2.1. Nanofluid preparation and methods

Pristine MWCNTs is a hydrophobic material and it cannot be dispersed in any solvent which is polar like distilled water. The suitable way to make MWCNTs hydrophilic is by introducing functional groups of carboxyl (-COOH) and hydroxyl (-OH) on its surface via acid treatment. This process also known as functionalization process, as shown in figure 1. This process was achieved by dispersing pristine MWCNTs in a solution of HNO_3 and H_2SO_4 at ratio of (1:3) (strong acid medium) through three steps. In the first step, the pristine MWCNTs was added into H_2SO_4 solution and mixed properly using magnetic stirrer for 30 minutes. In the second step, HNO_3 was gradually added to the mixture while the mixture container was placed in an ice bath to control the reaction temperature and avoid solution evaporation. The mixture solution was then stirred for 30 minutes followed by 3 h in ultra-sonication bath at (60 °C). In the final step, CF-MWCNTs were collected and washed thoroughly for 3 times with distilled water then centrifuged at 6000 rpm for 15 minutes to remove excess acid and then dried under drying oven for 24 h at 80 °C. Samples with different concentrations of CF-MWCNTs (0.02, 0.05, 0.08, and, 0.1wt %) were prepared by adding the CF-MWNTs into distilled water and mix them properly using probe sonicator for 60 minutes.

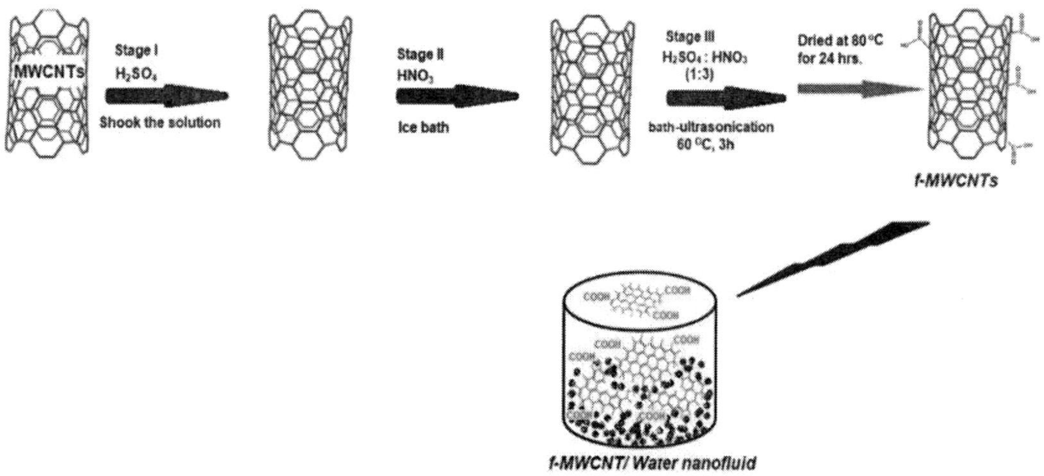

Figure 1. Schematic of functionalization and sonication process to making nanofluid.

2.2. Measurement devices

2.2.1. Evaluation of suspension stability and dispersibility. The evaluation of the FTIR spectra was done within a wavenumber range of 400-4000 cm^{-1} for FTIR spectroscopy. Ultraviolet-visible spectrophotometry analysis (UV-vis) was utilized to assess the dispersion quality of CF-MWCNTs in the aqueous suspension. To make available a quantifiable characterization of the stability can measurement of light absorbance of a suspension by UV-vis spectroscopy. The use of UV-vis is Lambda operating in the range of 200-800 nm wavelengths (UV-800/900, Lambda Company, USA). Light absorbance was measured by special quartz cuvettes suitable for UV region since all samples were at certain time intervals for an extra 30 days. A 1:20 ratio diluted in distilled water to allow suitable light transmission for all samples. The morphological characteristics were conducted by Field Emission Scanning Electron Microscopy (FESEM, SU8000, Hitachi), to characterize both the pristine-MWCNTs and the CF-MWCNTs nanofluids.

2.2.2. Thermal conductivity measurement. Thermal conductivity of the prepared nanofluids was measured using KD2 Pro thermal properties analyzer device (Decagon devices, Inc., USA), which has an accuracy of about 5%. Its operation principle is based on the transient hot wire method using the KS-1 probe which consists of a single needle sensor with 1.3 mm diameter and 60 mm long. To keep the samples at the preferred temperature during measurements, a water baths ''Memmert WNB 7-45" (Daihan Scientific Co., Ltd., Korea) with 1.4 kW and an accuracy of 0.1 °C was used.

2.2.3. Thermal conductivity measurement. Thermal conductivity of the prepared nanofluids was measured using KD2 Pro thermal properties analyzer device (Decagon devices, Inc., USA), which has an accuracy of about 5%. Its operation principle is based on the transient hot wire method using the KS-1 probe which consists of a single needle sensor with 1.3 mm diameter and 60 mm long. To keep the samples at the preferred temperature during measurements, a water baths ''Memmert WNB 7-45" (Daihan Scientific Co., Ltd., Korea) with 1.4 kW and an accuracy of 0.1 °C was used.

3. Results and discussion

3.1. Characterization of CF-MWCNTs nanoparticles

3.1.1. Fourier transform infrared spectroscopy. The FTIR spectra of pristine-MWCNTs and CF-MWCNTs are illustrated in figure 2. Figure 2 displays the absorbance peaks of CF-MWCNTs at 2850, 1650, 1383 and 1100 cm^{-1}, which are attributed to C-H, C=O, C-O and C-O-C bonds, respectively. The bonds reflect the attachment of acid treatment molecules on the surface and edge of MWCNT walls. Both bonds of C=O and C-O signify the successful direct esterification reaction. The CF-MWCNTs showed a broad absorbance peak at wavenumbers between 3450 cm^{-1}; attributable to -OH chains from acid treatment [20]. The CF-MWCNTs peaks indicate the success of CF procedure.

3.1.2. Scanning electron microscopy (SEM). FESEM images for CF-MWCNTs confirm the existence of broken tubes of MWCNTs after functionalization process as shown in figure 3 (a) and (b). Though, these tubes provide better dispersion stability, it may cause a small reduction in thermal conductivity [14]. On the other hand, the fictionalization treatment illuminated the upper surface roughness of the P-MWCNTs. The partial damage of graphitic carbon indicates higher roughness, which is mostly due to the effect of carboxyl groups.

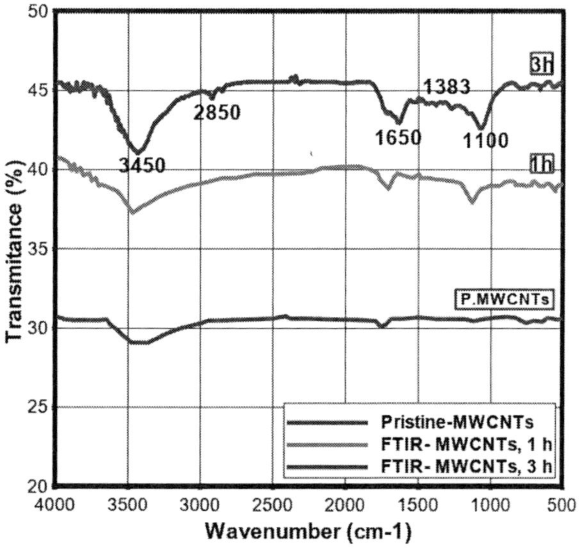

Figure 2. FTIR spectra of Pristine-MWCNTs and 1 and 3hr CF-MWCNTs.

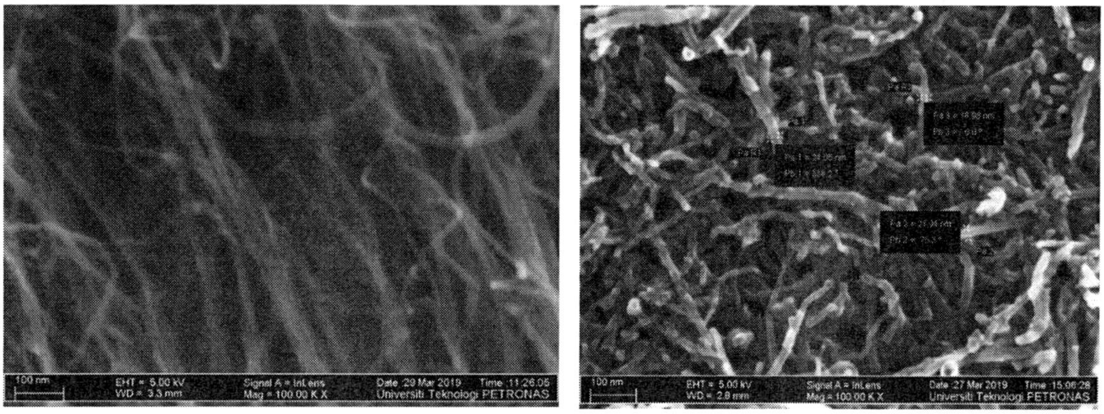

(a) (b)

Figure 3. FESEM images of (a) P-MWCNTs and (b) CF-MWCNTs.

3.1.3. Suspension stability and dispersibility. Individual MWCNT are active in the UV-vis region and exhibit characteristic bands corresponding to additional absorption due to ID van Hove singularities [21]. Further, it is possible to establish a relationship between different concentrations of CF-MWCNT dispersed in water and the intensity of the corresponding absorption spectrum. Moreover, UV-vis spectroscopy was used to monitor the dynamics of the dispersion process of MWCNT. After refluxation, the absorbance of MWCNT solutions show a maximum between 250 and 300 nm and gradually decreases from UV to near-IR, which is partly due to scattering, especially in the lower wavelength range. Initially, MWCNT exist as big aggregates and bundles in solution that are strongly and no absorption is evident in the UV-vis spectrum figure 4. After acid treatment, the provided mechanical energy overcomes the van der Waals interactions in the MWCNT bundles and leads to their disentanglement and improved dispersion. The increasing amount of dispersed CF-MWCNT results in an increasing area below the spectrum lines representing the absorbance.

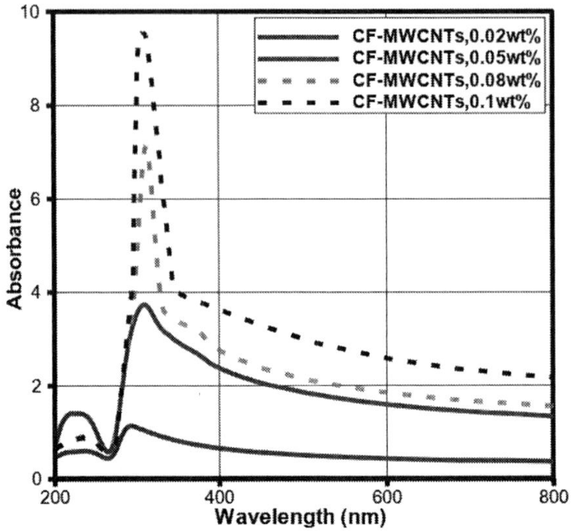

Figure 4. UV-vis absorption spectra of CF-MWCNTs nanofluids at different weight concentrations.

3.1.4. Thermal conductivity. Figure 5 shows the thermal conductivity of water-based nanofluids including CF-MWCNTs with different weight concentrations of 0.02, 0.05, 0.08 and 0.1wt% at different temperatures. It can be clearly seen that the thermal conductivity of water based CF-MWCNTs nanofluids with all; concentrations are higher than that of water. In addition, the thermal conductivity increases as the concentrations of MWCNTs in the water-based nanofluids increases. In, the main mechanism of thermal conductivity enhancement can be attributed to the Brownian motion of the suspended particles as it is one of the most dominating functions of temperature [22]. Figure 5 also shows adding 0.1wt% of CF-MWCNTs to the distilled water gives the highest the thermal conductivity of 0.83 W/m.K at 60 °C, which is 31% higher than that of distilled water (DW).

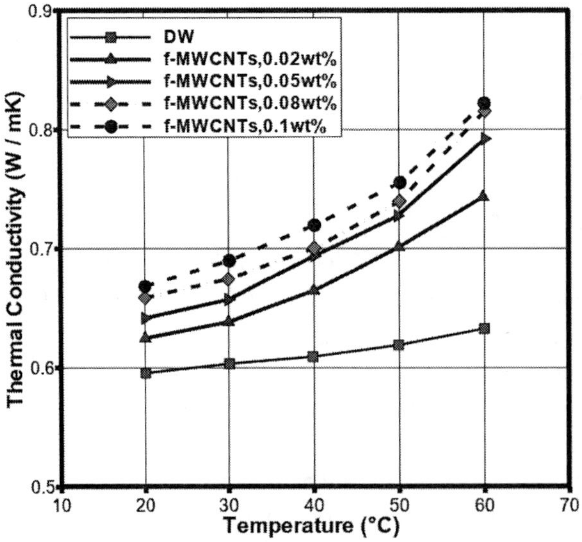

Figure 5. Thermal conductivity of CF-MWCNTs nanofluids as a function of temperature for different weight concentrations.

4. Conclusion

The preparation of homogenous and stabilized suspension of functionalized multi-walled carbon nanotube (CF-MWCNTs) in water based thermal fluid was achieved successfully. By proper chemical functionalization approach we managed to introduce some carboxyl and hydroxyl functional groups to pristine MWCNTs surfaces which significantly help to overcome the van der Waals interactions forces between pristine MWCNTs bundles and lead to their disentanglement and improved dispersion. These results were further confirmed using UV-vis, FTIR and FESEM characterization tools. Thermal conductivity measurement showed remarkable enhancement after functionalization treatment, whereas by adding 0.1 wt% of CF-MWCNTs to the distilled water, thermal conductivity increased up to 0.83 W/m.K at 60 °C, which is 31% higher than that of distilled water (DW).

Acknowledgments

The authors of current work wish to thank University Teknologi PETRONAS and SUNWAY University, for their facilities and support to conduct this research work.

References

[1] Choi S E, *Journal of Chemical and Engineering Data* **8:** 281–5.
[2] Solangi KH, Kazi SN, Luhur MR, Badarudin A, Amiri A, Sadri R, et al. 2015 *Energy*
[3] Hussain OA, Salem TK. 2017 *J Adv Technol Eng Res*.**2**.
[4] Zeinali H S, Fallahi M, Shanbedi M, Amiri A. 2016 Heat Mass Transf **52:**85–93.
[5] Murshed SMS, Leong KC, Yang C. 2005 *Int J Therm Sci*. **44**: 367–73.
[6] Yarmand H, Ahmadi G, Gharehkhani S, Kazi SN, Safaei MR, Alehashem MS, et al. 2014 *Entropy* **16**: 6116–32.
[7] Azizi Z, Alamdari A, Malayeri MR. 2015 *Energy Convers Manag* **101**: 515–24.
[8] Ding Y, Alias H, Wen D, Williams RA 2006 *Int J Heat Mass Transf*. **49**: 240–50.
[9] Askari S, Lotfi R, Seifkordi A, Rashidi AM, Koolivand H.2015 *Energy Convers Manag* **109**:10–8.
[10] Yu W, Xie H, Bao D. 2010 *Nanotechnology* **21:** 055705.
[11] Chang GL, Rhee K 2014 *J Mat Scie* **49**: 1506–1511.
[12] Amiri A, Sadri R, Shanbedi M, Ahmadi G, Chew BT, Kazi SN, et al. 2015 *Energy Convers Manag* **12:**051.
[13] Yarmand H, Gharehkhani S, Kazi SN, Sadeghinezhad E, Safaei MR 2014 *Sci World J*
[14] Khoshvaght-Aliabadi M. 2014 *Energy Convers Manag* **88**: 96–105.
[15] Behroyan I, Ganesan P, He S, Sivasankaran S 2015 *Int Commun Heat Mass Transf* **67**:163–72.
[16] Sabiha MA, Saidur R, Hassani S, Said Z, Mekhilef S 2015 *Energy Convers Manag* **09:**009
[17] Choi SUS, Zhang ZG, Yu W, Lockwood FE, Grulke EA 2001 *Appl Phys Lett* **79**:2252–4.
[18] Das SK, Putra N, Thiesen P, Roetzel W 2003 *J Heat Transfer* **125**:567.
[19] Solangi KH, Kazi SN, Luhur MR, Badarudin A, Amiri A, Sadri R, Zubir MN, Gharehkhani S, Teng KH 2015 *Energy* **89**:1065-86.
[20] Amiri A, Sadat Alehashem M, Kazi SN, Shanbedi M, Ahmadi G, Sadri R, et al. 2015 *RSC Adv* **5**:35425–34.
[21] Zare, Zardini H, Amiri A, Shanbedi M, Taheri, Kafrani A, Kazi SN, Chew BT, Razmjou A 2015 *Journal of Biomedical Materials Research Part A* **103**:2959-65.
[22] Halelfadl S, Maré T, Estellé P 2014 *Experimental Thermal and Fluid Scienc* **53**:104-10.

Urethane based nanocomposite coated proppants for improved crush resistance during hydraulic fracturing

T Qian[1], A S Muhsan[1,*], L Htwe[1], N Mohamed[2] and O Hussein[3]

[1] Petroleum Engineering Department, Universiti Teknologi PETRONAS, 32610 Bandar Seri Iskandar, Perak, Malaysia.
[2] Fundamental and Applied Sciences Department, Universiti Teknologi PETRONAS, 32610 Bandar Seri Iskandar, Perak, Malaysia.
[3] Mechanical Engineering Department, Universiti Teknologi PETRONAS, 32610 Bandar Seri Iskandar, Perak, Malaysia.

*E-mail: ali.samer@utp.edu.my

Abstract. Hydraulic fracturing is one of the stimulation techniques that has been used to increase the hydrocarbon flow in unconventional reservoirs. Proppants are normally used to maintain the fracture conductivity and increase the permeability to oil. Due to high closure stress of the fractured formation, the conventional or uncoated proppants are easily crushed. Therefore, this project aims to increase the compressive strength of proppants by coating it with urethane based nanocomposite. Glass beads and fracking sands are coated with urethane resin and mixed with reduced graphene oxide (rGO) and carbon nanotubes (CNTs) respectively. Analysis testing using optical microscopy (OM) was conducted to study the morphology and microstructures of the coated proppants before and after crush test. Results showed remarkable improvement in proppants compression strength up to 41% and 35% after adding only 0.5% of CNTs and 0.1% of rGO respectively to the urethane resin coating. Therefore, it is recommended to utilize this material for coating the proppant with the optimized concentration to prevent pore plugging and proppant from flowing back, achieving the objective and solving both problem statements.

1. Introduction

Hydraulic fracturing is a commonly used stimulation technique to further increase the hydrocarbon fluid production from unconventional reservoirs [1]. Unconventional reservoirs are reservoirs that have permeability of less than 1mD. Hydraulic fracturing helps to increase the reservoir permeability by inducing fractures around the near wellbore region. To prevent the fractures from closing, proppants are introduced to maintain the fracture width after the fracturing fluid has been removed [2]. Proppant is a solid particle that is pumped into the formation by suspending in the fracturing fluid. The proppant will then deposit in the fracture together with other proppants to form a "proppant pack" [3]. Proppants also comes in different sizes, shapes and types to provide an optimum production of hydrocarbon fluids.

Proppants can be divided to uncoated and coated proppants. Uncoated proppants are also known as traditional proppants. For instance, glass bead proppants. Glass bead proppants was once known as the strongest proppant as it is round in shape and are well sorted [4]. However, glass bead proppants were found to be easily crumbled in brine at high temperature and hence, not suitable to be used in formation with high temperature as it will create fines, reducing the permeability to oil [5]. Coated proppants on the other hand is mostly resin based coated proppants. Resin coated proppants is a further modification

IOP Conf. Series: Materials Science and Engineering **863** (2020) 012013   doi:10.1088/1757-899X/863/1/012013

of both sand and ceramic proppants by coating resin on the uncoated proppants. There are two types of resins can be used, which are curable resin and pre-cured resins. Curable resins are used to prevent proppant flowback by consolidating the proppants whereas pre-cured resins are used to trap the fines generated by the proppants [7]. According to Palisch [6], both resins have the function of encapsulating the fines generated by the proppant but doesn't improve the proppant compressive strength. The proppant may be crushed by high closure stress, but the fines will not be migrated to the wellbore. Resins are also easily degraded at low softening temperature, approximately above 121 °C [2]. In the case where resin is being heated above 121 °C, resin will change its surface to a rubbery state and migrate to the wellbore, leaving uncoated proppants with low compressive strength.

Therefore, this work aims to investigate the compressive strength of different concentration of CNTs and rGO reinforced resin/urethane based coated proppant. To mimic the actual proppants, glass beads were chosen as it has higher compressive strength of 5,000 psi compared to sand proppants 4,000 psi. The resulting compressive strength for both nanomaterials coated proppants with resin will then be compared and the optimum concentration of nanomaterials that gives the highest compressive strength will be reported. By incorporating resin with rGO and CNTs as a proppant coating, it is expected that the coated glass bead proppants will be able to withstand at higher temperature condition since both rGO and CNTs have great thermal properties [8]. As shown in figure 1, it is also expected that the fines generated by both resin with rGO and resin with CNTs coated glass bead proppants are low as rGO and CNTs are known to have high mechanical properties [8]. This allowing the proppants to withstand high closure stress as well as preventing fines from generating. The figure also shows the fines generated are captured by the resin with CNTs or resin with rGO coatings after applied the formation pressure.

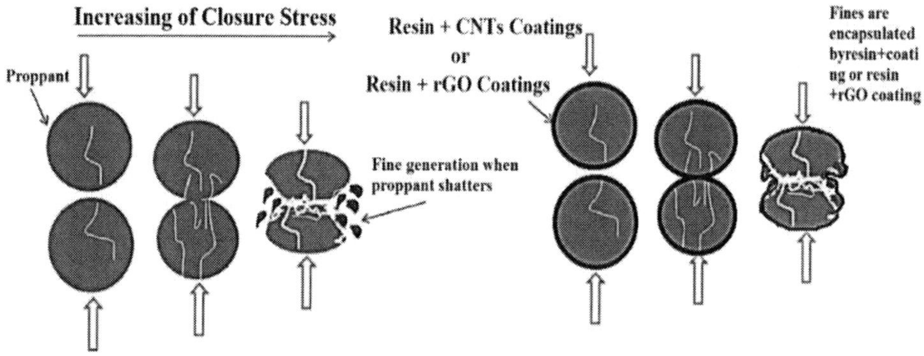

Figure 1. Fines are captured with resin, CNTs and GNPs coating.

2. Materials and methods

2.1. Materials

Different concentration of CNTs and rGO such as 0.1%, 0.5%, and 1%, resin, acetone and glass bead proppants are required in this experiment. Water also is required for the ultrasonic bath to ensure the uniform dispersion of nanomaterials in resin.

2.2. Apparatus and procedure

2.2.1. Carbon nanotubes coated proppants with resin. The CNTs is mixed with acetone and it is soaked into the ultrasonic bath to ensure the uniform dispersion of the CNTs in the solution. The purpose of using acetone is to dilute the resin or to reduce resin usage since resin has low thermal stability. The great thermal properties of CNTs will help to increase the overall thermal stability of proppants in the formation. The first step is then repeated by adding Urethane Part B and followed by Urethane Part A. The uncoated glass bead proppants are prepared by placing it on a siever. By using a pipette or a dropper, the well mixed CNTs and resin solution is then dripped onto the uncoated glass bead proppants. The excessive coatings will then flow to the bottom of the strainer and are wiped off. This is to ensure only a thin layer of coatings are coated on the proppants as well as to prevent any agglomeration on the proppants coating.

2.2.2. Reduced graphite oxide coated proppants with resin. The coating process is similar with CNTs but without the usage of ultrasonic bath as the reduced rGO is already well mixed in a solution form. The first step is to mix the rGO with the Urethane Part B. Then, another mixture of acetone with Urethane Part A is mixed. The purpose of adding acetone to Urethane Part A is to dilute the resin instead of mixing acetone with rGO as rGO is already well dispersed in a solution. Both mixtures are mixed together to form a rGO with resin coatings. The dropper will then drip the coatings onto the surface of the uncoated glass bead proppants and the excessive coatings are wiped off.

2.2.3. Compressive strength testing with Universal Testing Machine (UTM). Due to the limitation of compressive equipment, Universal Testing Machine (UTM) is used to perform crush test on all coated and uncoated glass bead proppants according to ISO 13503-2. The UTM machine exerts vertical stress onto the proppants to determine the compressive strength of the coated and uncoated proppants. A stainless-steel mold will be used together with the compression test. Once all of the proppants are fully crushed, the top plate is then released the load applied. A layer of proppants is prepared inside the mold and close it with a plunger. Then it is placed in between the top and bottom plate of the UTM machine. It is critical to ensure that only one even layer of proppants is placed in the mold as extra proppants will affect the compressive strength of the proppants. The results were presented in a stress vs deformation graph, with the maximum load (kN) and compression strength (MPa).

2.3. Measurement of fines Percentage (%) generated using 400 mesh siever

A 400 mesh siever is used to determine the amount of fines generated by each coated and uncoated glass bead proppants after the crush test. Any crushed proppants that are smaller than 40 mesh size are considered as fines. The % of fines generated here refers to the mass percentage and it is determined using equation (1):

$$\%\, of\ fines\ generated = \frac{Mass\ of\ crused\ proppants\ (g)}{Total\ Mass\ (g)} \times 100 \quad (1)$$

where the total mass is the mass of proppants used before the crush test and the mass of crushed proppants are the mass of fines that are smaller than 40 mesh size. After sieving the fines, the fines that are smaller than 40 mesh size are weighted, and the % of fines generated for each coated glass bead proppants and uncoated glass bead proppants is obtained.

2.4. Optical microscope on all coated proppants

Each coated proppant will undergo Optical Microscopy to observe on any agglomeration on the proppant coatings. According to figure 2, it is noticed that there is agglomeration of CNTs on the surface of the 1% CNTs coated proppant. Also, from figure 3, it is noticed that after 0.1% rGO coated proppants agglomeration of rGO is noticed on the proppant's coatings.

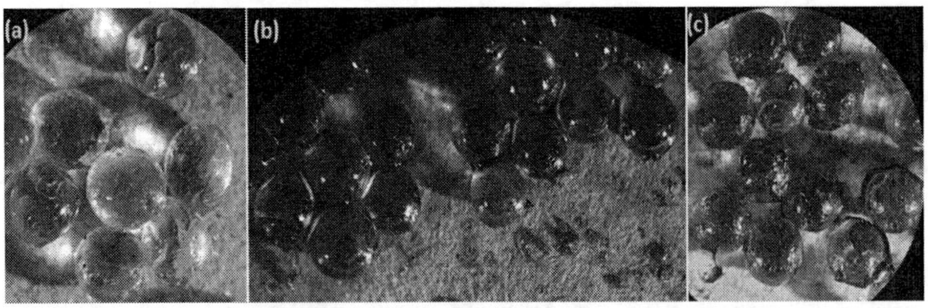

Figure 2. Optical microscope of (a) 0.1% CNTs, (b) 0.5% CNTs and (c) 1% CNTs.

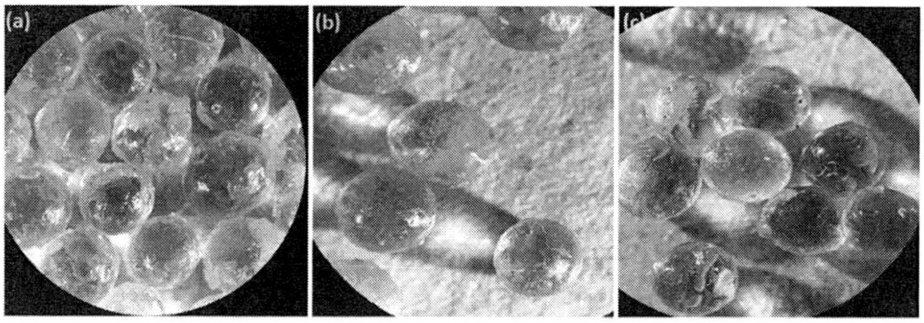

Figure 3. Optical microscope of (a) 0.1% rGO, (b) 0.5% rGO and (c) 1% rGO.

3. Results and discussion

Universal Testing Machine (UTM) was used to carry out crush test on all coated and uncoated glass bead proppants. Table 1 and figure 4 show the results from the UTM crush test showed that the uncoated glass bead proppants give the lowest compressive strength compared to the other CNTs coated proppants. This is because of the uncoated proppants are not protected by any coatings and hence, it is easily crushed with low load applied. The 0.5% CNTs coated proppants shows the highest compressive strength compared to the 0.1% CNTs and 1% CNTs coated proppants. In the beginning, the compressive strength of proppants increases as the concentration of CNTs increase. However, after 0.5% CNTs, it is noticed that the compressive strength of CNTs coated proppants starts to decline due to the agglomeration on the surface of the proppant coating. Agglomeration means there are uneven surface of coatings on the proppants and when pressure is applied on the proppants, the surface that has thinner coating than the agglomerated side tends to break first. This will cause the proppants have lower compressive strength. Thus, the most suitable CNTs concentration for proppant coatings is 0.5%.

Table 1. Compressive strength of each concentration of CNTs coated glass bead proppants.

Concentration of CNTs (%)	Fines generated (%)	Compressive Strength (MPa)
Uncoated	9.183	2923.125
0.1	12.847	4089.387
0.5	16.868	5369.251
1	13.536	4308.56

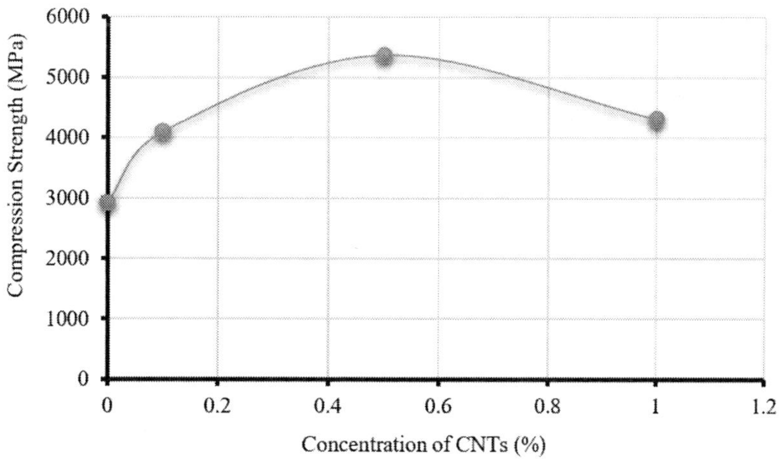

Figure 4. Compressive strength of uncoated proppants with different concentration of CNTs coated proppants.

For rGO coated proppants, it is noticed that 0.1% rGO gives the highest compressive strength compared to other concentration of rGO coated proppants as shown in Table 2 and Figure 5. It is also noticed that the uncoated proppants still have the lowest compressive strength among all the rGO coated proppants. Furthermore, the compressive strength of rGO coated proppants starts to decrease after 0.1% rGO coated proppants also due to the agglomeration of rGO on the proppants coating, which causes uneven stress exerted around the proppants surface. Usually, the side of the proppants that are agglomerated have protruding ends, which could increase the pressure applied on the agglomerated side and crush faster.

Table 2. Compressive strength of each concentration of rGO coated glass bead proppants.

Concentration of rGO (%)	Fines generated (%)	Compressive Strength (MPa)
Uncoated	9.183	2923.125
0.1	12.957	4124.395
0.5	11.832	3766.278
1	11.285	3592.18

4. Conclusion

In conclusion, 0.5% Carbon Nanotubes (CNTs) is the most suitable for proppants coating because it provides the highest compressive strength of 5369.25 MPa compared to 0.1% reduce Graphite Oxide (rGO) coated proppants as well as other concentration of CNTs coated proppants. This 0.5% CNTs coated proppants is able to improve the compressive strength of the uncoated glass bead proppants by almost 84%, which is twice the percentage of compressive strength improvement for 0.1% rGO coated proppants. Moreover, 0.5% CNTs coated proppants generate fines slightly more than the fines generated by 0.1% rGO coated proppants. With similar amount of fines generated but with different compressive strength, it is preferable to choose the coatings that are able to provide the highest compressive strength. By using 0.5% CNTs coated proppants, it is able to maintain the fracture width and fracture conductivity with its high compressive strength and prevent fines from migrating towards the near wellbore region, allowing more hydrocarbons to be produced. Therefore, it helps to prevent pore plugging and proppant from flowing back, achieving the objective and solving both problem statements.

References

[1] Bestaoui S N, Qu Q and Stephenson C J 2015 *Google Patents.*
[2] Liang F, Sayed M, Al-Muntasheri G A, Chang F F and Li L 2016 *Petroleum* **2** 26-39.
[3] Raysoni N, Salla R 2017 *Google Patents.*
[4] Robertson J Chilingarian G and Kumar S 1989 *Surface operations in petroleum production, II* **19B** (Elsevier Science) p 561.
[5] Geothermal fracture stimulation technology. Volume 1. Fracturing proppants and their properties 1980 *Report* 66-100 (USA) (https://digital.library.unt.edu/ark:/67531/metadc1182191/m1/66/).
[6] Palisch T T, Duenckel R., Chapman M A, Woolfolk S and Vincent M C 2009 *SPE Hydraulic Fracturing Technology Conference* (The Woodlands, Texas).
[7] DeCampos V, Sansone E and Silva G 2018 *Cerâmica* **64** 219-229.
[8] Meehan D N and Baker H 2011 *The way ahead* **7** 18-19.
[9] Sondergeld C, Simo H and Pournik M 2013 *SPE Production and Operations Symposium* (Oklahoma City, Oklahoma, USA).

IOP Conf. Series: Materials Science and Engineering **863** (2020) 012014 doi:10.1088/1757-899X/863/1/012014

Study of effect of moisture and temperature on thermal and physical properties of limestone using a transient plane source technique

F Anjum, F Rasool, F Mohyudin, A Ameen, M Y Naz*, A Ghaffar and S Shukrullah

Department of Physics, University of Agriculture, 38040 Faisalabad, Pakistan.

*E-mail: yasin603@yahoo.com

Abstract. This study investigated the thermophysical properties of limestone in its dry and moist states. A thermal constant analyser, based on Transient Plane Source (TPS) technique, was used to study the thermophysical properties. Temperature dependence of thermophysical properties of limestone was investigated by elevating the temperature up to 433 K. The chemical and mineral composition of samples were studied using SEM-EDS and XRD analysis. The thermal conductivity of dry and moist limestone showed that in situ measurements are necessary for the precise determination of thermal properties. The variations in thermal properties of limestones as a function of temperature were described by mineral composition and heat transfer mechanism. The thermal conductivity of limestone was measured in the range of 2.586 to 3.054 $Wm^{-1}K^{-1}$. The thermal diffusivity of samples was ranged from 1.390 to 1.601 mm^2s^{-1} while specific heat capacity of limestone was in the range of 1.847 to 2.066 $MJm^{-3}K^{-1}$. A noticeable change in thermal properties of moist limestone was observed as compared to the dry limestone.

1. Introduction

Porous stones are used in strengthen building structures due to their durability and high compressive strength. Natural stones as building materials are being used for thousands of years in the construction of structures and monuments. High volume of past literature is focused on sustainable energy resources and energy conservation techniques [1, 2]. The most efficient source of energy conservation is the use of thermal insulation to provide thermal comfort inside the buildings. In recent years, heat transfer properties of rocks are widely used to construct underground chambers for nuclear waste dumping, measurement of heat dissipation, determination of the rate of heat loss from earth due to the geothermal gradient, retrieving the history of earth's evolution, and construction of buildings on permafrost earth's zone [3, 4]. All these applications need accurate determinations of thermophysical properties of rocks.

In Pakistan, three types of stone are widely used in construction of monument and repairing the historic buildings. These stones include sandstone, limestone and marble. They can easily be cut into different shapes. For this study, limestone was chosen which is a sedimentary rock. The limestone consists of calcium carbonate (calcite) with small amount of magnesium carbonate (dolomite), silica, clay, iron oxide and carbonaceous material. Limestone is also used as raw material for cement [5, 6]. Calcite in limestone is sensitive to temperature and may affect the building environment.

The objective of this study was to investigate the influence of calcite content, moisture and temperature on the thermal properties of limestones. A thermal constant analyzer, based on transient plane source (TPS) method, was used to generate the data on thermal properties of limestone [5]. The

IOP Conf. Series: Materials Science and Engineering **863** (2020) 012014 doi:10.1088/1757-899X/863/1/012014

goal of this research was to compare the accuracy of in-situ thermal properties of dry and moist limestone samples with the existing theoretical thermal conductivity models. Furthermore, the temperature dependence of thermal conductivity, thermal diffusivity and specific heat capacity of limestone was measured to get important information about lithology and structure of the limestone

2. Materials and methods

The limestone samples were collected from a mountain located in the Murree Salt Range of Pakistan. The sampling was done at the height of 566 m to 799 m. From each bulk stone, three samples were cut in cubic blocks of dimensions 2.8 cm^3. The surface of all stones was polished with silicon carbide paper. After washing in distilled water, stone samples were dried in a heating oven at 110 °C for least 24 hours. The water absorption, water accessible porosity and bulk density of blocks were determined using Archimedes principle. The following equations were used in these calculations

$$WA(\%) = \frac{W_{wet}-W_{dry}}{W_{dry}} \times 100 \tag{1}$$

$$Porosity\ (\%) = \frac{W_{wet}-W_{dry}}{W_{dry}-W_{suspended}} \times 100 \tag{2}$$

$$Bulk\ density\ (g/cm^3) = \frac{W_{dry}}{W_{dry}-W_{suspended}} \tag{3}$$

where W_{dry} is the dry weight of samples after drying at 110°C, W_{wet} is the water-saturated weight of samples after boiling in distilled water for 4 h and cooling to room temperature. In case of building stones, W_{net} was determined by immersing stones in distilled water for 96 hours. $W_{suspended}$ is the suspended weight of water saturated samples, which was measured using equal arm balance (whose one pan was removed and a wire of equal weight was attached to suspend samples in water). During measurements, it was made sure that the sample does not touch the base or sides of the water container.

The chemical analysis of limestones was carried out through SEM based EDS. Thermal conductivity, thermal diffusivity and specific heat capacity of samples were measured using hot disk thermal constant analyzer based on TPS technique. Schematic design of humidity box attached with TPS setup for measurement of moist limestones is shown in figure 1.

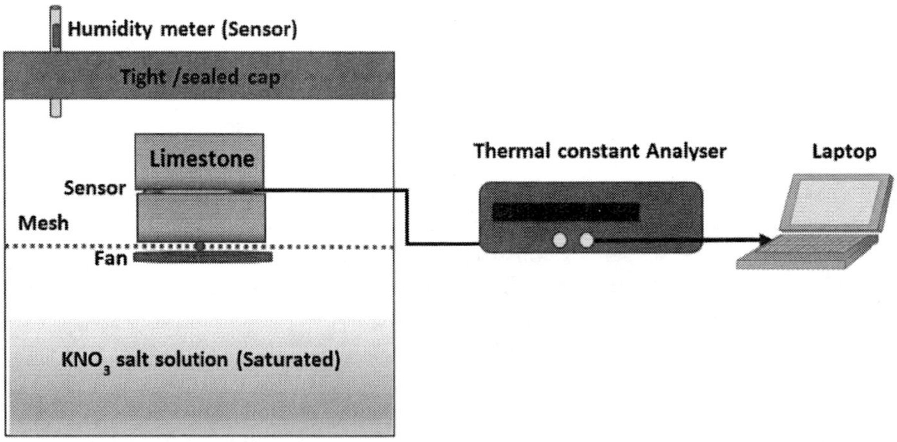

Figure 1. Schematic of humidity box attached with TPS setup.

3. Results and discussion

The results on physical properties, chemical and mineral composition and thermal properties of the limestone samples are discussed in this section.

3.1. Physical properties

Physical characterization of limestone samples is shown in Table 1. WA (%) of samples was found in a range of 0.8 to 1.4 and porosity (%) varied from 2.1% to 3.5%. Porosity (%) decreases with elevation from the baseline. The density of limestone also followed an increasing trend with elevation from the baseline. Overall, density increased from 2.62 g.cm^{-3} to 2.69 g.cm^{-3}. Figure 2 shows graphically the porosity at different positions above the baseline [7-9]. The trend in figure 2 satisfies a linear fit of the equation of $6.68 - 0.0061x$ and R^2 was determines as 0.91.

The compressive strength of limestones lied between 36 to 60.3 MPa and it is noticeable that the compressive strength of limestone decreased by 40% with moving from 566 m to 799 m above the baseline.

Table 1. Thermal properties of limestone samples.

Sample	Thermal conductivity ($Wm^{-1}K^{-1}$)	Thermal diffusivity (mm^2s^{-1})	Specific heat capacity ($MJm^{-3}K^{-1}$)
CC-1	2.586 ± 0.01	1.390 ± 0.09	1.865 ± 0.01
CC-2	2.860 ± 0.07	1.579 ± 0.11	1.847 ± 0.04
CC-3	2.903 ± 0.08	1.402 ± 0.03	2.066 ± 0.03
CC-4	3.054 ± 0.03	1.601 ± 0.05	2.029 ± 0.09

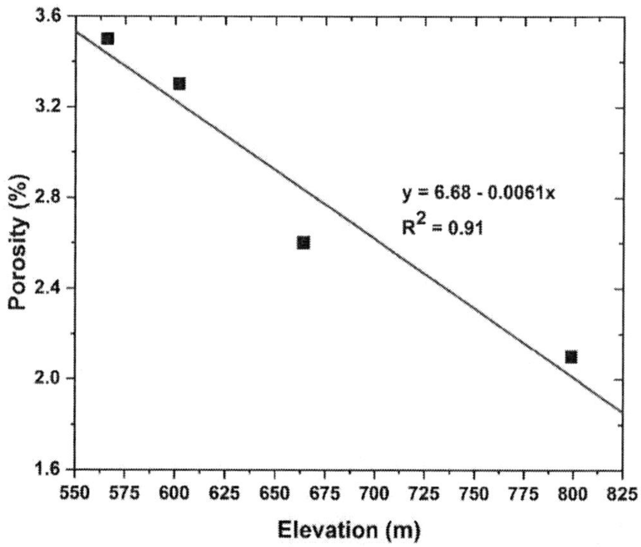

Figure 2. Variation in porosity (%) due to elevation from baseline.

3.2. Physical properties

EDS analysis was carried out on three data points for each limestone sample and the test data listed in Table 2 shows the average value of weight (%) with a standard deviation of each chemical element. The major elements were C, O, Ca. Some wt.% of Na, Mg, Al, Si, and K was also seen in the samples. Fe was found in CC-2, CC-3, and CC-4 samples while Cl was found in CC-1, CC-2, and CC-3 samples. EDS analysis confirmed that the major mineral phase in limestone is $CaCO_3$ (calcite).

Table 2. SEM-EDS results of limestone samples.

Elements	CC-1 Weight (%)	CC-2 Weight (%)	CC-3 Weight (%)	CC-4 Weight (%)	Chemical compound Standards
C K	12.60 ± 6.55	27.02 ± 3.76	21.01 ± 6.84	21.80 ± 1.25	$CaCO_3$ 1-Jun-99
O K	39.73 ± 8.49	55.05 ± 1.28	47.60 ± 0.66	47.80 ± 3.40	SiO_2 1-Jun-99
Na K	0.41 ± 0.49	0.22 ± 0.12	0.30 ± 0.25	0.28 ± 0.12	Albite 1-Jun-99
Mg K	0.19 ± 0.09	3.18 ± 3.56	0.18 ± 0.25	0.34 ± 0.23	MgO 1-Jun-99
Al K	0.20 ± 0.14	0.21 ± 0.05	0.19 ± 0.23	0.15 ± 0.04	Al_2O_3 1-Jun-99
Si K	0.08 ± 0.07	1.68 ± 2.59	1.07 ± 0.01	0.19 ± 0.01	SiO_2 1-Jun-99
Cl K	0.39 ± 0.39	0.09 ± 0.05	0.23 ± 0.13	-	
K K	0.11± 0.02	0.04 ± 0.04	0.27 ± 0.18	0.12 ± 0.01	MAD-10 Feldspar 1-Jun-99
Ca K	46.29 ± 14.58	12.30 ± 1.30	28.90 ± 7.88	29.02 ± 2.45	Wollastonite 1-Jun-99
Fe K	-	0.21 ± 0.01	0.25 ± 0.31	0.30 ± 0.22	Fe 1-Jun-99

3.3. Thermal properties

Table 3 shows the thermal properties of dry limestone samples at ambient temperature and pressure. The thermal conductivity of limestone samples was observed in the range of 2.586 to 3.054 $Wm^{-1}K^{-1}$. The porosity (%) and water absorption was decreased with an increase in height from baseline and density. However, the mineral composition remained unchanged by revealing that the thermal properties of the same type of stone characteristically dependent on physical properties [8, 9]. The thermal diffusivity of samples was ranged from 1.390 to 1.601 mm^2s^{-1} whereas specific heat capacity of limestone samples was in the range of 1.847 to 2.066 $MJm^{-3}K^{-1}$.

In rocks, porosity is generally very low and consequently, thermal properties primarily depend upon rock forming mineralogical constituents. Therefore, an increase in porosity can relate with an increase in pore fillings and as a result, the apparent density of limestone samples decreases (Table 1).

Table 3. Thermal properties of limestone samples.

Sample	Thermal conductivity ($Wm^{-1}K^{-1}$)	Thermal diffusivity (mm^2s^{-1})	Specific heat capacity ($MJm^{-3}K^{-1}$)
CC-1	2.586 ± 0.01	1.390 ± 0.09	1.865 ± 0.01
CC-2	2.860 ± 0.07	1.579 ± 0.11	1.847 ± 0.04
CC-3	2.903 ± 0.08	1.402 ± 0.03	2.066 ± 0.03
CC-4	3.054 ± 0.03	1.601 ± 0.05	2.029 ± 0.09

The thermal properties of water-saturated limestone samples are shown in table 4. In moist state, the thermal conductivity of limestone remained in the range of 2.984 to 3.153 $Wm^{-1}K^{-1}$, thermal diffusivity in the range of 2.079 to 2.289 mm^2s^{-1} and specific heat capacity of 1.406 to 1.598 $MJm^{-3}K^{-1}$. It is quite understandable that thermal conductivity of water-saturated sample is quite high: 30 times higher than the thermal conductivity of air [6]. The % increase in thermal conductivity of moist limestone however decreases with a decrease in porosity of limestone. The highest increase in thermal conductivity was 15.4% for moist CC-1 and the lowest increase in thermal conductivity was 8.5% for CC-3 and 8.7% for CC-4.

Table 4. Thermal properties of water saturated limestone samples.

Sample	Thermal conductivity $Wm^{-1}K^{-1}$	%increase from dry state	Thermal diffusivity mm^2s^{-1}	%increase from dry state	Specific heat capacity $MJm^{-3}K^{-1}$	%decrease from dry state
CC-1	2.984 ± 0.02	15.4	2.090 ± 0.11	49.9	1.431 ± 0.06	23.3
CC-2	3.153 ± 0.22	10.2	2.079 ± 0.14	31.6	1.598 ± 0.17	13.5
CC-3	3.149 ± 0.05	8.5	2.255 ± 0.20	60.8	1.406 ± 0.11	31.9
CC-4	3.321 ± 0.06	8.7	2.289 ± 0.24	43.0	1.465 ± 0.14	27.8

4. Conclusion

In this paper, four limestone samples were taken from a mountain at different heights from the baseline (566-799 m). The samples were characterized for their thermophysical properties under ambient conditions. The chemical and mineral composition of samples was measured by SEM-EDS analysis, which confirmed that the major composition of samples was calcite. The thermal conductivity of samples was measured about 2.586 to 3.054 $Wm^{-1}K^{-1}$. The thermal diffusivity of samples was ranged from 1.390 to 1.601 mm^2s^{-1} while specific heat capacity of limestone samples was in the range of 1.847 to 2.066 $MJm^{-3}K^{-1}$. A noticeable change in thermal properties of moist limestone was observed as compared to the dry limestone.

References

[1] Nardi I, De-Rubeis T, Buzzi E, Sfarra S, Ambrosini D and Paoletti D 2016 *Energies* **9** 677
[2] Raheem A, Abbasi S A, Memon A, Samo S R, Taufiq-Yap Y H, Danquah M K and Harun R 2016 *Energy Sustain. Soc.* **6** 1
[3] Maqsood A A 2007 *Int. J. Thermophys.* **28** 1371
[4] Suleiman B M 2006 *J. Phys. D. Appl. Phys.* **39** 547
[5] Horai K 1971 *J. Geophys. Res.* **76** 1278
[6] Clauser C and Huenges E 1995 *Handbook of Physical Constants*, ed T J Ahrens, American Geophysical Union, Washington DC USA 105
[7] Hall C and Hamilton A 2013 *Mater. Struct.* **48** 1265
[8] Anjum F, Ghaffar A, Jamil Y and Majeed M I *J. Mater. Cycles Waste Manag.* **0** doi:10.1007/s10163-018-0810-x.
[9] Somerton W H 1992 *Thermal properties and temperature related behaviour of rock/fluid systems*, Elsevier science publisher New York, USA

Viscosity impact investigation on erosion of carbon steel in sand-mixed-polivinyl alcohol

M Y Naz[1,*], N A Khan[2], M Naeem[2], F Sharif[2], F A Heral[2], A Ghaffar[2] and S Shukrullah[1]

[1] Department of Physics, University of Agriculture, 38040 Faisalabad, Pakistan.
[2] Department of Physics, Riphah International University Faisalabad, 38040 Faisalabad Pakistan.

*E-mail: yasin603@yahoo.com

Abstract. Erosion in oil and gas production sector is predominantly caused by fine particles' migration into the processing facilities. Fine particles are mostly transported to the facilities via carrier fluid. Other than sand, the viscosity of the carrier fluid also contributes to the sand erosion phenomenon. However, the impact of viscosity on sand erosion is not fully discovered by previous researchers. Therefore, this paper investigated the impact of fluid viscosity on erosion of carbon steel surface. In the presented work, an experimental rig was fabricated and tested for the study of impact of fluid viscosity on erosion of the carbon steel by considering the water mixed-polyvinyl alcohol as a carrier fluid. The metal coupons were impacted with sand slurries of different viscosities. Scanning electron microscopy along with energy dispersive spectroscopy and Universal scanning probe microscopy were used to analyse the centred and adjunct areas of the coupons. These analyses revealed that the viscous slurry is least effective in removal of metal surface in comparison with the pitting erosion. The obtained results reflected a decrease in erosion rate as the fluid viscosity upsurges and the type of metal surface erosion changes from uniform erosion to pitting erosion by fluid viscosity increment.

1. Introduction

Erosion is one of the most usual problems in oil and gas industry. The problems caused by sand increases as the wells and pipelines get older and more sand is produced. In addition to the environmental concerns and restrictions, the sand creates a variety of problems for the oil and gas industry such as solid particles interfere with control and instruments, reducing the reliability of the system significantly [1, 2]. The possibility to remove sand in a controlled and automated manner can grant the process systems with better control and safety. It can increase the lifetime of the equipment by allowing more sand to enter the production train and speed up the production rate [3].

Many sand management systems have been implemented over a period of time to exclude sand and its source down hole of the well [4]. These sand exclusion techniques involve gravel packing at the well head and/or using screens to exclude sand from entering the pipeline [5]. The sand exclusion methods along with continuous sand monitoring and control have been successful in cutting down the production of sand in pipelines to a large extent and are extensively used in oil and gas production wells. In spite of these measures, based on the geography of the well location, fine sands (less than 50-75 microns) find its way to the sand exclusion system and travels along with the fluid through the pipeline causing erosion and damage to the piping components [6]. In erosion phenomenon, tiny solid particles are carried through the pipelines of production well. Many parameters can exacerbate the erosion rate such as sand particle shape and size, impact angle, flow rate and fluid properties (viscosity and density) [7, 8]. It is beneficial to have clear understanding of how great the impact of each of these parameters on sand erosion; this understanding will lead us to have a better analysis and evaluation of how each of these parameters contribute to the magnitude of the sand erosion phenomenon.

Fluid properties such as density and viscosity play considerable role in the sand erosion rate and consequently damaging pipelines and other processing facilities [9]. The impact of fluid viscosity on erosion rate is not fully understood and still in its infancy [10]. However, significant research has been made to investigate the erosion phenomenon and to develop erosion models for predicting the erosion rate in various pipelines and different production situations [11, 12].

The effect of viscosity on the local thickness loss and total erosion ratio is crucial [9]. Sand particles are carried via fluid through the pipeline; the velocity of the flow in the pipeline will determine how big the erosion will be if the particle size assumed to be constant. It is well known that viscous fluids have lower rate of flow in comparison with fluid with lower viscosity. For many existing fields, the erosion potential of solid particles exists in any production well that involves the motion of solid particles within the carrier fluid. Erosion rate in processing facilities is highly related to several parameters such as fluid properties, flow rate, sand size and rate, material type, geometry as well as many others [11]. However, trying to isolate the effect of one parameter and evaluating its role in the erosion phenomenon can help to bring out a better understanding of each parameter status and parameter that should be given main consideration in analysis.

2. Materials and methods

The used sand was sandstone: fine sand matching the reservoir dimensions that can be a good representative of the real conditions. The carbon steel plates were used as erosion n sample for representing the inner surface of the production line that was eroded by fine sand produced from the oil reservoir. A two-phase pump was utilized to pump the polymer-sand mixture from the container to hit the carbon plate. A metal container was used as a reservoir having a mixture of 12 litres and 750 g of water and sand, respectively. A stand was used to hold the carbon steel plate in the top of the container, so it can be exposed to the fluid flow, as shown in figure 1.

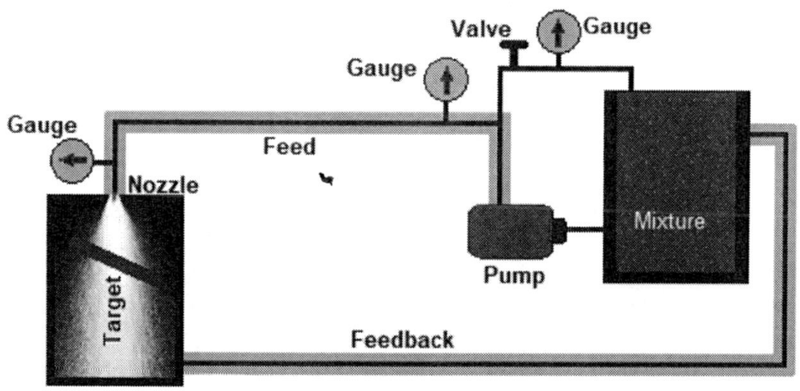

Figure 1. Schematic of fluid flow loop setup.

Polyvinyl alcohol was used as a viscosity increasing agent. Different amounts of polymer were added to the water-sand. These concentrations provided 1cP, 7 cP, 13 cP and 19 cP viscosity values to the fluid. Dry sieving method was used to characterise the sand particles. Sieve analysis is classic laboratory work implementation on a formation sand sample to determine grain/particle size distribution for sand control applications. The analysis was done by using a series of mesh having gradually decreasing screen sizes. The formation sample was placed on the top of the mesh series and seeped through the screens until it faces the screen which has smaller openings than the size of the grains. This method will used to obtain fine sands with dimension below 75 μm for the experiment.

Initially, 750 g of fine sand, with dimensions below 75 μm, was mixed in 12 l of tap water. This water to sand ratio remained constant for all experiments. The mixture of water and sand was pumped through a PVC pipe for duration of 8 hours and impinged onto the carbon steel plate. In the subsequent experiments, the water-sand mixture viscosity was increased by adding the polymer. Polyvinyl alcohol was used as the viscosity increasing agent and the effect of viscosity alteration on sand erosion phenomenon was measured using Universal scanning probe microscopy (USPM), energy dispersive spectrum (EDS), and scanning electron microscopy (SEM).

3. Results and discussion

The carbon steel coupons, eroded during the presented experimental work, were analysed through USPM, EDS, and SEM tools. For each experiment, two areas of interest were analysed in the treated sample, the area directly exposed to fluid flow and its adjunct area, which has not been directly eroded.

3.1. USPM analysis

After erosion experiments, the samples were washed and cleaned for topography analysis. The treated surfaces were scanned with a USPM probe to identify interactions in near-field regions. USPM analysis provided three dimensional (3D) models of the eroded plates plus erosion profile of the samples. The 3D models of directly eroded spots (central area) and the corresponding erosion profiles are shown in figure 2.

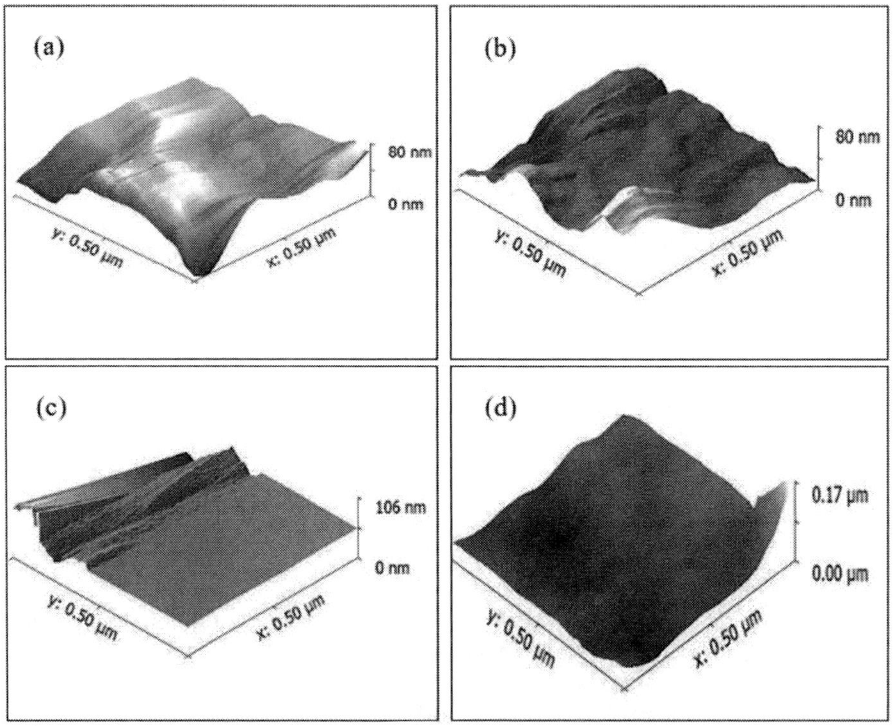

Figure 2. 3D models and corresponding erosion profiles of directly eroded coupons with: (a) 1cP solution, (b) 7 cP solution, (c) 13 cP solution and (d) 19 cP solution.

The results from 3D model show that the erosion severity decreases as the value of viscosity increases and the metal surface tend to be more unifrom. This is thought to be due to the high drag force and low fluid flow turbulance as the fluid viscosity increases [10]. Moreover, it can be observed from the erosion profile that although the severity of the sand erosion decreases, deep penetration by

IOP Conf. Series: Materials Science and Engineering **863** (2020) 012015 doi:10.1088/1757-899X/863/1/012015

the sand occures on the metal surface which inidcates that the type of erosion is changing from uniform surface erosion to pitting erosion.

The 3D models and corresponding erosion profiles of the adjunct areas are shown in figure 3. The analysis of USPM results related to the adjunct areas show that as it was expected, the erosion rate in the adjunct areas is lower than the directly eroded areas since the 3D models are showing a more homogeneous surface as the viscosity number increases. Moreover, the same phenomenon of erosion, as in case of directly eroded areas, was observed with an increase in viscosity of the fluid. The erosion was changing from uniform to pitting as the viscosity value increases. The number of deeper holes, resulted from pitting erosion, was higher in high viscosity values.

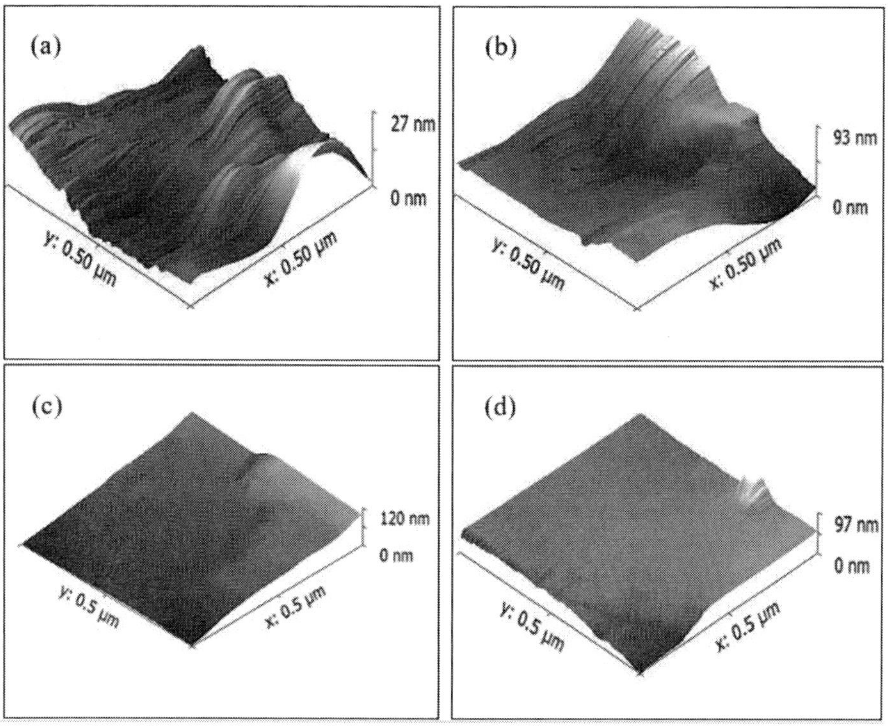

Figure 3. 3D models and corresponding erosion profiles of indirectly eroded coupons with: (a) 1cP solution, (b) 7 cP solution, (c) 13 cP solution and (d) 19 cP solution.

3.2. *SEM analysis*

SEM and EDS were used for surface and elemental analysis of the metal surface. For directly eroded spots, SEM results are shown in figure 4. Moreover, table 1 provides information regarding the detected iron and oxygen elements in the analysed spots. The pre-treatment samples were fine polished plates having no rough patches on the surface. However, after conducting sand erosion trials, the treated surface exhibited high degree of roughness. The viscosity of the carrier fluid governed the roughness of the eroded samples.

As the viscosity value increases, the congestion eroded area ratio to the not eroded one decreases since more untouched and untreated areas appear in samples treated in the experiment 3 and 4 with 13 cP and 19 cP solution viscosities. Moreover, it can be visually analysed that more dark holes were spotted in the SEM micrographs of the samples treated with high viscosity solutions. This can be a support to the earlier mentioned USPM analysis. Table 1 shows the percentage of iron and oxygen elements detected during the EDS analysis of the eroded samples. As it is tabulated, the percentage of detected iron is extremely higher in the samples treated with high viscosity solutions [7-10].

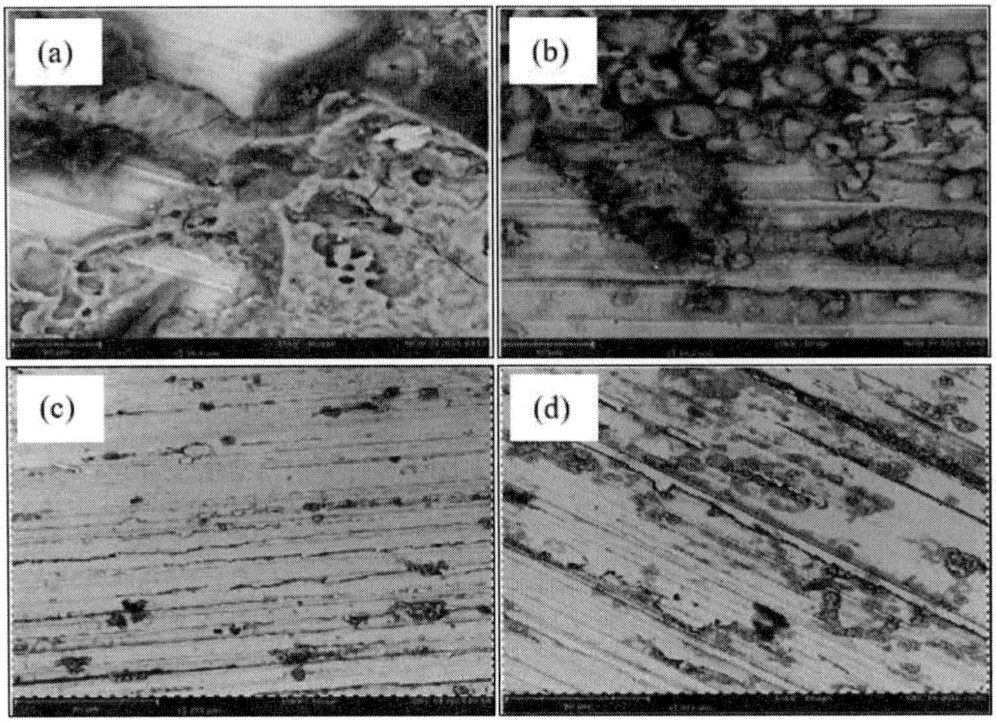

Figure 4. SEM illustration of eroded coupons with: (a) 1cP solution, (b) 7 cP solution, (c) 13 cP solution and (d) 19 cP solution.

Table 1. Iron and oxygen content of the eroded samples.

Viscosity (cP)	Element	Weight (%)
1	Oxygen	19
	Iron	25.2
7	Oxygen	33.1
	Iron	66.9
13	Oxygen	0
	Iron	100
19	Oxygen	0
	Iron	100

SEM of areas adjunct to the directly eroded spots are shown in figure 5. The erosion impact on the analysed surface starts to vanish after the second sample for viscosities higher than 7 cP. This observation is related to the higher viscosity values of the fluid, which resulted in less drag force and eventually less erosion values. The same analysis of the centred areas regarding the iron element percentage is valid for the adjunct areas. However, the iron percentage reduction in the centred areas as the value of viscosity increases remains lower than the iron percentage reduction in the adjunct areas. This behaviour evidently supports the conclusion made from the erosion profiles showing the uniform erosion in the adjunct areas which is more effective in surface metal removal in comparison with pitting erosion.

IOP Conf. Series: Materials Science and Engineering **863** (2020) 012015 doi:10.1088/1757-899X/863/1/012015

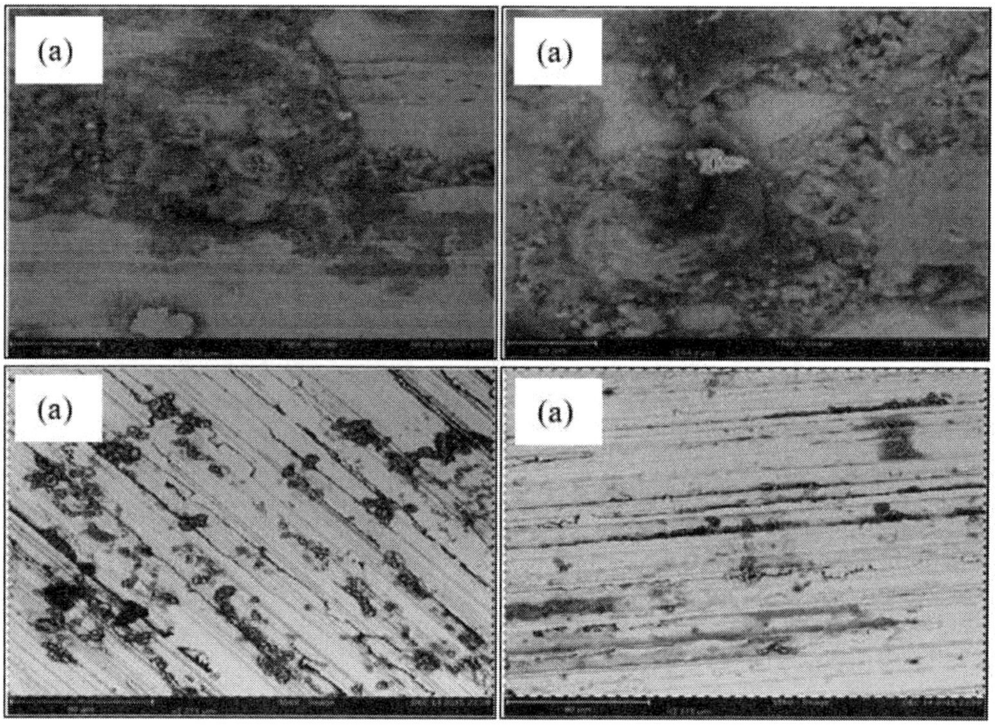

Figure 5. SEM illustration of indirectly eroded coupons with: (a) 1cP solution, (b) 7 cP solution, (c) 13 cP solution and (d) 19 cP solution.

4. Conclusion

Erosion of metal surface was investigated at four different viscosity values, 1 cP, 7 cP, 13 cP, and 19 cP. It can be concluded that the erosion velocity is more sensitive to the viscosity values below 8 cP as the erosion velocity plunges from about 120 cm/s to 40 cm/s. However, the effect of viscosity on erosion velocity losses its significance impact for higher viscosity values as the slope of the graph becomes more tedious.

The second observation is related to the erosion type. Overall there are two types of erosion in metal surface, one is uniform erosion where causes a uniform metal loss from the surface and the second is pitting erosion which causes deep holes in the metal surface. it can be observed from the erosion profiles that although the severity of sand erosion decreases, deeper penetrations caused by sand occurs on the metal surface which indicates that the type of erosion is changing from uniform erosion to pitting erosion in the centred areas. However, the governing type of erosion remains the uniform erosion, but with very low impact on surface. The obtained 3D models and EDS analysis also support this observation.

The erosion severity is indicated by the percentage of metal loss from the surface. For this purpose, the EDS and SEM analysis were used to support the conclusion. As discussed earlier and tabulated in the element detection table of each sample, as the viscosity number of the carrier fluid raises, the percentage of the detected metal on the surface upsurges. This suggest that the erosion impact in terms of its severity is restrained as the viscosity increases which is thought to be due to the lower drag force of the fluid and its lower turbulence stream. However, it is recommended to examine more viscosity values such as, 50 cP and 60 cP to have a holistic range of data points so they can be a better representative of the mere viscosity impact in extreme viscous oils. Furthermore, rerunning the same experiment procedure utilizing different common viscosity agents such as hydrolysed polyacrylamide can be fruitful due to the catalyst component difference of each polymer.

IOP Conf. Series: Materials Science and Engineering **863** (2020) 012015 doi:10.1088/1757-899X/863/1/012015

References

[1] Naz M Y, Ismail N I, Sulaiman S A and Shukrullah S 2016 *Instrum. Sci. Technol.* **45** 62
[2] Islam M A and Farhat Z N 2014 *Wear* **311** 180
[3] Naz M Y, Sulaiman S A, Shukrullah S, Ghaffar A, Ibrahim K A and AbdEl-Salam N M 2017 *Measurement* **106** 203
[4] Bellarby J 2009 *Develop. Petrol. Sci.* **56** 129
[5] Naz M Y, Ismail N I, Sulaiman S A and Shukrullah S 2015 *Sci. Rep.* **5** 16583
[6] Taslim M and Spring S 2010 *J. Am. Helicopter Soc.* **55** 22006.
[7] Parsi M, Najmi K, Najafifard F, Hassani S, McLaury B S and Shirazi S A 2014 *J. Natural Gas Sci. Eng.* **21** 850
[8] Chen X, McLaury B S and Shirazi S A 2004 *Comput. Fluids* **33** 1251
[9] Zhang Y, Reuterfors E, McLaury B S, Shirazi S and Rybicki E 2007 *Wear* **263** 330
[10] Mazumder Q H, Shirazi S A, McLaury B S, Shadley J R and Rybicki E F *Wear* **259** 203
[11] Naz M Y, Ismail N I, Sulaiman S A and Shukrullah S 2016 *Procedia Eng.* **148** 896
[12] Edwards J K, McLaury B S and Shirazi S A 2001 *J. Energ. Resourc Technol.* **123** 277

IOP Conf. Series: Materials Science and Engineering **863** (2020) 012016 doi:10.1088/1757-899X/863/1/012016

Experimental investigation of impact of dust accumulation on the performance of photovoltaic solar module

S Naz, M Nadeem, M Zain-ul-Abideen, S K Buttar, S Shukrullah*, M Y Naz and Y Jamil

Department of Physics, University of Agriculture Faisalabad, 38040 Faisalabad, Pakistan.

*E-mail: zshukrullah@gmail.com

Abstract. The effect of dust and sand on the surface of photovoltaic module has been investigated experimentally in this paper. The size distribution analysis of each sample was carried out using digital microscope. The chemical composition was investigated through EDX analysis. Experiments on photovoltaic solar module were conducted under controlled conditions by using artificial light source at different tilt angles and a dust blower. It has been concluded that there was only a small difference of degradation in maximum voltage and current when dust and sand was deposited on photovoltaic panel. The voltage decrement was 9.1%, 3.1%, 1.2% and 6.1% at tilt angles of 35°, 45°, 55° and 65°, respectively. These results reveal that the solar panel should be installed at angles between 45° and 55° in order to minimize the dust impact.

1. Introduction

The demand of sustainable and clean sources of energy is increasing every day. Solar photovoltaic (PV) energy has become the great concern of achieving electricity due to its several advantages over other energy sources. It proves green energy with no carbon emission, which is its main advantage and it is environment-friendly source of clean energy. According to a report, about 232 MWp energy from new solar modules was gained by the Chinese consumers until 2012. The Chinese government is planning to enhance this capacity to 1800 MWp in 2020. There are a lot of factors like humidity, soiling and wind velocity, tilt angle, mismatch of array and efficiency of invertor that effect the adeptness of solar modules. The impact of dust on performance of solar PV is of great concern. The significant impact of weather, dust and location may degrade the solar PV efficiency. Besides that, the size of dust and surrounding condition also have great influence on output power of modules. El-Shobokshy and Hussein [1] examined that smaller particles cover the solar cell more densely than larger particles. Many studies on this problem confirmed that high concentration of airborne dust on open-air PV installation degrade the cell glazing transmittance. This factor became the root of substantial reduction in efficiency of PV modules.

Nahar and Gupta [2] conducted some experiments in Thar dessert (India) to examine the influence of dust on different glazing surfaces and measure their transmittance. They observed that by increasing the tilt angles from horizontal position, the dust resolution decreases. They found 19.17%, 13.81% and 5.67% depletion in transmittance of glass surface at different tilt angles. The transmittance was also decreased significantly with increasing the tilt angle. El-Shobokshy and Hussein [1] measured energy output of solar cells under zero wind condition by polluting solar panel surfaces with different dust types. The result concluded that effect of dust might be not further associated to the exposure time and fine particles has greater impact than large particles [3]. It was revealed that dust type, dust size and density of dust accumulation cause a reduction in energy efficiency of PV solar panels. El-Nashar *et al.* [4] observed the impact of dust deposition on the efficiency of solar collectors in United Arab Emirates for different periods. The result revealed that the monthly glass transmissivity decline rate was between 6% to 10%, depending on season. Hottel and Wortz [5] conducted 90 days trial on performance of solar collectors at a fixed tilt angle of 30. The net performance of collector was decreased by 4.7%. The effect of dust was not so significant because collector remained cleaned due to frequent rainfall.

El Hamdani *et al.* [6] observed that the effect of dust deposition on solar PV modules is more pronounced for fine particles than large particles. Elminir *et al.* [7] examined the impact of dust on solar collectors having transparent cover by using hundred samples of glasses and concluded that at tilt angles of 0° to 90°, the transmittance decreases from 52.54% to 12.38%. The decline rate of transmittance is affected by dust deposition density. According to Mejia *et al.* [8], the decrease in performance of solar photovoltaic system is due to dust deposition is 7.4% for 145 days in summer season. Sulaiman *et al.* [9] studied the impact of dust by using different impediment materials on efficiency of photovoltaic panels and concluded that the outdoor elements decrease the solar performance up to 85%. Alwaeli *et al.* [10] observed almost 50% of reduction in performance of photovoltaic modules installed in Baghdad city of Iraq. They revealed that the thickness of dust increases and amount of transmittance decreases over specific period of time [10-14].

Therefore, the purpose of this research paper is to investigate current and voltage response to dust and sand which accumulated on surface of photovoltaic module for a specific time period. The experiment was conducted under controlled conditions at different tilt angles.

2. Materials and methods

The experiment was conducted under controlled conditions in plasma physics laboratory at university of Agriculture, Faisalabad, Pakistan. Figure 1 shows schematic of the experimental arrangement. A 30 W polycrystalline PV module was placed under an artificial lighting source, comprising of two filament tungsten bulbs each of 100 W. An artificial lighting setup was used to avoid inconsistency in solar radiation. A high precision digital multimeter was used to record current and voltage of PV module. Four digital thermometers were used to monitor the surface temperature at different points on the module. A dust generator was used to blow the module with dust and sand samples at different tilt angles.

The specification of solar PV module is summarized in table 1. Firstly, the solar panel was installed at four different angles of 35°, 45°, 55° and 65°. A stand was placed behind the module on which two filament tungsten bulbs were fixed by using nails. Two unit of 100 W bulbs were used as light source. A dust generator was fixed at 4 feet from the panel. The dust and sand were uniformly spread over module surface in separate experiments. The solar panel was divided into four sections and the temperature of each section was monitored using digital thermometers. Each experiment conducted for 60 minutes of duration and data were taken for five minutes interval. The flowrate was fixed at 500 g/h.

IOP Conf. Series: Materials Science and Engineering **863** (2020) 012016 doi:10.1088/1757-899X/863/1/012016

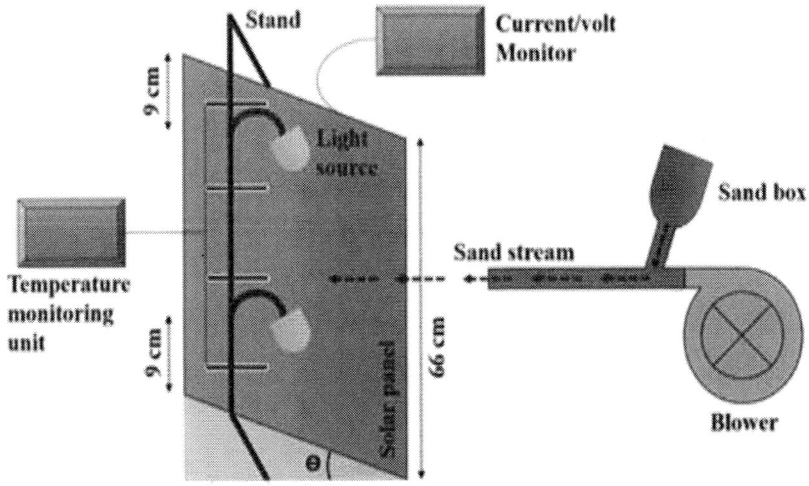

Figure 1. Schematic of experimental setup.

Table 1. Specifications of PV module.

Parameter	Value
Solar module type	BCT30-12
Peak power (pm)	30 W
Maximum power current (I_{mp})	1.83 A
Maximum power voltage (V_{mp})	16.4 V
Short circuit current (I_{sc})	1.98 A
Open circuit voltage (V_{oc})	21.6 V
Dimensions	660×408×25 mm^3

3. Results and discussion

The surface of panel was simulated for dusty surface by falling sample of dust and sand applied on the surface. In order to create the reference dust, the samples of dust and sand were prepared specifically for this work. Dust was collected from the agricultural field of Ghulam Muhammad Abad, Faisalabad and sand from the local building material shop. The samples were analysed for their surface morphology and particle size through a digital microscope. The obtained images were analysed to observe the nature and size distribution of sample particles. Chemical composition was also determined through EDX. The magnified images of dust and sand samples are shown in figure 2. The results show that the size of the samples remained between 62 μm and 128 μm.

Figure 3 shows the trend of temperature over time. The surface temperature remained unchanged over time. Since artificial light source was used to illuminate the PV module, the light intensity did not change over time and consequently the temperature. A comparison of average temperature for different tilt angles reveals no effect of angle on surface temperature under artificial lightening. Some variations in temperature were observed during early stage of the experiment, thereafter temperature become stable over time. Even dust accumulation did not affect the surface temperature. Similar temperature results were reported for both sand and dust accumulations.

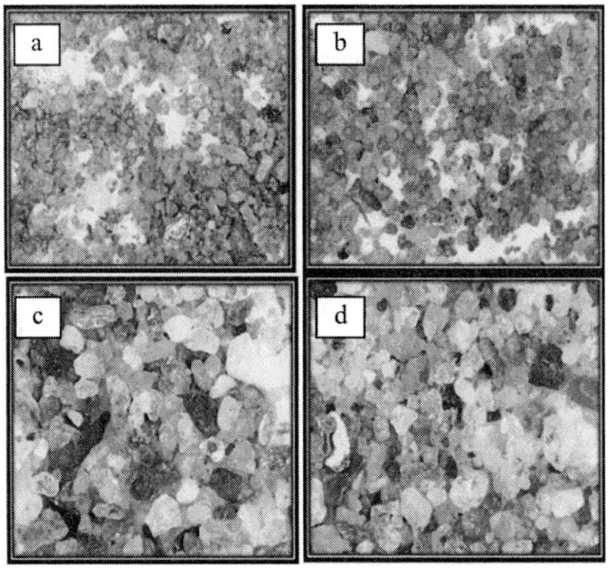

Figure 2. Image of dust (a, b) and sand (c, d) samples.

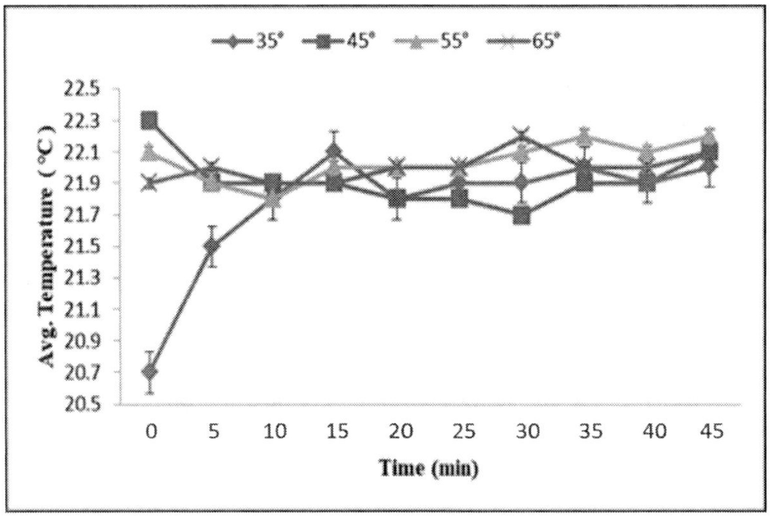

Figure 3. Average surface temperature as a function of time for different tilt angles.

Figure 4 and figure 5 show module voltage as a function of time at different tilt angles for both dust and sand. A degradation of the voltage was observed with time due to increament in dust layer thickness and tilt angle. For dust and sand, voltage was decreased by 2.5 V and 0.5 V respectively, with an increase in tilt angles. Similarly, voltage was decrease by 0.7 V and 2.5 V after 60 minutes of continuous dust and sand bombardment. It reveals that tilt angle is more important during dust accumulation and test time during sand accumulation.

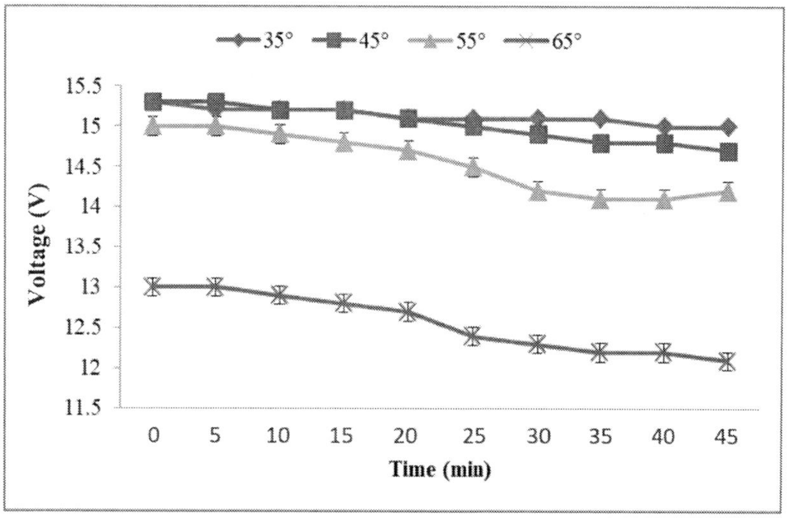

Figure 4. Voltage as a function of dust accumulation time for different tilt angles.

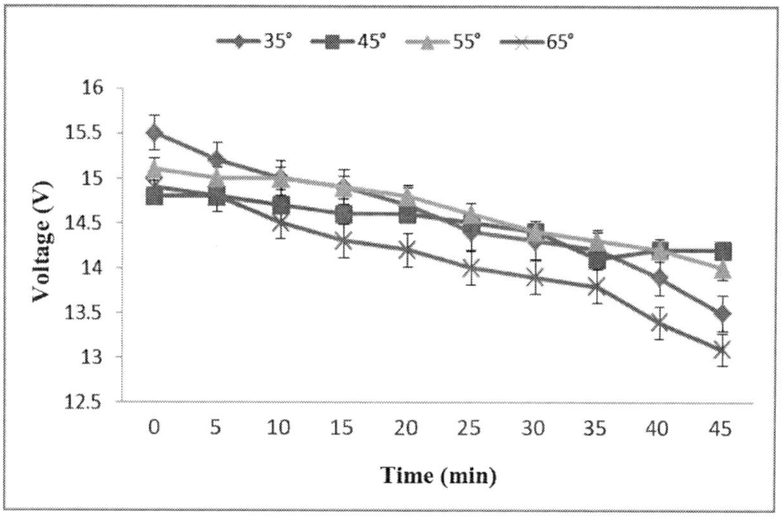

Figure 5. Voltage as a function of sand accumulation time for different tilt angles.

Figure 6 and 7 show current as a function of time at different tilt angles for both dust and sand. The current revealed a decreasing trend with tilt angle for both dust and sand. The current was decreased by 40 mA in dust environment and 35 mA in sand environment with an increase in tilt angle from at 35° to at 65°. For dust dominated environment, tilt angle of 35° to 45° is recommended for PV modules. For sand dominated environment, tilt angle of 45° to 55° is recommended for PV modules. There was no effect of exposure time on current was observed in the presented study. The current remained almost constant over time. These results were generated for fixed exposure time dust/sand flowrate. The presented results may vary with change exposure time and flowrate. The effect of these parameters on the performance of PV modules will be considered in future work.

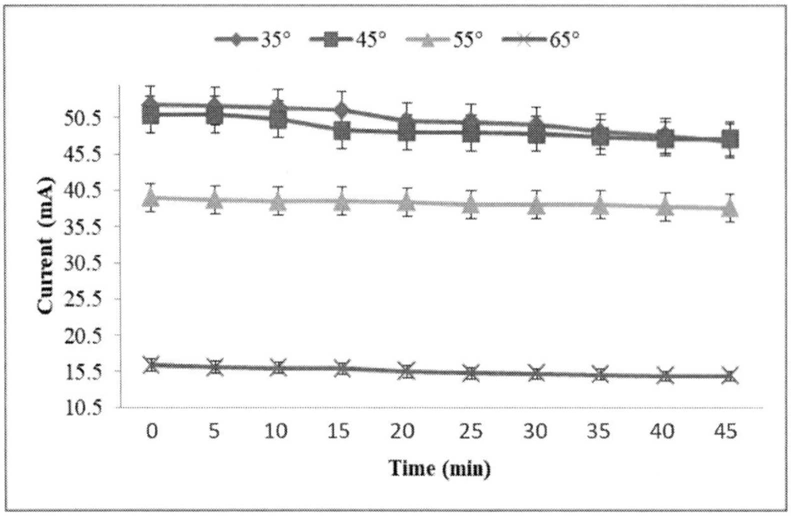

Figure 6. Current as a function of dust accumulation time for different tilt angles.

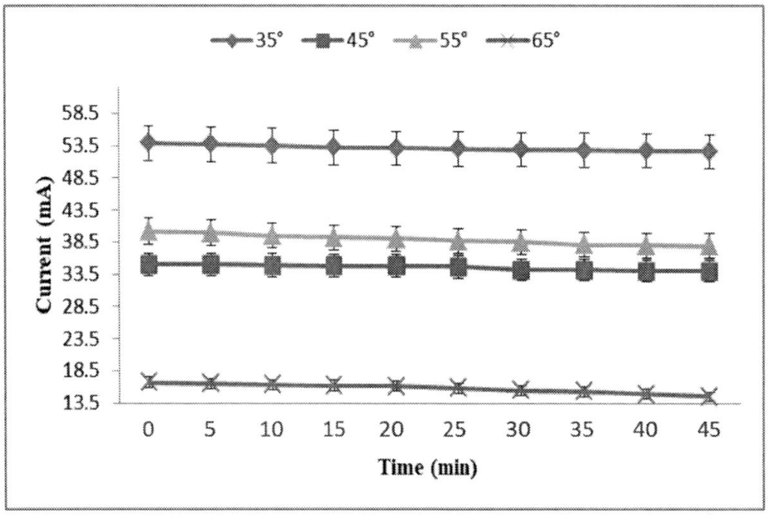

Figure 7. Current as a function of sand accumulation time for different tilt angles.

4. Conclusion

In this study, the impact of dust on performance of photovoltaic module has been investigated at different angles under constant artificial light source. Some variations in surface temperature were observed during early stage of the experiment, thereafter temperature become stable over time. The dust and sand accumulation did not affect the surface temperature. For condition of dust and sand, the voltage was decreased by 2.5 V and 0.5 V, respectively with an increase in the tilt angle. Similarly, voltage was decrease by 0.7 V and 2.5 V after 60 minutes of continuous dust and sand bombardment. It has been found that the tilt angle is more important during dust accumulation and test time during sand accumulation. For dust dominated environment, tilt angle of 35° to 45° is recommended for PV modules. Meanwhile, for sand dominated environment, tilt angle of 45° to 55° is recommended for PV modules.

References

[1] El-Shobokshy M S and Hussein F M 1993 *Sol. Energy* **51** 505.
[2] Nahar N M and Gupta P J 1990 *Sol. Wind Tech.* **7** 237.

[3] Mani M and Pillai R 2010 *Renew. Sust. Energ. Rev.* **14** 3124.
[4] Maghami M R, Hizam H, Gomes C, Radzi M A, Rezadad M I and Hajighorbani S 2016 *Renew. Sust. Energ. Rev.* **59** 1307.
[5] El-Nashar A M 1994 *Sol. Energy* **53** 105.
[6] Hottel H and Wortz B 1942 *Am. Soc. Mech. Eng.* **64** 91.
[7] El Hamdani S, Limam K, Abadie M O and Bendou A 2008 *Atmos. Environ.* **42** 8893.
[8] Elminir H K, Ghitas A E, Hamid R H, El-Hussainy F, Beheary M M and Abdel-Moneim K M 2006 *Energ. Convers. Manage.* **47** 3192.
[9] Mejia F A and Kleissl J 2013 *Sol. Energy* **95** 357.
[10] Sulaiman S A, Singh K A, Mokhtar M M M and Bou-Rabee M A 2014 *Energ Procedia* **50** 50.
[11] Alwaeli A, Chaichan M T and Kazem H A 2014 *Renew. Sust Energ. Rev.* **37** 734.
[12] Kazem H A, Chaichan T M, Saif S A, Dawood A A, Salim S A, Rashid A A and Alwaeli A A 2015 *Int. J. Sci. Eng.* **6** 293.
[13] Goossens D and Van-Kerschaever E 1999 *Sol. Energy* **66** 277.
[14] Hegazy A A 2001 *Renew. Energ.* **22** 525.

IOP Conf. Series: Materials Science and Engineering **863** (2020) 012017 doi:10.1088/1757-899X/863/1/012017

Photo-catalytic response of zinc-doped nickel ferrite nanoparticles for removal of chromium from industrial wastewater

M Shoaib-ur-Rehman, Z Javaid, I Asghar, M Azam, S Shukrullah*, F Iqbal, A Maqsood and M Y Naz

Department of Physics, University of Agriculture Faisalabad, 38040 Faisalabad, Pakistan.

*E-mail: zshukrullah@gmail.com

Abstract. In this study, Ni-$ZnFe_2O_4$ nanoparticles were synthesized by co-precipitation technique and used to remove the chromium (Cr) metal from the industrial wastewater. The synthesized nanoparticles were characterized for their structural properties and morphology by using X-ray diffraction (XRD) and scanning electron microscopy (SEM) techniques. The average particle size of synthesized nanoparticles was calculated by applying the Scherrer's formula on the characteristic peaks of XRD patterns. The large surface area, controlled size and well-defined structure of the synthesized Ni-$ZnFe_2O_4$ nanoparticles showed remarkable results for wastewater treatment. The characterized nanoparticles were distributed into the industrial wastewater under constant stirring for 20 minutes. The nanoparticles' suspension was subjected to UV–VIS spectrophotometry for the study of photo-catalytic behaviour of Ni-$ZnFe_2O_4$ nanoparticles. The effect of Ni-$ZnFe_2O_4$ was investigated under visible light irradiation. Zn doping in Ni-$ZnFe_2O_4$ showed remarkable effect on efficiency of Cr removal from the wastewater.

1. Introduction

Environmental pollution caused by industrial wastewater containing different chemicals especially synthetic dyes has become an area of special interest for scientists. Some of the industrial chemicals and synthetic dyes decompose aerobically and anaerobically resulting in the formation of carcinogenic compounds [1]. The classical techniques which were used in practice to decontaminate polluted water include adsorption, chlorination, and membrane process [2, 3]. In the past few decades, there has been great interest among scientists in developing different techniques with great prospective for environment protection applications such as water disinfection [4]. Photo-catalysts has been used successfully for the treatment of hazardous materials such as industrial effluents containing dyes. The ferrites are liked by scientists due to the properties which can be modified according to the need. The size and shape of nanoparticles are very important for adsorption process [5].

The method of adsorption with magnetic separation has been widely used in cleaning the harmful materials in the water which cause problem to human health as well as environment. Different ferrites have been used to separate different substances dissolved in the water which are not suitable [6].

Chromium is the most prior heavy metal pollutant which introduces in the human bodies from various industrials processes such as electroplating, tanning, processing of metals, fabrication of steel, paint manufacturing and photography. Chromium is mostly used in the ceramics, explosives and photography

[7]. Industries using chromium often discharge contaminated wastewater that may endanger human life and health. Chromium is a mutagenic and carcinogenic nature the presence of chromium in water is one of the most important issue.

This paper reports the efficiency of Ni-Zn ferrite nanoparticles in removing chromium form dye industrial wastewater. The adsorption technique and photo-catalytic behavior of Ni-$ZnFe_2O_4$ nanoparticles were investigated under visible light irradiation. UV- visible spectrometer was used to study the concentration of chromium in the solution. It was observed that chromium removal efficiency was increased with increasing of zinc doping.

2. Materials and methods

The $Ni_{1-x}Zn_xFe_2O_4$ with the selected "x" as 0.25, 0.50, 0.75, was prepared by co-precipitation method. In the co-precipitation method, calculated amounts of hexa-hydrated nickel chloride ($NiCl_2.6H_2O$), hexa-hydrated ferric chloride ($FeCl_3.6H_2O$), Zinc Chloride ($ZnCl_2$), sodium hydroxide (NaOH) and acetone (CH_3COCH_5) were weighted accurately and dissolved in doubly distilled water. The precipitation was carried out at a controlled pH of 12-13 using NaOH solution by constant stirring for 120 minutes. The beaker containing the dark brown precipitates was placed into pre-heated water bath containing water. The temperature of water bath was kept constantly at 75 °C for 2 hours for preparation of each samples. The precipitates were filtered and washed thoroughly with distilled water to make them free from sodium and chloride ions. This process continues until the pH of the residual solution reached 7.0. The final product was dried in an electric oven at heating temperature of 70 °C for 4 hours to remove the moisture contents. The fully dried samples were grinded to obtain the desired particle size. Before grinding the samples, pestle and mortar were washed carefully with acetone (CH_3COCH_5). Thereafter, the grinded precursor was annealed at 800 °C for 2 hours. The calcinated samples were set to cool slowly under ambient temperature conditions.

The ferrite phases were identified through X-ray diffraction (XRD) analysis and the morphology of the prepared ferrites was examined using scanning electron microscopy (SEM). Particle size of the ferrite samples was determined using Scherrer's formula.

The photocatalytic activity of the prepared $Ni_{1-x}Zn_xFe_2O_4$ ferrite samples was evaluated by absorption of Cr from dye industry under visible light irradiation. Quantity of samples S-1, S-2 and S-3 under visible light irradiation 0.05, 0.1, 0.15 and 0.20 of ferrite nanoparticles was added to 20 ml dye industry water. The solution was exposed to visible light under stirring for 20 minutes after addition of ferrite in dye water. At a given time interval, dye water was centrifuged to remove the Cr. The concentration of the Cr in solution was determined by using UV-Vis spectrophotometer.

3. Results and discussion

A typical XRD pattern of $Ni_{1-x}Zn_xFe_2O_4$ powder, prepared using co-precipitation method, is shown in Figure 1. This pattern confirms the formation of ferrite phases in the identified crystalline structures. The diffraction peaks also confirmed the good crystallization of the product. Sharp diffraction peaks, consistent with face centered cubic spinel structure, were noticed in XRD patterns of all the investigated samples. Peaks at 2θ of 31.9334, 34.6246, 35.6042, 36.4199 and 62.7154 were referred to XRD diffraction planes (220), (311), (311), (222) and (440) respectively. These peaks confirmed the formation of cubic spinel nickel copper ferrite when compared with JCPDS card 070-2674.

The lattice constant of $Ni_{1-x}Zn_xFe_2O_4$ ferrites was calculated using *d*-values, obtained from figure 1, in the following equation [8].

$$d = \frac{a}{(h^2 + k^2 + l^2)^{0.5}} \tag{1}$$

The lattice constant calculated for the cubic spinel structures is given in table 1. It is observed that the values of lattice constant for S-1, S-2 and S-3 are 8.36 Å, 8.35 Å and 8.33 Å, respectively. Lattice constant of $Ni_{1-x}Zn_xFe_2O_4$ was not significantly influenced by a change in concentration. Negligible effect of different concentrations on the lattice constant was noted in the presented work. The lattice constant values were in a close agreement with the results published by the American Society for Testing Materials [9].

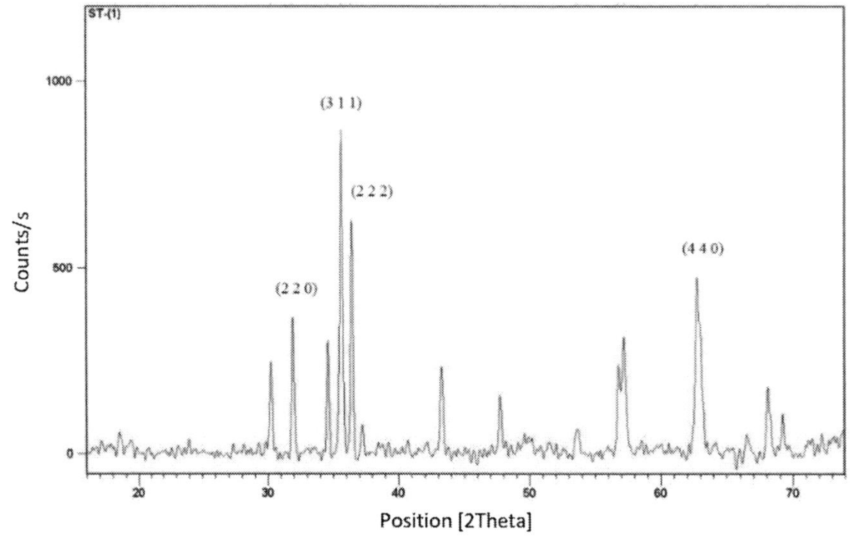

Figure 1. Typical XRD patterns of co-precipitated S-1 sample of $Ni_{0.75}Zn_{0.75}Fe_2O_4$.

Table 1. Estimated amount of Zn in Ni-Zn ferrite nanoparticles for the removal of Cr from industrial wastewater.

Zn Con.	Stirring time (min)	Cr in untreated samples (g/l)	Amount of Cr in treated samples (g/l)		
			S-1	S-2	S-3
0.05	20	0.65	0.42	0.40	0.36
0.10	20	0.65	0.40	0.30	0.30
0.15	20	0.65	0.32	0.36	0.30
0.20	20	0.65	030	0.35	0.26

The volume of the unit cell of the synthesized samples was calculated using the respective lattice constant values from table 1. The relation between the volume and the lattice constant exists in the form:

$$V = a^3 \tag{2}$$

The calculated volume of the samples is also given in table 2. It can be observed that the values of volume of the unit cell for S-1, S-2 and S-3 are 584.28 Å, 582.18 Å and 578.01 Å, respectively. Negligible effect of concentration on volume of the unit cell of $Ni_{1-x}Zn_xFe_2O_4$ was seen due to negligible variation in lattice constants.

The crystal size of $Ni_{1-x}Zn_xFe_2O_4$ ferrite samples was determined by considering the most intense diffraction peaks (311) in XRD patterns. In this study, the Debye-Scherrer formula was used to calculate the crystal size of tested samples as shown by equation (3):

$$T = \frac{0.9\lambda}{\beta \cos\theta} \tag{3}$$

Where 'θ' is the Bragg angle between incident ray and diffracted plane, 'λ' is the wavelength of X-rays used and 'β' is measured in radians (line broadening at intensity equal to the half maximum intensity). Table 2 shows that crystal size for S-1 ($Ni_{0.75}Zn_{0.25}Fe_2O_4$), S-2 ($Ni_{0.50}Zn_{0.50}Fe_2O_4$) and S-3

IOP Conf. Series: Materials Science and Engineering **863** (2020) 012017 doi:10.1088/1757-899X/863/1/012017

($Ni_{0.25}Zn_{0.75}Fe_2O_4$) calcinated at same temperature of 800 °C, are 51.56 nm, 39.92 nm and 33.01 nm, respectively. It has been found that the crystal size decreases with a decrease in the concentration of $Ni_{1-x}Zn_xFe_2O_4$ and increases with an increase in the ferrite concentration. The average crystal size decreases with the decreasing Ni content in $Ni_{1-x}Zn_xFe_2O_4$ nanoparticles. It may be due to small grain growth of Ni (Mote et al., 2013). Besides that, the crystal size increases with an increase in calcinating temperatures [10].

Table 2. Lattice constant a, volume *V*, crystalline size *T* and x-ray density of $Ni_{0.25}$-$Zn_{0.75}$ Fe_2O_4 calcinated at 800 °C.

Sample Name	a (A^0)	Volume	T (nm)	d_x(gm/cm^3)
S-1	8.36	584.28	51.56	5.36
S-2	8.35	582.18	39.92	5.43
S-3	8.33	578.01	33.01	5.41

The SEM images for $Ni_{1-x}Zn_xFe_2O_4$ (x= 0.25, 0.50 and 0.75) shown in figure 2. The grain sizes of S-1 and S-3 are same as 0.2 while S-2 is 0.1 μm. SEM graph of the synthesized samples shows their morphology. The graphs were obtained using secondary electron detector such as semi-inlens. Meanwhile, the images were obtained with the semi-inlens secondary electron detector. SIMS depth profiles were made using a ToF-SIMS 5 (IONTOF GmbH, Munster, Germany). Depth profiles of all the samples were carried out in the non-interlaced mode with instrument JSM-5910 at accelerated voltage of 20 volt.

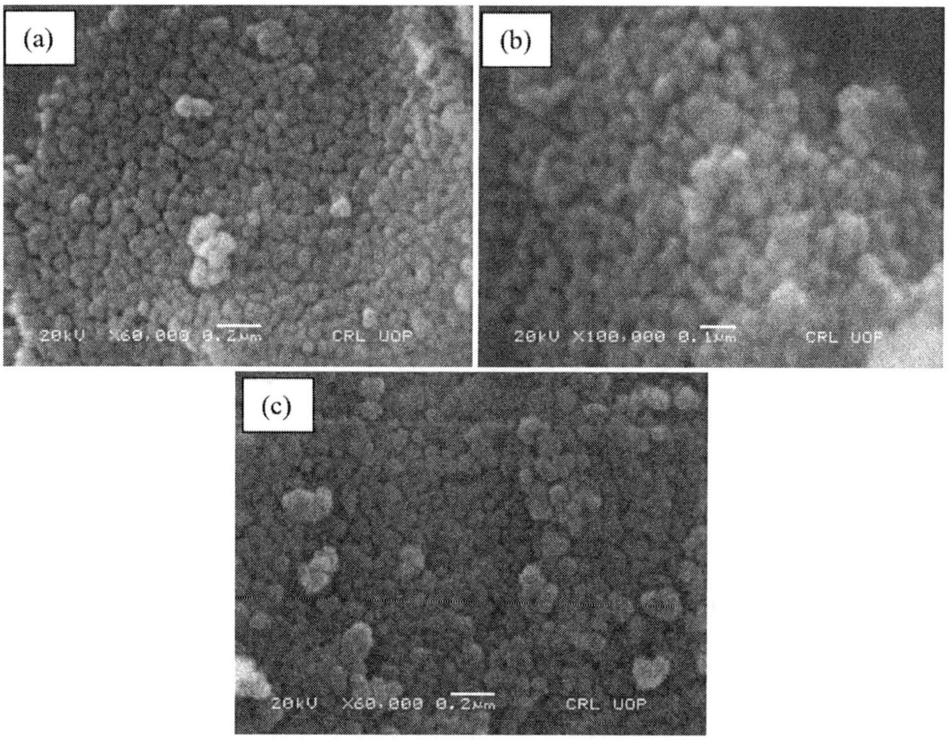

Figure 2. SEM micrographs of (a) S-1 ($Ni_{0.75}Zn_{0.25}Fe_2O_4$) (b) S-2 ($Ni_{0.50}Zn_{0.50}Fe_2O_4$) and (c) S-3 ($Ni_{0.25}Zn_{0.75}Fe_2O_4$).

The photoluminescence spectra was recorded for the study of optical properties and the recombination of energetic position of sub band gap defects states and photo generated charge carriers. Figure 3 shows the spectra of $Ni_{1-x}Zn_xFe_2O_4$ nanoparticles at excited wavelength of 430 nm. In this work, the emission peaks in the UV range at 371–493 nm results from the recombination of excited electron and holes. The emissions peaks between 415–450 nm correspond to the violet emission which might be due to the presence of lattice defects, oxygen vacancies and grain boundaries in $Ni_{1-x}Zn_xFe_2O_4$ nanoparticles [10]. The lowering of luminescence intensity of ferrites with increase in doping fraction, leads to the decrease in recombination rate of electrons-holes under light irradiation and the higher separation rate of photo generated electron-hole pairs efficiently.

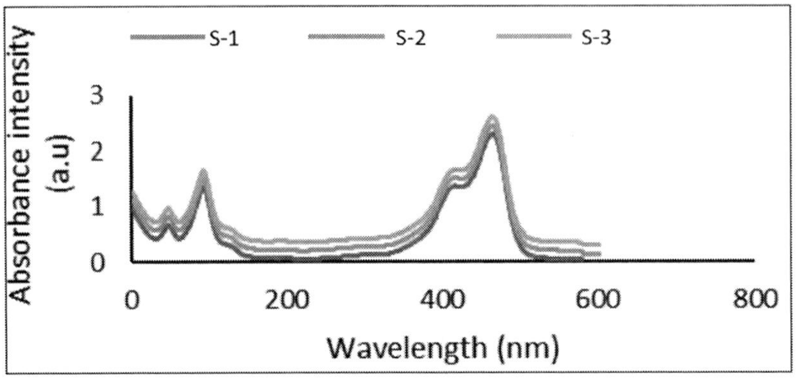

Figure 3. Photoluminescence spectra of S-1 ($Ni_{0.75}Zn_{0.25}Fe_2O_4$), S-2 ($Ni_{0.50}Zn_{0.50}Fe_2O_4$) and S-3 ($Ni_{0.25}Zn_{0.75}Fe_2O_4$).

Irradiation of industrial wastewater in the presence of $Ni_{1-x}Zn_xFe_2O_4$ (x= 0.25, 0.50 and 0.75) nanoparticles with two units of 200 Watt Hg lamp lead to a change in absorbance with 20 minutes time. Photocatalytic activity of NiZn Fe_2O_4 on dye industrial wastewater with different concentration of Zn is given in table 3. It was observed that the removal efficiency increases as the Zn concentration increase.

Table 3. Removal efficiency of co-precipitated Ni-Zn ferrite nanoparticles in industrial wastewater.

Sample Name	Concentration	Calcinated Temp. (°C)	Removal Efficiency (%)
S-1	0.25	800	40.22
S-2	0.50	800	48.18
S-3	0.75	800	60.09

It can be observed that for all three samples the highest removal efficiency from wastewater is obtained in 20 minutes of stirring. At higher stirring time, the increasing of the adsorption process efficiency is not significant. Therefore, it is not recommended to work under higher stirring time from the economic point of view.

4. Conclusion

The well single cubic structures of $Ni_{0.75}Zn_{0.25}Fe_2O_4$, $Ni_{0.5}Zn_{0.5}Fe_2O_4$ and $Ni_{0.25}Zn_{0.75}Fe_2O_4$ phases were synthesized and calcining all samples at 800 °C for 2 hours. The XRD patterns of co-precipitated samples showed the (3 1 1) peak at 2θ=35. With an increase in concentration of zinc in $Ni_{1-x}Zn_xFe_2O_4$ ferrite nanoparticles, the crystal size was decreased. The ranges of lattice constant for all the three samples are 8.36Å, 8.35Å and 8.33Å. The average crystal size for all the three samples of $Ni_{1-x}Zn_xFe_2O_4$ was remained in the range of 51.56 nm to 33.01 nm. The x-ray density of the samples increases with the increase in concentration of zinc. It was observed that for all three samples, the highest removal efficiency of the Cr from the wastewater was possible during the first 20 minutes of stirring and S-3 is found to be most effective among all other sample for Cr removal. At higher stirring times, the adsorption efficiency was not significantly by the time.

References

[1] Afkhami A, Sayari S, Moosavi R and Madrakian T 2015 *J. Ind. Eng. Chem.* **21** 920.
[2] Ali I 2012 *Chem. Rev.* **112** 5073.
[3] Chandamma N, Manohara B M, Ujjinappa B S, Shankarmurthy G J and Kumar M S 2017 *J. Alloy. Compd.* **702** 479.
[4] Cheng Z, Tan A L K, Tao Y, Shan D, Ting K E and Yin X J 2012 *Int. J. Photoenergy* **69** 1.
[5] Duan Q, Lee J, Chen H and Zheng Y 2017 *Water Sci. Technol.* **76** 3069.
[6] Erdem M and Tumen F 2004 *J. hazard. Mater. B* **109** 71.
[7] Fang B, Yan Y, Yang Y, Wang F, Chu Z, Sun X, Li J and Wang L 2016 *Water Sci. Technol.* **5** 1112.
[8] Toqeer I, Naz M Y, Khan Y, Azam M, Alkanhal M A S Meer R M M 2019 *Philos. Mag. Lett.* **99** 67.
[9] Ge F, Zhu L and Wang J 2008 *Desalination* **225** 156.
[10] Gul I H, Abbasi A Z, Amina F, Anis-ur-Rehman M and Maqsood A 2007 *J. Magn. Magn. Mater.* **311** 494.
[11] Gul I H, Ahmed W and Maqsood A 2008 *J. Magn. Magn. Mater.* **320** 270.

Computational fluid dynamics study of a single stage solar still for water treatment

M Hussain, B Shoukat, A A Wahid, S Shukrullah*, M Anwar, M Z Ashraf and M Y Naz

Department of Physics, University of Agriculture Faisalabad, 38040 Faisalabad, Pakistan.

*E-mail: zshukrullah@gmail.com

Abstract. The freshwater is only 2.5% of the total water on this earth and rest is the brackish. At present, the human being is facing a serious problem of availability of potable water. The amount of potable water can be enhanced by translating brackish water into drinkable through solar distillation in solar stills. Different geometries of solar stills are being designed to serve the distillation purpose. The performance of a solar still is mainly influenced by its design parameters and wall material. In the presented work, a 3-dimentional geometric modeling of a solar still was carried out by using k-epsilon model in ANSYS-FLUENT. The objective was to study the effect of the design parameters on performance of a solar still through CFD modelling followed by the experimental validation. ANSYS-FLUENT 14.0 package was used to simulate the solar still system. The simulated work was compared with the experimental data and a good agreement was seen among the experimental and simulated data.

1. Introduction

The depleting fresh water sources is a global issue. The uses of polluted water may result in numerous diseases. About thirty thousand individuals die daily because of water-borne diseases [1]. Population of the globe is increasing exponentially in the underdeveloped countries and so does the demand of pure water. For the existence of living organisms, the water is the most vital component. The most parts of Punjab and Sindh in Pakistan have saline to brackish ground water. By processing the saline water through solar distillation, it can be fairly converted into drinkable water [2]. Through convection, heat is transferred from the basin to the water, from the glass cowl to the atmosphere and from vapors to the glass cowl. From a solar still, heat flows to the atmosphere through the clear glass cowl and also the walls by physical phenomenon. But heat flows from the sun to the solar still through radiations [3, 4]. A better design of solar still can significantly improve the conversion process.

Passive and active solar stills are two major categories of solar distillation systems. In passive solar stills, lowering the depth of water in the basin may increase the conversion efficiency and the daily production rate of the single basin solar stills [5, 6]. To improve the simple basin stills, different techniques & modifications are reported in the published literature [5-8]. These modifications and techniques are helpful in increasing the efficiency and life of the solar stills.

Computational fluid dynamics (CFD) is a well know approach used to analyze and solve the fluid flow problems under different situations. Computing machines are responsible for this analysis, which is based on a simulation process [9]. For multipart configurations, due to large grid points, the computational time increases, which is difficult to manage. In this study, the water temperature and production rate of drinkable water were predicted and simulated results were compared with the experimental results. Further comparison has been made between simulation and experimental results of water temperature and glass cover temperature.

2. Materials and methods

This research deals with CFD study of single stage solar still by using ANSYS simulation software. The solar purification system on experimental side was consisted of a rectangular top glass-cover, wooden box with square shape base and a metal tray sheet. The surface area and volume of single stage solar still was 16.472 m^2 and 4 m^3, respectively. A 0.02 m thick glass sheet to cover the top surface of the solar still. A 0.05 m thick metal try sheet was used to increase the solar intensity and a wooden box was used to reduce the damages due to rising temperature. The glass-cover was fixed at an inclination angle of 24°. When the solar still having saline water was exposed to sunlight, the condensation and evaporation of saline water taken place. The experiments were performed during May 2018 in the University of Agriculture Faisalabad, Pakistan. From 9 am to 5 pm daily, the temperature of the basin water and glass cover were monitored continuously.

The geometry of the solar still was designed in ANSYS workbench, as shown in figure 1. The free stream temperature of upper and lower surfaces was set to 403 K and 293 K, respectively. The rest of the walls were considered thermally insulated. The maximum height of the rear and front walls of the cubic solar cell were set at 1.5 m and 2 m, respectively, and the glass lid was fixed at an inclination angle of 24°. The fine cell meshing of the geometry was performed, as shown in figure 1. The unstructured tetrahedral meshing was used for the reported solar still. The equations of continuity, energy and momentum for the fine meshed geometry were solved using ANSYS FLUENT 14.0 package. k-epsilon model was used to simulate the solar still process. The cell zone and boundary conditions were set according to the experimental design. *k*-epsilon model has been implemented in most general purpose CFD simulations and is considered as an industrial standard model. It is a stable and numerically robust simulation tool. It has well-established regime of predictive capability. For general purpose simulations, the model offers a good compromise in terms of accuracy and robustness. Within FLUENT, the turbulence model uses the scalable wall-function approach to improve robustness and accuracy when the near-wall mesh is very fine. The scalable wall functions enable solutions on arbitrarily fine near-wall grids, which is a significant improvement over standard wall functions.

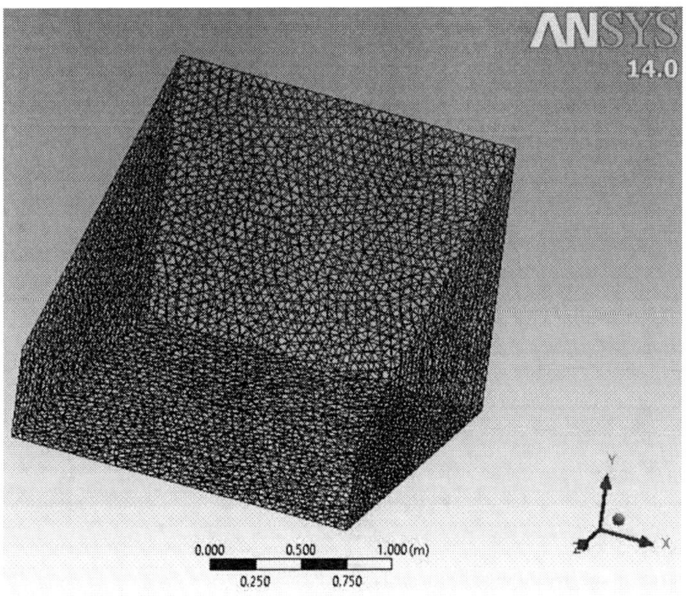

Figure 1. 3D meshed geometry of a solar still.

3. Results and discussion

The computation time per simulation, required to reach quasi steady-state conditions, was about 3-5 hour. Figure 2 shows temperature distribution on the glass cover of the solar still. This glass contour shows a range of colors according to the changing temperature. At bottom, the red color shows maximum temperature as compared to upper surface area. The lower part of glass cover was closer to the saline water. The vaporized water condenses in this glass area of maximum temperature. Kinetic energy of vapors increases at lower surface as compared to upper glass surface. A distance between upper and lower surfaces of glass plate and tilt angle was the reason of temperature difference between these surfaces. The absorber side plate was of Aluminum, which absorbs temperature from the coming rays through the transparent glass cover. Table 1 shows temperature of both absorber plate and transparent glass top [7-9].

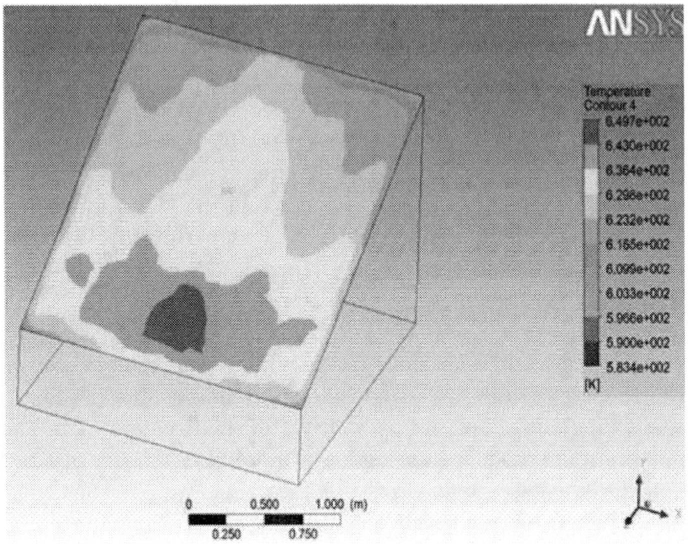

Figure 2. Temperature contour of glass plate cover.

Table 1. Temperature of glass and absorber plate.

X-axis	Temp. of glass plate (K)	Temp. of absorber plate (K)
1	6.31E+02	6.17E+02
2	6.29E+02	6.35E+02
3	6.26E+02	6.40E+02
4	6.38E+02	6.44E+02
5	6.38E+02	6.41E+02
6	6.34E+02	6.37E+02
7	6.29E+02	6.37E+02
8	6.23E+02	6.37E+02
9	6.10E+02	6.27E+02
10	6.20E+02	6.35E+02
11	6.32E+02	6.35E+02
12	6.34E+02	6.34E+02

The outer side walls ware made of wood to prevent heat loss. Figure 3 shows the contour plot of the wooden side walls. A comparison of temperatures of front glass wall and side walls is given in table 2. There is no big difference between side and glass plate temperatures, but difference gradually increases with time.

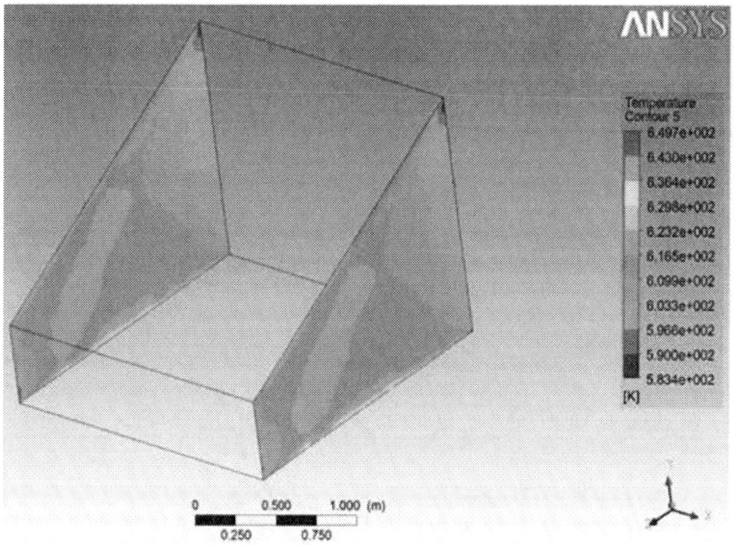

Figure 3. Contour plot of wooden side walls.

Table 2. Temperature of front glass cover and side walls.

X- axis	Temp. of Glass plate (K)	Temp. of side walls (K)
1	6.31E+02	6.02E+02
2	6.29E+02	6.11E+02
3	6.26E+02	6.15E+02
4	6.38E+02	6.15E+02
5	6.38E+02	6.15E+02
6	6.34E+02	6.19E+02
7	6.29E+02	6.06E+02
8	6.23E+02	6.08E+02
9	6.10E+02	6.10E+02
10	6.20E+02	6.12E+02
11	6.32E+02	6.14E+02
12	6.34E+02	6.18E+02

Figure 4 shows a change in temperature in upward direction. Due to fixed temperature of the lower surface and the upper surface, the temperature changes rapidly near the vertical front wall. In the glass centre and the front wall, the water enthalpy is also high in the absorption plate. Near a shorter height along the vertical wall, high heat transfer coefficient was observed due to high temperature gradient as shown in figure 5.

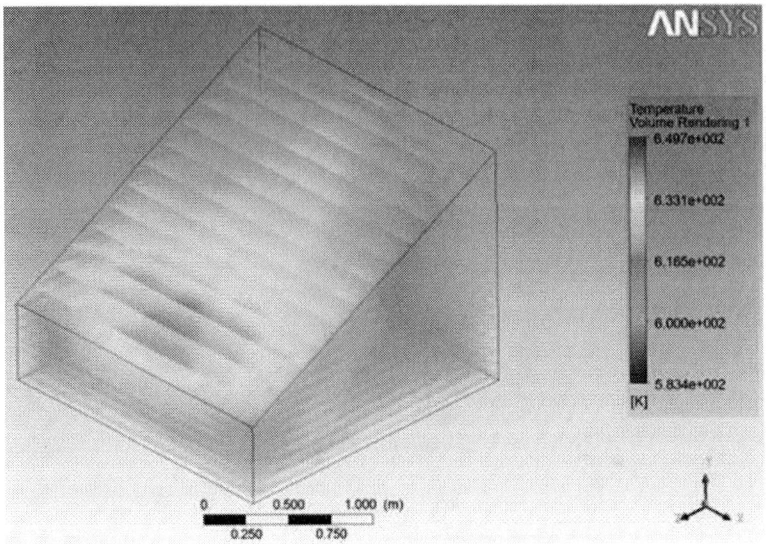

Figure 4. Vertical temperature rendering of the tested solar still.

Figure 5. Coefficient of heat transfer in the tested solar still.

On the experimental side, the fabricated solar still was tested during summer season. The data was generated for 30 days on daily basis. The experimental data was compared with simulated results. Figure 6 provides a comparison of experimental and simulated data. A good agreement was seen among the production rates of potable water predicted through experiments and simulation with the error as small as 8.7 percent.

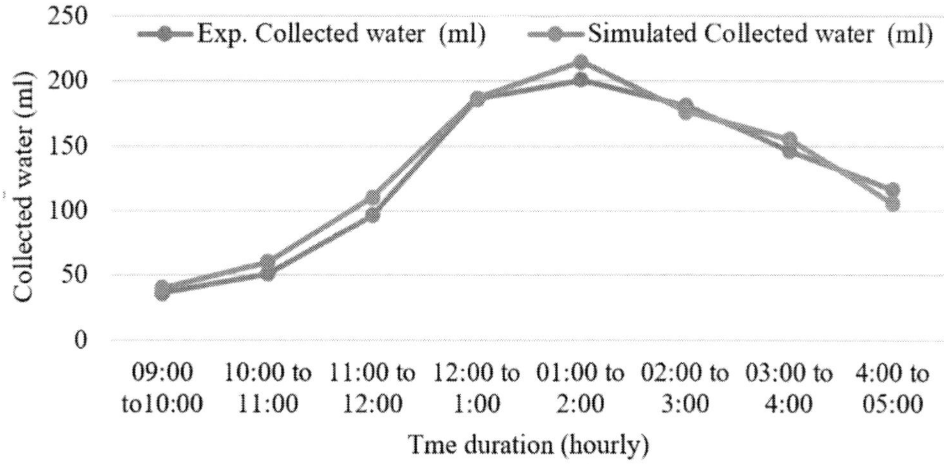

Figure 6. A comparison of water production rates achieved through experiments and simulation.

4. Conclusion

The evaporation and condensation happened at reasonably good speed in the fabricated solar still. The saline water was converted into drinkable water using the solar still. The experimental data was compared with simulated results. Comparison of experimental and simulated data as shown by figure 6 has a good agreement among the production rates of potable water through experiments and predicted through simulation with the error of 8.7 percent.

References

[1] Alaian W M, Elnegiry E A and Hamed A M 2016 *Desalination* **379** 10.
[2] Ali Samee M, Mirza U K, Majeed T and Ahmad N 2007 *Renew. Sust. Energ. Rev.* **11** 543.
[3] Boutriaa A and Rahmani A 2017 *Comput. Aided Chem. Eng.* **101** 31.
[4] Kabeel A E and El-Agouz S A 2011 *Desalination* **276** 1.
[5] Khare V R, Singh A P, Kumar H and Khatri R 2017 *Energy Procedia* **109** 447.
[6] Kumar D, Himanshu P and Ahmad Z 2013 INT J *ROBOT* RES **3** L 66–72.
[7] Mahmood T, Naz M Y, Sulaiman S A, Jamil Y, Shukrullah S, Zahid M and Inayat M 2017 CFD modeling and experimental validation of a solar still. *MATEC Web of Conferences* **131** 02010.
[8] Panchal H N and Patel N 2017 *Int. J. Ambient Energy* **1** 1.
[9] Rahbar N and Esfahani J A 2013 *Energy* **49** 289.

Activation energy of domestic wastewater sludge in combustion process by using thermogravimetric analysis

M S Zakaria*, S Hassan, and M N Faizairi

Department of Mechanical Engineering, Universiti Teknologi PETRONAS, 32610 Bandar Seri Iskandar, Perak, Malaysia.

*E-mail: mohd_16005855@utp.edu.my

Abstract. The increasing demand of energy and environmental problem due to the consumption of fossil fuel has led the exploration of biomass fuel as an alternative energy. One of the potential biomass feedstocks that is available in Malaysia is Domestic Wastewater Sludge (DWS). In this study, DWS fuel characterization and the effect of experimental condition such as gaseous atmosphere, heating rate and particle size to the thermal behaviour of DWS has been carried out. In addition, kinetic parameter such as activation energy was determined by using the iso-conversional model free kinetic method, Flynn Wall Ozawa (FWO) model. Furthermore, lower heating rate and smaller particle size of DWS produced more volatile gas (84.51%), resulted to low ignition temperature (209°C) and low burnout temperature (515°C). Higher volatile indicates that it can accelerate the ignition process. In addition, low ignition temperature and low burnout temperature reveal that DWS can be ignited easily and shorten the time for combustion process. Based on the finding, the average activation energy for combustion of DWS which is about 143.86 kJ/mol.

1. Introduction

Emissions of greenhouse gases from fossil fuels and an ever-growing demand for energy source and with a limited supply of fossil fuels are the current pressing global problems which need a sustainable solution. Biomass is one of the available sources of renewable energy that has become a great interest at the national and global level. Historically, biomass has been utilized by mankind since centuries ago when burned woods were used to cook and warm the environment [1]. The fossil fuel has become the world's main energy source for the past few decades [2-4]. This is due to the high demands from industrialization and population growth are the factors that increase the consumption of fossil fuel. Nonetheless, biomass is still believed to be a viable option to reduce the dependency on the fossil fuel. It is found that the biomass energy is a sustainable, abundant and environmentally friendly energy source that available to explore [5]. At present, biomass-based energy is used in developing countries for direct consumption such as cooking, and in industrialized countries for conversion into secondary energy such as heat and biofuel [6-8]. Among all the biomass available in Malaysia, the waste derived from wastewater treatment plant namely Domestic Wastewater Sludge (DWS) is one of the less explored biomass energy sources as it has a huge potential as solid fuel for energy recovery. This potential biomass source is classified in the industry waste group and require high operational cost to dispose. Currently, the main disposal method for DWS that has been practice in Malaysia is by landfilling. However, most of the landfill have reach its capacity whereas the volume of DWS keeps increasing annually [6]. Therefore, converting this waste to biofuel by Thermochemical Conversion (TCC) process will reduce the disposal problem as well as contribution to a new source of energy.

However, the details study on the combustion characteristics and kinetics study are very important. Through this study, it can provide a useful information and important data about the characteristic of the biomass as its degradation during the TCC process. This can be achieved using Thermogravimetric Analysis (TGA), which is the most popular technique in thermal analysis studies [9-11]. TGA can be applied to study the thermal degradation of biomass as well as the reaction kinetic determination. However, the mechanism occurred during the analysis relies on the controlled variable and experimental condition. For example, the heating rate, physical properties of material and the ambient gaseous in which the reaction takes place, always have significant effect to the degradation process [12-19]. Therefore, it is important to study the factors affecting thermal behaviour of DWS. In this study, TGA was carried out in oxygen and nitrogen atmosphere to study the behaviour of DWS combustion.

2. Methodology

In this study, the sample of DWS was collected at the Universiti Teknologi Petronas (UTP) wastewater treatment plant. The sample of DWS was fully dried in an oven at temperature of 105 ℃ for 24 hours. The dried sample was then crushed and sieved using to form a regular shape. The TGA analysis was performed by using Thermogravimetric Analyser (TGA) LABSYS Evo Setaram. In this research, non-isothermal experiment was practiced as the temperature in the TGA was increased from 30 °C to 800 °C. After 800 °C, there is no significant effect on the thermal degrading has been observed which determine the temperature limit on this experiment. There are four different heating rates were observed in this study which are 5, 10, 15 and 20 K/min. This type of heating rate may reduce the mass transfer and temperature gradient effect [15]. In addition, the particle size of DWS was fixed at 0.425 mm.

Based on the previous researcher, Flynn-Wall-Ozawa (FWO) iso-conversional model free kinetic method normally used to calculate the activation energy as contribute to reliable and accurate result [15]. FWO model is implemented by using several experiment conditions of non-isothermal experiments at varies of heating rate as stated in equation (1) [9,15,16]:

$$\ln(\beta) = \ln\left[\frac{AE_a}{Rg(\alpha)}\right] - 5.33 - 1.052\frac{E_a}{RT} \quad (1)$$

Hence, a graph of $\ln(\beta)$ vs $1/T$ is plotted and the determination of the activation energy can be calculated from the slope of the straight line produced in the plotted graph.

3. Results and discussion

The evaluation of activation energy of DWS combustion and pyrolysis were performed by kinetic analysis. This was done by using the FWO iso-conversional model free kinetic method. Based on this method, there are four different heating rates were studied. TG and DTG curves of DWS at four different heating rates in combustion process. Then, the temperature corresponding to conversion degrees at each heating rate, β was chosen. Subsequently, the curves of $\ln \beta$ versus $1/T$ were plotted. Figure 1 show the linear fitting for TGA combustion. Based on the curves, it can be seen that the nine straight lines are produced denote nine conversion degree, α was selected ranged from 0.1 until 0.9. In addition, the value of $1/T$ decreases with the increases the value of α as the value of temperature directly proportional to the value of α. The four dots at each line represent four different heating rate which are 5, 10, 15 and 20 K/min. Based on the graph, the value of $\ln \beta$ increase directly proportional to the value of the heating rates. However, there is different in gradient value for each α which can be explain in details in figure 2.

IOP Conf. Series: Materials Science and Engineering **863** (2020) 012019 doi:10.1088/1757-899X/863/1/012019

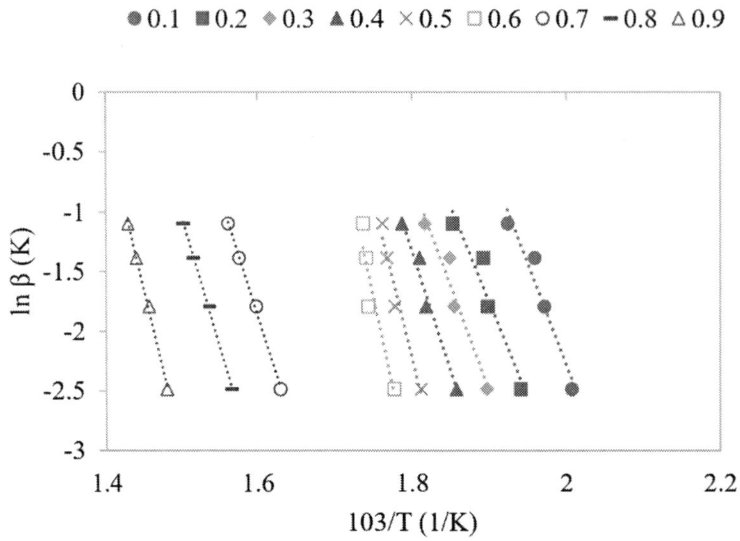

Figure 1. Plot of FWO linear fitting for TGA combustion.

Table 1 shows the gradient of FWO linear fitting, m, correlation coefficient, R^2 and the value of activation energy, E_a under oxidative atmosphere. It can be observed that the smooth linearity of data can be proved by the straight line which is achieves R^2 above 0.9 for combustion which is gives an accurate value for activation energy, E_a. Lastly, the activation energy is calculated by using the FWO aforementioned equation.

Table 1. Slope of FWO linear fitting, correlation coefficient, R^2 and calculated activation energy, E_a.

Combustion			
α	**Slope, *m***	R^2	E_a [kJ/mol]
0.1	-16.92	0.9461	133.77
0.2	-16.04	0.9245	126.92
0.3	-17.52	0.9431	138.35
0.4	-20.07	0.9728	158.62
0.5	-26.42	0.9604	208.71
0.6	-31.05	0.9031	245.42
0.7	-20.04	0.9983	158.67
0.8	-21.21	0.9976	167.74
0.9	-27.12	0.9965	214.43

Figure 2 displays the E_a obtained for every conversion. In general, the activation energy E_a obtained is different at each conversion. Therefore, a unique reaction mechanism at every degradation stage can be assumed. In addition, E_a value and conversion are correlated and dependent upon each other due to a complex multistep mechanism that occurred in a solid state [1]. Similarly, Cheng et al. [2] stated that distribution value of E_a is plausible during the whole decomposition process because thermal decomposition is a sequence reaction that occurred simultaneous and continuously. This resulted to an individual value of E_a.

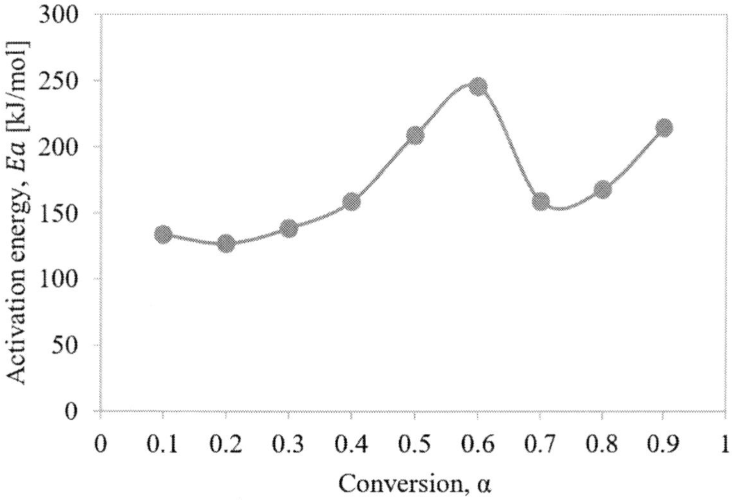

Figure 2. Comparison of E_a at each conversion for DWS TGA under O_2.

Activation energy value is slightly increase when the conversion increase for TGA pyrolysis. However, a slight drop of E_a at conversion of 0.2 can be seen before a sudden rise took place at conversion around 0.5 and 0.6. The activation energy, E_a, reached its maximum value at conversion 0.6 reveal that high activation energy is required for devolatilization stage which occurred at around $T = 300$ °C in TGA combustion. After a while, E_a drop drastically at conversion 0.7 due to the completion of devolatilization process at 400 °C. The E_a of DWS combustion slightly increased again because more energy is needed for char oxidation process. The average activation energy for combustion obtained in this study was 172.5 kJ/mol respectively.

4. Conclusion

Based on finding of this study, the average activation energy of DWS is low which is about 172.5 kJ/mol for combustion. Lower activation energy reveals that it has high reactivity and minimum energy required for the decomposition process is low.

References

[1] Pang Y, Shen S, and Chen Y 2017 *Waste and Biomass Valorization* **10** 127.

[2] Musabbikhah, Saptoadi H, Subarmono and Arif W M 2015 *International Journal of Engineering Transaction C: Aspects* **28** 3 pp 464-472.

[3] Asadullah M 2014 *Renew. Sustain. Energy Rev* **29** pp 201–215.

[4] Bidabadi M and Noroozi M J 2015 *International Journal of Engineering Transaction C: Aspects* **28** 3 pp 454-459.

[5] Manaf L A, Samah M A A and Zukki N I M 2009 *Waste Management* **29** 11 pp 2902–2906.

[6] Muyibi S A, Ambali A R and Eissa G S 2007 *Water Resource Management* **22** pp 485–508.

[7] Ahmad S, Kadir M Z A A and Shafie S 2011 *Renew. Sustain. Energy Rev* **15** pp 897–904.

[8] Shamsul N S, Kamarudin S K, Rahman N A and Kofli N T 2011 *Renew. Sustain. Energy Rev* **33** pp 578–588.

[9] Supatata N, Buates J and Hariyanont P 2013 *International Journal Environment Science* **4** pp 179–181.

[10] Tiara, Agustina T E and Faizal M 2018 *International Journal of Engineering Transaction C: Aspects* **31** 9 pp 1480-1486.

[11] Wojciechowska E 2015 *Water Resources* **39** pp 4749–4754.

[12] Jaafar M Z, Kheng W and Kamaruddin N 2003 *Energy Policy* **31** pp 1061–1072.

[13] Molla A H, Fakhru'l-Razi A, Abd-Aziz S, Hanafi M M, Roychoudhury P K and Alam M Z 2012 *Bioresources Technology* **85** pp 263–272.

[14] Planquart P, Bonin G, Prone A and Massiani C 1999 *Science Total Environment* pp 161–179.

[15] Manara P and Zabaniotou A 2012 *Renew. Sustain. Energy Rev.* **16** pp 2566–2582.

[16] Domínguez A, Menéndez J A, Inguanzo M and Pís J J 2006 *Biores. Techn.* **97** pp 1185–1193.

[17] Aniza N, Hassan S, Nor M F M and Fadhil M 2014 *Applied Mechanics Material* **695** pp 433–437.

[18] Cheng K, Winter W T and Stipanovic A J 2015 *Polym. Degrad. Stab.* **97** pp 1606–1615.

[19] Gil M V, Casal D, Pevida C, Pis J J and Rubiera F 2010 *Biores. Techn.* **101** pp 5601–5608.

IOP Conf. Series: Materials Science and Engineering **863** (2020) 012020 doi:10.1088/1757-899X/863/1/012020

Effect of calcination temperature on the structural and morphological properties of zinc sulphide nanomaterials

G Ahmad[1], A Batool[1], Q Iqbal[1], A Iqbal[1], F Ali[1], S Nawaz[1], N A Shad[2], M M Sajid[2] and Y Javed[1, *]

[1] Department of Physics, University of Agriculture Faisalabad, 38040 Faisalabad, Pakistan.
[2] Department of Physics, Government College University, 38000 Faisalabad, Pakistan.

*E-mail: myasi60@hotmail.com

Abstract. Metal sulfide nanomaterials have emerged an important material in recent times due to its potential applications in optical devices, nanoelectronics and nano-optoelectronics. In current research work, zinc sulfide (ZnS) nanoparticles were synthesis by simple chemical co-precipitation method and effect of calcination temperature on the structural and morphological properties of the nanomaterials have been investigated. It was observed that when temperature rises to 500 °C, cubic structure of ZnS transformed into hexagonal phase. Further increase in temperature oxidize the sulfide and ZnS changes into zinc oxide at 750 °C. This temperature rise was also evident in the SEM images where nanoparticles transformed into rod like morphology at 750 °C. Oxidation of the ZnS to ZnO also confirmed by FTIR spectrum which shows Zn-O bonds presence. These studies show that temperature variation can affect the structure and ultimately their relevant technological applications.

1. Introduction

Nanomaterials demonstrate distinguished characteristics as compared to those bulk materials. Nanomaterials based on semiconductor properties is an important class which shows excellent applications in many fields such as gas sensing [1], photocatalysis [2], LEDs [3], capacitors [4] etc. Metal sulfides become very popular inorganic nanomaterials during the last decade as a potential candidate to replace metal oxide-based materials due to their certain limitations [5]. ZnS is a type of metal sulfide belong to the II-IV group but with wider band gap from other materials of the family. Due to some exclusive and beneficial properties of ZnS, it becomes more popular as compared to ZnO. These properties include wide band gap, bond energy of exciton, high refractive index, luminescent characteristic, optical absorption and many others [6].

ZnS found in two crystalline forms i.e. zinc blende and wurtzite. Zinc blende, also known as sphalerite, has cubic crystal structure [7]. This phase shows more stability at low temperature than other form of ZnS. Sphalerite also occurs naturally in our environment. The range of energy band gap of Zinc blende is 3.5 eV to 3.7 eV. Wurtzite is the hexagonal structure of ZnS, stable at high temperature. Usually, this phase of ZnS obtained synthetically. The range of energy band gap of wurtzite is 3.7 eV to 3.9 eV [1].

It is essential task to prepare ZnS nanoparticles having small dimensions, specifically in the nanosized. The arrangement and size of nanoparticles affect the characteristics associated with various uses. This is the reason why size and structure of nanoparticles should be controlled during preparation of ZnS. In the production and development of nanotechnology, there are number of techniques used for the fabrication of ZnS nanoparticles such as green synthesis [8], microwave assisted [9], hydrothermal [10] and chemical co-precipitation method [11]. The chemical co-precipitation is very simple technique [12]. Co-precipitation is a technique in which solute is frequently dissolve in the solvent. Both solute and solvent are mixed together by adjusting pH of the mixture. These reactions can perform at normal temperature and pressure conditions. There are very few studies have been reported on the temperature effect on the ZnS nanomaterials. In this research work, ZnS nanoparticles have been prepared by using chemical co-precipitation method and effect of different calcination temperature has been investigated.

2. Experimental set up

All the chemicals were of analytical grade and used without any further purification. Experiment details are as follows: $ZnSO_4.5H_2O$ and $Na_2S.7H_2O$ were used as starting materials, dissolved separately in de-ionized water to make one molar solution and stirred at room temperature for 30 minutes. Then Na_2S solution was added in $ZnSO_4$ solution dropwise using burette while stirring. The reaction was completed in 2 hours. Then obtained solution was filtered. In order to remove the sodium impurities, the precipitates were washed with distilled water three times. The collected precipitates were dried at 80 °C for 16 hours in the oven. After drying, the resultant precipitates were ground in fine powder form with the help of pestle and mortar. Then sample was divided into four parts and treated at different temperatures i.e. as synthesized, 250, 500 and 750 °C. Schematic of the synthesis scheme is shown in figure 1. Samples were characterized by using X-ray diffraction, SEM, and FTIR.

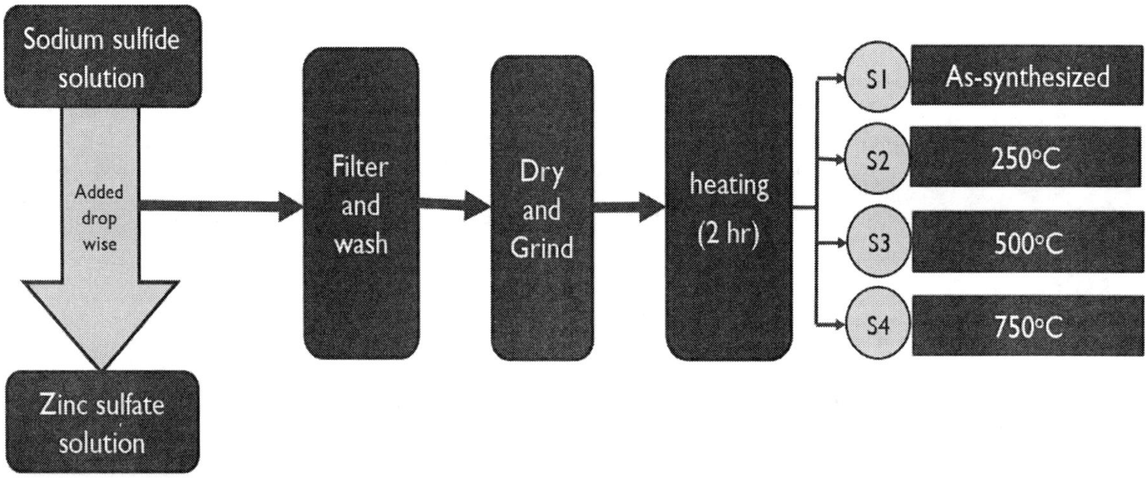

Figure 1. Schematic illustration of ZnS synthesis scheme.

3. Results and discussion

3.1. X-ray diffraction analysis

X-ray diffraction (XRD) is a useful technique for the determination of crystal structure. XRD pattern of the synthesized ZnS nanostructures treated at different temperature is shown in figure 2. As synthesized and treated at 250 °C nanostructures have presented typical XRD spectrum as it is observed in many studies with signature peaks present at 2 theta values of 28.53°, 47.45° and 56.3°. These 2 theta values correspond to (111), (222) and (311) planes of cubic zinc blend structure. XRD pattern of ZnS nanomaterials treated at 500 °C exhibits many additional peaks as compare to synthesize and heated 250 °C. The 2 theta values and corresponding diffraction planes are listed in table 1. While analysing and comparing with standard data, it is observed that ZnS at 500 °C actually has hexagonal crystal structure, indicating stretching in lattice parameter and angle. When nanomaterials are further heated up to 750 °C, 2 theta positions of the diffraction peaks are shifted and by matching with standard data. These peaks show ZnO structure with hexagonal crystal structure. The 2 theta values and corresponding diffraction planes are presented in table 1. This shows that at elevated temperature, oxidation of ZnS produce ZnO. In conclusion, current study shows that cubic phase of ZnS is stable till 250 °C. More heat changes the cubic phase into hexagonal phase with increased crystallinity. Further increase in temperature, may change the metal sulphide into metal oxide by oxidation of the sulphide nanomaterial.

Further investigation of the effect on crystallite size, Scherrer formula was used to calculate the crystalline size as tabulated in table 2. Scherrer equation can be written as equation (1).

$$Dp = \frac{K\lambda}{\beta \cos\theta} \tag{1}$$

Where, *Dp* is Average crystalline size, *K* is Shape factor (typically value about 0.94), λ is X-ray wavelength, β is Full width at half maximum of highest intensity diffraction peak, θ represents Bragg's angle.

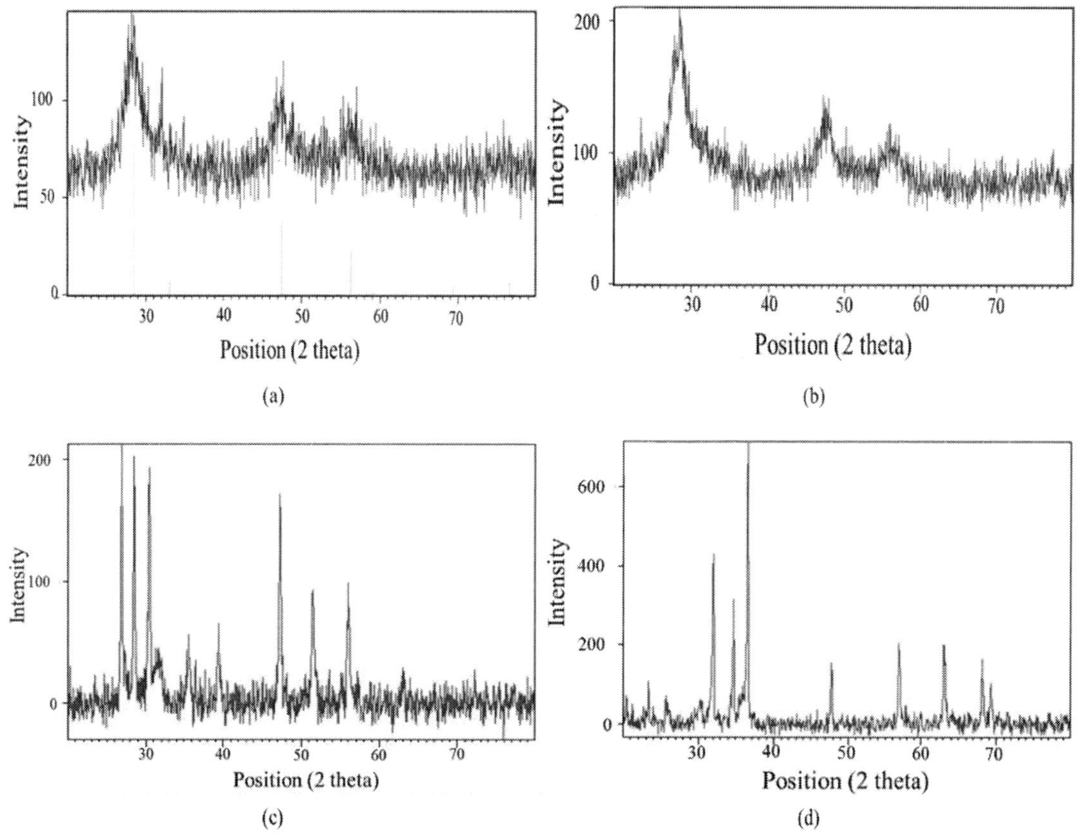

Figure 2. XRD pattern of ZnS nanomaterials calcinated at different temperatures (a) as synthesized, (b) 250 °C, (c) 500 °C and (d) 750 °C.

The crystalline size of as-synthesized and heated at 250 °C powder 3.72 nm broader peaks in the XRD pattern of these two samples indicates higher amorphous nature of material which result smaller grain size as it was calculated by Scherrer formula. Particle size at 500 °C is 42.47 nm which indicating heating effect and high crystalline order in the structure (also more diffraction peaks observation). Finally, at 750 °C, crystallite size increased to 47.47 nm, can be attributed to the heating effect.

IOP Conf. Series: Materials Science and Engineering **863** (2020) 012020 doi:10.1088/1757-899X/863/1/012020

Table 1. Peak analysis of samples.

Sample	Peak position (2θ)	Planes (hkl)	Compound
1	28.53	(111)	ZnS
	47.45	(220)	ZnS
	56.30	(311)	ZnS
2	28.53	(111)	ZnS
	47.45	(220)	ZnS
	56.30	(311)	ZnS
3	26.88	(100)	ZnS
	28.50	(002)	ZnS
	30.43	(101)	ZnS
	39.39	(102)	ZnS
	47.31	(110)	ZnS
	51.41	(103)	ZnS
	55.97	(200)	ZnS
4	27.16	(100)	ZnS
	32.02	(100)	ZnO
	34.50	(002)	ZnO
	36.51	(101)	ZnO
	47.80	(110)/(102)	ZnS/ZnO
	56.81	(112)/(110)	ZnS/ZnO
	63.06	(103)	ZnO
	69.27	(201)	ZnO

Table 2. Particle size calculated by Scherrer formula.

Sample	Temperature	Crystallite size (nm)
S-1	As-synthesized	3.72
S-2	250°C	3.72
S-3	500°C	42.47
S-4	750°C	47.47

3.2. SEM analysis

Scanning electron microscopy provides information about shape and morphology of the nanomaterials. SEM images have taken for samples calcinated at temperature of 500 °C and 750 °C as shown in figure 3 and figure 4, respectively. In figure 3, images are taken at magnification 300x, 495x, 718x and 1440x. Low magnification images present high yield of ZnS nanomaterials with hexagonal crystal structure. There are smaller spherical particles are observed along with large irregular aggregates of the nanomaterials. Whereas when sample is heated at 750 °C, these aggregates are reorganized and formed rod like morphologies as shown in figure 4. These results show that nanomaterials not only change from zinc sulphide to zinc oxide, there is also change in morphology with the increase in temperature.

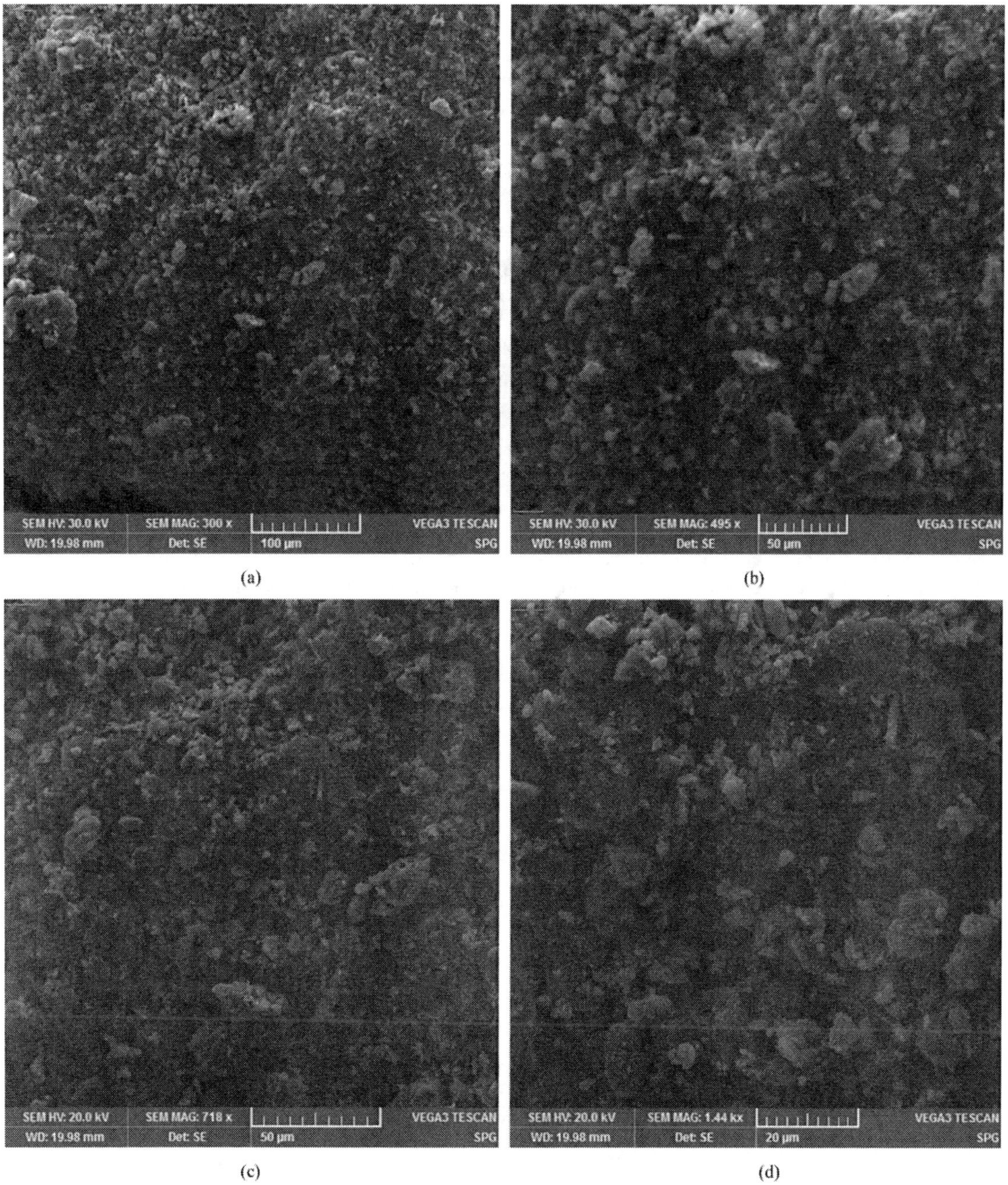

Figure 3. SEM images of ZnS nanoparticles calcinated at 500 °C at different magnifications. (a) 300x, (b) 495x, (c) 718x and (d) 1440x.

(a) (b) (c)

Figure 4. SEM images of nanomaterials heated at 750 °C at different magnifications (a) 2210x, (b) 3150x, (c) 4800x.

3.3. FTIR analysis

Fourier transform infrared spectroscopy provides information about impurity and functional groups attached with the nanomaterials. In this study, we have obtained FTIR spectrum of sample treated at 750 °C to confirm about the transformation of ZnS into ZnO form. FTIR spectrum of the sample is shown in figure 5. There is one visible band is observed at 605 cm^{-1} which is related to the stretching vibrations of ZnO bonds [13]. This spectrum confirms XRD results where we have observed ZnO diffraction peaks at 750 °C.

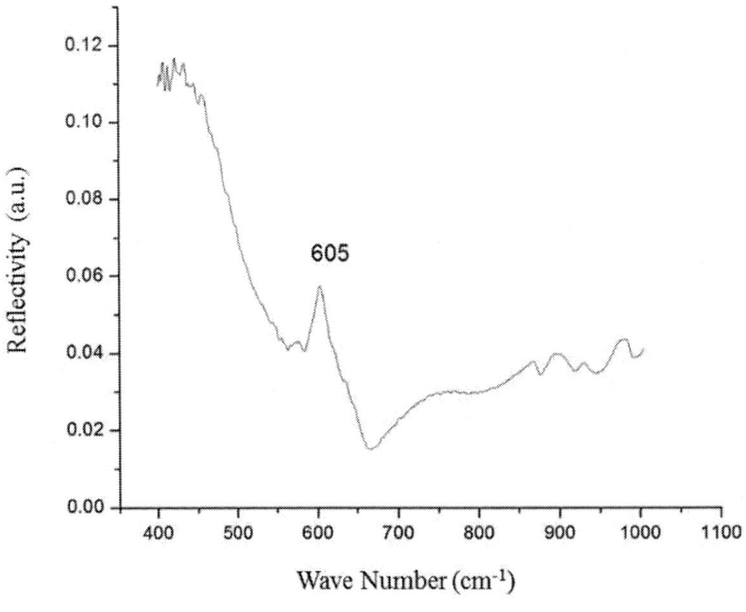

Figure 5. FTIR spectrum of ZnS nanomaterials heated at 750 °C.

4. Conclusion

ZnS nanomaterials have been synthesized successfully by using chemical co-precipitation method. Synthesized nanomaterials were subjected to different calcination temperatures i.e. 250 °C, 500 °C and 750 °C. Then these samples have been investigated by XRD, SEM and FTIR. XRD analysis shows changes in crystal structure with temperature and also showed oxidation of ZnS into ZnO nanomaterials at 750 °C. SEM images presented variation in morphology with temperature. FTIR spectrum confirms the formation of ZnO nanomaterials at elevated temperatures.

References

[1] Zhao Y, Zhang Y, Zhu H, Hadjipanayis G C and Xiao J Q 2004 *Journal of the American Chemical Society* **126** 6874
[2] Lai C-H, Lu M-Y and Chen L-J 2012 *Journal of Materials Chemistry* **22** 19
[3] Wang K-H, Zhu B-S, Yao J-S and Yao H-B 2018 *Science China Chemistry* **61** 1047
[4] Raj C J, Kim B C, Cho W-J, Lee W-G, Seo Y and Yu K-H 2014 *Journal of Alloys and Compounds* **586** 191
[5] Sadovnikov S I, Rempel A A and Gusev A I 2018 *Russian Chemical Reviews* **87** 303
[6] Shakil A, Das S, Rahman A, Akther U S, Hassan K and Rahman K 2018 *Materials Sciences and Applications* **9** 751
[7] Bai H-J, Zhang Z-M and Gong J 2006 *Biotechnology letters* **28** 1135
[8] Alijani H Q, Pourseyedi S, Mahani M T and Khatami M 2019 *Journal of Molecular Structure* **1175** 214
[9] Sousa D M, Alves L C, Marques A, Gaspar G, Lima J C and Ferreira I 2018 *Scientific Reports* **8** 15992
[10] Sabaghi V, Davar F and Fereshteh Z 2018 *Ceramics International* **44** 7545
[11] Parvaneh I, Samira S and Mohsen N 2015 *Chinese Physics B* **24** 046104
[12] Ayodhya D and Veerabhadram G 2018 *Materials today energy* **9** 83
[13] Bodke M, Purushotham Y and Dole B 2018 *Cerâmica* **64** 91

IOP Conf. Series: Materials Science and Engineering **863** (2020) 012021 doi:10.1088/1757-899X/863/1/012021

Characterization of manganese/cobalt oxide composites synthesized by chemical co-precipitation method

N-H Altaf[1], T Anjam[1], M M Sajid[2], N A Shad[2], S Shukrullah[1], M Y Naz[1] and Y Javed[1,*]

[1] Department of Physics, University of Agriculture Faisalabad, 38040 Faisalabad, Pakistan.
[2] Department of Physics, Government College University, 38000 Faisalabad, Pakistan.

*E-mail: myasi60@hotmail.com

Abstract. Composite materials have emerged as new class of materials which presented unique and enhanced applications in the nanotechnology field. In this study, Cobalt oxide-manganese oxide composites have been prepared by chemical co-precipitation method. Cobalt oxide particles were first prepared and used as seeds for making composites with manganese oxide. These composites have been characterized by XRD, SEM, UV-Vis spectroscopy and FTIR. Characterization tools confirmed the formation of Manganese/Cobalt oxide composites. XRD results presented the diffraction peaks for cobalt oxide, manganese oxide and $MnCo_2O_4$ compounds. FTIR spectrum also indicated the formation of bonds between Mn, Co and O especially at 461 cm^{-1} shows $MnCO_2O_4$ bonding. These composite materials absorb visible light of wavelength 682 nm and can be used in light activated photocatalytic activity for the degradation of toxic dyes.

1. Introduction

Nanoscale materials are the synthetic materials linked through noncovalent and covalent bonds. They may have the properties of organic, inorganic and bioactive materials which makes them versatile in nature. Their unique properties give opportunities to develop vast varieties of advanced materials of multiple functions and well-developed structures [1]. The properties of composites may be beneficial to many disciplines such as electronics, optical, catalysis, sensing, coating, energy storage and biomaterials [2]. Composites are attracting the attention of the researchers now a days due to the ever-growing interest, versatility and a path to the development of desired materials [3].

The composites are the atomic or molecular level composition of two different materials. These materials have chemical bonds between their different particles. Composites can be synthesized by variety of protocols including sol-gel methods, blending processes, coordination methods, intercalation, photo-polymerization, electrochemical preparation and synthetic routs of surface-grafting. Composite materials are providing multifunction in a single nano-system and consequently could be used to perform more than one function at a time [4]. Composites have potential application in biomedical field for combinational therapy e.g. gold-iron oxide nanocomposites are useful for both magnetic hyperthermia and plasma thermotherapy [5].

The bioinorganic hybrids are responsible for the development of biological and scientific materials. These are usually able to demonstrate the functions including drug delivery and molecular functioning of the drugs. The enzyme-clay nanohybrids and bio-materials related to the inorganic structures are used for the biosynthesis applications. The anionic exchange characteristics, versatile structure and wide chemical composition of nanocomposites makes the immobilization of biomolecules possible. For the immobilization of biomolecules, the double layered hydroxides are the most appropriate materials [6]. Manganese oxide nanoclusters with diameters of only 5 to10 nm have the tendency towards the ferromagnetic behavior. This behavior is shown towards the phase transition between the paramagnetic and other magnetic phases. The abnormal behavior shown by MnO may arise from the effects of cluster size of the material [7].

The anisotropy energy which is usually deformation dependent, overweighs the other anisotropy energies of Cobalt oxide. This property of CoO determines the direction and dimension of the magnetization which is to be directed towards the tetragonal deformation axis [8]. In this research, manganese-cobalt oxide composites were prepared by simple seed mediated co-precipitation method. The obtained composites have been characterized by x-ray diffraction, scanning electron microscopy, UV-Vis spectroscopy and FTIR for structural, morphological, optical and bonding properties.

2. Methodology

2.1. Chemical and components

All the chemicals were bought in pure form and used without any additional treatments. $CoCl_2.6H_2O$, $MnCl_2.4H_2O$ and NaOH were used for the preparation of the composites. Distilled water was used as solvent and for other reactions throughout the experiment. Reaction was based on two steps; first cobalt oxide was formed then used the cobalt oxide particles as seeds for the formation of composite.

2.2. Preparation of cobalt oxide powder

For cobalt oxide preparation, 0.4 M of each cobalt chloride hexahydrate and sodium hydroxide was taken and dissolved in 100 ml distilled water separately. NaOH solution was added dropwise into $CoCl_2.6H_2O$ solution with the help of burette. The colour of solution was first blue then turns into green and finally changes into dark brown. The pH of the solution was adjusted to 11. Then solution was magnetically stirred for two hours. The obtained precipitates were filtered, washed and dried in oven at 80 °C for 24 hours. The so formed cobalt oxide powder was used for further synthesis of composite materials.

2.3. Procedure for preparing composites

After measuring, the CoO (0.74g), $MnCl_2$ (7.91g) and NaOH (1.99g) were dissolved in 25 ml, 100 ml and 125 ml distilled water respectively. $MnCl_2.4H_2O$ solution were then mixed with CoO solution and sonicate it for 15 minutes. Then NaOH solution was added drop wise into the mixture of two solutions. The formed solution was magnetically stirred for 24 hours to produce composite materials. The obtained precipitates were separated by centrifugation, placed in a crucible for drying in hot air oven at 80 °C and ground into a fine powder using mortar and pestle to get the powdered form cobalt-manganese oxide composites. The obtained powder was then calcinated at 300 °C and 900 °C for 2 hours. The obtained composite materials have been characterized by XRD [Philips Xpert X-ray diffractometer with Cu Kα radiation (λ= 0.15406 nm)] for analysing phase purity of synthesized powder in the angle 2θ range of 20° to 70°. SEM (VEGA3 TESCAN) was used to observe the morphology of the structures. UV-vis spectrometer was used for optical properties and FTIR was used for bonding structures between different elements.

3. Results and discussion

3.1. X-ray diffraction analysis

The synthesized materials were characterized by X-ray diffraction (XRD). Figure 1 is showing XRD pattern of as synthesized composites and treated at 300 °C and 900 °C. There is only one prominent peak was observed at 29.5° which belongs to the (202) plane of cobalt oxide. As temperature rises up to 300 °C, two additional peaks were appeared at 25.3° and 35.8°. Diffraction peak at 35.8° belongs to (222) plane of Manganese oxide whereas 25.3° belong to pure Mn phase with JCPDS # 032-0634 as no such diffraction angle was observed in the different phases of cobalt oxide, manganese oxide and its composites. When temperature is elevated to 900 °C, number of diffraction peaks are increased at different diffraction angles. These peaks and their corresponding planes are presented in table 1. These diffraction peaks belong to both cobalt-manganese oxide composites having cubic crystal structure. From these observations, it can be concluded that rise in temperature improved the crystalline order in the materials and presented more stable composite behaviour at higher calcination temperature.

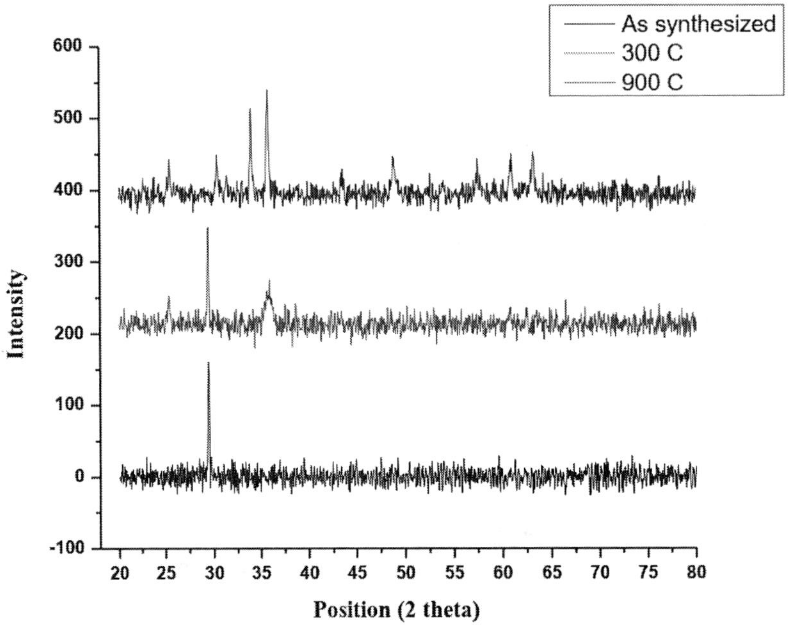

Figure 1. X-ray Diffraction pattern of CoO-MnO composites at different calcination temperature.

Table 1. Peaking matching of sample calcinated at 900 °C.

Sr. #	Peak (2θ)	Planes	Compound
1	30.3297	(220)	$MnCo_2O_4$
2	34.0208	(311)/(111)	MnO/CoO
3	35.7275	(222)/(311)	$MnO/MnCo_2O_4$
4	43.5245	(400)	$MnCo_2O_4$
5	57.4356	(511)/(220)	$MnCo_2O_4$/CoO
6	63.0822	(440)	$MnCo_2O_4$
7	66.695	(531)	$MnCo_2O_4$

3.2. Morphological analysis

Figure 2 shows SEM images of the samples calcinated at 300 °C and 900 °C at different magnifications. Low magnification image presents high yield of structures in both temperature values. SEM images of sample calcinated at 300 °C show larger aggregates and some flake like particles. Whereas, in case of 900 °C, more homogenous distribution of spherical particles is observed. High resolution images show that particles are present and existed in spherical shape. These particles have size in the range from 225 nm to 350 nm. SEM images at 900 °C shows better transformation of composite materials.

(a) (b) (c)

(d) (e) (f)

Figure 2. SEM images at different magnifications of composites calcinated at 300 °C (a-c) and 900 °C (d-f). Magnifications: (a) 2400x, (b) 5000x, (c) 8000x, (d) 8980x, (e) 13800x, (f) 23700x.

3.3. UV-visible spectroscopy

UV-vis spectroscopy is a useful technique to study the optical properties of the material. Visible light spectrum of the composites calcinated at 900 °C was recorded as shown in figure 3. The absorption edge of cobalt oxide-manganese oxide composites was obtained at 682 nm wavelength, indicating the potential use of these structures in visible light driven applications especially in photocatalytic degradation and thermotherapy.

3.4. Fast transform infrared spectroscopy

FTIR spectrum of CoO-MnO composites has been computed between the ranges 400-1100 cm^{-1} wave number region and composed of many close bands. Bands are observed at 461 cm^{-1}, 493 cm^{-1}, 553 cm^{-1}, 591 cm^{-1} and 611 cm^{-1}. The bands at 493 cm^{-1}, 591 cm^{-1} and 611 cm^{-1} attributes to Mn-O bonding for its different oxide types [9, 10]. The band at 553 cm^{-1} is related to the Co-O vibrations at the octahedral site [11]. Whereas band at 461 cm^{-1} corresponds to $MnCo_2O_4$ compound [12]. FTIR analysis confirms presence of cobalt oxide, manganese oxide and bond between cobalt and manganese indicating the formation of composite particles.

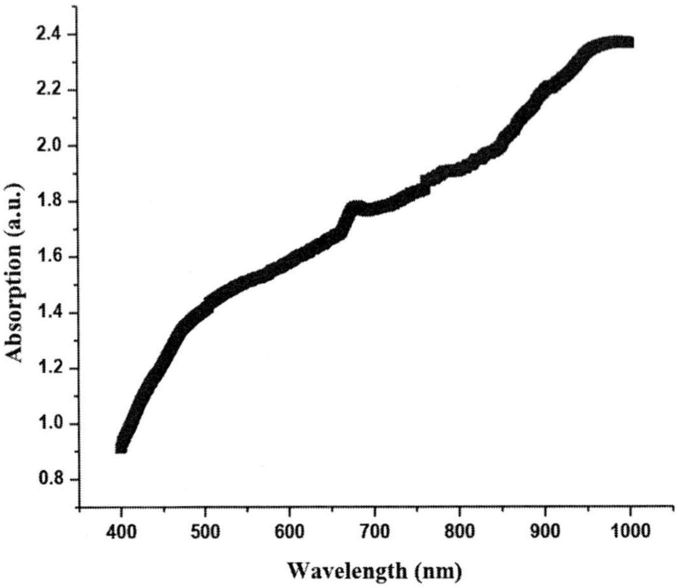

Figure 3. UV-Visible spectra of $CoMn_2O_4$ composites synthesized by co-precipitation method.

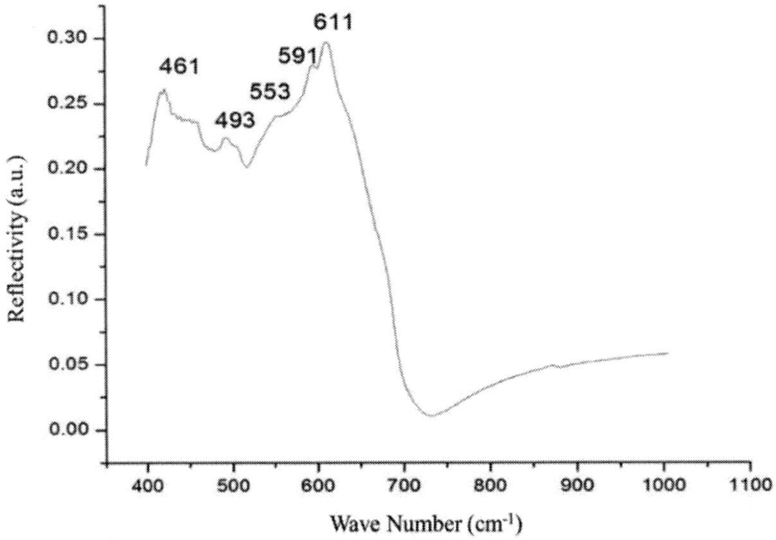

Figure 4. FTIR spectrum of $CoMn_2O_4$ composites calcinated at 900 °C.

4. Conclusion

The co-precipitation technique was effectively employed for the preparation of manganese cobalt oxide composites. Cobalt oxide powder was used as seed and manganese was allowed to grow on these seeded particles. XRD results confirm the formation of composites and present cubic crystal structure. SEM and UV-vis spectroscopy were used for morphological and optical properties. Particle size around 300 nm and absorption band at 682 nm wavelength was observed. Finally, FTIR spectrum also indicate the bonding between cobalt, manganese and oxygen which further asserted the results obtained from XRD analysis. These visible lights activated composites can be good candidate for photocatalytic degradation of organic dyes.

References

[1] Wu W, Jiang C and Roy V A 2015 *Nanoscale* **7** 38.
[2] Dahl M, Liu Y and Yin Y 2014 *Chemical reviews* **114** 9853.
[3] Liu L-H, Métivier R, Wang S and Wang H 2012 *Journal of Nanomaterials* **2012** 1.
[4] Treviso A, Van Genechten B, Mundo D and Tournour M 2015 *Composites Part B: Engineering* **78** 144.
[5] Leung K C-F, Xuan S, Zhu X, Wang D, Chak C-P, Lee S-F, Ho W K-W and Chung B C-T 2012 *Chemical Society Reviews* **41** 1911.
[6] Forano C, Vial S and Mousty C 2006 *Current Nanoscience* **2** 283.
[7] Lee G H, Huh S H, Jeong J W, Choi B J, Kim S H and Ri H-C 2002 *Journal of the American Chemical Society* **124** 12094.
[8] Farhadi S, Safabakhsh J and Zaringhadam P 2013 *Journal of Nanostructure in Chemistry* **3** 69.
[9] Mironova-Ulmane N, Kuzmin A, Skvortsova V, Chikvaidze G, Sildos I, Grabis J, Jankoviča D, Dindune A and Maiorov M 2017 In: *Proc. of the Int. Conf. on Oxide Materials for Electronic Engineering,* p 1013.
[10] Prasad K S and Patra A 2017 *Green Processing and Synthesis* **6** 549.
[11] Lima-Tenório M K, Pineda E A G, Ferreira C S, Rebelo Q H F, Souza R F B d, Passos R R and Pocrifka L A 2018 *Materials Research (Sao Carlos, Online)* **21** 1.
[12] Antunović V, Ilić M, Baošić R, Jelić D and Lolić A 2019 *PloS one* **14** e0210904.

Application of central composite design for optimization of microbial growth inhibition using pineapple leaves juice

A Ya'acob[1] and N Zainol[2,*]

[1] Faculty of Chemical & Natural Resources Engineering, University Malaysia Pahang, Lebuhraya Tun Razak, 26300 Gambang, Kuantan, Pahang, Malaysia.
[2] CARIFF, Universiti Malaysia Pahang, 26300 Gambang, Pahang, Malaysia.

*E-mail: azwina@ump.edu.my

Abstract. Microbial growth inhibitor is very important to prevent transmission of disease and infection. In this research paper, pineapple (Ananas comosus) leaves juice (PLJ) was used as microbial growth inhibitor agent. Microbe from infected pineapple leaves was used as microbe model. The selected factors in this study were temperature and microbial inhibition time with the range between 35 to 39 °C and 10 to 50 minutes respectively. In order to determine the optimum condition for microbial growth inhibition process, an experiment were designed according to a central composite design (CCD) in two variables following the response surface methodology (RSM). Microbial growth inhibition was measured from cell dry weight of microbe using dry weight measurement method. Result from RSM showed that microbial inhibition time have significant effect on the microbial growth inhibition. Model validation from experiments showed good correspondence between experimental and predicted values. The highest error was achieved at 8.03%. The optimal condition obtained for microbial growth inhibition was at 37 °C and 34.25 minutes. Minimum cell dry weight was found at 0.0508 gram which indicated the microbial growth inhibition at 94.73%. Optimal condition from this study would be useful for the application of PLJ in pineapple plantation.

1. Introduction

Microbial growth inhibitor is a type of agent which inhibits or prevents the growth of cells or can be known as static agent [1]. The control of microbial growth is needed in many practical situations and significant advances in agriculture, medicine, and food science. Therefore, microbial growth inhibitor is very important to prevent transmission of disease and infection, to prevent contamination by the growth of undesirable microorganisms and to prevent spoilage of materials by microorganisms. While several chemical agents are commercially available, these synthetic chemicals have undesirable side-effects and can cause disease to humans, plant and animals. Hence, there is an immediate need to search for the alternative natural products as microbial growth inhibitor agent. Pineapple leaves has been selected considering that pineapple has high phenolic compound compared to other fruits [2]. Since 1872, several phenolic compounds have been used to control microbial growth [1]. Phenolic compounds and its antioxidant biomolecules were the subject in research for many years which can inhibit the spread of an infectious agent. These antioxidants mainly come from plants in the form of phenolic compounds such as flavonoid, phenolic acids, ascorbic acid, isoflavones, flavones, anthocyanins, catechins, tannins, stilbenes, quinones and carotenoids [3, 4]. For a long period in history, plants have been valuable and essential sources of natural products for the health of human beings [5]. Plant phenolic have gained considerable interest in recent years for their potential effects against microorganisms. Seven phenolic compounds were isolated from the pineapple leaves which are ananasate, 1-O-caffeoylglycerol, 1-O-p-coumaroylglycerol, caffeic acid, p-coumaric acid, β -sitosterol and daucosterol [6-8]. Pernin et al. [6] reported that caffeic acid significantly affected growth of Listeria monocytogenes. β–sitosterol also inhibited the growth of Staphylococcus aureus and Escherichia coli respectively [9, 10]. Therefore, application of extracted pineapple leaves as natural microbial growth inhibitor considered a good alternative. Besides, pineapple leave is one of the abundantly available wastes materials in Malaysia and has not been utilized yet. This research aimed to optimize the microbial growth inhibition process using pineapple leaves juice extract. The obtained optimum condition from this study could improve the process of microbial growth inhibition.

2. Materials and methods

2.1. Materials
Pineapple (Ananas comosus) leaves and infected pineapple leaves were collected from pineapple plantation located at Pekan Pina Sdn Bhd, Pekan, Pahang. The pineapple leaves juices were extracted by using electrical sugarcane press machine. The microbe from infected pineapple leaves were streaked on the surface of the potato dextrose agar (PDA) in petri dish and was incubated at 37 °C for 24 h [11].

2.2. Total phenolic content (TPC) analysis
TPC analysis were conducted to determine the total phenolic content by using the Folin-Ciocalteu's reagent according to method described by Singleton and Rossi with minor modification with gallic acid (GA) as the standard [8]. Sample extract were centrifuged at 5000 rpm for 15 minutes. Then, 0.5 mL of the sample was mixed with 2.5 mL of 10-fold diluted Folin-Ciocalteu's reagent and allowed to react for 5 minutes. Then, 2 mL of 7.5% sodium carbonate (Na_2CO_3) was added and allowed to react for 1 hour at room temperature. The mixtures were measured using UV-visible spectrophotometer against the reagent blank. All determinations were carried out in triplicate. The sample was compared to a standard curve of prepared GA solution. The TPC is expressed as gallic acid equivalent (GAE), which reflects the phenol content as the amount of gallic acid units in 1 gram of extract (mg GAE g^{-1}) [12].

2.3. Analysis of cell dry weight
By using micropipette, the mixture of microbe broth and PLJ extract in 2.4 was spread onto the surface of new PDA plate and swabbed by using triangle stick. As for the control, only the microbe broth was spread onto the surface of new PDA plate without the PLJ extract. The PDA plate then was incubated in the incubator at 37 °C for 24 hours [11]. After 24 hours, all the microbe growth on the PDA plate were scraped out into the 10 mL nutrient broth in the centrifuge tube including the microbe from the control petri dish. All the empty centrifuge tube was weighed before proceeding to the next step. Next, the mixture was centrifuged at 5000 rpm for 15 minutes to separate the microbe and the broth respectively. The broth was discarded from the centrifuge tube. Then, the centrifuge tube was placed in the oven at 60 °C for overnight. Lastly, the dried centrifuge tube was weighed to determine the cell dry weight. The microbial growth inhibition (%) was measured in terms of cell dry weight. The cell dry weight and microbe growth inhibition were calculated using equation (1) and (2), respectively.

$$\text{Cell dry weight (g)} = \text{Dried centrifugal tube \& sample} - \text{Empty centrifugal tube} \quad (1)$$

$$\text{Microbial growth inhibition (\%)} = \frac{(\text{Cell dry weight}) - (\text{Cell dry weight of mixture})}{\text{Cell dry weight of control}} \times 100 \quad (2)$$

2.4. Experimental setup for optimization
Microbe broth was prepared by mixing the microbe with nutrient broth and incubated for 1 hour at 37 °C and 100 rpm. Then, the microbe broth was mixed with PLJ extract with volume ratio of 1:1. The mixture was incubated in the incubator shaker at 100 rpm at designed temperature and microbial inhibition time.

In order to determine the optimum condition for microbial growth inhibition process, the experiments were designed according to a central composite design (CCD) with two variables following the Response surface methodology (RSM) [13]. Table 1 showed the selected factors and level. The experiments were conducted according to variable setting as tabulated in table 2. All the data obtained were recorded in Design Expert software. The responses were analyzed using the Design Expert software to determine the optimum conditions of microbial growth inhibition. A total of 13 experiments were conducted and the outputs were analysed using Design Expert software. The response was analyzed using analysis of variance (ANOVA) based on the *p*-value with 95% of confidence level.

Table 1. Factors and ranges of CCD.

Factors	-alpha	-1	0	+1	+alpha
A: Microbial inhibition time (min)	10	20	30	40	50
B: Temperature (°C)	35	36	37	38	39

Validation experiment were performed based on the optimum conditions obtained from the data analysis of Design Expert software. The experiment was done to compare the predicted data from suggested conditions with experimental data from the experiment to determine the error between the experimental value and predicted value.

3. Results and discussion

3.1. Fitting the model

The experimental data is shown in table 2. The cell dry weight of Run 9 shows the lowest value of cell dry weight 0.0188 g which indicated the highest value of microbial growth inhibition at 94.09 %. The responses were obtained at microbial inhibition time of 30 minutes and temperature of 37 °C. Fitting of the data to various models (linear, two-factorial, quadratic and cubic) and their subsequent analysis of variance shows that cell cry weight is most properly described with a quadratic polynomial model. The adjusted R^2 of the quadratic model (0.8471) was higher than that of linear and two factorial models for all responses.

Table 2. Experimental data of CCD.

Run	Factor A: Microbial inhibition time (min)	Factor B: Temperature (°C)	Response 1: Cell dry weight (g)	Response 2: Microbial growth inhibition (%)
1	20.00	36.00	0.1327	58.26
2	40.00	36.00	0.1154	63.70
3	20.00	38.00	0.1053	66.88
4	40.00	38.00	0.0767	75.87
5	10.00	37.00	0.3765	18.43
6	50.00	37.00	0.1303	59.01
7	30.00	35.00	0.1905	40.08
8	30.00	39.00	0.1389	56.31
9	30.00	37.00	0.0188	94.09
10	30.00	37.00	0.0189	94.05
11	30.00	37.00	0.0485	84.74
12	30.00	37.00	0.0489	84.62
13	30.00	37.00	0.0693	78.20

3.2. Analysis of variance (ANOVA)

From analysis of TPC concentration using UV-spectrophotometer, at absorbance reading of 4.252, TPC in examined PLJ is 0.6010 mg GAE/mL. Table 3 summarize the ANOVA (F-test and p-value) to check the significance of model, each factor and to indicate the interaction strength of each factor. It was observed from the ANOVA that the confidence level was greater than 90% while the *p*-value of the model being 0.0015. The model with the *p*-value below 0.05 was statistically significant, which implied that the model was suitable for this experiment. The effect of temperature and the interaction of main effects on microbial growth inhibition in terms of cell dry weight are insignificant based on the value, 0.2254 and 0.8822 respectively which was greater than 0.05. Microbial inhibition time effect shows significant effect on microbial growth inhibition which explained by the *p*-value of 0.0039. The coefficient of determination (R^2) and adjusted coefficient of determination (R^2_{adj}) were 0.9108 and 0.8471, respectively which indicated that the estimated model fits the experimental data satisfactorily. Lee et al. (2010) suggested that for a good fit of a model, R^2 should be at least 0.80. The R^2 for these response variables was higher than 0.80, indicating that the regression models explained the mechanism well. Figure 1 shows the predicted versus actual microbial growth inhibition in terms of cell dry weight. A linear distribution is observed which is indicates the well-fitting model. The values predicted were close to the observed values of cell dry weight.

Table 3. ANOVA for response surface quadratic model.

Source of variation	Sum of Squares	Mean Square	F Value	P-value Prob > F	Remark
Model	0.097	0.019325	14.30	0.0015	significant
A-Microbial inhibition time	0.024	0.024147	17.87	0.0039	-
B-Temperature	0.002389	2.389E-03	1.77	0.2254	-
AB	3.19E-05	3.19E-05	0.024	0.8822	-
A^2	0.064	0.06395	47.31	0.0002	-
B^2	0.024	0.021532	15.93	0.0052	-
Residual	0.009461	0.001352	-	-	-
Lack of Fit	0.007561	0.00252	5.30	0.0704	not significant
Pure Error	0.001901	4.75E-04	-	-	-
Cor Total	0.11	-	-	-	-
R^2	0.9108	-	-	-	-
Adjusted R^2	0.8471	-	-	-	-

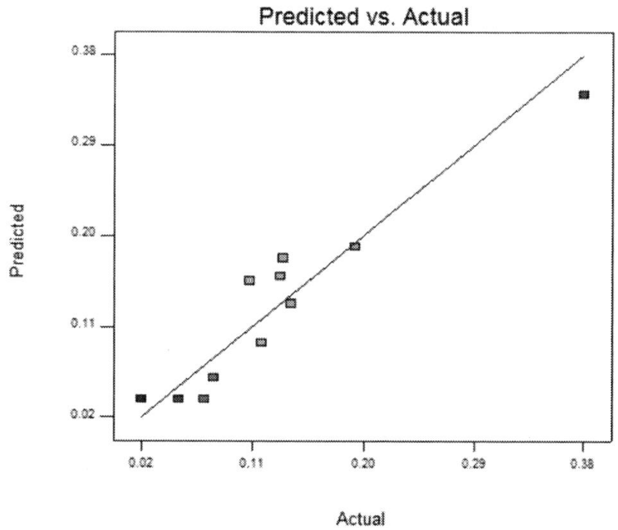

Figure 1. Correlation of actual and predicted values by the model.

3.3. Effect of independent and interactive parameters on cell dry weight

The effect of the two independent variables on the cell dry weight is shown in figures 2(a) and (b) respectively. Cell dry weight decrease with increasing microbial inhibition time from 20 min to 35 min as shown in figure 2(a). Meanwhile, upon increasing the microbial inhibition time from 35 min to 40 min, cell dry weight is increasing. This result indicates that increasing microbial inhibition time have a positive influence on cell dry weight. Its means that 35 minutes is adequate for microbial growth inhibition process as minimum cell dry weight indicates higher microbial growth inhibition. According to Maria et al [14], the phenol coefficient has potency of an agent against some microbe after 10 min. However, U.S Department of Health & Human Services (2016) reported that, 2 to 3 hours of microbial inhibition time between citral and *C. albicans* can also kill 99.9% of inoculum.

The effect of temperature is shown in figure 2(b). From the figure, it shows that the optimum temperature of microbial growth inhibition process is at 37 °C. The temperature did not give much effect to the cell dry weight as there was small difference between the cell dry weight from 35 to 39 °C. This is supported by the *p*-value, 0.2254 that obtained from ANOVA. The effect of temperature on cell dry weight are insignificant. Thus, the ambient temperature (37 °C) was selected for microbial growth inhibition process, this finding is similar to those reported by Yang et al. [15].

The effect of interaction between parameters, i.e. temperature and microbial inhibition time on cell dry weight was graphically represented by contour plot and 3D response surface generated by the model as shown in figure 3. An elliptical contour plot indicated the interaction between the variables were significant. A comparative study of the results showed that the cell dry weight at any temperature was affected by the microbial inhibition time. Decreasing of cell dry weight indicates the higher microbial growth inhibition. The plotted data depicts the effect of temperature and microbial inhibition time on cell dry weight. The optimum condition needed to achieve microbial growth inhibition of 91.65% with 0.0265 gram of cell dry weight are at 37 °C and 34.25 minutes. At this point, the cell dry weight is at their minimum rate which indicates the maximum microbial inhibition process. In addition, a study by Segundo [16] reported that by using bromelain extracted from pineapple stems as microbial growth inhibitor can inhibit 90% growth of *F. verticilioides*. It shows that the results are close to the result obtained in this study.

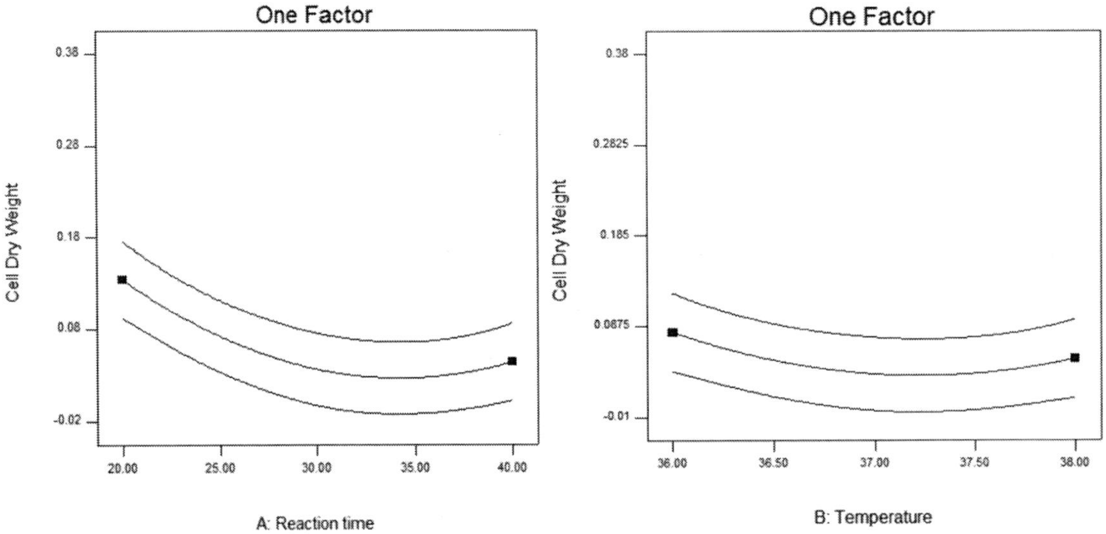

(a) (b)

Figure 2. (a) Effect of microbial inhibition time (reaction time) on cell dry weight and (b) Effect of temperature on cell dry weight.

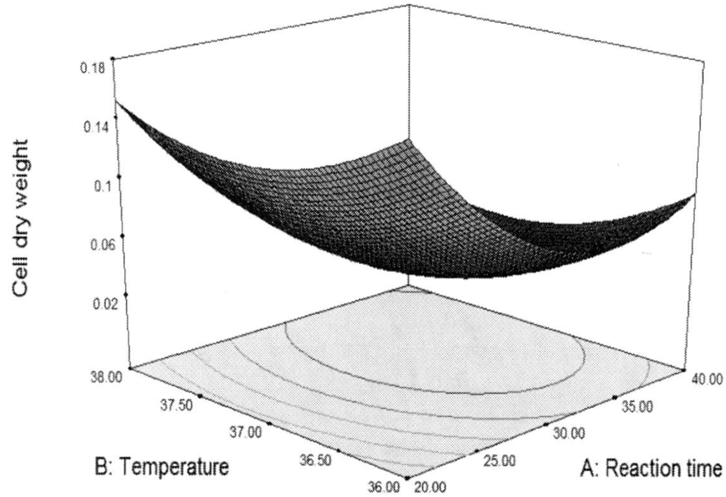

Figure 3. Contour plot and 3D surface plot of cell dry weight as function of microbial inhibition time (reaction time) and temperature.

3.4. Model validation

Table 4 shows the suggested optimum conditions derived from analysis of RSM designed by Design Expert software. It was suggested that at temperature of 37 °C and 34.25 minutes microbial inhibition time, the microbial growth inhibition obtained is 91.65%. The adequacy of the model equations was tested using the suggested optimum conditions. This validation experiments performed demonstrated that the experimental values were reasonably close to the predicted value. At the culture age and microbial incubation time of 24 hours, the highest experimental microbial growth inhibition obtained is 94.73%. The error from this experiment is calculated by using equation (3). The error obtained was between 1.73% to 8.03%. Hence, this concludes that the error was accepted since the highest error percentage was not greater than 10% for all responses [17].

$$\text{Error (\%)} = \frac{\text{Experimental} - \text{Predicted}}{\text{Predicted}} \times 100 \tag{3}$$

Table 4. Suggested optimum conditions.

Factor	**Value**
A: Microbial inhibition time	34.25 min
B: Temperature	37 °C
Cell dry weight	0.0265 g
Microbial growth inhibition (%)	91.65 %

4. Conclusion

This study focused on microbial growth inhibition process using pineapple leaves juice. Central composite design (CCD) of RSM was used to determine the optimum conditions for the process. The second order polynomial equation developed in this study shows a high correlation between experimented and predicted cell dry weight values. Response surface analysis was found to be a good approach for visualizing process parameter interaction. The optimum conditions obtained in this study are 37 °C and 34.25 minutes with maximum microbial growth inhibition of 94.73%. Results shows the potential of pineapple leave juice application as natural microbial growth inhibitor.

IOP Conf. Series: Materials Science and Engineering **863** (2020) 012022 doi:10.1088/1757-899X/863/1/012022

Acknowledgment
The author wishes to acknowledge the Universiti Malaysia Pahang for funding the project under grant RDU1803119.

References
[1] Carrillo-González R, Martínez-Gómez MA, González-Chávez MDCA and Mendoza Hernández JC 2016 *Sci. Total Environ.* **565** 872-881.
[2] Alias NH and Abbas Z 2017 *Malaysian Journal of Analytical Sciences* **21(3)** 690 – 699.
[3] Kasote DM, Katyare SS, Hegde MV and Bae H 2015 *Int J Biol Sci.* **11(8)** 982–991.
[4] Zhao HX, Zhang HS and Yang SF 2014 *Food Science and Human Wellness* **3** 183-190.
[5] Che C and Zhang H 2019 *Int J Mol Sci.* **20(4)** 830.
[6] Pernin A, Bosc V, Maillard MN and Dubois-Brissonnet F 2019 *Frontiers in Microbiology* **10** 137.
[7] Wang NS 2019 *Biochemical Engineering Laboratory Manual* (https://eng.umd.edu/~nsw/ench485/lab9c.htm, 2019).
[8] Singleton V and Rossi J 1965 *American Journal of Enology and Viticulture* **16** 144-158.
[9] Ododo MM, Choudhury MK and Dekebo AH 2016 *SpringerPlus* **5(1)** 1210.
[10] Andriana Y, Xuan TD, Quy TN, Minh TN, Van TM and Viet TD 2019 *Foods* **8(1)** 21.
[11] Wang K, Sipilä TP and Overmyer K 2016 *Scientific Reports* **6** 39403.
[12] Bizuayehu D, Atlabachew M and Ali MT 2016 *SpringerPlus* **5** 412.
[13] Zainol N and Rahim SR 2017 *Journal of Chemical Engineering and Industrial Biotechnology* **1**18-28.
[14] Maria CAL, Bezerra AP, Sousa JP, Guerra FQ and Lima E 2014 *Evidence-Based Complementary and Alternative Medicine* **2014** 1-9.
[15] Yang E, Fan L, Yan J, Jiang Y, Doucette C, Fillmore S and Walker B 2018 *AMB Express* **8(1)** 10.
[16] Segundo BS, López-García BI and Hernández M 2012 *Lett Appl Microbiol.* **55(1)** 62-67.
[17] Taylor JR 1997 *An Introduction to Error Analysis: The Study of Uncertainties in Physical Measurements, 2d Edition* (California:University Science Books).

IOP Conf. Series: Materials Science and Engineering **863** (2020) 012023 doi:10.1088/1757-899X/863/1/012023

Air co-gasification of palm kernel shell and polystyrene: Effect of different polystyrene content

M H Basha[1,2,*] and S A Sulaiman[1]

[1] Department of Mechanical Engineering, Universiti Teknologi PETRONAS, 32610 Bandar Seri Iskandar, Perak, Malaysia.
[2] Mechanical Engineering Programme, School of Mechatronic Engineering, Universiti Malaysia Perlis, Malaysia.

*E-mail:hafifbasha@unimap.edu.my

Abstract. Plastic waste management has been a problem as most of the plastics are not biodegradable. Although plastics are recyclable, only 21% of total plastic wastes are recycled. Using plastics as gasification feedstock/co-feedstock and converting it to usable energy is one of the solutions of removing plastics waste. Co-gasifying polystyrene (PS) with palm kernel shell (PKS) is promising as both of the feedstocks are abundant and have high energy content. However, the performance of air gasification of PKS and PS has not been studied previously. In this work, co-gasification of PKS and PS was carried out, focusing on the effect of PS content in the feedstock. The PS content was varied from 0 to 30 wt%. By increasing the PS content, the CH_4 volume percentage increases, while CO and H_2 volume percentage decreases. The high heating value of the producer gas increases with PS content, from 11.95 MJ/Nm^3 at 10 wt% PS, to 12.36 MJ/Nm^3 at 30 wt% PS. Higher PS content also increases the gas yield percentage.

1. Introduction

As the world's inhabitant increase, the use of plastic, a non-biodegradable material increases, causing a problem of plastic waste. In Malaysia, about a quarter of all solid waste consist of plastic waste [1]. The management of these plastic wastes remains a highly debated issue. Recycling is seen as a viable way to manage these wastes, but not with its challenges [2]. Polypropylene (PP), polyethylene terephthalate (PET) and polyethylene (PE) recycling used a widely common process. However, it cannot be said to same to polystyrene. The recycling of expanded polystyrene requires an additional process that are costly [3]. Thermochemical process is seen as a way to dispose of plastics. Gasification technology is one of them. Gasification uses heat to convert carbon materials to gas. The gas can be used directly as fuel, or as chemical feedstock for subsequent chemical process. Numerous gasification of plastics waste research has been done [4–8]. However, in an operation point of view, the gasification of plastics only can cause clogging problems when the plastic is melted and stick to the tubes. This problem can be solved by mixing plastics with biomass. Palm kernel shell (PKS) has a potential to be a co-feedstock as it has a high energy density.

In this work, co-gasification was performed using a batch type, externally heated gasifier with mixed blends of polystyrene (PS) and PKS as feedstocks, focusing on the effect of different polystyrene content of the blends. This study will analyze the gas composition and the heating value of the producer gas. The amount of tar produced from the gasification process will also be measured and analyzed. Information on the characteristics of the co-gasification reaction is essential before any real world use is considered.

2. Experimental

2.1. Feedstock and characterization
PKS is collected from FELCRA oil palm mill in Bota, Perak, Malaysia. The PKS was processed using a grinder, and sieved to obtain particle size of 2-4 mm. Prior to the experiment, the PKS was dried inside an oven at 105 °C for 24 hours. The polystyrene (PS) granule was supplied by Dow Plastics.

For mixed blends preparation, PKS and PS were weighted respectively according the desired percentages. The feedstock were mixed and stirred until it is uniformed. Table 1 shows the ultimate and proximate analysis of the feedstocks.

Table 1. Proximate analysis, ultimate analysis and calorific value of the feedstocks.

	Elemental analysis (wt%)				Proximate analysis (wt%)			
	C	**H**	**N**	**O[a]**	**Ash**	**Volatile Matter**	**Fixed Carbon**	**Calorific value (MJ/kg)**
PKS	41.95	5.49	1.43	50.97	8.27	60.78	30.94	22.59
PS	91.33	8.67	-	-	-	99.6	0.04	40.9

[a] by difference

2.2. Experimental set-up
In this work, the effect of different PS content was investigated. The electric heated batch downdraft gasifier used for the experiments is shown in figure 1. The diameter and vertical length of the cylinder reactor is 8 cm and 50 cm, respectively. Before the feedstock was fed, the reactor was heated to the desired temperature. The reaction temperature was set to 900 °C for all runs. Air was supplied by the air compressor, with the flow meter regulating the amount of air that was fed to the reactor. The air flow rate was set to 2.5 L/min. The reaction starts directly after the feedstocks were fed to the reactor. Condenser bottles condensed the liquid, and impinger bottles filled with solvents captures tars inside the producer gas. The gas product is directly analysed by the Emerson-Rosemount Analytical X-Stream X2GP online gas analyser, where the volume percentages of CO, CO_2, CH_4 and H_2 gas were measured. The experiment lasts about 10 minutes. The solid residue of the reaction was measured after the experiment ended. The polystyrene content that was tested are 0 wt% up to 30 wt%. Total mass of the feedstock was 100 grams.

2.3. Data analysis
The high heating value (HHV) of the producer gas was calculated using equation (1):

$$HHV_{gas} = HHV_{CO} \times CO\% + HHV_{H_2} \times H_2\% + HHV_{CH_4} \times CH_4\% \quad (1)$$

where HHV of CO, H_2 and CH_4 are 12.63 MJ/Nm^3, 12.74 MJ/Nm^3 and 39.82 MJ/Nm^3 respectively [9]. CO%, H_2%, and CH_4% are the average volumetric percentage of the respective gases. To calculate the mass gas yield of the produced gas, the solid, liquid and tar product of the reaction was measured respectively, and the mass gas yield was then calculated using the mass balance equations.

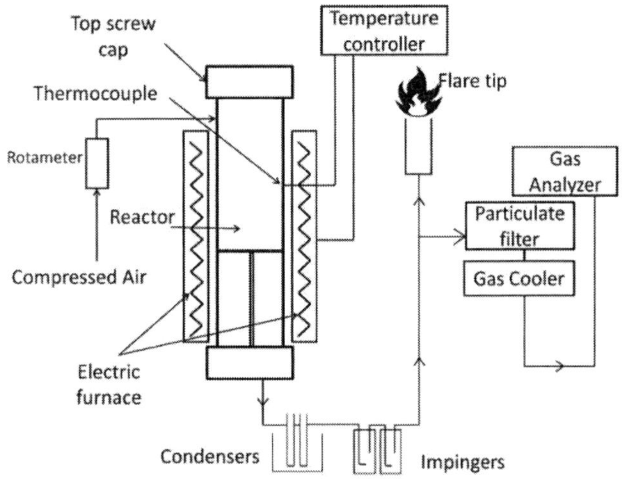

Figure 1. Schematic diagram of the electric heated downdraft gasifier.

3. Results and discussion

3.1. Producer gas composition

Figure 2 shows the relationship of the gas composition and the plastics percentage of the feedstock blend. With increasing PS content, CO and H_2 percentage decreases. The CO_2 percentage does not seems to affect much with increasing PS content. On the contrary, by adding PS as low as 10% to the feedstock the CH_4 percentage increases significantly from 15.6% to 19%. CH_4 shows and increasing trend with the increase of PS. This can be attributed to the degradation mechanism of PS which is a polymer. The polymeric hydrocarbon chain is cut into smaller piece of molecules, and ended up as the most simple form of hydrocarbon, which is methane [10].

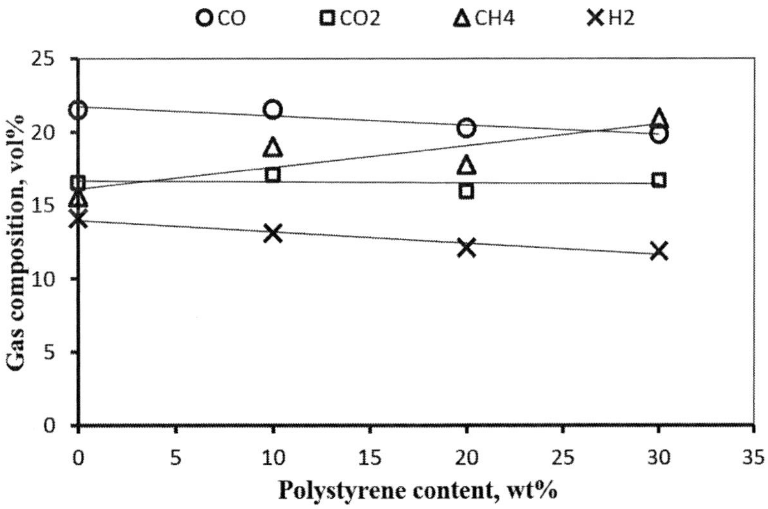

Figure 2. Effect of different polystyrene content on the gas composition of the producer gas of the co-gasification of PKS and PS at 900 °C.

3.2. High heating value (HHV) of producer gas and product distribution

Figure 3 shows the effect of PS content on the HHV of the producer gas. Feedstocks blend with PS yields producer gas with higher HHV, compared to the gas produced by the gasification of PKS only. Adding 10 wt% of PS to the feedstock blend yielded gas with HHV value of 11.95 MJ/Nm^3, an increase compared to the HHV value of gasification of PKS only, in which the HHV value of the producer gas is 10.72 MJ/Nm^3. This can be explained by observing the gas composition of the producer gas, in which adding plastic will increase the CH_4 content. CH_4 has high HHV, so the increase of CH_4 certainly affects the HHV value of the producer gas. Further increasing the PS content will show an increasing trend of HHV.

The product yield percentage is shown in also shown in figure 3. Increasing the PS content will increase the gas yield, while decreasing the solid yield and liquid yield. This can be attributed to the high volatile matter of PS. With increasing PS content in the feedstock, higher volatile content of the feedstock mixture will yield more gas, as more solids are converted to gas. The tar yield percentage shows a slight increment, with the increase of PS content.

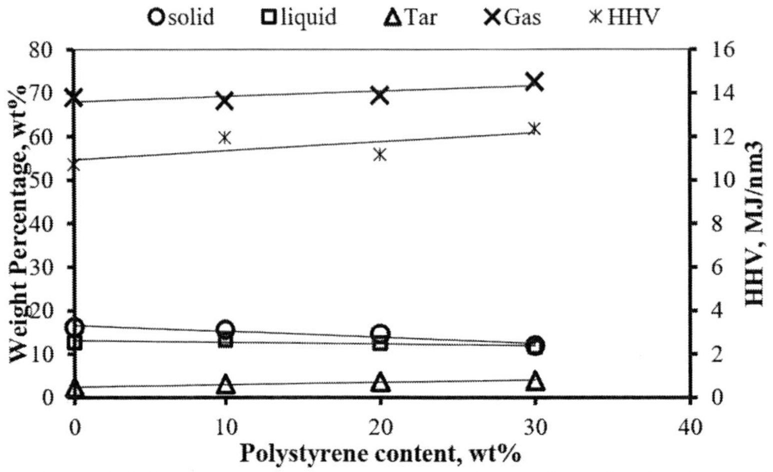

Figure 3. Effect of different polystyrene content on the product distribution and HHV of the producer gas of the co-gasification of PKS and PS at 900 °C.

4. Conclusion

In this work, the effect of PS content on the co-gasification of PKS and PS has been investigated. Increasing the PS content will increase the CH_4 percentage of the producer gas, while decreasing the CO and H_2 volume percentage. The HHV of the producer gas increased with PS content. With increasing PS content, the gas yield increases, while the solid and liquid yield decreases. Tar yield shows a slight increase with PS content. Co-gasification of PKS and PS is feasible for power generation as the produced gas has HHV in the range of 11.95 MJ/Nm^3 to 12.36 MJ/Nm^3, which is high enough for power generation. However, further study need to be done in order to investigate other parameters such as the reaction temperature and the equivalent ratio on the reaction.

Acknowledgment

The authors would like to thank Universiti Teknologi PETRONAS, Malaysia for the facilities and technical support for this work.

References

[1] Budhiarta I, Siwar C, and Basri H 2012 *Int. J. Adv. Sci. Eng. Inf. Technol.* **2** 129.
[2] Gershman B 2013 *Gasification of Non-Recycled Plastics From Municipal Solid Waste In the United States.*
[3] Noguchi T, Lnagaki Y, Miyashita M, and Watanabe H 1998 *Packag. Technol. Sci.* **11** 29–37.
[4] Arena U and Di Gregorio F 2014 *Energy* **68** 735–743.
[5] Tsuji T and Hatayama A 2009 *J. Mater. Cycles Waste Manag.* **11** 144–147.
[6] Xiao R, Jin B, Zhou H, Zhong Z, and Zhang M 2007 *Energy Convers. Manag.* **48** 778–786.
[7] Yang RX, Chuang K-H, and Wey M-Y 2018 *Energy & fuels.* **32** 5462–5470.

[8] Nunome Y, Suzuki T, Nedjalkov I, Ueki Y, Yoshiie R, and Naruse I 2019 *J. Mater. Cycles Waste Manag.* 1–11.
[9] P. Basu, *Biomass Gasification, Pyrolysis and Torrefaction: Practical Design and Theory*. 2013.
[10] Kantarelis E, Donaj P, Yang W, and Zabaniotou A 2009 *Journal of Hazardous Materials*. **167** 675–684.

IOP Conf. Series: Materials Science and Engineering **863** (2020) 012024 doi:10.1088/1757-899X/863/1/012024

Influence of non-covalently functionalized graphene nanoplatelets on viscosity and wettability for improved oil mobility at reservoir conditions

S S Hamdi[1], A S Muhsan[2,*] and H H Al-Kayiem[1]

[1] Mechanical Engineering Department, Universiti Teknologi PETRONAS, 32610 Bandar Seri Iskandar, Perak, Malaysia.
[2] Petroleum Engineering Department, Universiti Teknologi PETRONAS, 32610 Bandar Seri Iskandar, Perak, Malaysia.

*E-mail: ali.samer@utp.edu.my

Abstract. Noncovalent functionalized graphene nanoplatelets (GNPs) were successfully prepared with Gum Arabic as exfoliation solvent via an ultrasonication-assisted process to produce a highly dispersed material under high salinity and high temperature conditions. The implementation of the functionalization groups was validated with Fourier transform infrared spectroscopy and Transmission Electron Microscopy. The functionalized graphene nanoplatelets with Gum Arabic were dispersed in 3 wt.% high salinity brine with various concentrations to prepare Gum Arabic-GNP/brine nanofluids. The dynamic viscosity of the produced nanofluid at different concentrations was higher than that of crude oil at a high temperature (90 °C) and a high shear rate (1000 L/s). The functionalized graphene with Gum Arabic altered the wettability of sandstone from oil to water and changed the contact angle from 108° to 21° at 0.1 wt.%. Moreover, the value of contact angle reduced to 87% with presence 0.1 wt.% of GA-GNPs. These results are demonstrating that the GA-GNPs nanofluids could improve the mobility of the residual oil from the porous media to the production well with high efficiency of oil displacement.

1. Introduction

Since decades, nanofluids have been acquiring considerable concern with progressing of nanotechnology in various fields of industry. Nanofluids expression was proposed by Choi in 1995 [1], as an innovation of nanotechnology in heat transfer application, which consists of suspension exhibited high thermal conductivity by dispersion of a nanoscale materials, typical size less than (100 nm) in the base fluids [1]. Furthermore, the application of nanofluids in the oil and gas industry as a novel way more efficient than conventional fluids, economical and environment friendly [2]. Employment of nanofluids in chemical EOR have proven a significant potential to lower the interfacial energy of the colloidal system (between the oily and watery phase), a result of the high surface energy of nanoparticles. Also, it is used to modify the surface properties between the solid and liquid phases of the core with the presence of stable nanoparticles [3, 4]. In addition, the reduction in the viscosity ratio which are leading to increment the capillary number and eventually oil droplet is extracted and detached from the surface of rock pore by capillary force [5]. Most of literature have been reviewed the potentials of utilizing metal oxide nanoparticles like (SiO_2, Al_2O_3, ZnO, TiO_2, NiO, ZrO_2 and Fe_3O_4) for enhance oil recovery [6, 7].

However, the main issue of these nanofluids are unstable colloidal suspension and agglomeration of the nanoparticles at reservoir conditions high salinity and high temperature (HSHT). High concentration of salts would have high effect of ionic strength that reduce the electrostatic double layer forces between nanoparticles. However, nanoparticles have tendency to agglomerate, because of the high specific surface area to volume and the existence of mutual van der Waals forces of attraction among them. They caused clogging the pore throat and declining the porosity and the permeability of the porous media [8, 9]. Moreover, disordering of the nanoparticles at the wedge film which was created at the triple contact line between the oil droplet with the surface of the core rock and the stable nanoparticles according to the new concept by Wasan and Nikolov [10]. Thus, decrement the extent of disjoining pressure, this force was formed in the wedge film depend on the size and stability of the nanoparticles. Consequently, diminishing the recovery of the trapped oil and subsequently reduction the sweep efficiency [9]. These challenges could be reduced by preparing self-stabilize nanoparticles to overcome the agglomeration and sedimentation of nanoparticles under the harsh environmental condition and displacing more quantity of the oil from the pore than either unmodified nanoparticles or dispersant alone [11].

Graphene as new generation of carbon nanostructures have been attracting the vast interest of the scientists and researchers, due to its unique and remarkable properties [12, 13]. A graphene monolayer is usually stacked by an intense van der Waals force between layers to form graphene nanoplatelets (GNPs), which have better thermophysical and rheological properties, higher surface area, more highly ordered graphitic carbon, and lower cost than monolayer graphene [14]. The innate problem of GNPs is low dispersion, high hydrophobicity and strong van der Waals interactions. Numerous techniques have been developed to exfoliate graphene by physical and chemical process [15, 16]. In the previous studies, Chemical functionalization with organic hydrophilic group such as (hydroxyl, carbonyl, and carbonyl, etc.) can be added the radical at the edge and/ or the basal of graphene for improving the dispersibility of graphene [17]. Moreover, stable dispersion of graphene oxidized by common chemical process (Modified Hummer's method) [18], Also modified with metal oxide nanoparticles as hybrid nanofluid to stabilize the emulsion [19]. Some studies have utilized the polymer as surface modifier of graphene and graphene oxide (GO) to improve their dispersibility in high salinity of brine and high temperature. In this regard, Zuniga et al. [20], supposed to functionalize graphene oxide with a polyzwitterionic polymer to improve stability for a long period (140 days) at harsh conditions including high ionic strength of API and Arab-D brine and temperature. Luo et al. [21], prepared nanofluid of graphene amphiphilic Janus nanosheets by modified the surface of GO with alkylamine to investigate the colloidal stability of this nanofluid in unfriendly environments. Radnia et al. [22], synthesized amphiphilic graphene by direct functionalization of Nano porous graphene (NPG) prepared by chemical vapor deposition with sulfonic (-SO_3H) groups to enhance oil recovery. These functionalization techniques have used corrosive and harmful materials. Therefore, cost-effective and safe methods should be developed to functionalize graphene nanoplatelets.

Gum Arabic (GA) is considered the oldest one among the natural materials, which it is widely used in the food, soft drink, drug delivery and emulsion (oil/water) [23, 24] . However, GA is utilized with nanoparticles due to its ability to stabilize the dispersion of aqueous fluid over a wide range of pH. Meanwhile, GA has contained the hydroxyl and carboxyl bonds with long chains and complex molecules structure, which are absorbed on the surface of nanoparticles to provide by the steric repulsive [25].

This work aimed to synthesize hydrophilic graphene via an economical and environmentally green process with GA through direct exfoliation with assistance of ultrasonic process. To the best of our knowledge, few researches have reported the synthesis and stability of GNPs through noncovalent functionalization of surfactant at high salinity and high temperature. Thus, functionalized GNPs with GA were well prepared in this study. The quantity and the grafting of the functionalization groups on the graphene layers were evaluated. Moreover, the effect of GA-GNPs with various concentration on the oil wettability on surface of the sandstone and the dynamic viscosity under reservoir conditions were investigated.

2. Methodology

2.1. Materials

Pristine Graphene nanoplatelets grade (C) have average particle size less than 2μm and thickness of a few nanometers, with high specific area (750 m^2/g) were used in this study as raw material. Gum Arabic was utilized as exfoliation solvent with high purity were purchased from Sigma Aldrich. Sodium Chloride (NaCl) salt with purity ≥ 99.5% and purified deionized water (18.2 MΩ) at 25 °C were used to prepare high salinity brine. All the materials were used as received without any further purification.

2.2. Experimental set up

2.2.1. Preparation of GA-GNP nanofluids. In the first step of exfoliation process 0.5 gram of GA was mixed with 100 ml of deionized water (DW) at ambient temperature for 0.5 hour. Afterward, 1 gram of GNPs was added to the solution and stirred for 1 hour, to obtain the homogeneous suspension of nanoplatelets. Next, the mixture of suspension was dispersed by using probe sonicator (Ultrasonic processor, 750 Watt, and frequency 25 KHz) to reduce the van der Waals forces between the graphene layers. An ice bath was used to prevent the evaporation of the exfoliated solvent and to maintain its temperature near ambient temperature. The parameters for exfoliation the nanoplatelets were selected to successfully expel the layers of graphene with fewer defects at 60% amplitude for the apparatus and 2 hours of time for sonication. After exfoliation, the sample was centrifuged by speed (3000 rpm) for 15 minutes to remove the large layers of graphene and discarded. The supernatant was re-centrifuged to high speed 7500 rpm for 30 minutes. The final supernatant rinsed with ethanol and deionized water three times and filtrated to get rid of the extra amount of the dispersant (GA). The final solution of the exfoliated GNPs with GA denoted as GA-GNPs, dried in the oven at 60 °C for 48 hours.

2.2.2. Characterization of GA-GNPs. The functionalized GNPs with GA were characterized by using Fourier Transform infrared spectroscopy (Perkin Elmer- FTIR) and morphological characteristics was investigated by Field -Emission Transmission Electron Microscope (FETEM, Model; JEM-2100F).

2.2.3. Characterization of nanofluids. The final product of GA-GNPs was re-dispersed with various weight concentrations 0.01, 0.05 and 0.1 wt. % in high salinity brine (3 wt. % NaCl) as a base fluid by mild sonication for 1 hour. The contact angle measurement was used to evaluate the wettability alteration of glass slices as sandstone rock core. These slices of the substrate were aged in various fluids which are included a different concentration of the prepared nanofluids for 24 hours at 80 °C. The contact angle of oil droplet with the slices of glass was carried out by using Goniometer (Ramé-hart 260), the glass slice is immersed inside the visual cuvette filled with brine and droplet of crude oil is injected and released by an inverted needle to captured at the bottom of the glass slice. Afterward, magnified side view images of oil droplet were taken and analyzed to measure the contact angle by advance software to process the image of the droplet.

The dynamic viscosity of the nanofluids was investigated in rotational module by using HR-1, Discovery Hybrid Rheometer. The GA-GNPs NFs with different concentrations were examined at high temperatures. The prepared nanofluids (16 mL) were loaded between the concentric cylinders and allowed to attain temperature stabilization for 5 minutes. The test was carried out at various temperatures (25-90 °C) and a constant shear rate (1000 L/s).

3. Results and discussion

3.1. Characterization of GA-GNPs

The pristine and noncovalent functionalized graphene nanoplatelets were analysed through FTIR spectroscopy to evaluate the grafted GNPs with chemical bonds of the GA molecules as shown in figure 1. The FTIR results of Functionalized GNPs with GA exhibited different characteristic peaks with pristine, these involved a carboxylic band (C=O) at 2350 cm^{-1} [26]. Moreover, hydroxyl (COOH) deformation at the peak 1422 cm^{-1} which are agreement with the symmetric vibration of carboxylic acid group. Furthermore, 2923 cm^{-1}, 1067 cm^{-1}, 870 cm^{-1}, and 712 cm^{-1} which are correspond to the stretching of (C-H) alkanes band, epoxy absorbance band (C-O-C), (C-O) and (C-OH) bending in and out of plane deformation respectively [25], due to attachment of the GA molecules on the basal and edges of graphene. The hydroxyl group (-OH) appear with a strong and broad peak at 3400 cm^{-1}, which are related to absorption the (-OH) from ethanol alcohol and the chemical groups from GA [27]. Based on the results of FTIR with existing peaks compared to the pristine sample, which starkly confirms the successful grafting of Arabic gum functional groups to the GNPs plane.

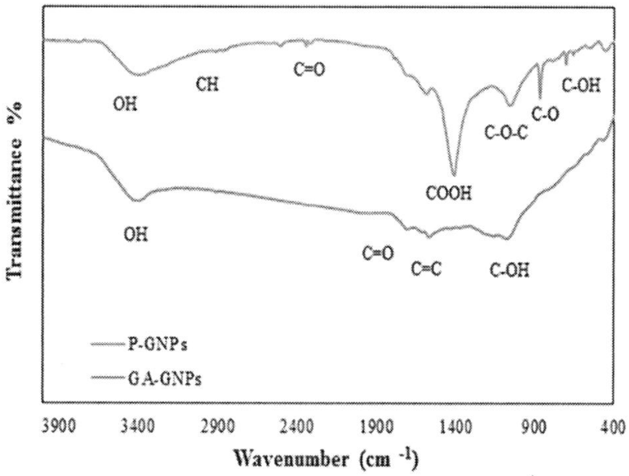

Figure 1. FTIR of (a) GA modified GNPs and (b) pristine GNPs.

Figure 2 shows the morphology analysis of pristine and functionalized GNPs with GA by high resolution TEM imaging which observed the disorder of the structure after the ultrasonication process. Meanwhile, the transparency and wrinkles of functionalized graphene nanosheets indicate to the successful process of noncovalent functionalization of graphene with Gum Arabic.

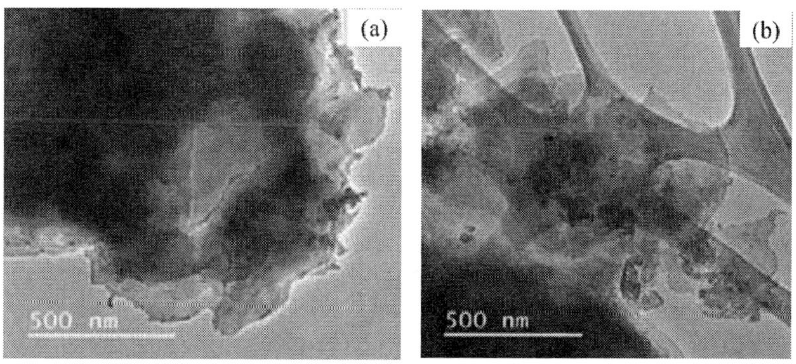

Figure 2. TEM image of (a) Pristine GNPs and (b) GA modified GNPs at high magnification.

3.2. Characterization of GA-GNPs nanofluids for improving oil mobility

Table 1 summarizes the dynamic viscosity of the different concentrations of prepared nanofluids and crude oil at ambient and high temperatures with a constant shear rate. It should be mentioned the dynamic viscosity of all nanofluids at high temperature with a high shear rate is lower than ambient temperature. However, the viscosity of crude oil declined from 8.13 mPa.s at 25 °C to 1.81 mPa.s at

IOP Conf. Series: Materials Science and Engineering **863** (2020) 012024 doi:10.1088/1757-899X/863/1/012024

90 °C and high shear rate of 1000 L/s. The viscosity ratio between the crude oil and various concentrations of GA-GNPs nanofluid at high temperature and high shear rate is higher than unity.

Table 1. The dynamic viscosity of the fluids.

Fluids	**Viscosity (mPa.s)**	
	25 °C	**90 °C**
Crude Oil	8.13	1.81
0.01 wt.% GA-GNPs	4.13	2.69
0.05 wt.% GA-GNPs	4.14	2.71
0.1 wt.% GA-GNPs	4.16	2.72

The contact angle of an oil droplet on the surface of treated slices with functionalized GNPs at 80 °C for 24 hours was illustrated in figure 3. The wettability of treated slices with three concentration of GA-GNPs nanofluids was altered to water wet. Eventually, the shape of the droplet changed, and the contact angle was recorded 21° at 0.1 wt. % of GA-GNPs nanofluid. In comparison, the contact angle with was 108° in the brine. In this regard, the contact angle declined to 87% with high concentration of GA-GNPs. These results were due to the functionalized graphene nanofluids that achieved the water wet with the presence a stable of nanoplatelets at a region of contact between the surface of the solid and liquid of (oil-brine). The functionalized graphene nanosheets adsorbed and spread along the glass slides to change the wettability from oil wet to water wet. Therefore, the oil droplet detached the surface of the porous and subsequently the oil simply to moved and displaced into the production well.

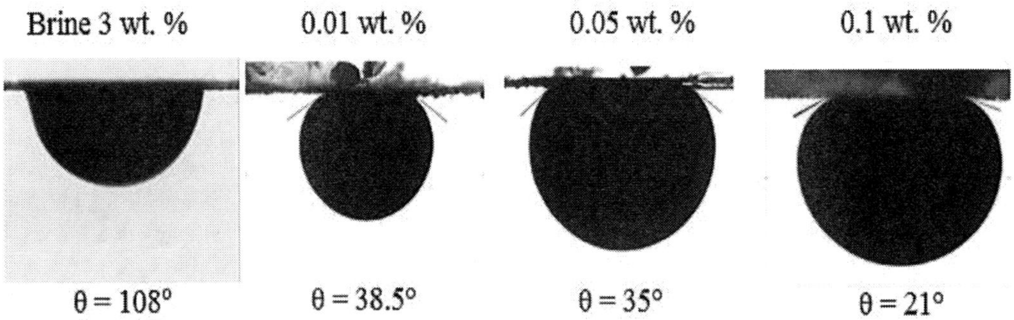

Figure 3. The contact angle of oil droplet below the untreated and treated glass slices with GA-GNPs at different concentrations.

4. Conclusion

Noncovalent functionalization of GNPs with GA has been successfully synthesized by economical and green environmentally process. The noncovalent functionalization of graphene layers was validated by using FTIR and FETEM. The adsorbing and spreading of GA-GNPs along the glass slides have led to altering the wettability from oil to water wet with reducing in the value of contact angle to 87%. However, the viscosity of the prepared nanofluid still higher than the viscosity of the crude oil at high temperature and high shear rate (90 °C and 1000 L/s) and the viscosity ratio higher than unity. Hence, the high stability of the GA functionalized GNPs at harsh conditions which is leading to high viscosity, altered the wettability to water wet could improve the oil mobility with increasing the sweep efficiency and resulting in favorable oil displacement.

References

[1] Choi S U and Eastman J A 1995.
[2] Mohd T A T, Baco J, Bakar N F A and Jaafar M Z 2016 *MATEC Web of Conferences*: *EDP Sciences*.

[3] Hendraningrat L, Li S and Torsater O 2013 *SPE International Symposium on Oilfield Chemistry*: (Society of Petroleum Engineers).
[4] Giraldo J, Benjumea P, Lopera S, Cortés F B and Ruiz M A J E 2013 *Fuels* **27**:3659-65.
[5] Seid M M, Moghadasi J and Naseri S 2014 Ira. J. Oil & Gas Sci. and Tech. **3**:18-26.
[6] Khalil M, Jan B M, Tong C W and Berawi M A 2017 *Applied Energy* **191**:287-310.
[7] Sun X, Zhang Y, Chen G and Gai Z 2017 *Energies* **10.**
[8] Hendraningrat L and Torsæter O 2015 *J. Transp. in Prous Media* **108**:679-96.
[9] Afolabi R O 2019 *Int. Nano Letters* **9**:1-15.
[10] Wasan D T and Nikolov A D 2003 *Nature* **423**:156.
[11] Behzadi A and Mohammadi A 2016 *J. Nanopart Res.* **18**:266.
[12] Ju M J, Kim J C, Choi H-J, Choi I T, Kim S G, Lim K, et al. 2013 *ACS Nano* **7**:5243-50.
[13] Wang Z, Luo J and Zhao G L 2014 *AIP Adva.* **4**:017139.
[14] Mehrali M, Sadeghinezhad E, Latibari S T, Kazi S N, Mehrali M, Zubir M N B M, et al. 2014 *Nanosc. Resea. Lett.* **9**:15.
[15] Geng Y, Wang S J and Kim J-K 2009 *J. Coll. and Intef. Sci.* **336**:592-8.
[16] Wang B, Hao J and Li H J D T 2013 *Dalton Transaction* **42**:5866-73.
[17] Amiri A, Sadri R, Shanbedi M, Ahmadi G, Chew B T, Kazi S N, et al. 2015 *Ener. Conv. and Manag.* **92**:322-30.
[18] Yoon K Y, An S J, Chen Y, Lee J H, Bryant S L, Ruoff R S, et al. 2013 *J. Coll. and Intef. Sci.* **403**:1-6.
[19] AfzaliTabar M, Alaei M, Bazmi M, Khojasteh R R, Koolivand-Salooki M, Motiee F, et al. 2017 *Fuel* **206**:453-66.
[20] Zuniga C A, Goods J B, Cox J R and Swager T M 2016 *ACS Appl. Mater. Interfaces* **8**:1780-5.
[21] Luo D, Wang F, Alam M K, Yu F, Mishra I K, Bao J, et al. 2017 *Chem. Mater.* **29**:3454-60.
[22] Radnia H, Rashidi A, Nazar A R S, Eskandari M M and Jalilian M 2018 *J. Molec. Liq.* **271**:795-806.
[23] Wang B, Wang L-j, Li D, Adhikari B and Shi J 2011 *Carbohy. polym.* **86**:343-51.
[24] Patel S and Goyal A 2015 *Inter. J. Food Prop.* **18**:986-98.
[25] Xu H and Suslick K S 2011 *J. the Amer. Chem. Soc.* **133**:9148-51.
[26] Moosa A A and Jaafar J N 2017 *Nanosci. and Nanotech.* **7**:38-47.
[27] Espinosa-Andrews H, Sandoval-Castilla O, Vázquez-Torres H, Vernon-Carter E J and Lobato-Calleros C 2010 *Carbohy. Polym.* **79**:541-6.

IOP Conf. Series: Materials Science and Engineering **863** (2020) 012025 doi:10.1088/1757-899X/863/1/012025

Correlation between microstructure and micro-hardness of 316L nitrided austenitic stainless steel

P Hussain*, H Mahmoud, S N Basha and A I Mohamad

Mechanical Engineering Department, Universiti Teknologi PETRONAS, 32610 Bandar Seri Iskandar, Perak, Malaysia

*E-mail: patthi_hussain@utp.edu.my

Abstract. This paper presents the effect of duration of solution nitriding towards microstructure and microhardness of AISI 316L stainless steel. The AISI 316L stainless steel was exposed to high temperature gas nitriding process at 1200 °C by using 50% ammonia gas and 50% nitrogen gas for different durations of 8, 15 and 20 hours. The nitrided austenitic stainless steel then underwent several analyses and tests to find the effect of different durations of nitriding towards its microstructures and microhardness. From the experiments conducted to the nitrided samples, it was found that the higher duration of nitriding resulted in the increase of microhardness of the samples due to the increase of nitrogen content in the stainless steel. However, there was no effect found on the microstructure of the nitrided samples. It was noted that the highest hardness was obtained from the nitrided sample of 20 hours that has 245.1 HV which is about 70% higher than the as-received sample.

1. Introduction

Stainless steel was developed in the early of 20th century and has become one of the most important materials in the world recently. The stainless steel has been applied in many types of industries such as medical equipment, textile, marine engineering, architecture, chemical engineering and other industries. Basically, any types of steels that contain more than 12 % chromium are considered as stainless steel. In order to increase the resistance towards localized corrosion, element like nickel is added into the stainless steel. Stainless steel has been classified into three groups according to its microstructure which are austenitic, martensitic and ferritic where each has a face-cantered-cubic, tetragonal crystal lattice and body-cantered cubic structures. Compared to the other two groups, austenitic stainless steel has been widely used in the industry due to its strength, corrosion resistance, weldability and ductility. However, the only drawback for austenitic stainless steel is its low hardness. So, in this paper, austenitic stainless-steel type AISI 316L was exposed to solution nitriding process which is a gas nitriding done at high temperature between 1000 °C to 1200 °C where the material was exposed to nitrogenous environment in a vacuum condition. This heat treatment method increased the hardness of stainless steel due to the diffusion of nitrogen into the steel. Recent studies have been carried out on austenitic stainless steel which focused on the corrosion resistance, effect of coating to its strength and effective duration of nitriding process. Hence, this paper investigates the correlation between microstructure and microhardness of AISI 316L austenitic stainless steel when subjected to solution nitriding process at 1200 °C using 50% ammonia and 50% nitrogen gas at different durations of 8, 15 and 20 hours.

IOP Conf. Series: Materials Science and Engineering **863** (2020) 012025 doi:10.1088/1757-899X/863/1/012025

Normally, the classification of stainless steel is according to its microstructure. One of the elements in stainless steel is chromium which function is to prevent the oxidation of the steel by forming a protective film on the surface of the stainless steel. Recent study showed that the amount of chromium in stainless steel is at least 11% compared to other elements [1]. At temperature of about 1050 °C the solubility limit of Cr is about 13% in the γ phase. Thus, the steel itself is in the γ phase where the austenite will change to martensite due to the high temperature condition for the martensite start temperature [2]. Hence, the steel is in martensitic condition at room temperature. The increase in the content of chromium in stainless-steel transforms austenite to ferrite. Thus, the stainless steel is in ferrite phase at room temperature. To prevent this from happening, nickel is added into stainless steel to keep austenite condition at higher content of chromium due to the ability of nickel which is a powerful former of austenite structure. So, the insertion of at least 9 % of nickel was enough to keep the stainless steel in the austenite form even at a high content of chromium. As a result, the temperature of martensite-finish and martensite-start will decrease to zero [3]. Hence, the stainless steel is in the form of austenite when cooled to ambient temperature. Austenitic stainless steel usually composed of the elements of Fe, Cr and Ni that have a face-cantered cubic (FCC) form due to the presence of some stabilizer elements for austenite like nitrogen, manganese and nickel [4]. Nitriding is a type of surface treatment which is thermo chemically conducted in order to increase the strength of the steel [5]. A perfect nitriding process results in the formation of nitrite in the steel structure. This can be proven by the formation of white layer comprising of $Fe_2(N, C)1-x$ and Fe_4N phases on the surface of the steel [6, 7]. The good properties like improvement in fatigue life, increased wear resistance, high surface hardness and anti-galling properties are achieved by the nitriding process which makes it suitable for case-hardening process of steel [8]. The low temperature used in the nitriding process had resulted in less deformation and distortion of the heat-treated parts. A low-temperature gas nitriding process is where the nitriding process is done at the temperatures ranging from 400 °C to 700 °C. In this process, the formation of nitrides occurred at the surface of the stainless steel due to the diffusion of nitrogen into the steel. As a result, the steel became hardened [9]. The drawback of this process is the decrease of the ability of the stainless steel against corrosion. The gas nitriding process that is done at the temperatures between 1000 °C to 1200 °C is called solution nitriding which is considered as the latest process to increase the strength of steel [10]. The solution nitriding process followed by quenching allows the nitrogen to diffuse interstitially into the steel which causes the increase of strength of stainless steel. This type of nitriding process is usually applied in the heat treatment process for austenitic, ferritic, martensitic and duplex stainless steels [11-13]. Recent studies had shown that high temperature gas nitriding causes the nitrogen to diffuse deeper into the stainless steel compared to low temperature gas nitriding. Another research had shown that the production of nitrides occurred at the surface of duplex stainless steel when it was subjected to high temperature gas nitriding at 1150 °C for 15 hours. Not only that, it was also found that the microstructure of austenitic stainless steel was not affected by high temperature gas nitriding [14]. For both steels, it was observed that the increase of pressure of nitrogen caused a higher absorption of nitrogen. Hence, the size of grain of steel increased along with the duration of high temperature gas nitriding when the steel had absorbed all the nitrogen.

Austenitic stainless steels possess high corrosion resistance; however, it has poor mechanical properties such as low hardness. Thus, high hardness is necessary so that it can be used in the high wear condition. The objective of this study is to improve the hardness of the austenitic stainless steel by nitriding technique. Nitriding was carried out at the elevated temperature.

2. Materials and methods

Stainless steel AISI 316L was used in this work. This type of stainless steel has a better corrosion resistance compared to type 304 stainless steel. The austenitic stainless steel is divided into three samples for different durations of solution nitriding which are 8, 15 and 20 hours. The stainless steel was ground and polished. The three samples of the AISI 316L then were then exposed to solution nitriding using 50% ammonia gas and 50% nitrogen gas for different durations of nitriding for 8, 15 and 20 hours for each sample, respectively. The nitrided samples were let to cool naturally. After that, all the nitrided samples were mounted, ground and polished. For metallography purpose, the nitrided samples were etched using Viella's Etchant. Then, the microstructure of the nitrided samples were observed under optical microscope. The chemical composition of the nitrided samples were analysed using EDS embedded with SEM machine to obtain the nitrogen content in the steel. Lastly, the as-received and nitrided samples were subjected to microhardness testing machine in order to acquire the hardness value for each sample along the cross-sectional area.

The data obtained through all the experiments and tests were analysed to investigate the correlation between the microstructure and microhardness of the nitrided austenitic stainless steel.

3. Results and discussion

3.1. Microstructure

Figure 1 shows the optical micrograph of as-received sample of AISI 316L conducted in this study. The image of this as-received sample was taken using optical microscope with the use of Marble's Etchant. Figure 1 also shows that the as-received sample had undergone cold working due to the rolling effect on its surface. The microstructure of the as received sample contains some recrystallization of twins and has a small grain size.

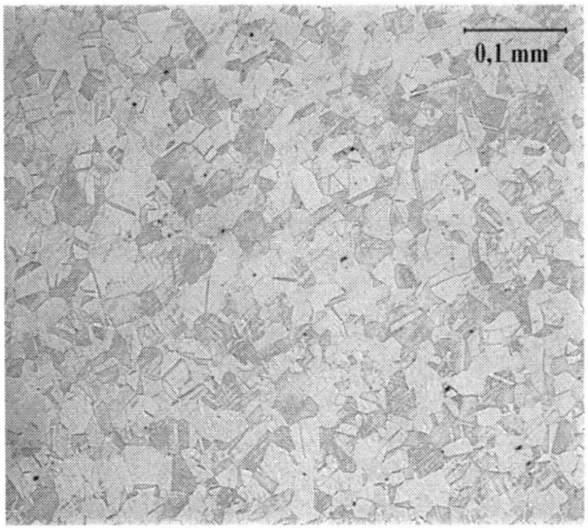

Figure 1. Optical Micrograph of as-received AISI 316L.

All the three samples of AISI 316L were undergone the solution nitriding for different durations of 8, 15 and 20 hours All the samples were let to cool naturally by means of air cooling. The nitrided samples were then ground, polished and etched with Viella's Etchant to increase the visibility of its microstructures to be observed by LECA Optical Microscope. Figures 2, 3 and 4 show the micrographs of the nitrided samples for 8, 15 and 20 hours, respectively, with the presence of twins on the microstructures. The presence of twins is considered normal for low stacking fault energy metals that have face-cantered-cubic structures like austenitic stainless steel. It is observed that the formations of twins in this paper consist of two types which are trans granular twin and suspended twin. These two types of twins are clearly shown in figures 2, 3 and 4 below. The trans granular twins are made up from four parts whereby two sides are in coherent twin planes and both ends are the grain boundary. For the suspended twins, they also have four parts, where the incoherent twin plane is at the head, the coherent twin planes are at both sides and the grain boundary is at the end.

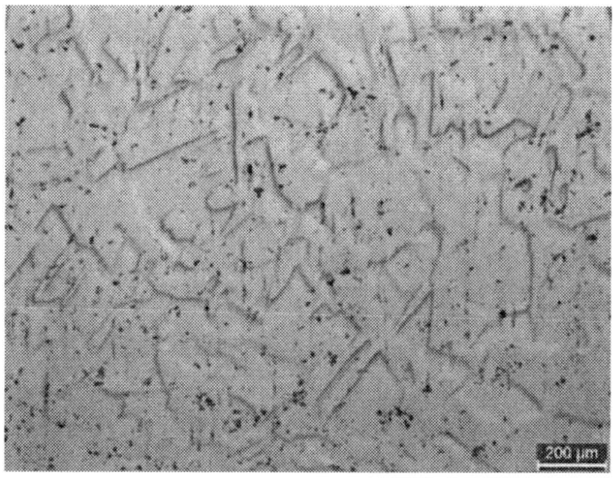

Figure 2. Optical micrograph of nitrided AISI 316L for 8 hours.

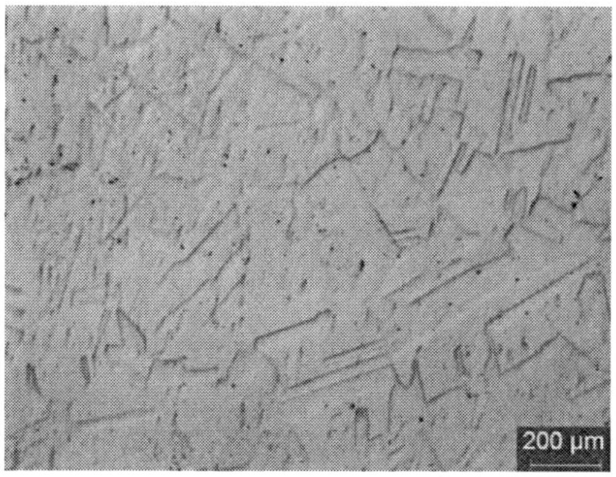

Figure 3. Optical micrograph of nitrided AISI 316L for 15 hours.

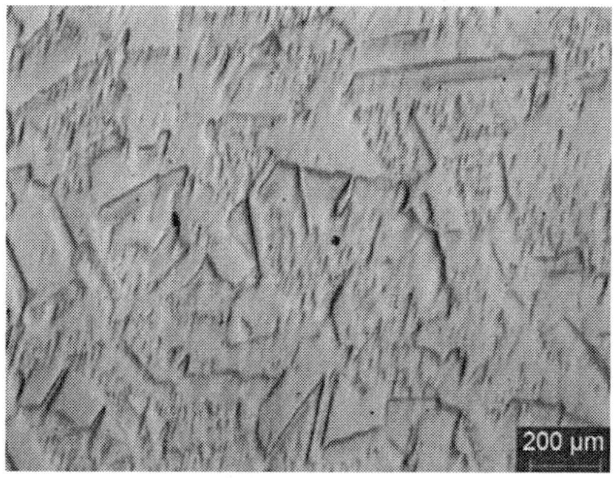

Figure 4. Optical micrograph of nitrided AISI 316L for 20 hours.

Moreover, it was also observed that the grain size of austenite in the samples increased with the increase of duration of solution nitriding because of slow cooling process. If the samples were quenched after nitriding process, the grain sizes became smaller and the toughness was higher. However, the duration of solution nitriding did not affect the single-phase structure of austenite. Thus, the diffusion of nitrogen into the steel is totally in solid solution austenite leaving no trace of nitrides behind. All the nitrided samples had experienced slow cooling process from temperatures of about 1200 °C. Next, the samples were exposed to the sensitization temperature of austenitic stainless steel which is in the range of 450 °C to 870 °C. At these temperatures, a normal austenitic stainless steel experienced the formation of chromium carbide precipitation on its grain boundaries. However, for all

IOP Conf. Series: Materials Science and Engineering **863** (2020) 012025 doi:10.1088/1757-899X/863/1/012025

the nitrided samples, it was observed that there was no formation of chromium carbide precipitation on its grain boundaries due to the low carbon content in the AISI 316L.

3.2. Chemical composition

The nitrided samples of AISI 316L were investigated using SEM with EDS to reveal the chemical composition of the nitrided austenitic stainless steel especially for the nitrogen content in the steel. The EDS analysis was done at random spots on the surface of the nitrided austenitic stainless steel. Table 1, table 2 and table 3 show the result captured from the EDS analysis by SEM at different durations of solution nitriding.

Table 1. EDS analysis by SEM of Nitrided AISI 316L for 8 hours.

Element Name	Weight Concentration (%)
Iron	63.0
Chromium	15.0
Oxygen	5.8
Nickel	9.1
Nitrogen	3.6
Molybdenum	2.9
Silicon	0.6

Table 2. EDS analysis by SEM of Nitrided AISI 316L for 15 hours.

Element Name	Weight Concentration (%)
Iron	60.7
Chromium	14.9
Oxygen	5.2
Nickel	8.2
Nitrogen	7.2
Molybdenum	3.1
Silicon	0.6

From the result of the EDS analysis in table 1, table 2 and table 3 above, it was clearly shown that the nitrogen content in the steel increases with the duration of solution nitriding even though the EDS analysis was executed at random spots on the surface of the steel not at the same spot. The lowest nitrogen content is 3.6 wt.% for 8 hours of solution nitriding whereas the highest nitrogen content is 8.5wt.% for 20 hours of solution nitriding. Hence, the higher the duration of nitriding, the higher the diffusion of nitrogen into the steel is.

3.3. Microhardness test

The Vickers hardness test was completed by using the LECO Microhardness Tester LM247AT. The Vickers hardness test was conducted at the cross-sectional surface of the as-received sample and the nitrided austenitic stainless-steel samples. The result obtained from the test was plotted in the charts as shown in figure 5 and figure 6. The error bar in figure 5 was estimated to be ± 5 HV due to some human error in conducting the hardness test.

Based on the chart in figure 5, hardness of nitrided austenitic stainless-steel increases with the increase in the duration of solution nitriding. It was also observed that the hardness is high at the surface of the as-received austenitic stainless steel and starts to decrease when approaching the centre of the steel due to the cold working of the as-received austenitic stainless steel. For nitrided AISI

316L, the hardness also is high at the surface and slowly decreases as it approached the centre of the nitrided AISI 316L due to higher concentration of nitrogen on the surface and the diffusion of nitrogen into the steel was less diffused to the centre.

Table 3. EDS analysis by SEM of Nitrided AISI 316L for 20 hours.

Element Name	Weight Concentration (%)
Iron	62.8
Chromium	15.4
Nickel	9.1
Nitrogen	8.5
Molybdenum	3.5
Silicon	0.7

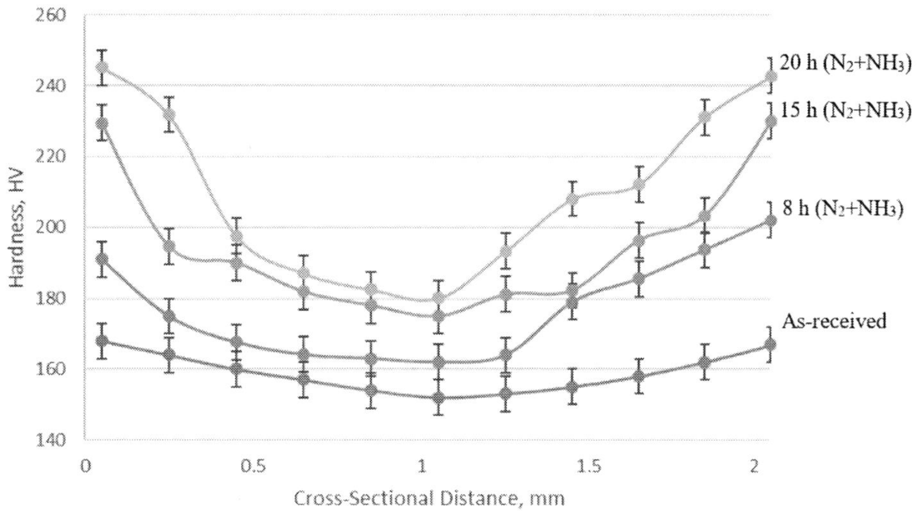

Figure 5. Chart of Hardness VS Cross-Sectional Distance for untreated and nitrided AISI 316L samples.

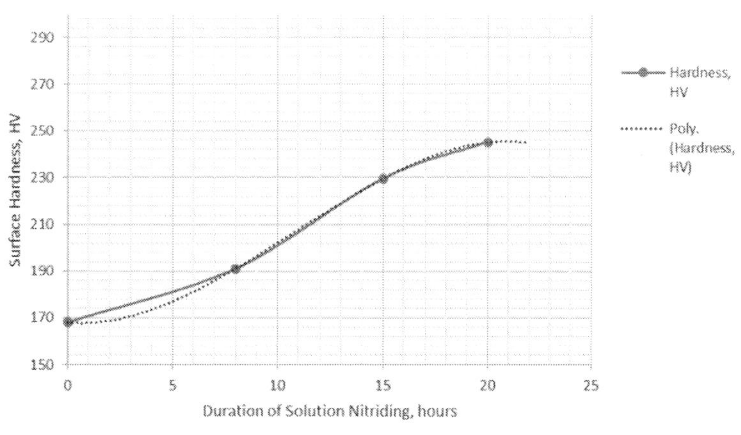

Figure 6. Plot of Surface Hardness VS Duration of Nitriding for AISI 316L.

The chart in figure 6 shows the effect of duration of solution nitriding on surface hardness of the AISI 316L. The surface hardness is expected to increase with the increase of the duration of solution nitriding until 22 hours before the surface hardness becomes constant. The highest surface hardness was obtained from the nitrided samples for 20 hours that has 245.1 HV which is 70% higher than the hardness of the as-received samples that only has 168 HV. Hence, this proves that micro-hardness increases with the duration of solution nitriding.

4. Conclusion

The addition of nitrogen into steels was indicated by the improvement of hardness on the steel. The highest hardness was acquired from the sample that was nitrided for 20 hours that have 245.1 HV which is about 70% higher compared to the hardness value of as-received sample. Longer duration of solution nitriding caused higher diffusion of nitrogen into steel. As a result, the hardness of the nitrided AISI 316L was increased by 46 % when compared to untreated sample and the grain size remained the same as received. From this study, it was found that the microstructure for both the as-received and nitrided samples were almost similar where both have the presence of twins. In conclusion, the increase of duration of solution nitriding has led to the increase of the micro-hardness of austenitic stainless steel but had no effect on its microstructure.

References

[1] Davis J 2001 ASM specialty handbook *ASM International* 385-418.
[2] Llewellyn D and Hudd R 1998 *Steels: metallurgy and applications* (Elsevier).
[3] Welsch G, Boyer R and Collings E 1993 *Materials properties handbook: titanium alloys* (ASM international) p 1169.
[4] Davis J R 1994 *ASM Specialty Handbook: Stainless steels* (ASM international) p 576.
[5] Firmanto H 2011 *Reaction layers in diffusion bonded of sialon to ferritic steel* (PhD Thesis: UTP).
[6] Hassani-Gangaraj S M and Guagliano M 2013 *Applied Surface Science* **271** 156-63.
[7] Aydin H, Bayram A and Topçu Ş 2013 *Materials Science* **19** 19-24.
[8] Baranowska J and Arnold B 2006 *Surface Coatings Technology* **200** 6623-8.
[9] Berns H and Siebert S 1996 *ISIJ international* **36** 927-31.
[10] Tschiptschin A P 2002 *Powder metallurgy aspects of high nitrogen steels* in *High Nitrogen Steels ,Stainless Steels: Manufacturing, Properties Applications.*
[11] Sung J, Kong J, Yoo D, On H, Lee D and Lee H J M S 2008 *Materials Science Engineering:* **489** 38-43.
[12] Garzon C and Tschiptschin A 2004 *Journal of materials science* **39** 7101-5.
[13] Tschiptschin A P, Garzón C M and Lopez D M 2006 *Tribology International* **39** 167-74.
[14] Mahmoud H R, Yusoff S A, Zainuddin A, Hussain P, Ismail M and Abidin K 2014 *MATEC Web of Conferences*: EDP Sciences) p 04021.

IOP Conf. Series: Materials Science and Engineering **863** (2020) 012026 doi:10.1088/1757-899X/863/1/012026

An investigation of thermal stability and heat capacity of imidazolium based ionic liquids at elevated temperature

B Bakthavatchalam[1,*], K Habib[1], R Saidur[2], K Irshad[3] and O A Hussein[1]

[1] Mechanical Engineering Department, Universiti Teknologi PETRONAS, 32610 Bandar Seri Iskandar, Perak, Malaysia.
[2] Research Centre for Nanomaterials and Energy Technology, School of Science and Technology, Sunway University, 47500 Selangor, Malaysia.
[3] Center of Research Excellence in Renewable Energy, King Fahd University of Petroleum & Minerals, Dhahran, 31261 Saudi Arabia.

*E-mail: balaji_17004028@utp.edu.my

Abstract. In this paper, the thermal stability and specific heat capacities of four imidazolium based ionic liquids namely 1-ethyl-3-methyl imidazolium octylsulfate, 1-ethyl-3-methylimidazolium diethyl phosphate, 1,3-dimethyl-imidazolium dimethyl phosphate and 1-butyl-3-methylimidazolium chloride have been experimentally analysed using Thermogravimetric Analysis (TGA) and Differential Scanning Calorimetry (DSC). The thermal stability was assessed by conventional and high resolution modulated TGA with the temperature range of 0 to 700 °C. The specific heat capacity was measured by using DSC within the temperature interval of 0 to 200 °C. The factors influencing the thermal stability and heat capacity of ionic liquids were studied. The thermal analysis results indicate that 1-ethyl-3-methylimidazolium octylsulfate exhibited excellent thermal stability even at 440 °C while 1-ethyl-3-methylimidazolium diethyl phosphate obtained a maximum specific heat capacity of 4.52 J/g.K.

1. Introduction

Ionic liquids (ILs) are an innovative class of salts that are liquid at room temperature. The most important characteristics of these liquids are as follows: high thermal stability, non-volatile, chemically stable, low melting point and also considered as green solvent. Ionic liquids consist of anions and cations where imidazolium is the most commonly used cation whereas anions include chloride, tetrafluroborate, hexafluorophosphate etc. Although early research works tended to assume that ILs were very similar to conventional liquids, it is now realized that ILs possess a wide variety of characteristics and the ubiquitous properties of ILs are thermal stability and specific heat capacity. One of the most interesting features of ionic liquids are their controllable anion and cation interactions, which can be a major factor in studying the characteristics of ILs at the molecular level. It was acknowledged from the outcomes of both experimental and theoretical studies that ionic liquids interionic interactions could determine the thermophysical properties. Studies on the effect of Imidazolium based ionic liquids on cellulose polymers, coal oxidation, conductivity and fluorination showed its capability to replace organic solvents in different applications [1-4].

Most researches have been done on analyzing the physiochemical properties of imidazolium based ionic liquid, but very few studies involving thermal stability and heat capacities on imidazolium based ionic liquids were recorded in the literature. In a recent study, Wust et al. [5] investigated the thermal stability of four pharmaceutical ionic liquids (Ranitidinium ibuprefenete, Bupivacainium ibuprofenate, Lidocainium ibuprofeate, Diphenydraminium ibuprofenate) and proved that these liquids were thermally stable up to 160 °C – 200 °C with an activation energy of 60-100 $Kjmol^{-1}$. Mora et al. [6] measured the density, thermal stability and heat capacity of fifteen ionic liquids (1-octyl-3-methylimidazolium, 1-octyl-2,3-dimethylimidazolium) based mixtures and demonstrated that ILs increased the thermal energy storage of the mixture. Bhattacharyya et al. [7] studied the thermal stability of thirteen ionic liquids (choline based amino acid ionic liquids) at 100 °C and 150 °C for 500 min. They revealed that ionic liquid with threonine as anion obtained higher thermal stability while α-amino acid anion produced less thermal stability. In addition, etherification on choline cation also contributes to a reduction in thermal stability. Gomez et al. [8] described the thermal behavior and heat capacities of pyrrolidinium, pyridinium, imidazolium based ionic liquids. Based on the thermal behavior of ionic liquids they found that ILs can overcome the energy of crystallization to attain more homogeneity and the heat capacities increased with respect to temperature and finally concluded that specific heat capacity of ILs can be increased by increasing the size of the anion. Zhang et al. [9] synthesized Br^- (Bromine), BF_4^- (Tetrafluoroborate), PF_6^- (Hexafluorophosphate), NTf_2^- (Bistriflimide) anions and incorporated dicationin ionic compounds to improve the thermal stability and thermal storage density of ionic liquids. They observed that high interaction energy resulted in high latent heat and thermal storage density. However, most of the researchers focused on analyzing the properties of ionic liquids only at low temperatures.

Researchers realized that ILs have several distinctive characteristics that enable new heat transfer fluids to be developed and synthesized by tuning the anion-cation structure to the required thermophysical properties and the targeted applications. The objective of this work is to determine the thermal stability and heat capacity of 1-ethyl-3-methyl imidazolium octylsulfate (EmimOSO4), 1-ethyl-3-methylimidazolium diethyl phosphate (EmimDEP), 1,3-dimethylimidazolium dimethyl phosphate (MmimDMP) and 1-Butyl-3 methylimidazolium chloride (BmimCl) ionic liquids as a function of temperature using TGA and DSC.

2. Methodology

2.1. Materials

The ionic liquids tested were bought from Sigma Aldrich, USA and their mass fraction purity were more than 99% for BmimCl ionic liquid and 98% for the remaining ionic liquids.

Table 1. Purity and water content of ionic liquids.

Ionic Liquids	Purity (mass fraction)	Water content
EmimOSO4	≥98%	≤1%
EmimDEP	≥98%	≤1%
MmimDMP	≥98%	≤1%
BmimCl	≥99%	≤0.2%

The water content of each ionic liquid is evaluated using a volumetric compact Karl Fischer Titrator, model V10S and its content was less than 250 ppm. The contents and structure of the ionic liquids utilized in this research are shown in table 1 and table 2, respectively. Wang et al. [10] clearly demonstrated the mechanisms and structures of Imidazolium based ionic liquids at molecular level.

Table 2. Structure and image of ionic liquids.

Ionic Liquids	**EmimOSO$_4$**	**EmimDEP**	**MmimDMP**	**BmimCl**
Structure	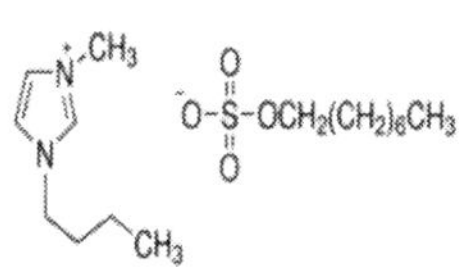	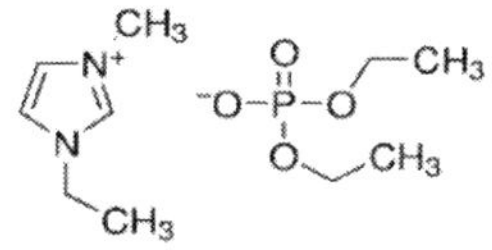	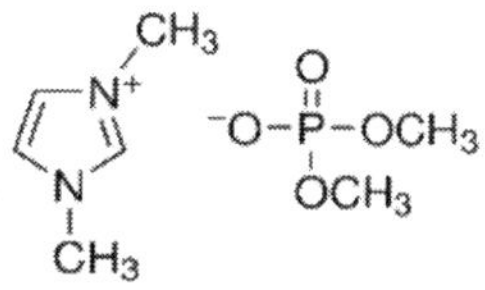	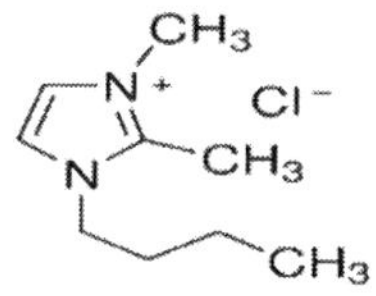
Formula	$C_{14}H_{26}N_2O_4S$	$C_{10}H_{21}N_2O_4P$	$C_7H_{15}N_2O_4P$	$C_8H_{15}ClN_2$
Image	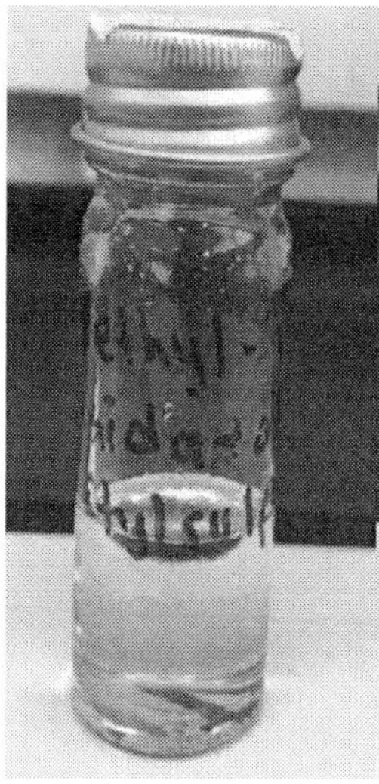			

2.2. *Apparatus and procedure*

All the four ionic liquids were evaluated by PerkinElmer TGA 400 instrument for Thermogravimetric analysis. A temperature range from 30 °C to 600 °C were used to measure the thermal stability with 30 °C/min heating rate under nitrogen atmosphere of 20 ml/min flow rate. The sample weight used in aluminum crucibles was about 10 mg to 16 mg. The schematic setup of TGA is shown in figure 1.

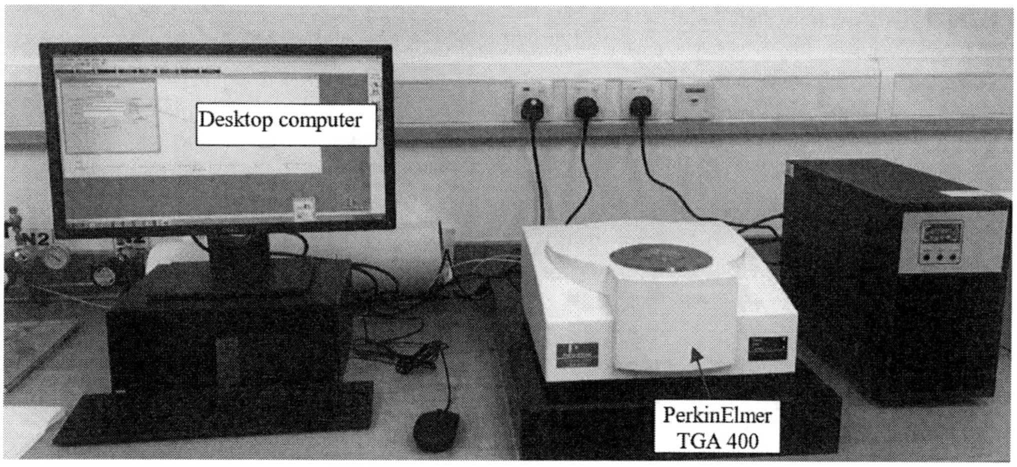

Figure 1. Photograph of TGA setup.

Specific heat capacity was assessed using a Differential Scanning Calorimeter (DSC 1000/C, LINSEIS Germany) with a heating rate of 10 °C/min and samples of weight between 10 and 15 mg in ceramic pans. The image of DSC setup is presented in figure 2.

Figure 2. Image of DSC setup.

3. Results and discussion

3.1. Thermal stability

The thermal stability of the four imidazolium based ionic liquids in a temperature range of 30 °C to 600 °C with a constant heating rate of 30 °C/min had been investigated using a Thermogravimetric Analyzer. This analysis was mainly performed to find the effect of temperature on the life of ionic liquids and to determine which ILs could be used in elevated temperature applications. The behavior of $EmimOSO_4$, EmimDEP and BmimCl ionic liquids were almost similar whereas MmimDMP ionic liquid showed a fluctuation in its gradual decrease of mass. The degradation of $EmimOSO_4$, EmimDEP, MmimDMP, and BmimCl ionic liquids with respect to temperature is represented in figure 3 to figure 6, respectively. It is observed that $EmimOSO_4$ started to degrade from ~150 °C and a sudden degradation curve is shown from 290 °C and ends at 430 °C while the complete decomposition occurred after 440 °C.

On the other hand, EmimDEP was thermally stable until 270 °C where the decomposition also started and resulted in an immediate degradation curve decline from ~295 °C and ended at 340 °C and a constant decomposition started at 345 °C. In addition, MmimDMP exhibited degradation at a low temperature of ~90 °C and revealed a gradual fall from 298 °C. Consequently, BmimCl displayed a similar trend of degradation behavior like EmimDEP with respect to time, which started to degrade at ~95 °C from where sudden fall of degradation occurred from 260 °C to 330 °C which are in good agreement with Huang et al. [11]. From the results, it is observed that the maximum change in the weight loss for all tested ionic liquids appeared after 200 °C which clearly indicated these liquids could be employed for elevated heat transfer applications. The TGA analysis can be considered as a helpful way for understanding the ionic liquids heat degradation. TGA evaluation of ionic liquids are accessible for restricted research, and further research is needed to fully comprehend the occurrence of temperature and time shift in degradation.

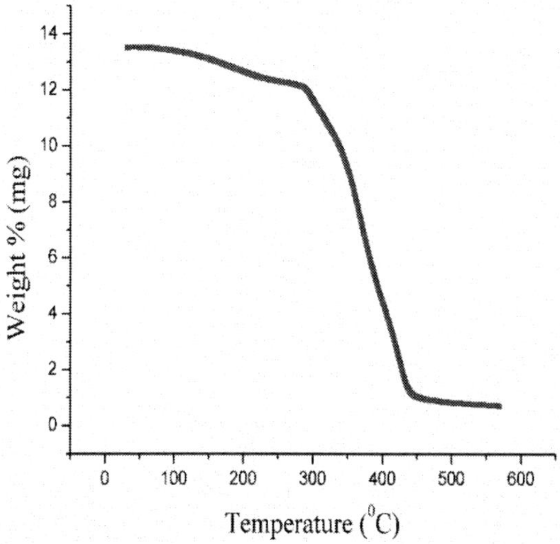

Figure 3. Effect of temperature on the mass of $EmimOSO_4$.

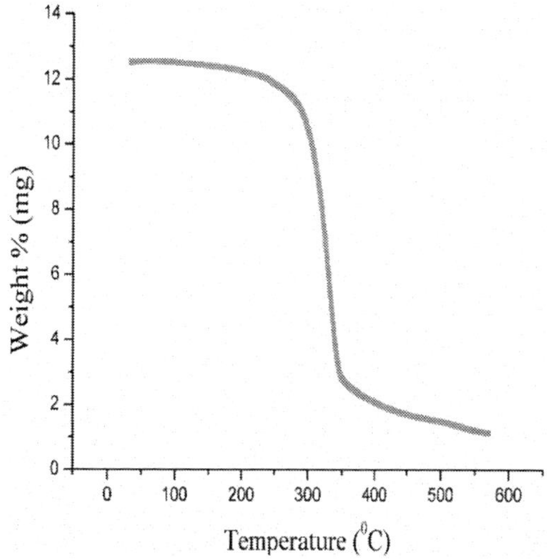

Figure 4. Effect of temperature on the mass of EmimDEP.

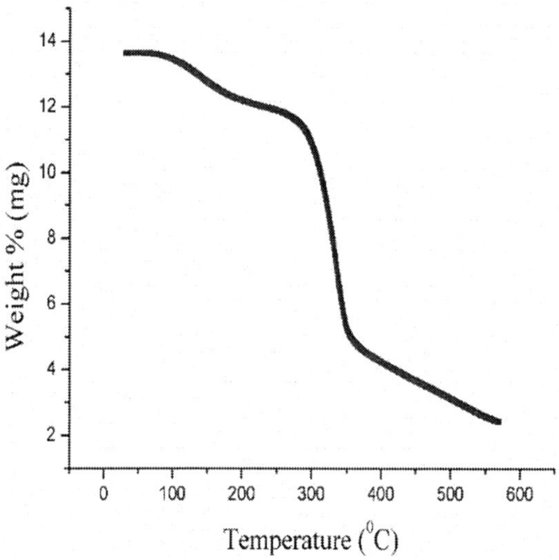

Figure 5. Effect of temperature on the mass Of MmimDMP.

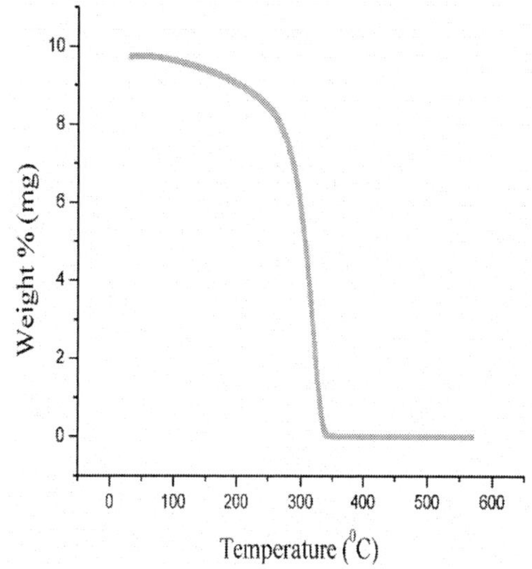

Figure 6. Effect of temperature on the mass of BmimCl.

3.2. Specific heat capacity

In this section, the specific heat capacity of EmimOSO4, EmimDEP, MmimDMP and BmimCl ionic liquids with a temperature ranging from 20 °C to 200 °C were investigated using a high precision Differential Scanning Calorimetry. Figure 7 to figure 10 show the specific heat capacity curves of all the studied ionic liquids with respect to temperature. As seen in figure 9 and figure 10 the specific heat capacity curves of MmimDMP and BmimCl ionic liquids showed similar trends whereas EmimDEP produced a steady rise in its curve while $EmimOSO_4$ displayed a gradual decrease which might be due to sulfate anion. As expected the specific heat capacity of all ionic liquids increased with respect to temperature except $EmimOSO_4$. The strong electrostatic interaction and Brownian motion of the molecules that occurred at elevated temperature made the specific heat capacity of ionic liquids increase with respect to high temperatures. The EmimDEP exhibited the highest value of specific heat capacity was obtained as 4.52 J/g.K in the tested ionic liquids, due to the proximity of the methylene group at the imidazolium ring, resulting in stronger anion-cation interaction. Consequently, the gradual decrease of specific heat capacity of EmimOSO4 was due to the formation of non-polar domain, leading to the decrease of electrostatic interaction.

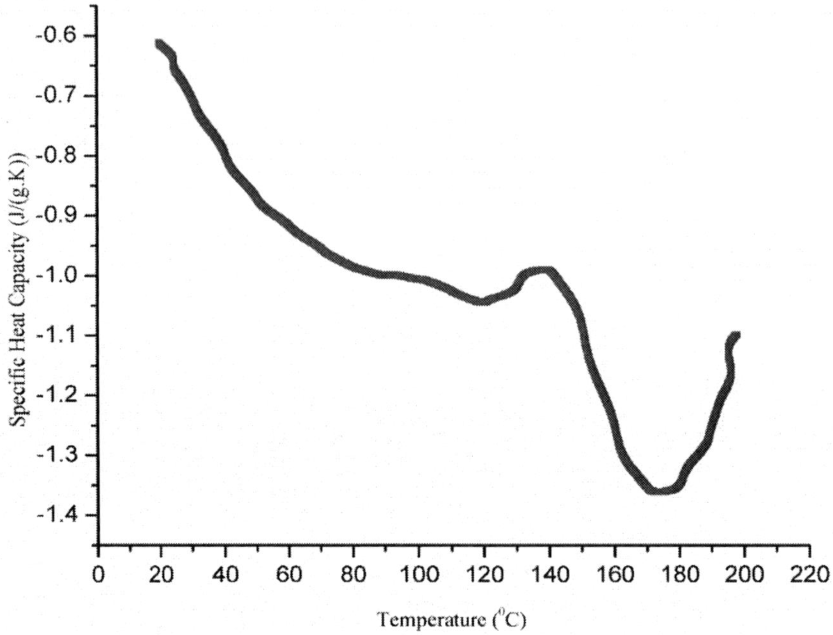

Figure 7. Specific heat capacity of EmimOSO4 with temperature.

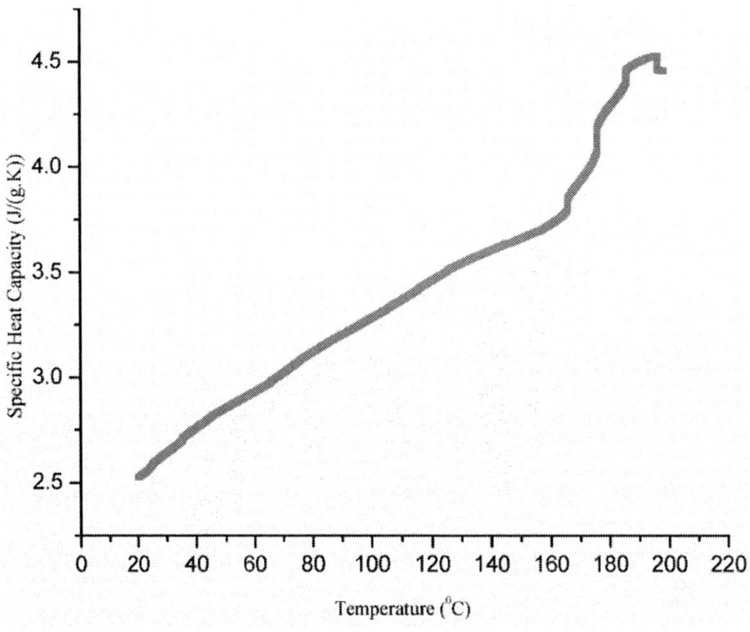

Figure 8. Specific heat capacity of EmimDEP with temperature.

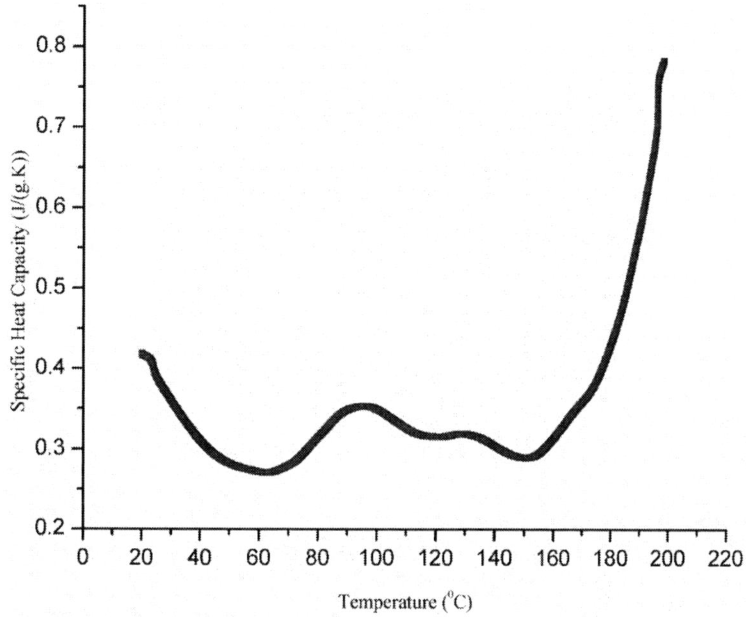

Figure 9. Specific heat capacity of MmimDMP with temperature.

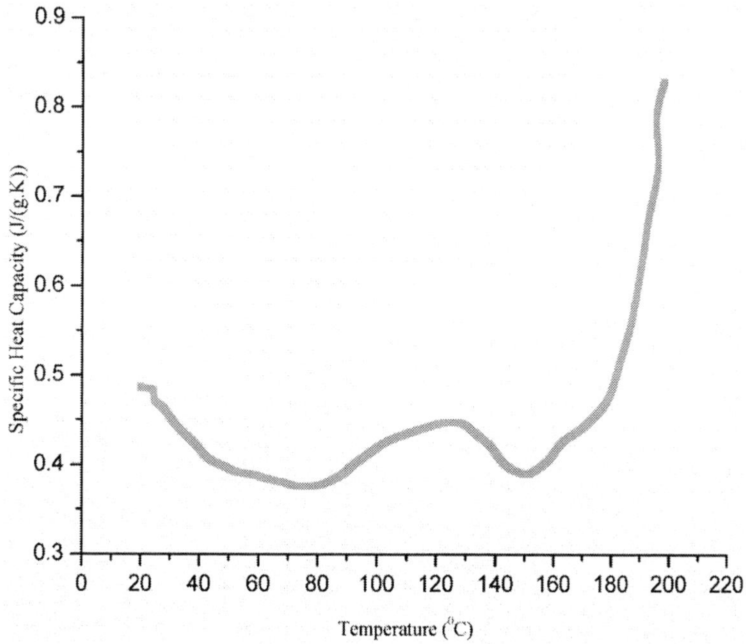

Figure 10. Specific heat capacity of BmimCl with temperature.

4. Conclusion

The evaluation of the thermal stability and the specific heat capacities of several imidazolium-based ionic liquids EmimOSO4, EmimDEP, MmimDMP and BmimCl have been carried out with TGA and DSC at very high temperatures. Overall, $EmimOSO_4$ exhibited a maximum thermal stability of 440 °C while EmimDEP resulted a maximum specific heat capacity of 4.52 J/g.K. The experimental results proved that $EmimOSO_4$ was more stable while EmimDEP showed more heat capacity detainment when compared to other ionic liquids. The ionic fluids utilized in this study showed remarkable characteristics that can be used as a potential application for heat transfer systems, where thermal stability and specific heat capacity plays a significant role.

Acknowledgements
We thank Centre of Research in Ionic Liquids (CORIL) in Universiti Teknologi PETRONAS and Research Centre for Nano-Materials and Energy Technology (RCNMET) in Sunway University for providing the necessary research resources. This work was supported by the YUTP grant (015LC0-118).

References

[1] Paiva T, Echeverria C, Godinho M H, Almeida P L and Corvo M C 2019 *European Polymer Journal* **114**:353-60.
[2] Zhang W, Jiang S, Wu Z, Wang K, Shao H, Qin T, et al. 2018 *Fuel* **217**:529-35.
[3] Arkhipova E A, Ivanov A S, Maslakov K I, Savilov S V and Lunin V V 2019 *Electrochimica Acta* **297**:842-9.
[4] Rodrigues A S M C, Fernandes A M, Dévemy J, Costa Gomes M and Santos L M N B F 2019 *Journal of Molecular Liquids* **282**:385-91.
[5] Wust K M, Beck T S, Hennemann B L, Villetti M A and Frizzo C P 2019 *Journal of Molecular Liquids* **284**:647-57.
[6] Mora S, Neculqueo G, Tapia R A and Urzúa J I 2019 *Journal of Molecular Liquids* **282**:221-5.
[7] Bhattacharyya S and Shah F U 2018 *Journal of Molecular Liquids* **266**:597-602.
[8] Gómez E, Calvar N, Domínguez Á and Macedo E A 2018 *Fluid Phase Equilibria* **470**:51-9.
[9] Zhang H, Xu W, Liu J, Li M and Yang B 2019 *Journal of Molecular Liquids* **282**:474-83.
[10] Wang X, Fu F, Peng K, Yang Z, Hu N, Chen X, et al. 2019 *Chemical Physics* **525**:110391.
[11] Huang G, Lin W-C, He P, Pan Y and Shu C-M 2018 *Journal of Molecular Liquids* **272**:37-42.

Multi-objective model optimisation using genetic algorithms for pleurotus sp. cultivation

N Zainol[1], A S Fakharudin[2,*], N A Dzulkefli[3] and M F A Bakar[2]

[1] CARIFF, Universiti Malaysia Pahang, 26300 Gambang, Pahang, Malaysia.
[2] Faculty of Computer Systems & Software Engineering, Universiti Malaysia Pahang, 26300 Gambang, Pahang, Malaysia.
[3] Faculty of Chemical & Natural Resources Engineering, Universiti Malaysia Pahang, 26300 Gambang, Pahang, Malaysia.

*E-mail: sahli@ump.edu.my

Abstract. Malaysia is the largest exporter of Elaeis Guineensis (Palm oil) in the international market. Oil palm cultivation generates a significant amount of lignocellulosic biomass derived from empty fruit bunches (EFB) as waste product. This research focused on optimizing the mycelium growth in Pleurotus sp. cultivation by using EFB as a culture medium. The EFB was cut into the range of size of substrate (S) from 1.5 cm to 3.0 cm, soaked in water for overnight, applied steam treatment and incubated at the selected range of temperature (T) from 29 °C to 32 °C. The responses were mycelium extension rate (M) and nitrogen concentration in mycelium (N). The multi-objective optimisation of M and N requires the objective functions which represent both processes. For this type of problem, multi-objective genetic algorithm was chosen as the methodology, specifically using NSGA-II algorithm. Through the implementation of selected multi-objective genetic algorithm, it was able to produce the pareto front for optimising both nitrogen concentration and the extension rate of the mycelium. The highest nitrogen concentration and mycelium extension rate was from the result with crossover and mutation probability of 0.5 and 0.2. It produced 388.45 mg/L of nitrogen concentration and 0.370 cm/day of mycelium growth.

1. Introduction

In recent years, oyster mushroom (Pleurotus ostreatus) has become one of the most cultivated mushrooms in the world, mainly in Brazil. There are many factors involved in mushroom production such as type of substrate, pH of substrate, moisture content, temperature [1], size of substrates [2], mass ratio of spawns to substrates pre-treatment of substrate [3] etc. There are some mushroom grower uses EFB as their substrate for mushroom cultivation but still at the stage of trials. Looking forward to optimising the factors affecting of Pleurotus sp. cultivation, most of other research only focus on the factors of mushroom growth and there are only very few research practices the optimisation for the growth condition and substrate preparation [4].

Multi-objective optimisation had been used widely to solve multi-disciplinary field which used pareto to find the optimal solutions. Qu et al. [5] proposed a Self-organised Speciation based Multi-objective Particle Swarm Optimiser (SS-MOPSO) to solve the multimodal multi-objective optimisation problems. The proposed method was tested on fourteen test problems and produced superior performance when compared to other methods. A new method also being proposed by Wilding et al. [6] to optimise a nuclear power conversion system (PCS) and emergency cooling system. A new mixed-integer optimisation method (MI-NSGA) was developed and used on PCS problem. The PCS optima reached efficiency of 35.63% and cooling system performance was improved.

Researchers had been used multi-objective optimisation using Response Surface Methodology (RSM) and Non-dominated Sorting Genetic Algorithm (NSGA-II) to obtain the optimal performance. The pareto front for optimal solutions had been generated by NSGA-II to optimise corrugated tube inserted with multi-channel twisted tape [7]. Three regression response models represent numerical data had been used to find the optimised response. A study by Li et al. [8] had optimised warpage, volumetric shrinkage and residual stress using NSGA-II to optimise fibre-reinforced composite injection moulding process. NSGA-II also being applied by [9] to minimise NO_x emission on a small single cylinder 4-stroke engine. The lowest NO_x emissions while using the biodiesel and exhaust gas recirculation was 63.76% with B10 fuel blend and 30% exhaust gas recirculation rate.

2. Process description of optimization using response surface method

2.1. Statistical analysis

Levels of independent factors such as size of substrate (*S*) and temperature (*T*) were based on results obtained from previous study. A two-factor-two level face centred central composite design (CCRD) was applied where the values of the independent factors were coded as in the range of -2 and +2 as shown in table 1. The experimental set-up shown in table 2 was design and analysed by using Design Expert software version 7.0 to build and evaluate models and to plot the three-dimensional response surface curves. The responses were mycelium extension rate (*M*) and nitrogen concentration in mycelium (*N*). Thirteen experiments were performed which involved five replicates for centre point. The five replicates for the centre point were used to estimate the experimental error.

Table 1. Factors and their range.

Factors	Symbol	Coded	Actual level	Coded level
Size of substrates (cm)	*S*	A	1.5	-2
			2.0	-1
			2.5	0
			3.0	1
			3.5	2
Temperature during spawn running (°C)	*T*	B	28	-2
			29	-1
			30	0
			31	1
			32	2

2.2. Preparation of substrate

The substrate which were empty palm fruit bunch (EFB) were collected from palm oil plantation at Banting, Selangor. EBB substrate was cut into the range of size as in table 2, then soaked in water for overnight, filtered to drain excess water and weighed for 100 grams. Then, the substrates were pre-treated with steam treatment and inoculated with 1:14 mass ratio of spawn to substrate, by placing the spawn on the surface of substrate. The bottles were closed and incubated at temperature set in table 2 in the dark condition.

2.3. Sample analysis

Sample analysis was conducted after all run fully colonized with mycelium. There were two responses which were mycelium extension rate (*M*) and nitrogen concentration in mycelium (*N*).

2.3.1. Determination of mycelium growth. Spawn run (mycelium extension) was observed regularly until appears white colonization. Each experiment was determined the days for complete mycelium growth. Mycelium growth was measured in centimetres as the length of the mycelium spreading from the surface of substrates toward the bottom of bottles [10].

Table 2. Experimental setup that has been constructed by using *RSM* by Design Expert software.

Run	Factor 1	Factor 2
	A: Size of substrates (*S*) (cm)	B: Temperature (*T*) (°C)
12	2.0	29
11	3.0	29
4	2.0	31
13	3.0	31
10	1.5	30
5	3.5	30
2	2.5	28
7	2.5	32
6	2.5	30
3	2.5	30
8	2.5	30
9	2.5	30
1	2.5	30

2.3.2. Nitrogen concentration analysis by using HACH Spectrophotometer. *N* was determined by using Persulfate Digestion Method (Method 10072). The mycelium was collected from the substrates and diluted by using deionized water using 40 dilution factors. Then, *N* was analysed by using HACH Spectrophotometer.

3. Genetic algorithm optimisation methodology

3.1. Objective function

The objective functions were acquired from a study by [11] which did the modelling of mycelium growth in Pleurotus sp. cultivation by using agricultural wastes. Equation (1) shows the objective function for mycelium extension rate (*M*). The objective function *M* represented the mycelium extension rate in cm/day which this rate was affected by five variables A, B, C, D and E.

$$M = 0.60 - 0.023A + 0.019B - 0.029C + 0.029D - 0.14E + 0.041BD - 0.034BE - 0.044CE + 0.044DE \tag{1}$$

Variables A represent the type of substrate and variables B represented size of substrate. The factor C was the value of mass ratio of spawn to substrate (SP/SS). While the variables D was the cultivation temperature and the variables E represent the types of substrate pre-treatment. The objective function for nitrogen concentration (*N*) is expressed by equation (2) which represented the volume of nitrogen concentration in mg/L which was also affected by the same five variables A, B, C, D and E. These two fitness functions will be used together for multi-objective optimisation to find the maximum extension rate and nitrogen concentration. The optimisation algorithm should be able to propose a pareto front of maximum value for both objectives functions.

$$N = 352 + 241.5A + 43B - 0.50C + 50.50D + 21.50E + 59.50AB + 41.00AD - 35.50BC + 47.50BE - 33.00CD - 39.00CE + 32.00DE \tag{2}$$

IOP Conf. Series: Materials Science and Engineering **863** (2020) 012027 doi:10.1088/1757-899X/863/1/012027

3.2. Optimisation process

The optimisation process from start to finish is shown in figure 1 to obtain the optimise mycelium rate and nitrogen concentration values. Firstly, the optimization problem will be loaded into the project. The problem contains the objectives and factors that will be used to create solutions. The next step was to initialize the parameter of the genetic algorithm such as the index and the probability for the genetic crossover and mutation. The crossover was set to simulate the interchange of genetic between two chromosomes and mutation was used to simulate the genetic alteration of a chromosome after crossover. Next, the parent population was generated based on loaded problem. The parent population was used to generate an offspring population through crossover and mutation. It also being used to create new population together with its offspring population. After crossover and mutation operators were applied to the population, it was combined between the parent population and offspring population.

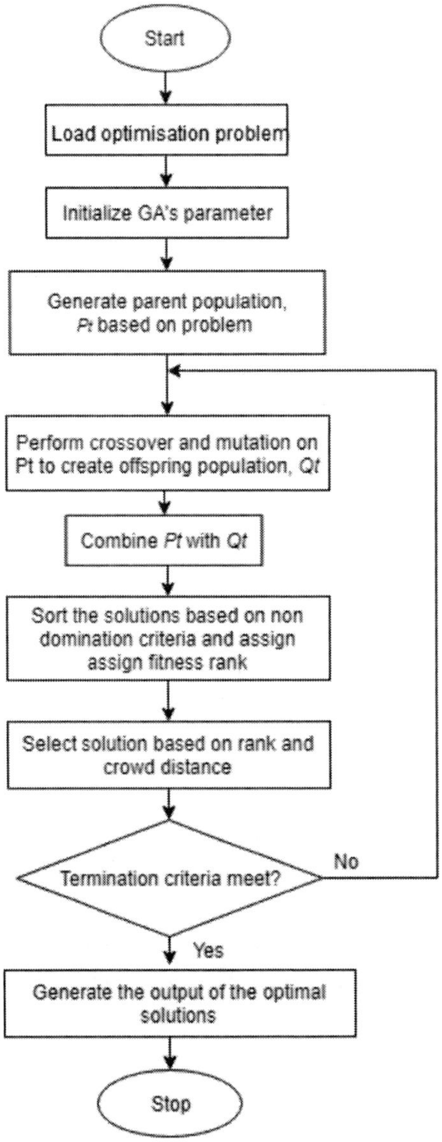

Figure 1. Flowchart of the optimisation process.

After the population was combined into one, it being sorted, and each chromosome was ranked accordingly. Non-dominate sorting method was being used during the sorting process and best solution was given higher ranking. The selection process was done next with the selection of the best solution. The selection process was based on the ranking given from previous step. If best solutions with similar rank exceed the population limit, then the solutions were selected using crowd distance to fit into the limit. The termination criteria checked was performed after the selection process. If the termination was not met, then the offspring will be generated again using crossover and mutation and selection process was applied next. The loop kept continued until the termination conditions had been satisfied. The final process was to generate the output for optimal *M* and *N* functions.

The optimisation process used the NSGA-II algorithm from jMetal [12] an object-oriented Java-based framework for multi-objective optimization with metaheuristics. NSGA-II is a very famous algorithm to use to solve multi- objective optimization problems [7-9]. The algorithm was slightly modified to ensure the it met the requirement to implement the optimisation process.

4. Results and discussion

Referring to figure 2, for highest value of nitrogen concentration was at 563.50 mg/L with its lowest value of mycelium extension rate was 0.357 cm/day. For the highest mycelium extension rate was at 0.387 cm/day but it was the lowest nitrogen concentration at 334.05 mg/L. The optimal values for both results were 385.57 mg/L of nitrogen concertation and 0.371 cm/day of mycelium extension.

Figure 3 shows the pareto front for the multi-objective optimisation with crossover of 0.7 and mutation of 0.1. The highest value of nitrogen concentration was at 689.93 mg/L and the lowest value of mycelium extension rate at 0.347 cm/day. The highest mycelium extension rate was at 0.387 cm/day, with the lowest nitrogen concentration at 334.13 mg/L. The optimal condition was at 387.38 mg/L of nitrogen concertation and 0.371 cm/day of mycelium extension.

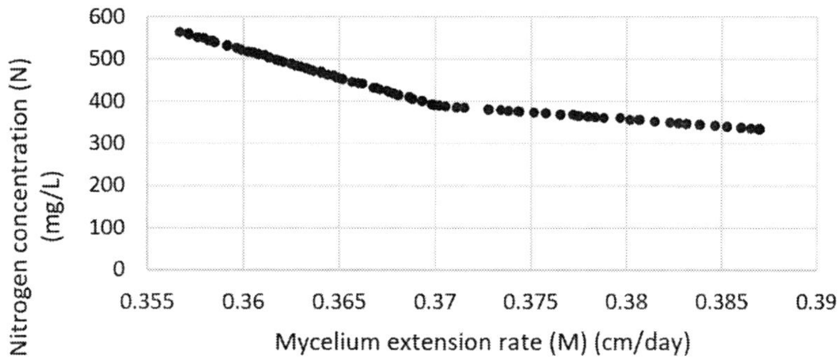

Figure 2. Pareto front for multi-objective optimisation with crossover of 0.5 and mutation of 0.1.

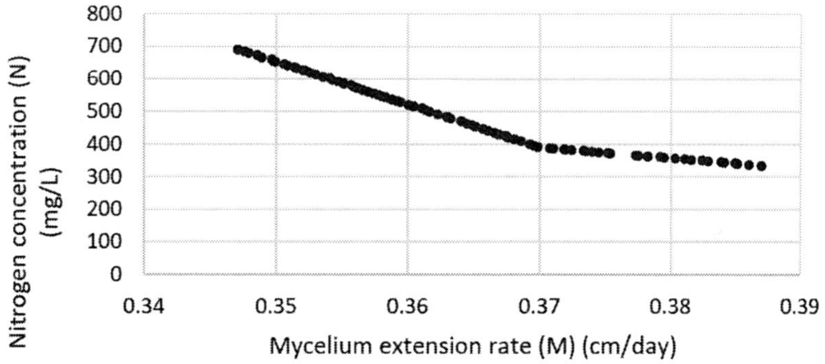

Figure 3. Pareto front for multi-objective optimisation with crossover of 0.7 and mutation of 0.1.

The results for optimisation with crossover of 0.5 and mutation of 0.2 is shown in figure 4. The highest value of nitrogen concentration was at 689.79 mg/L with the lowest value of mycelium extension rate 0.347 cm/day. The highest mycelium extension rate was 0.387 cm/day, with the lowest nitrogen concentration at 334.06 mg/L. The optimal values for both results were at 398.77 mg/L of nitrogen concentration and 0.369 cm/day of mycelium growth.

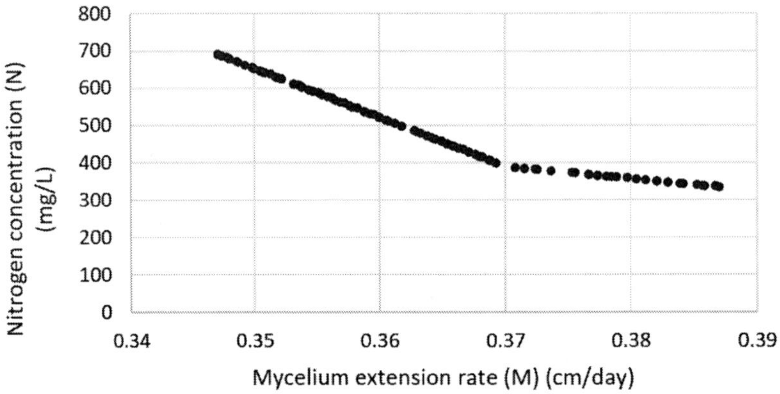

Figure 4. Pareto front for multi-objective optimisation with crossover of 0.5 and mutation of 0.2.

Figure 5 shows the pareto front for the multi-objective optimisation with crossover of 0.7 and mutation of 0.2. The highest value of nitrogen concentration was at 496.87 mg/L and the lowest value of mycelium extension rate at 0.362 cm/day. The highest mycelium extension rate was at 0.387 cm/day, with the lowest nitrogen concentration at 334.02 mg/L. The optimal condition was at 388.45 mg/L of nitrogen concertation and 0.370 cm/day of mycelium extension.

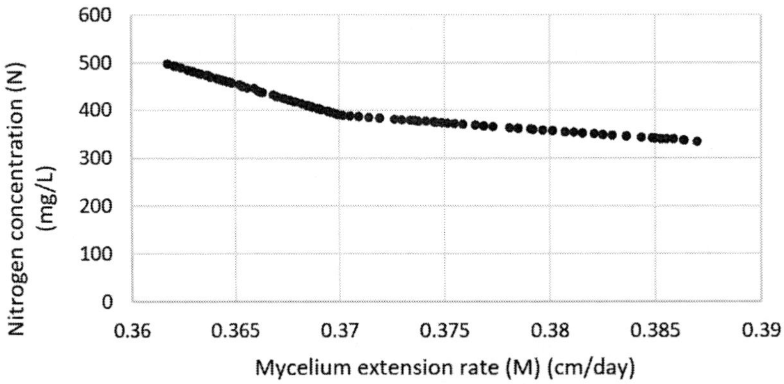

Figure 5. Pareto front for multi-objective optimisation with crossover of 0.7 and mutation of 0.2.

Referring to figure 6, for highest value of nitrogen concentration was at 689.98 mg/L with its lowest value of mycelium extension rate was 0.347 cm/day. For the highest mycelium extension rate was at 0.387 cm/day but it was the lowest nitrogen concentration at 334.05 mg/L. The optimal values for both results were 387.93 mg/L of nitrogen concertation and 0.370 cm/day of mycelium extension.

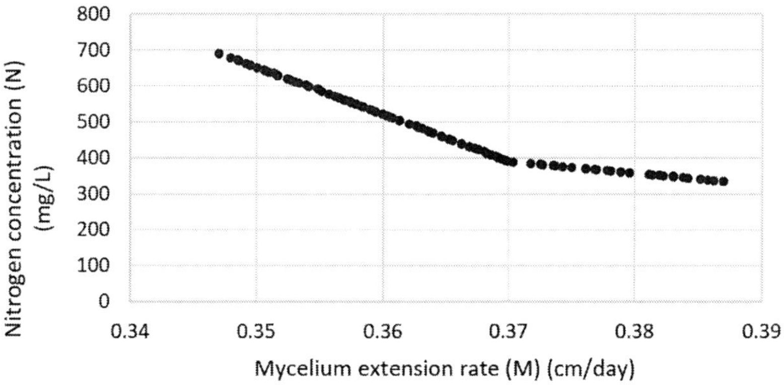

Figure 6. Pareto front for multi-objective optimisation with crossover of 0.7 and mutation of 0.3.

5. Conclusion

As observed on the plotted graphs, the distribution pattern looks quite similar, having a sharp curve in somewhere in middle. The maximum mycelium extension rate for all data set is similar, with the value of 0.387 cm/day. However, the maximum nitrogen concentration for all data set is different. The optimisation results with the parameters for crossover probability 0.7 and mutation probability 0.2 had produced the lowest maximum nitrogen concentration at 496.87 mg/L and the optimisation with crossover 0.7 and mutation 0.3 produced the highest nitrogen concentration at 689.98 mg/L compared to the other five optimisation results. The highest nitrogen concentration and mycelium extension rate was from the result with crossover and mutation probability of 0.5 and 0.2. It produced 388.45 mg/L of nitrogen concentration and 0.370 cm/day of mycelium growth.

Acknowledgments

The authors wish to acknowledge the Universiti Malaysia Pahang for funding the project under grant RDU1803119 and RDU1703295.

References

[1] Bellettini M B, Fiorda F A, Maieves H A, Teixeira G L, Ávila S, Hornung P S, Júnior A M and Ribani R H 2019 Factors affecting mushroom Pleurotus spp. *Saudi J. Biol. Sci.* **26** 633–46.

[2] Amirta R, Herawati E, Suwinarti W and Watanabe T 2016 Two-steps Utilization of Shorea Wood Waste Biomass for the Production of Oyster Mushroom and Biogas – A Zero Waste Approach *Agric. Agric. Sci. Procedia* **9** 202–8.

[3] Oliveira A P de and Naozuka J 2019 Preliminary results on the feasibility of producing selenium-enriched pink (Pleurotus djamor) and white (Pleurotus ostreatus) oyster mushrooms: Bioaccumulation, bioaccessibility, and Se-proteins distribution *Microchem. J.* **145** 1143–50.

[4] Vieira F R and de Andrade M C N 2016 Optimization of substrate preparation for oyster mushroom (Pleurotus ostreatus) cultivation by studying different raw materials and substrate preparation conditions (composting: phases I and II) *World J. Microbiol. Biotechnol.* **32** 190.

[5] Qu B, Li C, Liang J, Yan L, Yu K and Zhu Y 2019 A self-organized speciation based multi-objective particle swarm optimizer for multimodal multi-objective problems *Appl. Soft Comput.* 105886.

[6] Wilding P R, Murray N R and Memmott M J 2019 The use of multi-objective optimization to improve the design process of nuclear power plant systems *Ann. Nucl. Energy* 107079.

[7] Han H, Yu R, Li B and Zhang Y 2019 Multi-objective optimization of corrugated tube inserted with multi-channel twisted tape using RSM and NSGA-II *Appl. Therm. Eng.* **159** 113731.

[8] Li K, Yan S, Zhong Y, Pan W and Zhao G 2019 Multi-objective optimization of the fiber-reinforced composite injection molding process using Taguchi method, RSM, and NSGA-II

Simul. Model. Pract. Theory **91** 69–82.
[9] Jaliliantabar F, Ghobadian B, Najafi G, Mamat R and Carlucci A P 2019 Multi-objective NSGA-II optimization of a compression ignition engine parameters using biodiesel fuel and exhaust gas recirculation *Energy* **187** 115970.
[10] Pokhrel C P, Kalyan N, Budathoki U and Yadav R K P 2013 Cultivation of Pleurotus sajor-caju using different agricultural residues. *Int. J. Agric. Policy Res.* **1** 19–23.
[11] Dzulkefli N A and Zainol N 2018 Data on modeling mycelium growth in Pleurotus sp. cultivation by using agricultural wastes via two level factorial analysis *Data Br.* **20** 1710–20.
[12] Nebro A J, Durillo J J, Vergne M and Benítez A 2019 jMetal.

Review on effects of gasifying agents, temperature and equivalence ratio in biomass gasification process

F N Sidek, N A F Abdul Samad* and S Saleh

Faculty of Chemical & Process Engineering Technology, Universiti Malaysia Pahang, Lebuhraya Tun Razak, 26300 Gambang, Pahang, Malaysia.

*E-mail: asmafazli@ump.edu.my

Abstract. Biomass gasification is one of the promising technologies for converting biomass into gaseous fuels. It is crucial to fully understand the influence of operation parameters and types of gasifier on the performance of biomass gasification. It also can provide useful information for a better design and operation of gasification process. The general overview on each types of gasifier are discussed where it can be classified into fixed bed and fluidized bed gasifiers. Most of the review from literature focuses more on the effects of temperature and equivalence ratio rather than gasifying agent. However, the review on the effect of gasifying agent is still limited whereas the effect of gasifying agent is also important as it influences the yield and composition of product gas. In the present work, the effects of gasifying agent, temperature and equivalence ratio on the gasification process were reviewed. Firstly, this article highlights the advantages and disadvantages of each gasifiers which consists of fixed bed and fluidized bed. Based on review, a detailed comparison was made in terms of syngas composition obtained by using steam, air, oxygen and carbon dioxide as gasifying agents in order to provide basic knowledge regarding the selection of gasifying agents. The effects of temperature and equivalence ratio (ER) on the gas composition, tar content and reaction rate were discussed and analyzed. Finally, guidelines on the operation parameters in terms of gasifying agent, temperature and equivalence ratio are suggested in the summary review.

1. Introduction

The world is depending on fossil fuels for energy and about 80% of the world energy source is being fulfilled by these fuels which contributing to serious environmental problems such as global warming [1]. There has been a growing concern towards the environmental issues as well as depletion of fossil fuels in the future. Therefore, to secure the energy supply while reducing the effect of global warming, the shifting of energy from fossil fuel to renewables energy is essential. Among the renewables energy source, biomass is considered as promising source since it is widely available and it is derived from living matters. The utilization of biomasses for energy give positive impacts on the environment, as it is carbon neutral while the availability of biomass is abundance and uniform all over the world [2]. The examples of biomass are forest and agricultural wastes, municipal organic wastes and animal wastes [3].

Biomass can be converted into useful product in many different ways based on the characteristics and the condition of the product as well as its applications [4]. The most common application for biomass utilization is thermochemical conversion routes. Thermochemical conversion routes consist of three ways, which are gasification, liquefaction and pyrolysis [4]. The biomass gasification is an attractive process routes for energy conversion due to its conversion efficiency for various end product such as heat and electricity. In gasification process, the solid biomass is converted into combustible fuel through thermochemical conversion with the help of oxidant in a reactor known as gasifier [4]. The product of biomass gasification may undergo further processing step before it can be utilized for different applications such as power generation and cooking. Gasifier can be divided into two principle types namely fixed bed and fluidized bed [2]. Fixed bed gasifiers are the most common reactor use in gasification due to its simplicity and ease of operation in terms of design and operation. However, the fluidized bed is considered as the best due to its flexibility in terms of efficiency and type of fuels.

Many researchers have investigated the influence of operating parameters and types of feedstock on the gasification performance. Current reviews tend to focus more on the effect of temperature as well as equivalent ratio (ER) because it is easy to be varied and measured. However, the review on the effect of gasifying agent is still limited whereas the selection gasifying agent is also important. This is because different types of gasifying agent will influence the yield and composition of syngas. Therefore, it is necessary to understand how these factors influence the gasification reactions, which can provide useful information as it influence the composition of product gas [5]. In the present work, the effects of gasifying agents, temperature and equivalence ratio on the gasification process are reviewed. This information is selected due to the fact that gasification temperature significantly influences the composition of syngas, yield of tar content as well as reaction rate. The selection of gasifying agent is usually based on the desired quality or composition of syngas [3]. Furthermore, the equivalent ratio (ER) is considered one of the crucial variables that affect gasifier performance [6]. The correct ratio of ER will contribute to the good blending of feedstock over gasifying agent, which in turn affects the quality of syngas and the stability of the operation. This article also reviews the feasibility of fixed bed and fluidized bed by highlighting the advantages and disadvantages of each gasifier as well as the effects of different gasifying agents, temperature and ER on gasification performance.

2. Biomass gasification

A variety of biomass gasifier types have been performed in experimental and simulation works for representing the gasification process. It can be categorized into two major classifications which are fixed bed gasifier and fluidized bed gasifier [7].

2.1. *Fixed bed gasifier*

The most common type of gasifier used for biomass is the fixed bed gasifier. The fixed bed gasification is one of the earliest and most frequent reactor used for the production of syngas [2]. The fixed bed gasifier gain interest of researchers and industrial people due to its simple design and operation with no or few moving parts, but the syngas produced is a low gas calorific value with high tar content [4]. The fixed bed gasifier can be classified into three types which are updraft, downdraft and cross flow [8]. The first type of fixed bed gasifier is updraft gasifier. The major difference between the gasifier is depending upon the direction and the doorway of the airflow where for updraft gasifier the entry of the gasifying agent is from the bottom of the gasifier while the biomass is fed from the top of the gasifier [2]. However, the produced gas from the updraft gasifier contains huge amount of tar and moisture where the gas was vented at the top of the gasifier. In the case of downdraft gasifier, both biomass and gasifying agent were fed from the top of the gasifier while the direction is downward which is towards the lower section of gasifier unit while product gases leave just under the grate of gasifier [2]. This will allow the product gases to undergo partial cracking where the tar content is lower. The downdraft gasifier is divided into four zones, which are drying, pyrolysis, oxidation and reduction zone. The biomass with a high moisture and ash content will complicate the process of handling the biomass with a low overall thermal efficiency. For cross flow gasifier, the biomass moves downward from the top of the gasifier and the gasifying agent is supplied from the side of the gasifier [2]. The product gases exited from the top of the cross flow gasifier are expected to have high tar content with low overall energy efficiency.

2.2. *Fluidized bed gasifier*

Fluidized bed is considered as an attractive selection because of its flexibility in terms of fuel and efficiency [2]. The gasifiers are broadly used in coal gasification for many years. One of the advantages of fluidized bed over the fixed bed is the temperature in the reduction zone is uniform with the help of fine granular particle, which helps to fluidize the bed [9-10]. However, the downside of the fluidized bed gasifier is the complicated design with high operation and energy requirement. In addition, the major concern is the syngas from the gasifier has high tar content where gas cleaning is required. Fluidized bed is economical for 5 to 10 MW scale since the plant costs is high. Fluidized bed can be classified into two types, which are circulating fluidized bed and bubbling bed. The circulating fluidized bed is conducted based on the mechanism of continuous circulation of the bed material between the reaction vessel and cyclone separator. The advantages of a circulating fluidized bed are its ability to perform well with a high capacity of biomass throughputs and can be operated at high pressure [2]. In the bubbling bed, the gasifying agent enters the reactor from the bottom which is through the grate whereas the fine bed material is located above the grate where the biomass feed is fed. The bed material reacts with the high molecular weight tar to produce a syngas with lower tar content [2].

2.3. Advantages and disadvantages of different gasifiers

The advantages and disadvantages of each gasifier are highlighted in tables 1 and 2. There are numerous comparisons between fixed bed and fluidized bed gasifiers based on many aspects such as technology

Table 1. Fixed bed gasifier.

	Updraft Gasifier	**Downdraft Gasifier**	**Cross-flow gasifier**
Advantages	• Simple and low cost process • Suitable for high moisture and high inorganic content • Proven technology	• Required minimal or no tar cleanup as about 99.9% of tar formed is consumed [11] • Minerals remain with char or ash, reduce the need of cyclone • Proven technology, simple and low cost process.	• Ease of operation • Simple glass cleaning train (cyclone and bed filter)
Disadvantages	• Tar content is about 10 to 20% of tar by weight • Intensive cleaning is required	• The feed needs to be dry to lower the moisture (<20%) • The high temperature of syngas exits from gasifier require cooling before use • Unconverted carbon of 4 to 7% [12].	• High tar content • Low overall energy efficiency

Table 2. Fluidized bed gasifier.

	Circulating fluidized beds	**Bubbling beds**
Advantages	• Yields a uniform syngas • Uniform temperature distribution • Accept wide range of fuel • High rates of hear transfer between the inert material, fuel and gas • High conversion with low tar and unconverted carbon	• Suitable for fast reactions • High heat transport rates possible due to high heat capacity of bed material • High conversion rate with low tar and unconverted carbon
Disadvantages	• Gas bypass through the bed due to large bubble size	• Temperature gradients occur in direction of the solid flow • Size of fuel particles determine minimum transport velocity • Heat transfer less efficient than bubbling fluidized bed

and economy. The selection of gasifier will be depending on numbers of factors such as the feedstock, the product of producer gas required, and also operating condition. However, the complex design and operation as well as high energy consumption in fluidized bed is not favorable factor for gasifier selection. In addition, the tar content in a syngas of a fluidized bed is usually high where extra gas cleaning is required. In comparison to fluidized bed, the fixed bed gasifier appears to be more feasible for small scale plant application. This is due to the operation simplicity and cost-effective operation.

3. Effects of different gasifying agents

The most common gasifying agents used are air, steam, carbon dioxide and also pure oxygen. Usually the gasifying agent is selected based on the required quality of the syngas used in the downstream applications [3]. Ahmed et al. investigated woodchips char gasification by using steam and CO_2 as gasifying agent [13]. The gasification was conducted in a fixed reactor temperature and pressure of 900 °C and 2 bars. The variation of partial pressure of gasifying agents is from 0.6 to 1.5 bars with intervals of 0.3 bars. The steam and CO_2 flow rates were chosen in order to have an equal amount of oxygen in both of gasifying agents where the flowrates are 4.42 g/min of steam or 5.4 g/min of CO_2. The reactivity of steam is higher compared to CO_2 as the gasifying agent. The results show that the duration of gasification experiments for steam is approximately 22 minutes while the duration of CO_2 is approximately 60 minutes. Based on the findings, it shows that the reactivity of steam is approximately three times faster than that of CO_2.Therefore, as the reaction time faster, the time to produce syngas also shorter.

Xu et al. evaluated the gasification performance of municipal solid waste (MSW) by using thermodynamic equilibrium model [14]. Three type of gasifying agents were considered which are steam, hydrogen and air. In order to compare the performance of the gasifying agents, the temperature of the reactor and the mass flowrate of MSW are fixed at 1273 K and 100 kg/h. Based on the results, hydrogen gasifying agent contributed to highest mole fractions of CO and H_2 with smallest mole fractions of CO_2 and H_2O. As for steam gasifying agent, the mole fractions of H_2 and H_2O is the larger and smaller mole of fractions of CO_2 compared to air gasifying agent. This is due to the presence of oxygen content in the gasifying agent where hydrogen gasifying agent has the lowest oxygen content followed by steam gasifying agent and air gasifying agent. The presence of oxygen will contribute to the products composed of CO_2 rather than CO and H_2. Therefore, the selection of gasifying agent is depending on the product requirement. In this case, hydrogen gasifying agent is suitable to be used if H_2 and CO are needed with a lowest mass flowrate. On the other hand, air gasifying agent can be used in order to get high energy efficiency when there is no required quality and yield of the end products. However, the steam gasifying agent falls between the other two agents as the oxygen content of steam gasifying agent is higher than hydrogen gasifying agent but lower than air gasifying agent.

Prabawo et al. explored the high thermal efficiency in thermal gasification by replacing the steam gasifying agent by CO_2 gasifying agent [15]. The lab scale downdraft gasifier was used to study the effect of the gasifying agent on gas evolution and thermal efficiency. A list of biomass pyrolysis and CO_2 steam gasification experiments were conducted with or without the presence of O_2. The results show that CO_2-steam gasification with and without the present of O_2 produced a larger amount of combustible gas compared to pyrolysis. The yield of H_2 decreases while the yield of CO increases, as the steam is substituted by CO_2 gasifying agent. The advantage of CO_2 mixing ratio in terms of thermal efficiency was observed at the temperature greater than 850 °C. The efficiency of gasification process with CO_2 as a gasifying agent is higher if compared to the pyrolysis with N_2 in syngas production. Moreover, it is more efficient to use CO_2 than steam in the N_2 free syngas production.

Shayan et al. investigated and compared hydrogen production by using different gasifying agents [16]. Four gasifying agents were used which are air, oxygen-enriched air, oxygen and steam in gasification of wood and paper. The effect of key operating parameters on the yield of hydrogen, calorific value of producer gas, energy efficiencies of the process and energy destruction rate at different operating conditions. The gasification temperature and moisture content used are 1073 K and 10%. The highest amount of hydrogen produced when steam is used as the gasifying agent, which is higher than air gasifying agent. The reason for this finding is due to the fraction of nitrogen in steam gasification is zero where it leads to a shift in chemical equilibrium of the gasification reaction, which cause the yield of hydrogen to increase. Besides, molar fraction of hydrogen is higher by using oxygen enriched air and oxygen instead of environmental air due to lesser concentration of nitrogen to the gasifier. The results also show that the yield of CO in the producer gas is the highest when O_2 is used. The effect of gasification temperature on the yield of hydrogen of the producer gas for different gasifying agents also was investigated for both biomass wood and paper. The amount of hydrogen produced by using steam gasifying agent is the highest for all the gasification temperature ranges from 1000 to 1400 K for both biomass. On the other hand, the yield of hydrogen in the producer gas is the lowest for all the gasification temperature when air is used as the gasifying agent. The optimum gasification temperature by using steam gasifying agent for production of hydrogen is around 1050 K and 1075 K for wood and paper. As for the effect of gasification temperature on calorific value of producer gas, the results indicate that higher calorific value when using steam and O_2 gasifying agent. This is due to the amount of hydrogen and carbon monoxide content where it responsible for the calorific value of syngas. However, as the gasification temperature decreases, the calorific value of the syngas decreases for all types of gasifying agent.

4. Effects of gasification temperature

The most crucial parameters in biomass gasification are temperature where it will affect the gas composition, tar content, and reaction rate [3]. Xu et al. accessed the effect of steam temperature on the concentration of syngas, which are CH_4, CO, CO_2 and H_2 as well as the reactor temperature [14]. The steam temperature is varied from 973 to 2273 K and the steam to municipal solid waste ratio is set at 2. The effect of steam temperature on the concentration of syngas can be divided into two different regions. As the steam temperature is set below 1650 K, the concentration of H_2 increases up to 0.45 $Nm^3.kg^{-1}$ while the concentration of CO_2 increases from 0.09 to 0.19 $Nm^3.kg^{-1}$. On the other hand, the concentration of CH_4 decreases from 0.12 to 0 $Nm^3.kg^{-1}$ and the concentration of CO remains unchanged. However, as the temperature of the steam rises above 1650 K, the concentration of H_2 and CO_2 reduce slightly, while the concentration of CO slightly increases. On the other hand, the reactor temperature increases as the steam temperature increases. He et al. studied the production of hydrogen rich gas or syngas by using catalytic steam gasification of MSW in a bench-scale downstream fixed bed reactor [17]. The catalyst used is calcinated dolomite where the effect of catalyst and reactor temperature was studied. The effects of temperature on the product distribution (char, tar and gas) and gas fraction were investigated. The temperature of gasifier was varied from 700 to 950 °C. As the temperature increases, the conversion rate of MSW into hydrogen-rich gas increases from 27.01 to 53.29%. However, an increase in temperature results in decrease in the yield of char and tar while the composition of dry gas increase. The yield of char decreased from 21.68 to 8.12% whereas the composition of dry gas increased from 81.84 to 104.16%.

Pu et al. studied the biomass gasification of a pine with air and oxygen/steam atmosphere in a fixed bed reactor [18]. The effect of different reaction temperature on the gas composition and lower heating value (LHV) was investigated. The variation of reaction temperature is from 650 to 950 °C. The results show that, higher reaction temperature leads to higher H_2 content. The maximum value of reaction temperature is at 850 °C before the H_2 content started to decrease gradually. This is due to the water gas shift reaction and steam reforming reaction, which is endothermic. Therefore, the increase in reaction temperature is conducive for the yield of H_2 initially. However, some of H_2 may take part in oxidation reaction when the reaction temperature is further increased which contributes to decrease in the yield of H_2. Besides, the H_2 content also affect the LHV values. This is because, at reaction temperature range of 650 – 750 °C, as the H_2 content increases, the LHV also increases. Therefore, it can be concluded that, H_2 content plays an important role in LHV. Yan et al. carried out an experiment of biomass gasification in a fixed bed reactor [19]. The influence of temperature on the hydrogen yield was investigated. The variation of temperature used is from 600 to 850 °C. As the temperature increases, the yield of H_2 increases from 29.54 to 52.41%. It can be concluded that, an increase in temperature contributes to a sharp increase of hydrogen yield. Jamin et al. developed the thermodynamic equilibrium and predicted the composition of product gas based on the food and wood waste in a fluidized bed gasifier [20]. The gasification temperature used is in the range of 650 to 1000 °C. It was found that, as the temperature increases up to 1000 °C, the percentage of H_2 and CO gas increase which is from 29.58% to 34.03% and 31.85 to 45.78% for food waste. Meanwhile, the percentage H_2 and CO gas show the same trend where it increases from 31.25 to 39.87% and 26.33 to 34.81% for wood waste. Furthermore, LHV also exhibits the same trend as the product gas yield for both food and wood waste.

5. Effect of equivalence ratio (ER)

The equivalence ratio (ER) is defined as mass ratio of gasifying agent to fuel/biomass in any combustion unit [3]. Stoichiometric ratio is the minimum ratio of gasifying agent to fuel/biomass which is exactly enough to burn the fuel completely. The difference between combustion and gasification is combustion uses minimum stoichiometric ratio of gasifying agent to fuel whereas gasification uses gasifying agent-fuel ratio lower than stoichiometric ratio. On the other hand, the ratio between the gasifying agent-fuel ratio of the gasification process and the gasifying agent-fuel ratio for complete combustion is defined as equivalence ratio. Xu et al. investigated the effect of steam to MSW ratio on the concentration of syngas, which are CH_4, CO, CO_2 and H_2 and the reactor temperature [14]. The steam to MSW ratio (STMR) varied from 1 to 5 while the temperature is set to 1273 K. The result from the effect of the STMR is very identical to the effect of steam temperature. The results indicate that an increase of STMR will increase the reactor temperature that is cause by the energy balance. Guo et al. studied the effect of design and operating parameters on the gasification of corn straw [21]. The gasification was performed in a downdraft gasifier under atmospheric pressure and the gasifying agent used was air. Total of seven experiments were conducted by varying only the air feeding rate while the biomass feeding rate is kept constant. The equivalence ratio ranges from 0.18 to 0.37. The results show as the ER increases, the operating temperature in the gasifier increases, as high value of ER will enhance the combustion reactions to release heat. However, at the upper part of the gasifier, which is pyrolysis zone, the temperature is relatively low where the peak temperature only occurs at oxidation zone. The high temperature of gasifier caused by the high value of ER also led to higher tar cracking which in turn reduces tar yield. However, as ER increases above 0.32, the tar yield started to increase gradually. The lowest tar yield produced is 0.52 g/Nm^3 where the ER is at 0.32.

Pratik et al. explored biomass gasification of waste wood based on production of hydrogen energy [22]. The experiment was conducted in a downdraft gasifier where the effect of ER on the producer gas composition is evaluated. The ER was varied from 0.15 to 0.4. As the ER increases, the concentrations of H_2 and CO are also increased. However, as the ER is further increased up to 0.205, the concentrations of H_2 and CO are decreased. On the other hand, the CO_2 and N_2 show opposite trends compare to H_2 and CO concentrations. Furthermore, the result shows that the increase of ER will increase the production rate of producer gas. The optimum value of ER based on the production of producer gas point of view is 0.205, thus resulting to a maximum of yields of H_2 and CO. Gai et al. conducted a study on biomass gasification of corn straw in downdraft gasifier [23]. The study was carried out in downdraft fixed bed gasifier where air is used as gasifying agent. The influence of ER on the main gas composition is investigated where the range of ER is from 0.18 to 0.41. It was found as the ER increase, the temperature of gasifier increased. The results of ER can be divided into 2 categories, which is from 0.18 to 0.32 and above 0.32. The increase in ER, which is from 0.18 to 0.32 resulting the yield of CO_2 to decrease from 23.93 to 11.58% whereas the yield of CO increase from 11.35 to 19.81%. On the other hand, the yield of H_2 and CH_4 increases from 6.9 to 13.51% and from 1.27 to 3.96% respectively. When ER is greater than 0.32, the yield of CO_2 was observed to increase from 11.58 to 18.41% whereas the yield of CO gradual decrease from 19.81 to 15.16%. Moreover, the yield of H_2 and CH_4 reduced from 13.51 to 10.58% and from 3.72 to 1.57% respectively. Furthermore, as the ER increases from 0.18 to 0.32 the LHV increased from 2.69 to 5.39 MJ/m^3. However, as the ER increase above 0.32, the LHV started to reduce to 3.69 MJ/m^3.

6. Review summary

The summary of effects of gasifying agent, temperature and equivalence ratio (ER) and its remarks are described in the table 3. It can be concluded that the uses of different types or combination of gasifying agent have their own advantages and disadvantages. The gasifying agent should be selected based on application, required quality or composition product gas as well as the end use of the product gas. Based on the review, the usage of air as a gasifying agent will results in higher yield of CO_2 rather than CO or H_2. Therefore, making it suitable to be used for refrigeration and cooling application as the composition of CO_2 is high in the product gas. Next, in order to produce hydrogen rich gas, it is recommended to use steam as gasifying agent rather than air gasifying agent. The high amount of H_2 is suitable to be used in the fuel cell application. For example, power generation by using the H_2 as an input for proton exchange membrane fuel cell (PEMFC) [24]. Utilization of pure oxygen as gasifying agent will increase the composition of CO and H_2 while reduces the tar content in the product gas but the cost of the gasifying agent is high.

Table 3. Summary of parameters affecting the gasifier operation.

Parameters	Remarks	
Gasifying Agent	Air	• Higher composition of CO_2 rather than CO or H_2 rather than CO or H_2.
	Steam	• Favorable in hydrogen rich gas
	Pure Oxygen	• Higher composition of CO and H_2 • Reduce tar content • Cost of the gasifying agent high
Temperature	• For the production of hydrogen rich gas or syngas (above 850 °C) • Temperature ↑, LHV ↑, yield of char and tar ↓. • Temperature of biomass should not exceed 1000°C, as it will melt the ash • Reactor temperature ↑, ER ↑, tar yield ↓	
Equivalence Ratio (ER)	• ER ↑, yield of H_2, CO and CH_4 ↑ while yield of CO_2 ↓ • ER ↑ further to a certain ER value (ER>0.32), the yield of H_2, CO and CH_4 ↓, yield of CO_2 ↑	

For the effect of temperature on the composition of product gas, the findings indicate that for the production of hydrogen rich gas or syngas, high temperature is favorable. The high reactor temperature will help in reducing tar yield. However, as the temperature increase to a certain temperature above 850 °C, the composition of hydrogen decreases gradually. Therefore, it is important to study the optimal temperature in order to achieve high amount of hydrogen rich gas or syngas since the optimal temperature of every biomass is different. Furthermore, optimal temperature also contributes to optimal LHV and reduces the yield of char and tar. Despite the advantages of using high temperature, the temperature of biomass should not exceed 1000 °C as it will melt the ash especially the biomass with high ash content such as wheat straw. Besides, this requirement is to adhere to the reactor specification. On the other hand, as the ER increases, it will attribute to higher yield of H_2, CO and CH_4 and lower yield of CO_2. However, as the ER increase further to a certain ER value, the yield of H_2, CO and CH_4 will start to reduce gradually while the yield of CO2 will start to increase.

7. Conclusion

In this paper, the advantages and disadvantages of each gasifier were discussed. However, the selection of particular gasifier types and its design are depending on the properties of biomass feedstock and quality of product gas. The fixed bed gasifier is favorable for small-scale plant due to its design simplicity, easy to operate and economically feasible. On the other hand, the fluidized bed is more suitable for medium scale plant application due to its ability to achieve uniform yield and temperature distribution. The effects of gasifying agent, temperature and equivalence ratio on the performance of gasification process were also reviewed. As discussed above, the use of air gasifying agent will result in highest yield of CO_2 whereas steam gasifying agent will result in hydrogen rich syngas. On the other hand, the use of oxygen gasifying agent produced lower tar yield and higher composition of CO and H_2. In terms of gasification temperature and ER, it is favorable to use high reactor temperature in the range of 850 – 1000 °C and ER in the range of 0.2 – 0.32 in order to produce hydrogen rich syngas with a yield around 50%. The yield of H_2, CO and CH_4 increases while yield of CO decreases as the reactor temperature increases.

References

[1] D. Baruah and D. C. Baruah, 2014 "Modeling of biomass gasi fi cation : A review," vol. 39, pp. 806–815.

[2] T. K. Patra and P. N. Sheth, 2015 "Biomass gasi fi cation models for downdraft gasi fi er : A state-of-the-art review," vol. 50, pp. 583–593.

[3] M. Asadullah, 2014 "Barriers of commercial power generation using biomass gasi fi cation gas : A review," vol. 29, pp. 201–215.

[4] S. K. Sansaniwal, K. Pal, M. A. Rosen, and S. K. Tyagi, 2017 "Recent advances in the development of biomass gasi fi cation technology : A comprehensive review," vol. 72, no. December 2015, pp. 363–384.

[5] E. Madadian, V. Orsat, and M. Lefsrud, 2017 "Comparative Study of Temperature Impact on Air Gasification of Various Types of Biomass in a Research-Scale Down-draft Reactor," *Energy and Fuels*, vol. 31, no. 4, pp. 4045–4053.

[6] W. Ngamchompoo and K. Triratanasirichai, 2017 "Experimental investigation of high temperature air and steam biomass gasification in a fixed-bed downdraft gasifier," *Energy Sources, Part A: Recovery, Utilization and Environmental Effects*, vol. 39, no. 8. pp. 733–740.

[7] V. S. Sikarwar *et al.*, 2016 "An overview of advances in biomass gasification," *Energy Environ. Sci.*, vol. 9, no. 10, pp. 2939–2977.

[8] M. B. Muslim, S. Saleh, and N. A. F. A. Samad, 2017 "Effects of purification on the hydrogen production in biomass gasification process," *Chem. Eng. Trans.*, vol. 56, pp. 1495–1500.

[9] P. Lahijani and Z. A. Zainal, 2011 "Bioresource Technology Gasification of palm empty fruit bunch in a bubbling fluidized bed : A performance and agglomeration study," *Bioresour. Technol.*, vol. 102, no. 2, pp. 2068–2076.

[10] X. Ku, H. Jin, and J. Lin, 2017 "Comparison of gasification performances between raw and torrefied biomasses in an air-blown fluidized-bed gasifier," *Chem. Eng. Sci.*, vol. 168, pp. 235–249.

[11] M. B. Muslim, S. Saleh, and N. A. F. Abdul Samad, 2017 "Torrefied Biomass Gasification: A Simulation Study by Using Empty Fruit Bunch," *MATEC Web Conf.*, vol. 131, pp. 0–5.

[12] N. Couto, A. Rouboa, V. Silva, E. Monteiro, and K. Bouziane, 2013 "Influence of the biomass gasification processes on the final composition of syngas," *Energy Procedia*, vol. 36, pp. 596–606.

[13] I. I. Ahmed and A. K. Gupta, 2011 "Kinetics of woodchips char gasification with steam and carbon dioxide," *Appl. Energy*, vol. 88, no. 5, pp. 1613–1619.

[14] P. Xu, Y. Jin, and Y. Cheng, 2017 "Thermodynamic Analysis of the Gasification of Municipal Solid Waste," *Engineering*, vol. 3, no. 3, pp. 416–422.

[15] B. Prabowo, K. Umeki, M. Yan, M. R. Nakamura, M. J. Castaldi, and K. Yoshikawa, 2014 "CO 2 – steam mixture for direct and indirect gasification of rice straw in a downdraft gasifier : Laboratory-scale experiments and performance prediction," *Appl. Energy*, vol. 113, pp. 670–679.

[16] E. Shayan, V. Zare, and I. Mirzaee, 2018 "Hydrogen production from biomass gasi fi cation ; a theoretical comparison of using di ff erent gasi fi cation agents," vol. 159, no. December 2017, pp. 30–41.

[17] M. He *et al.*, 2009 "Hydrogen-rich gas from catalytic steam gasification of municipal solid waste (MSW): Influence of catalyst and temperature on yield and product composition," vol. 34, pp.

195–203.
[18] G. Pu, H. Zhou, and G. Hao, 2013 "Study on pine biomass air and oxygen / steam gasification in the fixed bed gasifier," vol. 8, pp. 1–7.
[19] F. Yan, S. Luo, Z. Hu, B. Xiao, and G. Cheng, 2010 "Bioresource Technology Hydrogen-rich gas production by steam gasification of char from biomass fast pyrolysis in a fixed-bed reactor : Influence of temperature and steam on hydrogen yield and syngas composition," *Bioresour. Technol.*, vol. 101, no. 14, pp. 5633–5637.
[20] N. A. Jamin, N. A. F. A. Samad, and S. Saleh, 2019 "Synthesis gas production of food and wood wastes in a fluidized bed gasifier using thermodynamic equilibrium model," *6TH Int. Conf. Environ. Empower. Environ. Sustain. Eng. Nexus Through Green Technol.*, vol. 2124, no. July, p. 020041.
[21] F. Guo, Y. Dong, L. Dong, and C. Guo, 2014 "ScienceDirect Effect of design and operating parameters on the gasification process of biomass in a downdraft fixed bed : An experimental study," *Int. J. Hydrogen Energy*, vol. 39, no. 11, pp. 5625–5633.
[22] P. N. Sheth and B. V Babu, 2010 "Production of hydrogen energy through biomass (waste wood) gasification," *Int. J. Hydrogen Energy*, vol. 35, no. 19, pp. 10803–10810.
[23] C. Gai and Y. Dong, 2012 "Experimental study on non-woody biomass gasification in a downdraft gasifier," *Int. J. Hydrogen Energy*, vol. 37, no. 6, pp. 4935–4944.
[24] F. R. A. Abdul Wahid, M. B. Muslim, S. Saleh, and N. A. F. Abdul Samad, 2016 "Integrated gasification and fuel cell framework: Biomass gasification case study," *ARPN J. Eng. Appl. Sci.*, vol. 11, no. 4, pp. 2673–2680.
[25] S. Safarian, R. Unnþórsson, and C. Richter, 2019 "A review of biomass gasification modelling," *Renew. Sustain. Energy Rev.*, vol. 110, no. May, pp. 378–391.

IOP Conf. Series: Materials Science and Engineering **863** (2020) 012029 doi:10.1088/1757-899X/863/1/012029

Effects of pure and mixed stabilizers on opto-electrical properties and morphology of TiO_2 nanoparticles synthesized by sol-gel method

M Ayyaz[1,*], N U Huda[1], F Rasool[1], H Sami-ur-Rehman[1], A Mehmood[2], M Y Naz[1], S Shukrullah[1] and A Ghaffar[1]

[1] Department of Physics, University of Agriculture Faisalabad, 38040 Faisalabad, Pakistan.
[2] Department of Chemistry, Government College University Faisalabad, Punjab 38000, Pakistan.

*E-mail: muhammadayyaz333@yahoo.com

Abstract. Titanium dioxide (TiO_2) is acknowledged as the most advanced nanomaterial. It has been used in certain application, such as semiconductor, photocatalyst, dye solar cells, paints, dyes, cosmetics, antifogging coatings, self-cleaning windows, etc. Although TiO_2 nanoparticles occur naturally in mineral form, but the particles are also being synthesized through different well-known techniques. The past literature reveals that sol-gel methodology is considered as the most attractive and suitable technique for the preparation of such kind of nanoparticles. This article reports the effects of three different stabilizers (HCl, HNO_3 and mixture of HCl+HNO_3) on the morphological and opto-electrical properties of TiO_2 nanoparticles. The nanoparticles of TiO_2 were synthesized through sol-gel process by using titanium tetra isopropoxide (TTIP) as precursor in a solution of deionized water and isopropanol at 80 °C under constant stirring. The synthesized TiO_2 nanoparticles were characterized through X-ray Diffractometry (XRD), Scanning electron microscopy (SEM) and UV-Visible spectrophotometry. The grain size with perfect crystallinity obtained through XRD were in good agreement with SEM results. The grain size of the prepared TiO_2 nanoparticles using HCl, HNO_3 and (HCl+HNO_3) were 3 nm, 2.8 nm and 3.3 nm, respectively. The optical characterization of TiO_2 revealed band gap energy values of 4.10 eV, 6.16 eV and 6.14 eV corresponding to absorption edges at 302 nm, 201.2 nm and 201.73 nm, respectively. The smaller grain sizes were having larger band gaps.

1. Introduction

TiO_2 is known as an inert and safest material and has been utilized for various applications in optical, electrical, magnetic and chemical fields [1, 2]. The particles of TiO_2 also known as ultrafine powder of titanium dioxide, are the particles with diameter lower than 100 nm [3]. This ultrafine powder of TiO_2 is also utilized as part of sunscreens due to its ability of blocking UV radiations coming from sun and it remains transparent to the skin. Its properties of photocatalytic sanitizing and applications in dye solar cells, paints, dyes, cosmetics, medicates, varnishes and nano fabrics make it more suitable for antifogging coatings and self-cleaning windows [3-5]. It known as the second most advanced nanomaterial in extremely using products after introduction of silver nanoparticles. Owing to its immense usage in commercial and technological applications, TiO_2 is known as "legacy nanomaterial"

[5-7]. The nanoparticles of TiO_2 are used on large scale for catalytical degradation of organic materials as catalyst, sensor technology and for chemical conversions in solar cells as photo catalyst [8-10].

A variety of techniques has been used to synthesize TiO_2 nanoparticles, such as wet chemical method, precursor method, coprecipitation method, solution route, microemulsion technique, sol-gel method and solvothermal method [11, 12]. Comparison of these methods shows that sol-gel approach is the most attractive and suitable technique for the preparation of nanoparticles due to its convenient features such as it is versatile in processing, it has extended composition ranges, stability, better homogeneity and low energy consumption for processing [13-15].

Sol gel technique is also termed as polymerization process because of its intrinsic nature [16], which involves conversion of monomers into a transparent colloidal solution known as "Sol". The sol is converted into viscous gel after further processing [17, 18] as shown in figure 1. A metal-oxo-polymer network (gel) is obtained from molecular precursors, such as metal salts or metal alkoxides [19]. Recently, Mahshid *et al.* [20] analysed the effects of pH on the particle size of TiO_2 nanoparticles prepared by the hydrolysis of TTIP. They observed that the particle shape was totally spherical because of the poor agglomeration due to acidic condition. Behnajady *et al.* studied the effect of different precursors and different solvents under different synthesis conditions utilizing the sol gel approach [21]. It was revealed that the grain size and rutile phase increase and anatase decreases with the change of solvent type along with increased photo-catalytical activities of TiO_2. They used TTIP instead of TBOT (Titanium butoxide). Nadzirah *et al.* [22] studied the role of three different stabilizers in preparation of TiO_2 films via sol gel method. It was predicted that Acetic acid stabilized TiO_2 films are more suitable for biosensor applications due to their surface roughness and electrical stability while HCl stabilized films were suggested to be used in photosensor applications due to their smooth surface. However, monoethalomine acid (MEA) stabilized TiO_2 films were suggested to be useless in electrical applications due to their low stability of electrical properties.

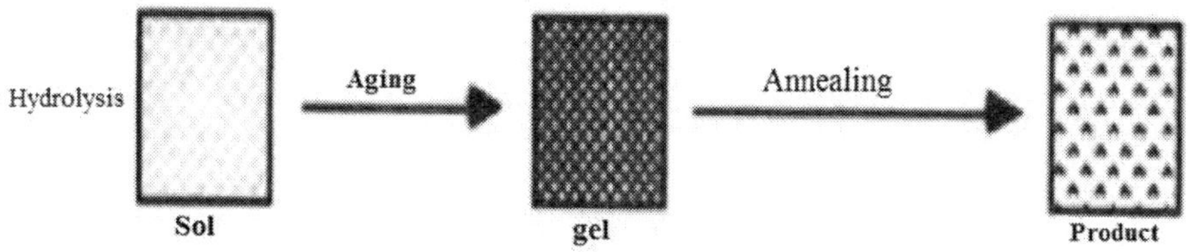

Figure 1. Schematic of sol-gel approach.

Although, the effect of some stabilizers on TiO_2 in sol gel method has been studied [22, 23] but the comparative study of the effects of pure and mixed stabilizers on TiO_2 nanoparticles under the sol gel method is still unclear. The main objective of this work was to synthesize the TiO_2 nanoparticles through sol gel approach by using three different stabilizers (HCL, HNO_3 and mixture of HCL and HNO_3) and titanium tetra isopropoxide as precursor.

2. Materials and methods

2.1. Chemicals

Chemicals used to prepare TiO_2 nanoparticles, were titanium tetra iso-propoxide (TTIP) $C_{12}H_{28}O_4Ti$ (98.0%, from Dae-Jung, Korea) as precursor, Isopropanol (C_3H_5OH, Riedel-deHaen, extra pure) as solvent, Hydrochloric acid fuming (HCL, Merck Germany, 37% extra pure) Nitric acid (HNO_3 68%, Huchems fine Chemical Corp. Korea) as stabilizers and deionized water as dispersing media [18, 24].

2.2. Synthesis of TiO_2 powder

To prepare TiO_2 using HCl as stabilizer, firstly, a 45 mL solution was prepared by mixing 30 mL of isopropanol as solvent with 15 mL of deionized water as dispersing media. The solution was kept under constant stirring at the temperature of 80 °C and 30 mL of TTIP was dissolved as a precursor drop wise to the solution [18]. After 1 hour, the temperature was reduced from 80 °C to 60 °C and a water-acid mixture (1.5 mL of HCL diluted with 50 mL deionized water) was mixed up with the TTIP dissolved solution. White thick precipitates were formed, which on continuous stirring turned into a transparent and stabilized white sol after 3 hours. The resulting sol was stirred continuously for 150 minutes to complete the process of hydrolysis and condensation. The obtained gel was annealed at 300 °C in open atmosphere for 2 hours to obtain TiO_2 nanocrystals. Finally, the resulted nanocrystals of TiO_2 were grinded for 45 minutes to obtain the fine powder of TiO_2 nanoparticles.

For synthesis of TiO_2 nanoparticles with HNO_3 as a stabilizer, 40 mL solution was prepared from 25 mL Isopropanol and 15 mL deionized water under constant stirring at 80 °C. Thereafter, 25 mL solution of TTIP was added drop wise to the prepared solution under continuous stirring. After 1 hour, the temperature was reduced from 80 °C to 60 °C and a water-acid mixture consisting of 1.5 mL of HNO_3 diluted with 60 mL deionized water, was added to the TTIP mixed solution under continuous stirring [22]. After 90 minutes of processing, the solution turned into a transparent white sol. The resulted sol was stirred continuously for 150 minutes to convert it into a viscous sol-gel. The obtained gel was dried at 300 °C in open atmosphere to obtain TiO_2 nanoparticles. The obtained nanoparticles were grinded using piston mortal for 45 minutes to obtain fine powder of TiO_2 nanoparticles.

Similarly, to obtain TiO_2 nanoparticles with a mixture of HCL and HNO_3 as stabilizer, 40 ml solution was prepared from 25 mL of Isopropanol and 15 mL of deionized water at 80 °C under continuous stirring. Thereafter 30 mL of TTIP was added drop wise to the solution under continuous stirring. After 1 hour of processing, the temperature was decreased from 80 °C to 60 °C and 61.4 mL of water acid mixture (0.6 mL HCL + 0.8 mL HNO_3) diluted with 60 mL of deionized water was added to TTIP mixed solution under continuous stirring. The solution turned into a transparent white sol after 90 minutes. The obtained sol was stirred continuously until it converts into a viscous gel after 2 hours. The resulted sol-gel was annealed at 300 °C for 2 hours in open air and nanocrystals of TiO_2 were obtained, which were grinded for 45 minutes to obtain fine powdered TiO_2 nanoparticles [25]. A flowchart of synthesis of TiO_2 nanoparticles is given in figure 2.

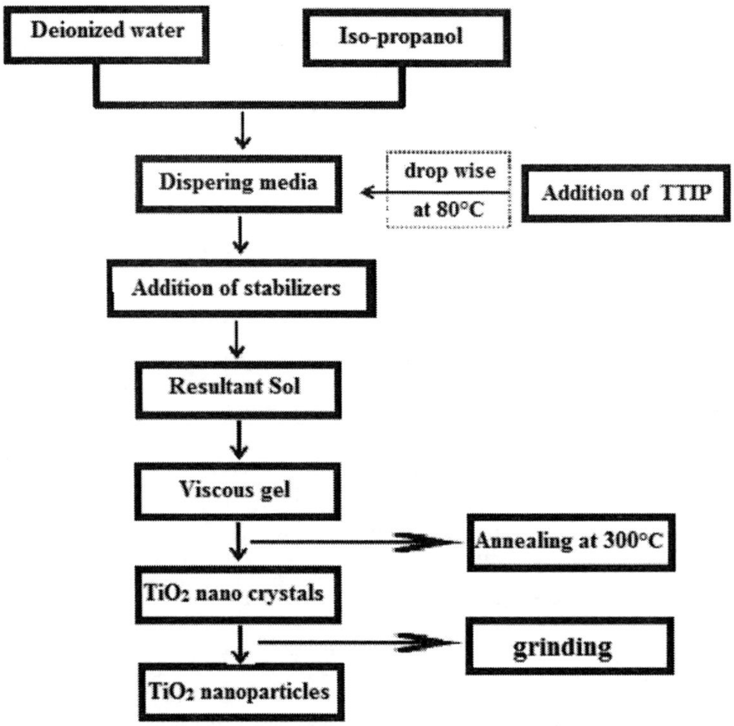

Figure 2. Flowchart of preparation of TiO_2 nanoparticles using sol-gel method.

3. Results and discussion

3.1. Structural properties

XRD was used to analyze TiO_2 samples synthesized with three different stabilizers. In figure 3, XRD spectra confirmed the formation of TiO_2. The diffraction peaks were observed in the angle range of $18° \leq 2\theta \leq 60°$. TiO_2 samples were crystalline in nature and composed of anatase and rutile phases. The HCl stabilized nanoparticles produced XRD peaks at 25.23°, 36.78°, 37.48°, 38.49°, 48.03°, 53.70° and 54.96° with corresponding anatase phases as (011), (013), (004), (112), (111), (020), (015) and (121), respectively. Similarly, XRD peaks at 27.29°, 35.80°, 38.97°, 40.91°, 43.68°, 53.68°, and 56.27° produced rutile phases as (110), (101), (200), (111), (120), (211) and (220), respectively.

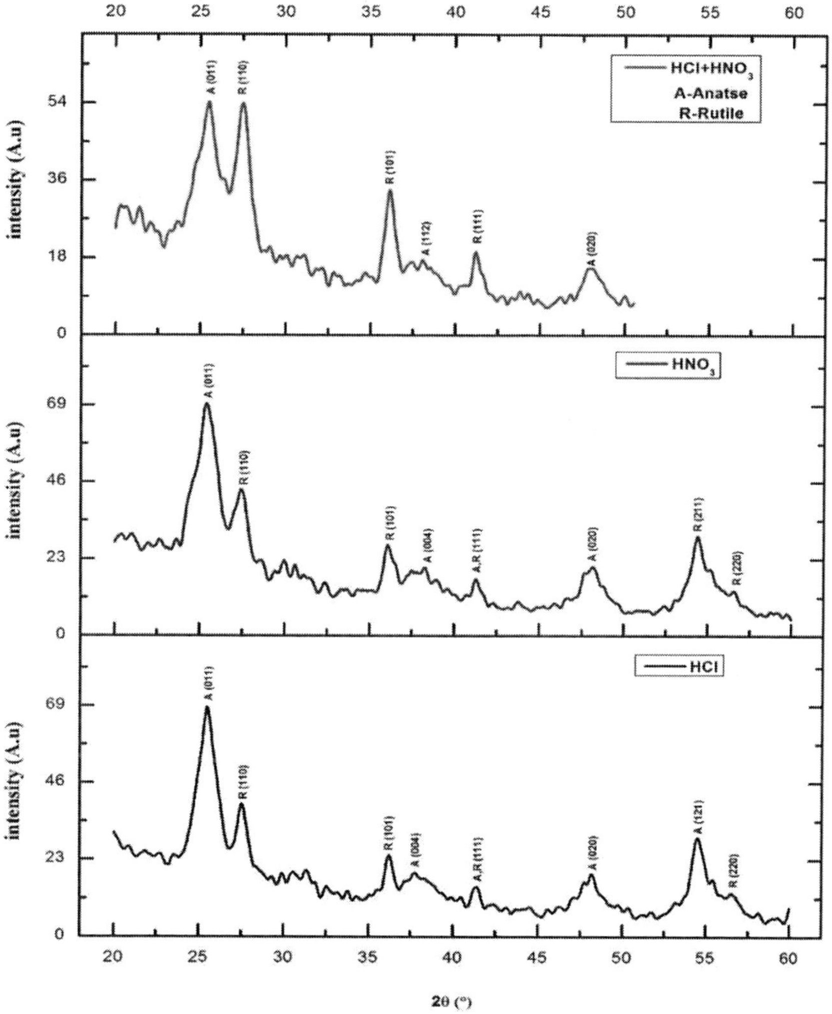

Figure 3. XRD spectra of TiO_2 synthesized with different stabilizers.

The HNO_3 stabilized nanoparticles produced XRD peaks at 25.24°, 36.77°, 37.63°, 38.53°, 40.88°, 47.95°, 53.77°, and 55.00° with corresponding anatase (h k l) values of (011), (013), (004), (112), (111), (020), (015) and (121), respectively. The XRD peaks at 27.22°, 35.81°, 38.94°, 40.88°, 43.73°, 53.89° and 56.24° represented (110), (101), (200), (111), (120), (211) and (220) phases of rutile TiO_2.

Similarly,(HCl+HNO_3 stabilized nanoparticles produced XRD peaks at 25.18°, 36.83°, 37.74°, 38.57° and 48.00° with corresponding (h k l) values (011), (013), (004), (112) and (020), respectively, which belong to anatase phase. The XRD peaks at 27.26°, 35.77°, 38.91°, 40.92°, and 43.71° presented the (110), (101), (200), (111) and (120) phases of rutile TiO_2. The most intense XRD peak was observed at 25.24°, which was used to calculate the particle size. This peak belongs to tetragonal anatase and rutile phase. The lowest peak intensity was observed at 25.18° by using mixed stabilizer. The particle size was calculated using Scherrer formula [24, 26]:

$$D = \frac{\lambda K}{\beta \cos \theta} \tag{1}$$

Where D is grain size, K is the constant of shape factor (K=0.89), β is full width half maximum (FWHM) in radian and λ is the X-ray wavelength for Cu target Kα radiations and θ is the Bragg's diffraction angle. Using above equation (1), the determined particle sizes are given in table 1.

Table 1. Influence of different stabilizers on crystallite size of TiO_2.

Stabilizer	Solvent	Precursor	Annealing Temperature	2θ (Deg.)	Particle size (nm)
HCl	Isopropanol	TTIP	300°C	37.48°	3.0
HNO_3	Isopropanol	TTIP	300°C	37.63°	2.8
HCl+HNO_3	Isopropanol	TTIP	300°C	25.18°	3.3

3.2. SEM analysis

SEM micrographs of TiO_2 nanoparticle samples are shown in figure 4. Irregular shapes of nanoparticles were seen in SEM micrographs, specifically for those synthesized with HCl and HCl+HNO_3 stabilizers. The nanoparticles, synthesized with HNO_3 stabilizer, were relatively smaller in size and but more agglomerated. However, particle size was uniform as compared to other two samples.

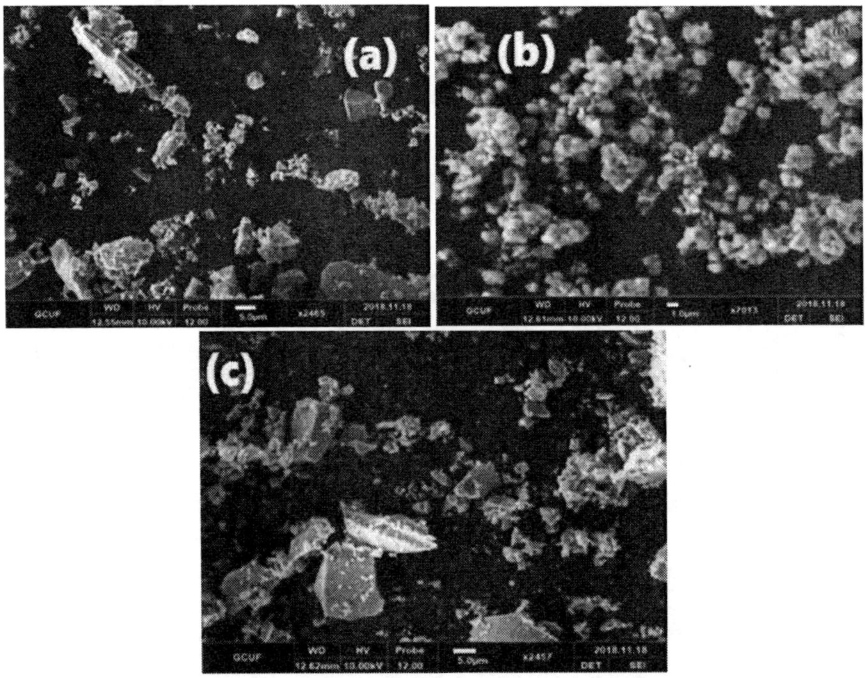

Figure 4. SEM micrographs of TiO_2 nanoparticles synthesized with: (a) HCl, (b) HNO_3, and (c) HCl+HNO_3 stabilizers.

IOP Conf. Series: Materials Science and Engineering **863** (2020) 012029 doi:10.1088/1757-899X/863/1/012029

The nanoparticles, synthesized with HCl and HCl+HNO_3 stabilizers, were poorly agglomerated. These particles revealed sharp rock like edges [20, 27]. Some soot or tiny buds were also seen on these nanoparticles. We could not study these buds due to limited magnification of SEM equipment, however, these buds might be oxidized form of the nanomaterial.

3.3. UV-Visible analysis

UV-Visible analysis was carried out to study the effect of different stabilizers on optical properties of TiO_2 nanoparticles. The band gap energy (E_g) was determined from the optical absorption spectra by using the equation:

$$E_g = \frac{1240}{\lambda}\ [\mathrm{eV}] \tag{2}$$

where E_g represents the band gap energy electron volts (eV) and λ is the wavelength of absorption edge obtained from the spectrum [28, 29]. Figure 5 shows the absorption spectra of nanoparticle samples obtained with 3 stabilizers.

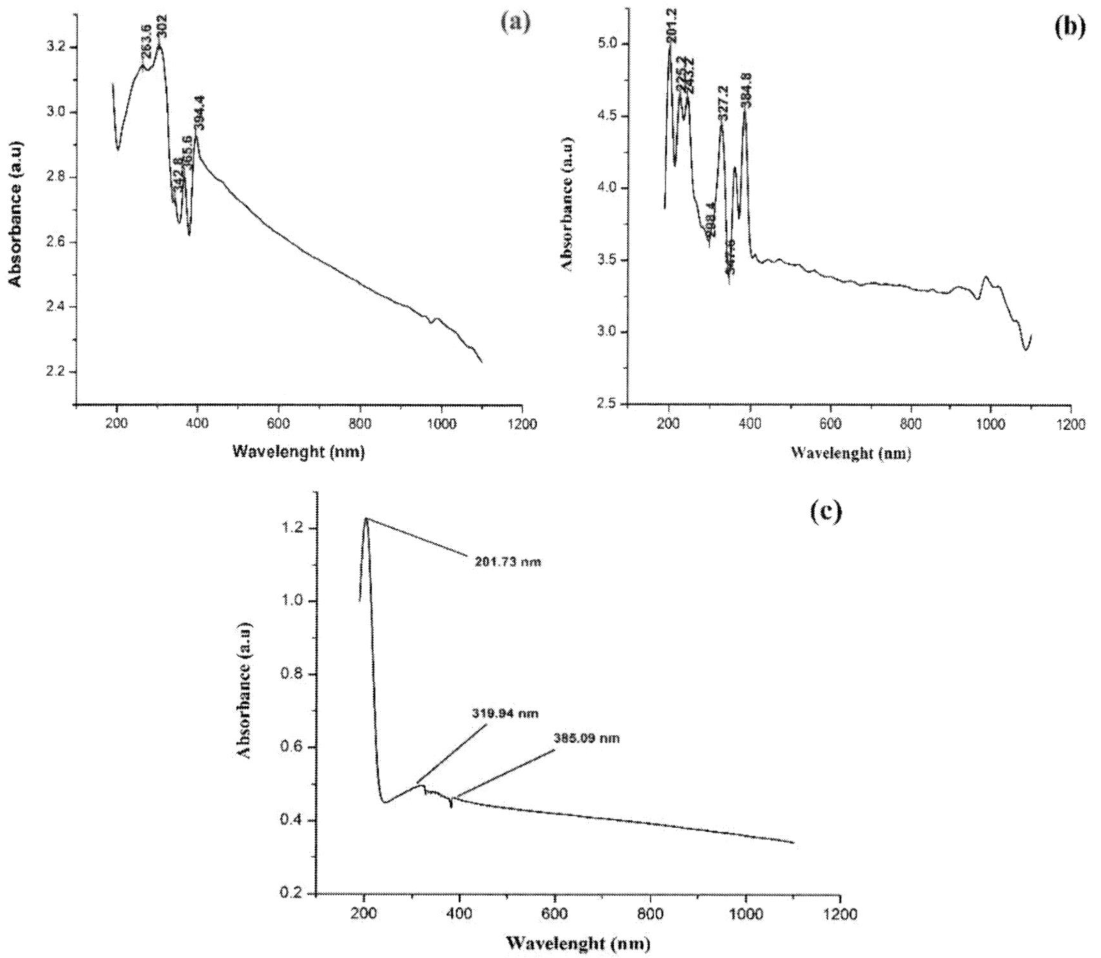

Figure 5. UV-Vis spectra of TiO_2 nanoparticles synthesized with: (a) HCl, (b) HNO_3, and (c) HCl+HNO_3 stabilizers.

The band gap energy of nanoparticles was calculated about 4.10 eV, 6.16 eV and 6.14 eV for HCl, HNO_3 and HCl+HNO_3 stabilizer, respectively. As the particle size decreases, the corresponding band gap increases and absorption edge seems to be shifted towards higher energy end (blue shift) [30].

The nanoparticles, stabilized with HCl, show absorbance up to wavelength of 302 nm. The nanoparticles, stabilized with HNO_3, show absorbance up to wavelength of 201.2 nm. Similarly, the nanoparticles stabilized with HCl+HNO_3 mixture exhibit absorbance up to wavelength of 201.73 nm.

4. Conclusion

In this work, TiO_2 nanoparticles were synthesized through sol-gel method by using three different stabilizers. Sol-gel is the most suitable technique for synthesis of such kind of nanomaterials due to its low cost and low process temperature. The minimum time to develop a sol was 90 minutes, which was converted into gel after 2 hours of processing with HCl+HNO_3 stabilizer. Irregular shapes of nanoparticles were seen in SEM micrographs, specifically for those synthesized with HCl and HCl+HNO_3 stabilizers. The nanoparticles, synthesized with HNO_3 stabilizer, were relatively smaller in size and but more agglomerated. However, particle size was uniform as compared to other two samples. The band gap energy of nanoparticles was calculated about 4.10 eV, 6.16 eV and 6.14 eV for HCl, HNO_3 and HCl+HNO_3 stabilizer, respectively. As the particle size decreases, the corresponding band gap increases and absorption edge shifts towards higher energy end (blue shift).

References

[1] Uboldi C, Urban P, Gilliland D, Bajak E, Valsami-Jones E, Ponti J and Rossi F 2016 *Toxicol. In Vitro* **31** 137.
[2] Zoccal J V M, Arouca F O and Goncalves J A S 2010 *Int. conf. on Materials Science Forum* vol 660 (Trans Tech Publ) pp 385-90.
[3] Zhang Y, Leu Y-R, Aitken R and Riediker M 2015 *Int. J. Environ. Res. Public Health* **12** 8717
[4] Hayle S T and Gonfa G G 2014 *Am. J. Nanosci. Nanotechnol.* **2** 1.
[5] Muniz E, Goes M, Silva J, Varela J A, Joanni E, Parra R and Bueno P R 2011 *Ceram. Int.* **37** 1017.
[6] Karami A 2010 *J. Iran. Chem. Soc.* **7** S154.
[7] Burunkaya E, Akarsu M, Camurlu H E, Kesmez O, Yesil Z, Asilturk M and Arpac E 2013 *Appl. Surf. Sci.* **265** 317.
[8] Shahab M, Tabish T, Zaman B, Tariq Z and Kamran M 2013 *ANNALS of F.E.H. – Int. J. Eng.* **11** 313.
[9] Phonkhokkong T, Thongtem T, Thongtem S, Phuruangrat A and Promnopas W 2016 *Dig. J. Nanomater. Bios.* **11** 81.
[10] Macwan D, Dave P N and Chaturvedi S 2011 *J. Mater. Sci.* **46** 3669.
[11] Gupta S M and Tripathi M 2012 *Cent. Eur. J. Chem.* **10** 279.
[12] Malekshahi Byranvand M, Nemati Kharat A, Fatholahi L and Malekshahi Beiranvand Z 2013 *J. Nanostruct.* **3** 1.
[13] Behnajady M, Eskandarloo H, Modirshahla N and Shokri M 2011 *Desalination* **278** 10.
[14] Gnanasekaran L, Hemamalini R and Ravichandran K 2015 *J. Saudi Chem. Soc.* **19** 589.
[15] Nikkanen J-P 2016 *Synthesis of TiO_2 by Various Methods: Structural Characteristics, Photocatalytic Activity and Usability of Powders and Coatings* vol 1431 (Finland: Tampere University of Technology Publication) p 114.
[16] Haider A J, Jameel Z N and Taha S Y 2015 *J. Eng. Technol.* **33** 761.
[17] Lee S, Cho I-S, Lee J H, Kim D H, Kim D W, Kim J Y, Shin H, Lee J-K, Jung H S and Park N-G 2010 *Chem. Mater.* **22** 1958.
[18] Nachit W, Touhtouh S, Ramzi Z, Zbair M, Eddiai A, Rguiti M, Bouchikhi A, Hajjaji A and Benkhouja K 2016 *Mol. Cryst. Liq. Cryst.* **627** 170.
[19] Wang C-C and Ying J Y 1999 *Chem. Mater.* **11** 3113.
[20] Mahshid S, Askari M and Ghamsari M S 2007 *J. Mater. Process. Technol.* **189** 296.

[21] Behnajady M A, Eskandarloo H, Modirshahla N and Shokri M 2011 *J. Photochem. Photobiol.* **87** 1002.
[22] Nadzirah S, Foo K and Hashim U 2015 *Int. J. Electrochem. Sci.* **10** 5498.
[23] Hak C R C, Fatanah D N E, Abdullah Y and Sulaiman M Y M 2018 *Int. J. Curr. Res. Sci. Eng. Technol.* **1** 172.
[24] Sharma A, Karn R and Pandiyan S 2014 *J. Basic Appl. Eng. Res* **1** 1.
[25] Kervee S E and Mohammadi M 2017 *J. Photochem. Photobiol. A: Chem.* **338** 104.
[26] Pookmanee P and Phanichphant S 2009 *J. Ceram. Process. Res.* **10** 167.
[27] Elbushra H, Ahmed M, Wardi H and Eassa N 2018 *MRS Adv.* **3** 2527.
[28] Binas V, Venieri D, Kotzias D and Kiriakidis G 2017 *J. Materiomics.* **3** 3.
[29] Wang B, Li L, Chen J, Duan C, Song J, Wang R and Zhang B 2018 *J. Nanopart. Res.* **20** 175.
[30] Yin G, Wang Y and Yuan Q 2018 *J. Nanopart. Res.* **20** 208.

Synthesis and characterization of silver nanorods using water coupled atmospheric pressure dc plasma

S Saleem[1], I Asghar[1], S Manzoor[1], A Abbas[1], S Shukrullah[1,*], M Y Naz[1] and Z Hussain[2]

[1] Department of Physics, University of Agriculture Faisalabad, 38040 Faisalabad, Pakistan.
[2] Department of Chemistry, Abdul Wali Khan University, 23200 Mardan, Pakistan.

*E-mail: zshukrullah@gmail.com

Abstract. Non-thermal microplasmas are gaining attention of research community due to their unique physical, chemical, optical and catalytic properties. Due to theses remarkable properties, microplasmas are being used in medicine, electronics, material processing and biotechnology. In this study, silver nanorods (AgNRs) were synthesized by coupling dc plasma with aqueous solution of silver nitrate ($AgNO_3$). Sucrose was added in the solution as a stabilizing agent. AgNRs were produced in aqueous solution by reducing $AgNO_3$ with dc plasma jet of argon at atmospheric pressure. No additional reducing agent was added used during production of AgNRs. Different molar ratios of $AgNO_3$ were used to control the size and dispersion of nanorods. Chemical composition and surface morphology of AgNRs were examined by using different techniques, such as scanning electron microscopy (SEM) and energy dispersive X-ray spectroscopy (EDX). The EDX spectrum confirmed that nanorods are AgNRs metallic silver with atomic weight of 16.86%. SEM micrographs revealed the formation of AgNRs. The rod length varied from 5 nm to 100 nm, depending on the molarity of solute.

1. Introduction

Synthesis of metal nanoparticles of extremely small dimensions, large surface area, controlled shapes and narrow size distribution is imperatively required for their use in cutting-edge technologies, such as optical sensing, catalysis, electronics and energy production devices [1-3]. Different methods for synthesis of silver nanoparticles (AgNPs) have been adopted, such as conventional biological, physical, photochemical and chemical methods. Each technique has benefits and drawbacks with cost, functionality, particle size and size distribution, etc [2, 3]. Among the available methods, chemical approach is regarded as simple and cost-effective method of synthesis of nanomaterials. Pure AgNPs can be produced through this method under mild conditions [4]. On the other hand, green synthesis method consists of photosynthesis and biosynthesis processes, which are usually carried out at high temperatures for longer periods under highly cleaning conditions [3, 5-8]. The chemical reduction approach is used to synthesize AgNPs with colloidal dispersions in water or organic solvents. Reducing silver ions in solvents produces distinct particle diameters of colloidal silver nanoparticles. Mostly, saccharides that include glucose, sucrose, soluble starch, lactose and fructose are used as reducing agents. Huang et al. [9] synthesized AgNPs by using an atmospheric pressure dc plasma jet and $AgNO_3$. A stabilizing agent was used to prevent aggregation of nanorods. Morphological features of synthesized AgNPs strongly depended on the parameters, such as process temperature, metal precursor, molar ratio of $AgNO_3$, reducing agents and stabilizing agents [9]. Bisht et al. [10] prepared AgNPs by using sucrose as a stabilizing agent. Filippo et al. [7] produced AgNPs under mild conditions by using sucrose and maltose as stabilizing agents and found the dramatic effect on the formation of nanoparticles. Water coupled atmospheric pressure dc plasma showed great authority on parameters of nanostructrues [11, 12].

Microplasma is a cost-effective single step procedure, which works at room temperature, low current, high voltage and low gas flow rate. Morphology and configuration of nanorods depends on the parameters, such as processing time, current, voltage, precursor and flow rate of plasma source. [9, 11-14]. Richmonds and Sankaran [15] explained the microplasma-assisted production of AgNPs by combining fructose with $AgNO_3$ in molar ratio of 0.1-1 M. In this study, AgNRs were synthesized by impinging dc plasma jet on a solution of $AgNO_3$ and sucrose. The distance of plasma jet, gas flow rate and amount of stabilizer were varied to control the nanorod properties.

2. Materials and methods

Argon gas, $AgNO_3$ and sucrose ($C_{12}H_{22}O_{11}$) were used as plasma source, precursor and stabilizing agent, respectively. Molar concentration of 0.06 M of $AgNO_3$ was made and sucrose was added to the solution to prevent agglomeration of AgNRs. 20 ml of solution was exposed to argon plasma jet under ambient conditions for synthesis of AgNRs. The experimental setup, used in the presented work, is shown schematically in figure 1.

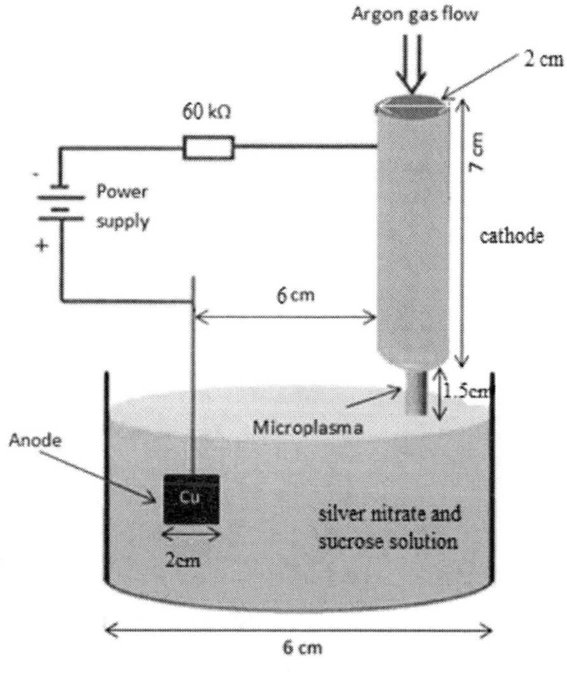

Figure 1. Schematic of experimental setup used for synthesis of AgNRs.

Two electrodes were used in this experiment. One electrode acted as cathode, which was a stainless-steel needle with outer diameter of 2 cm and inner diameter of 1 mm. The length of the needle electrode was 7 cm. The needle was placed just above the solution with 1 cm page. Second electrode was a tungsten electrode, which acted as anode. It was dipped in the solution to make a direct contact with the solution. The distance between electrodes was about 3 cm. Argon gas was coupled with cathode and gas was flowing through the capillary. An argon gas flowmeter was used to monitor and control the gas flow rate. Cathode was connected with negative terminal of dc power source and copper electrode was connected with the positive terminal. High dc voltage (2 to 10 kV, 20 to 40 mA) was provided to the electrodes. A 100 kΩ blast resistor was connected between the power supply and the cathode to avoid the breakdown. The argon flow rate was fixed to 1200 sccm. Figure 2 shows all the steps and materials used in this work are shown in figure 2. The final product was centrifuged to get AgNRs. These nanorods were dried in an over for 6 h at 80 °C.

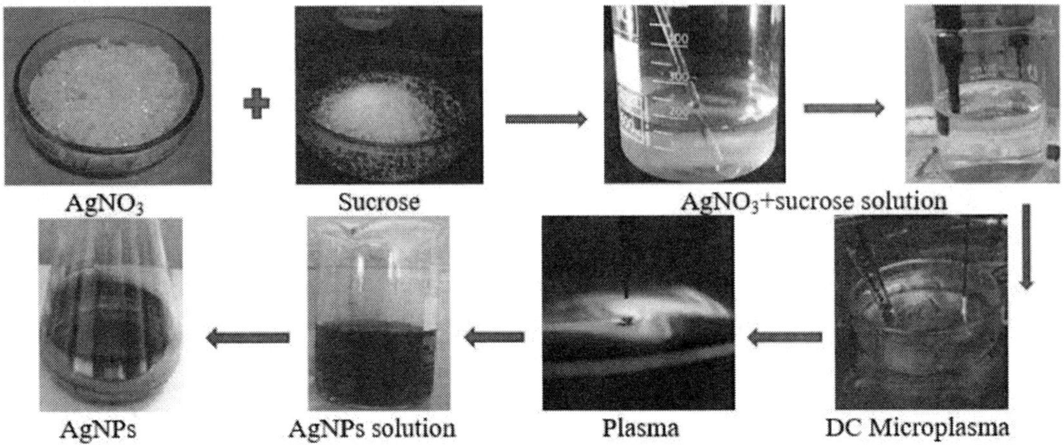

Figure 2. Different steps involved in synthesis of AgNPs.

3. Results and discussion

The chemical composition of AgNRs was analysed using energy dispersive X-ray spectroscopy (EDX). EDX analysis gives the information on elements present in the sample, but the exact content of each element in the sample cannot be determined through this technique because the elemental composition depends highly on the scan area. Figure 3 shows EDX pattern of the nanorods. The peaks around 3 keV, 1.8 keV, 0.5 keV and 0.3 keV correspond to Ag, Si, O and C, respectively. It was revealed that the nanorods are metallic silver. The weight of Ag was 16.86%, carbon 36.40% and oxygen 45.59%, as shown in table 1. Another peak, found in EDX spectrum, corresponds to Si. The Si peak in EDX appeared to the substrate on which AgNRs were deposited for characterization.

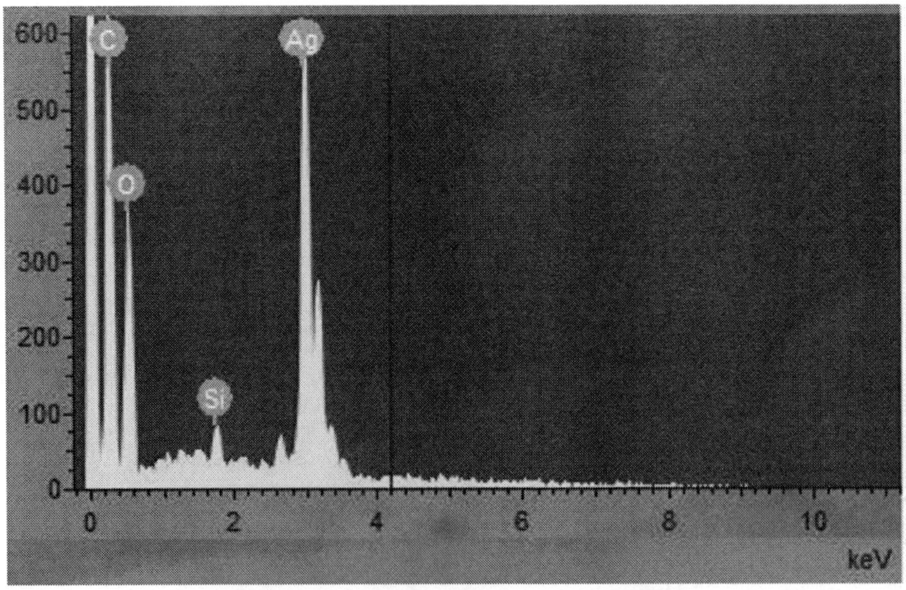

Figure 3. Typical EDX spectrum of AgNRs.

SEM micrographs were generated to study the size and surface morphology of AgNRs. SEM micrographs in figure 4 confirmed the formation of Ag nanorods rather Ag nanoparticles. These rods were more likely appeared in the form of nanostrips. The size and shape of the nanorods depended on the distance between solution and cathode needle. For gap of 0.5 mm between cathode and solution, the shape of nanorods was more spherical at 60% molar ratio of sucrose to $AgNO_3$ [10, 16].

Table 1. Chemical composition of as-grown AgNRs.

Element	Weight %	Atomic %
C K	0.06	36.40
O K	0.11	45.59
Si K	0.00	1.15
Ag L	0.26	16.86
Total	44	100

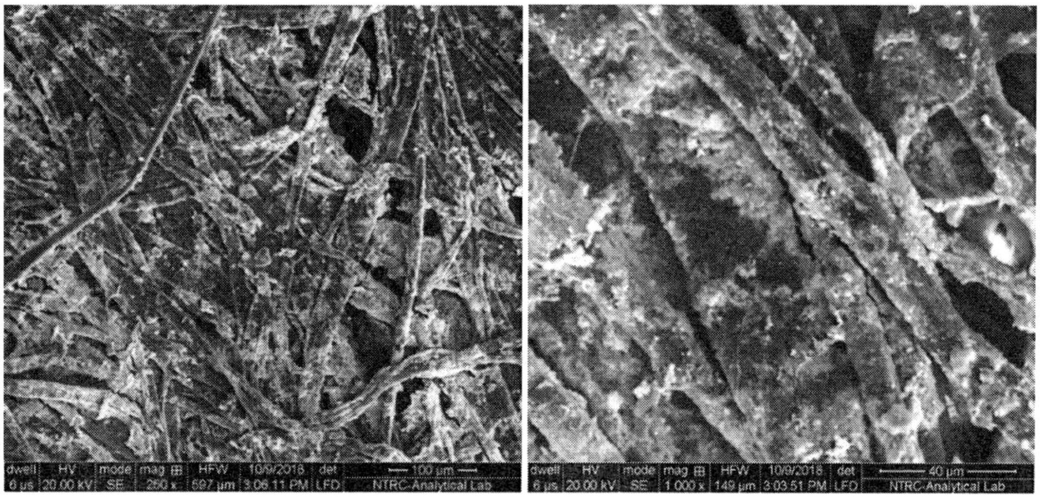

Figure 4. SEM micrographs of as-synthesized AgNRs using 20% molar ratio.

4. Conclusion

Silver nanorods were successfully synthesized through atmospheric pressure DC plasma jet of argon gas. The rod diameter remained between 5 nm and 100 nm, depending on the nozzle distance and solution molar ratio. Therefore, different molar ratios of $AgNO_3$ were used to control the size and dispersion of nanorods. The presented synthesis technique is cost-effective and environment friendly for synthesis of nanomaterials of different types.

References

[1] McKenna J, Patel J, Mitra S, Soin N, Švrček V and Maguire P 2011 *Eur. Phys. J: Appl. Phys.* **56**, 24020.
[2] Sahoo P, Kamal S, Kumar T J, Sreedhar B, Singh A and Srivastava S 2009 *Def. Sci. J.* **59** 447.
[3] Hussain J I, Kumar S, Hashmi A A and Khan Z 2011 *Adv. Mater. Lett.* **2** 188.
[4] El-Nour K M A, Eftaiha A A, Al-Warthan A and Ammar R A 2010 *Arab. J. Chem.* **3** 135.
[5] Bhaduri G A, Little R, Khomane R B, Lokhande S U, Kulkarni B D and Mendis B G 2013 *J. Photochem. Photobiol. A Chem.* **258** 1.
[6] Raveendran P, Fu J and Wallen S L 2003 *J. Am. Chem. Soc.* **125** 13940.

[7] Filippo E, Serra A, Buccolieri A and Manno D 2010 *J. Non-Cryst. Solids* **356** 344 [8] Hamedi S, Ghaseminezhad S. M, Shojaosadati S. A and Shokrollahzadeh S 2012.

[9] Huang X Z, Zhong X X, Lu Y, Li Y S, Rider A E and Furman S A 2013 *Nanotechnology* **24** 095604.

[10] Bisht A, Roshan D and Ilyas U 2008 *International Conference on Frontiers of Plasma Physics and Technology,* Bangkok,Thailand, 5-9 March 2007.

[11] Wang R, Zuo S, Zhu W, Wu S, Nian W and Zhang J 2014 *Plasma Process. Polym.* **11** 44.

[12] Mariotti D, Patel J, Švrček V and Maguire P 2012 *Plasma Process. Polym.* **9** 1074.

[13] Patel J, Němcová L, Maguire P, Graham W and Mariotti D 2013 *Nanotechnology* **24** 245604.

[14] Furusho H, Kitano K, Hamaguchi S and Nagasaki 2009 *Chem. Mater.* **21** 3526.

[15] Richmonds C and Sankaran R. M 2008 *Appl. Phys. Lett* **93** L 131501.

[16] Thong Y L, Chin O H, Ong B H and Huang N M 2015 *Jpn. J. Appl. Phys.* **55** 01AE19.

IOP Conf. Series: Materials Science and Engineering **863** (2020) 012031 doi:10.1088/1757-899X/863/1/012031

Effect of DC micro-plasma treatment on bacteria in its vegetative bacteria

M Naeem[1], W Bashir[1], N U H Altaf[1,*], M Y Naz[1], S Shukrullah[1], S U Rehman[2] and M Shahid[3]

[1] Department of Physics, University of Agriculture Faisalabad, 38040 Faisalabad, Pakistan.
[2] Institute of Microbiology, University of Agriculture Faisalabad, 38040 Faisalabad, Pakistan.
[3] Department of Biochemistry, University of Agriculture Faisalabad, 38040 Faisalabad, Pakistan.

*E-mail: noorulhuda100@yahoo.com

Abstract. Argon gas plasma consists of discharge pattern of micro hollow cathode. The pattern is used to inactivate the two main kinds of bacteria. The Culture of Escherichia Coli (gram positive) and Staphylococcus aureus (gram negative) are subjected to the plasma exposure. The micro-plasma jet (MPJ) discharge was generated in the voltage range of 2-10 kV with an operating current of 20-35 mA. Under such conditions, a MPJ of approximately 1 cm visible length was produced. The power efficiency of this device was approximately 80%. The efficacy of MPJ to inactivate the bacteria is determined by observing colony forming unit (CFU) counts before and after plasma exposure. The Colony counter is used here to find the CFU/plate. The effect of plasma exposure is observed by varying the time at 15, 30, 60, 90, 120, 150 and 180 seconds. An efficient reduction in CFU is observed after plasma exposure, particularly for 120 seconds. Approximate 50% decay of both cultures is observed after treatment time of 120 seconds.

1. Introduction

Plasma can be categorized into two classes, namely thermal plasma and non-equilibrium or low-temperature plasma (LTP) [1]. In the non-equilibrium plasma, electrons have higher temperatures and gas molecules have moderate temperatures without any local thermodynamic equilibrium, which renders lower temperatures to the whole system [2]. Cold plasma can be defined as "partially or wholly ionized gas carrying myriads of highly reactive species like electrons, negative ions, positive ions, free radicals, excited or non-excited atoms and photons at around ambient temperature". At atmospheric pressure, "cold plasma" is receptive to in-stabilities. By confining such plasmas to small dimensions has been given the reliability of atmospheric-pressure air plasma. Non-thermal micro plasma in DC discharges are generally created in closed discharge vessels using internal electrodes at relatively high potentials. These energies dissociate the gas molecules into a spark of ions, electrons, free radicals, charged/neutral gas molecules and other species which are active biologically and have potential to destroy microorganisms [3]. Low temperature DC-cold atmospheric pressure plasma consists of reactive species, which has attained considerable attention in biomedical application such as in wound healing, sterilization of heat sensitive reusable medical instruments or the surface modification of biocompatible materials [4]. Most of the "micro plasma" designs, such as plasma needles [9]. The atmospheric pressure plasma jets, several corona discharges and micro hollow cathode discharges are being developed for further applications in medical.

The properties of plasma can be altered by changing current, pulsed excitation time and voltage. A considerable improvement in tooth-whitening process, utilizing MPJ, is possible as compared to the conventional clinical methods [12]. The non-thermal RF-driven plasma jets of argon gas have been tested for antimicrobial treatment of heat sensitive materials. These jets can be operated at low powers, typical ranging from 10 W to 60 W. The highest reduction factor for viable bacteria is observed after direct plasma treatment of inoculated test strips [16]. It has been demonstrated by several authors that plasma treatment is an effective method due to production of UV radiations that can reduce the micro-organism at lower temperatures and even at ambient pressure up to 1 atm [8, 16].

In this study, non-thermal micro hollow cathode discharge (MHCD) design, based on clod plasma micro jet (PMJ), generated at atmospheric pressure [13]. Two main kinds of bacteria Escherichia Coli and Staphylococcus aureus were treated with PMJ. The samples were prepared and characterized in a petri dish for plasma exposure. A considerable reduction in the growth of these species is observed after the plasma exposure. The number of CFU were examined before and after plasma treatment.

2. Materials and methods

2.1. Production of micro-plasma jet

MHCD conception is used to generate an atmospheric pressure PMJ. The said structure comprises of a cathode with a micro hollow structure and an anode of copper rod. As the operating gas is propelled through this set up and a DC source is applied, a well-defined and concise PMJ can be produced in ambient air [6]. The appliance used here was developed built in Plasma Research Lab at the Department of Physics, University of Agriculture Faisalabad.

The opening of plasma jet nozzle was 2 mm. The copper anode was merged into bacterial media and powered with an external DC source. A hollow cathode was placed just above the media. A 2-20 kV DC supply was used generate PMJ for plasma exposure of bacteria. A blast resistance of 100 kΩ was placed between the power source and cathode to avoid circuit short. Compressed Argon gas was used as plasma precursor at a flowrate of approximately 2 SLM. The discharge current was measured in the range of 20-35 mA. Under such conditions, a PMJ of approximately 1 cm visible length was produced. The power efficiency of this device is approximately 80%. The temperature of the ionized gas was around 37 °C under atmospheric pressure.

2.2. Bacteria culture

Two most important bacteria cultures, namely Escherichia Coli and Staphylococcus aureus were treated with PMJ. The bacteria were grown and characterized in the Micro-biological lab at the University of Agriculture Faisalabad, Pakistan. These suspensions were grown in petri dishes and incubated for 24 hours to obtain the proper growth of bacteria. Plasma-induced deactivation of bacteria was observed for the specific stain of E.coli and Staphylococcus aureus [11]. The bacteria count, before and after plasma treatment, was carried out through CFU count. CFU is a way of finding colony forming units, as shown in figure 1. CFU/plate recorded for each petri dish before the treatment is given in table 1. The bacterial culture was carried out at room temperature. E.coli was gram positive whereas Staphylococcus aureus was gram negative.

Table 1. Bacteria cultures preparation for treatment.

Bacteria	Culture Temperature	Gram stain
E.coli	37°C	Positive
Staphylococcus aureus	37°C	Negative

Figure 1. Colony counter for CFU count.

2.3. Plasma treatment of bacteria

To avoid contamination, the treatment with plasma is carried out on a laminar flow clean bench. The jet nozzle was place just above the bacteria suspension. The bacterial samples were treated for different periods of time, as given in table 2. The efficacy of PMJ to inactivate the bacteria is determined by observing CFU count before and after plasma exposure [13]. The cultures treated with plasma were placed in petri dishes as well as incubated and sealed in a box at 37 °C temperature nearly 20-24 hours. CFU count for each plate before and after the treatment allowed us to get information about plasma efficiency.

Table 2. A manual description of inactivation of bacteria cultures.

Exposure time (s)	CFU/plate (E.coli)	CFU/plate (Staphylococcus)
0	81	185
15	77	180
30	79	168
90	65	132
120	45	92
150	44	90
180	42	87

3. Results and discussion

Initially, the impact of air on bacteria deactivation was observed without any plasma exposure. No particular decline of microbes was observed due to air flow. CFU numbers were assessed for both treated and untreated area of the petri dish. Due to the various coatings and layers, surrounding the genetic nucleus, spores were harder to inactivate [5]. The findings are reported in figure 2. All the microbes deactivated in the region exposed to plasma. With the passage of time, the bacteria within the unexposed parts also deactivate as shown in figure 2 (b).

While observing CFU for said two bacteria cultures, there was reasonable decrease in CFU count after plasma exposure. The plasma exposure is studied for cultures by using six different treatment time intervals ranging from 15 to 180 seconds. For smaller treatment times, E.coli CFU count decreases in small ratio but for treatment time of 120 seconds, the CFU count falls from 81 to 45. Similarly, Staphylococcus aureus CFU count decreased from 185 to 92 after treatment time of 120 seconds. There was a gradual decrease in CFU count of both types of bacterial below the exposure time of 120 seconds. A noticeable decay of CFU count was observed at the exposure time of 120 seconds [7]. It reveals that PMJ is effective technique for reducing CFU of bacteria if the exposure time is properly tuned. Table 2 provides information on CFU count over plasma exposure time. After 180 seconds CFU count of E.coli and Staphylococcus aureus was decreased to 42 and 87, respectively.

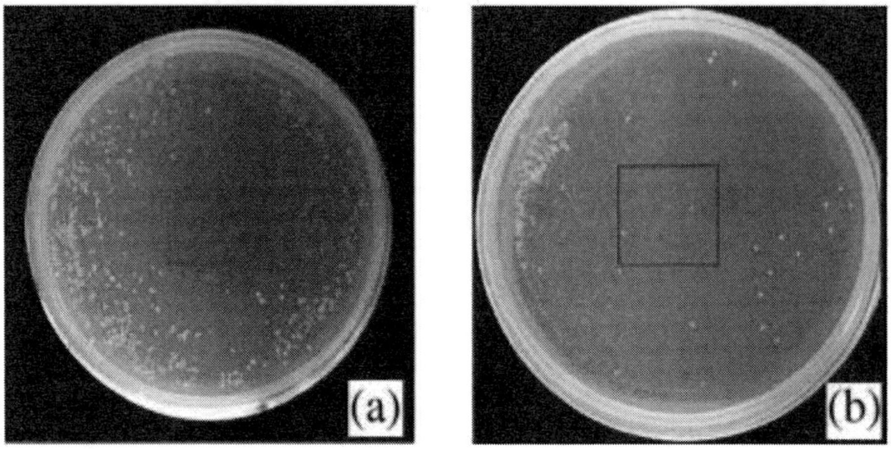

Figure 2. Scattering configurations of remaining CFUs on the petri dish after PMJ treatment for 180 s: (a) S. aureus and (b) E.coli.

Although CFU counting was carried out using a colony counter, the pictures of CFU in petri dishes can be taken by using a high resolving camera. Picture sin figure 3 reveal a gradual decrease in CFU count with increasing the treatment time [14, 15].

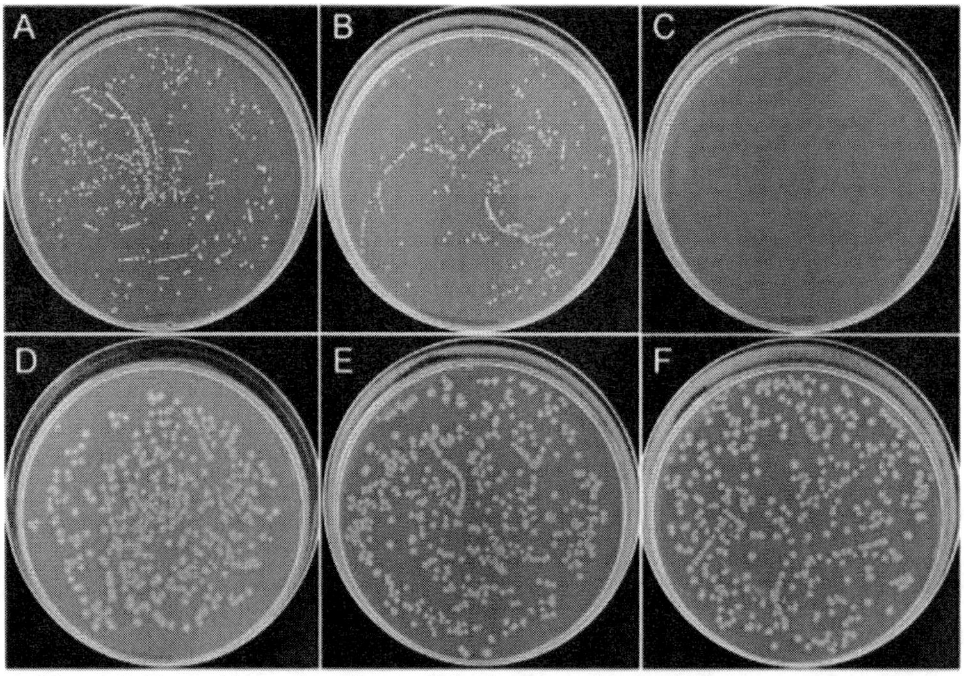

Figure 3. Treated bacteria cultures of the S. aureus for the different treatment time.

The bottom row in figure 3 shows the photographs of untreated bacteria and the top row shows the photographs of corresponding treated bacteria. The graphical representation of the observed response of CFU to the plasma exposure is given in figure 4 and figure 5. These plots clearly show a decreasing trend of CFU count over exposure time. For both bacteria, a sudden decrease in CFU count is observed after plasma exposure time of 120 seconds.

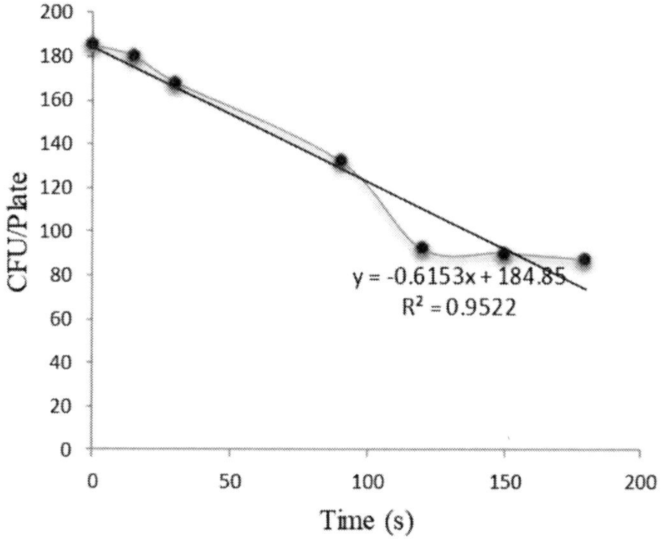

Figure 4. A graphical decay in CFU count of Staphylococcus aureus.

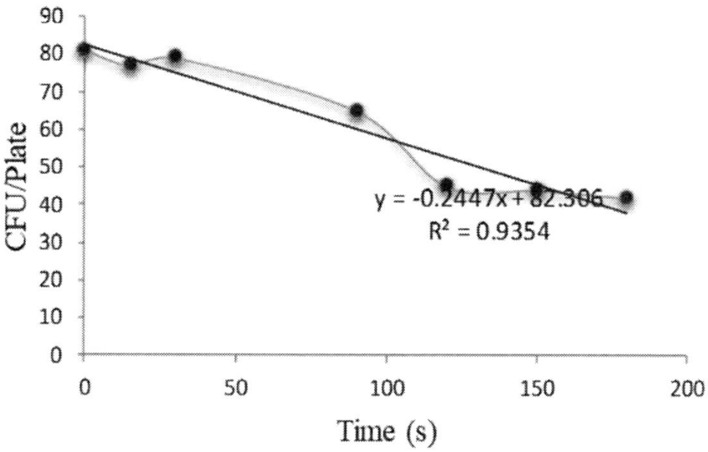

Figure 5. A graphical decay in CFU count of E.coli.

4. Conclusion

A dc atmospheric Argon gas cold plasma micro jet was used to deactivate two types of bacteria in their vegetative state. The presented plasma treatment model proved as an effective method for inactivation of bacteria in their vegetative state. Two main classes of bacteria culturesn namely E.coli and Staphylococcus aureus were treated with plasma to obtain the clear effect of plasma exposure on microbes. E.coli CFU count decreased from 81 to 45 after treatment time of 120 seconds. Similarly, Staphylococcus aureus CFU count decreased from 185 to 92 after same treatment time. A noticeable decay of CFU count was observed at the exposure time of 120 seconds. After 180 seconds, CFU count of E.coli and Staphylococcus aureus was decreased to 42 and 87, respectively. It reveals that PMJ is an effective technique for reducing CFU of bacteria if the exposure time is properly tuned.

References

[1] Conrads H and Schmidt M 2000 *Plasma Sourc. Sci. Tech.* 9 441.
[2] Haertel B, Von Woedtke T, Weltmann K D and Lindequist U 2014 *Biomol. Ther* **22** 477.
[3] Moselhy M, Petzenhauser I, Frank K and Schoenbach K H 2003 *J. Phys. D: Appl. Phys.* **36.**

[4] Ehlbeck J, Schnabel U, Polak M, Winter J, Von Woedtke T, Brandenburg R and Weltmann K D 2010 *J. Phys. D: Appl. Phys.* **44** 013002.
[5] Dobrynin, D, Fridman G, Mukhin Y V, Wynosky-Dolfi M A, Rieger J, Rest R F and Fridman A 2010 *IEEE T. Plasma Sci.* **38** 1878.
[6] Deng X L, Nikiforov A Y, Vanraes P and Leys C 2013 *J. Appl. Phys.* **113** 023305.
[7] Feng H, Sun P, Chai Y, Tong G, Zhang J, Zhu W and Fang J 2009 *IEEE T. Plasma Sci.* **37** 121
[8] Laroussi M and Leipold F 2004 *Int. J. Mass Spectrom.* **233)** 81.
[9] Laroussi M, Tendero C, Lu X, Alla S and Hynes W L 2006 *Plasma Process. Polym.* **3** 470.
[10] Shen J, Sun Q, Zhang Z, Cheng C, Lan Y, Zhang H and Chu P K 2015 *Plasma Process. Polym.* **12** 252.
[11] Liu F, Sun P, Bai N, Tian Y, Zhou H, Wei S and Fang J 2010 *Plasma Process. Polym.* **7** 231.
[12] Sun, P, Pan J, Tian Y, Bai N, Wu H , Wang L and Fang J 2010 *IEEE T. Plasma Sci.* **38** 1892
[13] Hsu D D and Graves D B 2003 *J. Phys. D: Appl. Phys.* **36** 2898.
[14] Shen J, Sun Q, Zhang Z, Cheng C, Lan Y, Zhang H and Chu P K 2015 *Plasma Process. Polym.* **12** 252.
[15] Brugger S D, Baumberger C, Jost M, Jenni W, Brugger U and Mühlemann K 2012 *Plos One* **7** e33695.
[16] Brandenburg R, Ehlbeck J, Stieber M, Woedtke T v, Zeymer J, Schlüter O and Weltmann K D 2007 *Contrib. Plasma Phys.* **47** 72.

IOP Conf. Series: Materials Science and Engineering **863** (2020) 012032 doi:10.1088/1757-899X/863/1/012032

Performance assessment of stratified chilled water thermal energy storage tank at district cooling plant

M A A Majid*, P S Hui and A A Soomro

Department of Mechanical Engineering, Universiti Teknologi PETRONAS, 32610 Bandar Seri Iskandar, Perak, Malaysia.

*E-mail: mamin_amajid@utp.edu.my

Abstract. Thermal energy storage (TES) is the key component of the district cooling (DC) plants. Its performance is important to be analysed. Various works have been carried out to analyse the TES tank performance. Among the methods, the most common are thermocline thickness and figure of merit. In this study nonlinear regression is used to analyse the temperature profiles of a stratified TES tank at DC plant at Universiti Teknologi PETRONAS. The TES performance was also evaluated using thermocline thickness and 1/2FOM. Results indicate curve fitting approach is enabled to plot temperature profiles during charging. The plots assist in understanding the development of thermocline within the TES tank during charging process. Results on the thermocline thickness and 1/2FOM during the understudy charging period indicate increased of thermocline thickness and decreased of 1/2FOM values respectively during the study period. Hence it can be concluded that deterioration occurred to the TES tank during the study period.

1. Introduction

Recently, the stratified thermal energy storage (STES) is getting more popular due to high energy demand. STES is the main component of district cooling (DC) plant and is used to store the cooling capacity during off-peak hours and during on-peak it is used to support the chilled water requirements to the buildings for space cooling [1]. The medium of the storage is water, due to its ease of availability, economic nature and high specific heat [2]. The volume of water having low temperature settle down and the volume of water having high temperature settle on top of the cold water due to density difference [3-5], separated by a transition layer known as thermocline (WTc) [3, 6] as shown in figure 1. Thermocline thickness is one of the accepted performance measures, it is a transition region of the temperature distribution [7]. Thicker thermocline indicates low performance whereas thinner thermocline is sign of high performance [2]. Various numerical and analytical approaches has been proposed and applied to evaluate the performance of TES tank in terms of thermocline thickness [8]. Among the major contributions of performance evaluation of TES tank have been reported by Yoo [8], Homan [9], and Musser and Behnfleth [10]. The definition of thermocline thickness proposed by Mussser and Behnfleth [10] remained the most dominant and widely applied. However, this approach still has one drawback, that is, in case of discrete temperature data available this method fails to predict the thermocline thickness accurately [11]. To overcome the issue of fail to predict the thermocline thickness accurately, this paper introduces non-linear regression [11] approach in order to evaluate thermocline thickness.

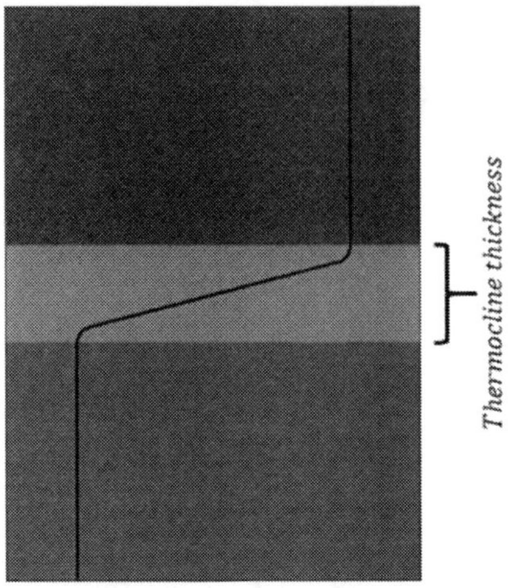

Figure 1. Schematic diagram of stratified thermal energy storage tank [3, 12].

2. Methodology

In this study, curve fitting method is used which representing mathematical relation between dependent and independent variable for a given set of data. The advantage of using curve fitting is that it enables visualization of data characteristics, to obtain important parameters and to summarize the relationships among variables [11, 13]. Conceptually, fitting a mathematical model on data set is to establish equation that defines the dependent variable as a function of an independent variable with one or more parameters. Temperature data of the TES system of Universiti Teknologi PETRONAS DC plant were acquired for the study, the TES system consists of two 1250 refrigeration ton (RT) of steam absorption chiller and four 325 RT electric chillers and one 54000 m^3 storage of TES tank with designed capacity of 10000 RTh. The electric chillers were used to charge the TES during night time. The inlet nozzle of the TES tank is made from 20 inch NPS located at elevation 3.4 meters height, while outlet nozzle is 12 inch NPS at elevation of 12.3 meters. Both nozzles are provided with diffuser on its end-connection in TES tank. Overflow line is connected at elevation of 14.025 meters. The entire tank is extensively insulated. The tank is equipped with 14 temperature sensors, installed at approximately 1 meter vertical interval, to measure the water temperature. The lowest temperature sensor is located at 0.51 meters height. All temperatures are hourly recorded with acquisition data system. Table 1 tabulates design specification of the TES tank. Meanwhile, table 2 shows the example of temperature readings acquired from TES tank by resistance temperature detector (RTD) vertically installed in the tank.

Table 1. Specifications of the TES tank.

No	Parameter	Value	Unit
1	Height	15	m
2	Diameter	22.3	m
3	Volume	5400	m^3
4	Mass flow rate	393	m^3/h
4	Material	Steel	-
5	Numbers of RTD	14	Nos

Table 2. Example of hourly data used as inputs.

		Temperature in each sensor data in °C													
Date	Hour	T1	T2	T3	T4	T5	T6	T7	T8	T9	T10	T11	T12	T13	T14
11-01-2016	18:00	14.3	14.3	14.2	14.2	14.1	14.1	13.8	13.6	13.1	9.7	8.0	7.7	7.8	7.5
	19:00	14.3	14.2	14.2	14.1	14.1	13.9	13.5	13.0	9.4	8.0	8.0	7.7	7.9	7.6
	20:00	14.2	14.2	14.1	14.1	13.9	13.6	12.8	9.2	8.0	7.8	7.9	7.7	7.9	7.7
	21:00	14.2	14.1	14.1	13.9	13.5	12.8	8.8	7.9	7.9	7.8	7.9	7.7	7.9	7.8
	22:00	14.1	14.1	13.9	13.6	12.6	8.7	7.7	7.8	7.9	7.8	7.9	7.8	8.0	7.8
	23:00	14.1	13.9	13.5	12.5	8.3	7.9	7.7	7.8	7.9	7.8	7.9	7.8	8.0	7.9
12-01-2016	00:00	13.8	13.5	12.0	8.3	7.9	7.8	7.7	7.8	7.9	7.8	7.9	7.8	8.0	7.9
	00:01	13.4	11.5	8.0	7.9	7.8	7.8	7.7	7.8	7.9	7.8	7.9	7.8	8.0	7.9
	00:02	10.5	8.1	7.9	7.9	7.8	7.8	7.7	7.8	7.9	7.8	7.9	7.8	8.0	7.8
	00:03	7.9	7.9	7.9	7.9	7.8	7.8	7.7	7.8	7.9	7.8	7.8	7.5	7.7	7.3
	00:04	7.9	7.9	7.9	7.8	7.8	7.8	7.7	7.8	7.8	7.6	7.4	7.0	7.1	7.0
	00:05	7.9	7.9	7.8	7.8	7.8	7.8	7.7	7.8	7.5	7.0	6.9	6.7	6.8	6.7
	00:06	7.9	7.9	7.8	7.8	7.8	7.8	7.6	7.3	6.9	6.7	6.7	6.5	6.7	6.6

Temperature distribution is an important parameter in performance analyses of TES tanks [14]. From the past researches it is confirmed that the temperature distribution in stratified TES tanks having S-curve profile [8-10]. In this study, the temperature distribution function is considered as the function of the average cold water (T_c), average hot water (T_h), position of thermocline (C) slope of the temperature gradient (S) and sensor elevation x as shown by equation (1). Since, the data follows the S-curve, the function from S-curve family i.e sigmoid function was chosen to fit the data as shown by equation (2).

$$T(x) = f(T_c, T_h, C, S, x) \tag{1}$$

$$T = T_c + \frac{T_h - T_c}{1 + 10^{(C-X)S}} \tag{2}$$

Formulation of thermocline thickness using SDR function was obtained by rearranging equation (2), as shown by equation (3).

$$\frac{T_h - T_c}{T - T_c} = 1 + 10^{(C-X)S} \tag{3}$$

The left term of equation term of equation (3) is re-arranged using dimension less cut-off temperature $\theta = \frac{T - T_c}{T_h - T_c}$ applying the approach of Musser (1998), describing the limit points of thermocline thickness.

$$\frac{1}{\Theta} = 1 + 10^{(C-X)S} \tag{4}$$

Distance from C to X express the half-thickness of the thermocline, and represented as follows;

$$C - X = \frac{\log(\frac{1}{\Theta} - 1)}{S} \tag{5}$$

Therefore thermocline thickness is defined as:

$$W_{TC} = \frac{Log(\frac{1}{\Theta} - 1)}{S} \tag{6}$$

Using the similar analysis, thermocline thickness for 4 parameters sigmoid (FPS) function is determined as:

$$W_{TC} = 2Ln(\frac{1}{\Theta} - 1).S \tag{7}$$

3. Results and discussion

Using historical data of Jan 2016 and Jan 2017 respectively, the data were used to plot temperature profiles for the case of without curve fitting and with curve fitting for three consecutive days in Jan 2016 and 2017. The historical data selected were 28th to 30th Jan for year 2016 and 2017 respectively. The plots were drawn using commercial software GraphPad prism [15]. The temperature profiles using curve fitting indicate clearly that the thermocline for the respective dates compared to the temperature profiles without curve fitting. This assists in visualization the development of thermocline during the charging stage. The charging period being studied was from 6 pm to 3 am. The plots as shown by figure 2(a) to figure 3(b) clearly indicate the stratification process occurred during charging. However, until 3 am the tank still did not achieve full charge. Figure 4(a) and (b) show plot of TES holding capacity (thermal energy status) status at different times during charging for 2016 and 2017.

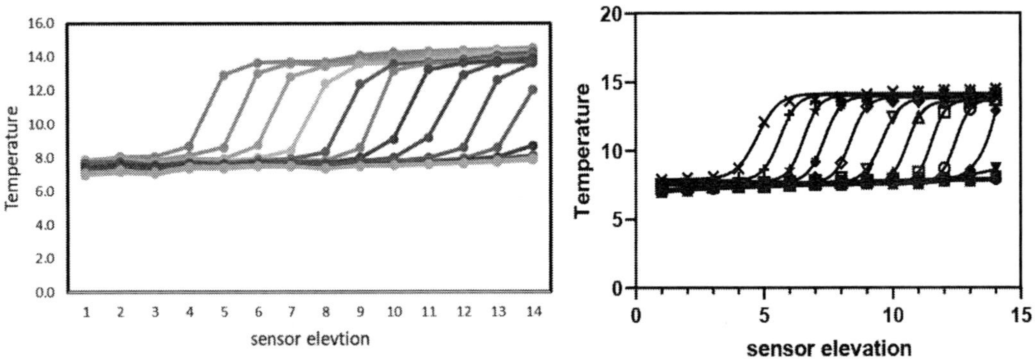

(a) (b)

Figure 2. Temperature profile for 28 Jan 2016 (a) without curve fitting, and (b) with curve fitting.

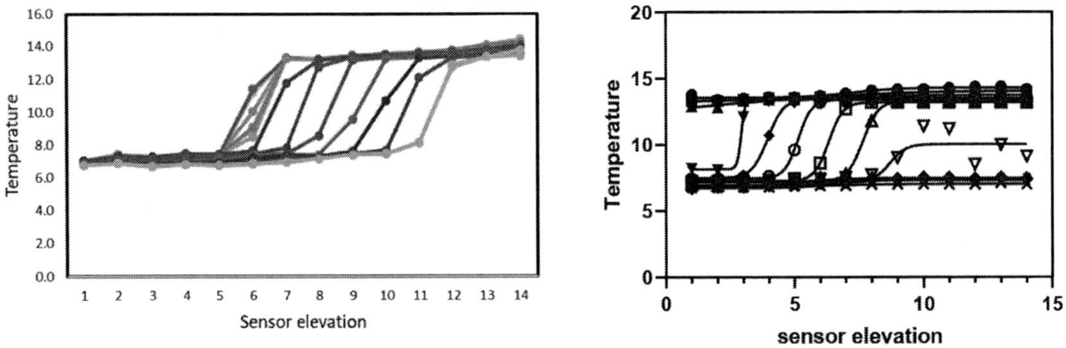

(a) (b)

Figure 3. Temperature profile for 28 Jan 2017 (a) without curve fitting (b) with curve fitting.

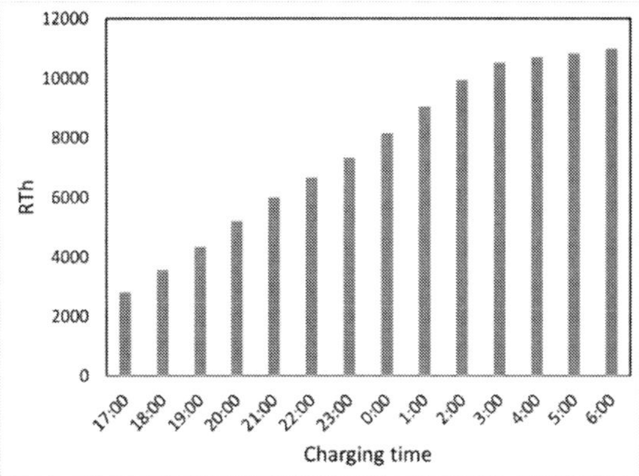

(a)

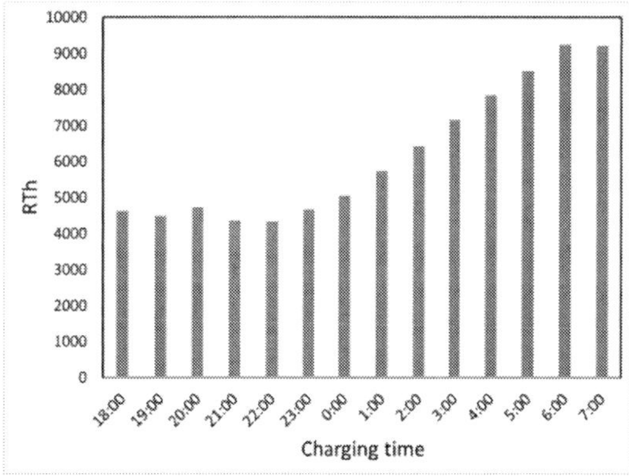

(b)

Figure 4. RTh vs charging time on (a) 28-29 Jan 2016 and (b) 29-30 Jan 2017.

Table 3 and table 4 indicate the thermocline thickness for month of Jan 2016 and 2017 respectively. Table 5 and 6 show the half cycle figure of merit (1/2FOM) for month of Jan 2016 and 2017. Figure 5(a) and (b) are the plots for the thermocline thickness for year 2016 and 2017 respectively. On the average for Jan 2016 the thermocline thickness varies from 2.929 to 4.809 meters. While for Jan 2017 the thermocline thickness varies from 2.338 to 9.076 meters. Year 2016 shows more stable results compared with year 2017. The average daily thermocline thickness was calculated by omitting the thermocline thickness value marked with '*'. This were done as the values seem to be out of bound. The evaluated 1/2FOM are included in table 5 and table 6 for 2016 and 2017 respectively. The daily average for 10 days during 2016 varies from 0.95 to 0.98. While for year 2017 the average varies from 0.82 to 0.98. The overall 10 day average for 2016 and 2017 are 0.97 and 0.94 respectively. The decrease in values of 1/2FOM proves deterioration of tank occurred.

The plot of thermocline thickness for year 2016 and 2017 are shown in figure 5(a) and (b) respectively. The 10 days over all average of thermocline thickness for 2016 is 4.192 meters and for 2017 is 4.655 meters. This indicates the performance of tank has deteriorated. It is also noted that the thermocline thickness and the behaviour during charging is similar as noted in the study of Musser [16].

Table 3. Tabulation of daily maximum thermocline thickness for year 2016.

Daily Maximum Thermocline Thickness (m)										
Time	Date									
	21-Jan	22-Jan	23-Jan	24-Jan	25-Jan	26-Jan	27-Jan	28-Jan	29-Jan	30-Jan
18:00	5.058	4.859	2.915	3.152	4.907	2.492	4.559	4.387	2.762	4.344
19:00	3.999	4.212	2.985	2.365	3.274	4.188	4.196	4.000	2.583	3.877
20:00	3.858	4.641	3.609	4.019	3.852	3.311	4.415	3.338	2.988	3.361
21:00	3.689	4.601	3.173	6.712	3.931	3.252	3.577	3.641	2.967	3.087
22:00	3.390	4.434	3.265	23.57*	4.361	5.619	3.317	3.585	2.867	2.766
23:00	3.786	4.519	2.452	18.36*	3.339	38.96*	2.664	2.827	2.392	2.922
0:00	3.725	4.882	2.376	14.63*	4.636	47.17*	3.495	3.559	2.669	1.765
1:00	4.360	4.725	3.281	48.03*	4.079	42.83*	3.457	3.117	3.189	2.355
2:00	5.228	6.403	2.750	28.38*	4.968	33.02*	4.392	3.721	2.890	2.345
3:00	19.50*	17.11*	3.258	30.52*	18.87*	45.53*	9.154	4.063	3.983	2.512
Average	4.121	4.809	3.006	4.062	4.150	3.772	4.323	3.624	2.929	2.934

* represents out of bound value

Table 4. Tabulation of daily maximum thermocline thickness for year 2017.

Time	Daily Maximum Thermocline Thickness (m) Date									
	21-Jan	22-Jan	23-Jan	24-Jan	25-Jan	26-Jan	27-Jan	28-Jan	29-Jan	30-Jan
18:00	3.31	11.1	12.86	3.85	2.22	2.66	3.50	2.87	2.68	4.12
19:00	3.10	10.8	13.42	3.50	2.25	3.28	3.06	2.36	3.15	5.28
20:00	3.02	10.0	12.94	2.21	2.31	2.95	3.50	2.34	2.32	4.32
21:00	2.93	11.2	11.51	6.88	2.25	2.27	2.76	2.32	2.80	3.48
22:00	2.03	9.59	9.29	49.40*	2.27	1.72	1.85	3.18	2.57	3.00
23:00	2.82	8.68	6.92	3.38	2.33	1.95	1.94	2.18	2.59	3.04
00:00	2.31	8.75	4.71	5.77	4.58	1.86	4.99	2.49	2.70	4.82
01:00	2.54	6.89	4.78	7.06	2.49	1.98	102.5*	3.20	2.25	3.87
02:00	2.06	6.85	4.46	2.51	35.7*	221.90*	3.65	2.60	2.33	4.95
03:00	17.90*	6.77	4.51	3.86	35.7*	23.24*	0.514	2.94	2.84	4.81
Average	2.685	9.076	8.544	4.340	2.592	2.338	2.866	2.652	2.627	4.171

* represents out of bound value.

Table 5. Tabulation of comparison of half cycle figure of merit (1/2FOM) for year 2016 and year

Time	Date											
	21 Jan	22 Jan	23 Jan	24 Jan	25 Jan	26 Jan	27 Jan	28 Jan	29 Jan	30 Jan	Average	Average of 10 days
18:00	0.90	0.94	0.95	0.96	0.97	0.97	0.97	0.97	0.97	0.91*	0.96	
19:00	0.89	0.93	0.94	0.95	0.96	0.96	0.97	0.97	0.96	0.92*	0.95	
20:00	0.95	0.96	0.96	0.97	0.97	0.98	0.98	0.98	0.98	0.98	0.97	
21:00	0.98	0.98	0.97	0.96	0.89*	0.91*	0.93*	0.72*	0.75*	0.74*	0.97	
22:00	0.95	0.97	0.97	0.97	0.97	0.98	0.97	0.98	0.97	0.92*	0.97	0.97
23:00	0.98	0.97	0.98	0.98	0.97	0.74*	0.65*	0.59*	0.62*	0.63*	0.97	
00:00	0.92	0.94	0.95	0.97	0.97	0.98	0.98	0.98	0.98	0.95	0.96	
00:01	0.90	0.93	0.95	0.96	0.96	0.97	0.97	0.98	0.97	0.97	0.96	
00:02	0.95	0.96	0.96	0.97	0.97	0.98	0.98	0.98	0.98	0.98	0.97	
00:03	0.95	0.96	0.97	0.98	0.98	0.98	0.99	0.99	0.99	0.99	0.98	

2017.

* represents out of bound value.

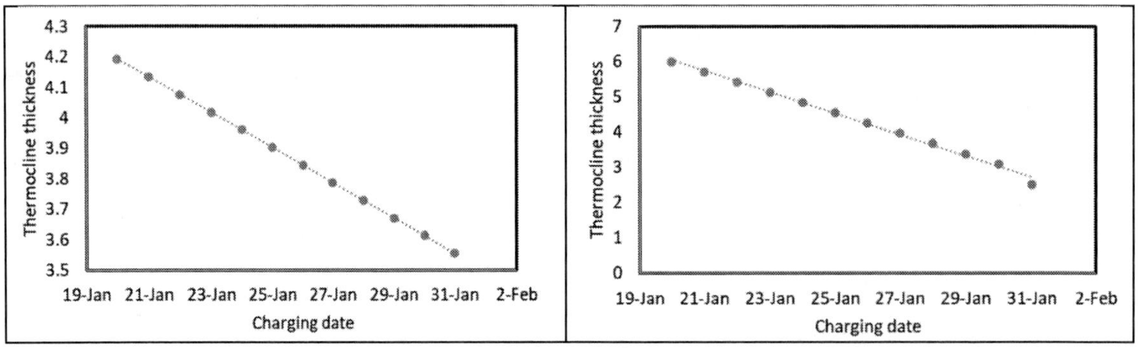

Figure 5. Average Daily Maximum Thermocline Thickness for (a) year 2016 and (b) year 2017.

Table 6. Tabulation of comparison of half cycle figure of merit (1/2FOM) for year 2016 and year 2017.

Time	Date											
	21 Jan	22 Jan	23 Jan	24 Jan	25 Jan	26 Jan	27 Jan	28 Jan	29 Jan	30 Jan	Average	Average of 10 days
18:00	0.96	0.97	0.98	0.98	0.99	0.98	0.99	0.99	0.99	0.92*	0.98	0.94
19:00	0.74	0.76	0.77	0.76	0.81	0.83	0.85	0.90	0.91	0.92	0.82	
20:00	0.81	0.80	0.82	0.86	0.90	0.93	0.96	0.96	0.96	0.97	0.90	
21:00	0.97	0.98	0.99	0.96	0.85*	0.75	0.75	0.78	0.94	0.93	0.89	
22:00	0.99	0.99	0.99	0.99	0.99	0.99	0.97	0.99	0.88*	0.88*	0.98	
23:00	0.98	0.98	0.98	0.99	0.99	0.99	0.99	0.99	0.72*	0.60*	0.99	
00:00	0.97	0.98	0.98	0.98	0.99	0.99	0.97	0.65*	0.90	0.99	0.97	
00:01	0.98	0.98	0.98	0.98	0.98	0.98	0.98	0.98	0.98	0.98	0.98	
00:02	0.96	0.96	0.97	0.96	0.96	0.97	0.97	0.98	0.98	0.98	0.97	
00:03	0.96	0.94	0.94	0.95	0.96	0.96	0.93	0.95	0.95	0.95	0.95	

* represents out of bound value.

4. Conclusion

The use of curve fitting assists in visualization of temperature profiles during charging which clearly indicate thermocline development at various stages of charging period. While from the evaluated values of thermocline thickness and 1/2FOM for year 2016 and 2017 indicates the decline in the value of thermocline from 4.19 to 4.65 meters average also 1/2FOM decreased from 0.97 to 0.94 thus deterioration of performance of the TES tank. Hence, maintenance of the tank is required in order to improve the tank performance. Due to escalating gas price the DC plant is current in the process of replacing the two steam absorption chillers with two 1250RT electric chillers and an additional thermal energy storage of same capacity as the current TES i.e. 10,000RTh.

Acknowledgments

The authors would like to acknowledge the support of Department of Mechanical Engineering at Universiti Teknologi PETRONAS for providing access to laboratories for the study. The authors would also like to acknowledge the financial support of YUTP (YUTP_0153AA-E51) for this study.

References

[1] I. Dincer, "On thermal energy storage systems and applications in buildings," *Energy and buildings,* vol. 34, pp. 377-388, 2002.

[2] I. Dincer and M. Rosen, *Thermal energy storage: systems and applications*: John Wiley & Sons, 2002.

[3] A. Karim, A. Burnett, and S. Fawzia, "Investigation of stratified thermal storage tank performance for heating and cooling applications," *Energies,* vol. 11, pp. Article number-1049, 2018.

[4] M. Karim, "Performance evaluation of a stratified chilled-water thermal storage system," *World Academy of Science, Engineering and Technology,* vol. 53, pp. 326-334, 2009.

[5] O. Abdelhak, H. Mhiri, and P. Bournot, "CFD analysis of thermal stratification in domestic hot water storage tank during dynamic mode," in *Building Simulation*, 2015, pp. 421-429.

[6] A. A. Soomro and A. A. Mokhtar, "Prediction of performance parameters of stratified TES tank using artificial neural network," in *AIP Conference Proceedings*, 2018, p. 080001.

[7] W. P. Bahnfleth, J. Song, and J. M. Cimbala, "Measured and modeled charging of a stratified chilled water thermal storage tank with slotted pipe diffusers," *HVAC&R Research,* vol. 9, pp. 467-491, 2003.

[8] J. Yoo, M. Wildin, and C. Truman, "Initial formation of a thermocline in stratified thermal storage tanks," *ASHRAE Trans.;(United States),* vol. 92, 1986.

[9] K. Homan, C. W. Sohn, and S. Soo, "Thermal performance of stratified chilled water storage tanks," *HVAC&R Research,* vol. 2, pp. 158-169, 1996.

[10] A. Musser and W. P. Bahnfleth, "Evolution of temperature distributions in a full-scale stratified chilled-water storage tank with radial diffusers," *ASHRAE Transactions,* vol. 104, p. 55, 1998.

[11] J. Waluyo, M. A. A. Majid, and M. Amin, "Temperature Profile and Thermocline Thickness Evaluation of a Stratified Thermal Energy Storage Tank," *Int J Mech Mechatron Eng,* vol. 1, pp. 7-12, 2010.

[12] A. A. Soomro, A. A. Mokhtar, A. Akbar, and A. Abbasi, "Modelling Techniques Used in The Analysis of Stratified Thermal Energy Storage: A Review," in *MATEC Web of Conferences*, 2018, p. 01015.

[13] S. V. Archontoulis and F. E. Miguez, "Nonlinear regression models and applications in agricultural research," *Agronomy Journal,* vol. 107, pp. 786-798, 2015.

[14] E. M. Kleinbach, W. Beckman, and S. Klein, "Performance study of one-dimensional models for stratified thermal storage tanks," *Solar energy,* vol. 50, pp. 155-166, 1993.

[15] H. Motulsky, "Prism 5 statistics guide, 2007," *GraphPad Software,* vol. 31, pp. 39-42, 2007.

[16] W. P. Bahnfleth and A. Musser, "Thermal performance of a full-scale stratified chilled-water thermal storage tank," *TRANSACTIONS-AMERICAN SOCIETY OF HEATING REFRIGERATING AND AIR CONDITIONING ENGINEERS,* vol. 104, pp. 377-388, 1998.

IOP Conf. Series: Materials Science and Engineering **863** (2020) 012033 doi:10.1088/1757-899X/863/1/012033

Investigation of the effect of annealing temperature on photophysical properties of manganese dioxide nanostructure prepared via co-precipitation route

M S Iqbal, M Iqbal, M Haseeb, M Aftab, A Amin and H Anwar*

Advanced Nanomaterials and Devices Laboratory, Department of Physics, University of Agriculture Faisalabad, 38040 Faisalabad, Pakistan.

*E-mail: Hafeez.anwar@gmail.com.

Abstract. Co-precipitation method was used to synthesize Manganese dioxide (MnO_2) nanostructures. The synthetization of Manganese dioxide (MnO_2) requires Manganese dichloride tetrahydrate ($MnCl_2.4H_2O$), Sodium hydroxide (NaOH) and distilled water. Manganese dioxide nanostructures were synthesized by taking suitable amount of manganese dichloride tetrahydrate and mixing it with 100 mL of distilled water. Sodium hydroxide (NaOH) was added into the precursor solution to adjust pH 12 and stirred for 18 hours. The prepared MnO_2 nanoparticles were characterized through various techniques such as X-ray diffraction (XRD), UV-visible spectroscopy and scanning electron microscopy (SEM). XRD results revealed that MnO_2 had tetragonal single phase. Average crystallite size for prepared samples is found in the range 39 to 45 nm. It is increased with increasing of annealing temperature. Volume and density of unit cell are found to be 258.83 $Å^3$ and 4.522 g/cm^3 respectively. SEM results showed that the annealing temperature has affected the shape of MnO_2. The morphology of MnO_2 changed from spherical to rod-like shape as annealing temperature increase. Absorption peaks are obtained between 362 nm to 366 nm of wavelength. These Manganese dioxide (MnO_2) nanoparticles can be applied as catalysts, permeable of toxic metals, ion/molecular-sieves, component of dry cell, inorganic pigment, electrodes for electrochemical batteries, electrodes for supercapacitors and cleaning of water.

1. Introduction

In recent years, research and development is mostly being done on nanotechnology that has delivered nanostructures (particles with size ≈ 100 nm or less) having physical properties that differ from those of the bulk counterparts and depend on polymorphism, morphology, size of particles, size distribution, coating, and the precursor used in the synthesis [1]. With the need for renewable energies, these nano-substances have undergone extensive research, to develop new systems that can be used for energy storage and conversion. Among them, transition-metal oxides including TiO_2, MnO_2, V_2O_5, etc. are stable and robust materials with tunable properties offering large surface areas.

Over the last three decades nanoparticles have received an increasing amount of research interest [2] Nanoscale materials have proved to have unique properties than its bulk due to large surface to volume ratio. MnO_2 nanoparticles are one of the most attractive oxides due to their unique properties as a low band gap, high optical constant semiconductor that exhibits ferroelectric and catalytic properties. MnO_2 is natural abundant, cheaper, and low toxic nanomaterials. In Ancient times, mineral-MnO_2 (MDOs) are utilized as blackish pigment for painting the rocks in Paleolithic pothole of the

IOP Conf. Series: Materials Science and Engineering **863** (2020) 012033 doi:10.1088/1757-899X/863/1/012033

Magdalenian traditions [3]. As a kind of important metal oxide, manganese dioxide is one of the most attractive inorganic materials because of its physical and chemical properties and wide applications in catalysis [4] molecular-sieves [5] ion exchange, biosensor, and especially energy storage.

Electrical properties of MnO_2 Nano-material basic research are the key to the development of future applications in electronics. MnO_2 has a considerable gap of approximately 1.3 eV at room temperature. MnO_2 exhibits specific capacitance within the range of 50-200 F/g in aqueous electrolyte, lesser than its $C_{theoretical}$ 1370 F/g [6]. MnO_2 based supercapacitors are ultimately restricted by poor electrical conductivity of MnO_2. The incorporation of other metal elements with MnO_2 improve their electrical conductivity and charge storage capacity. This can be achieved by mixing supplementary transition metal oxides such as RuO_2, NiO, Co_3O_4 with MnO_2 to make assorted metal oxides [7].

Manganese oxides with channel and layered crystal structures constitute a large family of porous materials. Manganese oxide (MnO_2) has a number of polymorph structures. The manganese dioxide found in different crystalline structures, including α-, β-, γ-, ε-, η-, δ-, λ -MnO_2, etc [8]. The MnO_6 structural unit is connected in diverse ways. MDOs are non-stoichiometric compounds, because of inevitable structural water molecules that are physisorbed. MDOs crystallize with various morphologies and crystallographic forms [9] including the α-, β-, γ-, δ- ε-, λ- and R-polymorphs [10] which are naturally occurring minerals such as hollandite (2×2), pyrolusite (1×1), nsutite (1×1)/(1×2) with hexagonal structure, birnessite (1×∞), akhtenkite (dense stack), spinel (1×1), and ramsdellite (1×2), respectively, where ($m\times n$) denotes the tunnel dimension [11].

In this research work, Manganese oxides (MnO_2) nanostructures were prepared by using co-precipitation technique and effect of annealing temperature on their different properties such as structural, morphological and optical properties were investigated.

2. Experimental setup

2.1. Chemicals and synthesis

All chemicals used throughout the research were of analytical regent grade and purchased from sigma-Aldrich. For the preparation of Manganese Dioxide nanoparticles schematic diagram is shown in figure 1. The Manganese Dioxide nanoparticles were synthesized by precipitation method using Manganese dichloride tetrahydrate [$MnCl_2.4H_2O$] and sodium hydroxide (NaOH) as precursors the $MnCl_2.4H_2O$ and NaOH in distilled water [12]. The obtained solution was stirred continuously using the magnetic stirrer for 8 hours.

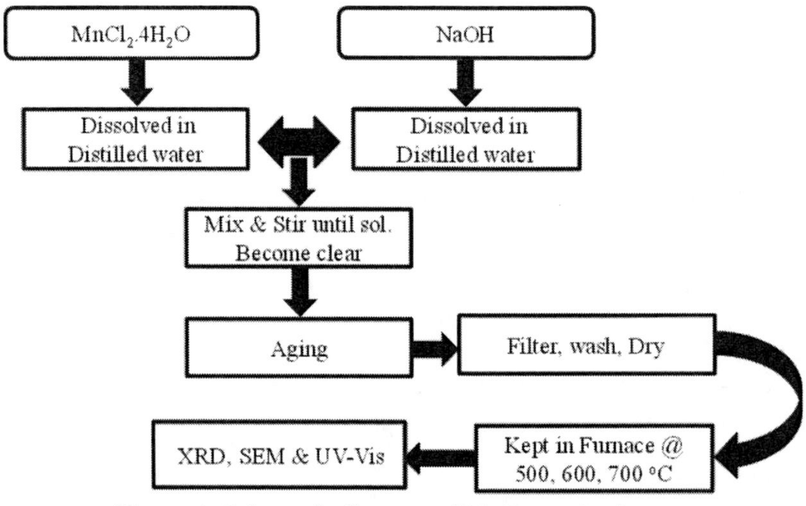

Figure 1. Schematic diagram of MnO_2 synthesis.

Sodium hydroxide solution (5 M) was added drop by drop to the $MnCl_2.4H_2O$ solution. The reaction could proceed for complete addition of sodium hydroxide solution, pH was maintained at 12.

The solution was kept overnight for aging. The solution was filtered and washed three times with distilled water to remove the by-products. The moisture free content was obtained after drying at 80 °C for 5 hours in an oven (Model: Shel-Lab). Then the grinded samples were annealed at different temperatures of 500 °C, 600 °C, and 700 °C in furnace (Model: SNOL-LHM01).

2.2. Characterization

The powder X-ray diffraction (XRD) was performed through XRD system Model (PANalytical) with Nickel filtered CuKα (λ=1.50405) radiation [13]. The particle size was calculated from broadening of line by using Scherrer's formula. The scanning electron microscope (SEM) studies were performed by using SEM Model (JEOL- JSM 5910) to study the morphology of the prepared nanoparticles.

3. Results and discussion

3.1. Structural analysis

The XRD pattern of NnO_2 nanoparticles prepared at 500 °C, 600 °C, 700 °C and untreated is shown in figure 2. The shown XRD pattern confirms the formation of nanoparticles. The MnO_2 nanoparticles are observed to be highly crystalline. Peaks at 2θ of 18.01°, 28.94°, 38.18°, 49.39°, 55.19°, 59.92°, 65.85° were referred to XRD planes (200), (310), (211), (411), (600), (521), (002). The XRD results showed that Manganese dioxide has the tetragonal structure. The average crystallite size of prepared MnO_2 nanoparticles is calculated by using Scherrer's relation as shown by equation (1).

$$L = \frac{0.9\lambda}{\beta \cos\theta} \tag{1}$$

where L is the crystal size, β is the line width (FWHM), θ is the diffraction angle and the λ is the wavelength of X-ray radiation. The lattice parameters, volume of unit cell and density of unit cell are presented in table 1. The values of lattice parameters for all the samples were calculated by using equation (2). The lattice parameter was not considerably influenced by the annealing temperature. These parameters are matching with the reported values of PDF file No. 44-0141.

$$\frac{1}{d^2} = \frac{h^2 + k^2}{a^2} + \frac{l^2}{c^2} \tag{2}$$

The volume of unit cell was calculated by using equation (3) by using the lattice parameter values from table 2. There is no significant effect of annealing temperature on the volume of unit cell.

$$V_{cell} = a^2 c \tag{3}$$

The X-ray density was calculated by using equation (4).

$$D = \frac{ZM}{N_A V_{cell}} \tag{4}$$

Where Z is the number of molecules per unit cell, M is the molar mass of the Zinc Oxide, N_A is Avogadro number, V_{cell} is the volume of the unit cell.

Table 1. Structural parameters extracted from XRD data of samples.

Sample	Lattice constant a=b (Å)	Lattice constant C (Å)	Unit cell Volume (Å)^3	Unit cell Density g/cm^3	Particle Size (nm)
A	9.98	2.687	267.62	4.3150	39.10
B	10.01	2.837	284.22	4.0630	29.25
C	9.9804	2.117	210.84	5.4772	39.75
D	9.836	2.818	272.671	4.2352	45.55

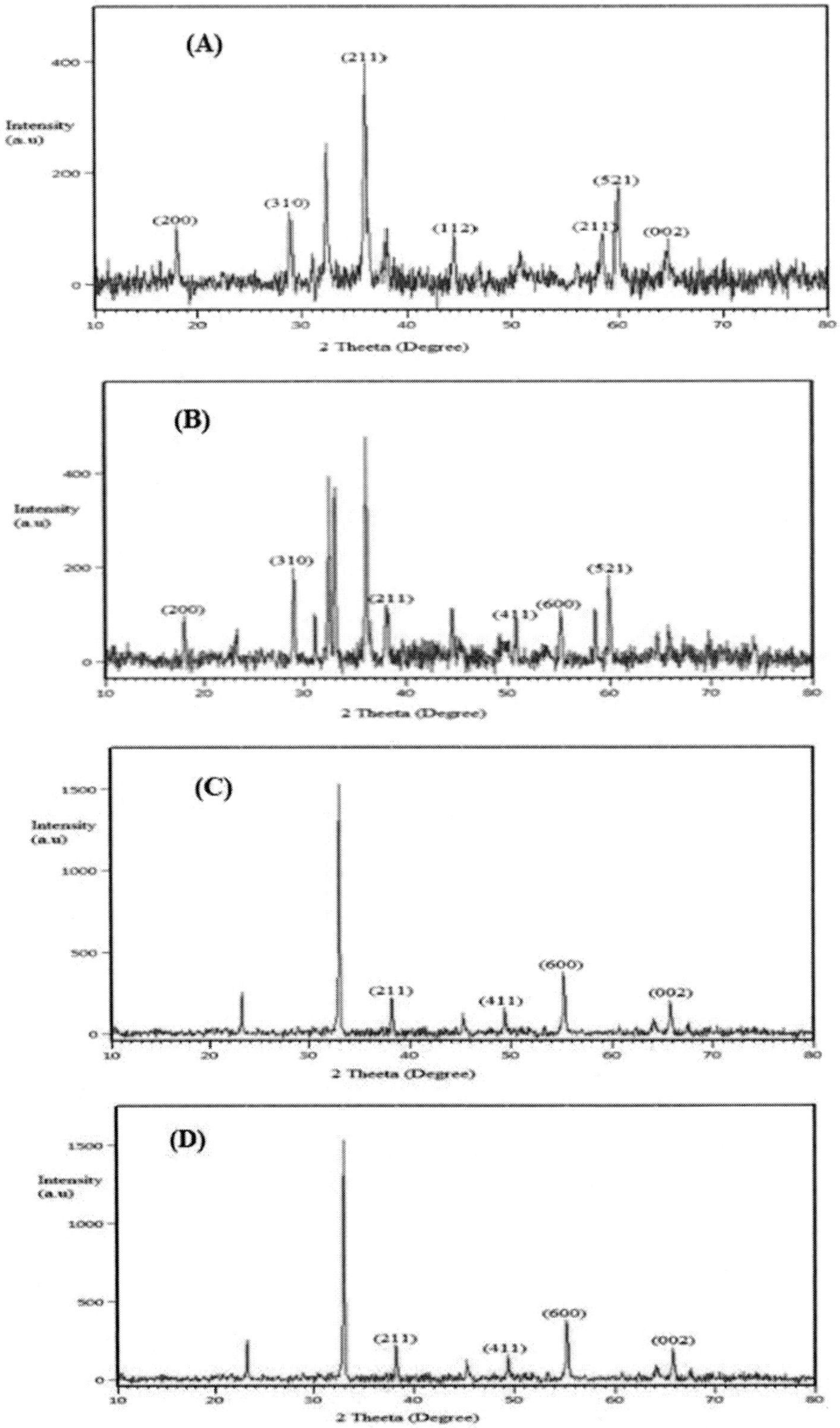

Figure 2. XRD pattern of samples A, B, C and D.

Average particles size is increased with increase in temperature treated as shown in table 1. This is because of change in growth rate between the different crystallographic planes.

3.2. SEM analysis

The scanning electron microscope images of prepared manganese dioxide nanoparticles of Sample A, Sample C and Sample D are shown in figure 3. The SEM images show the size and the morphology of the prepared nanoparticles. The SEM images show that the particles have random size and their morphology was changed from spherical to rod-like shape with increasing temperature.

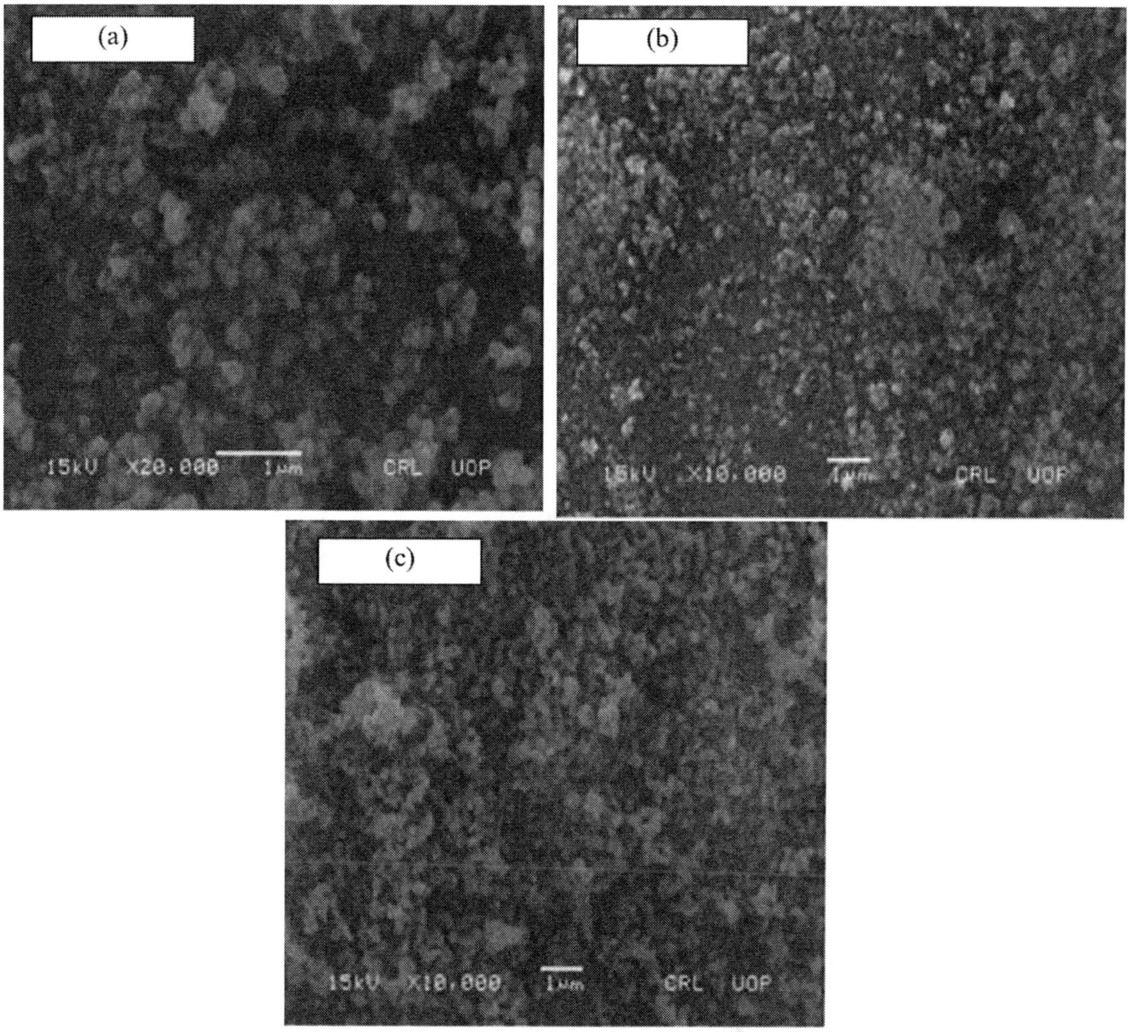

Figure 3. SEM micrographs for (a) Samples A, (b) Sample C and (c) Sample D.

3.3. UV visible analysis

Absorption spectra of prepared samples are obtained by UV Visible spectrometer. For UV-Vis studies a small amount of MnO_2 (0.02 g) for all samples is dissolved in 20 mL distilled water and stirred in sonicator. Graphs for absorption data for samples A, B, C and D are shown in figure 4.

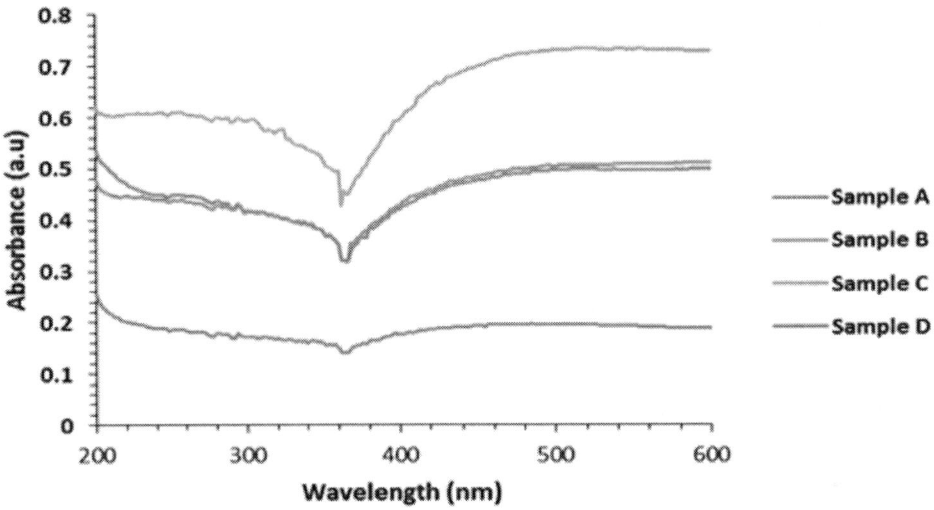

Figure 4. UV-vis spectra of all samples.

Optical properties of Manganese Dioxide Nanoparticles were analyzed using UV-Visible spectra. Absorption peak was observed at different wavelengths for all samples. Peaks were observed between 362 nm to 366 nm for all samples.

4. Conclusion

Manganese Dioxide nanoparticles samples are prepared by coprecipitation method. All samples are treated at different temperatures, i.e, 500 °C, 600 °C, 700 °C and a sample was kept untreated. Structured, morphological, and optical properties were characterized by XRD, SEM and UV visible. It is concluded that heat treatment on prepared samples of Manganese dioxide nanoparticle has affluence on particle size, volume and density of unit cell. Tetragonal structure and random particle size of MnO_2 was noted. Absorption peaks for prepared samples was observed between 362 nm to 366 nm. With increase in temperature particles agglomerated and rod-like shape is formed.

5. References

[1] Devan R S, Patil R A, Lin J H and Ma Y R 2012 *Advanced Functional Materials* **22** 3326.
[2] Bloom C J, Crew M D, Smithey D T, Miller W K and Morgen M M 2014 Nanoparticles comprising a cholesteryl ester transfer protein inhibitor and anon-ionizable polymer. Google Patents).
[3] Salomon H, Vignaud C, Lahlil S and Menguy N 2015 *Journal of Archaeological Science* 55 100
[4] Saputra E, Muhammad S, Sun H, Ang H M, Tade M and Wang S 2013 *Environmental science & technology* **47** 5882.
[5] Chen S-Y, Song W, Lin H-J, Wang S, Biswas S, Mollahosseini M, Kuo C-H, Gao P-X and Suib S L 2016 *ACS applied materials & interfaces* **8** 7834.
[6] Marceau H, Kim C-S, Paolella A, Ladouceur S, Lagacé M., Chaker M, Vijh A, Guerfi A, Julien C M and Mauger A 2016 *Journal of Power Sources* **319** 247.
[7] Birkner N R 2015 *Thermodynamics of Manganese Oxides at Bulk and Nanoscale: Phase Formation, Transformation, Oxidation-Reduction, and Hydration*: University of California, Davis).
[8] Birkner N and Navrotsky A 2012 *American Mineralogist* **97** 1291.
[9] Santos V, Soares O, Bakker J, Pereira M, Órfão J, Gascon J, Kapteijn F and Figueiredo J 2012 *Journal of catalysis* **293** 165.
[10] Sinopoli A, Othman Z, Rasool K and Mahmoud K A 2019 *Current Opinion in Solid State and Materials Science.*

[11] Julien C and Mauger A 2017 *Nanomaterials* **7** 396.
[12] Omar K A 2014 *Int. J. Res. Eng. Technol.* **2** 241.
[13] Li Y, Xie H, Wang J and Chen L 2011 *Materials Letters* **65** 403.

Performance and emission of turbocharger engine using gasoline and ethanol blends

S M Rosdi[1,*], R Mamat[1,3], A Alias[1], H Hamzah[1], K Sudhakar[1,2,3] and F Y Hagos[1,3]

[1] Faculty of Mechanical and Automotive Engineering Techology, Universiti Malaysia Pahang, 26600 Pekan, Pahang, Malaysia.
[2] Energy Centre, Maulana Azad National Institute of Technology Bhopal, M.P., India.
[3] Automotive Engineering Center, Universiti Malaysia Pahang, 26600 Pekan, Pahang, Malaysia.

*E-mail: mrosdi77@yahoo.com

Abstract. In this decade, alternative fuels such as ethanol were extensively studied as an additive or substitute for increasing engine performance and decreasing the release of harmful gases into the environment. The objective of the study is to assess engine performance and exhaust emission using Gasoline and Ethanol blends in a turbocharged spark-ignition engine. Gasoline was used as a reference fuel, and ethanol was blended into gasoline at 10%, 20% and 30% named as E10, E20, and E30, respectively. The turbocharger spark ignition (SI) engine has been coupled to eddy current 100 kW dynamometer to measure the engine efficiency. The engine test was carried out at an engine speed of 3000 rpm and 40% wide open throttle (WOT). As results, the performance parameters such as volumetric efficiency (VE), rate of heat release (ROHR), brake mean effective pressure (BMEP), brake specific fuel consumption (BSFC) of the engine increased with gasoline-ethanol blends compared to gasoline. Also, the complete combustion contributed to the reduction of the hydrocarbon (HC), carbon dioxide (CO_2) and nitrogen oxide (NOx) emissions. From the study, it can be concluded that ethanol is highly suited for turbocharged SI engines to reduce emissions and increase performance.

1. Introduction

With the rapid economic development, the problem of increase in pollution has become limitless [1]. Currently, the use of ethanol in gasoline has become a priority in the automotive industry due to the lack of fossil fuel reserves and environmental pollution issues. It is used in an internal combustion engine (ICE) to reduce the environmental effects of harmful emissions. Based on EU reports, CO_2 reduction targets are set between 20% and 30% for year 2020 and increase to 50% by the year 2050 [2] to endeavour sustainable growth. Some scientist predicts that if no action taken to reduce the emissions at present, greenhouse gas (GHG) concentration in the environment may double its range as early as 2035 [3] which may worsen the world's problems.

Biofuel such as ethanol are capable to reduce or control CO_2 emissions because they are renewable fuel. The use of ethanol in gasoline fuel is inevitable. Additionally, the use of ethanol in the combustion engine is cleaner than fossil fuels, thereby reducing environmental pollution and greenhouse emissions [4]. Ethanol has advantages as an alternative fuel because it is derived from renewable energy sources such as waste biomass, corn, soybeans, cassava, sweet potatoes, and so on. They have higher octane value, thus resulting in better anti-knock and allowing engine run under higher compression ratio [5].

Also, ethanol has low sulfur content, biodegradable and obtained from plant biomass [6]. Biofuels can reduce 80% greenhouse gas emission by replacing fossil diesel use in the internal combustion engines [7]. Hsieh et al. [8] concluded that the use of alcohol increases the volumetric efficiency (VE) due to the lower intake temperature. Zhuang & Hong [9] described that among various factor, charge cooling intake manifold with injecting ethanol into the cylinder significantly contributes to the increase in BMEP. The ethanol has a higher latent heat of vaporisation, which contributes to higher combustion velocity and higher energy content. Koç et al. [10] mentioned that to get the same energy output; the mass of alcohol fuel should be increased by 1.5-1.8 times compared to gasoline to increase the BSFC.

The formation of nitrogen oxide (NOx) is closely related to cylinder pressure. The NOx emission ideally takes place due to the Zeldovich mechanism reaction when the flame temperature is above 1850 K during the combustion. Three mechanisms are influencing the formation NOx: enrichment of oxygen, the reaction time of Nitrogen dioxide (NO) and higher cylinder temperature [11]. Carbon dioxide (CO_2) is one of the basic GHG, which is formed by the complete combustion of hydrocarbon fuel. The CO_2 formation is affected by the carbon-hydrogen (C/H) ratio in the fuel. Hydrocarbon is a product of incomplete combustion in the engine. It occurs in three condition: (1) absorption of fuel vapour into oil layer in cylinder block; (2) flame quenching due to incomplete combustion and (3) filling crevice volumes with the unburned mixture [12].

From the literature above, increasing alcohol fuel into the gasoline induces to vary suggestion in engine performance and emission. Furthermore, it is well known in the literature that using alcohol fuel as blends in gasoline contributes further substitution for gasoline feedstock. However, the effect of engine efficiency on different fuel blends in a turbocharged is not clear and less report. Considering this fact, this paper aims to investigate the engine efficiency and emissions characteristics of ethanol-gasoline blends in the turbocharged spark-ignition engine without any modification. In this study, the gasoline was utilized as a reference fuel. The ethanol was blended at concentrations of 10%, 20% and 30% with gasoline are named as E10, E20 and E30 respectively. These percentage of blended ethanol to gasoline fuel implemented in this study is a platform of knowledge and reduce dependency on fossil fuel. The experiments have been conducted at 40% wide open throttle (WOT) and 3000 rpm engine speed on the turbocharged SI engine.

2. Experimental methodology

The experiment was conducted on a turbocharged (1.8 Litre), spark ignition (SI), four-stroke, and a four-cylinder engine. The technical parameters of the engine are shown in table 1. The engine test rig, coupled with a 100 kW dynamometer (Dynalec Controls, India) is shown in figure 1.

Table 1. Specifications of the engine (Mitsubishi Motors Corporation, 1998).

Type	SOHC 16 V MPI
Number of cylinders	4
Type of chamber	Pent-roof Type
Displacement	1.8 Litre
Bore	8.1 cm
Stroke	8.9 cm
Compression ratio	9.5:1
Spark Timing	10° -35° BTDC
Injection Timing	715° BTDC
Power Maximum	118 kW
Torque Maximum	220 Nm

Ethanol is mixed with gasoline at 10%, 20%, and 30% ratios and labelled as E10, E20, and E30 respectively. A magnetic stirrer is used to blend ethanol and gasoline for about 10 minutes before testing the properties. The properties of the Fuel blends are assessed from the UMP Central laboratory. Table 2 shows the properties of gasoline, ethanol and ethanol blends used in the experiment. A gas analyzer from

Kane Auto-plus 5-2 is used to evaluate the engine emission, e.g. CO_2, HC, NOx and air-fuel ratio. The emission sensor is mounted at the exhaust tailpipe to obtain emission data.

Table 2. Properties of gasoline, ethanol and ethanol blends [11, 14, 15].

Properties	G	E	E10	E20	E30	Test standard
Chemical Formula	$C_{5-10}H_{12-22}$	C_2H_5OH	-	-	-	-
Density, (kg/m^3)	737	785	761	765	768	ASTM D4052
Boiling Temperature, °C	27-225	78.25	-	-	-	-
Octane Number	95	108.6	98.1	100.7	102.4	ASTM D2699
Motor Octane Number	82-92	92	99.9	101.6	-	ASTM D2700
Heat of Vaporization, kJ/kg	349-500	904-923	-	-	-	-
Lower Heating Value, MJ/Kg	44	26.9	41	39.4	37.8	ASTM D240
Flash Point, °C	45	12-20	-	-	-	-
Auto Ignition Temperature,°C	257	425	-	-	-	-
Water content (%)	0	7-11%	5.4	10.2	14.4	-
Air-fuel ratio	14.6	9	13.5	12.6	11.7	-

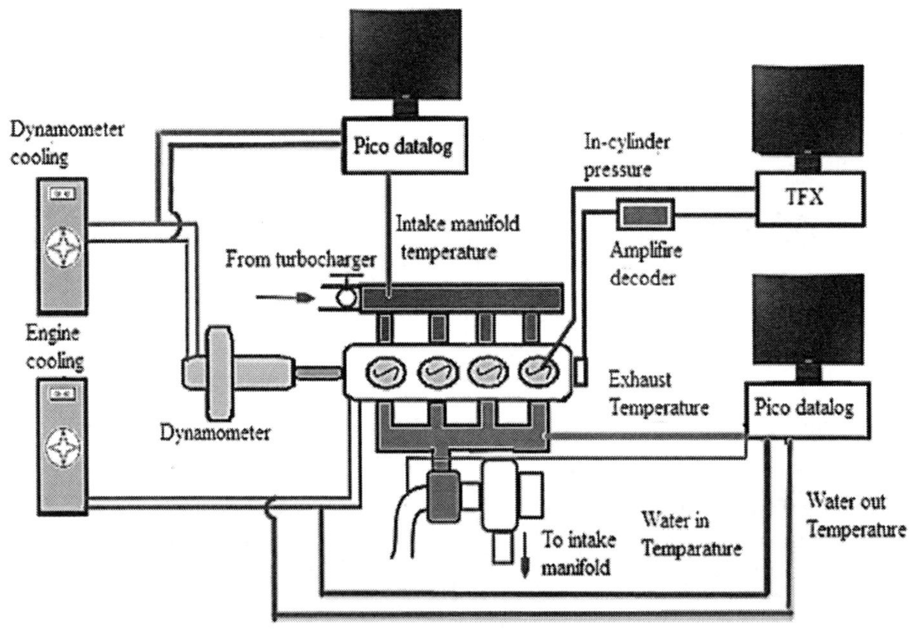

Figure 1. Schematic layout engine test rig.

The cylinder pressure sensor range between 0-200 bar with a sensitivity of 1.12 mV-Psi connected to the TFX Data Acquisition System (DAQ) used to record the in-cylinder pressure. Sensor-based optical-fibres and magnetic encoders are set up to detect crank angles and the combustion pressure in this study. An anemometer with a resolution 0.001 m/s (CENTER 330 model) is used to measure the air-flow rate into the intake manifold. The K-type thermocouple sensor with a data logger (TC-08 Picolog) was installed to measure coolant water, engine oil, air delivery in the manifold, manifold temperature and outlet exhaust gas temperature. The cooling tank circulates cold water to reduce the engine temperature. An electric fan is used to cool down the intercooler air temperature before entering the engine cylinder.

3. Results and discussion

The study was conducted to evaluate engine performance and exhaust emission on turbocharged SI engine at 3000 rpm for 40% WOT. The study was explored by comparing the test results of gasoline and gasoline-ethanol blends up to 30%. The testing was repeated three times to collect the data, and the average values are reported.

3.1. Volumetric efficiency

Figure 2 shows the trend of volumetric efficiency (VE) for stated fuel blends at 3000 rpm and 40% engine load. It shows averagely 6.5% increasing trend when using E10, E20 and E30 compared to G. The VE for G, E10, E20 and E30 was recorded at 52%, 54%, 58% and 64% respectively. The increasing of VE is due to the charge cooling of the intake manifold while using ethanol-gasoline blends. Also, the water content in ethanol correlates to the octane number in fuel. As the octane number increases, the VE also increases. Similar results were observed by Alahmer & Aladayleh [15] for various ethanol blends in engines. Lie et al. [16] described that the higher latent heat of alcohol fuels caused lower intake manifold temperature and induced to increase VE. They added that reduced combustion by using alcohol fuel was beneficial for improving thermal efficiency. In addition, Simsek & Ozdalyan [17] declared that alcohol fuels may have a cooling effect on the drawn fresh mixture due to their heating value below than gasoline and higher enthalpy of vaporization. This phenomena introduces to increase VE of the engine. Due to higher VE, engine operations may have higher torque and horsepower. On the other hand, with fresh charge cooling temperature engine operation, it will be operating with poorer mixtures, there may be significant decline in CO and NOx emission.

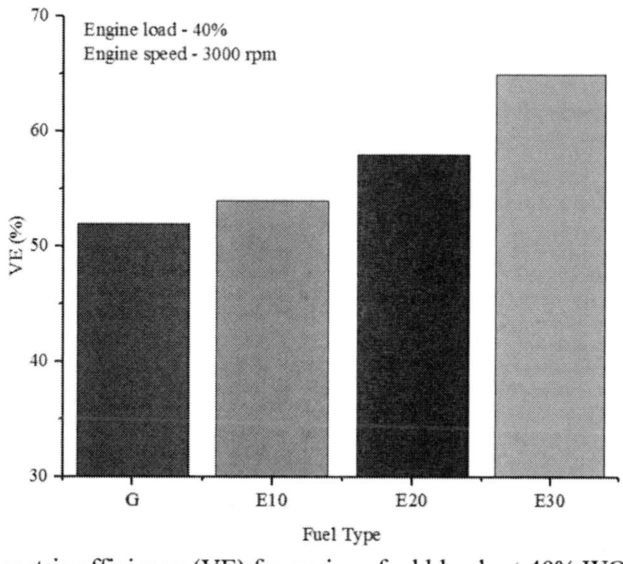

Figure 2. Volumetric efficiency (VE) for various fuel blends at 40% WOT at 3000 rpm.

3.2. Rate of heat release

The Rate of Heat Release (ROHR) is defined as the rate of chemical energy of the fuel released by combustion. The ROHR is calculated based on first law thermodynamic during the closed valve period from the average pressure. The cylinder pressure and ROHR profile were measured with crank angle (CA) for various blends gasoline-ethanol are shown in figure 3. The maximum cylinder pressure shows the range at 55-60 bar was obtained for all test fuels. The E30 shows higher pressure than others. When using ethanol blends, the ROHR was increased compared to gasoline. The corresponding ROHR for G, E10, E20 and E30 shows range at 60-70 J/degree. As a consequence, the use of ethanol blends in gasoline slightly increases and larger quantity of CA of cylinder pressure and ROHR. Thangavelu et al. [18] have similar result with this study. They described that the velocity of combustion by ethanol blend is faster, and higher peak heat release when using higher engine speed. In addition, ethanol blends show the higher laminar flow burning which promotes by formation of hydroxyl radical in combustion [19, 20].

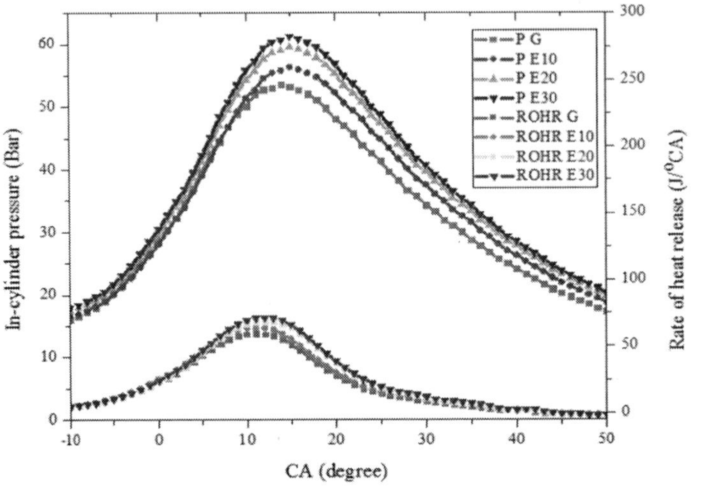

Figure 3. In-cylinder pressure and heat release rate for various fuel blends at 40% WOT at 3000 rpm.

3.3. *Break mean effective pressure*

Brake mean effective pressure (BMEP) is a parameter that reflects engine power. Figure 4 shows the comparison of the BMEP at 3000 rpm and 40% WOT for various ethanol-gasoline fuel blends. It was observed that ethanol fuel blends showed increases BMEP compared to gasoline. The latent heat of vaporization ethanol is higher than gasoline. Due to blends with gasoline, the latent heat of vaporization of ethanol-gasoline blends become higher compared to gasoline [21]. It offers an additional benefit of improved charge cooling and reduces the temperature of the intake manifold. With more air during suction stroke enter to the engine cylinders, fuel can be burnt properly to obtain high power output. In addition, effect of charge air cooling leads to higher volumetric efficiency and introduces higher BMEP. Furthermore, ethanol, as an oxygenated fuel introduces hydroxyl radical (-OH), enhances the efficiency of combustion in an engine which promotes complete combustion [18]. The finding also concordant with [5] that blends ethanol-gasoline increases the BMEP.

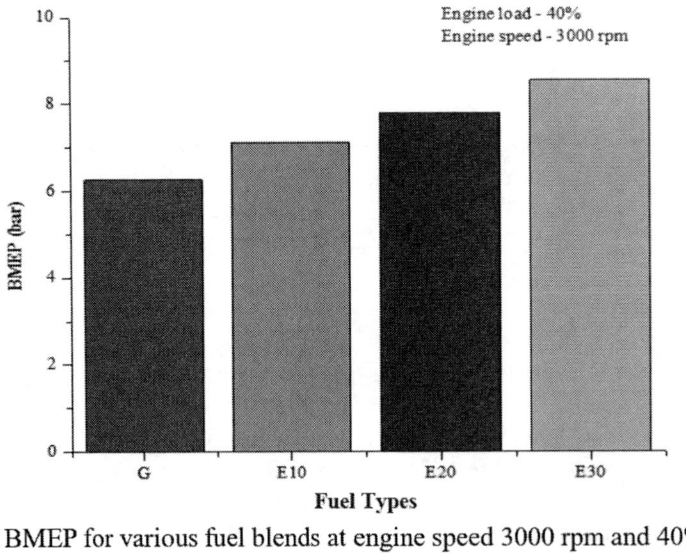

Figure 4. BMEP for various fuel blends at engine speed 3000 rpm and 40% WOT.

3.4. Brake specific fuel consumption

Brake specific fuel consumption (BSFC) defined as the amount of fuel consumed for each unit of brake power developed per power [18]. Figure 5 shows the BSFC with various blends of fuel at 3000 rpm and 40% WOT. From observation, BSFC increased with the increase percentage proportion blends. In these cases, the engine runs without any modification. The increase in BSFC is significantly related to the fuel properties such as lower heating value, high kinematic viscosity, density and the air-fuel ratio [22]. Najafi et al. [22] concluded that the increase of BSFC is due to higher density and octane number of ethanol-gasoline blends. Hassan et al. [23] described that due to lower heating value of ethanol, it should be 15% increased more than that of gasoline to achieve same power with gasoline. As consequently, the BSFC increases compared to gasoline.

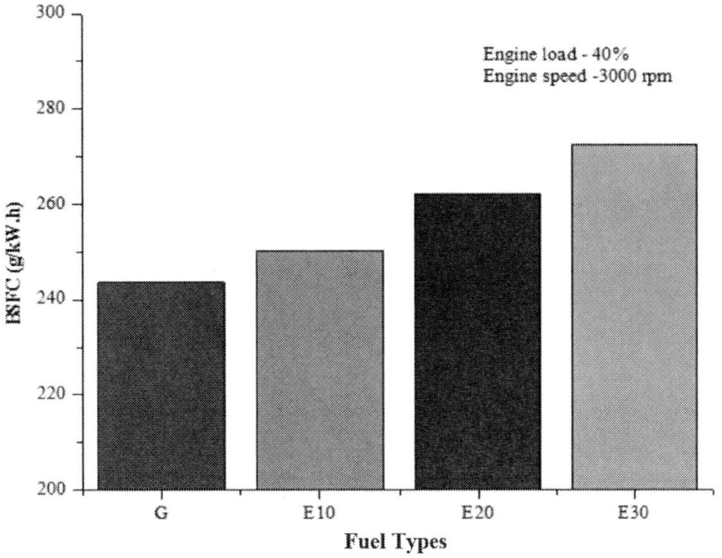

Figure 5. BSFC for various fuel blends at 40% WOT and 3000 rpm.

3.5. Nitrogen oxide emission

Nitrogen oxide (NOx) is harmful gaseous emission and generalized for NO and NO_2. Figure 6 shows the emission of NOx for various fuel blends at 3000 rpm and 40% WOT. From the figure, it is observed that the effect of using ethanol-gasoline blends shows decreasing trend of NOx emission compared to gasoline. The formation of NOx emission depends on the cylinder temperature, oxygen concentration and residence time for the reaction of gas temperature [24]. Increased cylinder pressure introduces to increase cylinder temperature [25]. With the ethanol-gasoline blends, the temperature of combustion is greatly reduced due to the high latent heat of vaporisation, lower heating value and oxygen content in alcohol, and these leads to the reduction in NOx emissions [11]. Some researchers identified that the reaction time of each engine cycle was reduced. The gas residence temperature within the cylinder was shortened [21, 23, 24]. This condition also led to lower NOx emissions [22].

3.6. Hydrocarbon emission

The Hydrocarbon (HC) emission for various fuel blends for at 40% WOT and 3000 rpm is shown in figure 7. Gasoline shows higher of HC compared to other fuel blends. Normally, HC emission presences when occurring incomplete combustion. HC emission shows decrease trend when engine uses ethanol-gasoline blends. Averagely, the HC reduced at 6.6% when proportion of percentage ethanol increase compared to gasoline. Park et al [27] mentioned that introducing ethanol in gasoline can reduce the HC emission because non-polar molecules cannot easily absorb the water molecules in the lubricating oil layer. Therefore, the probability of producing HC emission can be reduced. The reduction of HC is due to lean combustion and the presences of water content (about 7%) in the gasoline-ethanol blends which promotes complete combustion [5].

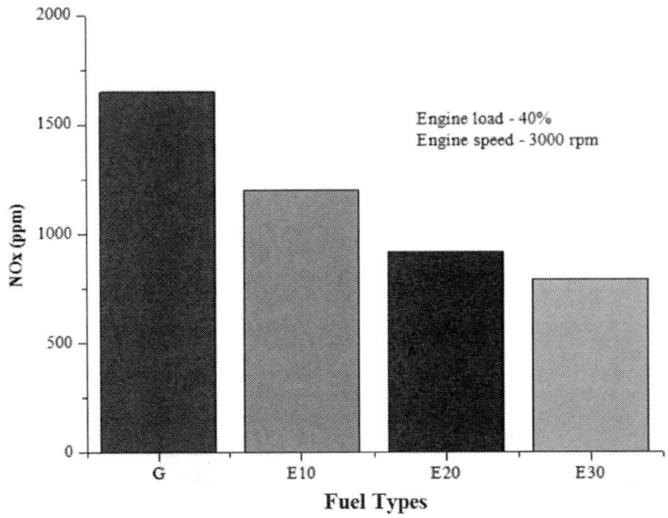

Figure 6. NOx emission for various fuel blends at 3000 rpm for 40% WOT.

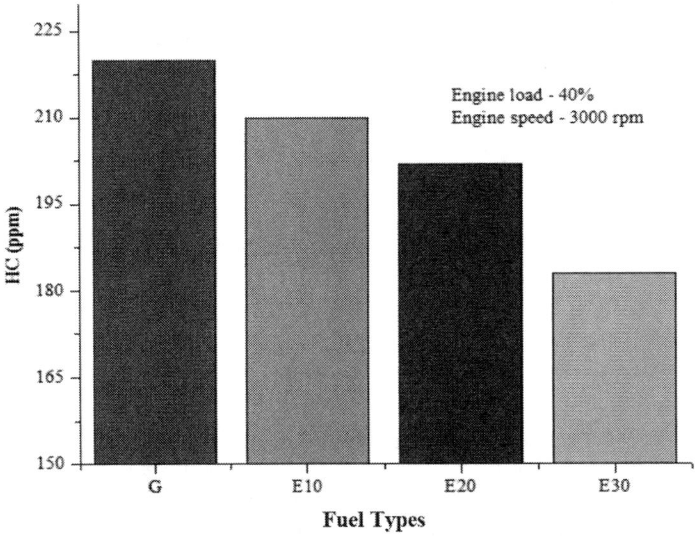

Figure 7. Hydrocarbon for various fuel blends at 40% WOT and 3000 rpm.

3.7. Carbon dioxide emission

Carbon dioxide (CO_2) is produced by the complete combustion of hydrocarbon fuel. Figure 8 shows the CO_2 emission for various fuel blends at 40% WOT and 3000 rpm. It showed that with the use of E30, the CO_2 emission decreased by 33% compared to gasoline. The CO_2 emission decreased with the increase in the proportion of ethanol. The engine has been running at lean condition when uses the ethanol-gasoline blends. The reduction in CO_2 is due to the hydroxyl radical (-OH) in ethanol, which encourages complete combustion in the engine cylinder. Also, the lower carbon content in ethanol compared to the gasoline lead to the reduction of CO_2 [11]. A similar result was obtained by [28], they mentioned CO_2 reduce extensively due to rapid evaporation and better mixing of air and fuel blends. The extra oxygen in gasoline fuel blends resulted excellent combustion. Due to the complete combustion of bioethanol in fuel, CO was transformed into CO_2 [22].

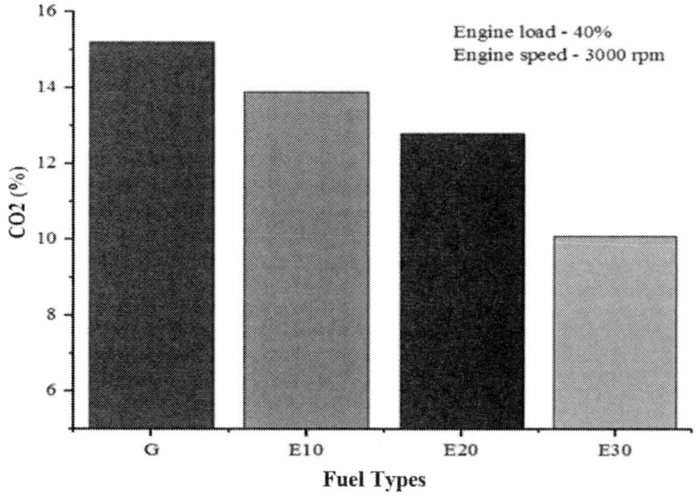

Figure 8. CO_2 emission for various fuel blends at 40% WOT and 3000 rpm.

4. Conclusion

The engine performance and emission characteristics of the 1.8L turbocharger engine fuelled with gasoline and ethanol blends was investigated at 3000 rpm and 40% WOT. The following conclusions are obtained from the study:

i) The use of ethanol up to 30% blends improves the performance of turbocharger engine due to high octane number and latent heat vaporization.
ii) The lower energy content and higher density of ethanol–gasoline blends increased the BSFC of the engine.
iii) The charge cooling effect the intake manifold introduces to increase the VE and also contributed to the lean combustion.
iv) The exhaust emission such as NOx, HC, and CO_2 showed a significant reduction due to the presence of oxygen in ethanol and combustion enhancement.
iv) Ethanol-gasoline blends show advances combustion leading to a higher in-cylinder pressure compared to gasoline. Also, the engine knock tendency is greatly reduced due to advance combustions.

Acknowledgements

The authors would like to thank the Faculty of Mechanical & Manufacturing Engineering, Universiti Malaysia Pahang (UMP) for providing the facilities to carry out this research, and Universiti Malaysia Pahang (UMP) for providing the research grant (RDU180391), and (RDU172204) to carry out the study.

References

[1] T. Venugopal and A. Ramesh, "Experimental studies on the effect of injection timing in a SI engine using dual injection of n-butanol and gasoline in the intake port," *Fuel*, vol. **115**, pp. 295–305, (2014).
[2] A. Alagumalai, "Internal combustion engines: Progress and prospects," *Renew. Sustain. Energy Rev.*, vol. **38**, pp. 561–571, Oct. (2014).
[3] H. A. Bekhet and N. S. Othman, "Impact of urbanization growth on Malaysia CO2emissions: Evidence from the dynamic relationship," *J. Clean. Prod.*, vol. **154**, pp. 374–388, (2017).
[4] B. M. Masum, H. H. Masjuki, M. A. Kalam, I. M. Rizwanul Fattah, S. M. Palash, and M. J. Abedin, "Effect of ethanol–gasoline blend on NOx emission in SI engine," *Renew. Sustain. Energy Rev.*, vol. **24**, pp. 209–222, Aug. (2013).
[5] X. Deng, Z. Chen, X. Wang, H. Zhen, and R. Xie, "Exhaust noise, performance and emission

characteristics of spark ignition engine fuelled with pure gasoline and hydrous ethanol gasoline blends," *Case Stud. Therm. Eng.*, vol. **12**, no. February, pp. 55–63, (2018).

[6] E. Galloni, G. Fontana, S. Staccone, and F. Scala, "Performance analyses of a spark-ignition engine firing with gasoline–butanol blends at partial load operation," *Energy Convers. Manag.*, vol. **110**, pp. 319–326, Feb. (2016).

[7] A. K. Hossain, D. I. Smith, and P. A. Davies, "Effects of Engine Cooling Water Temperature on Performance and Emission Characteristics of a Compression Ignition Engine Operated with Biofuel Blend," pp. 46–57, (2017).

[8] W. D. Hsieh, R. H. Chen, T. L. Wu, and T. H. Lin, "Engine performance and pollutant emission of an SI engine using ethanol-gasoline blended fuels," *Atmos. Environ.*, vol. **36**, no. 3, pp. 403–410, (2002).

[9] Y. Zhuang and G. Hong, "Primary investigation to leveraging effect of using ethanol fuel on reducing gasoline fuel consumption," *Fuel*, vol. 105, pp. 425–431, Mar. (2013).

[10] M. Koç, Y. Sekmen, T. Topgül, and H. S. Yücesu, "The effects of ethanol–unleaded gasoline blends on engine performance and exhaust emissions in a spark-ignition engine," *Renew. Energy*, vol. **34**, no. 10, pp. 2101–2106, Oct. (2009).

[11] M. Canakci, A. N. Ozsezen, E. Alptekin, and M. Eyidogan, "Impact of alcohol–gasoline fuel blends on the exhaust emission of an SI engine," *Renew. Energy*, vol. **52**, pp. 111–117, Apr. (2013).

[12] J. B. Heywood, *Internal combustion engine*, vol. **11**, no. 5. (1985).

[13] T. Topgül, H. S. Yücesu, C. Çinar, and A. Koca, "The effects of ethanol-unleaded gasoline blends and ignition timing on engine performance and exhaust emissions," *Renew. Energy*, vol. **31**, no. 15, pp. 2534–2542, (2006).

[14] M. A. Costagliola *et al.*, "Performances and emissions of a 4-stroke motorcycle fuelled with ethanol/gasoline blends," *Fuel*, vol. **183**, pp. 470–477, (2016).

[15] A. Alahmer and W. Aladayleh, "Effect two grades of octane numbers on the performance, exhaust and acoustic emissions of spark ignition engine," *Fuel*, vol. **180**, pp. 80–89, (2016).

[16] Y. Li *et al.*, "Effect of water-containing acetone-butanol-ethanol gasoline blends on combustion, performance, and emissions characteristics of a spark-ignition engine," *Energy Convers. Manag.*, vol. **117**, pp. 21–30, (2016).

[17] S. Simsek and B. Ozdalyan, "Improvements to the Composition of Fusel Oil and Analysis of the Effects of Fusel Oil–Gasoline Blends on a Spark-Ignited (SI) Engine's Performance and Emissions," *Energies*, vol. **11**, no. 3, p. 625, (2018).

[18] S. K. Thangavelu, A. S. Ahmed, and F. N. Ani, "Review on bioethanol as alternative fuel for spark ignition engines," *Renew. Sustain. Energy Rev.*, vol. **56**, pp. 820–835, (2016).

[19] T. D. M. Lanzanova, M. Dalla Nora, and H. Zhao, "Performance and economic analysis of a direct injection spark ignition engine fueled with wet ethanol," *Appl. Energy*, vol. **169**, pp. 230–239, (2016).

[20] X. Wang, Z. Chen, J. Ni, S. Liu, and H. Zhou, "The effects of hydrous ethanol gasoline on combustion and emission characteristics of a port injection gasoline engine," *Case Stud. Therm. Eng.*, vol. **6**, pp. 147–154, Sep. (2015).

[21] C. Wang, A. Janssen, A. Prakash, R. Cracknell, and H. Xu, "Splash blended ethanol in a spark ignition engine – Effect of RON, octane sensitivity and charge cooling," *Fuel*, vol. **196**, pp. 21–31, (2017).

[22] G. Najafi *et al.*, "SVM and ANFIS for prediction of performance and exhaust emissions of a SI engine with gasoline-ethanol blended fuels," *Appl. Therm. Eng.*, vol. **95**, pp. 186–203, (2016).

[23] A. O. Hasan, H. Al-Rawashdeh, A. H. Al-Muhtaseb, A. Abu-jrai, R. Ahmad, and J. Zeaiter, "Impact of changing combustion chamber geometry on emissions, and combustion characteristics of a single cylinder SI (spark ignition) engine fueled with ethanol/gasoline blends," *Fuel*, vol. **231**, no. May, pp. 197–203, (2018).

[24] W. N. M. Wan Ghazali, R. Mamat, H. H. Masjuki, and G. Najafi, "Effects of biodiesel from

different feedstocks on engine performance and emissions: A review," *Renew. Sustain. Energy Rev.*, vol. **51**, pp. 585–602, (2015).

[25] A. Calam, H. Solmaz, A. Uyumaz, E. Yilmaz, and Y. Içingür, "Investigation of usability of the fusel oil in a single cylinder spark ignition engine," vol. **88**, pp. 258–265, (2015).

[26] C. Ilkiliç and R. Behçet, "The reduction of exhaust emissions from a diesel engine by using biodiesel blend," *Energy Sources, Part A Recover. Util. Environ. Eff.*, vol. **32**, no. 9, pp. 839–850, (2010).

[27] C. Park, Y. Choi, C. Kim, S. Oh, G. Lim, and Y. Moriyoshi, "Performance and exhaust emission characteristics of a spark ignition engine using ethanol and ethanol-reformed gas," *Fuel*, vol. **89**, no. 8, pp. 2118–2125, Aug. (2010).

[28] M. Ghazikhani, M. Hatami, B. Safari, and D. Domiri Ganji, "Experimental investigation of exhaust temperature and delivery ratio effect on emissions and performance of a gasoline–ethanol two-stroke engine," *Case Stud. Therm. Eng.*, vol. **2**, pp. 82–90, Mar. (2014).

IOP Conf. Series: Materials Science and Engineering **863** (2020) 012035 doi:10.1088/1757-899X/863/1/012035

Microwave plasma assisted sol-gel technique for synthesis of TiO_2 nanoparticles

M Ayyaz[1,*], M Y Naz[1], A shoukat[1], M Marriam[1], M U Akram[1], Z Hussain[2], S Shukrullah[1] and I Toqeer[1]

[1] Department of Physics, University of Agriculture, 38040 Faisalabad, Pakistan.
[2] Department of Chemistry, Abdul Wali Khan University, 23200 Mardan, Pakistan.

*E-mail: muhammadayyaz333@yahoo.com

Abstract. Titanium dioxide (TiO_2) nanoparticles have attracted the attention of research community due to their novel functionalities as compared to the bulk material. TiO_2 is an excellent photocatalyst due to its high photosensitivity, nontoxicity, high refractive index, strong oxidizing ability, high stability, wide band gap and high resistance to photochemical. The main objective of this study was to investigate the influence of microwave (MW) plasma treatment on TiO_2 nanoparticles synthesized using sol-gel method. TiO_2 nanoparticles were obtained through sol-gel method at ambient temperature. The suspension was heated at 300 °C for 2 hours to evaporate the organic content. The obtained nanoparticles were placed in partially vacuumized chamber for MW plasma treatment. The plasma treatment is a promising technique for oxidation of nanomaterials. Both plasma treated and untreated samples were evaluated with X-ray diffraction (XRD), scanning electron microscopy (SEM) and UV-Visible spectroscopy for crystallite size, crystal phases, band gap energy and surface morphology. The obtained results confirmed the existence of anatase and rutile phases of TiO_2 with smaller particle size within the range of 0.2 to 14 nm. The particles were of aggregated and trigonal shapes. The MW Plasma treatment improved the photocatalytic activity of TiO_2 nanoparticles by raising their band gap energy and reducing the grain size.

1. Introduction

Titanium dioxide (TiO_2) is a white inorganic material. It is environmental friendly, cheap, easy to handle [1], resistive to photochemical and anti-chemical erosion and has high photocatalytic activity [2]. Due to unique electronic structure and light reduced redox phenomenon to act as sensitizers, TiO_2 and ZnO have attracted the great attention of the researchers. Among all the semiconductors, TiO_2 is widely used as n-type semiconductor [3]. TiO_2 nanopowders are also used in cosmetics, glazes, toothpaste, enamels, plastics, paper, fibers, pharmaceuticals, paints [4], antibacterial agents [5-7], foods, friction reducing agent, etc. [8].

TiO_2 is found in nature in polymorphs crystalline form, such as stable rutile, brookite and anatase [9]. Brookite phase is rarely synthesized in the laboratory, whereas the other two are widely synthesize in the laboratory. Anatase is a low temperature and chemically stable phase whereas rutile is a thermodynamically substantial form of TiO_2. Anatase can be converted in rutile at higher temperatures through phase transformation. This transformation from anatase to rutile happens at high temperatures higher than 400 °C [10]. However, this phase transformation can happen at ambient temperature under optimized synthesis parameters [11].

Physical and chemical properties of TiO_2 are related to its crystalline phase, size, and morphology [12]. TiO_2 nanoparticles adsorb enough dye molecules, consequently, produce large number of photons for current conversion efficiency. The photo-electrodes, coated with TiO_2, usually have higher transparency [13, 14]. Enough visible light can be transmitted through these materials. The smaller particle size results in only negligible scattering of light [15, 16]. Single phase TiO_2 has fewer application as compare to the mixed phase TiO_2 consisting of anatase and brookite phases.

Numerous methods have been reported during last few years to synthesize TiO_2. These methods include atomic layer deposition, chloride process, sol-gel route [17], direct oxidation of $TiCl_4$ [18, 19], chemical vapor deposition, hydrothermal method [20], liquid flame spray [21-23], sulfate process [24], and electrospinning method [25, 26]. Each technique has some merits and demerits [27]. Comparatively, sol-gel is a low cost and most commonly used method for synthesis of TiO_2 nanoparticles [28]. Different morphology of TiO_2 nanoparticles can be obtained with this technique. The morphologies include nanosheets, nanoparticles, nanowires, nanotubes, nanorods, mesoporous and aerogel [29]. The plasmas discharges are also being used to tailor the surface chemistry of such materials. A survey of the published literature shows scant information about the effect of plasma assisted calcination on the structural and optical properties of the metal oxide nanoparticles. Therefore, this research was carried out with the following objectives: (i) to synthesize the titanium dioxide nanoparticles through a sol-gel technique, (ii) to remove the dried gel, oxides and other impurities from the nanoparticles through MW plasma exposure and (iii) to study the effect of plasma treatment on the surface morphology, chemical composition and size distribution of the nanoparticles.

2. Materials and methods

2.1. Chemicals

The chemicals, used to prepare TiO_2 nanoparticles, were titanium tetra iso-propoxide (TTIP) $C_{12}H_{28}O_4Ti$ (98.0%, company Dae-Jung, Korea) as precursor, Isopropanol (C_3H_5OH, Riedel-deHaen, extra pure) as solvent, Nitric acid (HNO_3 68%, Huchems fine chemical corp. Korea) as stabilizers and deionized water as dispersing media [10, 30].

2.2. Synthesis of TiO_2 nanoparticles

TiO_2 nanoparticles were obtained by sol gel method in the presence of nitric acid (HNO_3) as a stabilizer. Titanium tetra isopropoxide (TTIP) and 2-propanol were used as starting materials. Solution A was prepared by mixing 24 mL of deionized water and 20 mL of iso-propanol [30, 31]. Solution B was obtained by dripping titanium tetra-isopropoxide in solution A and was continuously stirred on a magnetic hot plate at ambient temperature. After 1 hour of stirring, 0.5 ml of HNO_3 was added to the solution B and stirred continuously for 6 hours. During first 3 hours, a milky white solution was transformed into a transparent white solution known as sol. In next 3 hours, the viscous solution (sol) transformed into viscous gel. The obtained gel was subjected to continuous heating for 2 hours at 300°C. The organic content evaporates from gel during heating. Finally, TiO_2 was obtained in the form of nanocrystals, which were grinded using pestle mortar. The dried TiO_2 powder was placed inside a partially vacuumized chamber to perform MW plasma treatment. The plasma of oxygen gas was sustained with a 2.45 GHz microwave source. The TiO_2 sample was exposed to the plasma for fixed time of 5 minutes. The flowchart for the preparation of TiO_2 nanoparticles is shown in figure 1.

2.3. Characterization

SEM, XRD and UV-Visible techniques were used to evaluate the surface morphology, composition, structures, phases and size of MW plasma treated TiO_2 nanoparticles. The particle size was calculated using Scherrer formula:

$$D = \frac{\lambda K}{\beta \cos\theta} \quad (1)$$

where D is grain size, K is the constant of shape factor (K=0.89), β is full width half maximum (FWHM) in radian and λ is the X-ray wavelength for Cu target Kα radiations and θ is the Bragg's diffraction angle.

UV-Visible analysis was carried out to study the effect of different stabilizers on optical properties of TiO_2 nanoparticles. The band gap energy (E_g) was determined from the optical absorption spectra by using the equation (2).

$$E_g = \frac{1240}{\lambda}\, eV \tag{2}$$

where E_g represents the band gap energy electron volts (eV) and λ is the wavelength of absorption edge obtained from the spectrum.

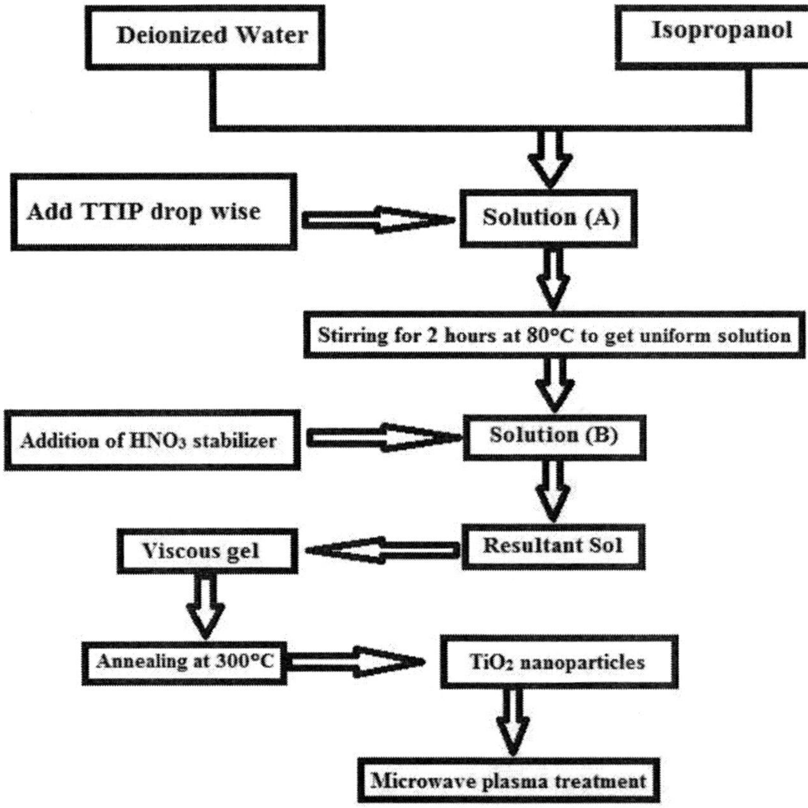

Figure 1. Flowchart of synthesis of TiO_2 nanoparticles through plasma assisted sol-gel method.

3. Results and discussion

3.1. Structural analysis

XRD spectra of TiO_2 samples was generated to study the particle size and crystalline phases. XRD spectra confirmed the formation of different phases of TiO_2. The major phases were anatase and rutile [32]. Figure 2 Shows XRD pattern of TiO_2 sample with peaks at 2θ= 25.28°, 27.19°, 36.06°, 36.94°, 37.66°, 38.60°, 41.33°, 47.94°, 54.30° and 56.70°. These peaks correspond to (011), (110), (101), (013), (004), (112), (111), (020), (211) and (220) planes of TiO_2. The peaks corresponding to planes (011), (013), (004), (112) and (020) indicates the anatase phase whereas (110), (101), (111), (211) and (220) planes represent the rutile phase of TiO_2. The observed peaks and planes were in good match with the standard (JCPDS no: 21-1272 and 21-1276) [33]. XRD analysis revealed 68% anatase phase and 32% rutile phase of TiO_2 [34].

The particle size was calculated using Scherrer's formula. The size was remained in the range of 0.2 nm to 14 nm, which is much smaller as compared to conventional sol-gel method. The conventional sol-gel approach produced the particle sizes almost 2 times larger than the plasma assisted sol-gel approach. The particle size distribution of plasma treated, and untreated nanoparticles is given in table 1.

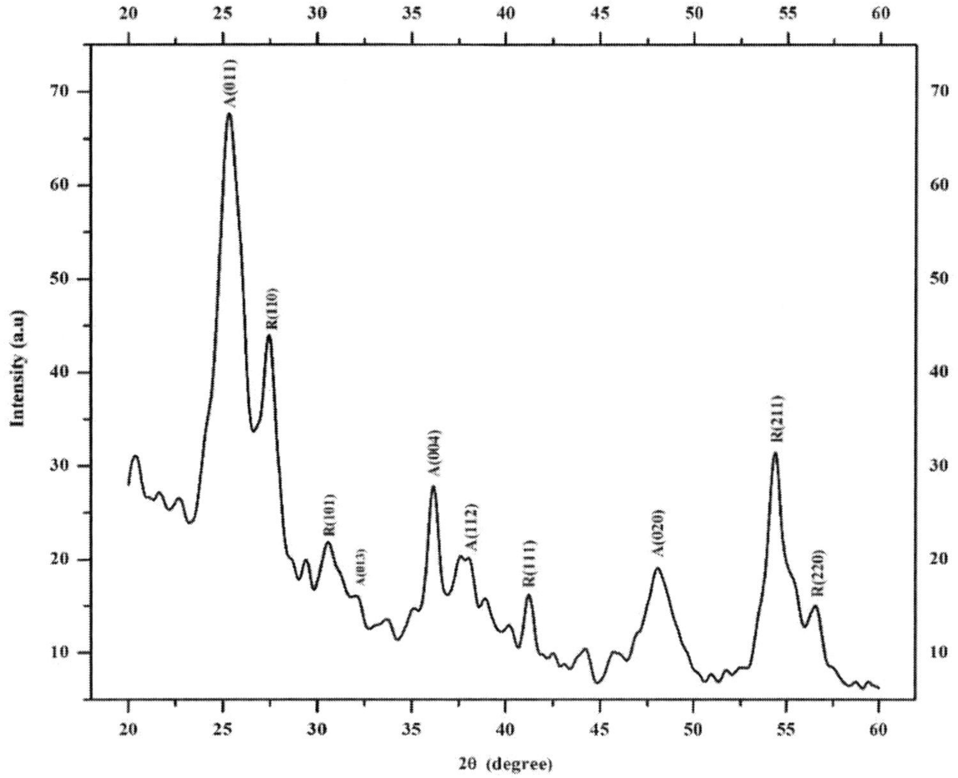

Figure 2. XRD spectrum of MW plasma treated TiO_2 nanoparticles.

Table 1. Particle size calculation for microwave assisted sol gel method.

2θ (deg.)	Particle size of untreated sample (nm)	Particle size of plasma treated sample (nm)
25.28	3.99	7.18
27.19	5.32	9.57
36.06	0.23	0.41
36.94	14.21	25.57
37.66	5.47	9.84
38.60	3.98	7.16
41.33	10.49	18.82
47.94	4.41	7.98
54.30	6.61	11.98
56.70	8.84	15.12

IOP Conf. Series: Materials Science and Engineering **863** (2020) 012035 doi:10.1088/1757-899X/863/1/012035

3.2. SEM analysis

Figure 3 shows SEM micrographs of untreated and plasma treated TiO_2 nanoparticles. Most of the nanoparticles in untreated sample exist in the form of nano-rocks. On plasma treatment, these nano-rocks break into nanoparticles. The size of nanoparticles almost reduces by half after plasma treatment. SEM Images clearly reveal a change in the surface morphology of plasma treated TiO_2 nanoparticles.

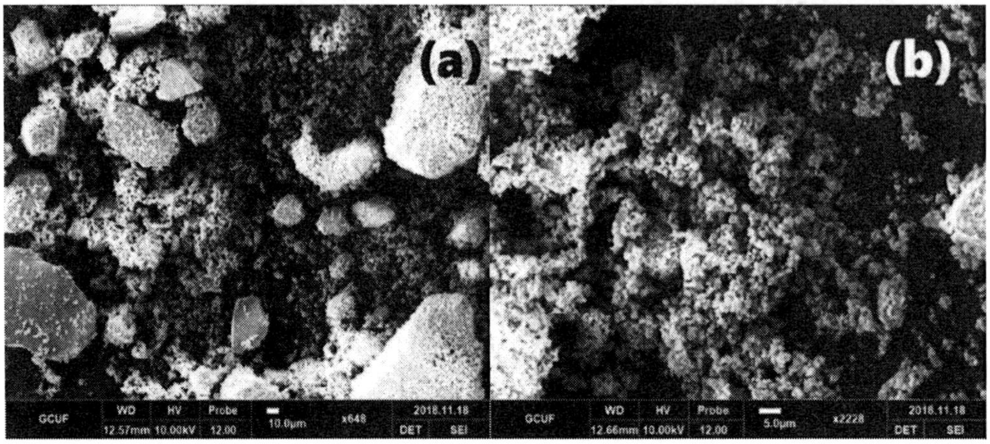

Figure 3. SEM micrographs of; (a) untreated and (b) plasma treated TiO_2 nanoparticles.

3.3. UV-Visible analysis

Figure 4 shows a typical absorption spectrum of plasma treated TiO_2 nanoparticles. UV-Visible study was conducted to evaluate the opto-electric properties of plasma treated TiO_2 nanoparticles. The band gap energy (E_g) was computed using equation (2) [35]. The plasma treated TiO_2 exhibited UV-resistance up to absorption edge wavelength of 200.55 nm. The band gap energy, corresponding to this wavelength, was calculated about 6.18 eV. This energy was almost 1.8 times higher than the band gap energy of untreated nanoparticles. The photocatalytic activity of nanomaterial semiconductors depends on the band gap energy, which is governed by the particle size. These results reveal that plasma treated TiO_2 nanoparticles have higher photocatalytic activity due to larger band gap energy and smaller grain size.

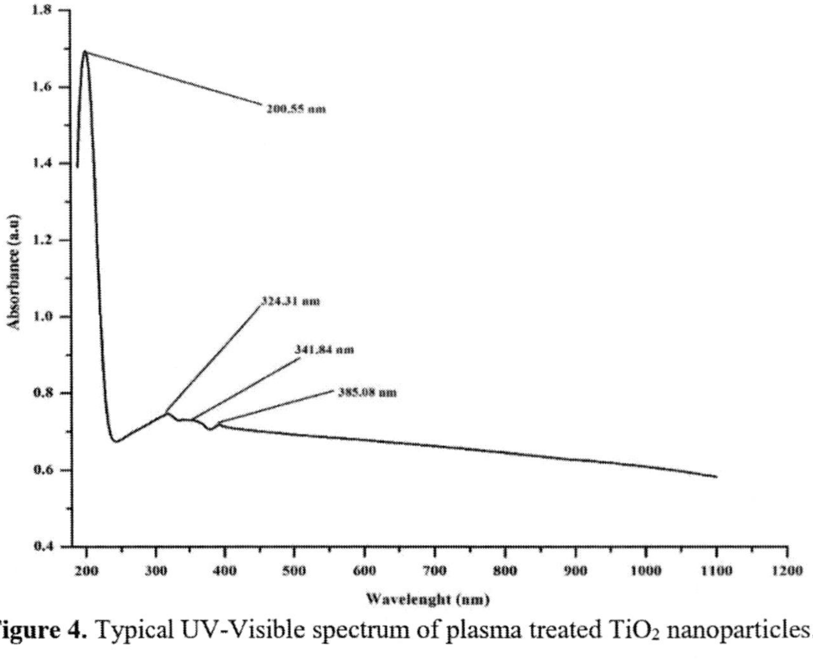

Figure 4. Typical UV-Visible spectrum of plasma treated TiO_2 nanoparticles.

4. Conclusion

Plasma assisted sol-gel is a promising technique for production of nanoparticles of controlled sizes. The size of nanoparticles, synthesized through plasma assisted sol-gel, remained in the range of 0.2 nm to 14 nm. The particle sizes obtained through conventional sol-gel method were two times larger than those obtained through plasma assisted sol-gel technique. Most of the nanoparticles in untreated sample existed in the form of nano-rocks. On plasma treatment, these nano-rocks broke into smaller nanoparticles. The size of nanoparticles almost reduces by half after plasma treatment. The band gap energy of plasma treated nanoparticles was calculated about 6.18 eV. This energy was almost 1.8 times higher than the band gap energy of untreated nanoparticles. The plasma treated TiO_2 nanoparticles showed higher photocatalytic activity due to larger band gap energy and smaller grain size.

References

[1] Addamo M, Augugliaro V, Di Paola A, Garcia-Lopez E, Loddo V, Marci G and Palmisano L 2008 *Thin Solid Films* **516** 3802.
[2] Haider A J, AL-Anbari R H, Kadhim G R and Salame C T J E P 2017 *Energy Procedia* **119** 332-45.
[3] Wang J, Mao B, Gole J L and Burda C 2010 *Nanoscale* **2** 2257.
[4] Sotiriou G, Gass S and Pratsinis S 2012 *MRS Online Proc. Library Arch.* **1386** 24.
[5] Heinonen S, Huttunen-Saarivirta E, Nikkanen J-P, Raulio M, Priha O, Laakso J, Storgards E and Levanen E 2014 *Colloids Surf. A: Physicochem. Eng. Asp.* **453** 149.
[6] Jiang G and Zeng J 2010 *J. Appl. Polym. Sci.* **116** 779.
[7] Yang J, Luo M, Tan Z, Dai M, Xie M, Lin J, Hua H, Ma Q, Zhao J and Liu A 2017 *Environ. Toxicol. Pharmacol.* **49** 112.
[8] Xue Q, Liu W and Zhang Z 1997 *Wear* **213** 29.
[9] Saleiro G, Cardoso S, Toledo R and Holanda J 2010 *Ceramica* **56** 162.
[10] Nachit W, Touhtouh S, Ramzi Z, Zbair M, Eddiai A, Rguiti M, Bouchikhi A, Hajjaji A and Benkhouja K 2016 *Mol. Cryst. Liq. Cryst. Sci.* **627** 170.
[11] Kim S J, Park S D, Jeong Y H and Park S 1999 *J. Am. Ceram. Soc.* **82** 927.
[12] Elbushra H, Ahmed M, Wardi H and Eassa N 2018 *MRS Adv.* **3** 2527.
[13] Ba-Abbad M M, Kadhum A A H, Mohamad A B, Takriff M S and Sopian K 2012 *Int. J. Electrochem. Sci* **7** 4871.
[14] Liang Y, Qiao B, Wang T-J, Gao H and Yu K 2016 *Appl. Surf. Sci.* **387** 581.
[15] Hore S, Vetter C, Kern R, Smit H and Hinsch A 2006 *Sol. Energy Mater. Sol. Cells* **90** 1176.
[16] Liu N, Chen X, Zhang J and Schwank J W 2014 *Catal. Today* **225** 34.
[17] Hayle S T and Gonfa G G 2014 *Am. J. Nanosci. Nanotechnol* **2** 1.
[18] Mahshid S, Askari M and Ghamsari M S 2007 *J. Mater. Process. Technol.* **189** 296.
[19] West R H, Celnik M S, Inderwildi O R, Kraft M, Beran G J and Green W H 2007 *Ind. Eng. Chem. Res.* **46** 6147.
[20] Liang Y, Sun S, Deng T, Ding H, Chen W and Chen Y J M 2018 *Materials* **11** 450.
[21] Aromaa M, Keskinen H and Makela J M 2007 *Biomol. Eng.* **24** 543.
[22] Gross K, Tikkanen J, Keskinen J, Pitkänen V, Eerola M, Siikamaki R and Rajala M 1999 *J. Therm. Spray Technol.* **8** 583.
[23] Keskinen H, Makela J M, Aromaa M, Keskinen J, Areva S, Teixeira C V, Rosenholm J B, Pore V, Ritala M and Leskela M 2006 *Catal. Lett.* **111** 127.
[24] Mohammadi M, Cordero-Cabrera M, Ghorbani M and Fray D 2006 *J. Sol-Gel Sci. Technol.* **40** 15.
[25] Heikkila P 2008 *Nanostructured fibre composites, and materials for air filtration* vol 749 (Finland: Tampere University of Technology Publication) p 88.
[26] Heikkila P, Olin N and Nikkanen J-P 2012 *J. Mod. Text. Sci. Eng.* **2** 1.
[27] Danks A E, Hall S R and Schnepp Z 2016 *Mater. Horiz.* **3** p 91.
[28] Chen X and Mao S S 2006 *J. Nanosci. Nanotechnol.* **6** 906.

[29] Gupta S M and Tripathi M 2012 *Cent. Eur. J. Chem.* **10** 279.
[30] Sharma A, Karn R and Pandiyan S 2014 *J. Basic Appl. Eng. Res* **1** 1.
[31] Kervee S E and Mohammadi M 2017 *J. Photochem. Photobiol. A: Chem.* **338** 104.
[32] Phonkhokkong T, Thongtem T, Thongtem S, Phuruangrat A and Promnopas W 2016 *Dig. J. Nanomater. Biostruct.* **11** 81.
[33] Ilkhechi N N, Alijani M and Kaleji B K 2016 *Opt. Quant. Electron.* **48** 148.
[34] Grishina M, Bolshakov O, Potemkin A and Potemkin V 2016 *Comput. Theor. Chem.* **1091** 122.
[35] Behnajady M, Eskandarloo H, Modirshahla N and Shokri M 2011 *Desalination* **278** 10.

IOP Conf. Series: Materials Science and Engineering **863** (2020) 012036 doi:10.1088/1757-899X/863/1/012036

Plasma-fabric interaction for surface activation and functionalization: A review

M Saleem*, N Kousar, B Shoukat, M Shoaib-ur-Rehman, F Batool, M Y Naz and A Ghaffar

Department of Physics, University of Agriculture, 38040 Faisalabad, Pakistan.

*E-mail: saleem_192@yahoo.com

Abstract. The applications and effects of plasma on textile fabrics are the main focus in this paper. We surveyed available literature on the plasma-fabric interaction and found that there are variation of parameters affected the plasma application. The parameters can be classified as weight loss, tensile strength, contact angle, aging effect and addition of contents of different chemical functional groups. The increase or decrease in weight of fabrics after oxygen, nitrogen, and atmospheric plasma treatment was obtained in this study. The weight loss increased with the exposure time of plasma but decreased with the pressure. The tensile strength decreases with plasma duration and increases with plasma power. The type of plasma also affects the wettability of fabric. Wettability of a fabric increases with plasma treatment time. The plasma treatment of fabric adds different elements such as oxygen, carbon and nitrogen on the surface. The amount of oxygen increases but carbon and sulphur decreases. The increase or decrease in contact angle depends on the conditions of plasma treatment. The aging time of plasma depends on different parameters of plasma treatment. The hydrophilicity character also affected by plasma treatment. The hydrophilicity decreased with treatment time.

1. Introduction

Plasma is a quasi-neutral gas of charged and neutral particles, which exhibit collective behaviour Nakasugi *et al* [1]. The photons, ions, electrons, atoms, free radicles, and excited molecules are present in the plasma [2]. A gas is an electrical insulator but when the electric potential is applied to the gas, it breaks down the gas and produces a conducting state of matter. The breakdown depends on several factors, such as gas species, gas pressure, gas flowrate, geometry of the setup and separation of the surfaces across which the voltage is sustained. When electric potential is applied, immediately a cloud of ions surrounds the negative ball and a cloud of electrons surrounds the positive ball. This phenomenon is called Debye shielding. If the thickness of charge particle layer is much smaller ($\lambda_D << L$) than the length of plasma dimensions, it becomes ionized gas known as a plasma and the number of charged particles in the Debye sphere must be greater than one ($N_D >>> 1$). Electrons in the plasmas are displaced from a uniform background of ions due to perturbation. Because of their inertia, the electron will overshoot and oscillate around their equilibrium position with their characteristic frequency known as plasma frequency. For an ionized gas to behave like a plasma, the condition $\omega_P > \upsilon_C$ should be fulfilled. Plasma is used in textile processing, where electrically charged particles interact with each other and surface of fabric to introduce unique physical and chemical properties at the surface.

The collective behaviour of plasma has the ability to screen out the local density perturbations and to produce a sheath between of external surface and plasma. In low-pressure discharge plasmas, the equilibrium is not reached between the electrons and heavy particles. The plasma has two types, one is low temperature plasma or weakly ionized plasma or gas discharge and other is high temperature plasma or fusion plasma. The low temperature plasma can be thermal or non-thermal. The temperature of all species in thermal plasma are same at thermal equilibrium, such as in stars and fusion plasma. In non-thermal plasma, all species do not have the same temperature. The electrons have temperature greater than the other species in plasma. There are many sources of creation of plasma, such as direct current, capacitively coupled plasma (CCP), inductively coupled plasma (ICP), microwave plasma, electron cyclotron resonance plasma, etc.

2. Characteristics of textile fabrics

A fibre is a unit of fabric, which is either natural or artificial. It is the basic element of fabrics and textile products. A fiber has a length at least 100 times of its diameter or width. This unit can be spun to made yarn or a fabric by methods including knitting, weaving, felting, braiding and twisting. The requirements for fibres to be spun include a length of at least 5 mm, cohesiveness, flexibility, elasticity, fineness, uniformity, durability, lustre and sufficient strength.

3. Effect of plasma on fabric properties

3.1. Weight loss

Trejbal *et al.* [3] determined the weight loss of plasma exposed fibre at different treatment times. The weight loss more familiar and frequently used for plasma treatment time over 60 seconds. After 480 seconds of treatment, the coated and uncoated fabrics lost their weight up to 2.4% and 4.7%, respectively. The loss in weight was proportional to the duration of plasma treatment. The loss in weight reduces the mechanical strength of fibres. Ling *et al.* [4] treated the grey cotton fabric with plasma. The pressure of inlet gas for plasma was in the range of 20 Pa to 100 Pa by keeping an interval of 20 Pa. The input discharge power was changed from 50 W to 350 W at interval of 50 W, and the exposure time was also changed from 3 min to 20 min. Lee *et al.* [5] treated the polyethylene terephthalate fibers with atmospheric pressure plasma. The weight loss increased by increasing the plasma treatment time but after the exposure time of 50 s, the increase in weight loss slowed down. It was due to the fact that the plasma etching rate was more rapid but after removing some etchable amorphous materials, the removal of crystalline materials was more difficult. The other reasons of slowing down the etching rate were of some molecular fragments in plasma re-deposited on the surface. Wang *et al.* [6] investigated the plasma treated grey cotton fabric by low pressure DC air plasma. The top layer of fabric consisted of oil, cotton wax and dirt particles. The low molecular weight layer introduced the hydrophobic properties at the surface. The high energy beams of electron collided with fabric and removed the contaminated layer from the fabric. The ablation of polymer chain started after plasma cleaning, which caused the weight loss. The weight loss was calculated using the following equation (1):

$$\text{Weight loss (\%)} = (w_f - w_i / w_i) \times 100 \tag{1}$$

Here W_f and W_i are the final and initial weight of the plasma treated fabric. Zhang *et al.* [7] treated the cotton fabrics with the atmospheric pressure plasma for different intervals of time. The loss in weight after plasma treatment is given table 1. The loss in weight increased with prolonged treatment time. It reveals that cleaning and etching affects were strong enough to degrade the sample surface.

3.2. Tensile strength

Trejbal *et al.* [3] studied the effect of plasma treatment time on surface properties of fabric. After 120 s of treatment time, 40% reduction in tensile strength was observed. The load bearing capacity of fiber was not affected during plasma treatment. The plasma only effected the fiber coating and did not affect the load bearing core. Similar results were reported by Ling *et al.* [4] who studied the effect of low-pressure nitrogen plasma on grey cotton fabric. The tensile strength reduced with nitrogen plasma pressure from 20 Pa to 100 Pa. The tensile strength was measured by strip method ASTM D5035. The dimensions of cotton fabric ware 5×35 cm^2 in warp direction. The tensile strength loss can be obtained using equation (2).

$$\text{Tensile strength loss} = \frac{T_0 - T_1}{T_0} \times 100 \quad (2)$$

where T_0 and T_1 represent the fabric tensile strength before and after plasma treatment. Stefano *et al.* [8] performed the atmospheric pressure plasma treatment of wool/cashmere (15/85%) by using dielectric barrier discharge plasma setup. Four samples (60×30 mm^2) were treated using Instron 5942 instrument. The traverse rate was maintained at 50 mm/min and the load was kept 200 N. The recording was performed at 800 point/min. Both extensibility and tensile energy of fabric increased but linearity and resilience decreased slightly.

Table 1. Comparison of different plasma treated fabrics by changing plasma parameters.

Plasma type	Processing parameters	Fabric property	Reference
Oxygen plasma on hydrophobic fiber	After 480 s, coated and uncoated fibers lost weight up to 2.4% and 4.7 %, respectively	Weight loss increased with treated time	Trejbal *et al.* [3]
Nitrogen plasma on grey cotton fabric	Plasma pressure range was (20.0 to 1000.0) Pa, power 200 W. Exposure time 5 min	Weight loss decreased with gas pressure	Ling *et al.* [4]
Atmospheric plasma on PET fabric	---	Weight loss increased with time	Lee *et al.* [5]
Atmospheric pressure plasma on cotton fabric	Pressure 20 Pa, treatment time 1 to 20 min	Weight loss increased with time	Zhang *et al.* [7]

Gaminian *et al.* [20] treated the fabrics with madder/TiO_2 and TiO_2 nanomaterial. The Alkaline hydrolysis of polyester fabric generated pits on the fabric surface that probably act as weak points when fiber is extended under the action of stress. The introduction of TiO_2 nanoparticles onto the surface of polyester fiber increased the tensile strength by filling the pits and interaction with hydroxyl and carboxyl groups of polyester chains. Chongqi *et al.* [9] investigated the effect of low temperature plasma on polyester fabrics. The tensile strength decreased significantly after plasma treatment, especially at voltages above 55 V. The large voltage induced rigorous etching on polyester surface and reduced the tensile strength. Sudhir *et al.* [10] studied the properties of carbon fiber after treating it with cold remote nitrogen oxygen plasma. The presence of oxygen in plasma induced some functional groups on the surface. The mixture of 1% O_2 in N_2 proved more effective than 0.5% O_2 in N_2 plasma. The effect of plasma treatment on tensile strength is summarized in table 2.

Trejbal *et al.* [12] investigated the effect of corona discharge and enzymatic treatment on the tensile strength of fabric. Wool fabrics were pre-treated with corona discharge at power of 100 W. These samples were then treated with enzyme by keeping humidity and temperature constant. Ming *et al.* [2] studied the effect of low temperature oxygen plasma on the ramie fabric. A decrease in tensile strength was observed with plasma treatment times due to etching and oxidation effects.

Table 2. Effect of plasma treatment on tensile strength of fabric.

Plasma type	Processing parameters	Property affected	Reference
Oxygen plasma on polymer macro fiber	Loading frame Web Tiv Ravestein FP100, loading rate 0.8 mm/min until load of 60 N and 0.6mm/min until ultimate strength	Tensile strength decreased with plasma duration	Trejbal *et al.* [3]
Nitrogen plasma on grey cotton fabric	Dimensions were 5.0 cm × 35.0 cm in warp direction, pressure 20.0 Pa to 100.0 Pa, power 200.0 W, time 5.0 min	Tensile strength loss reduced with pressure and enhanced with power and time	Ling *et al.* [4]
Atmospheric pressure plasma	Power 180 W,270 W, energy dose 1200, 3600 adsorption time 13 min, 1 min	Extensibility and tensile energy increased but linearity and resilience decreased	Stefano *et al.* [8]
Low temperature plasma on polyester fabric	Voltage from 40 to 60 V	Tensile strength decreased after plasma treatment	Chongqi *et al.* [9]
Cold remote nitrogen oxygen plasma on carbon fiber reinforced polymer composites	Pressure 10 Pa, time 5 min, nitrogen flow rate 2.5slm, working pressure 4hPa	Tensile strength increased by enhanced adhesion due to the presence of oxygen in nitrogen	Sudhir *et al.* [10]
Low temperature oxygen plasma on the ramie fabric	Gas pressure 20 to 100 Pa, power 150 W for time 3.0 min	Tensile strength decreased with time and discharge but increased with pressure	Ming *et al.* [2]
Corona discharge and enzymatic treatment with protease and cellulose on wool fabrics	Power 100 W, speed 0.5 cm/s, 5wt% enzyme at constant temperature and humidity	Strength decreased with plasma treatment time	Gaminian *et al.* [20]

3.3. Wettability

Stefano *et al.* [11] studied the effect of atmospheric pressure plasma on wool/cashmere (15/85%) textiles with dielectric barrier discharge in humid air (air water mixture). The adsorption time of de-ionized water droplets was decreased with plasma treatment. The untreated fiber did not adsorb the water droplets and droplets remained on the surface until evaporation. It was observed that wettability increases for both cashmere and wool/cashmere with energy dose and wheel speed. The cashmere fiber was influenced more than the wool/cashmere. In other experiments, power and energy dose were kept at 270 W and 4500 Wmin/m^2 and speed of mobile station was varied from 0.3 to 4.8 m/min and number of passes from 1 to 16. The wettability increased at higher number of passes. The wettability also revealed an increase with increasing the energy dose. The better results were obtained by providing energy in multiple steps. The multiple steps of energy caused more uniform surface modification.

Deterioration was observed at high dosage and low speed with the presence of burnt fibers, while there was no variation in organoleptic properties.

Shahidi *et al.* [21] carried out the drop test to investigate the hydrophobic and hydrophilic properties of the plasma treated samples. The drop absorption time of the untreated sample was larger than the plasma-treated sample. Moreover, with an increase in plasma treatment time, the roughness on the surface also increased. Stefano *et al.* [8] studied the effect of atmospheric pressure plasma on wool/cashmere (15/85%) textiles with dielectric barrier discharge arrangement. The energy was supplied in multiple steps. The treated samples lost their weight due to some dehydration during plasma treatment. The wettability of treated fibers increased with an increase of energy dose. The nitrogen plasma introduced some functional groups of cysteic acid (SO_3 groups), cystine monoxide (-SO-S- groups) and cystine dioxide (-SO_2-S-groups) on the treated surface. These groups provided polar characteristics and improved wettability to fabric. The effect of plasma treatment on wettability of fabrics is summarized in table 3.

Table 3. Effect of plasma treatment on wettability of fabric.

Plasma type	Processing parameters	Property affected	References
Atmospheric pressure DBD of nitrogen, wool/cashmere textile	Weight of drop 12 μL, 8 passes of mobile station. ATR spectra in 800-1800 cm^{-1}	Wettability increased by supplying energy dose	Stefano *et al* [11]
Low pressure nitrogen plasma on grey cotton fabric	Capillary rise height method, in warp direction, power 200 W, time 5.0 min, $K_2Cr_2O_7$ (1wt.%) for 1 cm	Capillary rise height decreased due to decrease of impurity removal efficiency	Ling *et al* [4]
Atmospheric pressure plasma with DBD in humid air on wool/cashmere textile	Experiment 1(power 270 W, speed 2.5 m/min), Experiment 2 (power 270 W, energy dose 4500 W min/m^2)	Wettability increased with supplying energy dose and number of passes	Stefano *et al* [8]
Diode plasma discharge device using DC Ar plasma on non-woven poly-hydroxy-butyrate (PHB)	Air flow rate 3 L/s, density of fiber= 1.25 g/cm^3, power 8W, time 240 s. Ar pressure 10 Pa, electrode area 48 cm^2, distance between electrode 5 cm	Plasma treatment decreased the contact angle and metallic sputtering increased the contact angle	Slepicka *et al* [18]
Dielectric barrier discharge on antimicrobial viscose fabric	Optimal treatment time was 120 s, energy density=14.4 J/cm^2	Plasma treatment increased the surface energy, wettability and capillary height	Kramar *et al* [21]

Ling *et al.* [4] investigated the effect of nitrogen plasma on grey cotton fabric. The fabric capillary rise height decreased as nitrogen pressure increased from 20 Pa to 100 Pa. The Capillary rise height decreased due to decrease of impurity removal efficiency during plasma treatment. Slepicka *et al.* [18] investigated the wettability and contact angle of non-woven poly-hydroxy-butyrate after treating with DC Argon plasma. The untreated sample was hydrophobic with surface contact angle of 120°. The plasma treated fiber exhibited a decrease in contact angle. The surface also became completely hydrophilic after treatment. The plasma treated samples were metalized through direct metal sputtering and immersing into Au/Ag nanoparticles' solution. The contact angle was increased after sputtering of gold and silver coated fabric due to formation of discontinuous metal clusters on the polymer. Kramar *et al.* [22] applied dielectric barrier discharge on antimicrobial viscose fabric to study the wettability and sorption properties. The capillary rise of sample, treated with air DBD, was increased appreciably. The plasma introduced the new polar groups on the surface through oxidation. such as hydroxyl functional groups and carboxylic groups. The increased number of polar groups improve the surface energy, wettability and capillary height. For longer plasma treatment times, some of the fibers get damaged. The optimal treatment time and energy density were revealed as 120 s and 14.4 J/cm^2, respectively.

3.4. Addition of functional groups

Trejbal *et al.* [3] studied the addition of oxygen groups on the surface of polymer fiber after the oxygen plasma treatment. The ion beam of plasma increased the surface roughness and activated the functional groups on the surface by reducing the surface energy. The X-ray photo electron spectroscopic analysis was carried out after one day of oxygen plasma treatment. XPS analysis showed that the prolonged plasma treatment leads to a bonding of 48 at% of O_2 on coated fiber and 21 at% on uncoated fiber. Both the coated and uncoated fibers were treated with oxygen plasma for 5 s, 30 s and 480 s. The plasma treatment significant enhanced the oxygen content in the fiber, consequently the hydrophilicity. These results were also supported by the contact angle measurements. After XPS analysis of deconvoluted high resolution C 1s XPS peaks, it was found that oxygen containing groups (C-O, C=O, and O-C=O) on the surface were increased by 17%, 21% and 34% after treated time of 5 s, 30 s and 480 s, respectively. Stefano *et al.* [11] studied the effect of atmospheric pressure plasma on wool/cashmere (15/85%) textiles using a dielectric barrier discharge arrangement. Nitrogen was used as a source gas for discharge plasma. Plasma treatment decreased the carbon content and increased the oxygen content of the fabric. The ratio of O to C increased due to etching of external layer of lipids and insertion of oxygenated species on surface after plasma treatment. The nitrogen plasma treated samples showed a better oxidation of lipid layer. Plasma treatment also reduces the amount of sulphur on the fiber.

Marija *et al.* [19] treated the cotton and rayon with low pressure water vapour plasma and an extract of japonica rhizome. The plasma treatment increased the adsorption capacity of cotton and rayon. Jian *et al.* [12] studied the effect of oxygen plasma on pure wool, chitosan deposited wool and polyaniline deposited wool. After plasma treatment, the carbon peak in XPS spectra became much lower and peaks of nitrogen and oxygen became stronger. The amount of nitrogen increased due to loss of carbon content. A decrease in C/N ratio and increase in O/N ratio was observed due to plasma etching of the surface and destruction of C-C bonds. The hydrophilic character of wool was increased due to attachment of groups, such as hydroxyl and carboxyl groups. The polyaniline/wool fabric was doped in acetic acid and hydrochloric acid. The doping enhanced C1s peak and decreased O1s peak. The C/N ratio was increased, and O/N ratio was decreased after plasma treatment. The effect of plasma treatment on surface chemistry of different fabrics is summarized in table 4.

Stefano *et al.* [8] investigated the effect of atmospheric plasma on pure cashmere and wool cashmere textiles. The XPS analysis showed a decrease in carbon content and increase in oxygen content in fiber with plasma treatment. Yongqing *et al.* [13] revealed a decrease in C-C/C-H component of C 1s peak and an enhancement of C- O, C= O and O- C= O components. Stefano *et al.* [8] and increased in C/O ratio due to etching of lipids and the insertion of oxygenated species. The insertion of oxygen by atomic oxygen was much higher than the nitrification induced by atomic nitrogen. The amount of nitrogen increased slightly due to removal of external hydrocarbon chains. Lee *et al.* [14] investigated the effect of diffused plasma on polyamide 66 (PA66). After the plasma treatment, the carbon content of PA66 fabric decreased and oxygen content increased. Nitrogen content also changed with treatment time. The plasma treatment added some polar functional groups to the surface of polyamide material. C1s and O1s spectra split into four peaks, which represent four carbon groups: C-C (285.0 eV), C-N (285.3 eV), C-O/C-OH (286.5 eV) and CONH (288.0 eV).

Table 4. Effect of different plasma treatments on surface chemistry of fabrics.

Plasma detail	Processing parameters	Property affected	References
Oxygen plasma on polymer fiber	Diameter of fiber 0.1mm, length 10 mm, treated time (5, 30, 480) s, power 100 W, pressure 56 Pa	Oxygen bonds increased on the coated and uncoated fiber	Trejbal *et al.* [3]
Nitrogen and humid air plasma on wool/cashmere (15/85%) textiles	Gas flow (20 L/min), power 180 W and 270 W, energy dose 1200 J and 3600 J	Percentage of oxygen increased but carbon and sulphur decreased	Stefano *et al.* [11]
Oxygen plasma on pure wool, low temperature plasma wool chitosan (CTS) deposited wool, polyaniline deposited wool,	Air pressure 20 Pa, radio frequency plasma for 1-6 min, power between 50 W and 400 W	Percentage of oxygen increased and of carbon decreased in LTP wool and vice versa in PANI wool, PANI/CTS wool	Jian *et al.* [12]
Atmospheric pressure plasma on wool/cashmere textiles in humid air	Treatment A (gas flow10 L/min, water vapour flow rate 3 g/h, station speed 2.5 m/min, power 450 W min/m^2), Treatment B with water flow rate 6 g/h.	Treatment A increased the percentage of O, N and S but treatment B increased O, N contents and decreased S	Stefano *et al.* [8]
Diffused plasma at atmospheric air pressure on polyamide 66 (PA66)	Treated for 10 s, 60 s. frequency (100 Hz,3 00 Hz, 500 Hz)	Amount of oxygen increased but carbon decreased	Lee *et al.* [14]

3.5. Contact angle

Trejbal *et al.* [3] investigated the effect of oxygen plasma on fiber reinforced composites material. The hydrophilicity was quantified by the contact angle measurements using an optical method. The fabrics were submerged in the demineralized water to evaluate the contact angles by using software CAMTIA. The angles were measured automatically at the outlines of menisci, formed due to adhering of water. Measurements were taken after 1, 7, and 30 days after plasma treatment. The uncoated fabric retained its hydrophilicity. The average contact angle of coated fabric was equal to 55.2°±2.3° and for uncoated fabric, the contact angle was 82.1± 2.8°. The plasma treatment reduced the contact angle by 24° to 29°.

Mehmet *et al.* [15] presented an atmospheric pressure plasma method to improve the pigment-based ink-jet printing performance on silk fabric. The contact angle was measured with DSA-100 drop shape analyzer (KRÜSS GmbH, Hamburg and Germany) to know the wettability of the silk fabrics. The probe liquid was distilled water and ethylene glycol. It was observed that the contact angle decreases by 20° and 15° for distilled water and ethylene glycol, respectively. Lee *et al.* [16] studied the effect of atmospheric-pressure air/He plasma on the surface of polyester fabrics. A mixture of air and 10% helium was used to generate plasma. The fabric was treated for 90 seconds. The effect of plasma treatment on hydrophilicity was investigated by measuring the dynamic contact angle. The probe liquids were the distilled water and ethylene glycol. It was observed that the contact angle of control fabric was about 85° and 72° for distilled water and ethylene glycol, respectively, which was decreased to 28° and 18°

after plasmas treatment. The effect of different plasma treatments on contact angle is summarized in table 5. Yongqiang *et al.* [13] studied the effect of helium plasma on cotton fabric. This work showed that coating a film of polymer on cotton fiber with plasma would result in high hydrophobicity. Park *et al.* [17] measured the water contact angle of the fabrics at ambient temperature by utilizing the sessile drop technique. The argon gas was used to produce plasma. The argon plasma caused sputtering of copper electrode. The copper was deposited on the surface of polyester/silk blended fabrics. 5 mL of distilled water was placed at five different points on the surface of fabric. The sputtering power was 100 W and distance between target and sample was 5.5 cm. The sputtering times were set as 4, 10, 20, 30 and 40 minutes. The contact angle of the copper coated fabric was increased many folds depending on the treatment time.

Table 5. Comparison of effect on fabric of different plasma treatment for contact angle.

Plasma detail	Processing parameters	Property affected	References
Oxygen plasma on fabrics reinforced composite materials	Measuring time 0, 1, 7 and 30 days after plasma treatment time of 0-480 s	Contact angle decreased after plasma treatment	Trejbal *et al.* [3]
Atmospheric pressure plasma on silk fabrics	Power 300 W, treatment time 90 s, dielectric space 3 mm, temperature 20 ± 2 °C and relative humidity 65 ± 5%	Contact angle decreased from 82° to 55° for distilled water and ethylene glycol	Mehmet *et al.* [15]
Atmospheric-pressure air/He plasma on surface of polyester plain weave fabric	Power 0 to 500 W, frequency 1kHz, pressure 100 kPa, air and 10% helium, dielectric space of 3 mm	Contact angle decreased to 28° and 18° for distilled water and ethylene glycol from 85° and 72°, respectively	Lee *et al.* [16]
Glow discharge of Helium low temperature plasma on the cotton fabrics	Frequency 13.56 MHz, power 0 to 500 W, process time 0-300 s	Contact angle increases with an increase in mass concentration of polymer	Yongqiang *et al.* [13]
Plasma sputtering process of copper on polyester/silk blended fabrics	Argon gas flowrate 14 mL/min, working pressure 6×10^{-6} mbar, power 100 W, sputtering time 4, 10, 20, 30 and 40 minutes	Contact angle increased from 0° to 146° after coating the fabric with copper	Park *et al.* [17]

4. Conclusion

In this paper, plasma-fabric interaction was reviewed pertaining to the effect of plasma treatment on physico-chemical properties of different fabrics. The post-plasma treatment weight loss, tensile strength, contact angle, aging and functional groups of the fabrics were reviewed. An increase or decrease in weight of fabrics generally happens after treatment depending on the type and conditions of plasma treatment. The weight loss increases with plasma exposure time and decreases with plasma pressure. The tensile strength decreases with exposure time but increases with plasma power. Plasma treatment also tailors the fabric wettability. Wettability of a fabric increases with plasma treatment time. The plasma treatment of fabric adds different functional groups to the fabric surface. The groups may contain oxygen, carbon and nitrogen, depending on the type of plasma.

References

[1] Nakasugi, Y., et al., *Experimental and Mathematical Analyses of Bio-electrochemical Conversion of Carbon Dioxide to Methane.* Energy Procedia, 2017. **114**: p. 7133-7140.

[2] Ming. S, et al., *Effect of low-temperature oxygen plasma on the degumming of ramie fabric.* Journal of Cleaner Production, 2015. **92**: p. 318-326.
[3] Trejbal. J, et al., *Deterioration of bonding capacity of plasma-treated polymer fiber reinforcement.* Cement and Concrete Composites, 2018. **89**: p. 205-215.
[4] Ling. W, et al., *A plasma aided process for grey cotton fabric pretreatment.* Journal of Cleaner Production, 2013. **54**: p. 323-331.
[5] Lee. L, et al., *Diffuse plasma treatment of polyamide 66 fabric in atmosphericpressure air.* Applied Surface Science, 2016. **362**: p. 348-354.
[6] Wang. C, et al., *Cotton fabric with plasma pretreatment and ZnO/Carboxymethylchitosan composite finishing for durable UV resistance andantibacterial property.* Carbohydrate Polymers, 2016. **138** p. 106–113.
[7] Zhang. C, et al., *Effect of atmospheric-pressure air/He plasma on the surface properties related to ink-jet printing polyester fabric.* Vacuum, 2017. **137**: p. 42-48.
[8] Stefano. Z, et al., *Characterization of atmospheric pressure plasma treatedpure cashmere and wool/cashmere textiles: Treatment inair/water vapor mixture.* Applied Surface Science, 2015. **349**: p. 235-240.
[9] Chongqi. M, Shulin. Z, and H. G, *Anti-static charge character of the plasma treated polyester filter fabric.* Journal of Electrostatics, 2010. **68**: p. 111–115.
[10] Sudhir. T, Bijwe. J, and P. S, *Adhesive wear performance of polyetherimide composites with plasma treated carbon fabric.* Tribology International, 2011. **44**: p. 782-788.
[11] Stefano. Z, a., et al., *Characterization of atmospheric pressure plasma treated wool/cashmere textiles: Treatment in nitrogen.* Applied Surface Science, 2017. **427**: p. 90-96.
[12] Jian. Yua, et al., *Conductivity and antibacterial properties of wool fabrics finished by polyaniline/chitosan.* Colloids and Surfaces A, 2018. **548**: p. 117-124.
[13] Yongqiang. L, et al., *Study of plasma-induced graft polymerization of stearyl methacrylateon cotton fabric substrates.* 2015. **357**: p. 2327–2332.
[14] Lee, H.J., et al., *Evaluation of a Dielectric Barrier Discharge Plasma System for Inactivating Pathogens on Cheese Slices.* Journal of Animal Science and Technology, 2012. **54**.
[15] Mehmet. K, et al., *The Evaluation with Statistical Analyses of the Effect of Different Storage Condition and Type of Gas on the Properties of Plasma Treated Cotton Fabrics.* Procedia - Social and Behavioral Sciences, 2015. **195**: p. 2170-2176.
[16] Lee. Li, et al., *Diffuse plasma treatment of polyamide 66 fabric in atmosphericpressure air.* Applied Surface Science, 2015. **362** p. 348–354.
[17] Park. M, et al., *Combined effect of corona discharge and enzymatic treatment on the mechanical and surface properties of wool.* Journal of Industrial and Engineering Chemistry, 2014. **20**: p. 179-183.
[18] Slepicka. P, et al., *Antibacterial properties of modified biodegradable PHB non-woven fabric.* Materials Science and Engineering C, 2016. **65**: p. 364-368.
[19] Marija.M, et al., *Dyeing of plasma treated cotton and bamboo rayon with Fallopia japonica extract.* Springer Science, 2016. **23:** pp 2221-2228.
[20] Gaminian. H, et al., *Simultaneous nano TiO2 sensitization, application and stabilization on polyester fabric using madder and NaOH producing enhanced self-cleaning with hydrophilic properties under visible light.* Journal of Photochemistry and Photobiology , 2017.**332:** pp 158-166.
[21] Shahidi.S,et al ., *Effect of plasma treatment on self-cleaning of textile fabric using titanium dioxide.* Micro & Nano Letters, 2015. **10:** pp. 408–413.
[22] Kramar. A, et al., *Antimicrobial viscose fabric prepared by treatment in DBD and subsequent deposition of silver and copper ions—Investigation of plasma aging effect.* Surface & Coatings Technology, 2013. **234**: p. 92-99.

IOP Conf. Series: Materials Science and Engineering **863** (2020) 012037 doi:10.1088/1757-899X/863/1/012037

A simple and low-cost purification method for microbial-free water using zinc oxide nanoparticles

S Muzaffar, M Naeem Aslam, S Rehman, S Umer, S Ahmed, S R Zahra, W Ahmad and H Anwar*

Advanced Nanomaterials and Devices Laboratory, Department of Physics, University of Agriculture Faisalabad, 38040 Faisalabad, Pakistan.

*E-mail: Hafeez.anwar@gmail.com

Abstract. Availability of standard quality drinking water is very essential for all living-beings. Availability of drinkable water has become a biggest challenge for the whole world. Water is a vital necessity for living-beings. Contamination of water has caused the millions of deaths in every year. Therefore, it is dire need to develop and explore low cost and simple methods to obtain quality drinking water. Zinc oxide (ZnO) is good reactive agent that had used to immobilize the toxic substance and destroyed the virus from contaminated water. Zinc oxide nanoparticles (ZnO-NPs) were prepared by using zinc sulfate heptahydrate ($ZnSO_4$. $7H_2O$) and sodium hydroxide (NaOH). As prepared zinc oxide nanoparticles (ZnO-NPs) were characterized by using different techniques such as X-ray diffraction (XRD), scanning electron microscopy (SEM) and UV-Vis spectroscopy. From XRD results, average crystallite size of prepared ZnO-NPs was calculated using Scherrer's equation and found to be 29 nm. SEM analysis showed that as prepared NPs has mixed morphology having hexagon and rod-like shapes. UV-Vis analysis showed that maximum absorbance range between 250 nm to 350 nm. After the characterization, zinc oxide nanoparticles (ZnO-NPs) were employed for purification of water. The treated water was then investigated by finding electrical conductivity (EC), analysis of pH, total dissolved solids (TDS), atomic absorption spectroscopy (AAS) for the study of heavy metals and colony forming units (CFU) for microbial count such as bacteria, virus and protozoa etc. It was found that water treated with ZnO killed 100 % of microorganism as compared to water without treatment of ZnO (63%).

1. Introduction

Sustainable Pure water is essential for all living-being. In developing countries, purified water has become a major problem. Water is a vital necessity for every life. Water has been used in daily routines work involving in cooking, drinking, washing, cleaning and agriculture [1]. Pakistan is a semi-parched area, and here the resources of purified water are shorted. In difficult situation, all living-beings have affected from lake of water, whose health has depended on pure water. Unhygienic water can cause infectious diseases as cholera, jaundice, paratyphoid fever, typhoid, dysentery, amoebiasis, and malaria [2]. Water pollution also caused by industrialization and environmental pollution. Residual dyes have a strong role played to contaminate water [3]. Dyes have produced by different sources as textile industries, pharmaceutical industries, paper and pulp industries, dye and dye intermediate industries and craft bleaching industries, etc. Available water distillation apparatus was in poor condition. High cost required to distill the water. Low cost apparatus was caused to contaminate the harmful substances, toxicity and pollution in water [4]. Therefore, many researchers have adopted different approaches including photocatalysis [Titanium dioxide (TiO_2), Zinc oxide (ZnO), Iron (III) oxide (Fe_2O_3), zirconia (ZrO_2), vanadium (V) oxide (V_2O_5), niobium pentoxide (Nb_2O_5) and tungsten trioxide (WO_3)] to purify water with efficiently. Among these, zinc oxide (ZnO) is most efficient and friendly candidate in environment to reduce the severe problem of water pollution because ZnO has unique properties such as strong oxidation, wide band gap (3.37 eV) near-UV spectral region, highly photocatalytic [5]. Zinc oxide nanoparticles (ZnO-NPs) are also well-known to destroy germs from water to make it drinkable water [6]. Bacteria are composed with cell membrane, cell wall and cytoplasm. The cell membrane occupies inside the cell wall. Cell wall has characterized by a homogeneous peptidoglycan layer. The function of cell wall organizes the osmotic-pressure of the cytoplasm and controls cell shape [7]. There are two types of bacteria namely Gram-positive bacteria and Gram-negative bacteria. The Gram-positive bacteria has consisted of one cytoplasmic membrane (have multilayers of peptidoglycan polymer) and cell wall which having thickness about 20 nm to 80 nm. Meanwhile, the Gram negative bacteria have composed with two cell membranes and plasma membrane which has consisted of thin layer of peptidoglycan and thickness about 7nm to 8nm. ZnO-NPs have comparable size with peptidoglycan, means that ZnO-NPs can be passed easily through the peptidoglycan to destroy the bacteria. Cytoplasm has consisted on fluid like as jelly, which has filled in all components of the cell excluding nucleus. Organisms have functions included the growth, reproduction, anabolism and metabolism [8]. Cytoplasm has contained amino acids, ion, salts, proteins, starches and water about 80%. Composition of cytoplasm leads to good electrical conductivity. Bacterial-cell wall has contained negative charge. Typically, bacterial cell structure has been shown in figure 1.

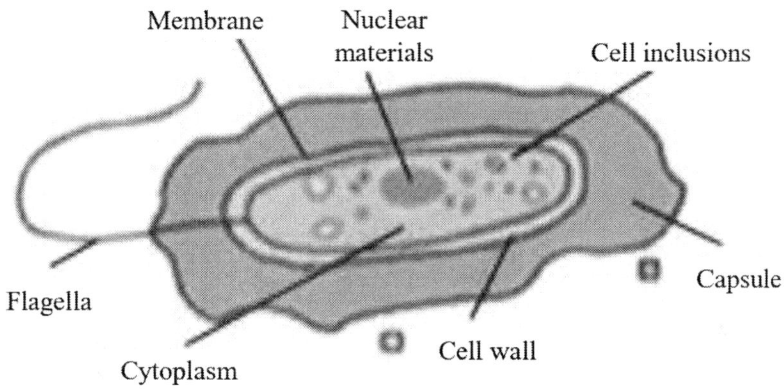

Figure 1. Bacterial cell structure.

There are several mechanisms for killing bacteria by using ZnO-NPs were conducted in previous studies. One of these mechanisms involved the direct contact of ZnO-NPs with cell walls, Reactive Oxygen Species (ROS) formation, liberation of antimicrobial ions mainly Zn, resulting in destruction of bacterial cell integrity. The property of anti-bacterial activity of ZnO-NPs has distinct for different medium such as if zinc (Zn) has reacted with species, which may cause to change toxicity mechanisms corresponding to the physico-chemical characteristics of ZnO-NPs [9].

ZnO possessed highly absorption factor to the ultra-visible light (UV). Highly absorption factor has been caused to increase conductivity dramatically, which leads to enhance the efficiency of ZnO to interact with bacteria. The morphology of ZnO has most efficient response to inactivate the micro-organisms. Rod shaped nanoparticles can be easily penetrating the bacterial cell wall and distort the peptidoglycan layer than the spherical shaped nanoparticles [10]. Zinc oxide has strong Annealing property Oxygen annealing (increase the oxygen atoms on the surface of ZnO) has been great effect to increase the antibacterial efficiency involving reactive oxygen spices (ROS), which provides an intensive strength for prohibit the bacteria. The size of particle has been much smaller then it provides efficient results to kill the bacteria, due to small particles have large surface area. Minimum Inhibitory Concentration (MIC) has been used to inhibit the bacterial growth and within perform of colony forming unit (CFU).Surface defect of ZnO has many corners for potential contact surface's sites [11]. In this research work, ZnO nanostructures were prepared by co-precipitation route and these as prepared samples were employed to contaminated water to get microbial-free drinking water. XRD, SEM, UV-vis spectroscopy were used to characterize as prepared ZnO-NPs whereas water analysis was done by using different techniques such as electrical conductivity, total dissolved solvent (TDS) and colony forming unit (CFU).

2. Experimental process

2.1. Chemicals and synthesis

All chemicals used throughout the research were of analytical regent grade and purchased from Sigma-Aldrich. For the preparation of zinc oxide nanoparticles schematic diagram is shown in figure 2. Zinc sulfate heptahydrate ($ZnSO_4$. $7H_2O$) and sodium hydroxide (NaOH) were used in this experiment. In order to aqueous solution of zinc sulfate, sodium hydroxide solution in a molar ratio of 1:2 was added slowly drop wise with vigorous stirring and the stirring was continued for 12 hours [12]. The obtained precipitate was filtered and washed three times thoroughly with deionized water. The precipitate was let to completely filtered and dried in an oven at 100 °C. The dried sample was grinded with agate pestle and mortar to obtain fine powder. The obtained powder was calcined at temperature 700 °C for 2 hours by using the muffle furnace. The as prepared ZnO-NPs were characterized using XRD, SEM and UV-vis spectroscopy. These prepared samples then employed to treat water for its purification. In this experiment the three samples were collected from different places such as home, field and factory. Sample (1, 2, 3) indicated uncoated and without sunlight samples. Sample (A, B, C) indicated uncoated and under sunlight samples and samples (A', B', C') was indicated coated and under sunlight. ZnO-NPs were prepared by co-precipitation and water purified method that was shown in figure 2.

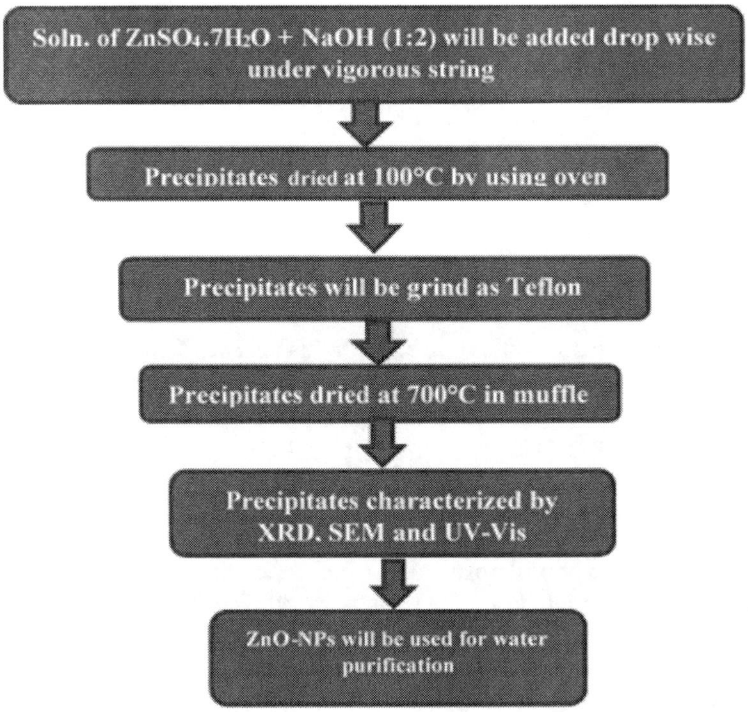

Figure 2. Co-precipitation method for synthesis of ZnO nanoparticles.

IOP Conf. Series: Materials Science and Engineering **863** (2020) 012037 doi:10.1088/1757-899X/863/1/012037

3. Characterization

The prepared samples were characterized by using different techniques such as X-ray diffraction (XRD), Scanning electron microscopy (SEM) and UV-vis spectroscopy for their structural, morphological and optical properties respectively. XRD was performed through XRD system Model (PANalytical) with CuKα (λ=1.54 Å) radiation. The particle size was calculated from broadening of line by using Scherrer's formula as given by equation (1).

$$T = \frac{0.9\lambda}{\beta \cos\theta} \tag{1}$$

The values of lattice parameters for all the samples were calculated by using equation (2).

$$\frac{1}{d^2} = \frac{4}{3}\left(\frac{h^2 + hk + k^2}{a^2}\right) + \frac{l^2}{c^2} \tag{2}$$

The volume of unit cell was calculated by using equation (3).

$$V_{cell} = a^2c \sin 60 \tag{3}$$

The density of prepared sample was calculated by using equation (4).

$$D = \frac{ZM}{N_A V_{cell}} \tag{4}$$

Where D is the crystal size, β is the line width (FWHM), θ is the diffraction angle and the λ is the wavelength of X-ray radiation. Z is the number of molecules per unit cell, M is the molar mass of the Zinc Oxide, N_A is Avogadro number, V_{cell} is the volume of the unit cell. The scanning electron microscope (SEM) studies were performed by using SEM Model (JEOL- JSM 5910) to study the morphology of the prepared nanoparticles. UV-vis spectroscopy was performed with UV-Vis spectrophotometer of Model (PG- T80).

4. Results and discussion

4.1. *Structural analysis*

4.1.1. XRD Analysis. The XRD pattern of ZnO nanoparticles prepared at 700 °C. Diffraction peaks has shown of ZnO in figure 4 that provides information particle size, volume and density. The diffraction peaks are good crystalline nature of ZnO, which will be seen in figure 3. The shown XRD pattern confirms the formation of nanoparticles.

The ZnO nanoparticles are observed to be highly crystalline. Peaks at 2θ of 31.80°, 32.66°, 34.47°, 36.28°, 47.58°, 56.64°, 62.93°, 68.00°, were referred to XRD planes (100), (002), (101), (102), (110), (103), (200) and (112), that shown in table 1. The XRD results showed that zinc oxide has the hexagonal wurtzite type structure. The average crystallite size of prepared ZnO nanoparticles was calculated by using Scherrer's relation and found to be 29 nm. The values of lattice parameters for all the samples were calculated by using equation 2 and found to be in accordance with the standard values. These parameters are matching with the reported values (JCPDS file No. 36-1451).The volume of unit cell was calculated is 47.604 $(Å)^3$,by using equation 3. There is no significant effect of annealing temperature on the volume of unit cell. The X-ray density (D) of prepared sample was found to be 5.0 g/cm^3 that was calculated by using equation (4).

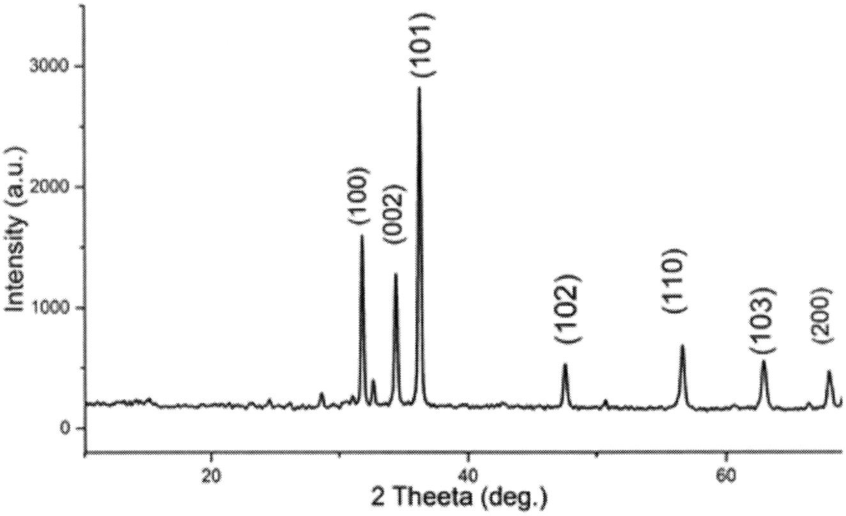

Figure 3. XRD pattern of prepared ZnO nanoparticles.

4.1.2 SEM analysis. SEM micrographs of the synthesized zinc oxide nanoparticles is shown in figure 4. It appears that the particles were agglomerated, clusters and hexagon along with some rod like nanostructures.

Figure 4. SEM micrograph of ZnO-NPs.

4.1.3. UV-Vis Spectroscopy. UV-Visible spectrometer is worked on the principle of Beer-lambert law. The concentration of ZnO (0.01 g) was dissolved in ethanol (30 ml). The UV-Visible transition of ZnO had maximum wavelength in range between 250 nm to 350 nm, which had shown in figure 5. UV-vis spectrum depends on different parameters such as method of synthesis, particle size and shape etc. This result has indicated that ZnO nanostructure had excellent UV-absorbing capability [13].

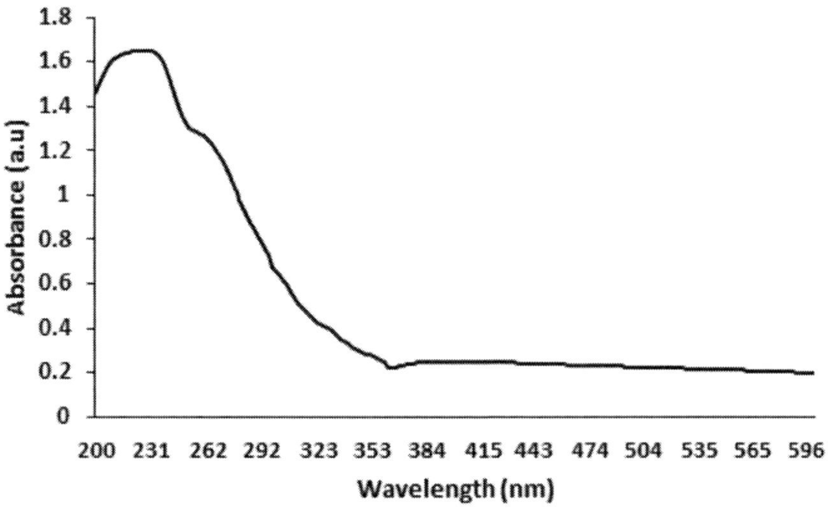

Figure 5. UV-visible Spectrum of ZnO-NPs.

4.2. Water analysis

In order to employ ZnO to kill bacteria, different samples of water from three different places were collected and gave those different named such as sample 1, sample 2 and sample 3. All the water samples were packed in three different containers and tested in laboratory. Different tests were conducted including pH, electrical conductivity (EC), total dissolved solids (TDS) and colony forming units (CFU) for microbial count such as bacteria, virus and protozoa etc. Again three water samples (sample A, sample B and sample C) had taken from the same place where they had already collected. All water samples (sample A, sample B and sample C) had placed under the sunlight for 6 hours daily, for four days (24 hours) and tested. Three containers had coated by zinc oxide nanoparticles. Again, water samples (sample A', sample B' and sample C') had taken at the same place from where they had already collected. Sample A', sample B' and sample C' had put into the coated three containers. Finally, all ZnO-NPs coated water samples had placed under the sunlight for 6 hours daily, for four days (24 hours). Therefore, water had purified by using ZnO nanoparticles.

Overall results of all three samples, which had collected from home, field and factory, and three uncoated samples had represented by sample 1, sample 2 and sample 3, respectively. Under sunlight, the samples were named as sample A, sample B and sample C. Coated sample with ZnO-NP's had indicated by sample A', sample B' and sample C'. From table 1, which had clearly conducted the results that coated water's sample had more efficient results than uncoated and under sunlight's samples. Comparison of all results of samples are shown in table 1.

Table 1. Overall results of uncoated, under sunlight and coated water samples.

Sources	**Sample name**	**pH**	**Electrical conductivity (µS/cm)**	**Total dissolved solvents (meq/L)**	**Colony forming units (Cfu/ml)**
Home	Sample 1	6.96	1207	12.07	2.80E+08
	Sample A	6.3	990	9.9	2.50E+07
	Sample A'	6.68	917	9.17	1.50E+07
Field	Sample 2	7.14	1790	17.9	3.00E+09
	Sample B	7	1044	10.44	1.76E+08
	Sample B'	7.02	895	10	2.87E+07
Factory	Sample 3	6.84	5100	510	2.61E+08
	Sample C	7.4	4480	44.8	1.20E+08
	Sample C'	6.98	580	5.8	1.90E+07

4.2.1. Percentage of CFU of uncoated and coated water samples under sunlight. The results of water treatment have been compared with uncoated water samples under the sunlight. The results have revealed that treated water has provided pure water. The results predicted that how many microbes were killed by using the zinc oxide nanoparticles (ZnO-NPs). The results of colony forming units (cfu/ml) killed in the coated and uncoated samples under the sunlight are given in the table 2.

Table 2. Cfu/ml killed in the uncoated and coated samples under sunlight.

Sample name	Cfu/ml killed in the uncoated samples under sunlight	Cfu/ml killed in the coated samples under sunlight
Home	$1.5x10^7$	$2.5x10^7$
Field	$1.76x10^8$	$2.87x10^7$
Factory	$1.2x10^8$	$1.9x10^7$

The percentage of Cfu/ml killed in the uncoated and coated samples under the sunlight is represented by in table 3.

Table 3. Percentage of cfu/ml killed in the uncoated and coated samples under sunlight.

Sample name	Cfu/ml killed in the uncoated samples under sunlight %	Cfu/ml killed in the coated samples under sunlight %
Home	60%	100%
Field	61%	100%
Factory	63%	100%

The Cfu/ml killed under sunlight, in the uncoated samples and coated samples of water named home, field and factory is shown in figure 6.

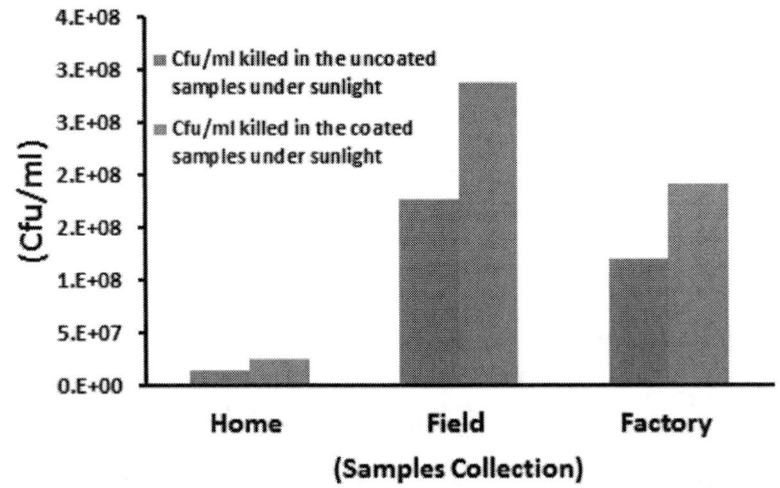

Figure 6. Cfu/ml killed under sunlight, in the uncoated and coated water samples.

Moreover, result was conducted that the microorganisms killed under the sunlight, in zinc oxide nanoparticles (ZnO-NPs) coated samples were 100% and in the uncoated samples were 63%. Finally, by coating zinc oxide nanoparticles (ZnO-NPs), all microorganisms were killed successfully and the water became totally pure form microorganisms, because all results are in the range of pure water shown above in the table 1.

5. Conclusion

Aim of this research was to synthesize and characterize the zinc oxide nanoparticles (ZnO-NPs) and its application for water purification. XRD technique was used to determine the crystalline size, density and volume of zinc oxide nanoparticles (ZnO-NPs) was 29 nm, 5.0 g/cm^3 and 47.604 $(Å)^3$, respectively. SEM technique determined that ZnO-NPs have hexagon with rod like shape. ZnO-NPs have shown highest UV-Visible absorbance range 320 nm to 380 nm, which has caused to anti-bacterial. Water had purified by coating zinc oxide nanoparticles (ZnO-NPs) inside the recycle plastic bottles of water and placed under the sunlight for 6 hours daily, for four days (24 hours). Finally, concluded that by coating zinc oxide nanoparticles (ZnO-NPs) under the activity of sunlight killed microbes effectively and made contaminated water as microbial-free water.

References

[1] Chalew T E A, Ajmani G S, Huang H and Schwab K J 2013 *Environmental health perspectives* **121** 1161.
[2] Lee K M, Lai C W, Ngai K S and Juan J C 2016 *Water research* **88** 428.
[3] Liu Y, Hua L, Li S and Xi G 2011 *Desalination and Water Treatment* **26** 297.
[4] Patil S S, Patil R H, Kale S B, Tamboli M S, Ambekar J D, Gade W N, Kolekar S S and Kale B B 2014 *Journal of nanoparticle research* **16** 2717.
[5] Bagabas A, Alshammari A, Aboud M F and Kosslick H 2013 *Nanoscale research letters* **8** 516.
[6] Baruah S, K Pal S and Dutta J 2012 *Nanoscience & Nanotechnology-Asia* **2** 90.
[7] Mirzaei H and Darroudi M 2017 *Ceramics International* **43** 907.
[8] Ganeev R A, Yang R, Hong J I, Cao B, Guo R, Okada T and El-Shall M S 2012 *Edited by Subhash Chandra Singh, Haibo Zeng, Chunlei Guo, and Weiping Cai* **4** 3.
[9] Srivastava D B, Leon K, Osmundson J, Garner A L, Weiss L A, Westblade L F, Glickman M S, Landick R, Darst S A and Stallings C L 2013 *Proceedings of the National Academy of Sciences* **110** 12619.
[10] Azizi S, Ahmad M B, Namvar F and Mohamad R 2014 *Materials Letters* **116** 275.
[11] Sirelkhatim A, Mahmud S, Seeni A, Kaus N H M, Ann L C, Bakhori S K M, Hasan H and Mohamad D 2015 *Nano-Micro Letters* **7** 219.
[12] Bedi P and Kaur A 2015 *World Journal of Pharmacy and Pharmaceutical Sciences* **4** 1177.
[13] Lanje A S, Sharma S J, Ningthoujam R S, Ahn J-S and Pode R B 2013 *Advanced Powder Technology* **24** 331.

IOP Conf. Series: Materials Science and Engineering **863** (2020) 012038 doi:10.1088/1757-899X/863/1/012038

Compressive behaviour of aluminium rectangular hollow tube using Digital Image Correlation (DIC) method

M Quanjin*, M N M Merzuki, M R M Rejab and M Z C Noh

Faculty of Mechanical and Automotive Engineering Technology, Universiti Malaysia Pahang, 26600 Pekan, Pahang, Malaysia.

*E-mail: neromaquanjin@gmail.com

Abstract. The lightweight structural application is commonly used aluminium material, which generally applied in automotive, military, aerospace and naval applications. The rectangular cross-section hollow tube of aluminium material performs the better compressive properties compared to other cross-section shapes. In order to investigate the compressive behaviour of aluminium tube, Digital Image Correlation (DIC) method is used to study the evolution of deformation and strain value. The quasi-static compressive test was performed, and the experimental result was compared with the DIC measurement results. It was obtained that the failure behaviour of the aluminium rectangular hollow tube was buckling mode with several inwards and outwards deformation. In addition, it was concluded that the DIC method results were lower than the experimental values on the maximum compressive stress and compressive modulus, which was considered as a good agreement between this two methods with 1.7 % percentage of error. It is suggested that DIC method is a sufficient method to measure the deformation and strain values with non-contact and non-destructive features.

1. Introduction

Nowadays, aluminium alloys have popularly been used in the automotive engineering applications due to its light-weight, corrosion resistance, recyclability and formability. Aluminium materials are largely used in light and crashworthy structural application compared to other metal materials such as aircraft, spacecraft and naval ship [1]. For instance, the aluminium crash box is commonly used to protect passengers and vehicle components by absorbing initial kinetic energy in the crash evidence [2]. Numerous works have studied to determine the cross-section shape of the crash box with aluminium alloys in automotive applications [3-7]. The rectangular cross-section of aluminium tube showed better compressive performance under quasi-static loading condition [8]. Therefore, the compressive behaviour of the rectangular cross-section aluminium tubes have been concerned and study by numerical researchers [9, 10].

Digital Image Correlation (DIC) is an innovative non-contact and non-destructive method to measure strain and displacement of crack propagation and material deformation [11]. In DIC method, a single illumination source and a single CCD (charged coupled device) are prepared to measure the specimen with the in-plane deformation [12]. 2D DIC method is not appropriate to obtain the measurement results from curved surfaces with three-dimensional deformation. In order to advance its limitation, 3D DIC method is developed to measure the three-dimensional deformation measurements from planar or curved surfaces [13, 14]. DIC technique is widely used in materials and engineering fields, which applies to measure the displacement and strain regarding as the structural analysis, finite element verification and quality assessment [15, 16]. Furthermore, several studies have been investigated the material properties using the DIC method [17-19]. This method could minimize the error of finite element simulation compared to experimental data, which is used to identify the material properties under quasi-static loading condition at high rates of strain [20].

Although there are several studies on the mechanical properties of aluminium material using DIC method, it is still limited works to investigate the compressive properties of aluminium material with a rectangular hollow tubular structure. Therefore, the aim of this paper is to deeply understand the compressive performance of aluminium tube with rectangular cross-section shape. The DIC method is used to measure the deformation evolution subjected to strain and stress values under compressive test. Experimental results and DIC results are analyzed and compared on compressive modulus and stress versus strain graph. In addition, the compressive behaviour of the aluminium tube is discussed as well.

2. Materials and Method

2.1. Specimen preparation
The tested specimen was selected aluminium 1100 type material with a rectangular hollow structure, and the dimension of aluminium rectangular hollow tube was as shown in figure 1. The parameter of the height to wall thickness ratio is 13.3. The properties of aluminium 1100 were mainly mentioned like tensile strength of 110 MPa, Poisson's ratio of 0.33 and an elongation of 12 %.

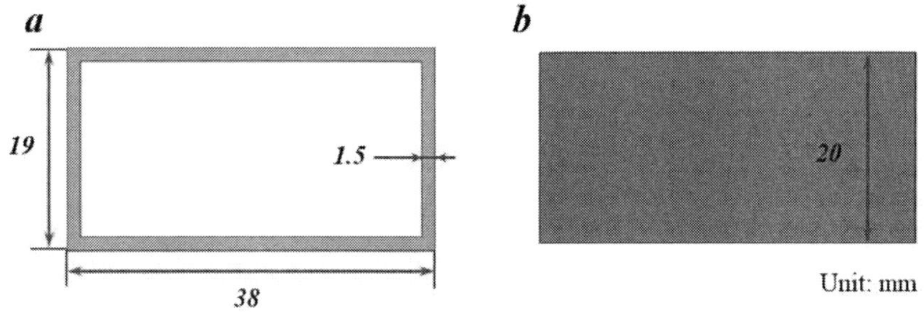

Figure 1. Dimension of aluminium rectangular hollow tube: (a) top view; (b) front view.

2.2. Quasi-static compression test
The quasi-static compressive test was performed using a standard universal testing machine Instron 3369 model with the maximum 50 kN load capability. The crosshead speed was set the 2 mm/minute. The load versus displacement curve was obtained by Bluehill software, and compressive behaviour of the tube was photographed during the test. The Nikon D3100 camera was used to recorded compressive behaviour images during the test, which was used to DIC method. However, the DIC software could only read the data from the movement of the random dots as on the surface of the tested specimen. Therefore, the aluminium rectangular tube was drawn the random dots using the black marker pen in a front surface of the specimen, which is shown in figure 2. In addition, in order to obtain better analysis results, the dots of the specimen should be cleared, and the recording zone must be accurate and adjusted.

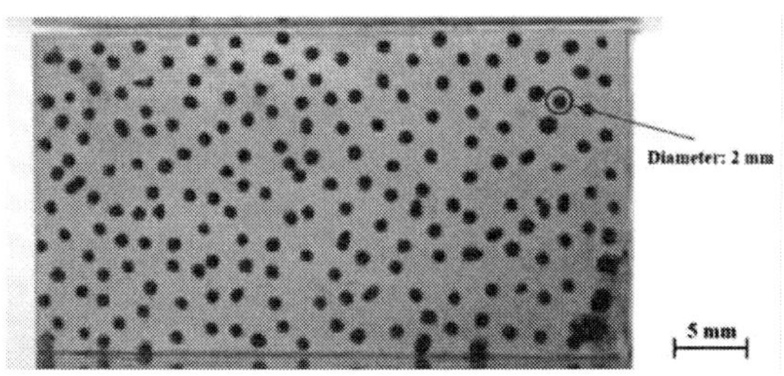

Figure 2. The front view of aluminium rectangular hollow tube with dots.

2.3. *Digital Image Correlation (DIC) method*

The flowchart of Digital Image Correlation (DIC) method procedure is briefly shown in figure 3, which involved video record, format convert, running Ncoor program from Matlab R2017a, strain method of Ncorr program, the setting of DIC parameters, DIC processing, calibration and scale. The testing video was recorded at different levels of quasi-static loading, and strains obtained in the region using DIC method. The accuracy of strain value and spatial resolution could be improved through increasing the resolution of target specimen area [21]. In this study, it is used the Nikon D3100 camera with a resolution of 4,608 3,702 pixels (14.2 megapixels). The recorded video was converted to digital images using free video to image converter software and saved into the new file folder. The value of frame per second (FPS) is calculated, and the total converted images are divided by the duration of the video time as the second unit.

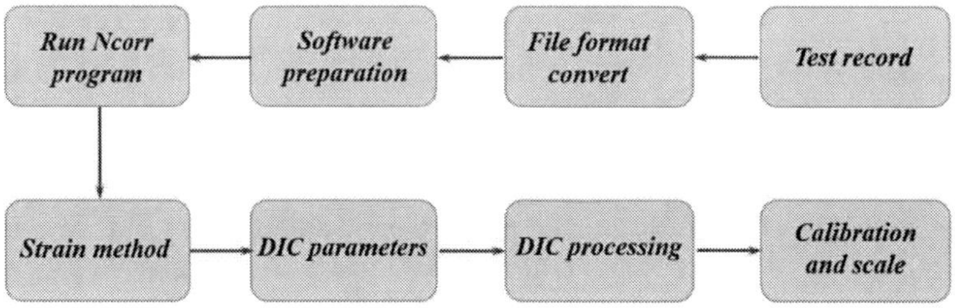

Figure 3. Flowchart of Digital Image Correlation (DIC) method procedure.

The Ncorr platform 1.1.2 version and Matlab R2016 version are downloaded and installed, which were the two software used in this study. Then, it is an important stage to determine and set the strain field obtained from the digital images. The Ncorr platform is suggested to use the toolboxes required such as Image Processing Toolbox and Statistics Toolbox [22]. In addition, it is also required to post-processing and correlation method. The procedure of image settings is used to correlate and process for displacement/strain field of tested specimen. The relevant images as the reference image were loaded when previous software is successfully setup. The current images were loaded based on the entire load used in this test, and the setting of the reference image is shown in figure 4.

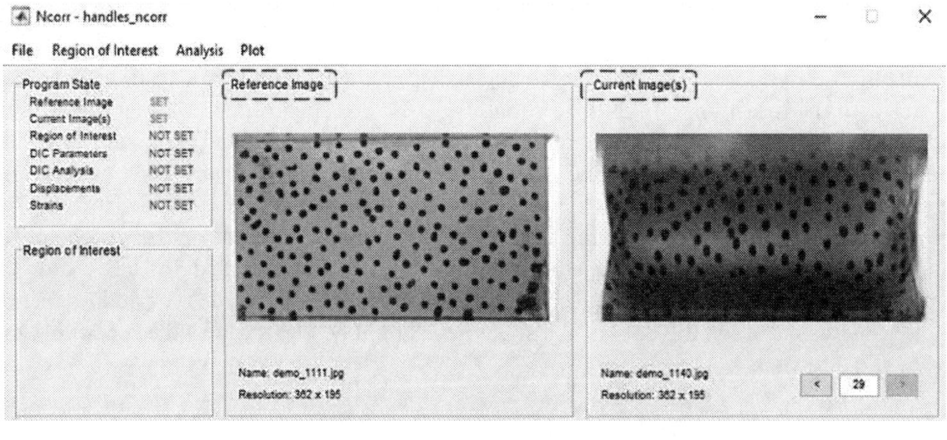

Figure 4. Loading of the reference and current images.

The reference image refers to the initial image recorded of the tested specimen before the compressive deformation occurred. The current image is used all the subsequent imaged to be correlated, which are compiled and ready to be processed via the correlation algorithm. The region of the specimen was drawn according to the area of the determined strain filed, and it was mainly depended on loading condition, failure behaviour and deformation mode. In this study, aluminium rectangular hollow tube was deformed in the middle of the total height, as shown in figure 5.

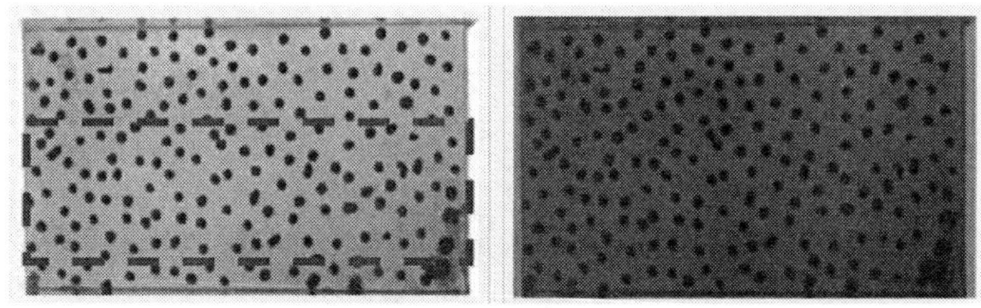

(a)

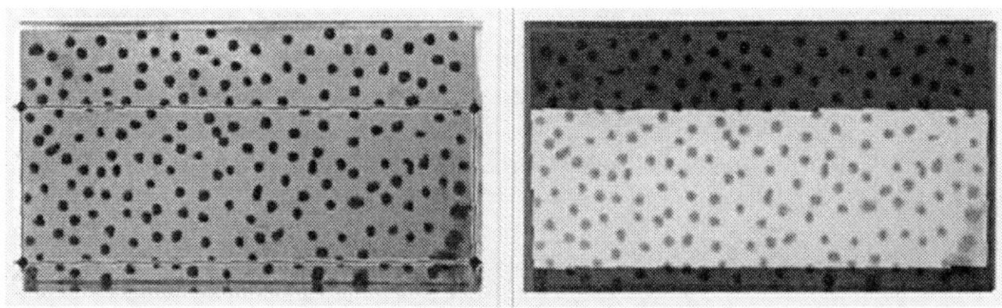

(b)

Figure 5. Setting of DIC parameters: (a) selecting the region of the tested specimen; (b) region selected condition.

Several parameters were provided on the left box on the software interface, and a green point was placed in the "Subset Location". For the subset option, the size of the subsets was dependent on spacing between the linked dots, which is used to reduce the computational load. Multithreading options are used for the speed up the computation process. The larger the number of threads, the more precise the results could obtain [23]. The parameter setting of the subset option is shown in figure 6, which involves the region interest and seed set in the region area. The seed region process provided the initial condition of the DIC method process, and the seed placement process was completed, as shown in figure 6(a). For instance, the single region was processed when it was set in the ROI interface as shown in figure 6(b).

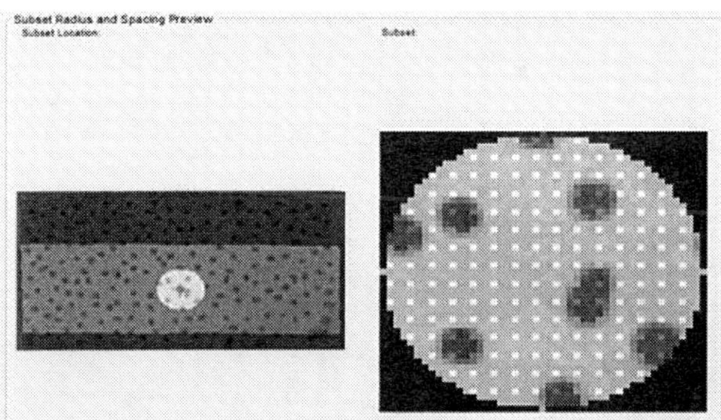

(a)

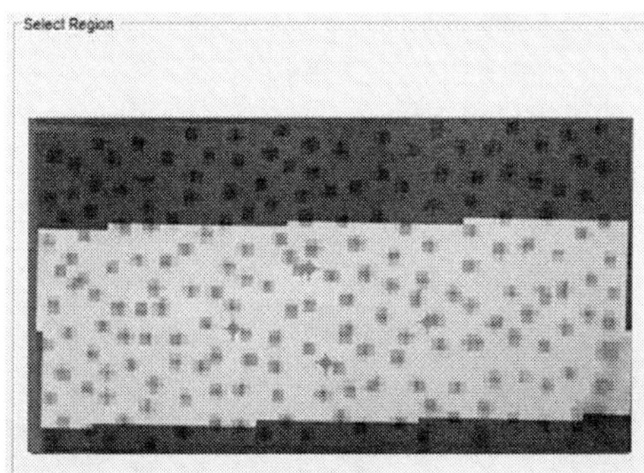

(b)

Figure 6. Subset option setting of DIC parameters: (a) subset option and region interest; (b) setting seed on region condition.

The seed region was checked to avoid its incorrect process, which showed the initial region for DIC method. For calibration and scale process, it was used to specify the corresponding pixels with the unit measurement according to its dimension. It converted the displacement from pixels to the true unit. The real value of the width from the tested specimen can insert at the "# of units", which was 38 mm in this case study. The relevant setting of calibration and scale is briefly presented in figure 7.

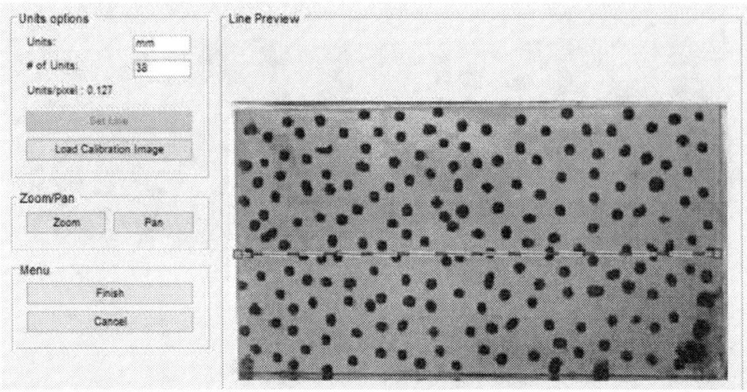

Figure 7. Calibration and scale process of record image using DIC method.

3. Results and discussion

3.1. Result of experimental test

Figure 8 showed the compressive behaviour of the aluminium rectangular hollow tube under the quasi-static compressive loading. It was observed that buckling mode occurred around the middle region of the aluminium rectangular tube. Therefore, the buckling mode was the main failure behaviour in the plastic deformation stage under quasi-static compressive loading, which agreed with similar studies on compressive properties of aluminium rectangular tube by other researchers [24-26]. Furthermore, the damaged specimen was illustrated in figure 9, which presented failure behaviour with buckling mode. There were some inwards and outwards deformation occurred on the boundary condition, which due to friction between the tube wall and crosshead.

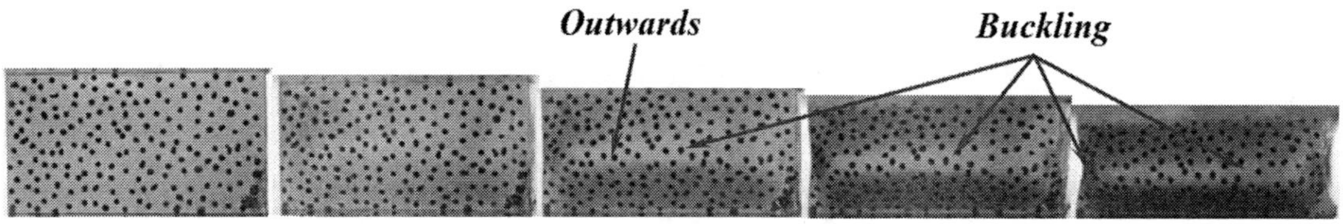

Figure 8. Compressive behaviour history of the aluminium rectangular tube under quasi-static compression test.

Figure 9. The damaged the aluminium rectangular tube.

The compressive properties of aluminium rectangular tube were investigated under the quasi-static loading condition, as shown in figure 10. The load markedly increased until it reached the maximum load of 40.74 kN in the elastic deformation stage. The load was slightly decreased during the plastic deformation stage, which mainly occurred the buckling mode in this stage. It was found that the maximum compressive stress occurred at the maximum load point of 245.48 MPa.

IOP Conf. Series: Materials Science and Engineering **863** (2020) 012038 doi:10.1088/1757-899X/863/1/012038

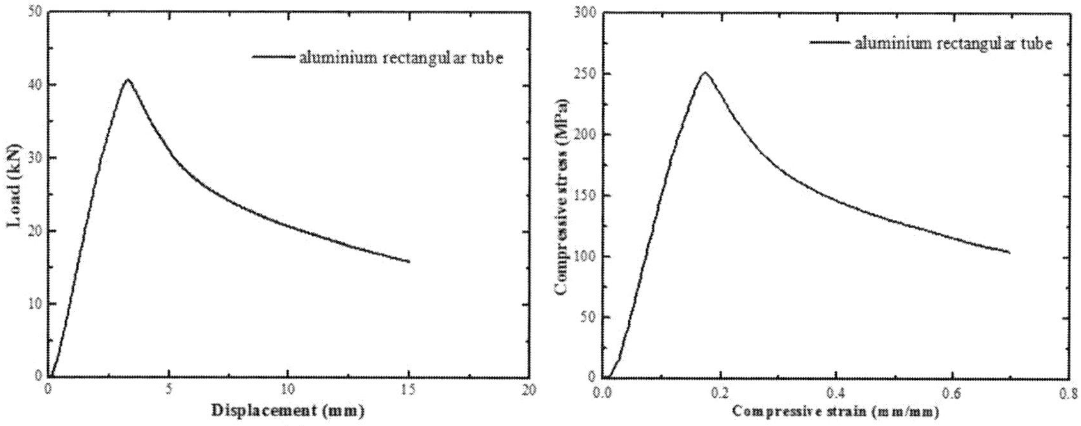

(a) (b)

Figure 10. The experimental traces of the aluminium rectangular hollow tube: (a) load versus displacement trace; (b) stress versus strain trace.

3.2. Result of Digital Image Correlation (DIC) method

Two methods are used to study the compressive properties of aluminium rectangular hollow tube. The stress versus strain graph is calculated according to its load versus displacement graph. The stress versus strain graph of DIC method is obtained from the DIC Ncorr software. There were three main conditions using DIC method, which were before compression, applied compression load and buckling at the middle of the specimen. Figure 11 presented the contour plot of measured strain before the load applied, which involved several parameters such as analysis type, image correspondences and correlation coefficient value. It obtained strain values with a negative value or 10-4 unit, which was recorded as an initial strain value before the load applied. Furthermore, the red contour plot is the area when the compression machine started to apply the load on the specimen. The maximum strain value in the area (red colour) with 0.052 and the minimum strain value (blue colour) is -0.173.

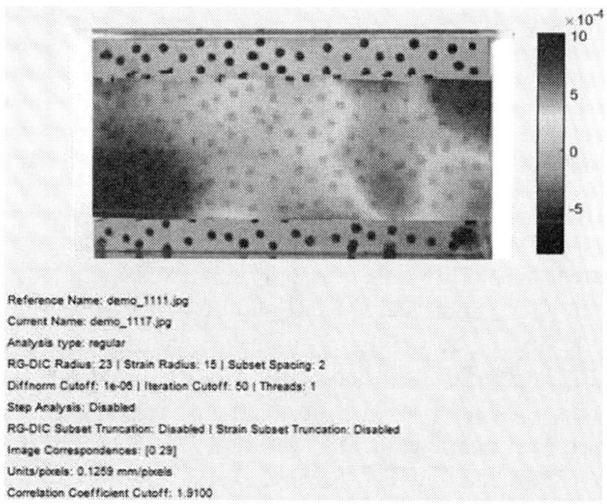

Figure 11. The result of DIC method before the load applied.

Figure 12 exhibited the contour plot of measure strain when the load applied, and it was observed the concentrated red colour around the middle of the specimen. It was shown that the strain region developed under the load applied, which indicated the deformation of buckling mode. Figure 13 showed the contour plot of strain value when buckling mode deformation initially occurred in the middle of the specimen. The reading of strain value was starting to become red colour around the

IOP Conf. Series: Materials Science and Engineering **863** (2020) 012038 doi:10.1088/1757-899X/863/1/012038

middle region of the specimen, as shown in figure 13. Each of the strain value could slowly change its measurable values until the compressive test is done. The reading of strain and stress values are completely measured and obtained at the selected point.

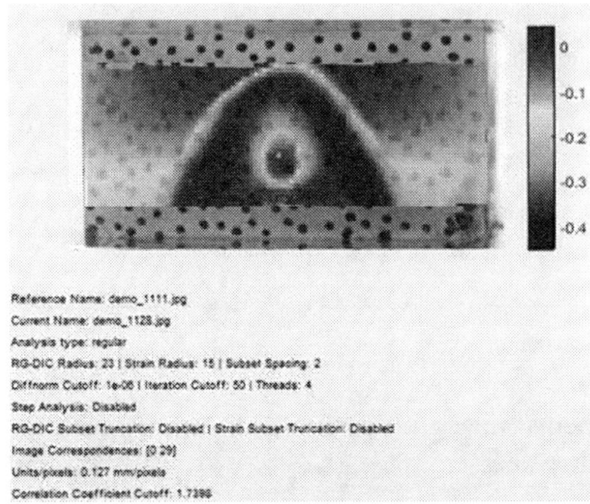

Figure 12. The result of DIC method when the load applies.

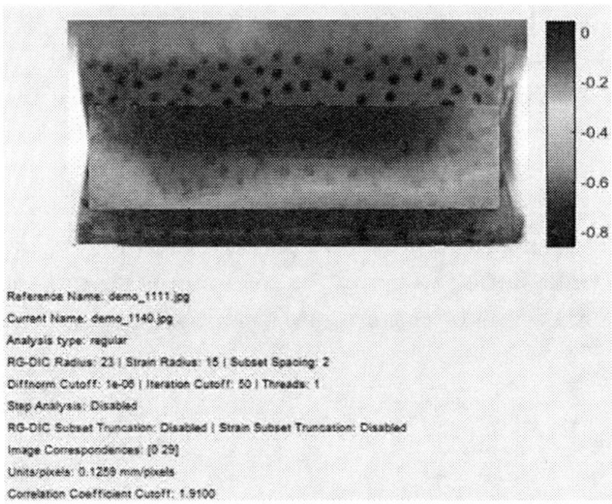

Figure 13. The result of DIC method when the buckling mode in the middle of the specimen.

The experimental data and DIC method data were compared and analyzed, which exhibited a similar trend pattern, as shown in figure 14. In order to obtain the compressive modulus value of this aluminium rectangular hollow tube, the data was only analyzed under the elastic deformation stage. For this case, the DIC method cannot detect the full pattern of the specimen due to the different value ranges between strain and stress. The maximum of stress value was 245.48 MPa for experimental method result and 241.12 MPa for DIC method result. The experimental result was a little bit higher than the result of DIC method, which provided the 1.7 % percentage of error. It was considered as a good agreement between experimental and DIC results. The compressive modulus was calculated according to stress versus strain trace, which obtained 1.22 GPa for the experimental case and 1.19 GPa for DIC case.

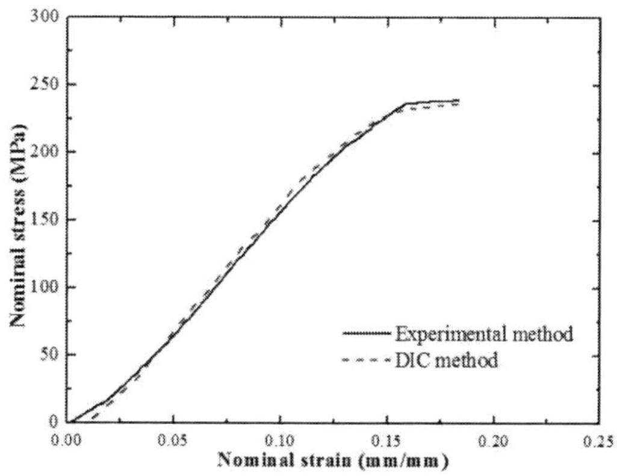

Figure 14. Results of the stress versus strain trace between experimental and DIC methods.

There are several factors that affect the results from the DIC method, which refer to the type of the material used, structural pattern of the tested specimen, size of the dots and the quality of video recorded [14, 27]. For the used material, it affects the range of error ratio with the provision of material type. For structural pattern, the suitability of the tested structure is another important parameter, which is due to the DIC software could only detect in the X and Y-axis. Therefore, the tested structure should select a two-dimension shape to analyze the strain value using DIC method.

For the size of the dots, it provides a high impact in order to obtain accurate results. The dots should follow the actual size, which affects the tracking procedure. The relevant displacement will be lost as the size dot is larger than the subset size, and it causes the dot cannot be recognized in the searching zone. Therefore, the dot should be smaller as the size of the subset and located in the search zone [28, 29]. Furthermore, it is also highlighted that the higher resolution image can generate better results by providing good lighting and closed distance between camera and specimen. The DIC software can only read the data from the movement of the dots on the surface of the specimen.

4. Conclusion

In recent years, Digital Image Correlation (DIC) method attracts much attention compared to other methods, which is generally used to measure the displacement mapping with cost-effective purpose. In this study, the compressive behaviour of aluminium rectangular hollow tube was studied, which mainly occurred the buckling mode of failure deformation. The load versus displacement and stress versus strain graphs were discussed. For DIC method, the contour plots of measured strain were summarized and highlighted on three conditions, which were before the load applied, applied load and buckling mode condition in the middle of the specimen. Furthermore, it was concluded the reasonably good agreement between the experimental test and DIC method, which obtained 1.7 % percentage of error. The experimental results were a little bit higher than the measurement values of DIC method. In addition, several factors affected the accuracy of DIC measurement results were discussed, which involved material type, specimen structural pattern, size of the dots and the quality of video recorded.

Acknowledgments

The authors are grateful to the Ministry of Education Malaysia: FRGS/1/2019/TK03/UMP/02/10 and Faculty of Mechanical and Automotive Engineering Technology, Universiti Malaysia Pahang for funding this research with PGRS180319.

References

[1] N. R. Council, *Application of lightweighting technology to military aircraft, vessels, and vehicles*. National Academies Press, 2012.

[2] H. Zarei and M. Kröger, "Optimization of the foam-filled aluminum tubes for crush box application," *Thin-Walled Structures,* vol. 46, no. 2, pp. 214-221, 2008.

[3] M. Ali, S. Jaffar, K. A. Abdullah, and Y. Aminanda, "Effect of Cross Sectional Shape on the Energy Absorbing Characteristics of a Tube under Quasi-Static Loading," in *Applied Mechanics and Materials*, 2013, vol. 315, pp. 334-338: Trans Tech Publ.
[4] L. Jun, X. Xiangsheng, and C. Qiang, "An investigation of the variable cross-section extrusion process," *The International Journal of Advanced Manufacturing Technology,* vol. 91, no. 1-4, pp. 453-461, 2017.
[5] M. Kamran, P. Xue, N. Ahmed, M. Zahran, and A. Hanif, "Axial Crushing of Uni-Sectional Bi-Tubular Inner Tubes with Multiple Outer Cross-Sections," *Latin American Journal of Solids and Structures,* vol. 14, no. 12, pp. 2198-2220, 2017.
[6] D. Bilston, D. Ruan, A. Candido, and Y. Durandet, "Parametric study of the cross-section shape of aluminium tubes in dynamic three-point bending," *Thin-Walled Structures,* vol. 136, pp. 315-322, 2019.
[7] R. M. Lima, Z. Ismarrubie, E. Zainudin, and S. Tang, "Effect of length on crashworthiness parameters and failure modes of steel and hybrid tube made by steel and GFRP under low velocity impact," *International journal of crashworthiness,* vol. 17, no. 3, pp. 319-325, 2012.
[8] S. B. Kim, H. Huh, G. Lee, J. Yoo, and M. Lee, "Design of the cross section shape of an aluminum crash box for crashworthiness enhancement of a car," *International Journal of Modern Physics B,* vol. 22, no. 31n32, pp. 5578-5583, 2008.
[9] P. K. Gupta, "An investigation into large deformation behaviour of metallic tubes," 2000.
[10] N. Gupta and S. Sinha, "Lateral compression of crossed layers of square-section tubes," *International Journal of Mechanical Sciences,* vol. 32, no. 7, pp. 565-580, 1990.
[11] M. Sutton and F. Hild, "Recent advances and perspectives in digital image correlation," *Experimental Mechanics,* vol. 55, no. 1, pp. 1-8, 2015.
[12] J. Blaber, B. Adair, and A. Antoniou, "Ncorr: open-source 2D digital image correlation matlab software," *Experimental Mechanics,* vol. 55, no. 6, pp. 1105-1122, 2015.
[13] K. Wu, X. Li, Y. Ge, and S. Ruan, "Determination of tubular material parameters in bulging test with three-dimensional digital image correlation method," *The International Journal of Advanced Manufacturing Technology,* pp. 1-9, 2018.
[14] S. Yoneyama, "Basic principle of digital image correlation for in-plane displacement and strain measurement," *Advanced Composite Materials,* vol. 25, no. 2, pp. 105-123, 2016.
[15] C. Niezrecki, J. Baqersad, and A. Sabato, "Digital image correlation techniques for NDE and SHM," *Handbook of Advanced Non-Destructive Evaluation,* pp. 1-46, 2018.
[16] S.-W. Khoo, S. Karuppanan, and C.-S. Tan, "A review of surface deformation and strain measurement using two-dimensional digital image correlation," *Metrology and Measurement Systems,* vol. 23, no. 3, pp. 461-480, 2016.
[17] H. Zeng and P. Bailly, "Experimental characterization of dynamic behaviour of gelatin-based material using DIC," *Polymer Testing,* vol. 63, pp. 298-306, 2017.
[18] C. Flament, M. Salvia, B. Berthel, and G. Crosland, "Local strain and damage measurements on a composite with digital image correlation and acoustic emission," *Journal of Composite Materials,* vol. 50, no. 14, pp. 1989-1996, 2016.
[19] M. Badaloni, M. Rossi, G. Chiappini, P. Lava, and D. Debruyne, "Impact of experimental uncertainties on the identification of mechanical material properties using DIC," *Experimental Mechanics,* vol. 55, no. 8, pp. 1411-1426, 2015.
[20] P. Wang, "Experimental Characterisation and Modelling of Polymer Foam Core Materials using the Virtual Fields Method and DIC," Aalborg Universitetsforlag, 2015.
[21] P. L. Reu *et al.*, "DIC challenge: developing images and guidelines for evaluating accuracy and resolution of 2D analyses," *Experimental Mechanics,* pp. 1-33, 2017.
[22] A. A. Justin Blaber. *Ncoor v1.2.* Available: http://www.ncorr.com/
[23] A. Ab Ghani, M. Ali, S. DharMalingam, and J. Mahmud, "Digital image correlation (DIC) technique in measuring strain using opensource platform Ncorr," *Journal of Advanced Research in Applied Mechanics,* vol. 26, no. 1, pp. 10-21, 2016.

[24] M.-N. Su, B. Young, and L. Gardner, "The continuous strength method for the design of aluminium alloy structural elements," *Engineering Structures,* vol. 122, pp. 338-348, 2016.
[25] A. Baroutaji, M. Sajjia, and A.-G. Olabi, "On the crashworthiness performance of thin-walled energy absorbers: recent advances and future developments," *Thin-Walled Structures,* vol. 118, pp. 137-163, 2017.
[26] P. C. Albert, A. Ghani, M. Z. Othman, and M. Zaidi, "Axial crushing behavior of aluminum square tube with origami pattern," *Modern Applied Science,* vol. 10, no. 2, pp. 90-108, 2016.
[27] Z. Wang, H. Kieu, H. Nguyen, and M. Le, "Digital image correlation in experimental mechanics and image registration in computer vision: similarities, differences and complements," *Optics and Lasers in Engineering,* vol. 65, pp. 18-27, 2015.
[28] P. Bing, X. Hui-Min, X. Bo-Qin, and D. Fu-Long, "Performance of sub-pixel registration algorithms in digital image correlation," *Measurement Science and Technology,* vol. 17, no. 6, p. 1615, 2006.
[29] J. D. Helm, "Digital image correlation for specimens with multiple growing cracks," *Experimental mechanics,* vol. 48, no. 6, pp. 753-762, 2008.

A review of synthesis roots of iron nanoparticles

T Anjam*, N U H Altaf, W Ahmad, M A Sajid, M A Munir and M Aslam

Department of Physics, University of Agriculture Faisalabad, 38040 Faisalabad, Pakistan.

*E-mail: tauseefanjam8@gmail.com

Abstract. Nanoscience is a standout amongst the most significant innovative work action in the outskirt of current science. From most recent two decades the nanotechnology and nanoscience have increased more significance in the field of research because of the enormous scale applications. The present work audits crafted by various researchers for the planning of iron nanoparticles by various procedures. Iron is a standout amongst the richest and generally utilized components on earth. Iron nanoparticles have a tremendous potential for various applications including magnetic liquids, magnetic resonance imaging (MRI) differentiate operators, catalyst for carbon nanotube development, nickel-iron batteries, and impetuses and sorbents for natural remediation. The preparation of iron nanoparticles by different methods, pulsed plasma, chemical reduction, chemical vapour condensation, one-step reduction, Green synthesis and Biosynthesis, its advantage over other processes are described. In this review article attempt has been made to compare each process by categorizing them specifically.

1. Introduction

Porous In the recent year, Nano science and technology are most important developing area of research. Nanotechnology implies the formation of nanoparticles in shape, size, controlled diversity, their potential and chemical composition uses for human benefits [1]. Today, nanotechnology is a developing area of investigation on planet. In this field the research explosion is because nanoparticles have remarkably better distinctiveness than their microscope counterpart [2]. The properties of materials change radically in their nano-form. In magnetic recording media (e.g. video tape, audio, data storage devices, magnetic fluid, and recording discs) the magnetic nanoparticles attract much attention as compared to other material due to their unique property. The various applications of magnetic nanoparticles such as bio-separation, targeted drug delivery, NMR imaging and cancer hyperthermia are uses in catalysis and other industries. To fabrication of electronic devices at nanoscale the application of magnetic nanoparticles is also used [3].

In nanoscale natural advance the innovation of iron nanoparticles is one of the first-generation of nanoparticles. In most recent couple of year, to deliver iron nanoparticles the various engineered strategies have been created, nanoparticles surface properties also adjust and improve the production for field transference and responses. Iron is plentiful and extensively used elements. Iron nanoparticles have potential for various application as well as catalysts for carbon nanotube configuration, nickel-iron batteries, sorbents for environmental remediation and magnetic fluid [4].

In the field of medicine, it also plays a vital role. For example, iron NPs are significant for absorbing material and waste water treatment. It also effectively controls the allotment of medicine powder which can targeted achieve liberation of chemotherapy drugs. It is important for absorbing material and waste water treatment [5]. It has high magnetization at room temperature as compared to another element. It has soft magnetic material because of cubic crystalline structure [6].

Due to their specific activity and large surface area in the field of magnetic material synthesis of iron NPs is one of the most attractive challenge. For preparation of magnetic nanoparticles, the common methods are one step reduction [5], Pulsed plasma [6], Chemical reduction [3], Biosynthesis [1], ball milling [7], green synthesis for green tea [8], microwave plasma [9] and chemical vapour deposition [10]. All these various methods take into account their environmental safety sufficiency, cost-effectiveness, simplicity and occasion for controlling the symphony and properties of iron NPs. The different methods are discussed for the synthesis of iron nanoparticles.

The aim of the study of pulsed plasma in liquid method is to synthesis the pure iron NPs. During the synthesis of iron NPs through pulsed plasma in liquid a number of factors medium applied voltages, temperature, pulse duration, applied liquid condition and medium type are affected by it. For size control and purity of nanoparticles are using the water-toluene. There are four different samples prepared from iron electrode submerge in liquid. The morphology, structure and magnetic properties were investigated obtained from these samples [6].

For the synthesis of iron nanoparticles in aqueous medium the different method has been reported. In this technique utilize the ascorbic corrosive for decreasing the $Fe(acac)_3$. The room temperature and saturation magnetization of iron is very high. The curie temperature of iron is also high, and this application is most important. It is simple, basic and reproducible method to incorporate iron nanoparticles. Chemically synthesize of iron nanoparticles using the ascorbic acid to reduce the $Fe(acac)_3$ in controlled condition. In the nonexistence of water yields this reaction were carried out. Iron nanoparticles were synthesized in the relative study [3]. The chemical vapour condensation is another method to synthesized iron nanoparticles by using the $Fe(CO)_5$ as a precursor. The effect of annealing process and processing parameter on the size and microstructure of iron nanoparticles were investigated [10]. The one step reduction method is also using to synthesize the iron nanoparticles [5]. Iron nanoparticles were removed from green tea leaves [8]. Because of their large quantity and low caste iron-based nanoparticles are also used in industry [11]. Iron nanoparticles are utilized to expel poisons from soil and water in nature and from modern waste streams [12]. Iron nanoparticles are utilized for a wide scope of uses including magnetic recording, media [13], ferrofluids [14], attractive cell separation [15], MRI differentiate agents [16], natural remediation [17], energy storage materials [18], bioimaging system [19] and genetic engineering [20].

2. Methods for the synthesis of iron nanoparticles

2.1. Synthesis of iron nanoparticles through pulsed plasma technique

In the year of 2014, Kelgenbaeve [6] studied that by using the method of pulsed plasma in liquid the iron nanoparticles were synthesized with 10 nm sizes. By using arc discharge method in which water toluene interface used as medium the iron nanoparticles were prepared. In very high potential difference the space between the electrodes were breakdown due to appearance of pulsed plasma from inter-electrode. The diameter of iron electrodes was 5 mm and purity 99.9%. These electrodes were immersed in 20 mm deionized water. The duration of electrical current between pulsed plasma discharges was 10 microsecond and gap between them 1 mm. For continue the discharge process one electrode was connected with vibrator. A magnetic stirring bar was used with rotational speed 900 rpm.

Different samples were ready to study the effect of toluene content by using pulsed plasma. The water-toluene (1) 95-5 (2) 20-80 (3) 70-30 (4) 60-40% were used respectively. In each experiment the (100 V, 6 A) of AC electric power was used for 1 hour. By using the evaporation and centrifuge the black precipitate was separated from liquid. The different ratio of toluene-water in emulsion was control the size and purity of iron nanoparticles. As the toluene fraction decrease from 40 to 5% the size of nanoparticles decrease from 21 to 9.1 nm and also increase the purity from 48 to 98%. The toluene-water ratio 95-5% was found for more purity formation of iron nanoparticles. The lattice parameters of iron sample calculated from XRD. The XRD found (a = 0.2927) and previously reported (a (BCCFe) =

0.2866). The surface morphology of iron nanoparticles were examined through SEM as shown in figure 1 at three different resolution of ×700, ×6,500 and ×100,000 [6]. The spherical agglomerates were evidently display. Because of chemical bond and high surface energy of iron nanoparticles, tendency to form agglomerate was rather strong. The magnetic dipole force was contribute to inter particulate forces.

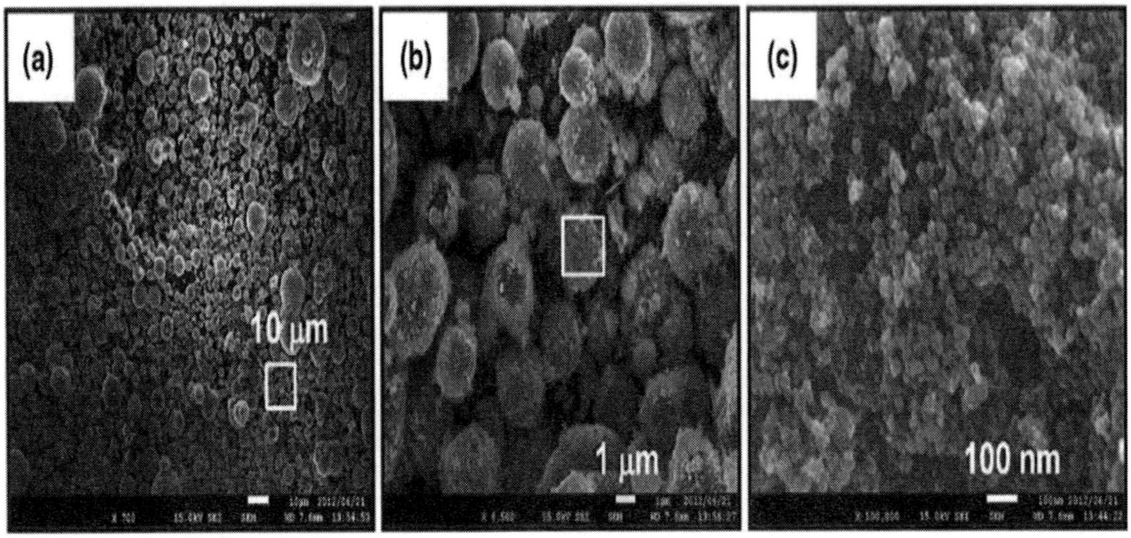

Figure 1. FESEM Iron nanoparticles images at different magnifications: (a) ×700; (b) ×6,500; (c) ×100,000 [6].

2.2. *Iron nanoparticles via chemical reduction of Fe(acac)₃ by ascorbic acid*

In the year 2014, Nene [3] studied that iron nanoparticles were chemically synthesized by using ascorbic acid for the reduction of $Fe(acac)_3$. For the synthesized of iron nanoparticles different method has been used but this method was used in first time. The chemical used in this method were diphenyl-ether, ascorbic acid, $Fe(acac)_3$ (purity 99%), dehydrated ethanol and ultrapure deionized water. The solution was prepared by 50Ml $Fe(acac)_3$ and 30 mM diphenyl-ether mixed. The solution of $Fe(acac)_3$ and diphenyl-ether was kept under stirring in atmosphere of nitrogen gas at 70 °C.

The 0.025 M solution was prepared of dehydrated ethanol and ascorbic acid. This solution was added drop by drop at rate of 2 mL/min when temperature reached at 70 °C. After adding this reducing acid solution, the solution heated to 190 °C. Then the solution was cooled down to 25 °C. The solution was washed and filtered 4-5 time by chloroform. Dry powder obtained by using reducing $Fe(acac)_3$ chemical method. Figure 2 shows the TEM images of iron nanoparticles at different magnifications. The iron nanoparticles size was observed by TEM to be 7 ± 1 nm.

2.3. *Synthesis of iron nanoparticles by chemical vapor condensation technique*

In the year 2002, Choi [10] studied the iron nanparticles were synthesis through chemical vapor condensation method. By using the pyrolysis of iron pentacarbonyl ($Fe(CO)_5$) the iron nanopaticles were synthesized. The influence of CVC parameter for size distribution and formation of nanoparticles was studied. The effect of annealing process and processing parameters on size and microstructure of iron nanoparticles were studies. The liquid iron pentacarbonyl was use as a precursor and helium or argon gas carried through heating bubble. The vaporization temperature was optimized at 150 °C. Flow gas was entered the precursor and pass through the tubular furnace to working chamber.

The precursor decayed in furnace and compress in particles. For conducting the experiment tubular furnace was heated at 400-1100 °C range. All particles were put on the surface of liquid nitrogen-cooled chiller. In working chamber, the powder was collected. The synthesized particles size was nearly equal to 6-25 nm in diameter. The structure of synthesized nanoparticles was core-shell type.

To examine the effect of corrosion temperature on particles size the different samples were synthesized at different temperature from 400-1100 °C by using helium or argon. Figure 3 shows the XRD pattern of samples formed at various disintegration temperatures. It also shows that through CVC iron NPs were also synthesized. With increasing the decomposition its width decreases and peaks intensity increase as shown in figure 3. In helium atmosphere there were no visible BCC peaks at decomposition temperature below 750 °C. BBC peaks was also not shown in X-ray diffraction pattern.

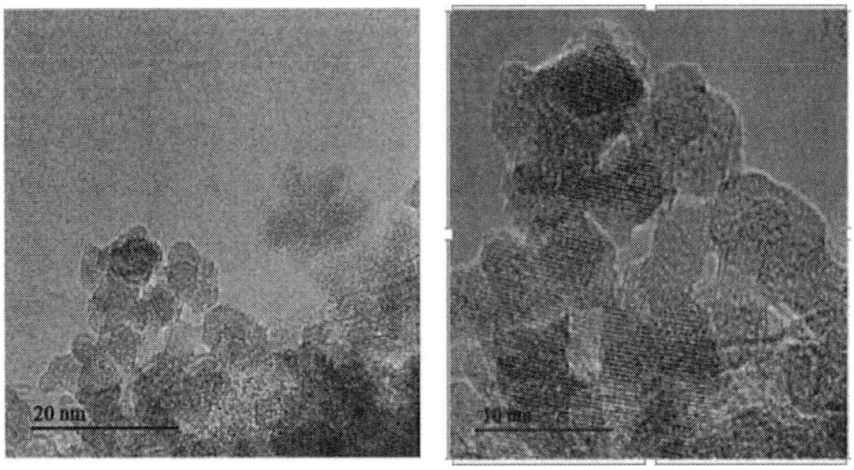

Figure 2. TEM images of iron nanoparticles [3].

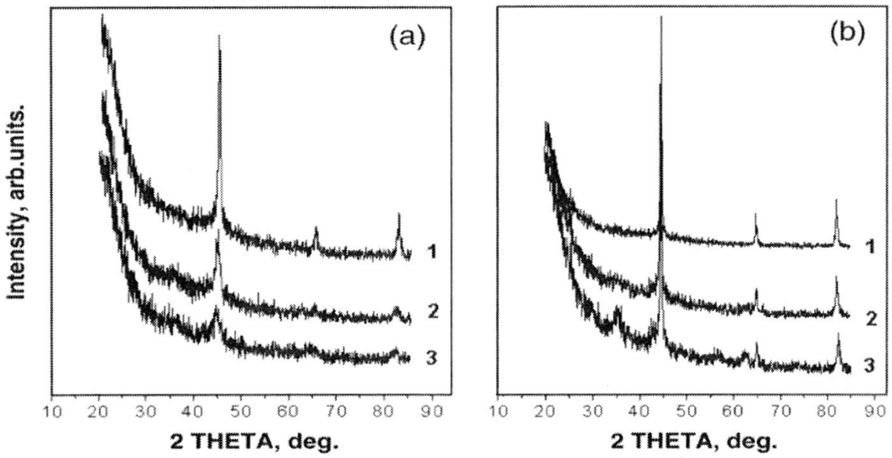

Figure 3. XRD patterns of iron NPs with diameter of 19 nm (1), 12 nm (2), and 8 nm (3) for (a) After heating 220 °C, and (b) After heating 320 °C with nanoparticles size of 25 nm in argon atmosphere [10].

2.4. Synthesis of iron nanoparticles via One-Step reduction process

In the year of 2016 Ping [5] studied the iron nanoparticles were synthesized through the method of one-step reduction. The citric acid ($C_6H_8O_7{\cdot}H_2O$) and ferric nitrate ($Fe(NO_3)_3{\cdot}9H_2O$) were used. With sol-gel method the gel was prepared. The 41.6 grams citric acid and 80 grams ferric nitrate were mixed in 122 mL deionized water. This solution was stirred at 17.5 r/s. The stirrer was continued for 30 minutes. After stirring the gel was prepared and again heated to 60 °C. The solution of ammonia was added by drop wise at 60 °C. This solution was controlling the PH at 7.10. This process was continued for 4 hours. After four-hour heating temperature was increased to 95 °C and addition of 10 mL of olefinic acid. The product was washed with ethanol three times after stirring and dried, milled. In the presence of argon gas atmosphere, the powder heated to 620 °C for 30 minutes. The black powders were obtained by one-step reduction method.

In one-step reduction method carbon source was used as carbon reduction that is provided by citric acid. For breaking of carbon in citric acid too much citric was added and also increase the temperature. Adding citric acid and increasing the temperature also results the formation of single matter C. This is used to condense the Fe_2O_3 to iron. Figure 4 shows the XRD pattern which showed all peaks was indexed to iron. The figure also shows the bcc structure from iron phase. The peaks or iron phases (110), (200), (211) at $2\theta \approx 44°$, 65°, 82°, respectively. This show the high crystallinity and iron nanoparticles.

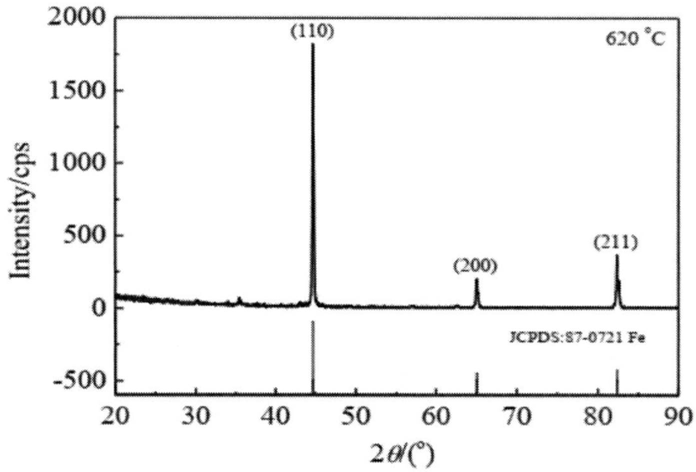

Figure 4. XRD patterns of iron nanoparticles [5].

2.5 Green synthesis of iron nanoparticles using green tea leaves extract

Gottimukkala in the year 2017 [8] extract the iron nanoparticles from the green tea. The weight of 20 grams tea was mixed with 1000 ml of water in order to extract the tea. For getting the tea extraction, the solution was heated in water bath at temperature of 80 °C. The extract of tea was collected and filtered. This solution was assembled and stored. This solution was dried in a beaker for additional use. In sterilized flask the 1:1 proportion of green tea extract and 0.01M ferric Chloride was added for synthesized of iron NPs. The colour of solution was changed when adding of green tea extract to ferric chloride solution. After changing the coloured the solution centrifuge wash with deionized water.

For synthesis of different the morphology of iron NPs green tea were used as reducing agent. The 4000 species have been found in polyphenols and $1/3^{rd}$ of total. This species was helps in reduction of salt precursor. The polyphenols consist of catechins and flavonoids. The reduction process was occurred due to the catechin. The standard potential of catechin was 0.57 V which decrease the Fe^{3+} to Fe^{0}. The iron ordinary potential was -0.036 V. When precursor was added in solution formed partial bonds due to breaking the -OH bond. This partial bond was made bond with metal ion. The partial bond was also breaking to reduce metal ion into nanoparticles.

In order to study diameter and functional groups of nanoparticles, the size and morphology were characterized through SEM and FTIR analysis. FTIR analysis send synthesis of iron NPs was ready to evaluate and analysis the attach biomolecules. Vibration stretch was also showed the FTIR at 3452 cm^{-1} for O-H and 1632 cm^{-1} for C=C. The C-N and C-H were observed to be 1383 and 2926 cm^{-1}, respectively. It was compared to FTIR results it showed broad stretch C=C at 1635 cm^{-1}, cm^{-1}, C-O-C at 1020 cm^{-1}, c-n 1379 cm^{-1} and O-H at 3419 cm^{-1} which match with extract. Polyphenols was function as capping agent as well as reducing agent in green tea extract. Figure 5 shows the average diameter of iron nanoparticles were about 116 nm found from SEM images. While comparing the all different methods for the synthesis of iron NPs, different particle sizes have been observed as shown in the table 1.

IOP Conf. Series: Materials Science and Engineering **863** (2020) 012039 doi:10.1088/1757-899X/863/1/012039

Table 1. Summary of the iron nanoparticles synthesized by different methods.

No.	Method	Study description	Procedure	Resulted iron nanoparticles size (nm)
1	Preparation of iron nanoparticles via pulsed plasma technique	The pulsed plasma in liquid is the suitable and cheap method for synthesis of nanoparticles. For increasing the demand of preparation of well dispersed nanostructure single-pulse plasma method is the low-cost conventional technique.	By using arc discharge method in which water toluene interface used as medium the iron nanoparticles were prepared.	≤10 nm
2	Preparation of Iron nanoparticles via chemical reduction method	It is a simple, reproducible and profitable method to make iron nanoparticles.	Nanoparticles of iron were chemically synthesized by reduction of Fe (acac) $_3$ using ascorbic acid in controlled condition.	7±1 nm
3	Iron nanoparticles via chemical vapour condensation	Iron nanoparticles by chemical vapour condensation (CVC) by the pyrolysis of organometallic Fe (CO) $_5$ as precursor. The impact of handling parameters and strengthening process on the microstructure and size of Fe nanoparticles were examined.	By using the pyrolysis of iron pentacarbonyl (Fe (CO) $_5$) the iron nanopaticles were synthesized.	6-25 nm
4	Synthesis of iron NPs via One-Step reduction technique	It is easy and simple method to synthesize the iron nanoparticles. Under moderate temperature iron nanoparticles were synthesized. Highly crystalline nanoparticles were obtained	To obtain high crystallinity and pure iron nanoparticle at appropriate temperature (620°C) is using excessive citric acid. The cttric acid ($C_6H_8O_7 \cdot H_2O$) and ferric nitrate ($Fe(NO_3)_3 \cdot 9H_2O$) were used.	30 nm
5	Iron nanoparticles via green tea leaves extract	The advancement over physical and chemical method is providing by green synthesizing. It is environment friendly, easy scaled up and cost-effective method to synthesis the iron nanoparticles. No need to use toxic chemical, high energy, pressure and temperature.	By take 20 g tea in 1000ml of water the extract of tea was prepared. For getting the tea extract the solution was heated in water bath at temperature of 80°C. The extract of tea was collected and filtered.	116 nm

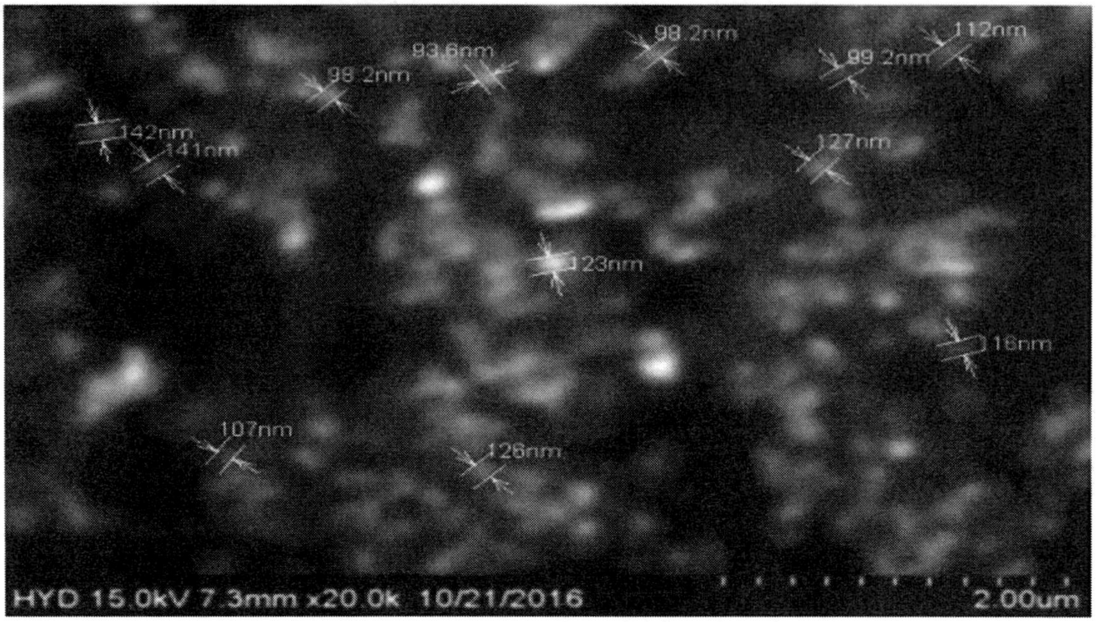

Figure 5. Zoomed SEM image of the iron nanoparticles [8].

3. Conclusion

Nanoscience is a champion among the most noteworthy imaginative work activity in the edge of current science. From latest two decades the nanotechnology and nanoscience have expanded more centrality in the field of research on account of the colossal scale applications. The present work reviews created by different specialists for the arranging of iron nanoparticles by different methods. Iron is a titleholder among the most extravagant and for the most part used segments on earth. Iron nanoparticles have an enormous potential for different applications including attractive fluids, attractive reverberation imaging (MRI) separate administrators, impetus for carbon nanotube advancement, nickel-iron batteries, and impulses and sorbents for regular remediation. The arrangement of iron nanoparticles by various techniques, beat plasma, concoction decrease, synthetic vapour build-up, one-advance decrease, Green combination and Biosynthesis, its bit of leeway over different procedures are portrayed. The resulted particles size was observed distinct through all different methods.

References

[1] Chandran M, Yuvaraj D, Christudhas L, Ramesh K V 2016 *Biotechnol. Ind. J.* **12**, 112.
[2] Okoth O J K 2016 (Doctoral dissertation, University of Nairobi).
[3] Nene A G, Takahashi M, Somani P R 2016 *World J. NanoSci. Eng.*, **6**, 20.
[4] Huang K C, Ehrman S H 2007 *Langmuir*, ***23***, 1419-1426.
[5] Ping H, Lu K, Jun Y, Fan F, Kuaishe W, Jinjing D, Dongxin L 2016 *Rare Metal Mat. Eng.*, **45**, 3112-3114.
[6] Kelgenbaeva Z, Omurzak E, Takebe S, Sulaimankulova S, Abdullaeva Z, Iwamoto C, Mashimo T 2014 *J. nanopart. res.*, **16**, 2603.
[7] Kerekes L, Hakl J, Meszaros S, Vad K, Gurin P, Kis-Varga M, Beke L 2002 *Czech. j. phys.*, **52**, A89-A92.
[8] Gottimukkala K S V, Harika R P, Zamare D 2017 *J. Nanomedine Biotherapeutic Discov.*, **7**, 151.
[9] Hoder T, Kudrle V, Frgala Z, Janca J, David B 2005 In *WDS* **5**, 300-305.
[10] Choi C J, Tolochko O, Kim B K 2002 *Mater. Lett.*, **56**, 289-294.
[11] Wang Ch, Xu L, Wang Q 2003 J. Nat. Gas. Chem. **12,** 10–16.

[12] Winkelmann K, German H, Hodes C, Li J, Price M, Termini C, Thiele C 2011 *J. Nano. Edu.* **3,** 62-81.
[13] Jorgensen F, *The Complete Handbook of Magnetic Recording*; McGraw-Hill: New York, 1995.
[14] Rosensweig R E 1966 *AIAA J.* **4,** 1751.
[15] Molday R S, Mackenzie D 1982 *J. Immunol. Methods* **52,** 353- 367.
[16] Tiefenauer L X, Tscgirky A, Kuhne G, Andres R Y 1996 *Magn. Reson. Imaging* **14,** 391 402.
[17] Elliott D W, Zhang W X 2001 *EnViron. Sci. Technol.* **35,** 4922-4926
[18] Zhang F, Braun G B, Shi Y, Zhang Y, Sun X, Reich N O, Zhao D, Stucky G, 2010 *J. Am. Chem. Soc.* **132,** 2850
[19] Fredin L A, Li Z, Ratner M A, Lanagan M T, Marks T J, 2012 *Adv. Mater.* **24,** 5946
[20] Shah B, Yin P T, Ghoshal S, Lee K B, 2013 *Angew. Chem.* **125,** 6310

IOP Conf. Series: Materials Science and Engineering **863** (2020) 012040 doi:10.1088/1757-899X/863/1/012040

A preliminary study on new cooling techniques for stingless bees hive

L A Halim[1], M F Basrawi[1,*], T M Yusof[1], M H Yusof[1], Z Mustafa[2] and T K Ibrahim[3]

[1] Meliponini Engineering Laboratory (MePEL), Faculty of Mechanical and Automotive Engineering Technology, Universiti Malaysia Pahang, 26600 Pekan, Pahang, Malaysia.
[2] Department of Neurosciences, School of Medical Sciences, Universiti Sains Malaysia, 16150 Kubang Kerian, Kelantan, Malaysia.
[3] Mechanical Engineering Departments, College of Applied Engineering, Tikrit University, Iraq.

*E-mail: mfirdausb@ump.edu.my

Abstract. Stingless bees are a type of insect which are very sensitive to the changes of their surroundings, especially to severe heat wave. A report stated that at temperature as high as 38 °C can cause death of the bees especially to the pupae. Therefore, the objective of this research is to evaluate new methods in regulating the temperature in the hive. To achieve this goal, two cooling methods were studied. One method is using greenroof, a type of roof which contains green vegetation and soil. It is used to cool down the space under the roof. Another method is using a photovoltaic (PV) powered thermoelectric cooler (TEC). Three units of MUSTAFA-hives were exposed under sunlight, one is without any temperature regulation and the other two hives used cooling techniques stated above. The temperatures inside each hive was measured at two points, and was compared with the hive without temperature regulation. It was found that, for the hive integrated with greenroof, the average hive temperature was about 3 °C and 6 °C lower in the honey cassette and brood-cells compartment, respectively. On the other hand, the hive with PV powered TEC recorded average temperatures of about 0.5 °C and 3 °C lower in honey cassette and brood-cells compartment, respectively. Therefore, it can be concluded that the implementation of greenroof or TEC could solve the problem of stingless bees' hive overheating. However, the greenroof has a better thermal performance, besides being economic and simpler solution.

1. Introduction

Stingless bees are fascinating creatures. They predate, *Apis* bees, the stinging honey bees, roaming the tropical earth for over 65 million years. They are diverse and although they share some similarities with *Apis* they are biologically distinct. Stingless bees' species are common to tropical and subtropical areas such as the Southeast Asia region. Diversity of stingless bees make classifying them a difficult task, especially in Indo-Malayan region [1]. Like honeybees, stingless bees produce honey in nest containing sterile workers and a queen. However, stingless bees produce less honey, and do not use water to cool their nest. [2]. Their hives usually consist of several sections including the brood cell area, as well as honey pot area. In tropical region, their nests are usually made inside trunk of trees.

Thermoregulation, or temperature control, of stingless bees' hives are crucial to ensure suitable condition for broodcell development. It has been suggested that the optimum range of temperature is between 31 °C to 32 °C [3]. In natural hives, thermostatic effect helps thermoregulation, where the wood insulation along with shades, are able to keep the temperature of the hive relatively constant [4]. Hive temperature of 38 °C and above can cause death in pupae development [5]. This is detrimental towards the growth of colony.

Typically, trees are cut down to obtain log containing the stingless bees' nests. The log is then attached with a topping compartment to facilitate the honey production. However, such practice can be damaging towards the natural ecosystem especially for tropical jungles and are not sustainable in long term. As stingless bees act as natural pollinator for tropical forests, the human disturbance to stingless bees community could potentially disrupt and alter the forest composition [6]. Hence, once a colony is obtained from a log, it can be put into artificial hives such as MUSTAFA-hives. Then, it can be split into two colonies. Once the colonies settled and become stable it can be divided again and again. The honey can be easily extracted as MUSTAFA-hives has separate honey cassette compartment. Thus, this minimizes the disturbance and intrusion into the colony.

Stingless bees' colonies in hives are often placed under trees or other suitable shaded areas. This is to prevent the hives from overheating especially in tropical hot and humid climate. However, this limited the area suited for beekeeping, leading to additional cost to build shading. Placing the hives under direct sunlight can cause temperature to build up and can be dangerous to the colony as brood cell cannot develop when subjected to extreme heat. Hence, suitable cooling method for hives may be developed to prevent overheating of hives when subjected to outdoor conditions without any shading.

One of the possible cooling technique is to use Thermoelectric Cooler (TEC). TECs are solid state heat pump based on Peltier effect. Essentially Peltier effect describes a phenomenon in a circuit with two different conductors, when current is applied, heat will be absorbed or rejected by the conductors depending on the direction of the current [7]. Although Peltier effect has been discovered back in 1834, only recently thermoelectric devices gained attraction. This is due to current development of semiconductor thermocouple materials. Such devices becoming more popular due to its miniature size and reliability with little or no need for maintenance as it has no moving parts.

Another possible cooling method is by using greenroof, a natural cooling technique that have been used for centuries [8]. Although the technique is centuries old, due to scarcity of planted surface in urban areas, greenroof become stabilising and efficient choice [9]. Essentially, a green roof is a type of roof made up of vegetation and soil as its outermost surface [10]. As vegetation have potential to provide cooling benefits [11], buildings using greenroof could reduce their inside temperatures as much as 3.4 °C [12]. There are also various literatures regarding analysis and improving the thermal performance of greenroof. These include using compost waste and inorganic substance to improve the physical properties of greenroof [13], and also some analysis on thermal conductivity of greenroof [14]. Therefore, greenroof have potential to be adapted to cool stingless bees' hive.

Currently, photovoltaic (PV) powered TEC system for thermoregulation has been used in rearing honey bees, and the loss of bees were minimal while production of honey increased [15]. Some automated control of temperature and other microclimate parameters for honeybees has also been investigated [16]. Hence, TEC system for thermoregulation is a promising cooling solution for stingless bees' hive. However, using TEC system for stingless bees' hives thermoregulation are unheard of.

Currently, researches on engineering application for meliponiculture especially in Malaysia are still scarce. At present, these include honey processing [17, 18] as well as engineering economic analysis for stingless bees keeping [19] and a few literature reported on thermal regulation [20, 21]. Thus, the main objectives of this study are to investigate the temperature profile of a hive exposed to outdoor conditions, and to evaluate the temperature regulation in the hive using greenroof and hive using PV powered TEC as cooling methods.

2. Methodology

An experiment was conducted at Universiti Malaysia Pahang, Pekan campus. Three identical MUSTAFA hives were exposed to outdoor conditions; without any shade. Two of the hives were fitted each with the cooling techniques respectively. Temperature readings were taken inside the broodcell compartment and honey cassette for each hives every five minutes during the day using thermocouples. The measurement was recorded by a temperature data logger (Lutron BTM-4208SD), which was then transferred into computer for further analysis.

Figure 1 shows the typical construction of a MUSTAFA-hive, where the hive is divided into three main components, which are the broodcell compartment, the honey cassette and the roof. In the brood cell compartment, a 25 mm diameter hole acts as the only entry and exit for stingless bees. The broodcell compartment is connected to the honey cassette with a 35 mm diameter hole. This provided access for the bees to store honey. These dimensions are optimised for rearing stingless bee of the *Heterotrigona Itama* species. Different species of stingless bees required different dimensions. The hives used in the experiment are all empty hives without colony, to ensure no thermal exchange occur with a living colony as well as to no disrupt them.

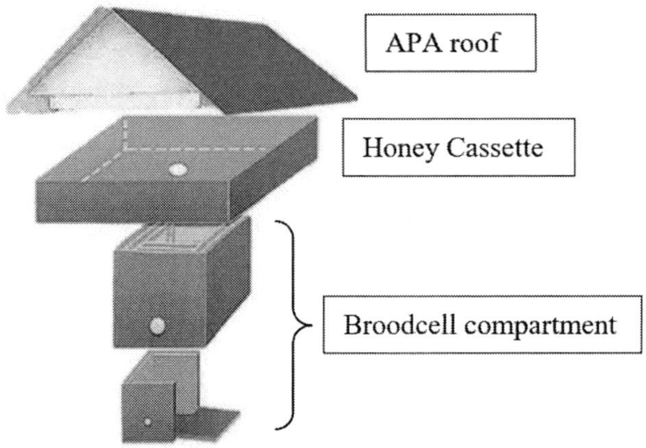

Figure 1. Typical construction of a Meliponiculture Using a Split-able Throne within Air-jacketed palace For Amplification-hive (MUSTAFA-hive).

Figure 2 shows the overall greenroof construction used in the experiment. The greenroof used was made up of four main components, which are the vegetation or the plant, soil and sand mixture in semi dry condition, a polyethylene layer, and wooden planks as roofing material and container.

Figure 2. Greenroof used in experiment.

Figure 3 shows the hive fitted with TEC setup, powered by PV panels. The cold side of TEC was positioned inside the honey cassette and the hot side was attached to a heatsink and fan outside. The inside of the honey cassette was also covered with aluminium sheet and some concrete to improve heat transfer of TEC inside the hive. The TEC was connected in parallel to two PV panels.

TEC fitted at the side of honey cassette, hot side coupled with a fan and the cold side attached to aluminium sheet inside honey cassette

PV panels position (parallel to the ground)

Figure 3. PV powered TEC system setup for experiment.

The TEC system used in the experiment consisted of one TEC1-12706 module with two PV 240 panels. The performance specification for TEC1-12706 are shown in Table 1. Meanwhile Table 2 shows the performance specification and dimensions of PV as provided by PV Hi-Tech Solar Sdn. Bhd.

Table 1. Performance specification and dimensions for TEC1-12706 [22].

Parameter	Minimum Value	Maximum Value
Hot Side Temperature (° C)	25	50
Q_{max} (Watts)	50	57
Delta T_{max} (° C)	66	75
I_{max} (Amps)	6.4	6.4
V_{max} (Volts)	14.4	16.4
Module Resistance (Ohms)	1.98	2.30
Dimensions	40mm x 40mm x 3.9mm	

Table 2. Performance specification and dimensions PV240 [23].

Parameter	Value
Model Name	PV 240
Number of Cells	60 cells in Series
Nominal Power @ STC (W)	240.00
Open Circuit Voltage (Voc)	37.95
Short Circuit Voltage (I_{sc})	8.67
Maximum Power Voltage (V_{mp})	30.51
Maximum Power Current (I_{mp})	7.88
Dimensions	1648mm x 991mm x 34mm
Weight	20 kg

Figure 4 shows the overall experimental setup. Three hives were exposed to outdoor conditions without any shades. The hive in the middle was fitted with a greenroof, the hive on the right was fitted with PV powered TEC system, and the hive on the left was not fitted with any cooling technique and acted as a control hive.

Figure 4. Overall setup, with thermocouples measuring temperatures inside honey cassette and brood cell compartment of each hives, connected to a temperature data logger.

3. Results and discussion

Table 3 shows the maximum, minimum, and average temperature inside the honey cassette and broodcell compartment of each hives. The control hive showed that, without cooling technique, the maximum temperature in the broodcell compartment could rise as high as 39.8 °C while in the honey cassette the maximum temperature was at 34.7 °C. This showed the need for cooling especially in the broodcell compartment.

Table 3. Minimum, maximum, and average temperature readings at honey cassette and broodcell compartment inside the hive.

	Honey cassette			**Broodcell**		
Temperature (°C)	**Control**	**Greenroof**	**TEC**	**Control**	**Greenroof**	**TEC**
Minimum	30.1	28.3	29.7	30.6	27.9	30.0
Maximum	34.7	30.8	34.6	39.8	31.6	39.2
Average	33.0	29.7	32.5	36.7	30.2	33.4

Between the two cooling methods tested; the greenroof and the PV powered TEC, the greenroof showed most significant cooling, where the average temperatures recorded are, 29.7 °C and 30.2 °C in the honey cassette and broodcell compartment respectively. This translated to an average of 3.3 °C and 6.5 °C lower than the control hive. On the other hand, the hive fitted with TEC system, experienced less cooling effect in both compartments. The average temperatures are 32.5 °C and 33.4 °C in honey cassette and the broodcell compartment respectively. This constituted to temperatures reduction of a mere 0.5 °C in the honey cassette and 3.3 °C in the broodcell compartment compared to the control hive.

The cooling performance of TEC are less significant compared to greenroof, where the minimum and maximum temperatures recorded are no more than 0.6 °C lower in both compartments. Meanwhile, the greenroof recorded 3.9 °C and 8.2 °C lower maximum temperatures for the honey cassette and broodcell compartment respectively. Figure 5 shows the temperature profile inside the honey cassette, represented by solid lines and broodcell compartment, represented by the pecked lines. The circle, triangle, and square marks represented the control hive, hive with greenroof and hive with PV powered TEC respectively.

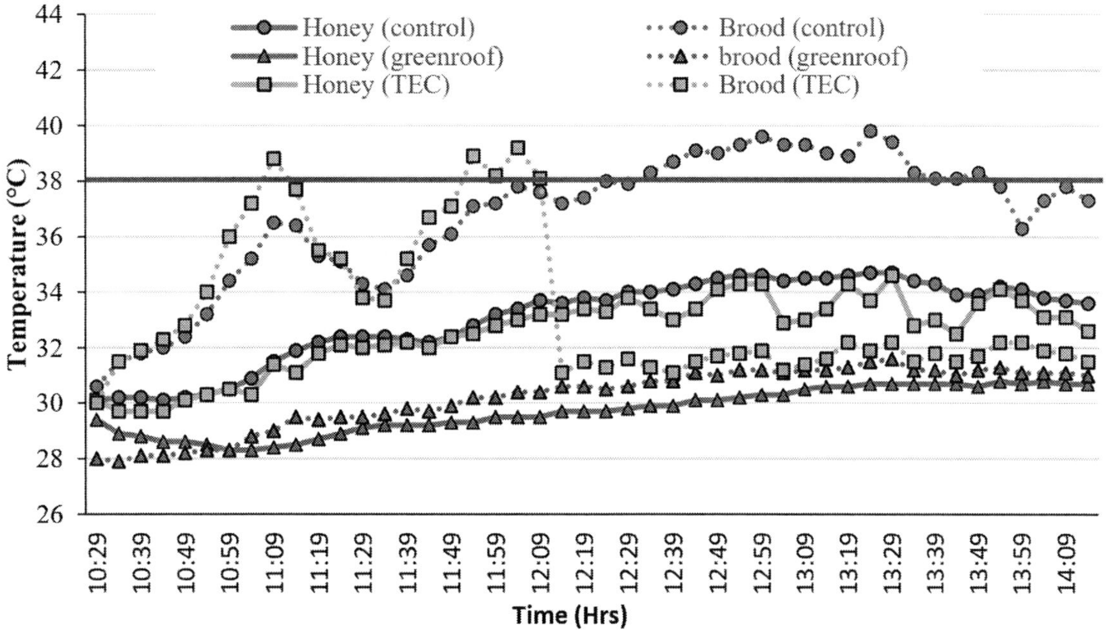

Figure 5. Temperature profile inside the broodcell compartment and honey cassette during the day for the control hive, hive with greenroof, and hive with TEC.

In the control hive, the broodcell compartment experienced a prolonged temperature rise above 38 °C, more than an hour, recorded around noon. For a colony, such extreme temperature are lethal to pupae in the broodcell compartment [5]. This can cause the colony to collapse and die out. The broodcell compartment also recorded significantly higher temperature compared to the honey cassette. This could partly be due to the roof covering most of the honey cassette from sun radiation while less covering of the broodcell compartment. As the hive was a closed system, the hot air may be trapped inside the broodcell compartment with only one exit, disrupting natural hot air circulation upwards.

Temperatures in hive with PV powered TEC followed similar trend to control hive, up until before noon. In fact, in broodcell compartment the PV powered TEC recorded higher temperatures at several points in time. However, around noon, the cooling effect of TEC system started to show with significant temperature drop in the broodcell compartment. This could be explained by the sun radiation started to peak around noon. This lead to the PV panels to produce enough power for the TEC to cool down the hive. This could be solved by adding more PV panels or adding battery packs. However, this could add more cost to the overall TEC system. On the other hand, in the honey cassette, the cooling effect was low. This could be because of air convection inside the honey cassette was low, resulted in lower heat transfer.

The hive with greenroof recorded more stable temperatures, with the broodcell compartment recorded slightly higher temperatures than the honey cassette. When compared to the control hive, the temperatures were consistently lower than the control hive, showing significant cooling effect around noon. The temperatures recorded were close to optimum temperature for stingless bees broodcells such as *Trigona denoiti*, which the broodcell area optimum temperatures are between 31 and 32 °C [3].

4. Conclusion

Three hives were exposed to outdoor conditions. Two of the hives were fitted with greenroof and PV powered TEC, respectively. Another hive remained without any cooling technique, acting as the control hive. Overall, the broodcell compartment in all hives recorded higher temperatures compared to the honey cassette up until around noon. This could be explained by low air convection inside the hive, which caused most of heat transfer to occur inside the broodcell compartment. However, the hive with PV powered TEC showed different temperature profile around. This could be explained by the PV only receiving enough power around afternoon. In the control hive, the broodcell compartment experienced prolonged temperature rise above 38 °C of more around noon. This could lead to death of pupae in broodcell. On the other hand, it was found that the hive fitted with greenroof has an impressive cooling performance. The temperatures inside the hive fitted with greenroof were significantly lower than the control hive when exposed to outdoor conditions without any shades. The average temperature reduction from the experiment were 3.3 °C and 6.5 °C in the honey cassette and broodcell compartment, respectively. Although the hive with PV powered TEC had the cooling potential as shown around afternoon where the temperatures dropped significantly, more PV panels might be needed. Alternatively, the PV panels could be supplemented with battery packs to provide enough power for the TEC system. Thus, it can be concluded that the greenroof is a feasible and economical solution towards cooling stingless bees' hives.

Acknowledgments

The authors would like to thank Universiti Malaysia Pahang for providing assistance for this research through the access of UMP Internal Grant RDU180313 and Ministry of Higher Education grant RDU190192 (FRGS/1/2018/TK03/UMP/02/25)

References

[1] Schwarz F 1939 The Indo-Malayan Species of Trigona *Am. Museum Nat. Hist.* **76** 83–141.
[2] Roubik D W 2006 Stingless bee nesting biology *Apidologie* **37** 124–43.
[3] Fletcher D J C and Crewe R M 1981 Nest structure and thermoregulation in the stingless bee, Trigona denoiti, Hymenoptera: Apidae *J. Entomol. Soc. South Africa* **44** 183–96.
[4] Sung I H, Yamane S and Hozumi S 2008 Thermal characteristics of nests of the Taiwanese stingless bee Trigona ventralis hoozana (Hymenoptera: Apidae) *Zool. Stud.* **47** 417–28.
[5] Vollet-Neto A, Menezes C and Imperatriz-Fonseca V L 2015 Behavioural and developmental responses of a stingless bee (Scaptotrigona depilis) to nest overheating *Apidologie* **46** 455–64
[6] Samejima H, Marzuki M, Nagamitsu T and Nakasizuka T 2004 The effects of human disturbance on a stingless bee community in a tropical rainforest *Biol. Conserv.* **120** 577–87.
[7] He W, Zhang G, Zhang X, Ji J, Li G and Zhao X 2015 Recent development and application of thermoelectric generator and cooler *Appl. Energy* **143** 1–25.
[8] Barrio E P Del 1998 Analysis of the green roofs cooling potential in buildings *Energy Build.* **27** 179–93.
[9] Niachou A, Papakonstantinou K, Santamouris M, Tsangrassoulis A and Mihalakakou G 2001 Analysis of the green roof thermal properties and investigation of its energy performance *Energy Build.* **33** 719–29.
[10] Sailor D J 2008 A green roof model for building energy simulation programs *Energy Build.* **40** 1466–78.
[11] Shiflett S A, Liang L L, Crum S M, Feyisa G L, Wang J and Jenerette G D 2017 Variation in the urban vegetation, surface temperature, air temperature nexus *Sci. Total Environ.* **579** 495–505.
[12] Tam V W Y, Wang J and Le K N 2016 Thermal insulation and cost effectiveness of green-roof systems: An empirical study in Hong Kong *Build. Environ.* **110** 46–54.
[13] Graceson A, Hare M, Hall N and Monaghan J 2014 Use of inorganic substrates and composted green waste in growing media for green roofs *Biosyst. Eng.* **124** 1–7.
[14] Ondimu S N and Murase H 2007 Combining Galerkin Methods and Neural Network Analysis to Inversely determine Thermal Conductivity of Living Green Roof Materials *Biosyst. Eng.* **96** 541–50.
[15] Altun A A 2012 Remote control of the temperature-humidity and climate in the beehives with solar-powered thermoelectric system *Control Eng. Appl. Informatics* **14** 93–9.

[16] Zacepins A, Brusbardis V, Meitalovs J and Stalidzans E 2015 Challenges in the development of Precision Beekeeping *Biosyst. Eng.* **130** 60–71.
[17] Ramli A S, Basrawi F, Daing Idris D M N, bin Yusof M H, Khalil Ibrahim T, Mustafa Z and Sulaiman S A 2017 A new dewatering technique for stingless bees honey *MATEC Web Conf.* **131** 03014.
[18] Ramli A S, Basrawi F, Ibrahim T K, Yusof M H Bin, Oumer A N, Johari N A, Muhamad A, Mamat M R and Habib K 2018 Experimental analysis on a novel low-temperature vacuum drying with induced nucleation technique for dewatering stingless bees honey *Dry. Technol.* **0** 1–7.
[19] Basrawi F, Ahmad A H, Daing Idris D M N, Maarof M R M, Chand M and Ramli A S 2017 Engineering economic analysis of meliponiculture in Malaysia considering current market price *MATEC Web Conf.* **131** 04005.
[20] Ramli A S, Luqman A H, Basrawi F, Oumer A N, Aziz A A and Mustafa Z 2017 A new cooling technique for stingless bees hive ed H Sakidin, M H Yusof, N Sa'ad, D Ling Chuan Ching and S A Abdul Karim *MATEC Web Conf.* **131** 03013.
[21] Basrawi F, A Hamid A R, Bahari R, Mohd Noordin M N and Yusof M H 2017 A Preliminary Study on the Thermal Performance of a Ventilated Honey Cassette for Stingless Bees *MATEC Web Conf.* **131** 2–7.
[22] Hebei IT (Shanghai) Co Ltd 2013 Thermoelectric Cooler TEC1-12706 1–3.
[23] PV Hi-Tech Solar Sdn. Bhd. 2015 PV Poly Series Photovoltaic Module 1–2.

Effect of superficial air velocity on the fluidized bed drying performance of stingless bee pot-pollen

L A Halim[1], M F Basrawi[1,*], S N Faizal[1], A S M Yudin[1] and T M Yusof[1]

[1] Meliponini Engineering Laboratory (MePEL), Faculty of Mechanical and Automotive Engineering Technology, Universiti Malaysia Pahang, 26600 Pekan, Pahang, Malaysia.

*E-mail: mfirdausb@ump.edu.my

Abstract. Stingless bees or "kelulut" also produce pot-pollen apart from honey. The pot-pollen is mixed with honey and bee secretion before stored in cerumen pots. It has high nutritive value and medicinal benefits. Pot-pollens are often neglected by the beekeepers due to difficulty in storing and preserving them due to high moisture content. Hence, fluidized bed dryer is proposed as a suitable method to dry and enable convenient storage and preservation of the pot-pollen. Pot-pollen sample of initial moisture content 30.5% is dried at three superficial air velocities, 1.0 m/s, 1.5 m/s, and 2.0 m/s for 30 minutes. Fluidized bed drying has managed to decrease the moisture content down to 23%, 20.5%, and 18.5%, respectively. Higher superficial air velocity lead to higher drying rate in of pot-pollen. Hence, using a fluidized bed dryer to dry stingless bee pot-pollen is a promising method for preserving them. Subsequently, the dried pot-pollen can be easily commercialized in the future.

1. Introduction

Stingless bees are native to Malaysia and locally known as "kelulut". They are important pollinator especially in tropical region such as Malaysia [1]. As stingless bee is usually for its honey, meliponiculture or beekeeping tend to neglect other bee products. Similar to honey bees, stingless bee collect pollen from variety of plant species [2]. The pollen is mixed with honey and bee secretion before stored in cerumen pots, coining the term pot-pollen. Although used interchangeably with bee bread, the term pot-pollen exclusively refer to pollen stored in stingless bees nest [3]. Pot-pollen have therapeutic and medicinal value [4]. Pot-pollen are usually neglected or simply thrown away by the bee keepers due to difficulty to preserve them as they contain high moisture content [5]. This can cause excessive fermentation and spoilage, deteriorating the quality of pot-pollen. Hence, preserving pot-pollen requires drying with suitable method. However, conventional methods may lead to loss of nutrient and useful compounds [6].

Currently, studies on engineering application for stingless bee focused either on thermal regulation for hives [7, 8], or honey processing [9, 10] and engineering economic analysis [11]. Fluidized bed dryer has shown promise as a viable method for drying of pot-pollen owing to its ability to operate at low temperature and short drying time [5]. Fluidized bed dryer employ the concept of fluidization, where the bed material is suspended in a fluidizing gas with air being the typical gas [12]. Fluidized bed dryer has several advantages compared to conventional dryers. Among the advantages of fluidized bed dryer are the system has homogeneous and rapid drying, suitable for heat sensitive products and efficient in drying process. Besides that, the system has simple design thus consume low cost of operation and maintenance. The system also can be designed with wide range of capacities [13].

In a fluidized bed, drying temperature is the most significant parameter towards drying rate [14–16]. However, it is not recommended to heat pot-pollen above 40 °C as its nutritional properties will be lost [6]. Hence, controlling other parameter such as superficial air velocity while maintaining low temperature is preferable towards drying of pot-pollen. This study aims to explore the effect of superficial air velocity on the fluidized bed drying of stingless bee pot-pollen.

2. Methodology

The drying experiments were conducted at the Meliponini Engineering Laboratory (MePEL), Faculty of Mechanical and Manufacturing Engineering, Universiti Malaysia Pahang (UMP). A schematic of laboratory scale fluidized bed dryer used in present study is shown in Figure 1 below. The cylindrical fluidized bed chamber is used to contain the bed material during drying. A ball valve is used to control the superficial air velocity. A perforated air distributor with diameter of 3 mm and 10 mm square pitch was used in this study.

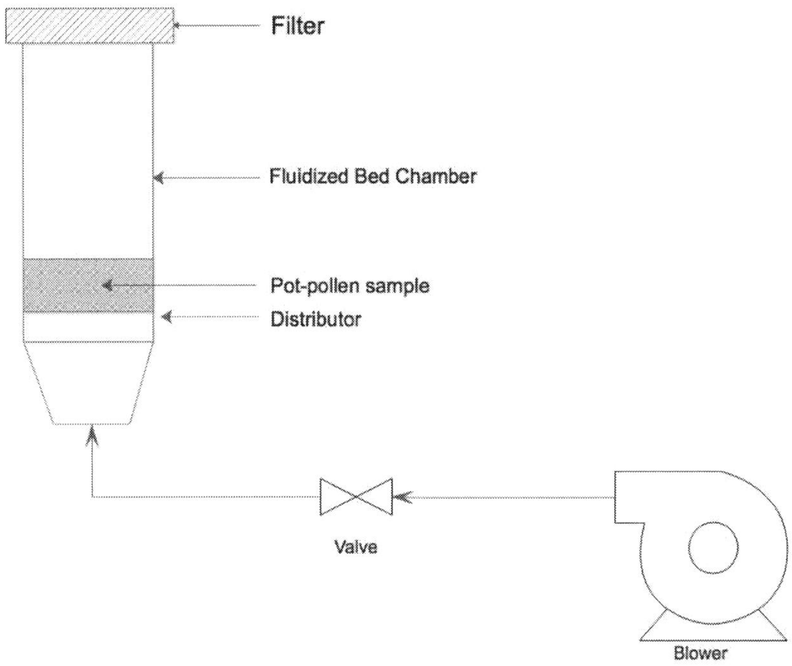

Figure 1. Schematic of laboratory scale fluidized bed dryer.

Raw stingless bee pot-pollen sample is sourced from a local farm in Kuantan, Pahang. The pot-pollen sample is cleaned and stored at -10 °C prior to the experiment. For each experiment, 50 g of pot-pollen is dried for 30 minutes in the fluidized bed dryer at ambient temperature conditions. The experiment is conducted at three superficial air velocities, 1.0 m/s, 1.5 m/s, and 2.0 m/s.

The moisture content of raw pot-pollen and dried pot-pollen sample is determined using hot air oven method. Each pot-pollen sample was weighed to 2 g before put in hot air oven at 100 °C for 24 hours. The moisture content will be the difference of the initial mass and the dried mass [17]. This is shown in equation (1).

$$Moisture\ content, MC\ [\%] = \frac{A - B}{B} \times 100 \quad (1)$$

where A is the initial mass in g of pot-pollen sample before oven drying and B is the final mass of pot-pollen after dried in the oven.

3. Results and discussion

Table 1 shows the moisture content for each pot-pollen sample dried in the fluidized bed dryer. The initial moisture content of pot-pollen sample is recorded at 30.5%. The moisture content level is within the typical range for stingless bee pot-pollen [3, 18–20].

Table 1. Moisture content level of pot-pollen sample.

Air velocity [m/s]	Before (g)	After (g)	Moisture Content [%]
Initial	2	1.39	30.5
1.0 m/s	2	1.54	23.0
1.5m/s	2	1.59	20.5
2.0 m/s	2	1.63	18.5

Figure 2 shows the initial and final moisture content of pot-pollen sample dried at each superficial air velocities. The final moisture content level are reduced down to 23%, 20.5% and 18.5%, respectively. The trend can be seen that increasing the superficial air velocity will reduce the moisture content further. This indicates that the drying rate is higher for higher superficial air velocity. Such observation can be explained by higher heat and mass transfer between fluidizing air and pot-pollen sample as the air velocity is increased. The best drying performance in this study is at 2.0 m/s air velocity, where the moisture content is reduced by 12%. This is a significant reduction for a short drying time of 30 minutes. Hence, using fluidized bed dryer is a viable and promising technique for pot-pollen drying.

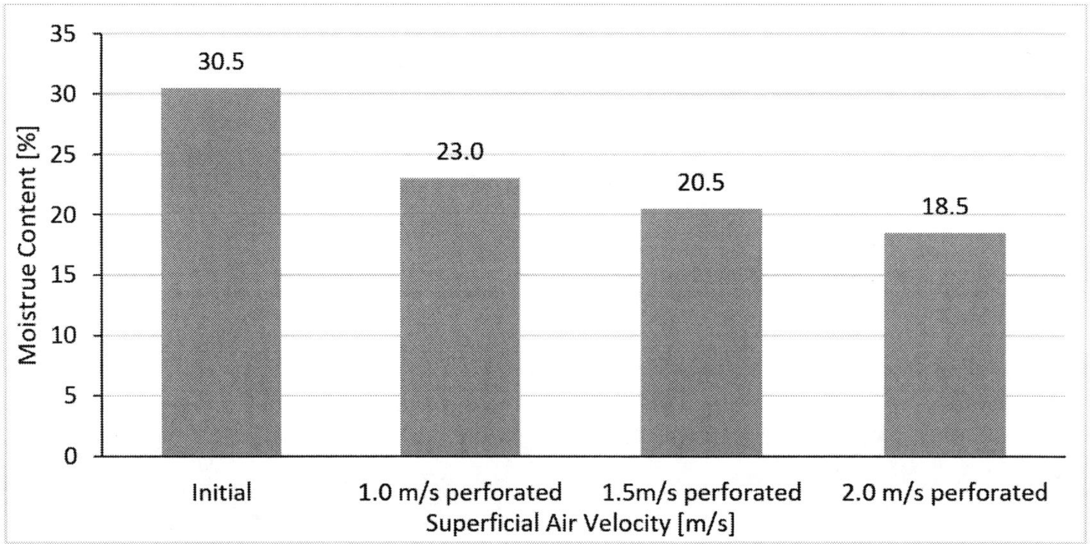

Figure 2. Moisture content for dried pot-pollen at different air velocities.

4. Conclusion

Raw stingless bee pot pollen sample is weighed to 50 g, and dried using fluidized bed dryer for 30 minutes at superficial air velocities of 1.0 m/s, 1.5 m/s, and 2.0 m/s. The initial moisture content level of pot-pollen sample is at 30.5%. As the superficial air velocities is increased, the drying rates also increase. The final moisture content at each superficial air velocities are at 23%, 20.5%, and 18.5%. The best drying performance from this study is at 2.0 m/s superficial air velocity, where the moisture content level reduced by 12%. In conclusion, fluidized bed dryer is able to dry stingless bee pot-pollen within a short period of time. Hence, dried pot-pollen can be easily stored and may be commercialized in the future.

Acknowledgments
The authors would like to thank Universiti Malaysia Pahang for providing assistance for this research through the access of Post Graduate Research Scheme PGRS190342 and Ministry of Higher Education grant RDU190192 (FRGS/1/2018/TK03/UMP/02/25)

References

[1] Mohd N, Mohd F, Sajap a. S, Rosliza J and Suri R 2010 Conservation and sustainable utilization of stingless bees for pollination services in agricultural ecosystems in Malaysia *Int. Semin. Enhanc. Funct. Biodivers. Relev. to Sustain. Food Prod. ASPAC* 1–11.

[2] Ghazi R, Zulqurnain N S and Azmi W A 2018 Melittopalynological Studies of Stingless Bees from the East Coast of Peninsular Malaysia *Pot-Pollen in Stingless Bee Melittology* (Cham: Springer International Publishing) pp 77–88.

[3] Vit P, D'Albore G R, Barth O M, Peña-Vera M and Pérez-Pérez E 2018 Characterization of Pot-Pollen from Southern Venezuela *Pot-Pollen in Stingless Bee Melittology* (Cham: Springer International Publishing) pp 361–75.

[4] Flavia Massaro C, Villa T F and Hauxwell C 2018 Metabolomics Analysis of Pot-Pollen from Three Species of Australian Stingless Bees (Meliponini) *Pot-Pollen in Stingless Bee Melittology* (Cham: Springer International Publishing) pp 401–17.

[5] Halim L A, Basrawi F, Chong J Z, Oumer A N, Yudin A S M, Yusof M H bin and Sulaiman S A 2018 Preliminary Study on Drying of Stingless Bee Pot Pollen Using Novel Fluidized Bed Dryer with Swirling Distributor ed S A Abdul Karim, N Zainuddin, M H Yusof and N Sa'ad *MATEC Web Conf.* **225** 04007.

[6] Akhmetova R, Sibgatullin J, Garmonov S and Akhmetova L 2012 Technology for extraction of bee-bread from the honeycomb *Procedia Eng.* **42** 1822–5.

[7] Ramli A S, Luqman A H, Basrawi F, Oumer A N, Aziz A A and Mustafa Z 2017 A new cooling technique for stingless bees hive ed H Sakidin, M H Yusof, N Sa'ad, D Ling Chuan Ching and S A Abdul Karim *MATEC Web Conf.* **131** 03013.

[8] Basrawi F, A Hamid A R, Bahari R, Mohd Noordin M N and Yusof M H 2017 A Preliminary Study on the Thermal Performance of a Ventilated Honey Cassette for Stingless Bees *MATEC Web Conf.* **131** 2–7.

[9] Ramli A S, Basrawi F, Daing Idris D M N, bin Yusof M H, Khalil Ibrahim T, Mustafa Z and Sulaiman S A 2017 A new dewatering technique for stingless bees honey *MATEC Web Conf.* **131** 03014.

[10] Ramli A S, Basrawi F, Ibrahim T K, Yusof M H Bin, Oumer A N, Johari N A, Muhamad A, Mamat M R and Habib K 2018 Experimental analysis on a novel low-temperature vacuum drying with induced nucleation technique for dewatering stingless bees honey *Dry. Technol.* **0** 1–7.

[11] Basrawi F, Ahmad A H, Daing Idris D M N, Maarof M R M, Chand M and Ramli A S 2017 Engineering economic analysis of meliponiculture in Malaysia considering current market price *MATEC Web Conf.* **131** 04005.

[12] Holdich R G 2002 *Fundamentals of particle technology* (Midland Information Technology and Publishing).

[13] Hall C W 1988 *Handbook of Industrial Drying* vol 6.

[14] Vizcarra-Mendoza M G, Ruiz-Martinez R S, Martinez-Vera C, Iruegas-Evaristo A and Carrillo-Guerreno J M 1998 Treatment of Pollen By Fluidization *Dry. Technol.* **16** 1843–53.

[15] Reyes A, Vega R and Garcia G 2008 Drying sawdust in a pulsed fluidized bed *Dry. Technol.* **26** 476–86.

[16] Li H W and Guo H 2017 Analysis of drying characteristics in mixed pulsed rectangle fluidized

beds *Powder Technol.* **308** 451–60.

[17] Ismail W I W, Hussin N N, Mazlan S N F, Hussin N H and Radzi M N F M 2018 Physicochemical Analysis, Antioxidant and Anti Proliferation Activities of Honey, Propolis and Beebread Harvested from Stingless Bee *IOP Conf. Ser. Mater. Sci. Eng.* **440**.

[18] Souza B, Roubik D, Barth O, Heard T, Enríquez E, Carvalho C, Villas-Bôas J, Marchini L, Locatelli J, Persano-Oddo L, Almeida-Muradian L, Bogdanov S F and Vit P 2006 Composition of stingless bee honey: Setting quality standards *Interciencia* **31** 867–75.

[19] Chuttong B, Phongphisutthinant R, Sringarm K, Burgett M and Barth O M 2018 Nutritional Composition of Pot-Pollen from Four Species of Stingless Bees (Meliponini) in Southeast Asia *Pot-Pollen in Stingless Bee Melittology* (Cham: Springer International Publishing) pp 313–24.

[20] de Oliveira Alves R M, da Silva Sodré G and Carvalho C A L 2018 Chemical, Microbiological, and Palynological Composition of the "Samburá" Melipona scutellaris Pot-Pollen *Pot-Pollen in Stingless Bee Melittology* (Cham: Springer International Publishing) pp 349–60.

IOP Conf. Series: Materials Science and Engineering **863** (2020) 012042 doi:10.1088/1757-899X/863/1/012042

Low voltage reliability equivalent using monte-carlo simulation technique

N F M Fauzi, N N R Roslan and M I M Ridzuan*

Faculty of Electrical & Electronics Engineering Technology, Universiti Malaysia Pahang, 26600 Pekan, Pahang, Malaysia.

*E-mail: ikhwanr@ump.edu.my

Abstract. Reliability is the ability of a system to supply continuous electricity to customer which ends with zero fault that occurs under a specific period of time. Most of the literature focus more on medium voltage (MV) and high voltage (HV) compared to the low voltage (LV) due to the general absence of exact data in LV network and sizing of LV network. In addition, an increament in size of the LV network makes the network more complex and difficult to assess. Therefore, in this paper, the performance of reliability in LV network will be evaluated in detailed network model. To reduce simulation time, methodology of reducing detailed network into an equivalent network is introduced. This equivalent network is obtained by simplifying the complex network using Monte-Carlo Simulation technique. The results in this research are quantified and compared between these detailed and equivalent networks in reliability indices; SAIFI, SAIDI and CAIDI. The values of SAIFI, SAIDI and CAIDI in detailed network are slightly higher than in equivalent network.

1. Introduction

Power systems are perhaps the most complex large-scale engineered systems today, and it is predicted to have the highest level of reliability. The interruption always occurs in the system, yet their customers expect more reliability and affordability [1]. Over decades, the reliability of power system evaluation was focused more on the generation and transmission compared to the distribution system especially on the low-voltage distribution although this low voltage will also affect the performance of power system. The goal of the present planned in distribution system is to ensure that the performance of power system, especially in LV and MV will give better effect to the customers. Thus, it is important to ensure that the customer will get the continuity of supply with minimum interruption occurring. In most of the power system, both MV and LV are represented in a lumped model due to the complexity of calculation and the volumes of LV and MV [2]. Hence, the simulation will take time to compute the result for the reliability analysis in such a complex and large network. Thus, the representation of an equivalent network will simplify the complex network, consequently reducing simulation time.

Since distribution network is currently supplying the most customers, it is tied up by the target or minimum customer satisfaction level imposed by Energy Regular, which in Malaysia is Energy Commission (EC). These targets mostly involve the frequency and duration of interruption. To attain that target, distribution network operators (DNOs) must correctly assess their reliability performance. Therefore, it is critical to have accurate distribution network configurations and parameters. However, due to the size of distribution network, low voltage (LV) network is often represented by aggregate model [3-8].

In this research, the main intention is to represent the whole LV network with a single equivalent component. This simplified equivalent component is implemented in MV system specifically at the downstream of the aggregation of MV point. The performance of reliability in a detailed network should be assessed first before an assessment in an equivalent network can be done. The detailed network will give more details and specific information of the system such as the interruption on the specific location and occurrence of fault while the equivalent network consists of simplified information of the network.

2. Monte-Carlo simulation technique

In reliability assessment, there are two types of methods that can be used to evaluate the reliability performance of the network, which are analytical and probability assessment. The analytical method uses a mathematical based approach which evaluates the performance of reliability in power system using mathematical solution while the probability method uses random nature process. In terms of contingency, basically the analytical approach will choose the states in increasing order of which each state is evaluated in just one time. The reliability indices are then calculated using mathematic solution based on the statistical data related to each state [9]. As power system consists of a large and complex network, Monte-Carlo Simulation (MCS) will be used to equivalate and assess the performance of networks [10, 12]. Figure 1 shows the flowchart of how the performance of reliability in the system is assessed. The sequential MCS technique is used which simulates the system chronological behaviour by sampling the system state sequences for several period of time. For this method, two basic inputs which are fault rate and repair time, need to be identified first before it can be randomly generated [13, 14].

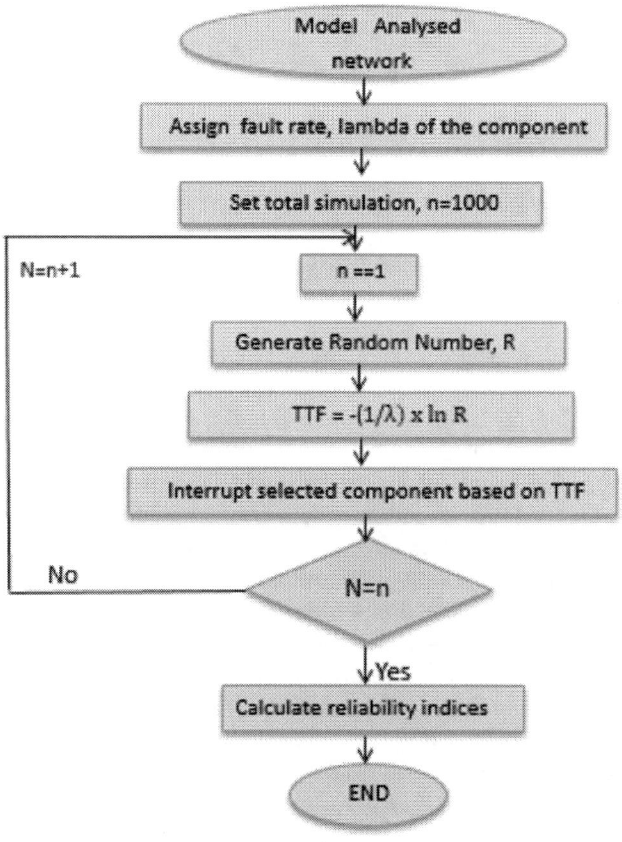

Figure 1. Flowchart of Monte-Carlo simulation.

In this research, two models of bus system, which are Network 14 and Network 4gs (networks from Matpower) are used to represent the distribution network. Network 14 represents LV distribution network for Case 1 as shown in figure 2. Meanwhile Network 4gs represents MV distribution network for Case 2 as shown in figure 3. Both of the networks are being modified by only one generator to analyse the reliability performance of the systems. Before an equivalent network can be assessed, a few analysis needs to be done in both LV and MV networks. In distribution system, the LV network is always located to the downstream of MV network. Even though the objective of this research is to evaluate the performance of LV network, the assessment of MV network is also included to quantify and justify the importance of detailing the distribution network. Hence, there are about 4 different networks that need to be analysed differently tabulated in table 1.

Table 1. Description of cases.

Case	Description
1	LV network consists of 14 buses and 20 branches (figure 2)
2	MV network consists of 4 buses & 4 branches (figure 3)
3	Detailed (a combination of MV and LV) network consists of 56 buses & 84 branches (figure 4)
4	Equivalent network (a combination of MV and LV equivalent) network (figure 5)

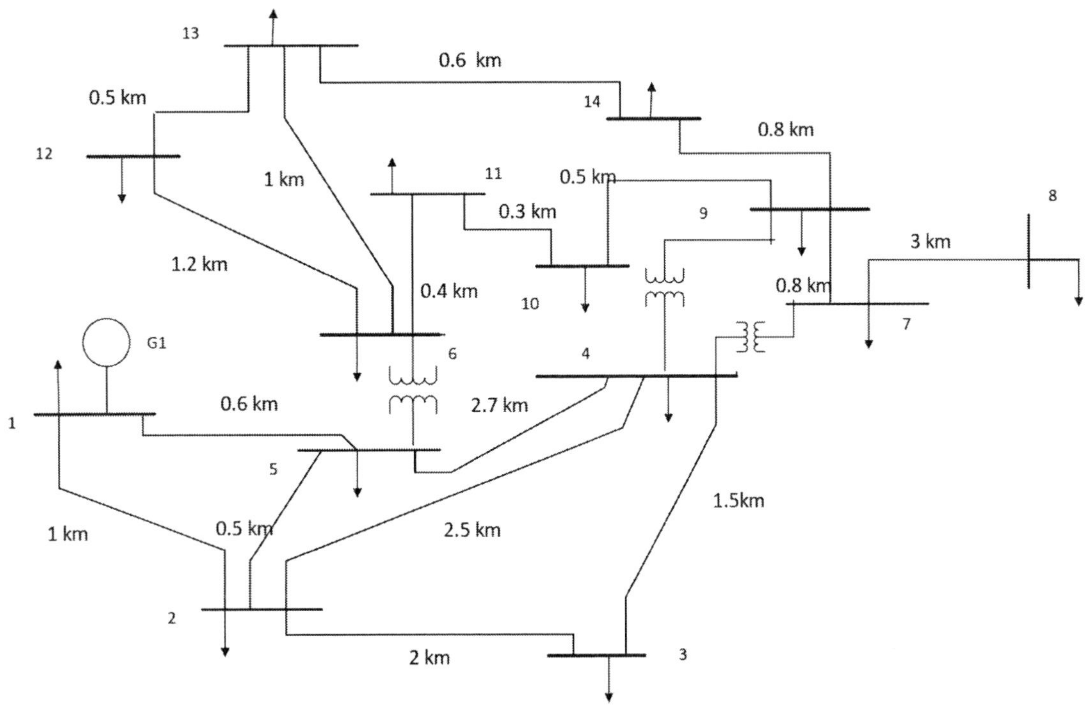

Figure 2. LV Network for case 1.

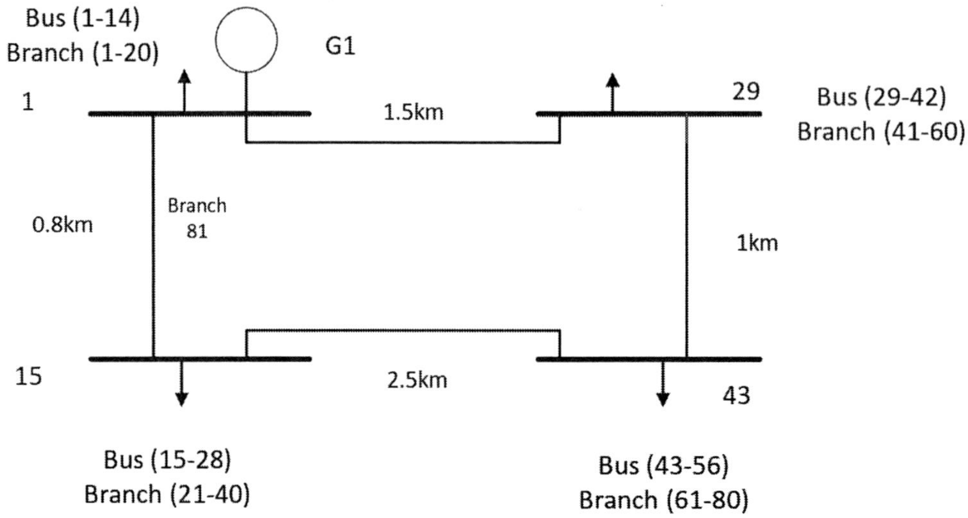

Figure 3. MV Network for case 2.

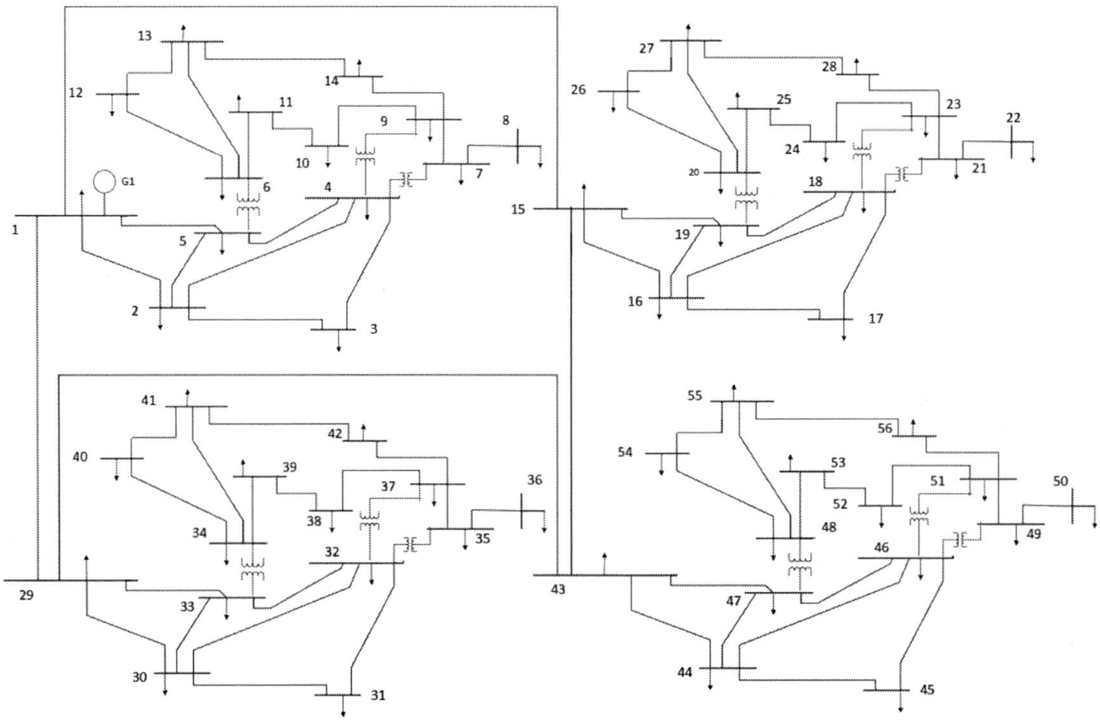

Figure 4. MV and detailed LV Networks for case 3.

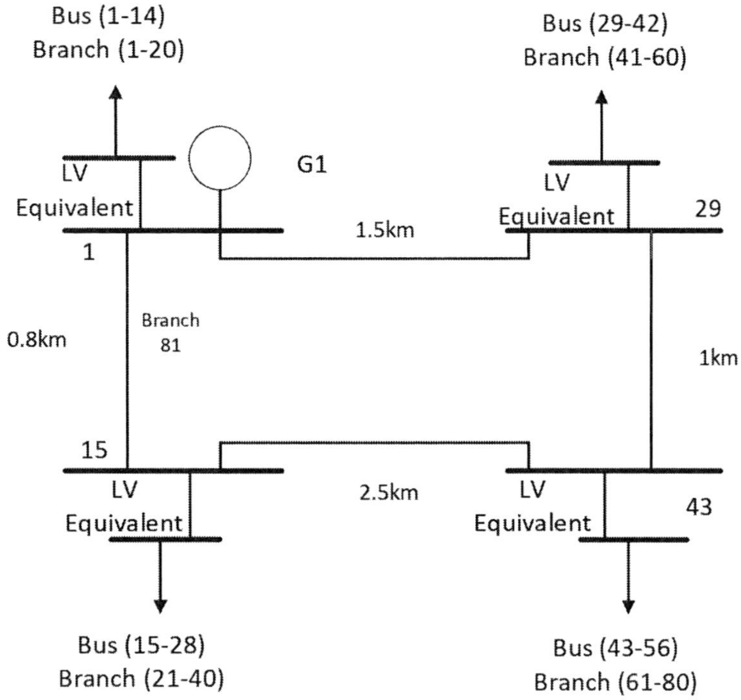

Figure 5. MV and Equivalent LV Networks for case 4.

2.1. Input data, fault rate and repair time

As stated above, two basic inputs in the reliability assessment are fault rate and repair time. It is very important to select the accurate value for both of these inputs since it will indirectly affect reliability performance. For each of failure rate and repair time, it will consider the reliability performance of every component in the network such as transformer, circuit breaker, etc as tabulated in table 2. In this research, two components of power systems are included, but the main focus is overhead line since it is the most dominant component in the network.

Table 2. Parameter for reliability analysis in LV and MV network [15,16].

Component	Voltage (kV)	Fault rates (failure/year)	Repair times (hours/fault)
Overhead lines	11	0.123	5
	0.4	0.168	6.44
Transformer	11/0.4	0.015	5

2.2. Reliability Indices

In distribution system, the assessment of the reliability can be divided into two different groups which are load indices and system indices [17]. There are a few reliability indices used as a parameter to evaluate the performance of reliability in the system which are *SAIDI, SAIFI, MAIFI, CAIDI, ENS* and *AENS* [18]. In this analysis, only three common indices are considered, which are *SAIDI, SAIFI* and *CAIDI* as defined by equation (1) to equation (3). These indices are very important especially to the service provider to record the performance of reliability in power system in order to ensure better quality of services received by the customers [19].

$$SAIFI = \text{Total number of customers interrupted } (LI) \,/\, \text{Total number of customers served} \quad (1)$$

$$SAIDI = \text{Total number of interruption durations } (LI) \,/\, \text{Total number of customers served} \quad (2)$$

$$CAIDI = SAIDI \,/\, SAIFI \quad (3)$$

3. Results and discussion

Figure 6 below illustrates the average indices of 4 different cases. Based on figure 6, the average of *SAIFI* in Case 1 network is slightly lower than in Case 2. This is because the total interruption in LV is lower than in MV, which is directly related to failure rates from table 2. Another contributing factor is the number of component in Case 1 is higher than Case 2. Since the formula of *SAIFI* is related the total interruption, hence an increase in the occurrence of interruptions in the system will also increase the value of *SAIFI*. Since LV network is located to the downstream of MV; thus the interruptions in LV will affect the total interruptions in MV for Case 3 and 4. Hence, the interruptions in MV will be higher than in LV. This is one of the reasons why the average of *SAIFI* is lower than in MV.

The repair time used in Case 1 and Case 2 are 5 hours/fault and 6.44 hours/fault, respectively. Since the average of interruption hours (*CAIDI*) is inversely proportional to the average failure rate (*SAIFI*), hence the higher the value of *SAIFI*, the lower the value of *CAIDI*. Figure 6 illustrates Case 2, which has higher *SAIFI* and the lowest *CAIDI*. While for Case 3 and Case 4, the average values of *SAIFI, SAIDI* and *CAIDI* are close each other. The result of the average reliability indices obtained in both Case 3 and Case 4 are acceptable since these network models need to be the same or almost the same for all the indices. This is because the representation of the equivalent network (Case 4) is to simplify the large/complex network (Case 3) without changing any parameters of the network.

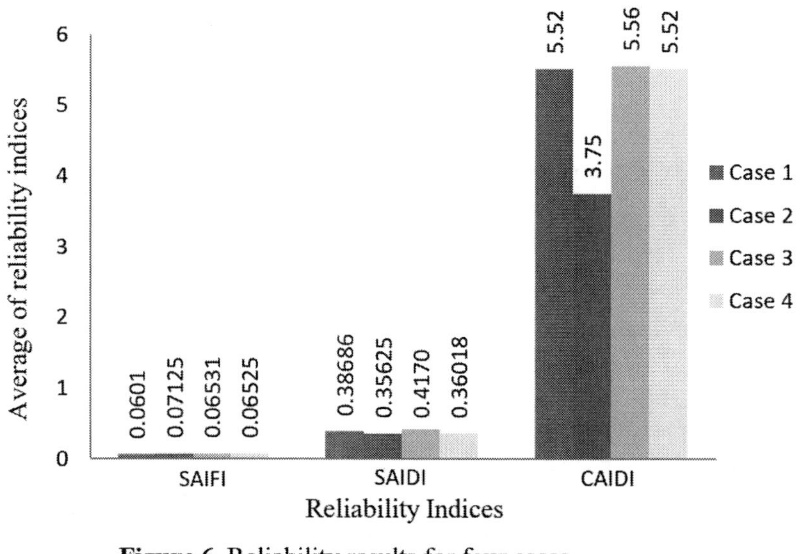

Figure 6. Reliability results for four cases.

Table 3 below shows the percentage error between detailed (Case 3) and equivalent network (Case 4). The average of SAIFI between Case 3 and Case 4 are close to each other. Hence, the percentage error between these two is the lowest. Since Case 3 is the combination of LV and MV networks, therefore, the repair time of the components is different according to the types of networks. Thus, the percentage error in the average of *CAIDI* between Case 3 and Case 4 is about 0.72 % which is higher than percentage error in *SAIFI*. Lastly, the percentage of error in *SAIDI* is the highest compared to the others. The *SAIDI* index is the total duration of interruption over the total number of customers. The total duration of interruptions is related to the repair time and interruptions of components. Since the interruptions in detailed network varies and there are a few customers who are not interrupted at all, hence it will affect the overall average of *SAIDI* in detailed network. Thus, the percentage error of *SAIDI* between these two networks are the highest.This percentage error of *SAIDI* can be reduced by increasing the simulation time.

Table 3. The percentage error between detailed (Case 3) and equivalent network (Case 4).

Average Index	Case 3	Case 4	Percentage Error (%)
SAIFI	0.06531	0.06525	0.09
SAIDI	0.4170	0.36018	13.63
CAIDI	5.56	5.52	0.72

3.1. Detailed network

The detailed network (Case 3) represents the combination of both MV with LV detailed networks. All the parameters of components in the network must be configured and analysed. These parameters such as resistance, R and reactance, X of components in the network must be represented by equivalent values, which are Req and Xeq. All the information of parameters used in this analysis are obtained from MatPower. In Case 3, detailing the network model required more time to model the network and higher simulation time compared to Case 4. The positive side of Case 4 is it can provide more detailed information, especially on the specific location/component of interruption and duration of interruption in the system.

Since the failure rate of the lines depends on the length of lines, hence increasing the length of the line will increase the failure rate. The data of interruptions of a specific customer will facilitate the service provider to detect the location of failure in a short time, hence reducing the duration of interruptions experienced by the customers. Figure 7 shows the reliability indices for each of the customers. This graph displays that the reliability indices for every customer varied among them. This means that each of the customer will experience a different total number of interruptions. Based on the graph, for customers 1, 31, 32, 33, 44, 46 and 47, there is no reliability indices recorded. It means that these customers did not experience interruptions at all. This is due to many combinations of electrical path from source to load, which increase the security level for these customers.

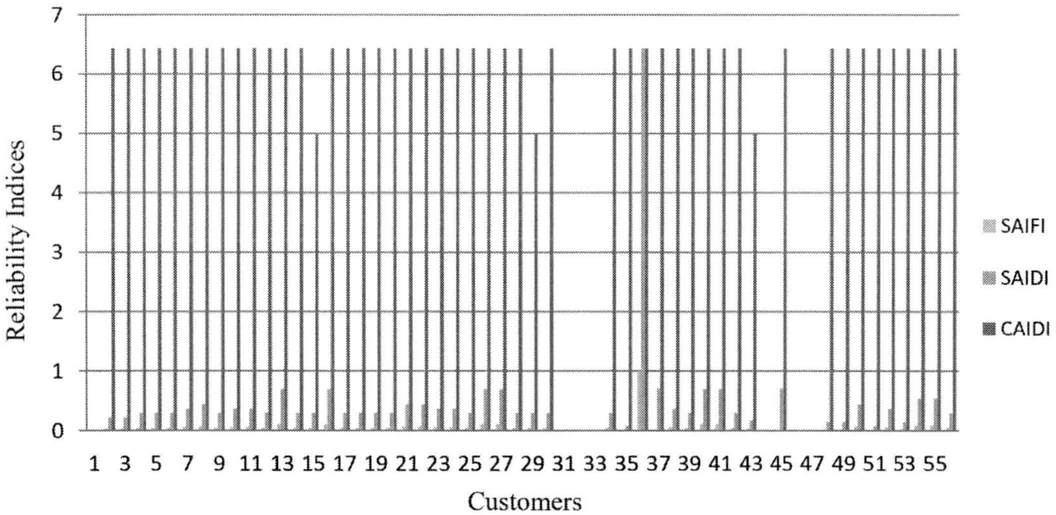

Figure 7. Detail reliability indices for case 3.

3.2. Equivalent network

This equivalent representation (Case 4) will not change the parameter of the components in the network because the total number of the same parameter is represented with one equivalent value. There are about 56 customers in the detailed network (Case 3); thus, the parameters of LV network (Case 1) at every 14 customers will be represented with one equivalent customer. In this case, the reliability indices (*SAIFI, SAIDI, CAIDI*) are used to justify the representation of detailed network (Case 3) with an equivalent network (Case 4).

The equivalent network (Case 4) has benefits, especially in reducing the simulation time, but it is really difficult to detect the interruption and location of interruption occurring in the network. This is because one equivalent value represents a numerous values of components and configurations in the LV network. If the type of fault component and location of fault is detected, the service provider will be able to provide mitigation plan to overcome this interruption by re-routing the electrical path from source to customers. Hence, it is crucial, especially to the service provider to decide, either to detect the specific location of the failure in the network or save detailed network modelling time and simulation time.

Figure 8 shows the reliability indices obtained for each customer in Case 4. Based on the graph, the value of *CAIDI* obtained is constant for each of the customers. The result is acceptable since the values for Case 3, and Case 4 are almost the same. The value differences between Case 3 and Case 4 are able to be reduced by increasing the simulation value.

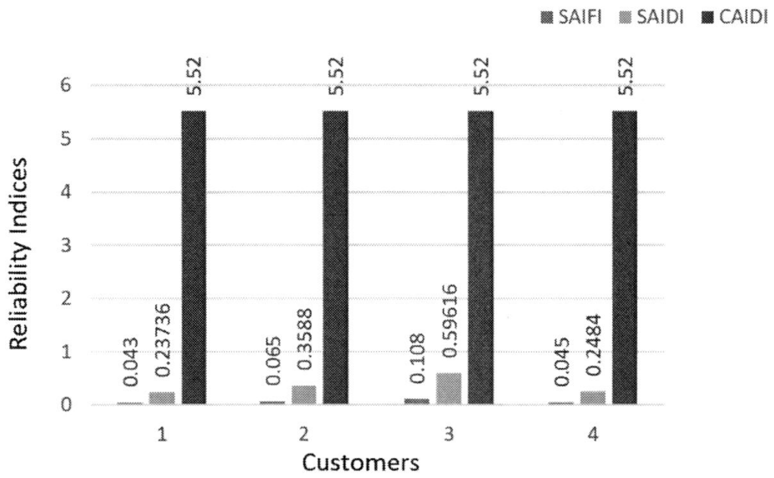

Figure 8. Detail reliability indices for case 4.

4. Conclusion

This paper has introduced the methodology of reducing large/complex network into a single equivalent network. The complexity of the network is represented by one equivalent network in which the parameter of reliability indices of these networks will have the same value or close to each other depending on the number of simulations. The percentage error of reliability indices can be reduced by increasing the number of simulations (years). Although the equivalent of a network can simplify the network and reduce simulation time, the disadvantage of this network is the difficulty to determine the location of fault and faulty component.

Acknowledgements

Universiti Malaysia Pahang Internal Grant RDU1703260 supports this research. The authors would also like to thank the Faculty of Electrical & Electronics Engineering Universiti Malaysia Pahang for providing facilities to conduct this research and financial support throughout the process.

Reference

[1] I. Hernando-Gil, B. Hayes, A. Collin, and S. Djokic, "Distribution network equivalents for reliability analysis. Part 2: Storage and demand-side resources," *IEEE PES ISGT Europe*, IEEE, pp. 1–5, Oct-2013.

[2] I. S. Ilie, I. Hernando-Gil, and S. Z. Djokic, "Reliability Equivalents of LV and MV Distribution

Networks," *IEEE International Energy Conference and Exhibition, ENERGYCON 2012*, pp. 343–348, 2012.

[3] S. Kazemi, "Reliability evaluation of smart distribution grids," PhD thesis, Aalto University, Espoo, Finland, 2011.

[4] O. Siirto, M. Loukkalahti, M. Hyvarinen, P. Heine, and M. Lehtonen, "Neutral point treatment and earth fault suppression," in *Electric Power Quality and Supply Reliability Conference (PQ)*, 2012, pp. 1–6.

[5] M. Katsanevakis, R. A. Stewart, and L. Junwei, "A novel voltage stability and quality index demonstrated on a low voltage distribution network with multifunctional energy storage systems," *Electr. Power Syst. Res.*, vol. 171, pp. 264–282, Jun. 2019.

[6] M.-G. Jeong *et al.*, "Optimal Voltage Control Using an Equivalent Model of a Low-Voltage Network Accommodating Inverter-Interfaced Distributed Generators," *Energies*, vol. 10, no. 8, p. 1180, Aug. 2017.

[7] I. Afandi, P. Ciufo, A. Agalgaonkar, and S. Perera, "A holistic approach for integrated volt/var control in MV and LV networks," *Electr. Power Syst. Res.*, vol. 165, pp. 9–17, Dec. 2018.

[8] A. Di Fazio, M. Russo, M. De Santis, A. R. Di Fazio, M. Russo, and M. De Santis, "Zoning Evaluation for Voltage Optimization in Distribution Networks with Distributed Energy Resources," *Energies*, vol. 12, no. 3, p. 390, Jan. 2019.

[9] O. G. I. Okwe Gerald Ibe, "Adequacy Analysis and Security Reliability Evaluation of Bulk Power System," *IOSR J. Comput. Eng.*, vol. 11, no. 2, pp. 26–35, 2013.

[10] R. Billinton and R. Allan, *Reliability Evaluation of Power Systems*, 2nd ed. New York, 1996.

[11] D. Urgun and C. Singh, "A Hybrid Monte Carlo Simulation and Multi Label Classification Method for Composite System Reliability Evaluation," *IEEE Trans. Power Syst.*, vol. 34, no. 2, pp. 908–917, Mar. 2019.

[12] L. Peng, B. Hu, K. Xie, H.-M. Tai, and K. Ashenayi, "Analytical model for fast reliability evaluation of composite generation and transmission system based on sequential Monte Carlo simulation," *Int. J. Electr. Power Energy Syst.*, vol. 109, pp. 548–557, Jul. 2019.

[13] M. Muhammad Ridzuan, S. Djokic, M. I. Muhammad Ridzuan, and S. Z. Djokic, "Energy Regulator Supply Restoration Time," *Energies*, vol. 12, no. 6, p. 1051, Mar. 2019.

[14] M. I. Muhammad Ridzuan, I. Hernando-gil, and S. Djokic, "Reliability Analysis on Protection Devices Inclusion in LV Residential Distribution Network," *J. Telecommun. Electron. Comput. Eng.*, vol. 10, no. 1, pp. 137–141, 2018.

[15] M. I. Muhammad Ridzuan, "Reliability Assessment of Distribution Networks Incorporating Regulator Requirements , Generic Network Equivalents and Smart Grid Functionalities," The University of Edinburgh, 2017.

[16] I. Hernando Gil, "Integrated assessment of quality of supply in future electricity networks," *PhD's thesis, University of Edinburgh*, 2014.

[17] M. Anumaka, "Fundamentals of Reliability of Electric Power System and Equipment," *Int. J. Eng. Sci. Technol.*, vol. 3, 2011.

[18] "IEEE guide for electric power distribution reliability indices," *IEEE 1366*, 2004.

[19] Energy Commision Malaysia, "Performance and Statistical Information in Malaysia 2016," *Suruhanjaya Tenaga*, p. 103, 2016.

Experimental analysis on the effect of cooling surface area and flow rate for water cooled photovoltaic module

M F Basrawi[1,*], M N A F Anuar[1], T K Ibrahim[2] and A A Razak[3]

[1] Energy Sustainability Focus Group (ESFG), Faculty of Mechanical and Automotive Engineering Technology, Universiti Malaysia Pahang, 26600 Pekan, Pahang, Malaysia.
[2] Department of Mechanical Engineering, College of Applied Engineering, University of Tikrit, Tikrit, Iraq.
[3] Faculty of Engineering Technology, Universiti Malaysia Pahang, Lebuhraya Tun Razak, 26300 Gambang, Pahang, Malaysia.

*E-mail: mfirdausb@ump.edu.my

Abstract. Application of water spray or water flow on the surface of photovoltaic (PV) modules is one of the techniques used to increase efficiency. Main parameter that affect the performance by this technique is water flow rate and cooling surface are. However, there is less study focus on those parameters. Thus, the objective of this paper is to investigate the effect of water-cooled surface area and water flow rate on the temperature and power output of the PV. Orifices were used to create half-cooled and fully-cooled surface area for water to flow as cooling techniques while the hand valve was used to control the flow rate of water at 120 L/h, 180 L/h and 240 L/h flowing onto the panel. A solar simulator was constructed and used to provide 600 W/m^2, 1,000 W/m^2, and 1,200 W/m^2 irradiance for the panel. The testing methodology consists of three different experiments for each irradiance level. It was found that more cooling surface area covered could significantly reduce temperature in any irradiance level, and fully-cooled module could keep the temperature at below 40 °C eventhough the irradiance was at 1,200 W/m^2. In addition, the optimum flow rate also depends on the cooling surface area. Thus, there is a unique relation between cooling surface area and optimum flow rate. Thus, further investigation is needed on this relation.

1. Introduction

Renewable energy is a source of energy which exists in this world, and solar energy is one of them. In the year 2017 alone, the capacity of electrical energy produced by PV panel is more than any other type of power generating with total of 402GW [1]. However, The PV panel efficiency in converting light into electricity has a decreasing rate from 0.25 % to 0.5 % per Degree Celsius, depending on the material of PV cell used [2]. Many methods were investigated to reduce the temperature of module, and one of it is by using water. It could be by spraying or flowing water on the surface of modules.

As the water covers the surface of the module, the temperature of module reduced the power output increased. However, this effect may depend on the possible heat transfer parameters which are the area of module covered with water or water cooling surface area, and the water flow rate. There are many studies on water cooled by spray method as reported in [3-6], and by water flow method as reported in [7-11]. However, all studies only focus on the possibility of water cooling, but less study in detail on the effect of surface and flow rate. Only studies in [5] and [6] considered flow rate as parameters, but

IOP Conf. Series: Materials Science and Engineering **863** (2020) 012043 doi:10.1088/1757-899X/863/1/012043

only Basrawi et al. focused on the effect of cooling surface area on the performance of PV module [12], and only one flow rate was considered in the report. Thus, the objective of this study is to investigate the effect of cooling surface area and flow rate for water cooled PV module.

2. Methodology

Figure 1 shows the experimental setup. The setup had a sun simulator using six 150 W halogen lamps. Voltage, current, temperature and flow rate were measured from the experiment. PV module placed for the setup was at 10° inclination angle to help the water used for cooling could easily flowed on the surface of the module. The module used for this experiment is a brand of Venus Solar and table 1 shows the specifications. It has peak power output of 30 W, height of 66 cm and width of 39 cm.

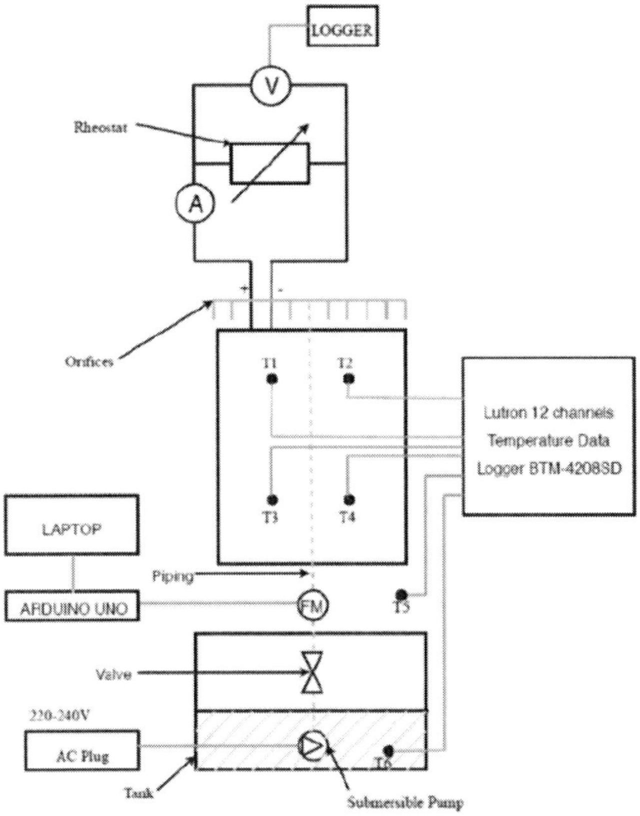

Figure 1. Schematic layout of the experimental setup.

Table 1. Technical specification of the PV panel.

Item	Unit	Value
Model		Venus Solar KL-30W-36P
Pmax	[W]	30
Voc	[V]	21.6
Isc	[A]	1.8
Vmpp	[V]	17.28
Impp	[A]	1.73
No. of cells		36
Effective panel area	[m^2]	0.173

The module was connected to a rheostat 50W 4R7J that have a range of resistance between 1Ω and 1000Ω to act as a load. The voltage and current were measured by using Fluke 317 clamp meter and

Lutron CM-9930 Clamp meter. Lutron CM-9930 clamp meter was connected with Lutron DL-9602SD logger to record the voltage. The temperature was measured by using K-type thermocouples. They were placed at 4 different positions at the back side of the PV panel as shown in the figure 2. A two thermocouples were used to measure the ambient temperature and temperature of water in reservoir. All the thermocouples used in this experiment were connected to Lutron BTM-4208SD 12 channel temperature logger. A 10 W submersible pump was used to pump the water to the top of the PV panel to be run on the surface of PV panel. The pump has a head value of 1.33 m, it is more than enough to provide flow rate of water needed for the experiment. It was connected to a 240V AC power supply. The water flow rate was measured by using a flow meter and a valve was used to control the flow rate that was chosen to be 100 L/h and 250 L/h. These values were selected to be in the range of flow rate that being used by other research based on the literature. The flow rates were chosen also influenced by the ability of the submersible pump used for this experiment set up. The water was pumped from the reservoir to the top of the PV panel and flowed on top of the PV panel. It will recirculate back to the reservoir using the water harvesting pipeline. This was a recycle method of water usage.

The flow rate chosen was 120 L/h, 180L/h and 240 L/h as shown in table 2. The surface of cooling area was either uncooled, half-cooled, fully-cooled, and the condition of that is shown in figure 2 and 3. For each surface area three values of irradiance were investigated as shown in table 2.

Table 2. Parameters for the experiment.

Flow rate (L/h)	Cooling Surface Area	Irradiance (W/m^2)		
120	Half	600	1,000	1,200
	Full	600	1,000	1,200
180	Half	600	1,000	1,200
	Full	600	1,000	1,200
120	Half	600	1,000	1,200
	Full	600	1,000	1,200
Uncooled		600	1,000	1,200

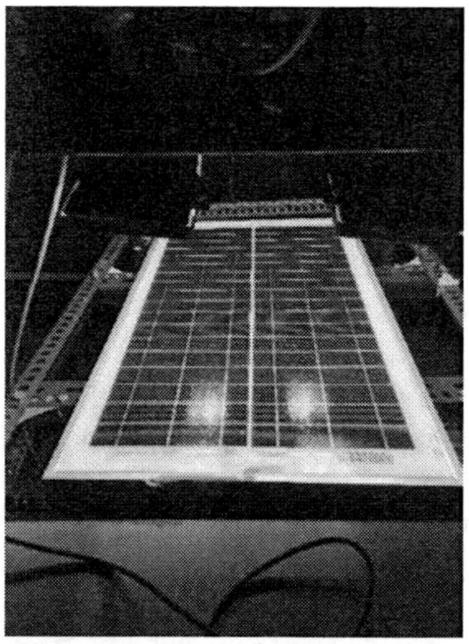

Figure 2. Uncooled PV panel.

Figure 3. Half-cooled (left) and fully-cooled (right) PV module.

To calculate power output, the following formulas is used.

$$P = V \times I \quad (1)$$

where P is the power output [W], V is the instantaneous voltage [V], and I is the instantaneous current [A].

3. Results and discussion

3.1. Analysis data for the effect of cooling surface area on the efficiency of the PV panel

3.1.1. Temperature (T). Figure 4 shows the result of average temperature of the module throughout the experiment for 30 minutes at 120 L/h. The colours differentiate the amount of irradiance used for different sets of experiments. The dashed lines, dotted lines and normal lines show the temperature of uncooled PV module, half-cooled, and fully-cooled.

It was found that a clear trend in which all uncooled PV module recorded the highest temperature for each irradiance. The highest temperature was at the irradiance of 1,200 W/m^2 and the lowest was at the irradiance of 600 W/m^2. Following the uncooled PV panel was all the half-cooled PV module. This set recorded temperature values in between the uncooled and fully-cooled module. The fully-cooled recorded the lowest temperature with all three different irradiance below 40 °C which is very good for the performance of the PV module.

The reason for the uncooled PV panel with irradiance of 1,200 W/m^2 had the highest temperature value is because it gained the highest amount of irradiance compare with the other two irradiances, 1,000W/m^2 and 600 W/m^2. The set for half cooled PV panel recorded a lower temperatures values compare to the set of uncooled PV panel. This is because the presence of the water on the surface of the PV panel that carried away the heat from the PV panel. However, the set of fully cooled PV panel recorded lower temperature values.

It can be concluded that more cooling surface area covered could significantly reduce temperature in any irradiance level. In addition, fully-cooled module could keep the temperature of module at below 40 °C eventhough the irradiance was at 1,200 W/m^2.

There were two more flow rates that have been experimented which is 180 L/h and 240 L/h. However, both flow rates show the same pattern with the results shown in figure 4. Thus, only the final data at 30 minutes are shown in table 4. Table 4 shows the values of the final temperature for all sets of experiment. There were 3 sets of experiment consist of uncooled, half-cooled and fully-cooled module. Each set have 3 subsets based on irradiance which were 600 W/m^2, 1,000 W/m^2 and 1,200 W/m^2. The water-cooled PV panel with flow rate of 120 L/h has been the bench mark to make comparison for the other two flow rates, 180 L/h and 240 L/h. The percentage of the difference of temperature were also shown.

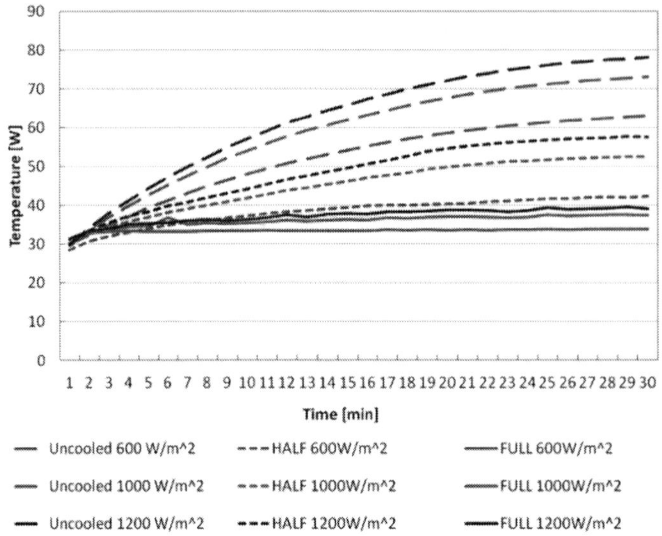

Figure 4. Results of temperature changes.

Table 3. Results of temperature for different flow rate.

Setting Condition		120 L/h (°C)	180 L/h T_3(°C)	ΔT_3(%)	240 L/h T_4(°C)	ΔT_4(%)
600 W/m^2	Uncooled	63.3	63.3	0	63.3	0
	Half-cooled	42.3	36.2	14.4	36.8	13.1
	Fully-cooled	33.8	33.1	2.1	33.1	2.2
1,000 W/m^2	Uncooled	73.3	73.3	0	73.3	0
	Half-cooled	52.5	40.1	23.4	39.8	24.1
	Fully-cooled	37.4	34.7	7.2	34.1	8.7
1,200 W/m^2	Uncooled	78.3	78.3	0	78.3	0
	Half-cooled	57.5	42.2	26.6	41.3	28.2
	Fully-cooled	39.2	34.1	12.9	34.7	11.4

It was found that for half-cooled module, the difference percentage of ΔT_3 of half-cooled for all the 3 different irradiances have values between 14.41 % and 26.73 %. This shows that the temperature of the module when water flowed onto it at 180 L/h has significant differences compare to the 120 L/h flow rate. However, there are only small temperature differences on the percentage difference between the 180 L/h and 240 L/h flow rate. This shows that the optimum flow rate of water-cooled technique for the module lies in between these flow rates, 180 L/h and 240 L/h. However, for fully-cooled module there is no significant difference between every flow rate, and therefore the optimum flow rate lies at the flow rate less than 120 L/h. Thus, the optimum flow rate depends on the cooling surface area.

3.1.2. Power (P). Figure 5 shows the result of power generated by the PV panel throughout the experiment for 30 minutes at 120 L/h. The pattern of power generation for the PV panel is related to the pattern of the temperature shown in figure 4. Since all fully-cooled module could maintain temperature below 40 °C, the power generation of module also was kept at its initial power output. Then, half-cooled module at the middle range of power output, and finally the uncooled module at the lower range of power output.

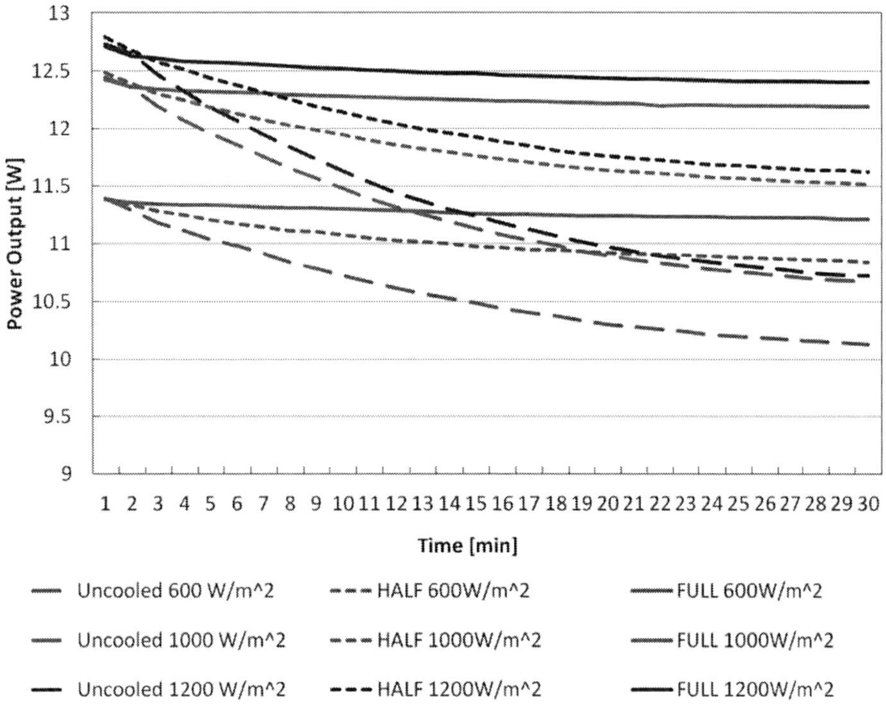

Figure 5. Results of power output changes.

Table 4 shows the values of the final power generated for flow rate at 120 L/h, 180 L/h and 240 L/h. Similar trend was found with table 3 for temperature. However, referring to the set of experiment with 1,000 W/m² irradiance, the percentage of the increment of the power generation for the module did not improve as much as the percentage of the reduced temperature. This was probably due to the power generation for the module did not depends on the temperature of the module only.

Table 4. Results of power output for different flow rate.

Setting Condition		120 L/h	180 L/h		240 L/h	
		(W)	T_3(W)	ΔT_3(%)	T_4(W)	ΔT_4(%)
600 W/m^2	Uncooled	10.13	10.13	0	10.13	0
	Half-cooled	10.84	10.84	2.95	11.19	3.23
	Fully-cooled	11.21	11.22	0.09	11.29	0.71
1,000 W/m^2	Uncooled	10.67	10.67	0	10.67	0
	Half-cooled	11.51	11.61	0.09	11.64	1.17
	Fully-cooled	12.19	12.20	0.08	12.31	0.98
1,200 W/m^2	Uncooled	10.72	10.72	0	10.72	0
	Half-cooled	11.62	12.12	4.3	12.14	4.48
	Fully-cooled	12.40	12.45	0.04	12.59	1.53

3.2. The effect of water flow rate and cooling surface area on water cooled PV module

Figure 6 shows the relation between cooling surface area and temperature at different flow rates. The colours refer to different irradiance in which red line is for 600 W/m², blue line is for 1,000 W/m², and grey line is for 1,200 W/m². The dotted lines, dashed lines and normal lines show water flow rate 120 L/h, 180 L/h, and 240 L/h.

It was found that when cooling surface area increases from 0% to 100%, temperature of the module decreases in any irradiance, and difference of temperature between different flow rates is insignificant. However, there is a large difference of temperature at 50% of surface area between flow rate of

120 L/h and 180-240 L/h. This suggests that there is a unique relation between cooling surface area and optimum flow rate. Thus, further investigation is needed on this relation.

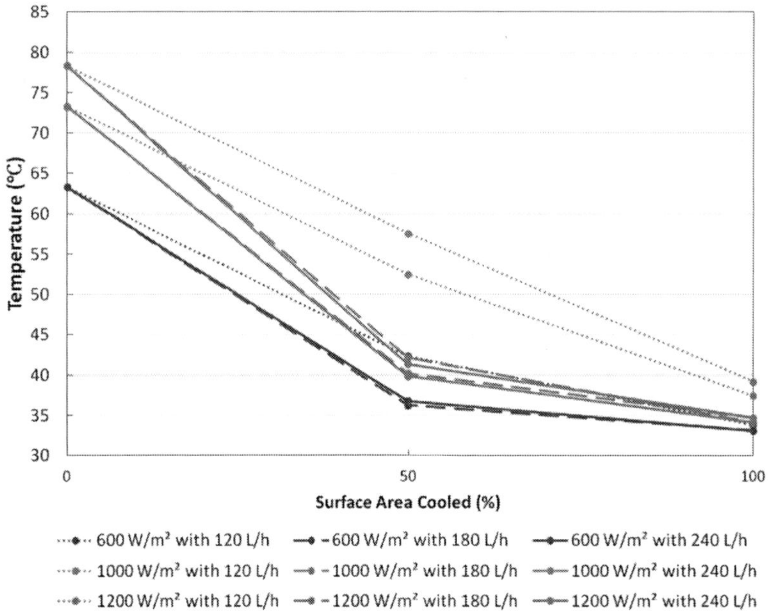

Figure 6. Relation between water-cooled surface area and temperature for different irradiance.

4. Conclusion

Effect of water flow rate and cooling surface area were investigated by a 30 W PV module. It was found that:

i) More cooling surface area covered could significantly reduce temperature in any irradiance level. In addition, fully-cooled module could keep the temperature of module at below 40 °C eventhough the irradiance was at 1,200 W/m^2.
ii) The optimum flow rate of water-cooled technique for the module lies in between 180 L/h and 240 L/h. However, for fully-cooled module there is no significant difference between every flow rate, and therefore the optimum flow rate lies at the flow rate less than 120 L/h. Thus, the optimum flow rate depends on the cooling surface area.
iii) All fully-cooled module could maintain temperature below 40 °C the power generation of module also was kept at its initial power output.
iv) There is a large difference of temperature at 50% of surface area between flow rate of 120 L/h and 180-240 L/h. This suggests that there is a unique relation between cooling surface area and optimum flow rate. Thus, further investigation is needed on this relation.

Acknowledgment

The financial support offered by the Universiti Malaysia Pahang under RDU190192 (FRGS/1/2018/TK03/UMP/02/25) are gratefully acknowledged.

References

[1] Braid, J.L. and R.H. French, *1 - Overview: Power Degradation and Failure of PV Systems1*, in *Durability and Reliability of Polymers and Other Materials in Photovoltaic Modules*, H.E. Yang, R.H. French, and L.S. Bruckman, Editors. 2019, William Andrew Publishing. p. 3-21.
[2] Filip Grubišić-Čabo, S.N.a.T.G.M., & Giuseppe Marco, T., *Photovoltaic panels: A review of the cooling techniques*. Transactions of FAMENA XL, 2016. **40**(SI-1): p. 63-74.

[3] Salih, S.M., Abd, O. I., & Abid, K. W., *Performance enhancement of PV array based on water spraying technique.* International Journal of Sustainable and Green Energy, 2015. **4**(16): p. 8-13.

[4] Irwan, Y.M., et al., *Indoor Test Performance of PV Panel through Water Cooling Method.* Energy Procedia, 2015. **79**: p. 604-611.

[5] S. Nižetić, D.Č., A. Yadav, and F. Grubišić-Čabo, *Water spray cooling technique applied on a photovoltaic panel: The performance response.* Energy Conversion and Management, 2016. **108**: p. 287-296.

[6] Bendra, O., *OPTIMIZING SOLAR CELLS EFFICIENCY BY COOLING TECHNIQUES.* 2017, AL AKHAWAYN UNIVERSITY IN IFRANE p. 32.

[7] K. A. A. Moharram, M.S.S.A.-E., H. A. A. Kandil, and H. El-Sherif, *Enhancing the performance of photovoltaic panels by water cooling.* AinShams Engineering, 2013. **4**(4): p. 869-877.

[8] Matias, C.A., Santos, L. M., Alves, A. J., & Calixto, W. P, *Increasing photovoltaic panel power through water cooling technique.* Transactions on Environment and Electrical Engineering, 2017. **2**(1).

[9] Tabaei, H. and M. Ameri, *Improving The Effectiveness Of A Photovoltaic Water Pumping System By Using Booster Reflector And Cooling Array Surface By A Film Of Water.* Iranian Journal of Science and Technology Transactions of Mechanical Engineering, 2015. **39**(M1): p. 51-60.

[10] Odeh, S. and M. Behnia, *Improving Photovoltaic Module Efficiency Using Water Cooling.* Heat Transfer Engineering, 2009. **30**(6): p. 499-505.

[11] Nair Milind, M.A., Febin Francis, Jithu Francis, Joson Varghese, Sajith U K, *Enhancing the Efficiency of Solar Panel Using Cooling Systems.* Int. Journal of Engineering Research and Application 2017. **7**(3): p. 5-7.

[12] Basrawi, F., et al., *Experimental Analysis on the Effect of Area of Surface Cooling for a Water-Cooled Photovoltaic.* MATEC Web Conf., 2018. **225**: p. 01011.

Experimental study on the effect of bed aspect ratio to the drying rate of chilli for swirling fluidized bed dryer

M F Basrawi[1]*, F Redzlan[1], T K Ibrahim[2] and A S M Yudin[1]

[1] Energy Sustainability Focus Group (ESFG), Faculty of Mechanical and Automotive Engineering Technology, Universiti Malaysia Pahang, 26600 Pekan, Pahang, Malaysia.
[2] Department of Mechanical Engineering, College of Applied Engineering, University of Tikrit, Tikrit, Iraq.

*E-mail: mfirdausb@ump.edu.my

Abstract. Swirling Fluidized Bed Dryer (FBD) has better performance because of the improvement of air flow with less pressure drop. However less study focus on the performance of swirling FBD especially at small-scale. Thus, the objective of this study is to investigate drying characteristics of green chillies at different ratio of chilli, R_{chilli} and bed aspect ratio, R_{Bed} by the swirling fluidized bed dryer. An electrical blower with 600 W of power was used as a main source to supply air at flow rate of 0.324 kg/s which equals to the velocity at the distributor of 8.25 m/s. The chamber has 400 mm height and 150 mm diameter with maximum capacity of 2 kg. The air distributor used has 67% inclination angle to provide air swirling inside the FBD. It was found that R_{Bed} increased from 0.8 to 1.6, the drying performance decreased in which the final moisture ratio was higher at 0.72-0.79 for R_{Bed} =1.6 as compared to better moisture ratio at 0.18-0.24 for R_{Bed} = 0.8. The drying rate was at 0.225-0.275 %/min for R_{Bed} = 0.8, and 0.075-0.10 %/min for R_{Bed} = 1.6. When the R_{chilli} increased from 1:1 to 2:1, the drying performance showed slightly decreased. Moisture ratio also slightly increased from 0.18 to 0.24 for R_{Bed} = 0.8, whereas from 0.72 to 0.79 for R_{Bed} = 1.6. This finding suggest that the R_{Bed} has more significant effect than R_{chilli}.

1. Introduction

The drying process is a process intended to reduce the humidity content of most products, whether industrialized or not, to ensure their storage and transport preservation. Fluidization process is highly attractive for powder drying and wet granular materials. This method has been used industrially it is very common in the drying of coarse materials, pharmaceuticals and food products in various countries around the world [1, 2]. The major parts of the fluidized bed reactors are: the fluidization vessel, the solids discharge, dust separator, the air supply and heat supply. A fluidized bed is called a vessel that contains a bed of solid particles, e.g. sand or catalytically active particles, and a method to introduce gas. It is a machine that dries a continuous flow of moist granular material [3].

Among parameters that affect the drying performance in fluidized bed dryer are the bed aspect ratio, R_{Bed} and the chilli ratio, R_{chilli}. The bed respect ration, R_{Bed} is define as the ratio of high of the bed to the diameter of the bed. Meanwhile, the chilli ratio, R_{chilli}.is define as the ratio of chilli length to the chilli diameter. Lower R_{Bed} and smaller R_{chilli} of particles improved the drying rate of the samples. However, there are no studies in drying chilli at different R_{Bed} together with R_{chilli} for Swirling FBD.

Thus, the objective of this study is to investigate drying characteristics of green chillies at different ratio of chilli, R_{chilli} and bed aspect ratio, R_{Bed} by the swirling fluidized bed dryer.

2. Methodology

Figure 1 shows the experimental setup of swirling fluidized bed dryer. For the blower, 600 W blower was used as a main source to supply air at flow rate of 0.324 kg/s and the velocity at the distributor was 8.25 m/s. The chamber has 400 mm height and 150 mm diameter with maximum capacity of 2 kg. The air distributor used has 67% inclination angle to provide air swirling inside the FBD.

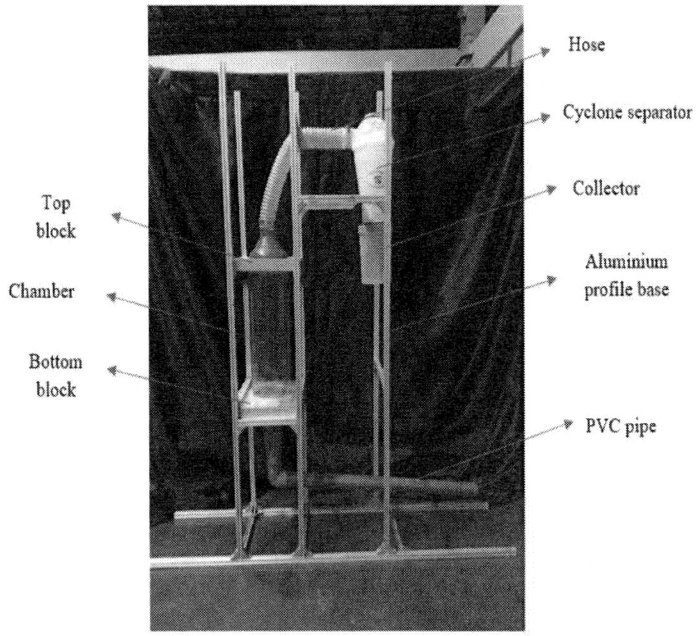

Figure 1. Schematic layout of the experimental setup.

Table 1 shows the variables in conducting the experiment for drying the samples using the fluidized bed dryer. Approximately, 2 kilograms of sample green chillies were used to determine the rate of drying by the fluidized bed dryer. 2 kilograms of green chillies have been purchased. Both ends of the chillies were removed and only the middle portions, which resemble a cylindrical shape, were used to produce the required samples. Samples were prepared at two R_{chilli} of 1:1 and 2:1, at 1 kg and 2 kg respectively. Figure 2 shows the sample of chilli with R_{chilli} of 2:1. After preparation, chillies were kept in a plastic container and washed in order to make uniform moisture distribution before commencement of experimentation.

Table 1. The variables of the experiment in drying the chilli.

Weight	Length to diameter ratio of chilli (R_{chilli})	Length to diameter ratio of chilli (R_{chilli})	Bed aspect ratio (R_{Bed})
1 kg	1:1	2:1	0.8
2 kg	1:1	2:1	1.6

The experiment was conducted to investigate the rate of drying of the samples and performance of the fluidized bed dryer that have been fabricated. Figure 3 Shows approximately 1 kg of sample was placed in the fluidized bed column. The swirling of the bed was observed and recorded.

The air flow of 8.25 m/s from the blower was on during the drying process so that it will evaporate the moisture content of the samples. Clamper was use to clamp the chamber tightly so that no airflow moves out from the chamber because there was a small gap between the chamber and the aluminium blocks. In this experiment, high weight differences were required to indicate the weight loss of the samples by the FBD. Other than that, low moisture ratio gives good swirling effect to the bed. Lastly the graph of moisture ratio and weight loss vs time was plotted for every ratio of the samples.

Figure 2. Sample of chilli with length to diameter ratio of 2:1.

Figure 3. Drying of chilli sample.

After preparation, chillies were kept in a plastic container and washed in order to make uniform moisture distribution before commencement of experimentation. Equation (1) is the formula on how to get the value for the length to diameter ratio of chilli (L_{chilli} / D_{chilli}),

$$R_{chilli} = \frac{L_{chilli}}{D_{chilli}} \tag{1}$$

where L_{chilli} is length of chilli (m), and D_{chilli} is diameter of chilli (m). For the bed height, 1 kg of chilli consist of 12 cm shown in figure 4 of bed height. For 2 kg of chilli, the bed height was 24 cm shown in figure 5. The diameter of the bed was 15 cm. The bed aspect ratio, (R_{Bed}) need to calculate and the value is shown in table 1. R_{Bed} is shown in equation (2).

$$R_{Bed} = \frac{H_{Bed}}{D_{Bed}} \tag{2}$$

where H_{Bed} is height of the bed (m), and D_{bed} is diameter of the bed (m). Other than that, low moisture ratio gives good swirling air flow motion to the bed. The weight losses, W_l are calculated by subtracting the weight of the samples before and after drying for every 5 minutes throughout 40 minutes shown in equation (3). Lastly the graph of moisture ratio (M_r) and weight loss (W_l) vs time was plotted for every ratio of the samples.

$$W_l = W_n - W_{n+1} \tag{3}$$

Figure 4. 12cm bed height with aspect ratio of 0.8

Figure 5. 24cm bed height with aspect ratio of 1.6

where W_n is weight of chilli at specific time (kg), and W_{n+1} is weight of chilli at subsequent time in kg. M_r is calculated based on equation (4).

$$M_r = M_{r_n} - \frac{W_n - W_{ds}}{W_{ds}} \tag{4}$$

where, W_n is weight of sample at specific time in kg, W_{ds} is weight of dried sample, $M_{r,n}$ is moisture ratio at specific time (%). The drying rate (D_r) is dividing the difference of the moisture ratio of two consecutive samples by the drying duration. It is calculated based on equation (5).

$$D_r = \frac{M_{r_i} - M_{r_{n+i}}}{t} x\ 100 \tag{5}$$

where, D_r is drying rate (%), M_{rn} is moisture ratio at specific time, M_{rn+i} is moisture ratio at subsequent time, and t is drying duration (s).

3. Results and discussion

3.1. Effect of R_{chilli} and R_{Bed} on the drying rate of fluidized bed dryer

Figure 6 shows the results of weight loss and moisture ratio for different R_{chilli} and R_{Bed}. Red lines show the moisture ratio, whereas the blue bars show the weight loss. It should be noted that the blue bars are weight loss in each interval, and not the overall weight of the sample. On the top side of the figure is the result for R_{Bed} = 0.8, and at the bottom side of the figure is R_{Bed} = 1.6. On the right side is R_{chilli} at 1:1, and on the left side is R_{chilli} at 1:2.

It was found that when R_{Bed} increased from 0.8 as shown in figure 6(a) and (b) to 1.6 as shown in figure 6(c) and (d), the drying performance decreased in which the moisture ratio at 40 minutes was higher at 0.72-0.79 for R_{Bed} =1.6 as compared to better moisture ratio at 0.18-0.24 % for R_{Bed} = 0.8. This is because higher R_{Bed} means the bed has higher bed pressure drop that makes frictional shear forces within the bed loading increases. In term of drying rate, the drying rate was at 0.225-0.275 %/min for R_{Bed} = 0.8, and 0.075-0.10 %/min for R_{Bed} = 1.6. This result is consistent with the study by Sadaka et. al, which concluded that the overall drying rate decreases with increases in aspect ratio or R_{Bed} [4].This suggests that R_{Bed} has significant effect on the drying performance even for swirling FBD at small-scale.

When R_{chilli} increased from 1:1 as shown in figure 6(a) and (c) to 2:1 as shown in figure 6(b) and (d), the drying performance slightly decreased. Moisture ratio for slightly increased from 0.18 to 0.24 for R_{Bed} = 0.8, whereas from 0.72 to 0.79 for R_{Bed} = 1.6. This is because R_{chilli} = 1:1 has smaller surface area and therefore particle has more contact surface are with the air, and particles dry faster as compared to R_{chilli} =2:1. This result is consistent with the study by Oluleye et al., smaller R_{chilli} will dry faster because smaller grains show high resistance to airflow therefore reducing the fan outputs and also increasing the amount of moisture reduced in the grains [5]. However, R_{chilli} has less effect as compared to R_{Bed}.

Thus, R_{Bed} has more significant effect than R_{chilli} on the drying performance of swirling FBD at small-scale for chilli. When R_{Bed} increased two times, moisture ratio decreased 53-55%, whereas when R_{chilli} increased two times, moisture ratio only decreased 6-7%.

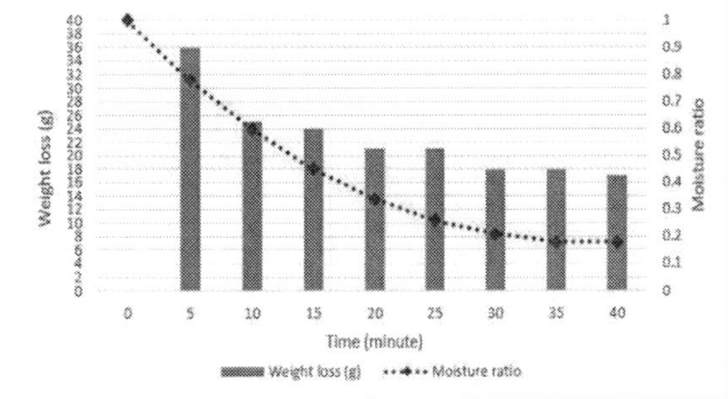

(a) R_{chilli} =1:1 and R_{Bed} =0.8 chilli

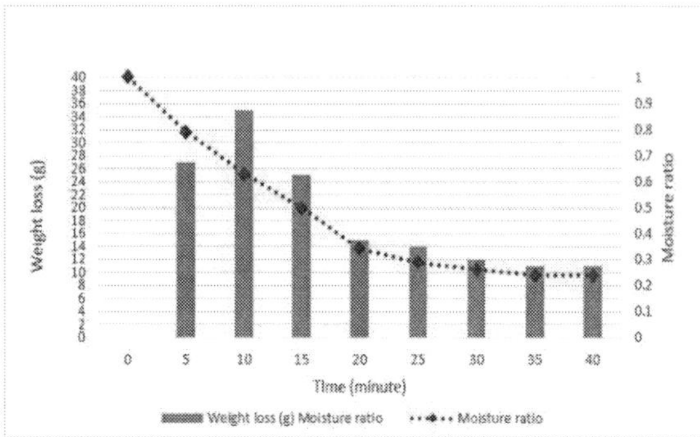

(b) R_{chilli} =2:1 and R_{Bed} =0.8 chilli

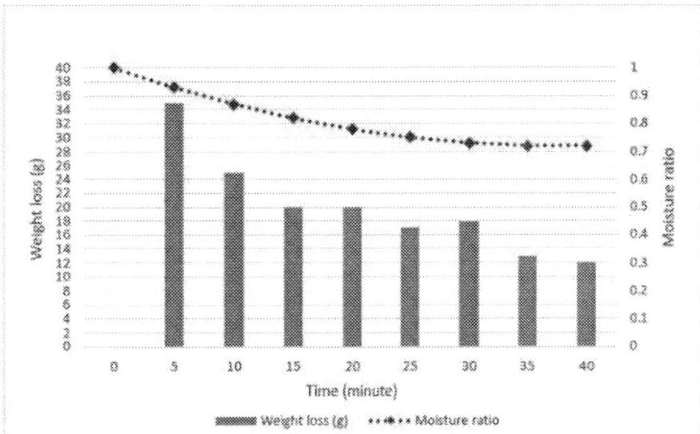

(c) R_{chilli} =1:1 and R_{Bed} =1.6 chilli

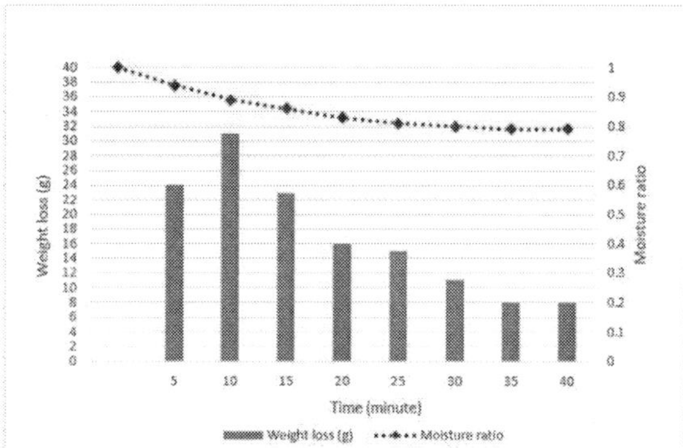

(d) R_{chilli} =2:1 and R_{Bed} =1.6 chilli

Figure 6. Results of weight loss and moisture ratio.

4. Conclusion

i) When R_{Bed} increased from 0.8 to 1.6, the drying performance decreased in which the final moisture ratio was higher at 0.72-0.79 for R_{Bed} =1.6 as compared to better moisture ratio at 0.18-0.24 for R_{Bed} = 0.8. This is because higher R_{Bed} means the bed has higher bed pressure drop that makes frictional shear forces within the bed loading increases.

ii) The drying rate was at 0.225-0.275 %/min for R_{Bed} = 0.8, and 0.075-0.10 %/min for R_{Bed} = 1.6. This suggests that R_{Bed} has significant effect on the drying performance even for swirling FBD at small-scale.

iii) When R_{chilli} increased from 1:1 to 2:1, the drying performance slightly decreased. Moisture ratio slightly increased from 0.18 to 0.24 for R_{Bed} = 0.8, whereas from 0.72 to 0.79 for R_{Bed} =

1.6. This is because R_{chilli} = 1:1 has smaller surface area and therefore particle has more contact surface are with the air, and could dry faster.

iv) However, R_{Bed} has more significant effect than R_{chilli}. When R_{Bed} increased two times, moisture ratio decreased 53-55%, whereas when R_{chilli} increased two times, moisture ratio only decreased 6-7%.

Acknowledgment
The financial support offered by the Universiti Malaysia Pahang under RDU190192 (FRGS/1/2018/TK03/UMP/02/25) are gratefully acknowledged.

References

[1] F. ABERUAGBA, J.O.O., K.R. ADEBOYE, M.A. OLUTOYE, *Fluidization Characteristics of a Prototype Fluidized Bed Reactor* Leonardo Journal of Sciences, 2005(6): p. 29-41.
[2] Halim, L.A., et al., *Preliminary Study on Drying of Stingless Bee Pot Pollen Using Novel Fluidized Bed Dryer with Swirling Distributor.* MATEC Web Conf., 2018. **225**: p. 04007.
[3] Sivakumar, R., et al., *Fluidized bed drying of some agro products – A review.* Renewable and Sustainable Energy Reviews, 2016. **61**: p. 280-301.
[4] Sadaka, S., K. Luthra, and G.G. Atungulu, *Effects of Aspect Ratio, Drying Temperature and Drying Duration on the Performance of a Custom-Made Fluidized Bed Drying System*, in *2018 ASABE Annual International Meeting.* 2018, ASABE: St. Joseph, MI. p. 1.
[5] Oluleye A. E, O.A.A., Anyaeche C.O., *Design and Fabrication of a Low Cost Fluidized Bed Reactor.* 2012. **3**: p. 24-35.

IOP Conf. Series: Materials Science and Engineering **863** (2020) 012045 doi:10.1088/1757-899X/863/1/012045

Stabilizing solar farm power output by micro gas turbine integration: Emission performance

A Al-Anati[1], M F Basrawi[1,*], M H Yusof[1], T K Ibrahim[1,2], A A Razak[3] and M R R Chand[1]

[1] Energy Sustainability Focus Group (ESFG), Faculty of Mechanical and Automotive Engineering Technology, Universiti Malaysia Pahang, 26600 Pekan, Pahang, Malaysia.
[2] Department of Mechanical Engineering, College of Applied Engineering, University of Tikrit, Tikrit, Iraq.
[3] Energy & Environmental Engineering Technology, Faculty of Technology, Universiti Malaysia Pahang, Lebuhraya Tun Razak, 26300 Gambang, Pahang, Malaysia.

*E-mail: mfirdausb@ump.edu.my

Abstract. Worldwide growth of Photovoltaics (PV) technology has been close to exponential rate in the past decades. However, the increase of power generation by PV imposes new challenges on the existing electrical grid due to the unpredictability of PV power output that depends mainly on solar irradiance. One of the solutions to stabilize solar farm power output fluctuation by hybridizing Micro Gas Turbines (MGTs) that is controlled to meet the total power production target. Thus, the objective of this paper is to investigate the effect of different operation strategies on the environmental performance on MGT integrated solar farm. Different operation strategies for MGT integrated solar farm have been implemented to constantly produce a total power output of 800 kW throughout the year. The three strategies then compared by analysing emission produced by MGTs. A simulation was conducted in Simulink environment using real ambient temperature and global sun irradiance data. It was found that Operation Strategy 3 reduced carbon dioxide (CO_2) by 13.8%, nitrogen oxides (NOx) emission by 3.2%, and carbon monoxide (CO) by 97.9%. This shows that the proposed strategy could stabilize the total power output of the solar farm at minimum emission level.

1. Introduction

Fossil fuels including coal, oil and natural gas are the primary energy sources [1]. Yet the steady increment of energy consumption is always coupled with pollution and climate change that could harm human life and the environment [2]. Fortunately, renewable energy specially PV power generation could provide an alternative [3]. PV power generation noticed significant growth in recent years. However, PV penetration to the grid could causes various impacts to electrical grid systems [4]. Large-scale growth of PV causes concerns in keeping supply-and-demand balance as the power output of PV shows considerable fluctuations due to weather conditions.

One of the current issues reported by CAISO is the well-known "Duck Curve," which represents the net load on the grid (e.g., total demand minus wind and solar generation) [5]. The duck curve illustrates two areas of concern. First is the possibility of excess generation in the middle of the day due to the inability of the thermal fleet to integrate a large amount in the solar generation, resulting in solar generation being curtailed and over-generation. The second area of concern is the need for resources to ramp up quickly enough to meet the evening peak.

Different solutions have been considered by researchers such as employing various storage technologies, batteries, electric capacitors and fuel cells [6]. The cost of energy storage systems remains the major challenge besides the sizing problems [7, 8]. Furthermore, methods like control strategies based on moving average, low pass filters and ramp rate limitation have also been proposed for mitigating PV power output fluctuation [7, 9–11]. The main drawback with the control strategies-based on low pass filter and the moving average is that they lag PV power output and therefore would not provide the same level of smoothing as the ramp rat limiting methods.

Although several studies have been conducted to study the impact of solar farm power quality delivered to the grid and ways to minimize it, there is a lack of studies on the performance of hybridizing a PV with MGT to keep the total power output of the plant constant. Concerns on using hybrid system using MGT is the lack of studies on hybridizing PV with MGT using new concept operation strategies that could solve the problem of power output fluctuation. The impact on the grid due to the unpredictable changes of weather which in return reduces PV overall efficiency. Besides, MGT performance depends on the load factor. It is well known that when MGT operates at partial loads negatively affects efficiency while emission and fuel consumption increase. Therefore, a new method is proposed and simulated to stabilize PV power output fluctuation. Thus, the objective is to investigate the effect of different operation strategies on the environmental performance of MGT integrated solar farm.

2. Materials and methods

Environmental data has been collected from a workstation at district of Pekan, Malaysia. Sun irradiance I_G and ambient temperature T_{amb} [12]. Data were collected for one year. The simulation of the system is conducted based on the real profile of environmental data. The system components are selected based on the capacity of the plant objective as it must produce constant output of 800 kWh throughout the year. That is achieved by 5 MGT units and 3080 PV modules. The overall system is demonstrated as shown in figure 1.

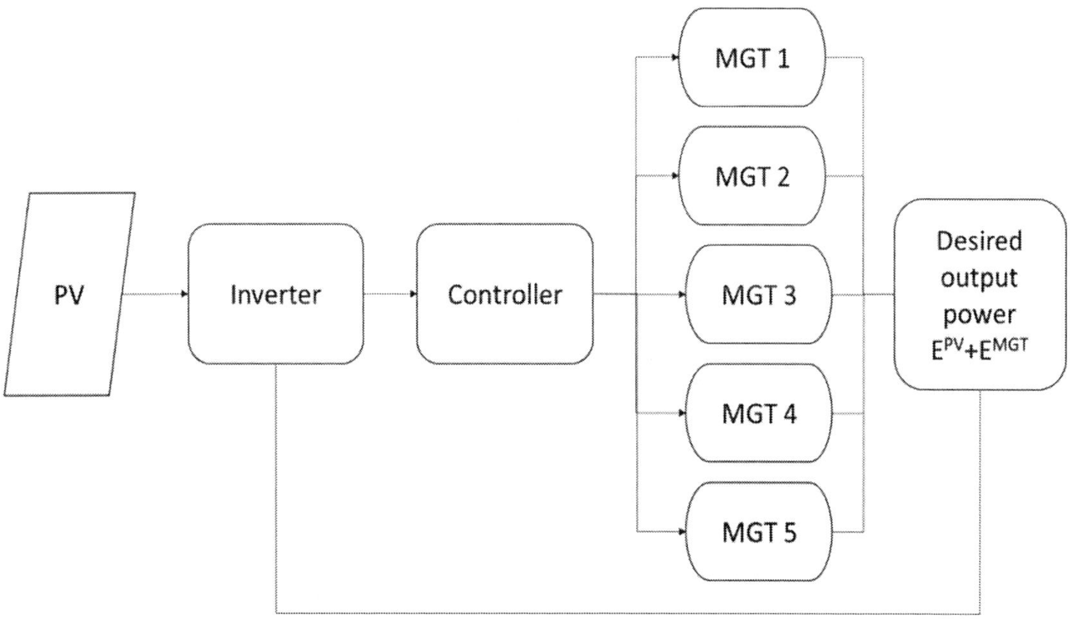

Figure 1. Overall system.

2.1. *PV system modelling*

Sizing and modulation of the solar farm is based on methods proposed by [12–14]. The model basically requires two inputs which are the T_{amb} and $\bar{I}_G$. Based on the hourly inputs, the model calculates the amount of energy generated by the PV module, E_{PV} [kWh] as the output in that hour. Table 1 presents technical data of chosen PV module and matched inverter during sizing process.

Table 1. PV and inverter specifications.

Inverter Input (DC)	Value	PV Parameter	Value
Type designation	PVS800-57-1000kW-C	Peak power (*Pmax*)	315 W
DC voltage range, mpp ($U_{DC,\ mpp}$)	600 to 850 V	Rated voltage (*Vmpp*)	54.7 V
Maximum DC voltage (U_{max} ($_{DC}$))	1100 V	Rated current (*Impp*)	5.76 A
Maximum DC current (I_{max} ($_{DC}$))	1710 A	Open circuit voltage (*Voc*)	64.6 V
Efficiency	98.8%	Maximum system voltage (*UL*)	600

2.2. *MGT modelling*

The mathematical model of MGT units was developed based on the C-200 micro turbine model manufactured by Capstone Turbine Cooperation [15,16]. A 200 kW MGT unit was chosen based on the availability of commercial models and demand need. The performances are reported at ISO conditions, by definition, it is at standard sea level pressure of 101.325 kPa, 15 °C of T_{amb} and at 60% of relative humidity. As shown in table 2, at the rated power output, $E_{MGT\text{-}FL}$ which is 200 kW, the MGT unit can achieve electrical efficiency, $\eta_{MGT\text{-}FL}$ up to 33%.

Table 2. Capstone Turbine specifications.

Parameter	Value
Net power output	200(+0/-4) kW
Net efficiency (LHV)	33(+-2) %
Nominal steady fuel flow (HHV)	2400000 kJ/hr

To mathematically model the MGT unit, the method formulated by [13] was implemented. The modelling method are analogy to the backward facing quasi-static approach in which the actual input of the MGT unit were treated as outputs. As shown, the inputs of the model were T_{amb} and *LF* respectively, while, the output of the model were power output, E_{MGT}, fuel energy consumption, $Q_{f\text{-}MGT}$, exhaust temperature, T_{Ex} and the exhaust mass flow rate, $\dot{m}_{Ex}$. The data on T_{amb} were used as input parameters to calculate the temperature derating of the output parameters at full load operation.

2.3. *Operation strategies*

The three operation strategies had the same set up of equipment, and therefore only the method of operation is elaborated. All operation strategies controlled MGT units on desired LF in order to keep the total power output of the plant constant through the year at 800 kW. Figures shown below are example cases when solar farm generates 300 kW. Figure 2 (a)-(c) illustrates MGT unit's interaction in each operation strategy.

Strategy 1 keeps the plant total output at 800 kW by subtracting the power produced by PV plant from the 800 kW to get the amount of power (kW) needed to be generated by MGT units. The needed value then is divided equally into 5 MGT units. Thus, all MGT units will run at the same LF at all cases.

Figure 2(b) shows strategy 2 with all units operate sequentially. MGT 1 will operate in the full load, once the demand is higher than the MGT capacity, the MGT 2 will on the operation. In strategy 3, MGT performance and efficiency were taken into consideration. As high LF ensure that MGT will require less fuel ration and low emission, the aim of this control strategy is to ether keep the MGT units operating at LF larger than 0.6 all the time, or 0 which is off. The strategy operates the units according to PV output by dividing the needed power by the max number of MGT units that can operate it by considering LF larger than 0.6. The strategy is shown in figure 2(c).

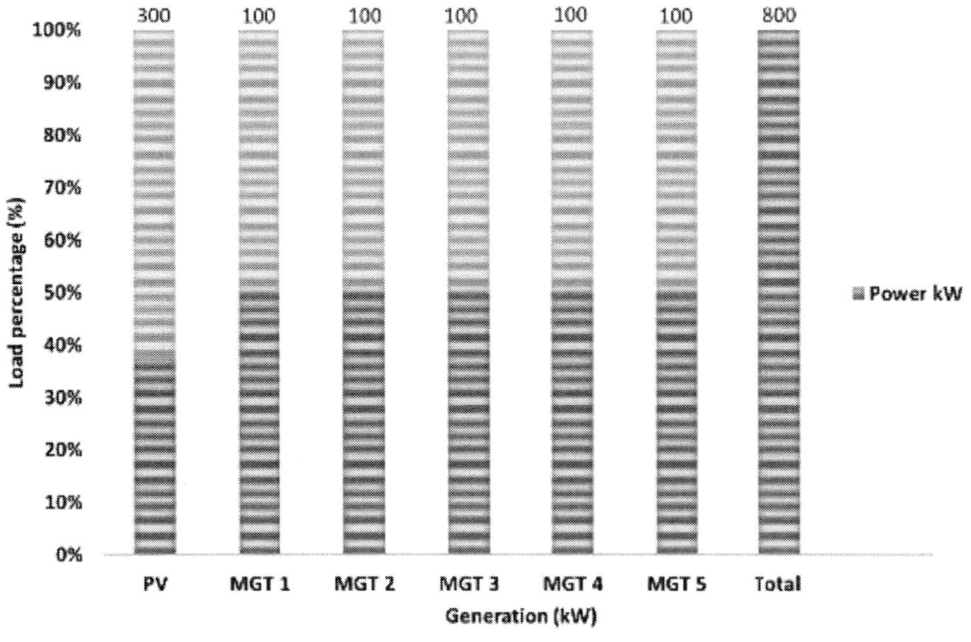

(a)

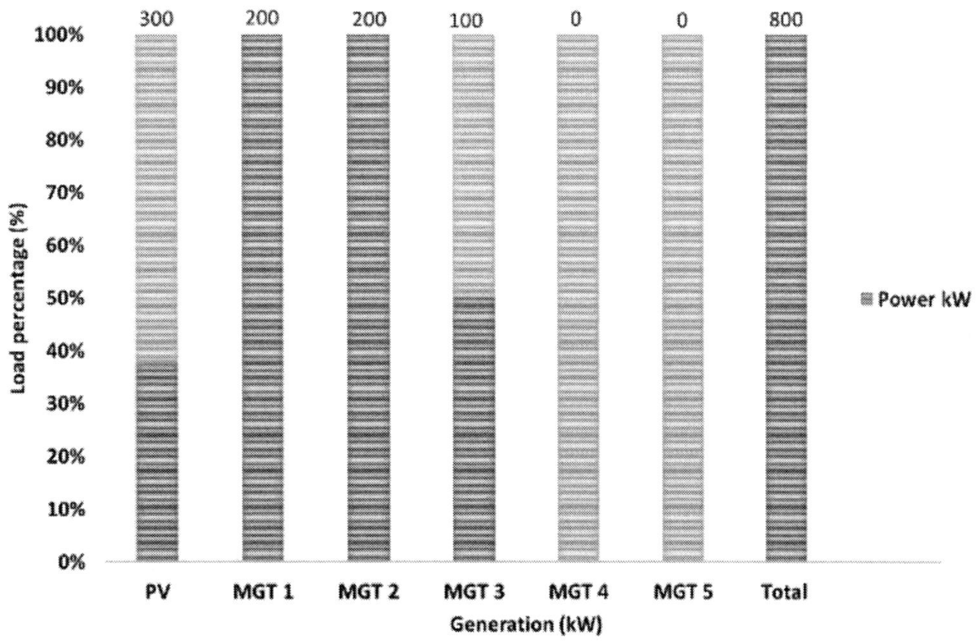

(b)

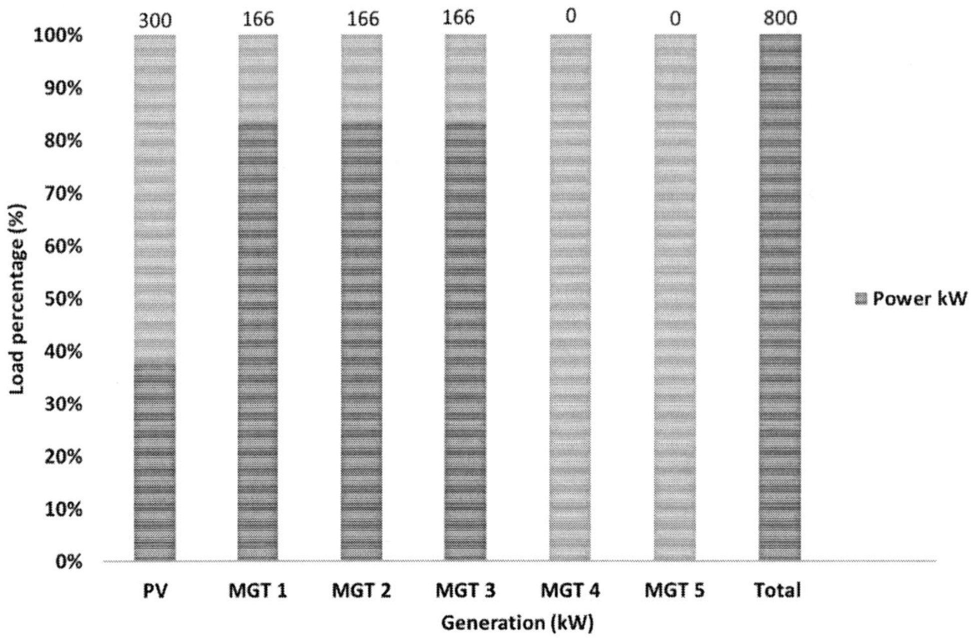

(c)

Figure 2. Operation strategy of the MGT unit (a) Strategy 1, (b) Strategy 2, and (c) Strategy 3.

2.4. Environmental analysis

To evaluate emission performances of the hybrid system under each of the strategies, the Emission Reduction Index (ERI) analysis was utilized as a benchmark. The ERI is an indicator that reflects the amount of emission reduction that can be achieved by the hybrid system as compared to the conventional system. The type of emissions acknowledge in the ERI analysis were narrowed down to only CO_2, NO_x as well as the *CO* emission. In the case of MGT, the Emission Factor (EF) were calculated based on [15] and [16] method using EF. The calculated EF of the MGT unit were then substituted into equation 1to obtain the total amount of CO_2, NOx and CO emission produced were n is type of emission

$$E_n = EF_n \times (P_{net}) \tag{1}$$

3. Results and discussion

Figure 3 shows operation of the plant delivering constant power output of 800 kWh to grid on a randomly chosen day. As PV generation increases, MGTs generation will decrease to meet the total power output of 800 kWh throughout the day. MGTs are running according to the solar farm generation with respect to demand.

3.1. Environmental performances

Emission Reduction Index was used to obtain the total amount of CO_2, NO_x and CO emission produced by each MGT unit under all strategies. Figure 4 presents CO_2 emitted by MGT. The values are closely equal where that is justified by the equal levels of MGT Emission factor for CO_2 while running at partial loads.

NO_x emission as show in figure 5 was expected to be at nearly equal ranges for all MGT units using all strategies because the emission factor value was very low. Operating MGT at partial loads do not affect the amount of NO_x as compared to CO2 that has higher emission factor at lower LF.

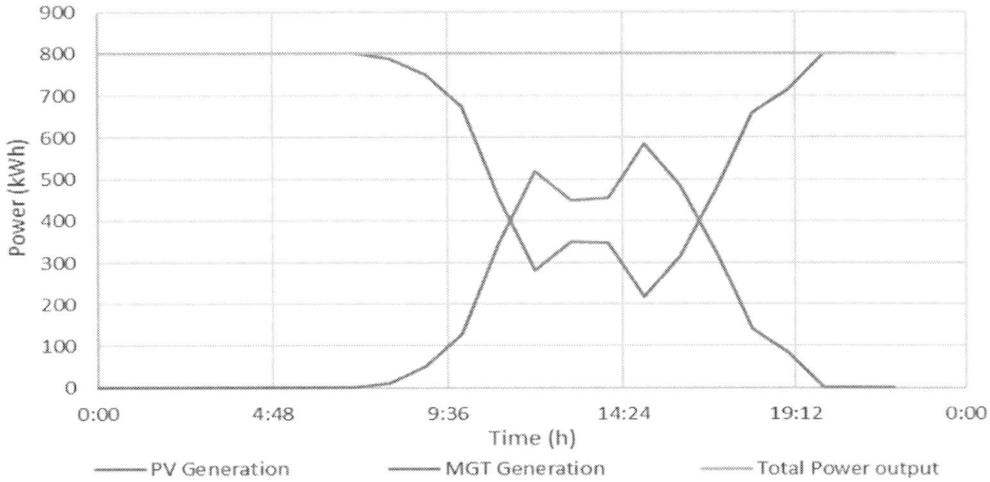

Figure 3. Condition of all MGTs with PV output when controlled by any of the three strategies.

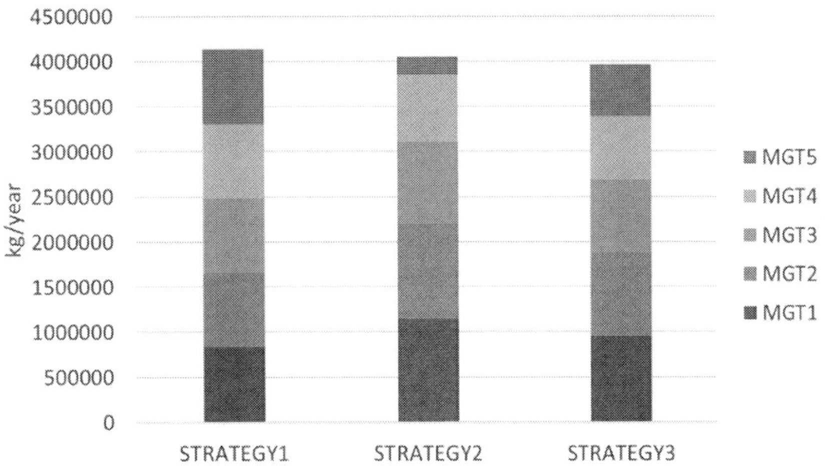

Figure 4. CO_2 emission by the three operation strategies.

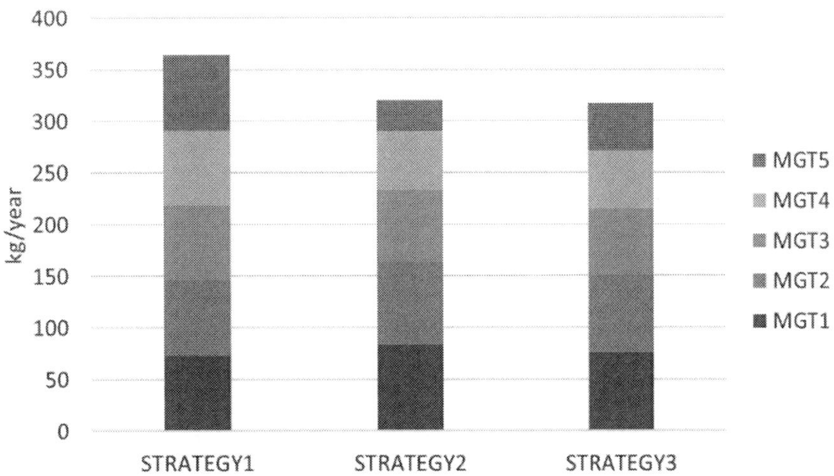

Figure 5. NO_x emission by the three operation strategies.

Figure 6 shows that the highest MGT units in emitting CO was in Strategy 1. This is because the continuous occurrence of LF that lays between LF ranges from 0 to 0.4. The first four units in strategy 2 emitted 1,219.2 kg of CO, were MGT5 alone emitted 1,523.45 kg of CO, and this is because the low LF that MGT5 was operated at, in which the first four MGT were capable of satisfying the demand for 84% of the time, and MGT5 was operated only when the surrounding temperature reduced the performance of MGT 1, MGT 2, MGT 3 and MGT 4. The maximum LF recorded for MGT 5 was 0.32. Strategy 3 was expected to emit the lowest CO emission as the algorithm keeps MGT running at high LF.

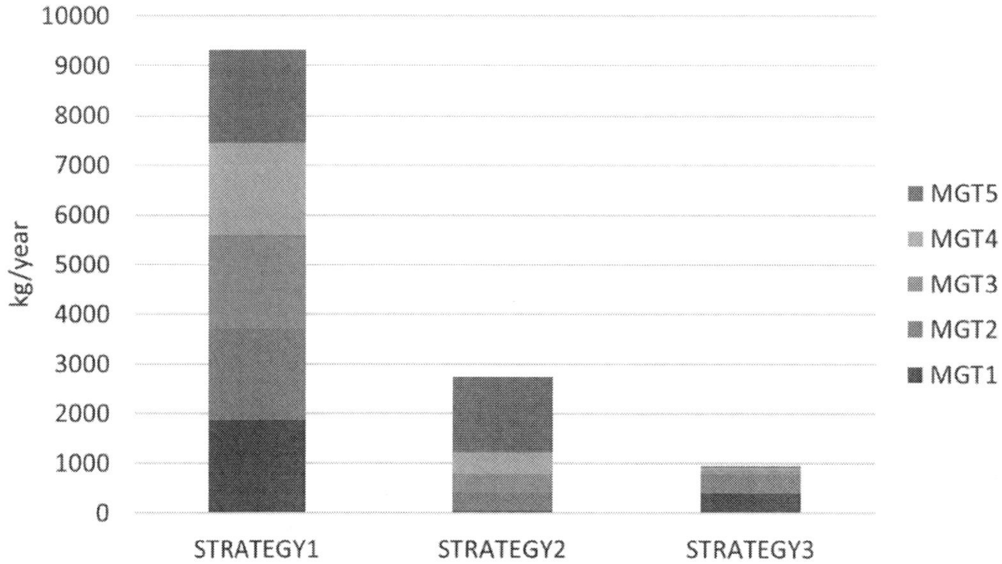

Figure 6. CO emission by the three operation strategies.

4. Conclusion

The hybrid MGT-PV solution succeed in stabilizing power output fluctuation throughout the year at constant power output. It was found that different operation strategies had different emission levels. Operation Strategy 3 was chosen to be the optimum way for operating the plant because it has the lowest levels of CO_2, CO and NO_X emission. Therefore, operating MGTs at the highest LF possible as in Operation strategy 3 was the key of emitting the least emission percentages in result. Further studies should be conducted to reduce percentage of MGT emission production at higher levels.

Acknowledgments

The financial support offered by Universiti Malaysia Pahang under (FRGS/1/2018/TK03/UMP/02/25), RDU180313 and RDU190192 are gratefully acknowledged.

References

[1] Yang, H., Lu, L., & Zhou, W. (2007). A novel optimization sizing model for hybrid solar-wind power generation system. Solar energy, 81(1), 76-84.

[2] F. Basrawi, T. Yamada, and S. Obara, "Economic and environmental based operation strategies of a hybrid photovoltaic-microgas turbine trigeneration system," *Appl. Energy*, vol. 121, pp. 174–183, 2014.

[3] N. Strachan and A. Farrell, "Emissions from distributed vs . centralized generation : The importance of system performance," vol. 34, pp. 2677–2689, 2006.

[4] J. M. Pearce, "Expanding photovoltaic penetration with residential distributed generation from hybrid solar photovoltaic and combined heat and power systems," *Energy*, vol. 34, no. 11, pp.

1947–1954, 2009.

[5] D. Curve and A. W. Ahead, "Electricity Currents : A Survey of Current Industry News and Developments California ' s ' Duck Curve ' Arrives Well Ahead of Schedule Never Mind Oil and Water :," vol. 29, pp. 71–72, 2016.

[6] S. Morozumi, N. Energy, and I. T. Development, "NEDO Research Related to Battery Storage Applications for Integration of Renewable Energy," *Trains*.

[7] S. P. Dimakis, G. K. Papagiannis, and G. C. Christoforidis, "Control strategy to alleviate PV power output fluctuations through a superconducting coil," *Proc. Univ. Power Eng. Conf.*, 2014.

[8] X. Li, Y. Li, X. Han, and D. Hui, "Application of fuzzy wavelet transform to smooth wind/PV hybrid power system output with battery energy storage system," *Energy Procedia*, vol. 12, pp. 994–1001, 2011.

[9] Y. Noro *et al.*, "Power fluctuation suppression system for large scale PV," *Proc. IEEE Power Eng. Soc. Transm. Distrib. Conf.*, pp. 1–6, 2012.

[10] M. Y. Worku, M. A. Abido, and R. Iravani, "Power fluctuation minimization in grid connected photovoltaic using supercapacitor energy storage system," *J. Renew. Sustain. Energy*, vol. 8, no. 1, p. 13501, 2016.

[11] T. Senjyu, M. Datta, A. Yona, H. Sekine, and T. Funabashi, "A new method for smoothing output power fluctuations of PV system connected to small power utility," *7th Internatonal Conf. Power Electron. ICPE'07*, pp. 829–834, 2008.

[12] F. Basrawi and T. Yamada, "Theoretical analysis of performance of a micro gas turbine co / trigeneration system for residential buildings in a tropical region," *Energy Build.*, vol. 67, pp. 108–117, 2013.

[13] K. H. Koo, M. R. R. Chand, H. Ibrahim, A. A. Aziz, and F. Basrawi, "Theoretical Analysis on the Effect of Operation Strategies on the Environmental Performance of Micro Gas Turbine Trigeneration System in Tropical Region," vol. 5, pp. 19–25, 2015.

[14] F. Basrawi, T. K. Ibrahim, K. Habib, and T. Yamada, "Effect of operation strategies on the economic and environmental performance of a micro gas turbine trigeneration system in a tropical region," *Energy*, vol. 97, pp. 262–272, 2016.

[15] Capstone. (2016). Product Catalog Applications and Features. In C. T. Corporation (Ed.). Capuder, T., & Mancarella, P. (2014). Techno-economic and environmental modelling and optimization of flexible distributed multi-generation options. Energy, 71, 516-5.

[16] Zimmerman, B. (2011). Capstone MicroTurbine for the Oil and Gas Industry. Zonen, K. (2016). Instruction Manual : CMP and CMA Series Kipp & Zonen.

Investigation of the role of pH on structural and morphological properties of titanium dioxide nanoparticles

S A Sadiq, Z Waseem, S Hanif, S Riaz, K Hayat, A I F Batool, Y Jamil, M Y Naz and H Anwar*

Advanced Nanomaterials and Devices Laboratory, Department of Physics, University of Agriculture Faisalabad, 38040 Faisalabad, Pakistan.

*E-mail: Hafeez.anwar@gmail.com

Abstract. In the present research work, titanium dioxide (TiO_2) nanoparticles were synthesized by sol-gel method in which titanium tetraisopropoxide (TTIP) used as a precursor and the effect of pH on their structural and morphological properties was investigated systematically. X-ray diffraction (XRD) and scanning electron microscopy (SEM) were used to characterize the prepared samples. The acidity-alkalinity of the gel was adjusted to different pH values (1, 4, 7 and 10) by using respective media, i.e. HCl and NH_4OH. The results showed that from strong acidic media to strong basic media, the average particle size was found to increase from 14 nm at pH 1 to 20 nm at pH 10, whereas the average particle size was 19 nm at pH 7. All peaks in XRD patterns perfectly indexed to the stable anatase tetragonal phase except sample at pH 1, where some traces of rutile phase was also present. XRD results showed the formation of single phase high purity nanoparticles. TiO_2 nanoparticles showed homogeneous morphology with low agglomeration at pH 7 and pH 10.

1. Introduction

The last four decades have witnessed a great interest in the characterization of titanium dioxide nanoparticles for their beneficial use in a large number of applications such as cosmetics, ceramics, solar cells, pollution-reduction and analytical devices. Titanium dioxide is one of the most promising nanomaterials because it is highly stable, low-cost and is believed to be non-toxic. Titanium dioxide has been synthesized by different methods including hydrothermal method [1] direct oxidation, sol-gel method [2, 3] and chemical vapour deposition [4]. Among them sol-gel method is one of the common method because of its good chemical homogeneity. It is a well-known fact that the specific surface area and the surface-to-volume ratio is increased remarkably with the decrease in the size of the material. High surface area caused by the small size of the particle is highly advantageous and beneficial to many TiO_2 based applications and devices [5]. Zhang and Banfield presented a useful and productive report on structural characteristics, mechanical and thermodynamic properties of nanocrystalline TiO_2 [6]. Ibrahim and Sreekantan [7] identified that in the formation of titanium dioxide structure, pH plays an important role. Results showed that anatase structure was formed at lower acidic condition and rutile phase was formed at higher acidic condition. They explained from SEM images that at pH 1 TiO_2 powder showed rod like shape while sample prepared at pH 3 and 5 were spherical in shape. They also observed that particle size was smaller at higher pH. Degree of crystallinity of anatase is pH dependent and lower acidity enhanced the crystallinity also promotes formation of big crystallite size. They prepared titanium dioxide (TiO_2) nanoparticles through sol-gel method using titanium tetraisopropoxide as a precursor.

They investigated the effect of pH and reaction morphology on the preparation of TiO_2 nanoparticles. The experimental results showed that morphology of titanium dioxide was in nanopowder form when NH_4OH catalyst was used and in case of HCl, it showed a granular or bulk form. The pH of the prepared solution has great influence on the particle size of titanium dioxide nanoparticles [8, 9].

In this research work, a simple and cost effective method for the preparation of titanium dioxide nanoparticles at different pH through sol-gel method was used in which titanium tetraisopropoxide used as a precursor. All samples were characterized by using different techniques i.e. scanning electron microscopy (SEM) and X-ray diffraction (XRD). Then effect of pH on their structural and morphological properties was investigated.

2. Experimental procedure

Titanium dioxide nanoparticles were synthesized by sol-gel method at different pH conditions and their structural and morphological properties were observed. Titanium dioxide nanoparticles were prepared from titanium tetraisopropoxide, ethanol, hydrochloric acid and distilled water. In this work, titanium tetraisopropoxide used as a precursor and hydrolysis of titanium tetraisopropoxide was carried out at different pH values through a respective media i.e. NH_4OH and HCl. In First step, solution A was prepared when 5.5 ml titanium tetraisopropoxide (TTIP) and 54.8 ml ethanol were mixed together and placed on magnetic stirrer for continuous stirring (300 rpm). In second step, solution B contained a homogenous mixture of 1.68 ml hydrochloric acid and 19 ml of distilled water. Then the solution B was allowed to mix with already prepared solution A of titanium tetraisopropoxide (TTIP) in ethanol under continuous stirring. The resultant solution of titanium tetraisopropoxide (TTIP), ethanol, hydrochloric acid and distilled water were kept under constant stirring for further 20 hours. At this stage, pH of solution was found to be 1 and was measured by using Eutech pH meter. The prepared solution was filtered and after filtration, the material was dried in an oven Model (Shel-Lab) at 80 °C for 1 hours. The dried sample was ground to fine powder using agate pestle and mortar. The particles obtained after grinding were calcined in furnaceModel (SNOL-LHM01) at 500 °C for 4 hours.

Thus, first sample A of titanium dioxide nanoparticles having pH value 1 was ready for characterization. The next three samples namely sample B, sample C and sample D of titanium dioxide nanoparticles with pH values of 4, 7 and 10, respectively. These samples were prepared in the similar manner with only difference that their pH was adjusted by gradual addition of suitable amount of NH_4OH or HCl. Synthesis process is shown in figure 1.

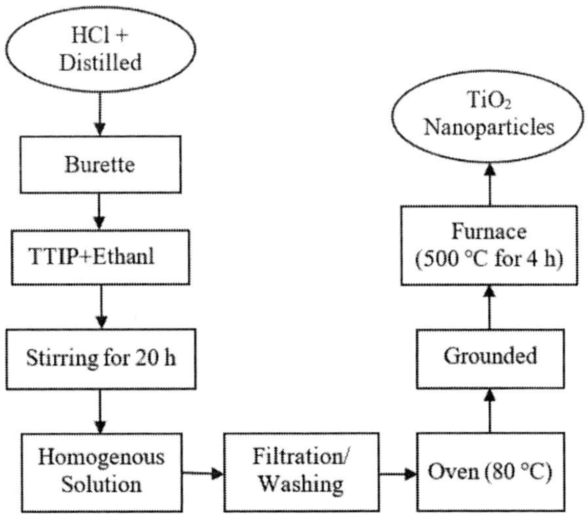

Figure 1. Schematic of preparation of TiO_2 nanoparticles.

3. Characterization

Powder X-ray diffraction was used to determine the average particle size and crystal phase of titanium dioxide nanoparticles. XRD patterns of TiO_2 nanoparticles was recorded using PANanalytic XPert PRO with λ =1.54056 Å which is the wavelength characteristic of the Cu- K_α radiation operated at 15 kV. The average crystallite size (Scherrer's Equation), lattice parameters, volume of unit cell and X-ray density were measured by using equation (1), equation (2), equation (3) and equation (4).

$$T = \frac{0.9\lambda}{\beta \cos\theta} \quad (1)$$

$$\frac{1}{d^2} = \frac{4}{3}\left(\frac{h^2 + hk + k^2}{a^2}\right) + \frac{l^2}{c^2} \quad (2)$$

$$V_{cell} = a^2 c \sin 60 \quad (3)$$

$$D = \frac{ZM}{N_A V_{cell}} \quad (4)$$

where T is the crystal size, β is the line width (FWHM), θ is the diffraction angle and the λ is the wavelength of X-ray radiation. Z is the number of molecules per unit cell, M is the molar mass of the titanium oxide, N_A is Avogadro number, V_{cell} is the volume of the unit cell. The morphological and structural properties of titanium dioxide nanoparticles were determined by JEOL JSM 5910 scanning electron microscope. The samples were mounted on specimen stub by using double sided carbon tape and a thin film of gold was deposited in order to have good results.

4. Results and discussion

XRD patterns of the all samples (A, B, C, D) synthesized by sol-gel method at different pH values (1, 4, 7 and 10) and calcined at 500 °C for 4 hours were showed in figure 2. All the peaks can be perfectly indexed to the stable anatase tetragonal phase of TiO_2 single crystal structure with the lattice constant a=b=3.8 Å and c=9.50 Å, indicating the high purity and crystallinity of the obtained samples. For example, sample B of titanium dioxide prepared at pH 4 exhibit diffraction peaks in the range of 2θ, 25.43°, 37.88°, 48.08°, 54.05°, 55.05°, 62.7°, 70.37° and 75.27°, indexed to (hkl) planes (101), (004), (200), (105), (211), (204), (220) and (215), showing that the sample comprises of pure anatase phase of TiO_2. Concerning the presence of the anatase phase, the structure is body-centered-tetragonal. Moreover, it can be seen from table 1 that the values of densities of all the four prepared samples were nearly equal to 3.9 g/cm^3, which can be taken another obvious evidence of the fact that prepared samples have anatase phase. It is also noticed that pH affects particles size and degree of crystallinity. A trace of rutile was found in sample prepared at pH 1 at 27° corresponding to rutile phase (110). In this case, it is found that high acidity in medium solution favours the formation of rutile phase while lower acidity favours anatase formation [10]. This mechanism could be explained using the concept of partial charge model [11].

4.1. Effects of pH on the particle size of TiO_2

Figure 2 shows the XRD patterns of the particles prepared at different pH conditions. The sharp peaks of crystalline anatase appeared in all the XRD patterns of the all specimens. The most intense peak was (101) at the value of 2θ round about 25° peak of anatase in all samples, whereas, for sample prepared at pH = 1, (110) peak for rutile phase at 2θ = 27.67° was present. The average particle size of sample A at pH 1 (strong acidic end) was 14 nm. At neutral condition pH 7, particle size of sample C was 19 nm. Similarly, at highly basic end (pH =10) particle size of sample D was found to be 20 nm. The variation in crystallite size as per increase in pH values is contrary to literature [12]. According to previously reported literature, average particle size decreased with increasing pH values, i.e., from acidity to alkalinity. These observed and calculated facts suggested that there is a strong correlation between pH and average particle size [13].

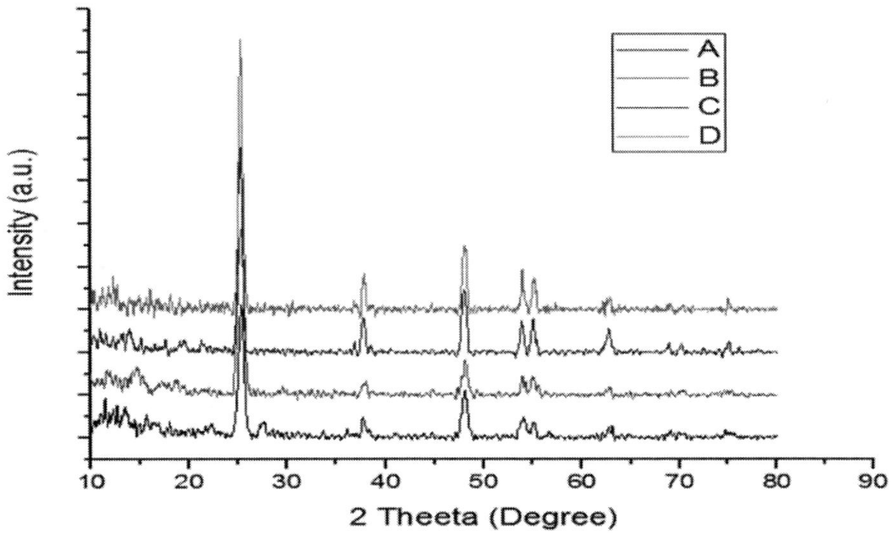

Figure 2. XRD patterns of titanium dioxide nanoparticles synthesized through sol-gel method with different pH values for sample with (A) pH 1, (B) pH 4, (C) pH 7 and (D) pH 10.

The lattice parameters did not show any significant change at different pH value, indicating that the interaction of hydrochloric acid with the TiO_2 nanoparticles has been occurred only at the surface of the nanoparticles. Various structural parameters calculated from XRD data were shown in table 1.

Table 1. Various structural parameters extracted from XRD data at different pH values.

Samples	**pH**	**Average particle size (nm)**	**Lattice constant (Å)**		**Volume of titanium dioxide (Å^3)**	**Density of titanium dioxide (g/cm^3)**	**Phase**
			a=b	c			
A	1	14	3.79	9.50	136.46	3.89	Anatase with traces of rutile
B	4	12.5	3.78	9.50	135.74	3.91	Anatase
C	7	19	3.81	9.50	138.0	3.84	Anatase
D	10	20	3.80	9.50	137.18	3.87	Anatase

4.2. Effect of pH on the morphology of TiO_2

The surface morphology of TiO_2 nanoparticles is shown in figure 3.Titanium dioxide nanoparticles prepared at acidic conditions showed an agglomerated and aggregated nanoparticles of irregular shape as shown in figure 3(a) and (b). TiO_2 nanoparticles showed uniform morphology with low agglomeration at neutral and higher basic condition in figure 3(c) and figure 3(d). It is found that the morphology markedly depends on the pH of precursor solution.

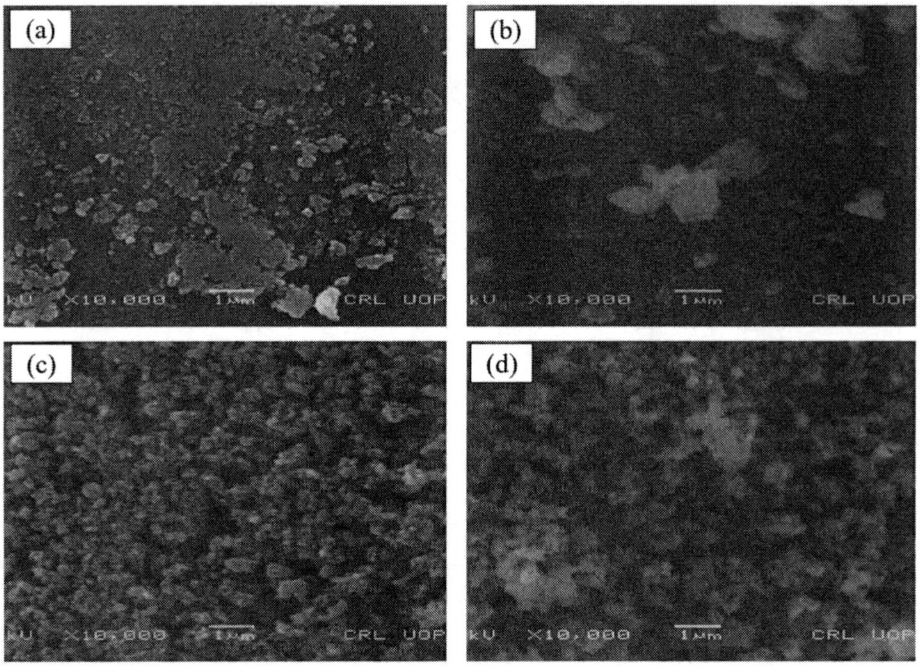

Figure 3. SEM micrographs of titanium dioxide nanoparticle: (a) pH 1 (b) pH 4 (c) pH 7 and (d) pH 10.

5. Conclusion

Effect of pH on morphological and structural properties of titanium dioxide (TiO_2) nanoparticles were investigated through scanning electron microscopy (SEM) and X-ray diffraction (XRD) techniques. From these investigations, it can be concluded that TiO_2were prepared at different pH conditions (1, 4, 7 and 10) and calcined at 500 °C had different average particle size and morphologies. The average particle size was found to be increase with increasing pH values ranging from 14 nm to 20 nm. All peaks in XRD patterns can be perfectly indexed to the stable anatase tetragonal phase; except sample prepared at pH 1 that showed some rutile traces. The results show the high purity and crystallinity of the obtained samples. Agglomerated and irregular to uniform morphology was found with increasing pH values.

Acknowledgement

Hafeez Anwar is grateful to Higher Education Commission (HEC) of Pakistan for providing funding under SRGP 266.

References

[1] Rasalingam S, Wu C-M and Koodali R T 2015 *ACS applied materials & interfaces* **7** 4368.
[2] Nabi G, Khalid N, Tahir M B, Rafique M, Rizwan M, Hussain S, Iqbal T and Majid A 2018 *Journal of Inorganic and Organometallic Polymers and Materials* **28** 1552.
[3] Senić Ž, Bauk S, Vitorović-Todorović M, Pajić N, Samolov A and Rajić D 2011 *Scientific Technical Review* **61** 63.
[4] Lopez L, Daoud W A, Dutta D, Panther B C and Turney T W 2013 *Applied Surface Science* **265** 162.
[5] Ahmed M, El-Katori E E and Gharni Z H 2013 *Journal of Alloys and Compounds* **553** 19.
[6] Zhang J, Yu J, Zhang Y, Li Q and Gong J R 2011 *Nano Letters* **11** 4774.
[7] Ibrahim S A and Sreekantan S 2011 Effect of pH on TiO2 nanoparticles via sol-gel method. In: *Advanced Materials Research*: Trans Tech Publ) pp 184.

[8] Lim C S, Ryu J H, Kim D-H, Cho S-Y and Oh W-C 2010 *Journal of Ceramic Processing Research* **11** 736.
[9] Prasad S and Madhavan J 2013 *Synthesis* **5** 2970.
[10] Chen Z, Wang F, Balachandran S, Li G, Liu P, Ding Y, Zhang S and Yang M 2018 *Nanotechnology* **29** 125602.
[11] Li L, Yan J, Wang T, Zhao Z-J, Zhang J, Gong J and Guan N 2015 *Nature communications* **6** 5881.
[12] Tsega M and Dejene F 2017 *Heliyon* **3** e00246.
[13] Devi G S, Kumar K S and Reddy K S 2015 *Particulate Science and Technology* **33** 219.

IOP Conf. Series: Materials Science and Engineering **863** (2020) 012047 doi:10.1088/1757-899X/863/1/012047

Drying process of black pepper in a swirling fluidized bed dryer using experimental method

N F M Roslan and A S M Yudin*

Energy and Sustainability Focus Group, Faculty of Mechanical and Automotive Engineering Technology, Universiti Malaysia Pahang, 26600 Pekan, Pahang, Malaysia.

*E-mail: shukrie@ump.edu.my

Abstract. Black pepper (piper nigrum) is one of the well-known spices extensively used worldwide. Drying is the most popular methods of preserving the black pepper. Unfortunately, this operation may negatively influence product quality. Moreover, due to the high heat capacity of water, drying is usually a long-lasting and energy-intensive process, thus new drying techniques are continuously being sought. Fluidized bed dryers have been applied to dry raw materials due to the advantages of good mixing efficiency and high heat and mass transfer rate. In order to control and optimize the drying process of fluidized bed dryers, it is necessary to develop reliable methods to measure the moisture content of solid particles in the bed. For this project we dried the black pepper using two methods which is drying under sun drying and swirling fluidized bed dryer using heater. The temperature for both drying will be the same. Then, the characteristic and the drying moisture ratio of the black pepper were compared between these two drying methods. The performance of the swirling fluidized bed dryer was evaluated based on the percentage of moisture content of loss reduction. The time taken for this drying is 10 hours, which considered as 1 day of drying process. The percentage of moisture content loss reduction is 57.9%. The moisture ratio after 10 hours of drying is 0.3420. In conclusion, drying using swirling fluidized bed dryer is the new technology that can be used in drying industry.

1. Introduction

Drying is the process of the removal of water (moisture) from hygroscopic materials at low to medium moisture contents (normally <30% wet basis) by means of evaporation. When the moisture content of the agricultural products is high (usually >50% wet basis) the process of removal of moisture is referred to as dehydration [1]. The examples of the products that are dried include cereals, oilseeds, legumes, and some processed foods; and examples of the products that are dehydrated include fruits, meats, and vegetables.

Conventional method is drying under sunlight or solar drying. Natural method is drying under sunlight on sunny day. The material to be dried is placed directly under hostile climate conditions like solar radiation, ambient air temperature, relative humidity and wind speed to achieve drying. In direct solar dryer, the materials to be dried is placed in an enclosure, with transparent covers or side panels. Heat is generated by absorption of solar radiation on the product itself as well as the internal surfaces of the drying chamber. This heat evaporates the moisture from the drying product and promotes the natural circulation of drying air. In the open sun drying, the products are spread on the concrete floor and are raised platform which are contaminated by dust, dirt, insect's damaged and other pollutants, thereby, degrading the quality of the products [2]. Other than that, sun drying method involves multiple works such as heaped and covered the product at night or during rainy weather that can cause re-wetting [3].

The quality the products will further be degraded by the delays drying when a rainy day or long drying time and the moisture content of drying products are not uniform, which causes the mould growth [4]. Furthermore, time taken to remove the moisture content by using conventional method is to long in around one to six hours one to ten days. Since Malaysia has monsoon season between March to August, it is also will disturb the drying process and the drying rate become longer. Swirling fluidized bed drying is proposed to improve the drying of black pepper.

Fluidized bed dryer is a new method that were used nowadays for a better quality of the output products. It offer significant advantage like high heat and mass transfer. Undergo uniform moisture reduction with less drying time and high drying rate. The loss of moisture weight reduction from the wet solid per unit time (drying rate) is the method to measure the dryer performance. Swirling fluidized bed dryer has four stage of swirling patterns. Firstly, packed regime: in the beginning, there are none movement in the bed. At this condition, resistance of bed towards air flow is at its peak due to the particles are stick together. Secondly, two layer: then, bottom layer particle begin to swirl while upper layer having vibration condition. At this regime, tangential velocity component was decayed and at certain height, superficial velocity is insufficient to create any swirling in the bed. Thirdly, wave motion: at this region, the bed having partial swirl. Swirling motion extends over a certain area of this bed, while the remaining area is static. The swirling area moved black paper to the static area and the motion stopped, while triggered second swirling motion at static area. Lastly, fully swirling: as the black paper become drier, stable swirling pattern was finally reached [4]. Both wave motion and fully swirling condition occurred due to the swirling velocity finally becomes higher as a result of lower bed weight as a result of continuous moisture transfer.

Considering the disadvantages of the open sun drying, an attempt has been made in this study to investigate the drying characteristic of black pepper by using conventional method of sun drying and also to investigate the drying characteristic of black pepper inside the swirling fluidized bed dryer. Lastly, to compare the drying of black pepper by using conventional method and swirling fluidized bed drying.

2. Drying technology of black pepper

2.1. Drying under solar

The drying process under sun drying is a conventional method for harvesting and processing for black pepper which is the biggest spices in the world. Black pepper are from unripe green berries from pepper plant. Matured green spikes with 1 or 2 berries in the spike starting to yellow are harvested for the preparation of black pepper. In a traditional pepper garden where vines are about 3m high, harvesting of spikes on the upper canopy is done using a ladder. Blanching is sometimes practised by immersing the berries in hot water, of about 90 °C, for 2-3 minutes prior to drying. This reduces microbial load, washes off dust particles, and helps in drying pepper faster. Blanched, dried black pepper has a glossy appearance and a better, more uniform colour [5]. Good sun-dried black pepper should not contain more than 12% moisture. Generally, the conversion rate is about 33% of the green berries, but it can vary from 29 to 36% depending upon the variety and maturity at time of harvest. After blanching, the product can be dried to a uniform moisture level (8–10%) within three days in full sunlight. The end product has a shiny black colour. First, determine the PPO (polyphenol oxidise) activity, reduced weight percentage, and the changes in colour and appearance of green pepper berries after blanching and sun drying, and we found that PPO activity decreased dramatically after blanching, whereas the rate of browning and water loss increased [5]. The fresh pepper berries was then subjected to heat treatment with hot water for 1, 2, and 3 minutes at 80, 90, and 100 °C, respectively. The fresh pepper berries was then compared with the pepper berries soaked in boiled water for 10 minutes.

Blanching is known to activate the phenolase, which is responsible for producing the black colour of pepper berries. The black colour that the pepper acquires upon drying is due to the oxidation of colourless phenolic compounds present in the skin. The lower the blanching temperature and the shorter the sun drying, the lighter the sample colour. The unblanched pepper berries remained yellow-green during sun drying for 16 hours.

2.2. Tunnel dryer

The Tunnel dryer is a direct continuous type of dryer. It is a largest scale dryer. In this dryer, the materials that to be dried are sent to the air heated tunnel for drying purpose. The material is entered at one end and the dried material is collected at the other end of the tunnel. The outgoing material met the incoming air to ensure maximum drying and the outgoing air contacted the wettest material so that the air was as nearly saturated as possible. One of the doors of the tunnel is opened and the materials to be dried are placed to the trolleys and trucks are pushed slowly in the tunnel and then door is closed. Hot air is circulated and passed through the rail truck and perforated trolleys. The hot air then followed is recirculated with the help of fans and the material becoming dried. The moist air is passed out through the exhaust after completion of drying. The door is opened and the trolleys are taken out of the funnel and some new trolleys with the wet materials are introduced into the trucks and the process is repeated [6].

2.3. Swirling fluidized bed dryer

The optimal method for controlled, gentle and even drying of wet solids. Higher throughput with better quality for a range of products (having different shape, bulk density, physical and chemical properties) can be achieved in FBD drying by using medium and low grade thermal energies. The drying process with fluidized bed drying reduces the drying time in the drying oven by approximately twenty times. In addition, fluidized bed drying provides controlled and uniform drying conditions compared to the uneven drying in trays [7]. It must be ensured that drying always takes place in a thermodynamic equilibrium.

Acts as a high mixing ability and improved solid-gas contact to shorten the drying time of products. Hydrodynamic study was conducted for three beds loadings of 1.0 kg, 1.4 kg at a drying temperature of 90 °C. The SFBD has shown excellent potential to dry the pepper with a relatively short drying time compared to the conventional method [8]. It was found that bed higher bed loading of wet pepper requires longer drying time due to higher amount of moisture content in the bed. In SFBD method, when air flow was introduced through the bottom of a bed in SFBD, solid particles become suspended (at least at drag force equal to gravitational force) and behave fluid-like and high solid-gas contact in the bed allows vigorous mixing [4]. As the drying cycle time increase, the moisture content decrease. When the moisture content is high, there is pre-heating period [9].

2.4. Drying parameter

The drying parameter that used to calculate the drying result is moisture content and moisture ratio. It is important to calculate the moisture content reduction loss to get the result of the experiment. Drying curves obtained under controlled conditions provide important information regarding the water transportation mechanisms, and they are used in the determination of the effective diffusion coefficient. The moisture content (M) at any time of drying (%), was calculated using equation (1):

$$M = \frac{W_i - W_d}{W_i} \times 100\% \qquad (1)$$

The reduction of moisture ratio with drying time was used to analyse the experimental drying data. Moisture ratio (MR) represents the amount of moisture remaining in the samples reported to the initial moisture content. It was calculated using equation (2):

$$MR = \frac{M - M_e}{M_0 - M_e} \qquad (2)$$

The equilibrium moisture content (M_e) were determined by drying until no further change in weight was observed for the samples in each treatment and drying conditions [10].

3. Experimental setup and procedure

3.1. Conventional drying experiment

The experiment was carried out under the sunlight or solar. The black pepper weight is 1008 grams was dried on the drying mat that placed on the table that exposed to the sunlight without any disturbance such as building. The drying black pepper was conducted to investigate the drying time and the moisture ratio content that dried from the black pepper. A fresh black pepper was used to ensure the moisture content are neglected as shown in figure 1.

Figure 1. Fresh black pepper that were used in experiment.

The drying mat basically from the wire gauze so that the air surrounding can propagate around the black pepper as shown in figure 2.

Figure 2. The black pepper were placed on wire gauze.

The temperature of the surrounding was recorded using temperature data logger and the weight of the black pepper before and after the drying was recorded to determine the moisture ratio that were loss using electronic weighing scale.

3.2. Swirling fluidized bed drying experiment and setup

The fluidized bed apparatus is located at Energy and Sustainability Focus Group laboratory, Universiti Malaysia Pahang. The photograph of the experimental setup is shown in figure 3. In order to scientifically investigate the performance of the distributor configurations and its effect to the bed material, the following requirements are specified such as fluidized bed system that allows interchangeability of the distributor, clear silica glass cylinder to enable visualization of studies, pressure drop measurement and the velocity of the air flow.

The fluidization column consists of transparent cylindrical silica glass which has 108 mm internal diameter, 5 mm thickness and 310 mm length. An acrylic plate that has the same diameter with the glass column's flange was mounted at the bottom. The air flow was controlled by the controller that controls the speed of the blower between 0 and 50 Hz speed range. Fluidization air was generated from a blower and it was heated to a desirable temperature using an electrical heater. The temperature and humidity sensor was installed at the outlet of the bed to measure the temperature and relatively humidity of the exhaust air.

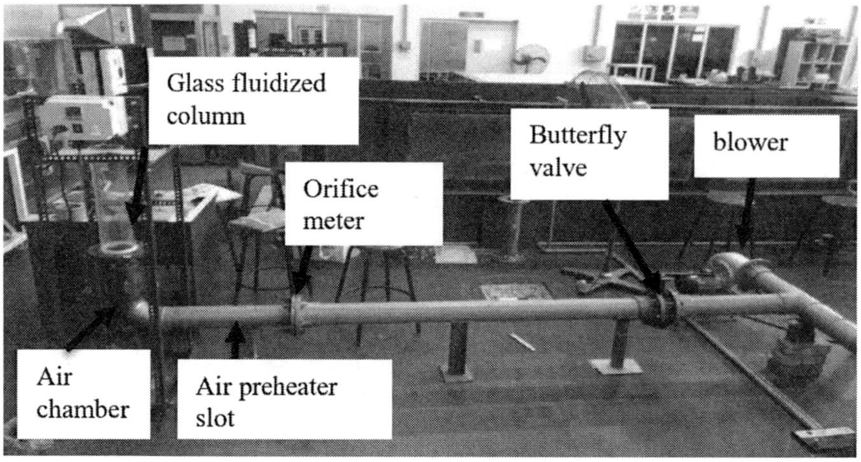

Figure 3. The fluidized bed dryer setup.

The black pepper weight of 1008 grams was dried in the swirling fluidized bed dryer. The weight of black pepper was measured using electronic weighing scale before put in the swirling fluidized bed dryer to calculate its moisture content. The blower was turn on and the speed of the blower was 4.26 m/s. The heater was turn on but the temperature on the heater must not higher than 40 °C, so that the temperature of the conventional drying is the same with the temperature of the swirling fluidized bed dryer. After the black pepper was dried, the black pepper was weigh again to calculate the moisture ratio and to calculate moisture content loss reduction.

3.3. Distributor

A design of air distributor made of aluminium with 115 mm diameter and 8 mm thickness was tested in the experiments. The distributor geometry was modified to introduce novel swirling distributor by making the edge opening with 45 inclined air intakes as shown in figure 4.

Figure 4. The 45° distributor that used in the fluidized bed dryer.

4. Results and discussion

4.1. Conventional drying (sun drying)

By considering previous sun drying that was done by other researcher, the moisture content that should be removed is 12%. For conventional drying, the black pepper was weight first and 1008 grams black pepper was dried and the moisture content of the black pepper that should be loss (x) were calculated as shown in equation (3):

$$x = \frac{12}{100} \times 1008\, grams = 120.96\, grams \tag{3}$$

The drying of black pepper was also affected by surrounding temperature. The average temperature that recorded was 35 °C. Figure 5 shows that the condition of black pepper before conventional drying. The temperature of the surrounding was recorded using temperature data logger. The drying time taken for conventional drying is 7 days. Figure 6 shows the condition of black pepper after 7 days of drying. The remaining weight of the dried black pepper is 476 grams. Table 1 shows the weight of the black pepper for each day drying and its moisture ratio using the calculation of moisture ratio.

Figure 5. Condition of black pepper before drying.

Figure 6. Condition of black pepper after 7 days of drying.

Table 1. the weight before, after and moisture ratio of conventional drying for each day.

Time (Day)	Weight before (g)	Weight after (g)	Moisture ratio
1	1008	879	0.8546
2	879	764	0.7249
3	764	615	0.5570
4	615	547	0.4803
5	547	499	0.4262
6	499	482	0.4070
7	482	476	0.4003

Figure 7 shows the graph for the moisture ratio and temperature vs the time taken for conventional drying. From the graph, the highest temperature recorded was on day 3 which is 37 °C and the moisture ratio is decreasing as the time (day) of drying is increasing.

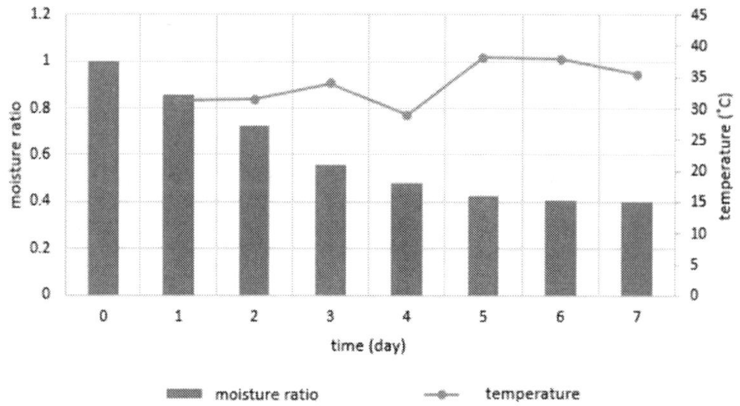

Figure 7. The moisture ratio and temperature vs time taken for conventional drying.

The percentage of moisture content of loss reduction of drying black pepper using conventional drying is 52.7%. The visualisation of the black pepper after drying is the colour does not shiny because the product was spoilt because was dried on the surrounding so there may be some impurity on the black pepper. The colour of black pepper also blackish grey.

4.2. Swirling fluidized bed drying

The weight of the black pepper used in the swirling fluidized bed dryer setup is 1008 grams. Figure 8 shows the condition of black pepper in swirling fluidized bed before drying process took place. The time taken for drying in a swirling fluidized bed only 10 hours. For the swirling fluidized bed dryer, the temperature of the hot air from the heater was in the range of 33 °C and 38 °C. Figure 9 shows the condition of black pepper in a swirling fluidized bed dryer during drying process.

Figure 8. Condition of black pepper before swirling.

Figure 9. Condition of black pepper when swirling.

During the drying process in a swirling fluidized bed, the black pepper swirls in a spiral motion in a clock-wise direction. Air entrainment at 45° inclination angles provides swirling flow thus improve drying time to only 10 hours. The mechanism of swirling motion that occurs rapidly at the bottom of the bed is due to the non-uniform air formation from the tangential air flow through the inclined slots. The chaotic air formation induced the black pepper particles in clock-wise swirling motion which the pattern is similar to blenders. However, due the black pepper weight, the swirling motion is weakened as the air bubble progressed upward through the dense black pepper. Instead, the air erupted and the occurrence is randomized across the bed surface. Table 2 shows the weight and moisture ratio of the black pepper during for 2, 4, 6, 8 and 10 hours in the swirling fluidized bed drye. Figure 10 shows the moisture ratio and temperature of the heater versus time taken during drying in the swirling fluidized bed.

Table 2. The weight and moisture ratio of black pepper for 2, 4, 6, 8 and 10 hours in swirling fluidized bed dryer.

Time (hour)	Weight before (g)	Weight after (g)	Moisture ratio
2	1000	920	0.9091
4	920	783	0.7534
6	783	547	0.4852
8	547	451	0.3761
10	451	421	0.3420

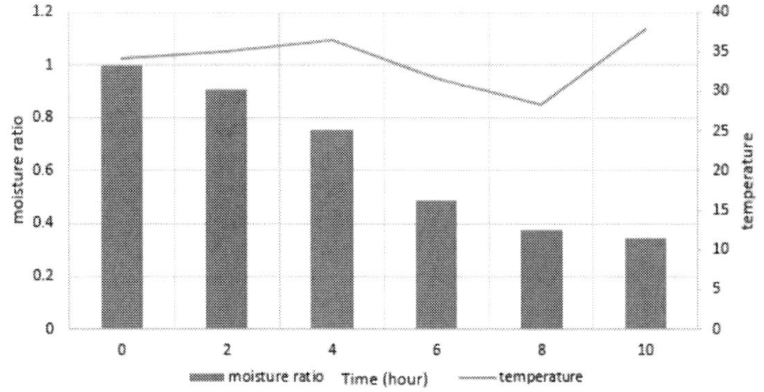

Figure 10. The moisture ratio and the temperature of the heater vs time taken for drying black paper in swirling fluidized bed dryer.

The percentage of moisture content reduction at the end of drying is calculated at 57.9%. Figure 11 shows the visualisation black pepper using swirling fluidized bed dryer at the end of drying time. The colour of the black pepper is pure black. The shape also round and shrink and the black pepper also shiny shows that the quality is better because there is no impurities.

Figure 11. The visualisation of result of drying black pepper using swirling fluidized bed dryer.

4.3. The comparison between conventional drying and swirling FBD

Table 3 shows the comparison between conventional drying and swirling fluidized bed drying. The conventional drying takes more time because the drying using the sunlight. It can be done effectively from 9 am to 4 pm. The rest of the time is idle. The weather also affect the drying especially in raining day, the drying process is impossible to be conducted due to high moisture content in the air. The conventional drying also should be done in an open field. Trees, buildings and other structures may block sunlight and hinder the drying process. So, the drying using conventional takes more time than swirling fluidized bed drying.

Table 3. The comparison of the conventional drying and the swirling fluidized bed dryer.

Comparison	Conventional drying	Swirling FBD
Remaining weight after drying(g)	476 g	421 g
The moisture content loss reduction	52.7%	57.9%
Drying time	7 days	1 day (10 hours)

5. Conclusion

In short, it can be concluded that the time taken for drying the black pepper by using conventional drying is 7 days, while drying using swirling fluidized bed dryer is reduced to only 10 hours. Its shows that the drying time using swirling fluidized bed is faster than drying using conventional method. The quality of the black pepper after drying using fluidized bed dryer also shows better quality because the black pepper colour is dark black and it look shiny shows that no impurities in the black pepper. The remaining weight for the dried black pepper by conventional drying is 476 grams while by swirling fluidized bed dryer is 421 grams. The percentage of the moisture content reduction for swirling fluidized bed dryer which is 57.9% meanwhile the percentage of moisture content reduction for conventional drying which is only 52.7%.

In general, the drying of black pepper in a swirling fluidized bed dryer is more efficient and high quality than conventional drying. Hence, the three objectives of this project which is to investigate the drying characteristic of black pepper by using conventional method sun drying, also to investigate the characteristic of black pepper by using swirling fluidized bed dryer and to compare drying of black pepper performance by using conventional method and swirling fluidized bed drying was achieved.

Acknowledgments

The authors wish to express their gratitude to the Faculty of Mechanical and Manufacturing Engineering, Universiti Malaysia Pahang (UMP), Pekan, Pahang, Malaysia for providing the research facilities and supporting the research under University Research Grant RDU171110, RDU180320, RDU190383 and Malaysia of Higher Education Research Grant (FRGS/1/2019/TK10/UMP/02/26).

References

[1] Scherer G W 1990 *J. Am. Ceram. Soc.* **73(1)** pp. 3–14.
[2] Codex Alimentarius 2003 *Code of Practice for the Prevention and Reduction of Mycotoxin contamination in cereals* pp. 1–6.
[3] Yahya M 2016 *Contemporary Engineering Sciences* **9(7)** pp. 325–336.
[4] Ozbey M and Soylemez MS 2005 *Energy Convers. Manag.* **46** pp. 1495–1512.
[5] Gu F, Tan L, Wu H, Fang Y and Wang Q 2013 *Food Chem.* **138(2–3)** pp. 797–801.
[6] Prakash O and Kumar A *Renew. Sustain. Energy Rev.* **(29)** pp. 905–910.
[7] Daud W R W 2008 *Adv. Powder Technol.* **19(5)** pp. 403–418.
[8] Haron N S, Zakaria J H and Batcha M F M 2017 *IOP Conf. Ser. Mater. Sci. Eng.* **226(1)**.
[9] Sivakumar R, Saravanan R, Perumal A E and Iniyan S 2016 *Renew. Sustain. Energy Rev.* **61** pp. 280–301.
[10] Seremet L, Botez E, Nistor O V, Andronoiu D G and Mocanu G D 2016 *Food Chem.* **195** pp. 104–109.

IOP Conf. Series: Materials Science and Engineering **863** (2020) 012048 doi:10.1088/1757-899X/863/1/012048

Experimental study of drying characteristics of cocoa bean in a swirling fluidized bed dryer

M A Zulkarnain, M K Shahriman, and A S M Yudin*

Energy and Sustainability Focus Group, Faculty of Mechanical and Automotive Engineering Technology, Universiti Malaysia Pahang, 26600 Pekan, Pahang, Malaysia.

*E-mail: shukrie@ump.edu.my

Abstract. Drying is a process to reduce the weight for easy transportation and enhance the shelf life of product without quality losses over an extended period of time. Cocoa bean is an agricultural product that contains high moisture. The drying process of cocoa bean was conducted by two methods which are conventional method and swirling fluidized bed dryer. The conventional method is by drying under the sunlight. By using swirling fluidized bed dryer, it can shorten the time where heater is used to supply heat in this experiment. This fluidized bed dryer functions by the addition of design of air distributor used in the experiment. The distributor installed to improve mixing inside the bed. Therefore, in order to minimize the cost of using high capacity blower as well as to reduce energy, viable design of air distributor which can contribute to low pressure drop and improved particulate mixing in fluidized bed is essential. A fluidized bed column of 108 mm in diameter with one slotted distributor with inclination angles of 45° is used in the experiment. We find the results by using 1kg sample for conventional method are 11 days drying time, 58.55% weight loss reduction, and 0.4145 moisture ratio. Meanwhile for swirling fluidized bed dryer the time required was 2 days, 59.16% weight loss reduction, and 0.4083 moisture ratio. Based on the results obtained, the drying by using swirling fluidized bed dryer is more efficient compare to conventional method.

1. Introduction

Cocoa (Theobroma cacao) is one of the economic tree crops grown in Malaysia. It is grown in tropical countries like Nigeria, Ghana, Ivory Coast, Brazil, Venezuela and Indonesia [1]. The cocoa pod which is botanically a berry, attains a height of 129 to 159 mm and diameter of 72 to 83 mm when fully ripe [2]. It normally contains 20 to 40 seeds surrounded by a mucilaginous pulp when the pod is ripe. By weight, cocoa pod is composed of about 74.4% husk, 22.5% wet beans and 3.1% placenta. Cocoa is a perennial cash crop with three important varieties viz. criollio, foresterio and trinitario. Cocoa beans is mainly consumed as chocolate and widely used in beverages, cosmetics, pharmaceuticals and health benefits such as anti-carcinogenic, antiulcer, anti-inflammatory, anti-microbial and analgesic [3,4].

Processing of cocoa beans involves the harvesting of cocoa pods, breaking of the harvested pods, fermentation of wet mass of cocoa beans obtained from broken pods, drying of fermented cocoa beans to moisture content of about 6 to 8% and storage of the dried beans till the time of use. Figure 1 shows a flow chart of the preparation processes. Out of these steps of processing cocoa pods into beans, the two major steps are fermentation and drying. Fermentation of cocoa beans involves keeping a mass of cocoa beans well insulated while at the same time air is allowed to pass through. This process is carried out to develop the chocolate flavour and aroma in the beans [5].

During fermentation, the pulp surrounding the bean is removed and the sugar in the pulp is converted into acetic acid. The different methods of fermentation are box, basket and heap and tray fermentation. The duration of fermentation varies from 4 to 7 days depending on the method of fermentation employed. Fermentation is the initial step needed in the development of various flavour precursors in the beans [6]. After fermentation, the moisture content of the beans is about 55% and this must be reduced to 6 to 8% for safe storage [7]. The drying process apart from reducing moisture content of the beans, aids in the completion of the chemical reactions that were started during the fermentation process and in the development of the chocolate brown colour of well fermented beans [8]. Drying of cocoa beans on most farms is carried out naturally by making use of the sun rays while few large farms use artificial dryers to achieve the drying operations [5]. Drying is essentially a process of simultaneous heat and mass transfer. In most drying operations, water is the liquid evaporated, and air is the drying medium. Heat, necessary for evaporation, is supplied to the particles of the material and moisture vapor, evaporated liquid, is removed from the material into the drying medium. Fluidized drying of granular products of solids can be either batch wise or continuous. Batch operation is preferred for small scale production and for heat sensitive materials. Fluidized bed dryers are widely used in a number of industry sectors to dry finely divided 50–5000lm particulate materials. Compared with other drying techniques, fluidized bed drying offers many advantages. High heat and mass transfer rates between the gas and the particles are possible because of good contact or large contact area between the particles and gas, good and rapid mixing of solids, nearly uniform moisture content distribution throughout the bed, closely controllable temperature in the bed, ease in transport and handling of particles, and simplicity in construction. On the other hand, the disadvantages include high pressure drop, attrition of the solids and erosion of the containing surfaces [9].

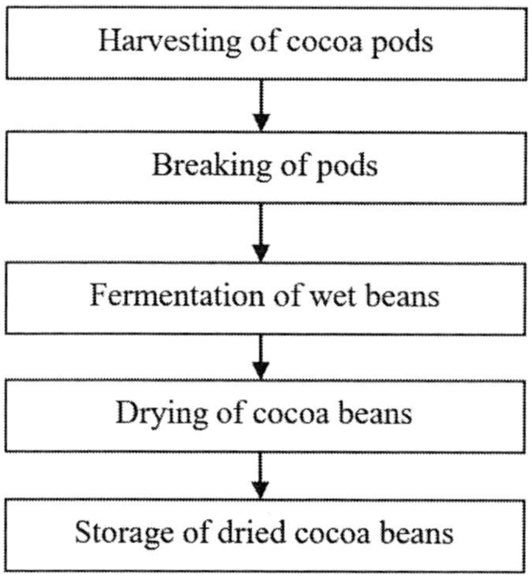

Figure 1. Preparation processes of cocoa bean.

2. Drying technology of cocoa bean

2.1. Sun drying

The natural way to dry the cocoa beans is in the sun on drying mats, which is a simple and cheap method. The mats are usually placed on a raised platform to protect the cocoa beans against animals and foreign matters. Sun drying can be carried out in a drying 'autobus', which is a shelter and ramps at different heights allowing trays to slide from the sun under the roof. Sliding roofs and immobile drying platforms are also developed [11].

The drying area should be located away from contaminant sources and should receive maximum sun exposure and air circulation during most times of the day, to speed up the drying process of cocoa beans. Thermometer was used to measure the peak temperature of drying [5]. Shady areas should be avoided. In rainy or wet regions, cocoa beans must be covered and re-spread once the drying surface has dried. Ensure that the drying surface is clean and located away from contaminants sources. The layer of drying cocoa beans should not exceed 6 cm thick to avoid slow or inadequate drying that may lead to mould growth. Beans must be turned several times each day to ensure uniformly dried beans. Rake over the cocoa bean layer frequently during the day time to allow faster drying and reduce the risk of fungal growth (5 - 10 times per day). Protect cocoa beans during drying from rain and dew. The cocoa beans should be heaped and covered at night or during rainy weather to avoid re-wetting. [10].

2.2. Plastic roof solar

The dryer has a width of 2.0 m, length of 5.0 m and height 2.2 m. The walls and roof are covered by a polyethylene film with 2.0 mm of thickness and supported by corrugated rods (diameter 0.9525 cm) separated 1.0 m. A platform (rack), which contained the product to be dried, was installed inside the dryer. The platform has a width of 1.0 m, length of 5.0 m and height 0.9 m. The main purpose of the plastic roof solar dryers is to make the solar radiation incident more efficient and to provide protection from rain and insects [12].

2.3. Artificial drying

Artificial drying is necessary in weather conditions that are cold and rainy or at plantations that are so big that sun drying would take too much space. Ovens are simple dryers that use heat to dry the cocoa beans. They can be heated either by the base of the oven, or by hot air from an external fire. The beans are spread in trays, allowing the air to permeate through a ladder system.

Using this method, the cocoa beans are not contaminated with smoke from the fire since dry beans easily absorb flavours and aromas from the environment. To avoid this, it must be ensured that the smoke ducts are impermeable. This method is also used in family plantations. Mechanical dryers are occasionally used in larger plantations, where big volumes of cocoa beans are to be dried.

The cocoa beans are placed on rotating platforms that circulate in a tunnel through which hot air is blown [11, 13, 14]. Heat was generated by the heater integrated into the side walls of the oven and the hot air flowed through the samples [15]. This method takes 10 – 20 hours, depending on the initial moisture-content of the cocoa bean. And the beans were left to rest at room temperature overnight. The tempering step is a common routine in cocoa drying, and the purpose is to redistribute the internal moisture to the outer beans layer after each drying cycle [15]. This is to reduce bean breakage associated with drying without tempering [16].

2.4. Swirling fluidized bed dryer

Fluidized bed dryers have been widely used for drying various agricultural products such as apple [17], olive pom ace [18], canola [19], soybeans [20], castor oil seeds [21], bird's eye chilli, red bell pepper, carrots, black tea, baker's yeast, coconut and hazelnut. Fluidized bed drying, a technique that was originally adapted for catalytic cracking of crude oil, offers several advantages over other types of drying such as solar drying, freeze drying, osmotic dehydration, spray drying, and vacuum drying. The main advantage is the thorough mixing of solids in this drying process which results in efficient mass and heat transfer, thus leading to rapid and economic drying [22].

Furthermore, inherent characteristics such as temperature uniformity and ease of control make fluidized bed dryers highly suitable for drying heat-sensitive products [23][24]. Also, fluidized bed dryers allow for easy handling and transport of the dried products which makes them appropriate for industrial purposes [25]. The main limitations of fluidized bed drying include loss of product qualities such as colour, texture, flavour, and nutrients [26]. However, such drawbacks can be avoided by using appropriate drying conditions [27].

2.5. Drying parameter

The drying parameter that used to calculate the drying result is moisture content and moisture ratio. It is important to calculate the moisture content reduction loss to get the result of the experiment. Drying curves obtained under controlled conditions provide important information regarding the water transportation mechanisms, and it used in the determination of the effective diffusion coefficient. The moisture content (M) at any time of drying (%), was calculated according to equation (1).

$$M = \frac{W_i - W_d}{W_i} \times 100\% \quad (1)$$

The reduction of moisture ratio with drying time was used to analyze the experimental drying data. Moisture ratio (MR) represents the amount of moisture remaining in the samples reported to the initial moisture content. It was calculated using equation (2).

$$MR = \frac{M - M_e}{M_0 - M_e} \quad (2)$$

The equilibrium moisture content (M_e) were determined by drying until no further change in weight was observed for the samples in each treatment and drying conditions [11].

3. Experimental set up and procedure

3.1. Conventional sun drying experiment

The conventional drying under direct sunlight was conducted at Kampung Beruas, Kuala Pahang. The cocoa pods were selected according to their ripeness and maturity levels. There were 13 pods spread on the net and dried under direct sunlight for periods of 11 days (12 March - 22 March) from 10 am to 5 pm. The range of ambient temperature is at Kuala Pahang around 28 °C (minimum) to 34 °C (maximum). The ambient temperature was recorded using Temperature Data Logger. The temperature data logger used was a 4 channel K Type thermocouple with SD card data logger.

The sun drying experimental set up is shown in figure 2. The spreaded cocoa bean is covered with net to minimize the insect damages, dirt, dust and other pollutants. The purpose of using the net and using basket as their base is to smoothen the air ventilation. The weights of the cocoa bean were recorded every day after 5 pm within 11 days to do the analysis of the weight losses and the moisture ratio.

Figure 2. Conventional sun drying experimental set up.

3.2. Swirling fluidized bed dryer experimental set up

The fluidized bed apparatus is located at Energy and Sustainability Focus Group laboratory, Universiti Malaysia Pahang. In order to scientifically investigate the performance of the distributor configurations and its effect to the bed material, the following requirements are specified such as fluidized bed system that allows interchangeability of the distributor, clear silica glass cylinder to enable visualization of studies, pressure drop measurement and the velocity of the air flow. The fluidization column consists of transparent cylindrical silica glass which has 108 mm internal diameter, 5 mm thickness and 310 mm length as shown in figure 3 and figure 4. An acrylic plate that has the same diameter with the glass column's flange was mounted at the bottom. The air flow was controlled by the controller that controls the speed of the blower between 0 and 50 Hz speed range. Fluidization air was generated from a blower and it was heated to a desirable temperature using an electrical heater. The temperature and humidity sensor was installed at the outlet of the bed to measure the temperature and relatively humidity of the exhaust air.

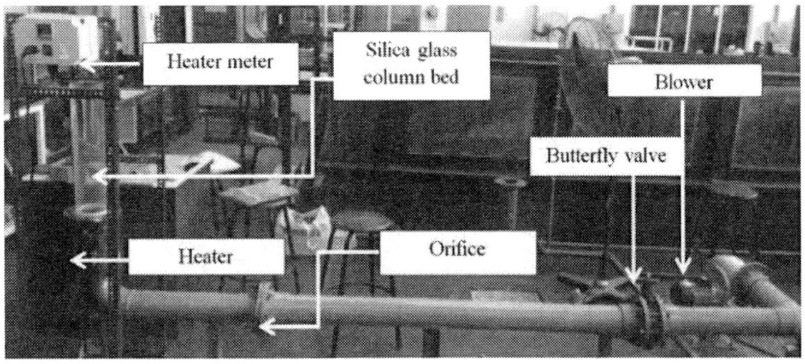

Figure 3. Swirling Fluidized Bed Dryer Set Up.

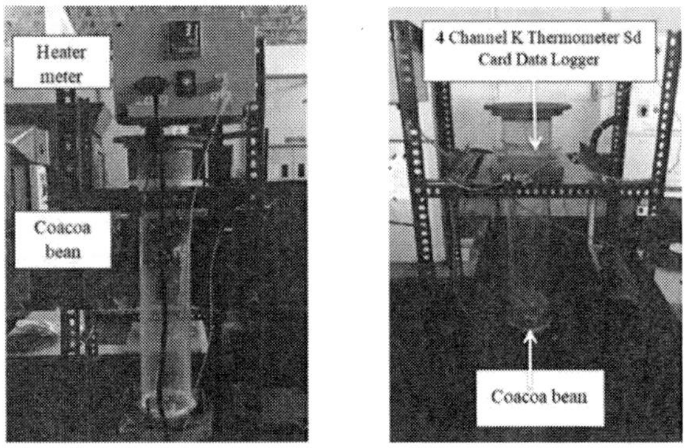

Figure 4. Initial set up before the experiment started.

3.3. Distributor

A design of air distributor made of aluminium with 115 mm diameter and 8 mm thickness was tested in the experiments. The distributor geometry was modified to introduce novel swirling distributor by making the edge opening with 45 inclined air intakes as shown in figure 5.

Figure 5. The 45° distributor that were used in the fluidized bed dryer.

4. Results and discussion

4.1. Conventional drying

The sample of 1.023 kilogram of cocoa bean was dried under direct sunlight within 11 days starting from day 1 to day 11. Based on figure 6 for moisture ratio versus time, the average ambient temperature for 11 days of drying is around 29 to 32 °C. The time taken for cocoa bean to dry completely is until the value of moisture ratio is constant. Table 1 also shows the weight of cocoa bean before and after the drying and the moisture ratio of the cocoa bean in 11 days.

Table 1. The data for conventional drying under direct sunlight for cocoa bean.

Time (Day)	Weight before (g)	Weight after (g)	Moisture ratio
1	1023	667	0.6520
2	667	560	0.5474
3	560	500	0.4888
4	500	497	0.4858
5	497	483	0.4721
6	482	464	0.4536
7	464	451	0.4409
8	451	441	0.4311
9	441	432	0.4223
10	432	424	0.4145
11	424	424	0.4145

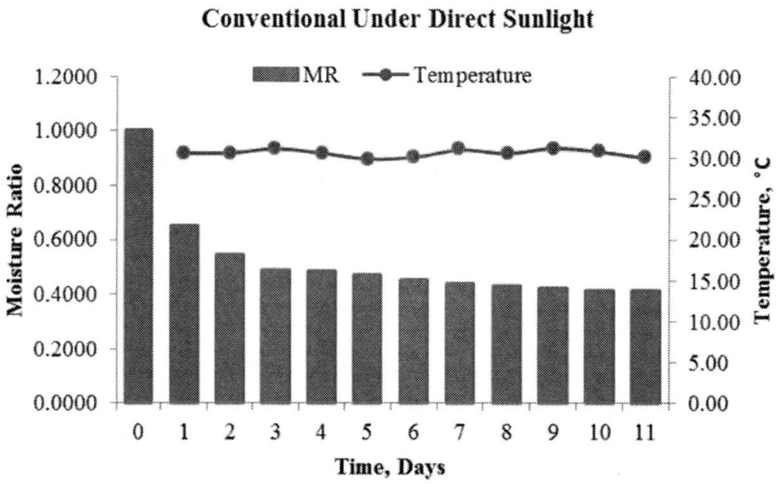

Figure 6. Graph for moisture ratio and temperature versus time (days) of the cocoa bean.

4.2. Swirling fluidized bed drying

A sample of 0.502 kilogram of cocoa bean was dried using Swirling fluidized bed dryer within 2 days. Table 2 tabulates the average temperature for 2 days using the heater of drying is around 34 to 35 °C. Meanwhile, moisture ratio and temperature versus time (days) of the cocoa bean is shown in figure 7. Based on data in figure 7, it shows that the percent of weight loss for swirling fluidized bed in day 1 and day 2 is 42.34% and 59.16%. While using conventional method in day 1 and day 2, the percent of weight loss is 34.8% and 45.26%. It shows that the drying performance by using swirling fluidized bed is better than conventional method. In addition the final moisture ratio for swirling fluidized bed dryer is lower than conventional method which is 0.4083 and 0.4145.

Table 2. The data by using swirling fluidized bed dryer for cocoa bean.

Time (Day)	Weight before (g)	Weight after (g)	Moisture ratio
1	502	290	0.5777
2	290	210	0.4083

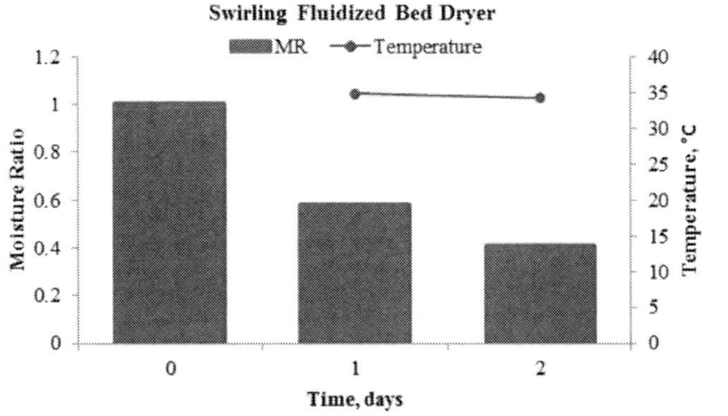

Figure 7. Graph for moisture ratio and temperature versus time (days) of the cocoa bean.

The drying characteristics of cocoa bean have been experimentally studied both by conventional drying method and by swirling fluidized bed dryer. It can be concluded that for conventional method, the drying time red required to dry cocoa bean took too much time which 11 days for it to fully dried because the factor of weather need to be considered in order to conduct this experiment as. However, drying by using fluidized bed dryer only took 2 days to evaporate the moisture content of the cocoa bean. It shows that fluidized bed dryer could save more time compared to conventional method.

This result shown that the performance of fluidized bed dryer in drying the cocoa bean was more efficient compared to conventional method. Furthermore, the visual characteristic of cocoa bean can be observed by using both methods. In short, it is proved that the performance of swirling fluidized bed dryer is more efficient in terms of save time, percentage of moisture content reduced and the quality is maintained compared to conventional method. Hence, the objectives for this project were achieved.

4.3. The comparison between conventional drying and swirling FBD

Figure 8 shows the comparison of moisture ratio by using conventional method and swirling fluidized bed. From the graph, table 1 and table 2, the moisture ratio by using swirling fluidized bed dryer is lower compared to conventional method which is the final moisture ratio by using swirling fluidized bed dryer is 0.4083 and for conventional method the final moisture ratio is 0.4145. The difference of weight loss between these methods within day 1 and day 2 is 7.2% and 12.7% respectively.

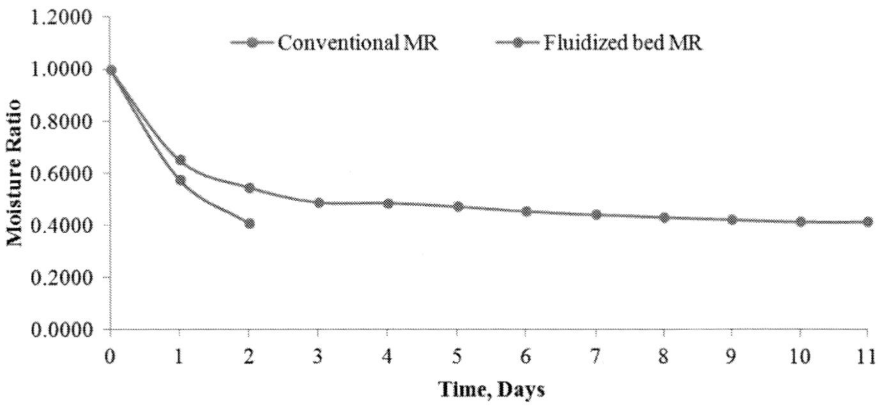

Figure 8. Comparison of moisture ratio by using conventional method and swirling fluidized bed.

5. Conclusion

As conclusion, we find the results by using 1kg sample for conventional method are 11 days drying time, 58.55% weight loss reduction, and 0.4145 moisture ratio. Meanwhile for swirling fluidized bed dryer the time required was 2 days, 59.16% weight loss reduction, and 0.4083 moisture ratio. Swirling fluidized bed drying is proven experimentally to be more efficient than conventional drying for cocoa bean. From the study, it has resulted that the characteristics of cocoa bean by swirling fluidized bed drying is better than the conventional drying. The study can be improved by introducing different design of air distributors which can contribute to low pressure drop and improved particulate mixing in fluidized bed dryer. This swirling fluidized bed drying also can be extended to other agricultural products such as black pepper, etc. which the has characteristics of low in moisture and feasible shape to be tested in the swirling fluidized bed dryer.

Acknowledgments

The authors wish to express their gratitude to the Faculty of Mechanical and Manufacturing Engineering, Universiti Malaysia Pahang (UMP), Pekan, Pahang, Malaysia for providing the research facilities and supporting the research under University Research Grant RDU171110, RDU180320, RDU190383 and Malaysia of Higher Education Research Grant (FRGS/1/2019/TK10/UMP/02/26).

References

[1] Beckett. SI 1994 *Industrial Chocolate Manufacture and Use*, 2nd Ed. (New Delhi: Springer Science)
[2] Huizhen L and Morey V 1984 *Trans. of the ASAE*, **27** pp 581–585.
[3] Aprotosoaie AC, Miron A, Trifan A, Luca VS, Costache II 2016 *Diseases* **4(4)** pp 39.
[4] Taubert D, Roesen R. and Schomig E 2007 *Arch Intern. Med* **167(7)** pp. 626 – 634.
[5] Lasisi D 2014 Int. J. of Eng. Res. & Tech. 3(1) pp. 991–996.
[6] Hii CL, Law CL, Cloke M. and Suzannah S 2009 *Biosystem Eng.* **102(2)** pp 153-161.
[7] Opeke LK 1987 *Tropical tree crops Chihester* (Ibadan Nigeria: Spectrum Books Ltd.)
[8] Wood GAR. and Lass RA 1986 *Cocoa(Tropical Agricultural Series).* (New York: Longman's Scientific and Technical. Co-published with John Wiley and Sons, Inc.).
[9] Ozbey M and Soylemez MS 2005 *Energy Convers. Manag.* **46** pp. 1495–1512.
[10] Codex Alimentarius 2003 *Code of Practice for the Prevention and Reduction of Mycotoxin contamination in cereals* pp. 1–6.
[11] Mikkelsen L 2010 *Quality Assurance along the Primary Processing Chain of Cocoa Beans from Harvesting to Export in Ghana,* Student in Food Science at the University of Copenhagen, Faculty of Life Sciences, Frederiksberg, pp. 2-37.
[12] Pierucci S *et al.*, 2017 *Chem. Eng. Trans.* **57(2012)** pp. 8–13.
[13] Guehi TS, Zahouli IB, Ban-Koffi L, Fae MA and Nemlin JG 2010 *Int. J. Food Sci. Technol.*, **45(8)**, pp. 1564–1571,
[14] Musa NA 2012 *Journal of Engineering and Applied Sciences* **7(2)** pp. 194–197.
[15] Guehi ST *et al.* 2017 *Powder Technol.* **45(8)** pp. 1564–1571.
[16] Oke DO and Omotayo KF 2012 *J. Cereals and Oil seeds* **3(1)** pp. 1–5.
[17] Kaleta A, G´ornicki K, Winiczenko R and Chojnacka A 2013 *Energy Convers. Manag.* **67** p. 179–185.
[18] Meziane S 2011 *Energy Convers. Manag.* **52(3)** pp. 1644–1649.
[19] Gazor HR and Mohsenimanesh A 2010 *Czech J. Food Sci.* **28(6)** pp. 531–537.
[20] Soponronnarit S, Swasdisevi T, Wetchacama S and Wutiwiwatchai W 2001 *J. Stored Prod. Res.* **37(2)** pp. 133–151.
[21] Perea-Flores MJ *et al.*, 2010 *Ind. Crops Prod.* **38(1)** pp. 64–71.
[22] Lozano-Acevedo A *et al.,* 2011 *American J. Potato Res.* **88(4)** pp. 360–366.
[23] Murthy ZVP and Joshi D 2007 *Dry. Technol.* **25(5)** pp. 883– 889.
[24] Senadeera WW, Bhandari BR, Young G and Wijesinghe B 2003 *J. Food Eng.* **58(3)** pp. 277–

283.
[25] Srinivasakananm C and Balasubramanian N 2009 *Adv. Powder Technol.* **20(4)** pp. 390–394.
[26] Sagar VR and Kumar PS 2010 *J. Food Sci. Technol.* **47(1)** pp. 15–26.
[27] ElKhodiry MA *et al.,* 2015 *J. Food Process.* **2015**, pp. 1–10.

IOP Conf. Series: Materials Science and Engineering **863** (2020) 012049 doi:10.1088/1757-899X/863/1/012049

R1234yf vs R134a in automotive air conditioning system: A comparison of the performance

M Z Sharif[1,*], W H Azmi[1,2], N N M Zawawi[1], R Mamat[1] and A H Hamisa[1,3]

[1] Department of Mechanical Engineering, College of Engineering, Universiti Malaysia Pahang, Lebuhraya Tun Razak, 26300 Gambang, Kuantan, Pahang, Malaysia.
[2] Automotive Engineering Centre, Universiti Malaysia Pahang, Pekan, Pahang 26600, Malaysia.
[3] Faculty of Engineering, University College of Yayasan Pahang (UCYP), Taman Gelora, Jln Dato' Abdullah, Kuantan Pahang, 25050, Malaysia.

*E-mail: sharif5865@yahoo.com

Abstract. The main objective of this study is to evaluate the performance of a compact automotive air conditioning (AAC) system operating with R1234yf and R134a refrigerant. The different ranges of initial refrigerant charge and compressor speed have been tested on the AAC to evaluate the effect of different refrigerant on the thermodynamic performance. The results of this study have shown that the cooling capacity of R1234yf AAC system is significantly lower (between 5 to 25 %) than R134a system. In the experiment, the power of the compressor is slightly lower (up to 11 %) for refrigerant R1234yf compared to R134a inside the AAC system. Finally, the experimental results reveal that, under the same operating conditions, the COP of R1234yf AAC system is always lower than with R134a (14.5 % on average). In order to maximize the efficiency and performance of AAC system, further optimization needs to be done in order to reduce the gap in term of performance of the AAC system operating with R1234yf refrigerant.

1. Introduction

In the past, numerous types of refrigerant are being used inside vapor compression system. However, in recent years, researcher has understood that not all refrigerant is created equal. Some of the refrigerant blends is not safe and durable, while some of them contributes to ozone depletion and also global warming [1]. In 1992, the Kyoto protocol which is an international treaty has been held to discuss measures taken to reduce global warming by limiting greenhouse emission. At that time, R134a, R22, and R12 which is widely used in automotive should be phased out as the result of the treaty due to high global warming potential (GWP) [2]. The researcher has done various research to replace these high GWP and ozone depletion potential (ODP) refrigerant [1]. The promising replacement for the phased-out refrigerant is R-1234yf, R-290 (propane), R-600a (isobutane), R-1270 (propylene), R-774 (CO_2), and R-717 (ammonia). All these refrigerants have zero ODP and low GWP.

As such, the R1234yf refrigerant identified as the best alternative to replace the current R134a refrigerant. At present, most of the European car is equipped with R1234yf refrigerant in their automotive air conditioning (AAC) system and is used without restriction. Moreover, the R1234yf refrigerant has thermodynamic properties almost similar to R134a refrigerant. Thus it can straightforwardly be used in the current AAC system without much modification to the original system.

The performance of the vapor compression system employing R1234yf refrigerant has been a debatable subject among researchers in these past several years. From the conducted literature survey, it has been found that several experimental studies have been performed to investigate the potential of direct substitution of R134a refrigerant with the R1234yf refrigerant system with minimum modifications to the existing system.

Several investigation findings showed that R1234yf refrigerant has significant environmental benefit compared to R134a refrigerant in AAC [3]. However, Park, Lee, Choe and Jung [4] reported a reduction of COP up to 2.7 % and the cooling capacity up to 4.0 % in the R1234yf refrigerant system compared to R134a. Hence, to overcome the performance reduction concern, the author proposed a mix ratio of R1234yf/R134a refrigerant. This newly proposed mixture has resulted in the reduction of the compressor discharge temperature up to 6.7 °C while the amount of refrigerant charge is decreased up to 11 %. Chen, Zhao and Qi [5] performed extensive experimental studies of R1234yf refrigerant MAC system at the different condition of driving mode. The studies show even under the various driving mode conditions, the system with R1234yf refrigerant proved to give out a lower performance of COP and cooling capacity compared to R134a system. Hence, further optimization and new energy-saving technologies are required to overcome this concern as suggested by the author. Later, Lee and Jung [6] conducted an experimental analysis to compare the performance of R1234yf and R134a in a MAC bench tester system integrated with the heat pump. The performance of these systems in summer and winter conditions using R134a and R1234yf refrigerant was done on an open type compressor for MACs bench tester. Interestingly, even though the experiment shows the reduction in cooling capacity and COP of the system with R1234yf compared to R134a refrigerant, the amount of refrigerant charge is 10 % lower in the R1234yf. This is because R1234yf has a lower density, so the use of the refrigerant can be saved [7]. Also, the compressor discharge temperature is 6.5 °C lower than those of the R134a system.

To date, experimental data for AAC systems using R1234yf are still limited. This paper is intended to update the knowledge regarding the behaviour of R1234yf inside the AAC system for the compact car. The experimental test setup used in this investigation used the real component from vehicles. In this paper, performance analysis of an AAC system using R134a and R1234yf is carried out experimentally. A comparative study of AAC performance will be performed for refrigerants R134a and R1234yf. The study will be conducted to analyse the effect of initial refrigerant charge and compressor speed on refrigerant mas flow rate, cooling performance, compressor performance, and Coefficient of Performance (COP).

2. Methodology

2.1. AAC experimental setup

The AAC test setup is build based on the principle of vapor compression refrigeration system. The four main components of the AAC system are compressors, condensers, expansion valves and evaporators (cooling coils). In the air-conditioning system, the primary working fluid used in the system is the refrigerant and compressor lubricants. This experimental setup was developed for the R134a as the refrigerant. For this experiment, the performance of the AAC system using refrigerants R134a and R1234yf will be compared. The role of the lubricant in the compressor is to protect and reduce the friction between two surfaces that slide together [8]. The PAG ND12 lubricant is used as a lubricant compressor as recommended by the manufacturer. All major components are taken from real vehicles including pipes and expansion valves to ensure the authenticity of the experiment. The main driver of the AAC system which is the internal combustion engine has been replaced with an electric induction motor to drive the compressor. Figure 1 shows the schematic diagram of the experimental test setup. The experimental test setup is connected with the water-cooling system where the evaporator is put inside the insulated calometric water bath. The circulation of the water inside the calometric will act as the heating load to the evaporator of the AAC system. The difference of water inlet and the outlet is used to calculate refrigeration mass flow rate and cooling capacity in accordance to the ASHRAE standard of Standard 41.9-2000: Calorimeter Test Methods for Mass Flow Measurements of Volatile

Refrigerants. The experimental setup will be placed in a special room with constant ambient temperature, which 25 °C with an increment of ± 0.1 °C and constant humidity.

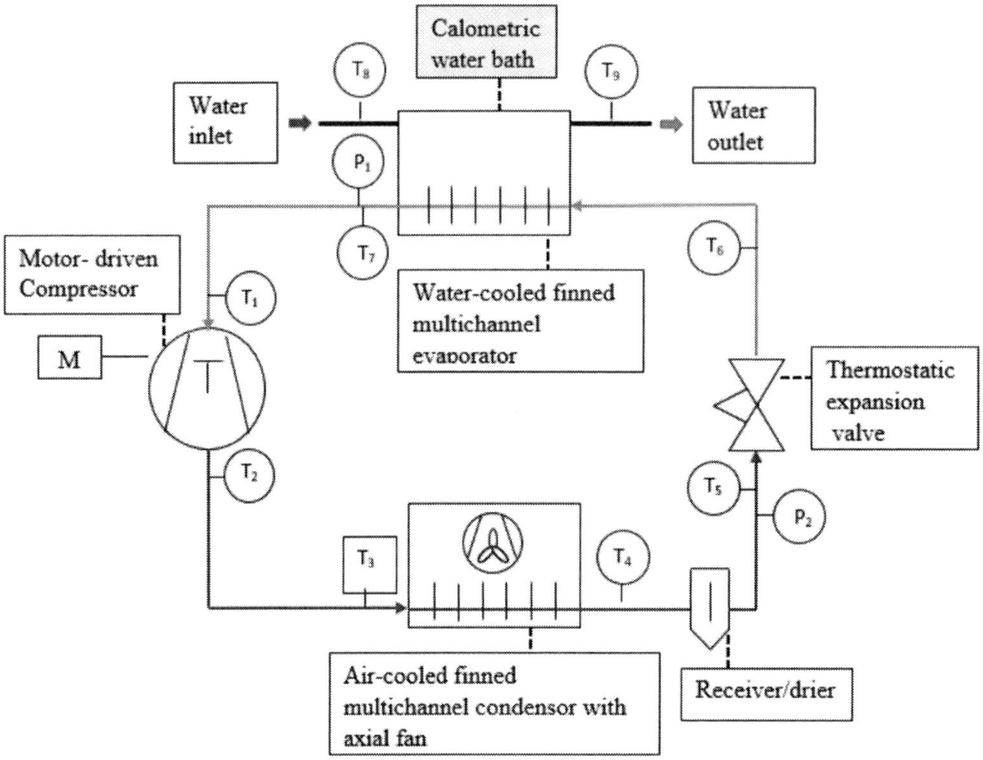

Figure 1. Schematic diagram of Automotive Air Conditioning experimental set-up.

2.2. Instrumentation and uncertainties

Multiple instruments and sensors are used for measurement purposes in AAC test rig investigations under various working conditions. In the AAC system investigation, important parameters such as pressure and temperature will be recorded. The location of all sensors are shown in figure 1 (Section 2.1). The measurement of temperature and pressure will be taken using T-type thermocouple and pressure sensor. The accuracy of the measuring instrument is illustrated in table 1. Uncertainty for each measuring device is summarized in the table. The temperature and pressure readings will be retrieved and stored in PC using data acquisition tool for analysis purposes.

Table 1. The summary for the uncertainties of the experimental parameters.

Parameters	Full scale	Uncertainty
K-type thermocouples, K	233.15 to 648.15	±1.5
pressure gauge, psi	0- 200	±0.1
Tachometer, rpm	0-20,000	±2
water flow meter, LPM	0-100	±0.1
Weighing scale, kg	0-25	±0.001

2.3. Performance analysis of the AAC system

The vapor compression refrigeration system (VCRS) in the air conditioning system works by absorbing heat from refrigerated space and releasing to the surrounding. The compressor will compress the refrigerant to high-pressured gas. Then, the condenser will condense this high-pressure gas refrigerant to high-pressure saturated liquid refrigerant. Later, the expansion valve will change the refrigerant pressure via the throttling process. The pressure reduction process changes the state of the refrigerant from liquid to the gas in which the refrigerant starts to boil. The boiling process happens inside the evaporator and this condition causes the refrigerant to absorb a lot of heat from the cooling medium. The VCRS process include an isentropic compression (reversible adiabatic) in the compressor, heat rejection of constant pressure in the condenser, extension (isenthalpic) expansion in the capillary tubes and the constant pressure heat absorption in the evaporator.

The process of heat absorbs from the cooling medium via the evaporator is calculated from the enthalpy difference of the evaporator inlet and outlet. The heat absorb in the evaporator (Q_L) can be calculated as:

$$Q_L = h_1 - h_5 \tag{1}$$

The cooling capacity in the evaporator is the product of the enthalpy difference and the refrigerant mass flow which is calculated as:

$$\dot{Q}_L = \dot{m}_r(h_1 - h_5) \tag{2}$$

The refrigerant mass flow rate, $\dot{m}_r$ are obtained by using the following relation [9, 10]:

$$\dot{m}_r = \frac{\dot{Q}_{water}}{(h_1 - h_5)} \tag{3}$$

where,

$$\dot{Q}_{water} = \dot{m}_{water}(h_8 - h_9) \tag{4}$$

The indicated work in the compressor is calculated from the enthalpy difference of the compressor inlet and outlet [11]. The compressor work (W_{IN}) can be calculated as:

$$W_{IN} = \dot{m}_r(h_2 - h_1) \tag{5}$$

The indicated power in the compressor is the product of the enthalpy difference and the refrigerant mass flow which is calculated as:

$$\dot{W}_{IN} = \dot{m}_r(h_2 - h_1) \tag{6}$$

The coefficient of performance (COP), which is the ratio of the cooling capacity and the compressor power input which is expressed by:

$$COP = \frac{\dot{m}_r(h_1 - h_5)}{\dot{m}_r(h_2 - h_1)} \tag{7}$$

2.4. Experimental procedures

Before starting the experiment, all the conditions of the equipment used and the conditions of the test setup must be checked and ensured that all safety precautions were followed before the experiment started. The experimental procedure of the AAC performance investigation will be conducted by following the regulations and recommendations from the standard of SAEJ2765 [12]. The AAC experimental test set-up was vacuum first by using a vacuum pump to remove the moisture and to determine if there are leaks in the system. The compressor lubricant is filled into the compressor with 110 ml. The compressor lubricant used in this experiment is PAG ND12 which specially formulated for use in R1234yf and R134a AAC system. According to the manufacturer, PAG ND12 are specially formulated to solve several issues regarding the solubility and the higher moisture content of R1234yf refrigerant. The charging machine will be utilized to charge the refrigerant R1234yf and R134a into the AAC system. The desired amount of refrigerant charge is charged into the system. The refrigerant tank will be weighed on a weighing scale to determine the amount of refrigerant charge. The water in the calorimetric water tank was heated up until the temperatures for the inlets and outlets were the same. Then, the experiment started by starting the induction motor with a speed of 900 rpm, which is adjusted by the frequency inverter. The experiment was kept running for 20 minutes and the data reading of temperatures, pressure, power analyzers, and water mass flow rates were taken for 10 minutes after that. The data will be recorded and analyzed. The variables of the following parameters were used to evaluate the response of the AAC system performance which is the compressor speed (n) and initial refrigerant charge (m_{rc}).

3. Results and discussion

3.1. Refrigerant mass flow rate performance of R1234yf vs R134a

Figure 2 presents the experimental data of the refrigerant mass flow rate as a function of the initial refrigerant charge for both R134a and R1234yf AAC system at a various compressor rotational speed. From the data in figure 2, it is apparent that the refrigerant mass flow rate increases with the increment of the initial refrigerant charge. In addition, the compressor speed has a clear effect on the refrigerant mass flow rate. The mass flow rate increases with the increasing compressor speed. These findings further support the theoretical predictions from previous research [13], where they also reported the same trend for the refrigerant mass flow rate in the vapor compression system experiment. A comparison of the two results reveals that the mass flow rate values for R1234yf are up to 12 % higher compared R134a, especially at a higher initial refrigerant charge. The higher refrigerant mass flow rate in R1234yf AAC system is related to the refrigeration system pressure. The compression ratio in inlet and outlet through the compressor for AAC system with R1234yf is smaller than the AAC system with R134a. Therefore, the number of refrigerant specific volume in the compressor inlet is small under the same saturated temperature, thus more refrigerant is released by the compressor in order to achieve equivalent cooling performance [14].

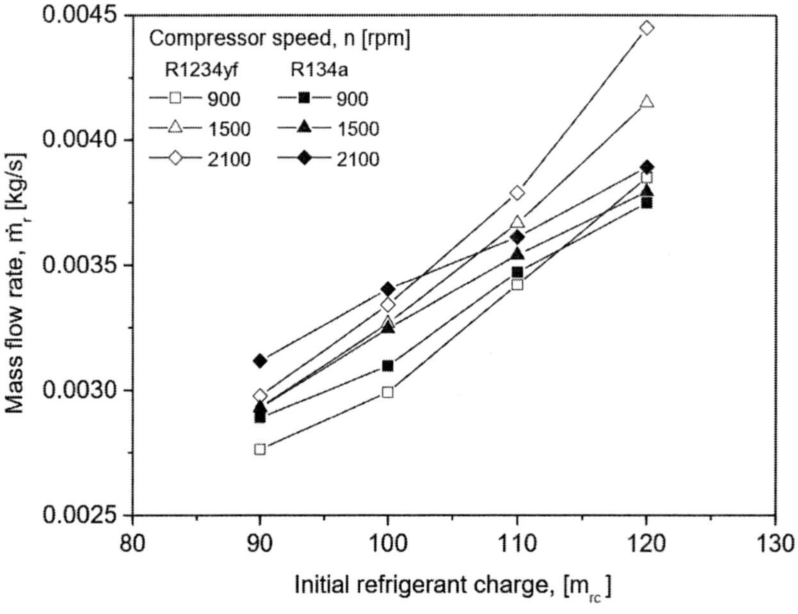

Figure 2. Variation of refrigerant flow rate against various initial refrigerant charge.

3.2. Cooling performance of R1234yf vs R134a

Figure 3 present the variation of heat absorb against the initial refrigerant charge for different compressor speed. The heat energy differences for the evaporator inlet and outlet show decreasing trend with the initial refrigerant charge and the compressor speed for the both refrigerant. The compressor of the AAC system with a lower initial refrigerant charge undergo starvation due to the lack amount of refrigerant. Due to this condition, the AAC system with an expansion valve system increase the superheat as a feedback mechanism to satisfy the same cooling requirement. Interestingly, the AAC system with R1234yf has a significantly lower heat absorb as compared to R134a AAC system. The possible explanation of this result is related to the properties of latent heat of vaporization. The R1234yf have a slightly lower latent heat of vaporization thus it can absorb less heat energy compared to R134a [15].

Figure 4 shows the experimental results of the cooling capacity against the function of the initial refrigerant charge for refrigerants R134a and R1234yf at various compressor rotational speed. As shown in figure 4, the experimental results show the cooling capacity is in the increasing trend with the initial refrigerant charge. For any initial refrigerant charge, the cooling capacity increases with the compressor speed for both refrigerants. Contrary to the finding in figure 3, the cooling capacity show increment with initial refrigerant charge and compressor speed despite having a reduction in heat absorb. This is due to the mass flow rate, which is shown in figure 2, where the increment in the refrigerant mass flow rate overweight the effect of the evaporator heat absorb reduction. It is also seen that for any refrigerant charge and compressor speed value, the cooling capacity of the R134a AAC system has a significantly higher cooling capacity (from 5 to 25 %) than the R1234yf AAC system. This result may be due to the higher heat absorb value as shown in figure 3 which help the evaporator absorb more heat and provides a better cooling effect in the R134a AAC system.

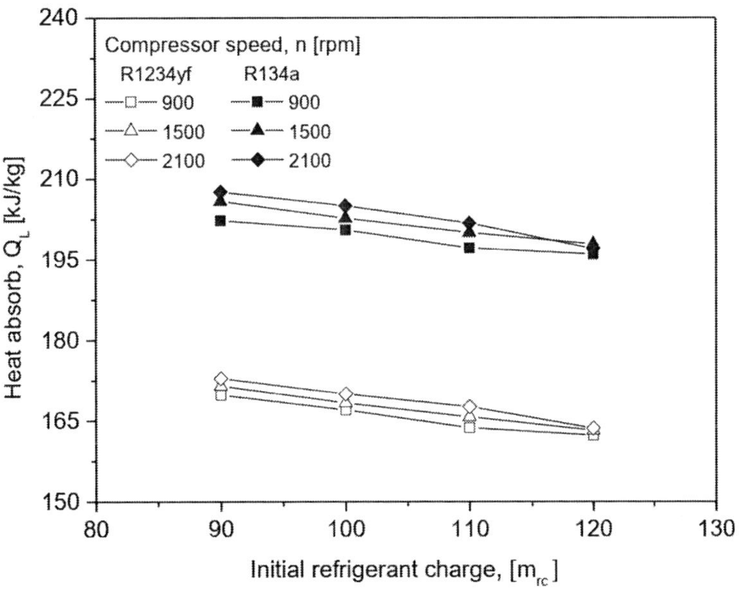

Figure 3. Heat absorb per unit mass of R1234yf and R134a AAC system as a function of initial refrigerant charge.

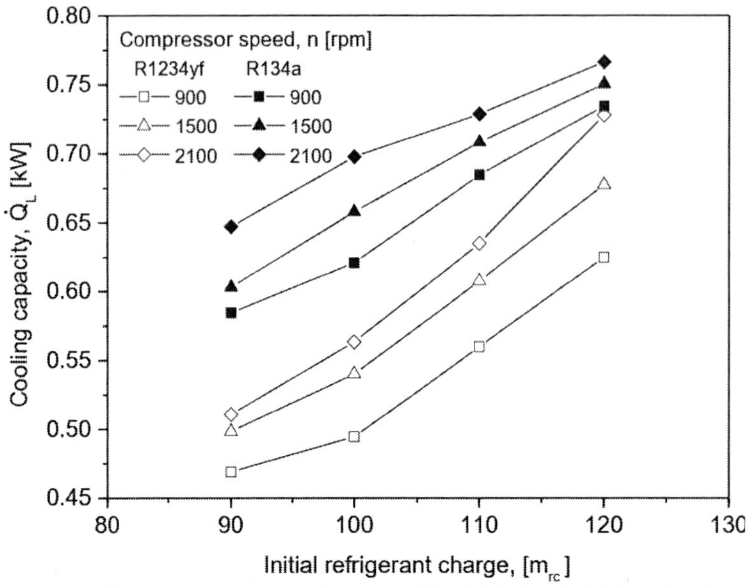

Figure 4. Cooling capacity per unit mass of R1234yf and R134a AAC system as a function of initial refrigerant charge.

3.3. Compressor performance of R1234yf vs R134a

The indicated compressor work and compressor powers for the same set of experiments are shown in figure 5 and figure 6, respectively. Figure 5 shows the compressor work is in an increasing trend with the compressor speed, for both refrigerants. The compressor at high speeds tends to work much harder than at low speeds, and subsequently reduces its efficiency. However, the compressor work shows no obvious changes with the increment of the initial refrigerant charge. It is interesting to note that the R1234yf AAC system requires lower compressor work compared to R134a AAC system. On the compressor indicated power data analysis of the AAC experiment, this study found that compressor indicated powers increase with increasing values the initial refrigerant charge and compressor speed as shown in figure 6. Previously, theoretical analysis from the previous study has already shown that the refrigerant cycle predicts the isentropic compressor power to increase with the initial refrigerant charge and compressor speed [13]. The increment of the compressor power is due to the rise in the refrigerant mass flow rate, which is shown in figure 2. Finally, a comparison between the experimental results for R134a and R1234yf show a clear effect on the compressor indicated power. The experimental values for the compressor indicated power is slightly lower (up to 11 %) for the AAC system with R1234yf compared to R134a. As noted in another study, R1234yf AAC system has a lower compressor ratio due to the influence if the thermodynamic characteristic thus the compressor work and compressor power are lower compare to R1234yf AAC system [14].

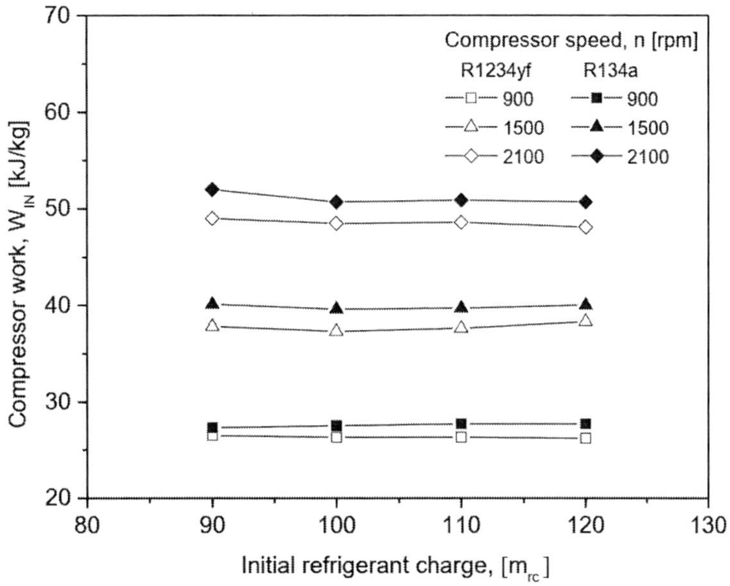

Figure 5. Compressor per unit mass of R1234yf and R134a AAC system as a function of initial refrigerant charge.

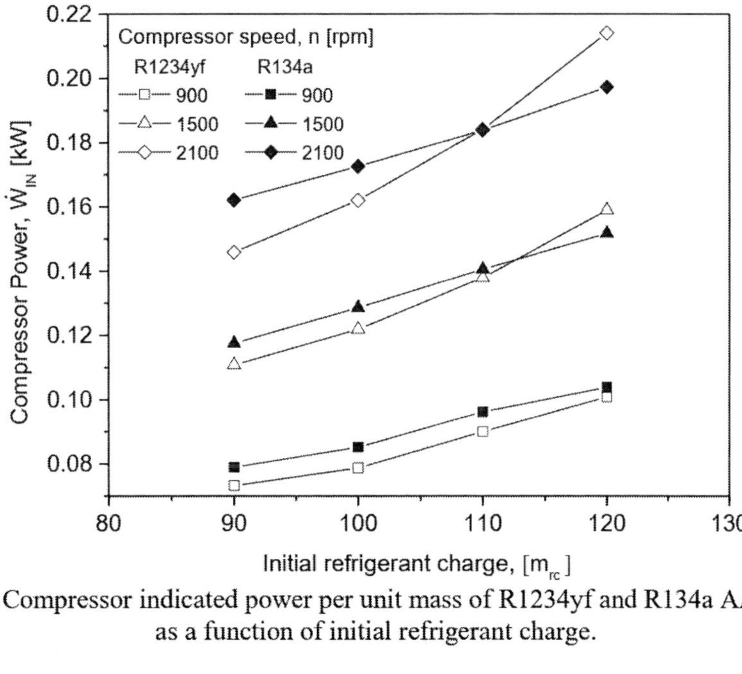

Figure 6. Compressor indicated power per unit mass of R1234yf and R134a AAC system as a function of initial refrigerant charge.

3.4. Coefficient of performance of R1234yf vs R134a

Figure 7 compares the experimental data of COP for the same cases considered in the previous figures. From the graph, there are no obvious changes in COP increment of the initial refrigerant charge. As expected, there is a decrement trend in the COP value with the increasing compressor speed. This is due to the high compressor work and compressor power (as shown in figure 5 and figure 6) with the compressor speed, that outweighs the increment of the heat absorb and cooling capacity in COP calculation. Finally, the experimental results reveal that, under the same operating conditions, the COP of R1234yf AAC system is always lower than with R134a (14.5 % on average). The findings of the current study are consistent with the previous study where the author found the same finding regarding the trends of COP inside the refrigeration system operating with R1234yf [6, 13, 14].

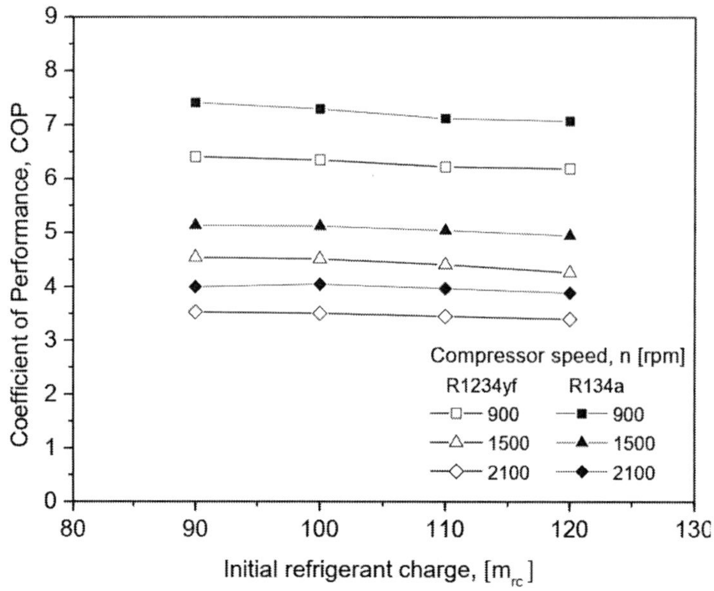

Figure 7. COP of R1234yf and R134a AAC system as a function of initial refrigerant charge.

4. Conclusion

The performance of the AAC system operating with R1234yf and R134a refrigerant were analysed and compared. The study was carried out by analysing the effect of initial refrigerant charge and compressor speed on the heat absorb, cooling capacity, compressor work, compressor power, and coefficient of performance. In this study, there is a notable difference in term of performance between R1234yf and R134a operating inside AAC system. A comparison of the two results reveals the mass flow rate values for refrigerant R1234yf are up to 12 % higher than the R134a at a higher initial refrigerant charge. The AAC system with the R1234yf has a significantly lower heat absorb as compared to the R134a AAC system. It is also noticed that for any refrigerant charge and compressor speed value, the cooling capacity of the R1234yf AAC system has a significantly lower cooling capacity (from 5 to 25 %) than the R134a system. However, the compressor work shows no obvious changes with the increment of the initial refrigerant charge. Interestingly, the experimental values for the compressor indicated power is slightly lower (up to 11 %) for the R1234yf compared to the R134a operating inside the AAC system. Finally, the experimental results reveal that, under the same operating conditions, the COP of the R1234yf AAC system is always lower (14.5 % on average) compared to the R134a AAC system. Nevertheless, a detailed optimization procedure is recommended in order to reduce the gap in term of performance of the AAC system operating with R1234yf refrigerant.

Acknowledgments

The authors are grateful to the Universiti Malaysia Pahang (UMP) for financial supports given under PGRS190335.

References

[1] Calm J M 2008 The next generation of refrigerants–Historical review, considerations, and outlook *international Journal of Refrigeration* **31** 1123-33

[2] Manoj V, Balamurugan V and Kumar B S 2016 A Review on selecting an eco-friendly refrigerant alternate to R134a in domestic refrigerators *International Journal of Science, Engineering and Technology Research* **5** 2992-9

[3] Koban M 2009 HFO-1234yf low GWP refrigerant LCCP analysis *SAE Technical Paper* **2009** 1-5

[4] Park K-J, Lee Y-H, Choe D-S and Jung D-S 2010 Performance of R1234yf and R1234yf/R134a mixture under mobile air-conditioner operating conditions *Korean Journal of Air-Conditioning and Refrigeration Engineering* **22** 837-44

[5] Chen J, Zhao Y and Qi Z 2011 New developments in mobile air conditioning systems in China *Frontiers of Energy and Power Engineering in China* **5** 53-8

[6] Lee Y and Jung D 2012 A brief performance comparison of R1234yf and R134a in a bench tester for automobile applications *Applied Thermal Engineering* **35** 240-2

[7] Minor B and Spatz M 2008 HFO-1234yf low GWP refrigerant update

[8] Zawawi N N M, Azmi W H, Redhwan A A M, Sharif M Z and Samykano M 2018 Experimental investigation on thermo-physical properties of metal oxide composite nanolubricants *International Journal of Refrigeration* **89** 11-21

[9] Sharif M Z, Azmi W H, Zawawi N N M, Mamat R and Shaiful A I M 2019 Energy and exergy analysis of compact automotive air conditioning (AAC) system. In: *IOP Conference Series: Materials Science and Engineering*: IOP Publishing) p 012042

[10] Redhwan A A M, Azmi W H, Sharif M Z, Mamat R, Samykano M and Najafi G 2019 Performance improvement in mobile air conditioning system using Al 2 O 3/PAG nanolubricant *Journal of Thermal Analysis and Calorimetry* **135** 1299-310

[11] Redhwan A A M, Azmi W H, Sharif M Z and Hagos F Y 2017 Development of nanolubricant automotive air conditioning (AAC) test rig. In: *MATEC web of conferences*: EDP Sciences) p 01050

[12] SAEJ2765 S I S 2008 SAE J2765 - Procedure for Measuring System COP [Coefficient of Performance] of a Mobile Air Conditioning System on a Test Bench. SAE internationals) p 20

[13] Sieres J and Santos J M 2018 Experimental analysis of R1234yf as a drop-in replacement for R134a in a small power refrigerating system *International Journal of Refrigeration* **91** 230-8

[14] Cho H, Lee H and Park C 2013 Performance characteristics of an automobile air conditioning system with internal heat exchanger using refrigerant R1234yf *Applied Thermal Engineering* **61** 563-9

[15] Wang C-C 2014 System performance of R-1234yf refrigerant in air-conditioning and heat pump system–An overview of current status *Applied Thermal Engineering* **73** 1412-20

IOP Conf. Series: Materials Science and Engineering **863** (2020) 012050 doi:10.1088/1757-899X/863/1/012050

The stability of TiO_2/POE nanolubricant for automotive air-conditioning system of hybrid electric vehicles

A H Hamisa[1,4], T M Yusof[1,2], W H Azmi[1,2,3,*], R Mamat[1,2,3] and M Z Sharif[1]

[1] Department of Mechanical Engineering, College of Engineering, Universiti Malaysia Pahang, Lebuhraya Tun Razak, 26300 Gambang, Kuantan, Pahang, Malaysia.
[2] Automotive Engineering Centre, Universiti Malaysia Pahang, 26600 Pekan, Pahang, Malaysia.
[3] Centre of Excellence for Advanced Research in Fluid Flow, Lebuhraya Tun Razak, 26300 Kuantan, Pahang, Malaysia.
[4] School of Engineering, Faculty of Science and Technology, University College of Yayasan Pahang (UCYP), Taman Gelora, Jln Dato' Abdullah, Kuantan 25050 Pahang, Malaysia.

*E-mail: wanazmi2010@gmail.com

Abstract. Nanolubricant are containing nanoparticles that are mixed with the base lubricant. Nanoparticle consists of ultra-fine size of particles from 1 to 100 nm. It was widely used as additional material to enhance properties of lubricant. The thermo-physical properties of the nanolubricant enhanced and thus able to improve the performance of vapour compression refrigeration system (VCRS) and automotive air-conditioning (AAC) system. The stability condition is one of the priority in formulation of nanolubricant for new application in AAC system of hybrid electric vehicles (HEV). The main objective of this study is to investigate the stability of TiO_2 nanoparticles dispersed in Polyol-Ester (POE) lubricant. The TiO_2/POE nanolubricant was prepared at volume concentration of 0.01 to 0.1% using the two-step method without any surfactant. The stability investigations were conducted by using visual sedimentation observation, micrograph observation, UV-Vis Spectrophotometer measurement and zeta potential measurement. The findings by visual sedimentation observation showed the best stability condition for more than 14 days at 0.01% and 0.03% volume concentration. Meanwhile the optimum sonication time is observed at 5 hours by visual and supported by UV-Vis evaluation. Further, the concentration ratio from UV-Vis evaluation was recorded above 95% for 5 hours sonication time and more than 30 days observation. Finally, the Zeta potential for the present nanolubricant was measured with 81.1 mV and obtained within the range of very good stability condition. Hence, this can confirm a good stability condition for the present TiO_2/POE nanolubricant. Further investigation is required for the properties evaluation and performance of the nanolubricant in AAC system of HEV.

1. Introduction

Nanoparticle was widely used as value added in producing new materials. Nanoparticles are also increasingly widespread in the production of enhanced material properties. Nanoparticle is a kind of material that has a very small and fine particle size of less than 100 nm. It has a variety of properties that can be applied to the production of materials to be used on various applications. Dispersion of nanoparticle is used in sectors such as biomedical, health care, food agriculture, industrial, electronics, environment, renewable energy and textile [1, 2]. Application of nanoparticle in liquid can be divided into two types that were commonly used in industry. There are nanolubricant and nanofluid and others type is nanorefrigerant in application of refrigerator or air-conditioning system. Choi et al. [3] has applied the first concept of nanoparticles in a base fluid called nanofluids. Thermo-physical properties of the base fluid can be increased and it was proven by using this method.

The element of thermo-physical properties like thermal conductivity, viscosity, density and stability have been one of the main criteria to investigate since many years ago by many researchers [4-8]. Nanolubricant stability is one of the first elements need to be conducted before proceed to the next step. The stability investigation also has been done by the previous researches [9-11]. The application of nanoparticles in lubricants is widely been used in the automotive industry to improve the AAC system performance. As technology rapidly developed, nanolubricant technology has evolved from time to time. The first generation of nanolubricant was more focused on one type of nanoparticle which is mixed into the base lubricant. Sharif et al. [12, 13] and Redhwan et al. [14] investigated the effect of Al_2O_3 and SiO_2 nanoparticles additives into PAG lubricant. While Krishnan et al. [15] and Nabil et al. [16] reviewed that nanoparticles can enhance the performance in vapour compression refrigeration system.

Operation of AAC system is one of the most significant factors affecting real-world fuel consumption [17]. Fuel consumption is associated to the AAC system by up to 27% [18]. Dahlan et al. [19] have performed the study about performance of electric compressor in non-electric vehicle. This researcher also give opinion that the DC compressor has high potential to be implemented to the conventional non-electric vehicle to have lower fuel consumption because of lower load with better performance. Other researcher conclude that using the direct current compressor has high possibility in reducing the load of engine system and result the higher coefficient of performance and lowering the fuel consumption. The result shows about 15% to 54% increment of electrically driven compressor (EDC) compressor performance comparing with belt-driven compressor. It also cause the fuel reduction percentage is about 5% to 14% [20].) The novelty of the SiO_2/PAG nanolubricant in term of energy saving have been proven by the findings of the compressor work reduced up to 16.5% and the reduction in the compressor directly associated with the energy saving characteristic of the vapour compression system [21].

It is essential to find and innovative AAC solution to HEVs or EVs [22]. Due to that, many studies and approaches have been carried out to find out how to improve and enhance the performance and affect the energy-saving from the AAC system especially in new technology of HEV. In this present study, this nanolubricant will be use in hybrid electric vehicle (HEV) air-conditioning system. The compressor in AAC system for HEV is the most different component compared to conventional AAC system. Common vehicle use belting driven compressor and HEV used electrically driven compressor (EDC). The lubricant inside the compressor is also different in terms of the properties. PAG lubricant is primarily used in common vehicle AAC system while POE lubricant commonly is used in HEV AAC system. Therefore, the goal of this study is to investigate the stability of the TiO_2/POE nanolubricant. An experimental study will be conducting through visual sedimentation observation, UV-Vis Spectrophotometer and zeta potential to find the best processing concentration and sonication time of TiO_2/POE nanolubricant. After that, HEV AAC system performance can be proceeded for further investigations.

2. Methodology

2.1. Material properties

In producing a sample of the nanolubricant, it involved two main types of materials namely TiO_2 nanoparticle and lubricant base type Polyol-Ester (POE) RL68H. This TiO_2 nanoparticle will be dispersed in the base lubricant of POE RL68H. The TiO_2 nanoparticle purchased from US Research Nanomaterials, Inc. The size is 50 nm with a purity of 99%. POE has been widely and successfully used with electric compressors for over 20 years and its good dielectric properties make it the popular choice for electric compressors in hybrid cars and other vehicles with electric compressors. The details properties of nanoparticle are given in table 1 and the properties of Polyol-Ester (POE) RL68H are presented in table 2. The nanoparticle size characterization of the TiO_2/POE nanolubricant by field scanning electron microscope (FESEM) technique carried out using JEOL JSM 7800F machine. The FESEM image for nanoparticle is shown in figure 1.

Table 1. Properties of TiO_2 nanoparticle [10, 23].

Properties	TiO_2
Thermal Conductivity (W m^{-1} K^{-1})	8.4
Specific heat/J kg^{-1} K^{-1}	692
Density, kg.m^{-3} @ 293 K	4230
Molecular mass, g mol-1	79.86
Average particle diameter, nm	50

Table 2. Properties of POE RL68H lubricant [24].

Properties	POE RL68H
Viscosity @ 40◦C (cSt)	66.6
Viscosity @ 100◦C (cSt)	9.4
Pour Point (◦C)	-39
Density @ 20◦C (g/ml)	0.977
Flash Point COC (◦C)	270

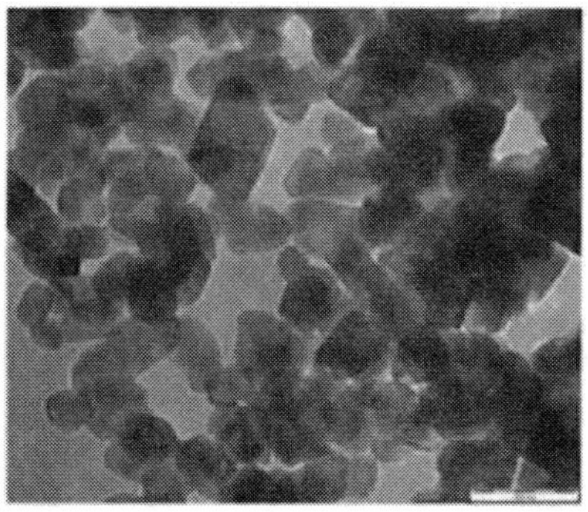

Figure 1. TEM Image for TiO_2 nanoparticles at 100 nm scale [10, 23].

2.2. Preparation of TiO_2/POE nanolubricants

The process of two step-method have been used in preparing the nanolubricant as recommended [25]. It also been used by the previous studies [26, 27]. TiO_2 nanoparticle and POE base lubricant have been mixed. The preparation of the nanolubricant starts by measuring the mass of TiO_2 nanoparticles. Equation (1) [26, 28] has been used to determine the volume concentration of the nanolubricant required.

$$\emptyset = \frac{m_p/\rho_p}{m_p/\rho_p + m_L/\rho_L} \text{ X } 100 \quad (1)$$

The nanoparticle was filtered with strainer in the process of weighing the nanoparticle. Then, the nanoparticle dispersed in POE and stirred using magnetic stirrer about half an hour and undergo with sonication. The sonication process using the ultrasonic bath Fisherbrand (model: FB15015) to enhance the stability. This nanolubricant was prepared at a volume concentration of 0.01, 0.03, 0.05, 0.07 and 0.1%. The nanolubricant was first prepared at the highest concentration followed to the lower concentration.

2.3. The stability of nanolubricant

The stability studies need to be conducted using qualitative and quantitative experiments. Qualitative experiment involved visual observation of sedimentation. This observation was carried out at least for 14 days [23, 26]. The stability of the nanolubricant considered stable when the concentration is constant [29]. Then, the quantitative experiments involved the UV-Vis Spectrophotometer measurement and zeta potential measurement was conducted. The UV–Vis was used to determine the peak absorption wavelength and absorbance value of the sample by comparing intensity level with the base lubricant and zeta potential is the potential difference between the dispersion medium and the stationary layer of fluid attached to the particles [30]. The UV-Vis measurement was conducted for at least 14 days (340 hours) by varying the sonication time. The wavelength of the UV-Vis Spectrophotometer is set at 530 nm after the nanolubricant was scanned to determine the peak absorption wavelength for 0.01% of concentration. The stability evaluation by UV-Vis was also used by previous studies by Sharif et al. [21] and Yu & Zie [25]. After that, the measurement of zeta potential was conducted using Anton-Paar Litesizer 500 to support and prove the stability of the nanolubricant.

3. Results and discussion

3.1. Analysis method with UV-Vis Spectrophotometer

The observation of absorbance for volume concentration from 0.01 to 0.1% is shown in figure 2. The linear relationship between absorbance, A and concentration, ϕ was found and proven to follow the linear relation of the Beer-Lambert law. Beer-Lambert law is about a linear relationship between the absorbance intensity and the nanolubricant concentration [21].

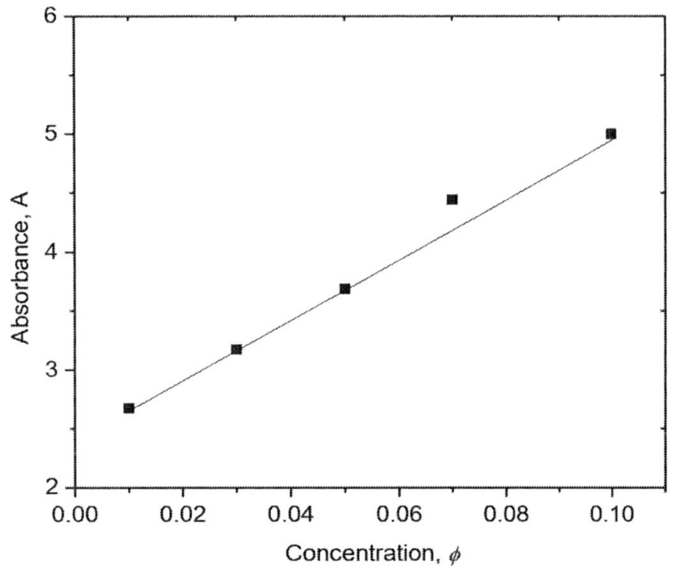

Figure 2. UV-Vis Spectrophotometer linear relation graph between absorbance and TiO_2/POE nanolubricant concentration.

Volume concentration of 0.01% of the nanolubricant for five different sonication hours was demonstrated in figure 3. The ideal absorbance ratio is one (100%) which present the ideal stability of the nanolubricant [31]. From the findings, all concentrations ranged from 0 to 7 hours showed that it started to decrease after 24 hours. It showed a downward trend of 0 hour sonication, and continue up to 15 days. Meanwhile, after 2 days (48 hours), concentration of 1, 3, 5 and 7 hours remain stable until 31 days (744 hours). The concentration ratio for 1, 3, 5 and 7 hours were kept above 95% after 30 days although its decrease a little bit day by day. From this figure, it was found that 5 hours sonication time indicated the best absorbance ratio compared to others sonication time. Therefore, to produce this TiO_2/POE nanolubricant, the 5 hours sonication time was the best selection.

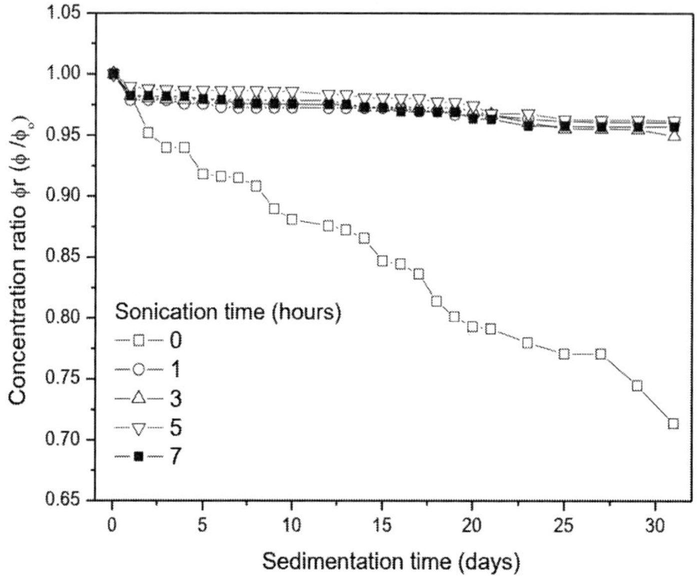

Figure 3. The concentration ratio of TiO_2/POE for different sonication time as a function of time.

3.2. Concentration ratio comparison with literature

The comparison of the current study among the best nanolubricant stability from previous study showed in figure 4 [21, 29]. SiO_2/PAG nanolubricant gave the best absorbance ratio about 70% in 2 weeks time with the 2 hours of sonication time and 0.2% concentration [29] and Al_2O_3/PAG nanolubricant give the best absorbance ratio about 90% also in 2 weeks time with 1.5 hours of sonication time with 0.2% of concentration [21]. Based on this comparison, it can be concluded that the stability for the TiO_2/POE nanolubricant shows a better and stable absorbance ratio in same range of 2 weeks time.

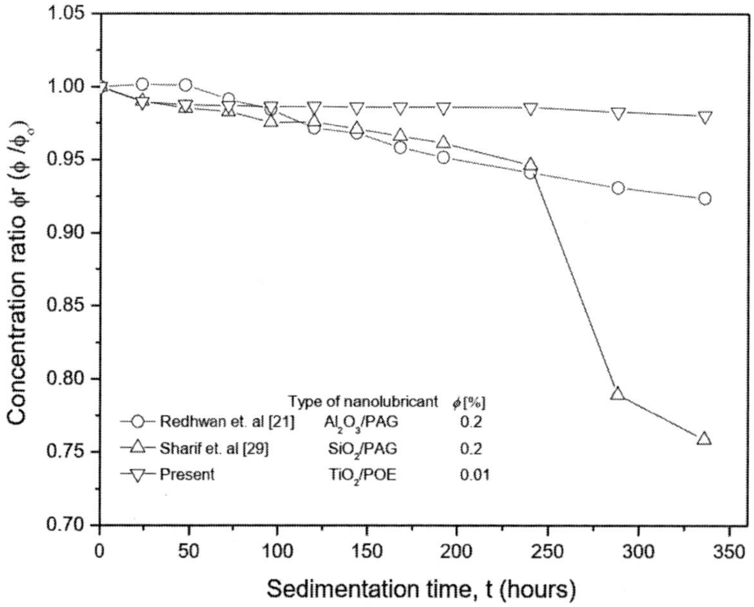

Figure 4. Comparison of concentration ratio with sedimentation ratio with literature.

3.3. Zeta Potential evaluation

Zeta potential is another quantitative experiment to evaluate and support the stability of the nanolubricant. It validate the quality of nanofluid or nanolubricant stability through a study of its electrophoretic behaviour [25]. In this case, TiO_2/POE nanolubricant has been tested using the Anton-Paar Litesizer 500. The higher of absolute value, the better dispersion of the particles into the base lubricant shows the better stability [21]. Table 3 show the zeta potential and the classification of stability.

Table 3. Zeta potential and classification stability [32]

Zeta potential (mV)	Stability
0	Little or no stability
15	Some stability but settling lightly
30	Moderate stability
45	Good stability, possible settling
60	Very good stability, little settling likely

The result of zeta potential measurement for TiO_2/POE nanolubricant is 81.1 mV. The result has then been compared with the classification of nanofluid stability based on the zeta potential of very good stability as shown in figure 5. The absolute value above 60 mV is desirable for very good stability. It proved that the TiO_2/POE is beyond the stable limit of 60 mV. From this finding, it can be concluded that the stability of this TiO_2/POE nanolubricant confirmed in the rate of very good stability.

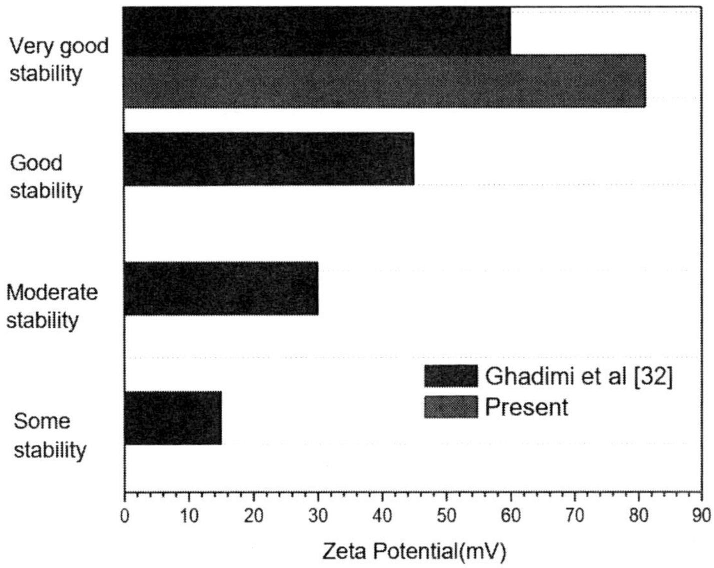

Figure 5. Zeta potential value of TiO_2/POE nanolubricant.

3.4. Visual observation and the effect

The images of TiO_2/POE nanolubricant for a volume concentration of 0.01% with different sonication time 0, 1, 3, 5 and 7 hours are shown in figure 6 and the images of TiO_2/POE nanolubricant for a volume concentration of 0.01, 0.03, 0.05, 0.07 and 0.1% with 3 hours sonication time are shown in figure 7. These images were taken after preparation and continue every day until at least 14 days or as long as it can hold to see the sedimentation and deposition of the nanoparticle in the liquid. From figure 6, no sedimentation of particles observed after the nanolubricant is prepared. The sedimentation of particles started after a day for 0 hour and others by day 7 afterward but in minimum sedimentation. After 14 days, the sedimentation can be seen especially for 0 and 7 hours compared to 3 and 5 hours which are more stable with a fewer sedimentation. Figure 7 also shows the same trend as figure 6 but this time, concentration of 0.05, 0.07 and 0.1 clearly can be seen the sediment after 3 days. Compared to 0.01 and 0.03, the naolubricant are still in the stable condition with very small sediment.

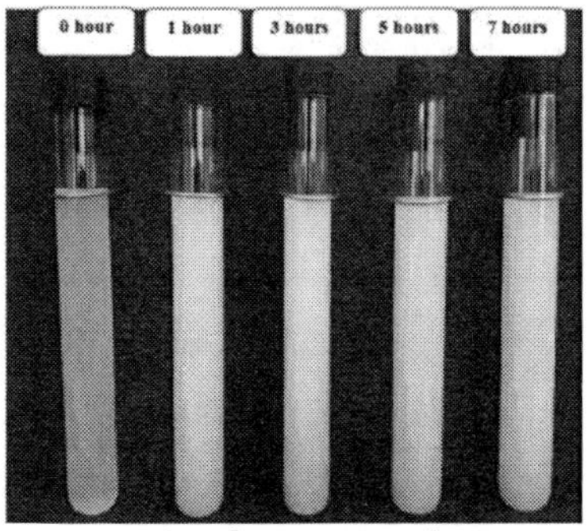

a) After preparation

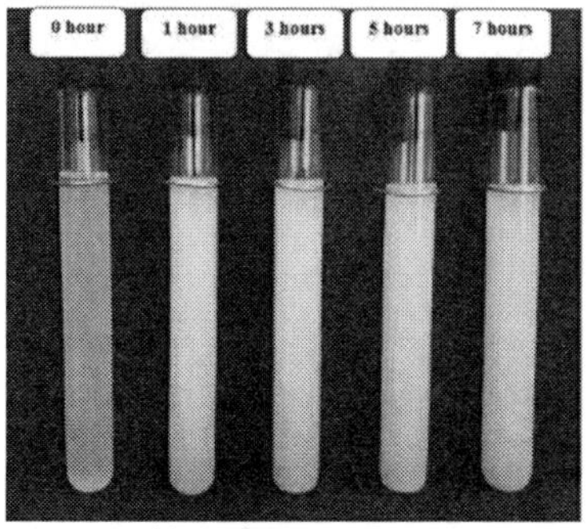

b) 3 days

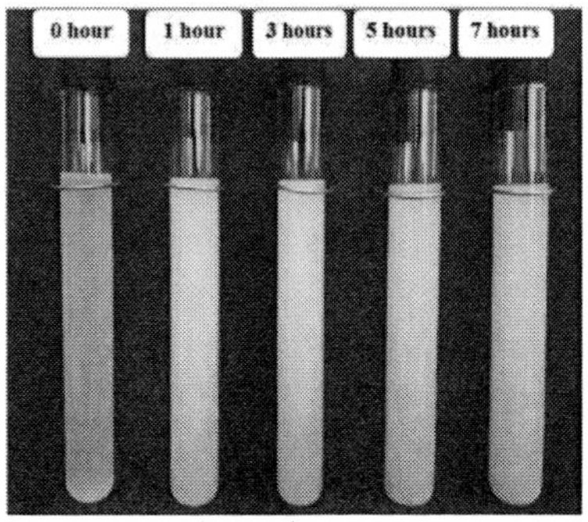

c) 11 days

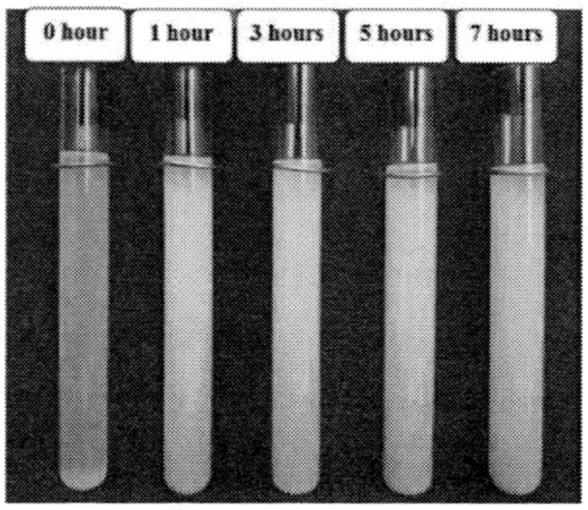

d) 14 days

Figure 6. Sedimentation observation of TiO_2/POE nanolubricant with different sonication time.

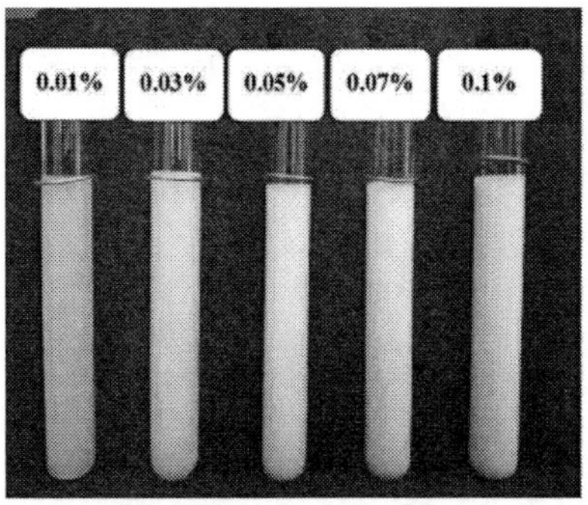

(a) After preparation

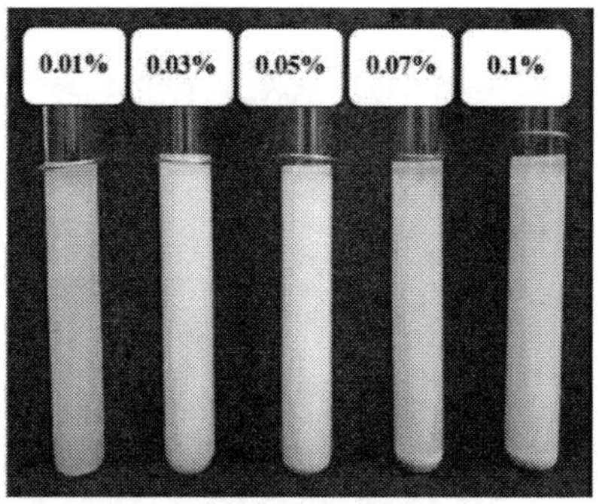

b) 3 days

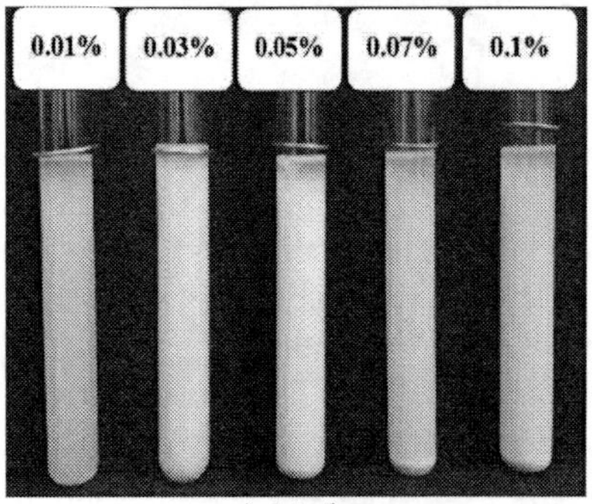

c) 11 days

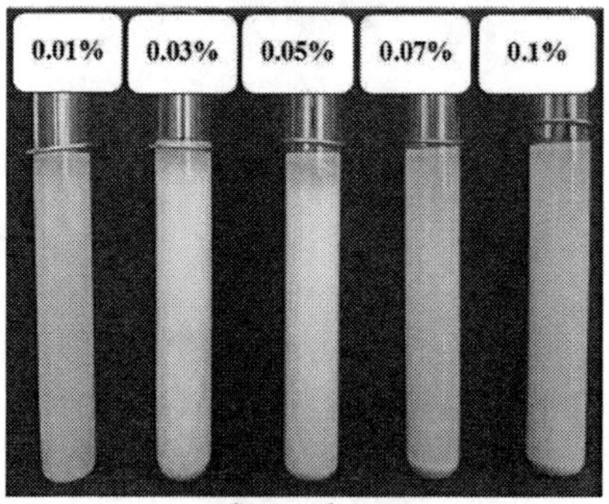

d) 14 days

Figure 7. Sedimentation observation of TiO_2/POE nanolubricant with different concentration.

4. Conclusion

TiO_2 was successfully dispersed into the POE lubricant which normally utilize in the refrigeration system by using two-steps method without any surface stabilizer or surfactants. The optimum sonication time process is 3 hours with 0.01% of concentration was confirmed by using the UV-Vis spectrophotometer analysis. It was done by the evaluation of various concentration and sonication time with both qualitative and quantitative method involving visual observation and UV-Vis spectrophotometer analysis. From the visual observation of the sedimentation, it can be concluded that this nanolubricant was in stable condition even after 14 days even some of the mixing show drastically occurred the sediment. It was shown that the peak absorbance of the TiO_2/POE nanolubricant lies at 530 nm wavelengths. After 14 weeks, the result from the UV-Vis showing that the relative concentration of these nanolubricant was maintained at over 95% compared to the initial concentration. Supported by Zeta Potential finding which is 81.1 mV, it has clearly shown that TiO_2/POE nanolubricant was in a very good stable condition after 14 weeks.

Acknowledgments
The authors are grateful to the Ministry of Education Malaysia and Universiti Malaysia Pahang (www.ump.edu.my) for financial supports given under FRGS/1/2018/TK10/UMP/02/10 and RDU190146. The authors also thank the research team from Automotive Engineering Centre (AEC), Centre of Excellence for Advanced Research in Fluid Flow (CARIFF) and Advanced Automotive Liquids Laboratory (AALL), who provided insight and expertise that greatly assisted in the present research work.

References

[1] Tsuzuki T 2009 *International Journal of Nanotechnology* **6** 567-78.
[2] Azmi W H, Zainon S N M, Hamid K A, Mamat R 2019 *Journal of Mechanical Engineering and Sciences* **13** 5182-211.
[3] Choi S U S 1995 *Developments and Applications of Non-Newtonian Flows* (New York: American Society of Mechanical Engineers (ASME)) p 99–105.
[4] Chandrasekar M, Suresh S 2009 *Heat Transfer Engineering* **30** 1136-50.
[5] Lee S, Choi S U S, Li S, Eastman J A 1999 *Journal of Heat transfer* **121** 280-9.
[6] Hatwar A S, Kriplani V M 2014 *Int J Adv Res Sci Eng* **3** 175-83.
[7] Sharif M Z, Azmi W H, Mamat R, Shaiful A I M 2018 *International Communications in Heat and Mass Transfer* **92** 56-63.
[8] Zawawi N N M, Azmi W H, Redhwan A A M, Sharif M Z, Samykano M 2018 *International Journal of Refrigeration* **89** 11-21.
[9] Zawawi N N M, Azmi W H, Redhwan A A M, Sharif M Z 2018 *J Mech Eng* 28-38.
[10] Hamid K A, Azmi W H, Nabil M F, Mamat R 2018 *International Journal of Heat and Mass Transfer* **118** 617-27.
[11] Redhwan A A M, Azmi W H, Sharif M Z, Zawawi N N M, Mamat R 2018 *J Mech Eng* 14-27.
[12] Sharif M Z, Azmi W H, Redhwan A A M, Mamat R 2016 *International Journal of Refrigeration* **70** 93-102.
[13] Sharif M Z, Azmi W H, Redhwan A A M, Mamat R, Yusof T M 2017 *International Journal of Refrigeration* **75** 204-16.
[14] Redhwan A A M, Azmi W H, Sharif M Z, Mamat R, Zawawi N N M 2017 *Applied Thermal Engineering* **116** 823-32.
[15] Krishnan R S, Arulprakasajothi M, Logesh K, Raja N D, Rajendra M 2018 *Materials Today: Proceedings* **5** 20580-7.
[16] Nabil M F, Azmi W H, Hamid K A, Zawawi N N M, Priyandoko G, Mamat R 2017 *International Communications in Heat and Mass Transfer* **83** 30-9.
[17] Carlson R B, Wishart J, Stutenberg K 2016 *SAE International Journal of Fuels and Lubricants* **9** 260-8.

[18] Fontaras G, Zacharof N G, Ciuffo B 2017 *Progress in Energy and Combustion Science* **60** 97-131.

[19] Dahlan A A, Zulkifli A H, Nasution H, Aziz A A, Perang M R M, Jamil H M 2014 *Applied Mechanics and Materials* **554** 464-8.

[20] Zulkifli A A, Dahlan A A, Zulkifli A H, Nasution H, Aziz A A, Perang M R M, et al. 2015 *IOP Conference Series: Materials Science and Engineering* **100** 012028.

[21] Redhwan A A M, Azmi W H, Sharif M Z, Mamat R, Samykano M, Najafi G 2019 *Journal of Thermal Analysis and Calorimetry* **135** 1299-310.

[22] Wyczalek F A 1993 *IEEE Aerospace and Electronic Systems Magazine* **8** 9-14.

[23] Hamid K A, Azmi W H, Mamat R, Sharma K V 2016 *International Communications in Heat and Mass Transfer* **73** 16-24.

[24] Emkarate. RL68H. Typical Properties Data Sheet. CPI Engineering Services: Lubrizol; 2016.

[25] Yu W, Xie H 2012 *Journal of nanomaterials* **2012** 1.

[26] Zawawi N N M, Azmi W H, Sharif M Z, Najafi G 2019 *Journal of Thermal Analysis and Calorimetry* **135** 1243-55.

[27] Sharif M Z, Azmi W H, Redhwan A A M, Zawawi N N M, Mamat R 2017 *Journal of Mechanical Engineering* 233-47.

[28] Usri N A, Azmi W H, Mamat R, Hamid K A, Najafi G 2015 *Energy Procedia* **79** 397-402.

[29] Sharif M Z, Azmi W H, Redhwan A A M, Zawawi N M M 2017 *MATEC web of conferences* **90** 01049.

[30] Mukherjee S, Paria S 2013 *IOSR Journal of Mechanical and Civil Engineering* **9** 63-9.

[31] Ramadhan A I, Azmi W H, Mamat R, Hamid K A, Norsakinah S 2019 *IOP Conference Series: Materials Science and Engineering* **469** 012068.

[32] Ghadimi A, Saidur R, Metselaar H S C 2011 *International journal of heat and mass transfer* **54** 4051-68.

IOP Conf. Series: Materials Science and Engineering **863** (2020) 012051 doi:10.1088/1757-899X/863/1/012051

Forced convection heat transfer and friction factor of water/bio-glycol mixture based TiO_2-SiO_2 nanofluids

S N M Zainon[1] **and W H Azmi**[1,2,3,*]

[1] Department of Mechanical Engineering, College of Engineering, Universiti Malaysia Pahang, Lebuhraya Tun Razak, 26300 Gambang, Kuantan, Pahang, Malaysia.
[2] Automotive Engineering Centre, Universiti Malaysia Pahang, 26600 Pekan, Pahang, Malaysia.
[3] Centre of Excellence for Advanced Research in Fluid Flow, Lebuhraya Tun Razak, 26300 Kuantan, Pahang, Malaysia.

*E-mail: wanazmi2010@gmail.com

Abstract. The research on nanofluids as heat transfer fluid has been done for many years and the manipulation of nanoparticles kept evolving as time passes by. After an exceptional heat transfer performance enhancement was found in single nanofluids, combining two types of nanoparticles in a base fluid has garnered researcher's attention all over the globe. The study on performance of heat transfer for a combination of two types of nanoparticle in a mixture of water and Bio-glycol has yet to be established; hence, the present study was conducted to investigate the heat transfer performance of TiO_2-SiO_2 nanofluids in 60:40 mixture of water and Bio-Glycol. The TiO_2-SiO_2 nanofluids was prepared by using the two-step method at three different particle concentration, namely 0.5%, 1.0% and 1.5%. Forced convection heat transfer experiment was carried out at bulk temperature 30°C and Reynolds number in the range of 1000 to 8000. A constant heat flux was supplied to the test section throughout the experiment. The Nusselt number of TiO_2-SiO_2 nanofluids was found at particle concentration 1.5%, 70.11% enhanced from the base fluid. The average enhancement of heat transfer coefficient for each particle concentration 0.5%, 1.0% and 1.5% are 40.3%, 63.4% and 70.3%, respectively. Heat transfer coefficient enhancement for 1.0% and 1.5% particle concentration almost similar at low concentration, however, as Reynolds number increase, the augmentation gap increases. The results of friction factor displayed a decrease trend with the increase of Reynolds number. While the friction factor of TiO_2-SiO_2 nanofluids insignificantly increased from the base fluid, the increment between particles concentration almost negligible.

1. Introduction

In the past few years, the evolution of research on heat transfer fluid has grown rapidly due to the limitation of conventional heat transfer fluids like water, oil, and ethylene glycol that stunted the enhancement of the heat transfer [1, 2]. Demand to improve efficiency of industrial applications of heating and cooling such as heat exchange has become the focus of researchers in designing a compact heat exchanger while reducing its cost, basically by increasing heat transfer to accommodate heat transfer coefficient [3, 4]. As a result, fuel consumption will be significantly reduced, and the service life of machinery could be extended. The act of dispersing ultra-fine particles in a base fluid pioneered by Masuda et al. [5] to investigate the thermal conductivity enhancement triggered many experimental investigations in the heat transfer performance of heat transfer fluid due to its potential in enhancing the thermo-physical properties of fluids. However, heat transfer fluid dispersed with miniature metallic oxide leads to an increase in viscosity, thereby contributing to high pumping penalty. Hence, to fully utilize the colloidal suspension as heat transfer fluid, selection of type of particles, their size and shape, thermal conductivity, and particle concentration should be wisely chosen. A comprehensive review on thermo-physical properties of hybrid nanofluids was presented by Nabil et al. [6] and Azmi et al. [7]. In addition, various studies were undertaken to investigate the thermal conductivity and dynamic viscosity of nanofluids [8-11].

A previous study on heat transfer and pressure drop by Pak and Cho [3] in circular tube under turbulence flow revealed how the colloidal suspension improves the heat transportation in fluids. The authors used Al_2O_3 and TiO_2 nanoparticles dispersed in water as working fluids at different concentrations and observed the increment in Nusselt number when volume concentration of fluids and Reynolds number increased. At fixed Reynolds number, the researcher compared the Al_2O_3 nanofluids at different concentrations 1.34% and 2.78% and found 45% and 75% of convective heat transfer coefficient being augmented, respectively. Despite the anomaly improvement in convective heat transfer coefficient, the increase in pumping penalty cannot be avoided as viscosity increased at high volume concentration. The friction factor and pumping penalty for Al_2O_3 nanofluids at 2.78% volume concentration was reported to be 28% and 31% higher than water, respectively.

Azmi et al. [12] utilized Al_2O_3 nanofluids that was dispersed in a mixture of water and ethylene glycol to study the heat transfer performance under turbulence flow regime using forced convection setup. As a result, they found the highest increase in convective heat transfer coefficient at 1.0% volume concentration. This enhancement was found at 60:40 mixture ratio of water and ethylene glycol, which is 24.6%. In addition, the increase in friction factor is not apparent which 5% only, influenced by the low volume concentration of nanoparticles applied in this study. Further heat transfer performance studies were investigated in the recent works [13, 14]. Another study conducted by Abdolbaqi et al. [15] for properties evaluation and Abdolbaqi et al. [16] using water/Bio-glycol based TiO_2 nanofluids also found about 28.2% enhancement of the Nusselt number from the base fluid. Since there is no study on combination of two particles dispersed in a mixture of water and Bio-Glycol, the present study aimed to investigate the heat transfer performance of water/Bio-Glycol based TiO_2-SiO_2 nanofluids at low concentration range from 0.5% to 1.5% with working bulk temperature 30 °C.

2. Methodology

2.1. TiO_2-SiO_2 nanofluids preparation

TiO_2 and SiO_2 water-based nanofluids were considered in the present heat transfer analysis. There were commercially produced by US Research Nanomaterials, Inc for research purposes in the form of liquid suspension with 40 wt% concentration sized range from 30-50nm while SiO_2 nanopowder dispersion also acquired from US Research Nanomaterials, Inc. (USA) with 25 wt% concentration sized 30nm. The procured TiO_2 and SiO_2 nanofluids will then be prepared to a new concentration by the dilution technique using equation (1) and equation (2). In order to prepare the nanofluids by dispersing the nanoparticles in a water/Bio-glycol based fluid, proper mixing and stabilization of the particles are required.

$$\varphi = \frac{\omega \rho_{bf}}{\left(1-\frac{\omega}{100}\right)\rho_p + \frac{\omega}{100}\rho_{bf}} \tag{1}$$

$$\Delta V = (V_2 - V_1) = V_1\left(\frac{\varphi_1}{\varphi_2} - 1\right) \tag{2}$$

In the present study, TiO_2 and SiO_2 nanoparticles were mixed with water/Bio-glycol based fluid then were sonicated continuously by an ultrasonic vibrator generating ultrasonic pulses of 80 W at 36 ± 3 kHz for 3 h to break down agglomeration of the nanoparticles prior to being used as the working fluid.

2.2. *Thermo-physical properties*

Thermo-physical properties like thermal conductivity, dynamic viscosity, specific heat, and density are fluid properties that change with temperature. These properties have long been studied by various researchers in different kind fields to evaluate the characteristic of heat transfer fluid and determine the optimum condition for heat transfer fluid to work effectively in the system. In the present study, Thermal conductivity of TiO_2-SiO_2 nanofluids was measured using C-Therm Thermal Conductivity Analyzer with 5% accuracy and 1% precision, supplied by C-Therm Technologies (Ltd.). While the measurement for dynamic viscosity was carried out using Anton Paar RheolabQC Rotational Rheometer. Density Meter was used to measure the density of the nanofluids, while specific heat was estimated using equation (3) or also known as mixture relation equation.

$$C_{hnf} = \frac{(1-\phi_p)\rho_{bf}C_{bf} + (0.2\times\phi)\rho_{TiO_2}C_{TiO_2} + (0.8\times\phi)\rho_{SiO_2}C_{SiO_2}}{\rho_{hnf}} \quad (3)$$

All these properties were evaluated at temperature 30 °C and particles concentration 0.5% to 1.5%. Summary of the thermo-physical properties are presented in table 1.

Table 1. Thermo-physical properties of TiO_2-SiO_2 nanofluids at 30 °C.

Volume Concentration, ϕ *(%)*	**Thermal Conductivity,** *K (W/m. K)*	**Dynamic Viscosity,** *μ (Pa.s)*	**Density, ρ** *(kg/m³)*	**Specific Heat, *c*** *(J/kg.K)*
0.5	0.430589	0.0039054	1053	3753.84
1.0	0.432905	0.0041633	1058	3715.763
1.5	0.438889	0.0040707	1068	3678.255

2.3. *Experimental setup*

The experimental setup used for forced-convection experiment is shown in figure 1. Forced-convection test rig built with control panel, collecting tank capable of accommodating a maximum of 0.08 m^3 working fluids, 1.0 HP circulating pump, 1.5 m long test section with inner diameter (ID) 16 mm, outer diameter (OD) 19 mm, and thickness of 3 mm. The test section is made of copper tube and was well insulated with fibre glass to minimize heat loss to the surroundings. 600 W heat was supplied using a voltage regulator by attaching two nichrome to the outer wall of the test section to supply a constant heat flux. Eight K-type thermocouples were wrapped at 6 different locations of the section and the other two were fixed at test section inlet and outlet, labelled from T_1 to T_8. A data logger type ADAM View Advanced Data Acquisition was connected to the test rig to record and display data measured from the experiment. The outlet of the test section is connected to a 2.8kW chiller to cool down the hot working fluid before circulating it back to the collecting tank. A battery-powered flow meter measured in LPM is installed between the pump and inlet test section. A pressure transducer is placed between the inlet and outlet of the test section and displayed as T_9 at the data logger. Validation of experimental setup was carried out by comparing the Nusselt number data of base fluid, water/Bio-Glycol calculated using equation (4):

$$Nu_{exp} = \frac{hD}{k} \quad (4)$$

where h is heat transfer coefficient, D stands for inner diameter of test section, and k is thermal conductivity of nanofluids, with estimated Nusselt number using Dittus-Boelter equation, shown in equation (5).

$$Nu_{DB} = 0.023\mathrm{Re}^{0.8}\,\mathrm{Pr}^{n} \tag{5}$$

where Re is Reynolds number and Pr represents Prandtl number. Since the nanofluids in the present study is heated, value of n=0.4 is used.

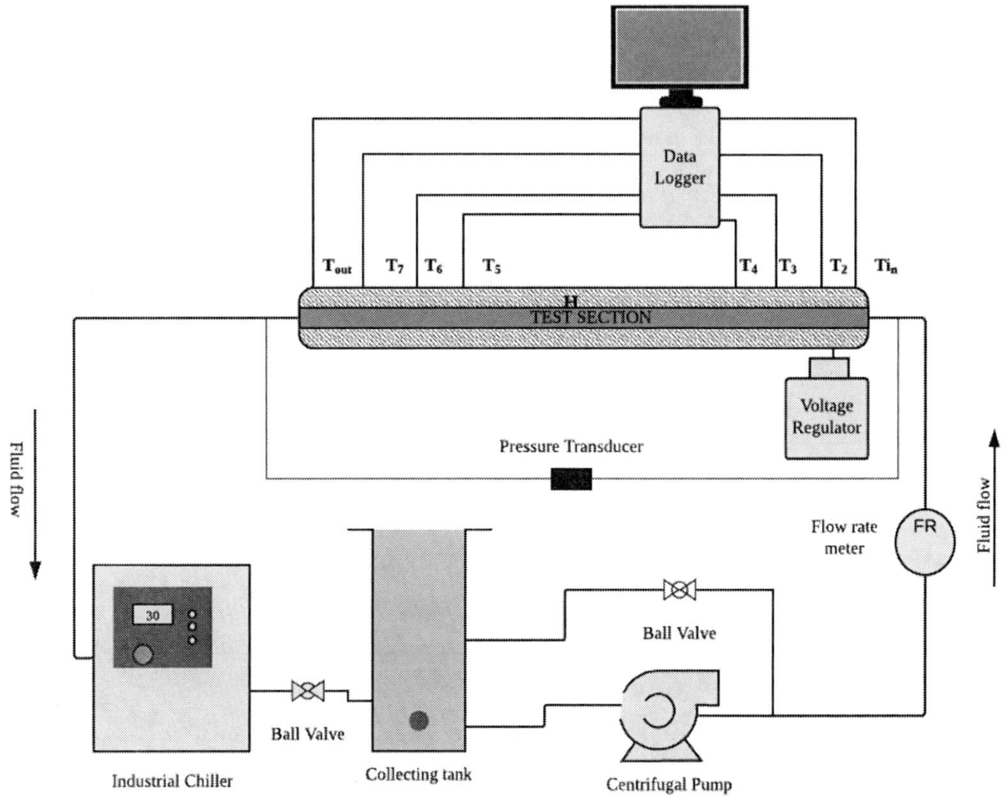

Figure 1. Schematic diagram of forced convection experimental setup.

3. Results and discussion

To establish an accurate experimental method, a validation on the experimental setup was done by using a base fluid water/Bio-Glycol (W/BG) at bulk temperature 30 °C. The results presented in figure 2 confirmed the reliability of the experiment by comparing the Nusselt number data of W/BG with estimation from Dittus-Boelter [17] with ±2% average deviation. The forced convection experiment in this study was conducted using TiO_2-SiO_2 nanofluids at three different concentrations (0.5%, 1.0%, 1.5%) and operating bulk temperature of 30°C. The variation of the Nusselt number in Reynolds number ranges from 1000 to 8000 is shown in figure 3. From the figure, it evident that the Nusselt number is influenced by the nanofluids concentration and Reynolds number. The Nusselt number shows an increment as concentration and the Reynolds number increases, which the highest Nusselt number can be obtained at 1.5 vol.%, about 70.11% higher than the base fluid. The possible explanation behind this enhancement is due to the fact the higher volume concentration possessed higher thermal conductivity compared to the lower volume concentration [18]. Several investigators like Torii [19], Chandra Sekhara Reddy and Vasudeva Rao [20] and Kayhani et al. [21] also found similar trend in their researches.

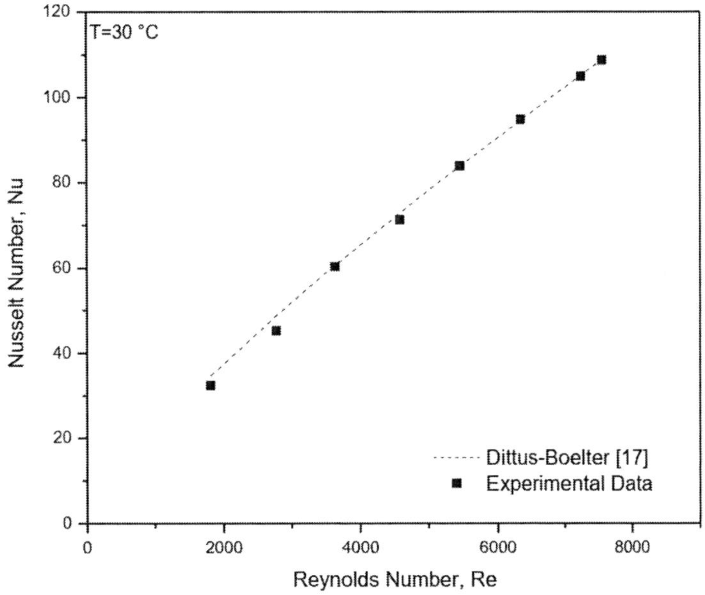

Figure 2. Validation of Nusselt Number of W/BG with Dittus-Boelter [17].

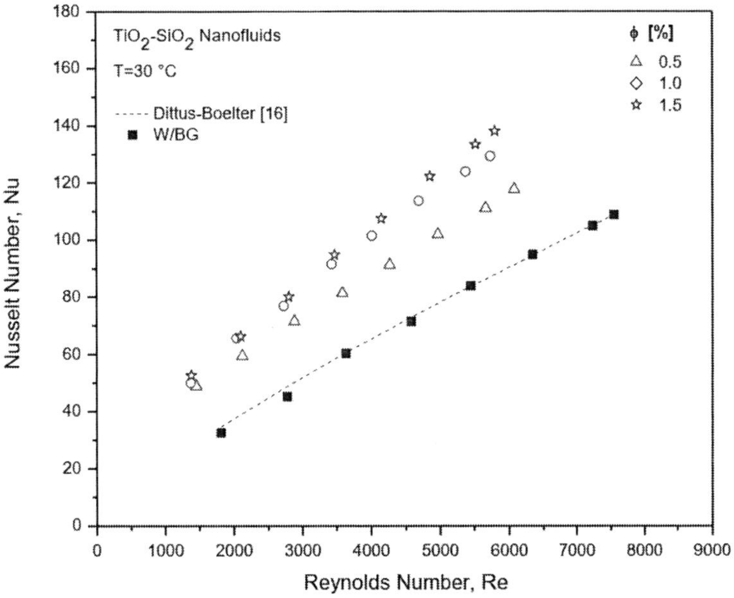

Figure 3. Nusselt number of TiO_2-SiO_2 nanofluids at different particle concentration.

Variation of heat transfer coefficient of TiO_2-SiO_2 nanofluids against Reynolds number at different concentration is plotted in figure 4. An increase trend was found in heat transfer coefficient as the volume concentration increase. From the pattern observed in figure 3, it is obvious that small addition of nanoparticles can greatly influence the heat transfer coefficient of the heat transfer fluid. This statement was supported by average enhancement achieved at 1.5%, which is approximately 70.3%, following by 1.0% volume concentration that was enhanced up to 63.4% and 40.2% for 0.5% volume concentration. Enhancement of heat transfer concentration in the function of Reynolds number at different volume concentration is presented in figure 5. Based on the data showed in this figure, the

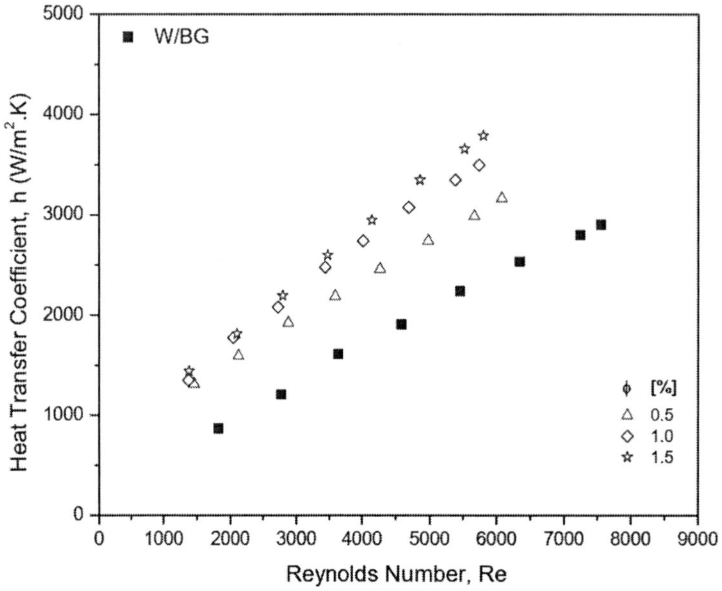

Figure 4. Heat transfer coefficient of TiO_2-SiO_2 nanofluids at different particle concentration.

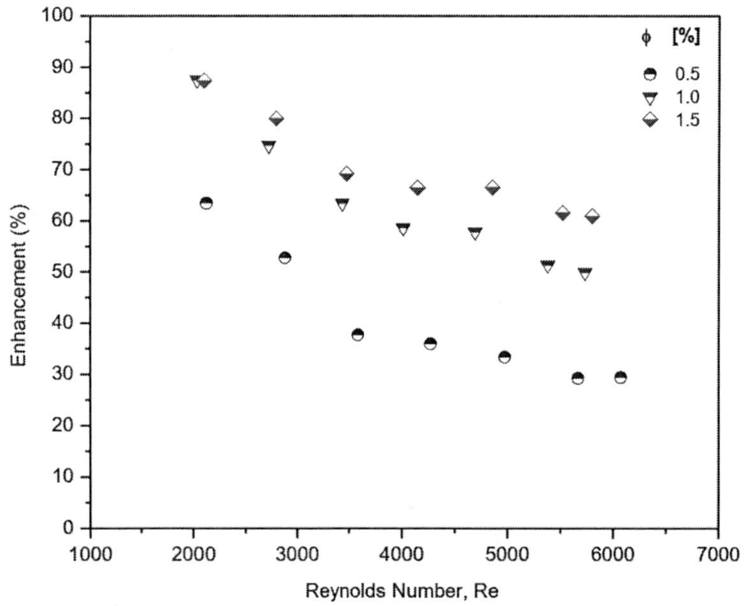

Figure 5. Enhancement of heat transfer coefficient of different particle concentration with Reynolds number.

enhancement TiO_2-SiO_2 nanofluids for 1.0% and 1.5% particle concentration almost similar at low Reynolds number, however the difference become obvious as Reynolds number increase. The outcomes from this experiment showed that the enhancement of heat transfer coefficient increases with particle concentration. Similar findings also can be seen in the researches performed by Bhanvase et al. [22] that achieved maximum of 105% augmentation in heat transfer coefficient using water/ethylene glycol based TiO_2 nanofluids at 0.5% volume fraction. Azmi et al. [23] encountered same results where the enhancement of heat transfer coefficient increases as particle concentration increase up to

3%, however, further increase in particle concentration caused the enhancement to decrease. Another researcher that encountered similar results is Hamid et al. [24].

The friction factor of W/BG and TiO_2-SiO_2 nanofluids in the function of Reynolds number is provided in figure 6. It shows that the friction factor decreases as the Reynolds number increases. The estimation of friction factor for base W/BG using Blasius [25] and Petukhov [26] seems in good agreement especially at higher Reynolds number. A little increase in the friction factor of TiO_2-SiO_2 nanofluids from the base fluid was observed. However, the increase almost insignificant at high Reynolds number. In comparison, between particle concentration, results revealed a negligible increment of friction factor. Friction factors of all concentration fall closely to Blasius [25] and Petukhov [26], proved the TiO_2-SiO_2 particles are not a strong influencer to the friction factor. Similar findings also reported in prior researches [12, 14, 27, 28].

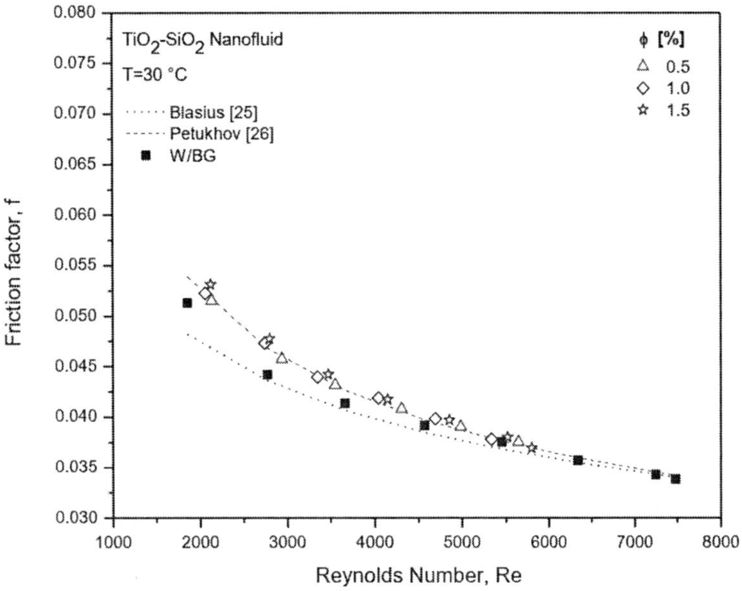

Figure 6. Variation of friction factor of TiO_2-SiO_2 nanofluids at different particle concentration with Reynolds number.

4. Conclusion

Conventional heat transfer fluids have reached its limit in improving heat transfer performance of heating and cooling system. Thus, the employment of small amounts of nanoparticles dispersed in base fluid has become a trend in research due to its anomalous heat transfer enhancement. Water/Bio-Glycol based TiO_2-SiO_2 nanofluids at low concentrations was employed in this experimental study on forced convection heat transfer at bulk temperature 30 °C. The study leads to the following conclusions:

i. Nusselt number of nanofluids increases with particle concentration and Reynolds number, where the highest Nusselt number was observed at 1.5 vol.%, which is 70.11% higher than base fluid.
ii. Small addition of nanoparticles led to great heat transfer enhancement. Average enhancement for 0.5%, 1.0% and 1.5% are 40.3%, 63.4% and 70.3%, respectively. At high Reynolds number, maximum of 60.9% heat transfer coefficient augmentation was observed at 1.5 vol.%.
iii. Friction factor TiO_2-SiO_2 nanofluids decrease as the Reynolds number increases. Dispersing TiO_2 and SiO_2 in a mixture of water and Bio-Glycol did not significantly influence the friction

factor especially at high Reynolds number, and there is no apparent increment of friction factor in term of concentration.

Acknowledgments
The authors are grateful to the Ministry of Education Malaysia and Universiti Malaysia Pahang (www.ump.edu.my) for financial supports given under FRGS/1/2018/TK10/UMP/02/10, RDU190146 and PGRS1903110. The authors also thank the research team from Automotive Engineering Centre (AEC), Centre of Excellence for Advanced Research in Fluid Flow (CARIFF) and Advanced Automotive Liquids Laboratory (AALL), who provided insight and expertise that greatly assisted in the present research work.

References

[1] Duangthongsuk W, Wongwises S 2010 *International Journal of Heat and Mass Transfer* **53** 334-44.
[2] Angayarkanni S A, Philip J 2015 *Advances in Colloid and Interface Science* **225** 146-76.
[3] Pak B C, Cho Y I 1998 *Experimental Heat Transfer* **11** 151-70.
[4] Gugulothu R, Reddy K V K, Somanchi N S, Adithya E L 2017 *Materials Today: Proceedings* **4** 1051-6.
[5] Masuda H, Ebata A, Teramae K 1993.
[6] Nabil M F, Azmi W H, Hamid K A, Zawawi N N M, Priyandoko G, Mamat R 2017 *International Communications in Heat and Mass Transfer* **83** 30-9.
[7] Azmi W H, Zainon S N M, Hamid K A, Mamat R 2019 *Journal of Mechanical Engineering and Sciences* **13** 5182-211.
[8] Khdher A M, Sidik N A C, Hamzah W A W, Mamat R 2016 *International Communications in Heat and Mass Transfer* **73** 75-83.
[9] Hamid K A, Azmi W H, Nabil M F, Mamat R, Sharma K V 2018 *International Journal of Heat and Mass Transfer* **116** 1143-52.
[10] Redhwan A A M, Azmi W H, Sharif M Z, Mamat R, Zawawi N N M 2017 *Applied Thermal Engineering* **116** 823-32.
[11] Usri N A, Azmi W H, Mamat R, Hamid K A, Najafi G 2015 *Energy Procedia* **79** 397-402
[12] Azmi W H, Usri N A, Mamat R, Sharma K V, Noor M M 2017 *Applied Thermal Engineering* **112** 707-19.
[13] Hamid K A, Azmi W H, Mamat R, Mohamad M S, Sharma K V 2019 *Numerical Heat Transfer; Part A: Applications* **75** 271-89.
[14] Hamid K A, Azmi W H, Nabil M F, Mamat R 2018 *International Journal of Heat and Mass Transfer* **118** 617-27.
[15] Abdolbaqi M K, Sidik N A C, Aziz A, Mamat R, Azmi W H, Yazid M N A W M, et al. 2016 *International Communications in Heat and Mass Transfer* **77** 22-32.
[16] Abdolbaqi M K, Mamat R, Sidik N A C, Azmi W H, Selvakumar P 2017 *International Journal of Heat and Mass Transfer* **108** 1026-35.
[17] Dittus F W, Boelter L M K 1930 443-61.
[18] Hussein A M, Sharma K V, Bakar R A, Kadirgama K 2013 *Journal of Nanomaterials* **2013** 12.
[19] Torii S 2010 *Advances in Mechanical Engineering* **2** 917612.
[20] Chandra Sekhara Reddy M, Vasudeva Rao V 2014 *International Communications in Heat and Mass Transfer* **50** 68-76.
[21] Kayhani M H, Soltanzadeh H, Heyhat M M, Nazari M, Kowsary F 2012 *International Communications in Heat and Mass Transfer* **39** 456-62.
[22] Bhanvase B A, Sarode M R, Putterwar L A, K.A A, Deosarkar M P, Sonawane S H 2014 *Chemical Engineering and Processing: Process Intensification* **82** 123-31.
[23] Azmi W H, Sharma K V, Sarma P K, Mamat R, Anuar S, Dharma Rao V 2013 *Experimental Thermal and Fluid Science* **51** 103-11.

[24] Hamid K A, Azmi W H, Mamat R, Sharma K V 2016 *International Communications in Heat and Mass Transfer* **73** 16-24.
[25] Blasius H 1913 *Mitteilungen über Forschungsarbeiten auf dem Gebiete des Ingenieurwesens* Springer) p 1-41.
[26] Petukhov B, Irvine T, Hartnett J 1970 *Academic, New York* **6** 503-64.
[27] Azmi W H, Abdul Hamid K, Usri N A, Mamat R, Mohamad M S 2016 *International Communications in Heat and Mass Transfer* **76** 24-32.
[28] Sarma P K, Kedarnath C, Ramanarayanan C P, Kishore P S, Rao V D, Ramakrishna K, et al. 2008 *International Journal of Heat and Technology* **26** 75-83.

IOP Conf. Series: Materials Science and Engineering **863** (2020) 012052 doi:10.1088/1757-899X/863/1/012052

Performance of Al_2O_3-SiO_2/PAG employed composite nanolubricant in automotive air conditioning (AAC) system

N N M Zawawi[1] and W H Azmi[1,2,3,*]

[1] Department of Mechanical Engineering, College of Engineering, Universiti Malaysia Pahang, Lebuhraya Tun Razak, 26300 Gambang, Kuantan, Pahang, Malaysia.
[2] Automotive Engineering Centre, Universiti Malaysia Pahang, 26600 Pekan, Pahang, Malaysia.
[3] Centre of Excellence for Advanced Research in Fluid Flow, Lebuhraya Tun Razak, 26300 Kuantan, Pahang, Malaysia.

*E-mail: wanazmi2010@gmail.com

Abstract. Automotive Air conditioning (AAC) system is the contributor of harmful gasses emission and global warming. In order to solve the issues, investigation on potential improvement in lubrication used in air conditioning system is done to improve the system performance and efficiency. The performance of AAC system namely cooling capacity, compressor work, coefficient of performance (COP) and power consumption was investigated by comparing pure lubricant and Al_2O_3-SiO_2/PAG nanolubricants. Performance of AAC by using different ranges of refrigerant charges (95 to 155 g) and speeds (1200 and 1800 rpm) was investigated. The result shows that the cooling capacity and COP of pure lubricant were relatively lower than Al_2O_3-SiO_2/PAG nanolubricants. The compressor work and power consumptions of Al_2O_3-SiO_2/PAG nanolubricants were greatly reduced. Cooling capacity and COP are enhanced by 102.99% and 23% respectively. The compressor work and power consumption are reduced by 25.9% and 28.24% respectively. From the results, nanolubricants gives more advantages in AAC performance over pure lubricant. Therefore, Al_2O_3-SiO_2/PAG composite nanolubricants is recommended to be used as the compressor lubrication to enhance AAC performances system.

1. Introduction

Environmental issues around the world are increasing rapidly nowadays and immediate preventive measures and solutions are needed to solve the issues. Air conditioning system is the biggest energy consumptions and major contributor to global warming [1] in the world. Automotive Air Conditioning (AAC) system is one of best application of refrigeration system. AAC system is necessary to provide thermal comfort to the vehicle passengers [2] and remove the unwanted heat. However, to maintain desired level of comfort, more fuel is consumed to cool down the vehicle which causing harmful gasses emission into the atmosphere, eventually raising the average temperature on earth.

AAC system consists of the compressor, condenser, expansion valve and the evaporator. A working fluid is needed to be present to ensure the systems work efficiently. The working fluid involves in the AAC system is the refrigerant and the lubricant which will flow throughout the system and at the same time transfers heat along the system. In an ideal working environment, the working fluid involves in the AAC system is the combination of both refrigerant and the lubricant simultaneously flowing and transferring heat throughout the system. Thus, the combination of the lubricant and refrigerant is needed to be utilized and considered in an ACC system [3]. Hydrofluorocarbons (HFCs) refrigerant (R134a) is currently used as common refrigerant worldwide [4]. Lubricant is need for lubrication of the condensers in the air conditioning system in preventing excessive friction in the bearings [5], time reducing the refrigerant temperature during compressing process and in the same time sealing the refrigeration leakage path [6]. Moreover, lubricants are able to enhance the efficiency of the compressor by reducing the compressor friction loss [7]. Choosing the right lubricant is important to provide maximum coefficient of performance (COP) value thus reducing power consumption [8].

Therefore, studies are need to be done to investigate potential improvement in lubrication to increase the efficiency of the refrigeration system. The use of nanoparticles in refrigeration systems is considered by many scientists for its contribution in enhancement of thermo-physical properties and heat transfer capabilities [9-11]. Nanolubricants have the higher advantages over pure lubricant in terms of heat transfer, efficient in dispersion stability, clogging particle and pumping force reduction [1, 12]. Besides, nanolubricants also decrease the amount of energy needed to operate the systems. This helps in reducing energy consumption, emissions, global warming potential and greenhouse gas effects [13]. Studies have been done to prove the potential of nanolubricants in enhancing the thermodynamic and mechanical performance of the refrigeration system [14] and also in its capability to increase the performance of the refrigeration system [1] and systems efficiency. The study on addition of nanoparticle in refrigerant or lubricant mixture in improving performance for refrigeration system are done by few researchers [10, 15-20].

Researches on different metal oxides nanolubricants combinations in refrigeration systems [21, 22] and on the utilization of nanolubricants; mainly focusing on single-component based nanolubricant for AAC system performance[23, 24] are considered very limited even though extensive nanoparticles research in refrigeration system has been done. Zawawi et al. [14] studies on AAC performance of Al_2O_3-SiO_2/PAG nanolubricants with 50:50 composition ratios for refrigerant charges (95 to 155g) at compressor speed of 900, 1500 and 2100 rpm. So, as the continuation for the study reported by Zawawi et al. [14], study on AAC performance of Al_2O_3-SiO_2/PAG nanolubricants with 60:40 composition ratios is done. In the present study, the AAC system setup is developed for performance analysis of Al_2O_3-SiO_2/PAG nanolubricants. The performance (cooling capacity, compressor work, COP and power consumption) of the Al_2O_3-SiO_2/PAG nanolubricants for 0.06% volume concentrations has been investigated with a refrigerant charges range of 95 to 155 g and at compressor speed of 1200 and 1800 rpm. The performance investigation is conducted to compare the performance of the employment of pure lubricant and Al_2O_3-SiO_2/PAG in AAC system.

2. Methodology

2.1. Al_2O_3-SiO_2/PAG nanolubricants preparations

From previous study [22], Al_2O_3-SiO_2/PAG nanolubricants is selected as the best metal oxide combination for applications in refrigeration compressors. Thus, in this study, Al_2O_3-SiO_2/PAG nanolubricants is utilized in the AAC system study as the main lubricant. Table 1 shows the properties of the Al_2O_3 and SiO_2 nanoparticles and PAG 46 lubricant properties at the atmospheric pressure are given in table 2. Two-step method was used in the preparation of Al_2O_3-SiO_2/PAG nanolubricant for 0.06% volume concentration. Researchers widely used this method in nanofluids preparation [25-28]. Al_2O_3/PAG nanolubricant and SiO_2/PAG nanolubricants were separately prepared. These two single nanolubricants were then mixed and homogenized together by 60:40 composition ratio for 2 hours to maintain the stability of the nanolubricants. Dispersion and stability of the Al_2O_3-SiO_2/PAG is done with accordingly to previous researches [22] with method suggested by Sharif et. al. [29]. The volume concentrations of the nanolubricant were calculated using equation (1) [27, 29-31].

$$\phi = \frac{m_p/\rho_p}{m_p/\rho_p + m_l/\rho_l} \times 100 \tag{1}$$

Table 1. Properties of nanoparticles [32].

Properties	**Al_2O_3**	**SiO_2**
Molecular mass (g/mol)	101.96	60.08
Average particle diameter (nm)	13	30
Density (kg/m^3)	4000	2220
Thermal Conductivity (W/m.k)	36	1.4
Specific Heat (J/kg.K)	773	745

Table 2. Properties of polyalkylene glycol (PAG 46) [32].

Properties	**PAG 46**
Density (g/cm^3) @ 20°C	0.9954
Flash Point (°C)	174
Kinematic viscosity (cSt) @ 40°C	41.4-50.6
Pour Point (°C)	-51

2.2. AAC test setup

The AAC system setup in this study is utilized from SD Speciation of Perodua Kancil as one of representative compact car in Malaysia [30]. Development and specification of the setup was referred with Redhwan et. al [30]. The bench setup is divided into four main parts i.e. refrigeration system made from AAC system of the car, driver and control system where an electrical motor and an inverter frequency controller are utilized, water bath system for evaporators and piping system and complete instrumentation and data logger of the AAC system. The main setup of the system such as compressor, condenser, evaporator, expansion valve and piping system are included. The components are listed in table 3.

Table 3. List of components used in experimental setup.

No.	**Description**	**No**	**Description**
1	Compressor	6	Pressure Gauge
2	Three phase induction motor	7	Data Logger Module
3	Evaporator (inside Water bath)	8	Water Heater
4	Frequency Inverter Controller	9	Flow Transducer
5	Power Analyser	10	R134a gas

As stated in previous study [14], the experimental procedure of AAC performance investigation is conducted by following regulation and recommendation from SAEJ2765 standard. A 105 ml of Al_2O_3-SiO_2/PAG nanolubricants was prepared to replace the previous lubricant in the compressor. 95g refrigerant mass was weighed using weight balanced and is charged into the system. Water heater was then used to heat water in water tank until the inlet and outlet achieve equilibrium temperature. The controlled room ambient temperature and humidity are between 24.5 to 25.5°C and 50 to 60% respectively. Induction motor was started at 1200 rpm speed and the experiment was run for a 20 minutes time interval. The temperatures and water flow rate measurement were monitored and recorded through a data logger module. Next, the cycle of experiment was repeated for different refrigerant chargers and compressor speed of 1800 rpm. The schematic diagram of the AAC system test rig is shown in figure 1.

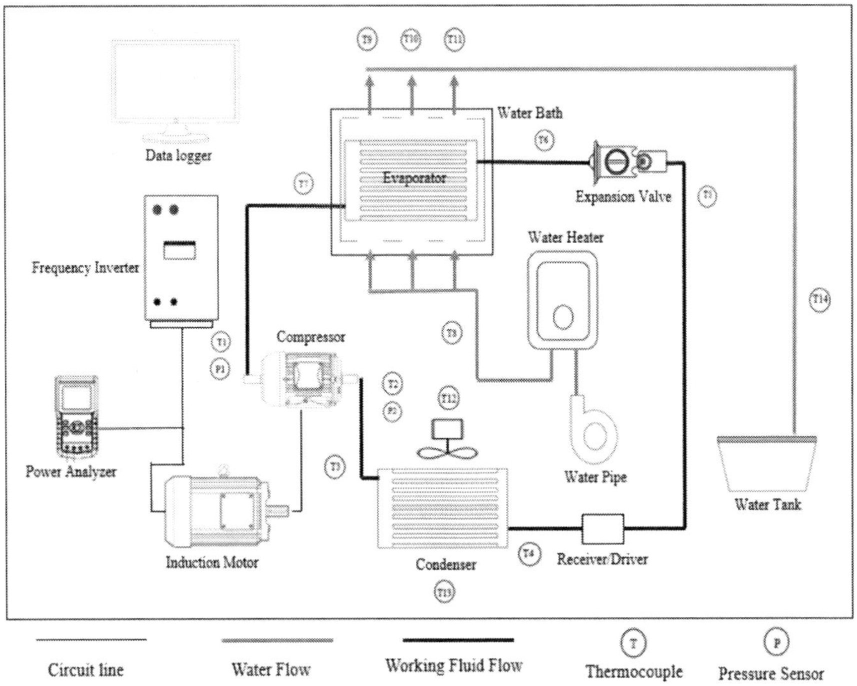

Figure 1. Schematic diagram of AAC test rig.

3. Results and discussion

AAC system performance namely cooling capacity, compressor work, COP and power consumption is investigated for various compressor speed at constant volume concentration of 0.06%. Figure 2 depicts the cooling capacity for Al_2O_3-SiO_2/PAG nanolubricant as a function of refrigerant charges. The cooling capacity of Al_2O_3-SiO_2/PAG nanolubricants increases along with increment of refrigerant. The trend is observed to agree well with the reported findings [1, 33]. It can be observed that the cooling capacity of the Al_2O_3-SiO_2/PAG nanolubricants at different compressor speed are higher than pure PAG lubricants. This is due to the nanoparticles in the nanolubricant contributes in the enhancement in thermal conductivity of nanolubricant to [34] and also enhancement of the heat transmission and causes the heat transfer to rise in the evaporator [31]. A high cooling capacity indicates that the system has an overall higher heat transfer and the system is able to cool down more efficiently. With the increasing of compressor speed, the mass flow rate tends to rise higher causing the heat transfer to rise in the evaporator [35] hence eventually improves the cooling capacity. Thus, the conservation of mass equation is proven. The highest cooling capacity is found to be 1.17 kW or 102.99% for compressor speed of 1800 rpm at 155 g refrigerant charge.

The compressor work of PAG and nanolubricants against refrigerant charges is shown in figure 3. It is clearly observed in the figure that the compressor work increases with compressor speeds but decreases as the refrigerant charges increases. Compressor at high speed tends to work much harder compare to lower speed [36]. This can also be deduced that the low compressor speed produces lower load on the compressor thus having a lower compressor work compare to a higher compressor speed. Besides, the compressor works for Al_2O_3-SiO_2/PAG nanolubricants are relatively lower compared to the PAG lubricants. The compressor work for AAC system utilized by Al_2O_3-SiO_2/PAG nanolubricants is maximumly reduced by 25.90% at 1800 rpm for 155g refrigerant charge meanwhile the minimum reduction is 8.11% also at 1800 rpm. Nanolubricants contributes in reducing pumping force of the compressor to achieve comparable heat transfer [37] and this helps in reducing the work required by the compressor to run the system [1].

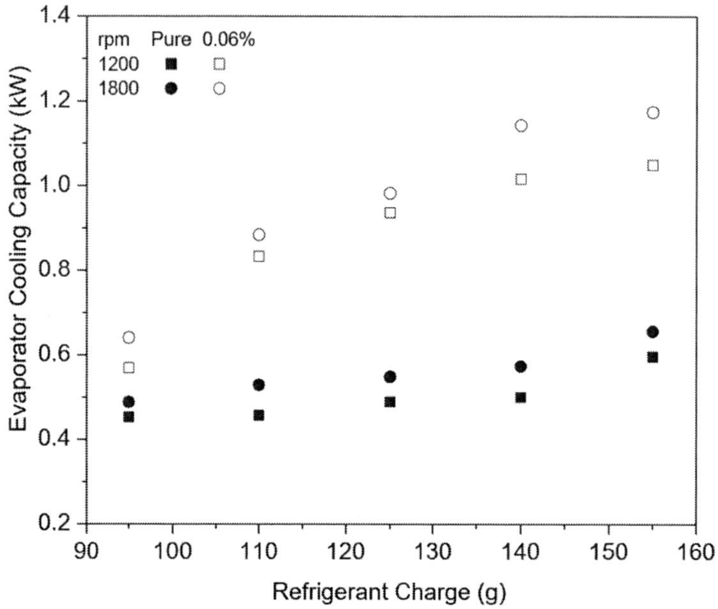

Figure 2. Evaporator cooling capacity against refrigerant charges.

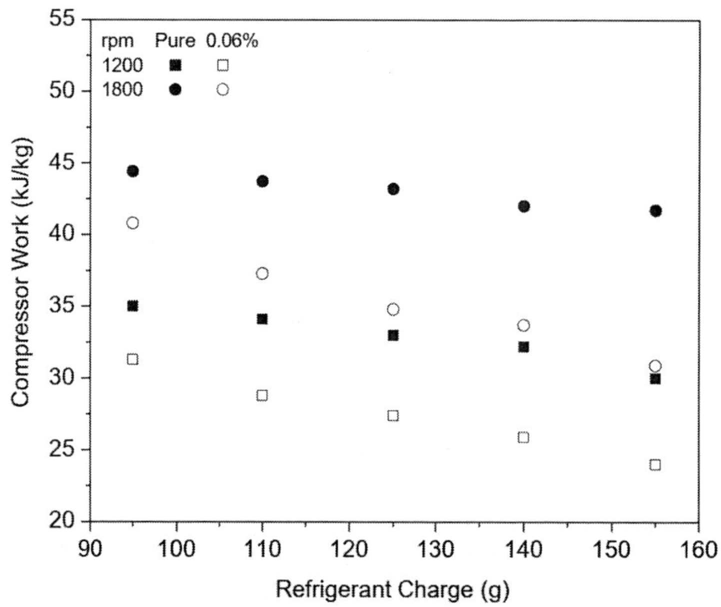

Figure 3. Compressor work against refrigerant charge.

Figure 4 depicts COP against the refrigerant charge of the system. COP of Al_2O_3-SiO_2/PAG nanolubricants is higher compared to pure lubricant. This is due to pure lubricant has low cooling capacity and high compressor work. However, Al_2O_3-SiO_2/PAG nanolubricants has higher increment of the cooling capacity reducing compressor work and eventually increases the COP of the system. Similar trend is also observed in previous researches reported in the literature [23, 24]. Al_2O_3-SiO_2/PAG nanolubricants gives the maximum COP of 7.16 for 1200 rpm compressor speed. For the same compressor speed, COP value is recorded at 5.82 for pure lubricant. COP value for Al_2O_3-SiO_2/PAG nanolubricants is increased by 23% compared to pure lubricant.

Power consumption of Al_2O_3-SiO_2/PAG against refrigerant charge is illustrated in figure 5. The power consumption of 0.06% volume concentration is lower compared to pure lubricant. The maximum reduction was recorded as 28.24% at 1200 rpm for 155g refrigerant charge. Meanwhile, the minimum reduction of 13.09% at 1800 rpm compressor speed for 95g refrigerant charge was obtained. The power consumption of Al_2O_3-SiO_2/PAG was compared with the results from previous study by Redhwan et. al [24]. Al_2O_3-SiO_2/PAG nanolubricant has a lower dynamic viscosity compared to Al_2O_3/PAG thus lower the power consumption of the system and promotes energy saving. At higher viscosity, the compressor tends to work harder to overcome the viscosity resistance and more pumping force is needed to pump the refrigerant at the suction side. Hence, Al_2O_3-SiO_2/PAG is recommended to be used in AAC system as nanolubricants gives more advantages in AAC performance over pure lubricant.

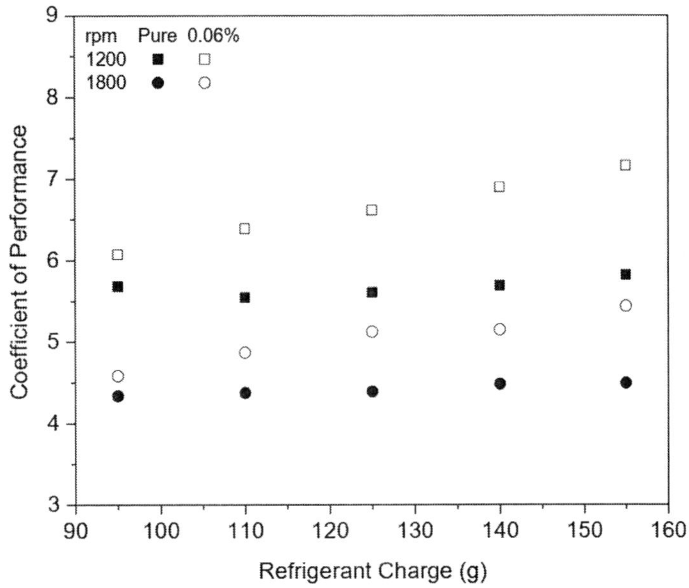

Figure 4. Coefficient of performance against refrigerant charges.

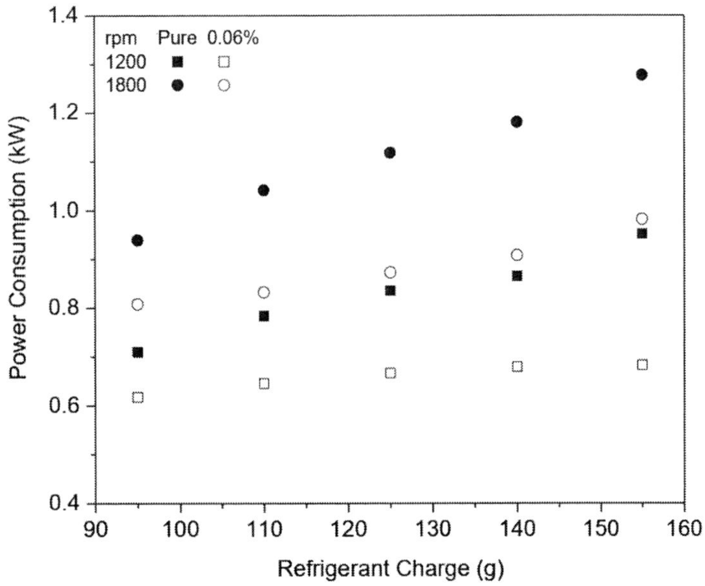

Figure 5. Power consumptions against refrigerant charges.

4. Conclusion

The performance of automotive air conditioning system operating with Al_2O_3-SiO_2/PAG nanolubricant was determined through the evaluation of cooling, compressor work, COP and power consumption. The highest cooling capacity and COP enhancement of 102.99% and 23% were recorded for AAC system utilized with 0.06% Al_2O_3-SiO_2/PAG nanolubricant. Meanwhile the compressor work and power consumptions were greatly reduced at maximum reduction of 25.9% and 28.24% respectively. Based on the findings, the author highly recommends the Al_2O_3-SiO_2/PAG nanolubricants for applications in automotive air systems as it yielded significant enhancements in system performance compared to PAG based lubricant. Further investigations on the performance of refrigeration system using the Al_2O_3-SiO_2/PAG nanolubricants are required to extend the present work.

Acknowledgments

The authors are grateful to the Ministry of Education Malaysia and Universiti Malaysia Pahang (www.ump.edu.my) for financial supports given under FRGS/1/2018/TK10/UMP/02/10 and PGRS190335. The authors also thank the research team from Automotive Engineering Centre (AEC), Centre of Excellence for Advanced Research in Fluid Flow (CARIFF) and Advanced Automotive Liquids Laboratory (AALL), who provided insight and expertise that greatly assisted in the present research work.

References

[1] Azmi W H, Sharif M Z, Yusof T M, Mamat R, Redhwan A A M 2017 *Renewable and Sustainable Energy Reviews* **69** 415-28.
[2] Atik K, Aktaş A 2011 *Isi Bilimi Ve Teknigi Dergisi/ Journal of Thermal Science and Technology* **31** 11-7.
[3] Hundy G F 2016 Elsevier Science.
[4] Dang C, Jia L, Xu M, Huang Q, Peng Q 2017 *International Journal of Heat and Mass Transfer* **104** 351-61.
[5] Aminullah A R M, Azmi W H, Redhwan A A M, Sharif M Z, Zawawi N N M, Kadirgama K, et al. 2018 *J Mech Eng* 49-61.
[6] Wu X, Xing Z, He Z, Wang X, Chen W 2017 *Applied Thermal Engineering* **112** 340-51.
[7] Dagilis V, Vaitkus L 2017 *International Journal of Refrigeration* **83** 99-107.
[8] Afshari F, Comakli O, Lesani A, Karagoz S 2017 *International Journal of Refrigeration* **74** 503-14.
[9] Sharif M Z, Azmi W H, Zawawi N N M, Mamat R, Shaiful A I M 2019 *IOP Conference Series: Materials Science and Engineering* **469** 012042.
[10] Kumar D S, Elansezhian R 2012 *International Journal of Modern Engineering Research* **2** 3927-9.
[11] Redhwan A A M, Azmi W H, Sharif M Z, Mamat R 2016 *International Communications in Heat and Mass Transfer* **76** 285-93.
[12] Nabil M F, Azmi W H, Hamid K A, Zawawi N N M, Priyandoko G, Mamat R 2017 *International Communications in Heat and Mass Transfer* **83** 30-9.
[13] Mahbubul I M, Saidur R, Amalina M A 2013 *International Communications in Heat and Mass Transfer* **43** 100-4.
[14] Zawawi N N M, Azmi W H, Sharif M Z, Shaiful A I M 2019 *IOP Conference Series: Materials Science and Engineering* **469** 012078.
[15] Bi S-s, Shi L, Zhang L-l 2008 *Applied Thermal Engineering* **28** 1834-43.
[16] Bi S, Guo K, Liu Z, Wu J 2011 *Energy Conversion and Management* **52** 733-7.
[17] Wang R, Wu Q, Wu Y 2010 *Energy and Buildings* **42** 2111-7.
[18] Jwo C S, Jeng L Y, Teng T P, Chang H 2009 *J Vac Sci Technol B* **27** 1473-7.
[19] Subramani N, Prakash M 2011 *International Journal of Engineering, Science and Technology* **3** 95-102.

[20] Sabareesh R K, Gobinath N, Sajith V, Das S, Sobhan C 2012 *international journal of refrigeration* **35** 1989-96.
[21] Hamisa A H, Azmi W H, Zawawi N N M, Sharif M Z, Shaiful A I M 2019 *IOP Conference Series: Materials Science and Engineering* **469** 012044.
[22] Zawawi N N M, Azmi W H, Redhwan A A M, Sharif M Z, Samykano M 2018 *International Journal of Refrigeration* **89** 11-21.
[23] Sharif M Z, Azmi W H, Redhwan A A M, Mamat R, Yusof T M 2017 *International Journal of Refrigeration* **75** 204-16.
[24] Redhwan A A M, Azmi W H, Sharif M Z, Mamat R, Samykano M, Najafi G 2019 *Journal of Thermal Analysis and Calorimetry* **135** 1299-310.
[25] Hamid K A, Azmi W H, Nabil M F, Mamat R, Sharma K V 2018 *International Journal of Heat and Mass Transfer* **116** 1143-52.
[26] Azmi W H, Hamid K A, Usri N A, Mamat R, Mohamad M S 2016 *International Communications in Heat and Mass Transfer* **76** 24-32.
[27] Usri N A, Azmi W H, Mamat R, Hamid K A, Najafi G 2015 *Energy Procedia* **79** 397-402.
[28] Azmi W H, Zainon S N M, Hamid K A, Mamat R 2019 *Journal of Mechanical Engineering and Sciences* **13** 5182-211.
[29] Sharif M Z, Azmi W H, Redhwan A A M, Zawawi N N M, Mamat R 2017 *J Mech Eng* **4** 233-47.
[30] Redhwan A A M, Azmi W H, Sharif M Z, Mamat R, Zawawi N N M 2017 *Applied Thermal Engineering* **116** 823-32.
[31] Zawawi N N M, Azmi W H, Redhwan A A M, Sharif M Z 2018 *J Mech Eng* **1** 28-38.
[32] Zawawi N N M, Azmi W H, Sharif M Z, Najafi G 2019 *Journal of Thermal Analysis and Calorimetry* **135** 1243-55.
[33] Kim J H, Cho J M, Lee I H, Lee J S, Kim M S 2007 *International journal of refrigeration* **30** 43-9.
[34] Pico D F M, da Silva L R R, Schneider P S, Bandarra Filho E P 2019 *International Journal of Refrigeration* **100** 104-12.
[35] Sabharwall P, Utgikar V, Gunnerson F 2009 *Nuclear Technology* **166** 197-200.
[36] Zawawi N N M, Azmi W H, Redhwan A A M, Sharif M Z, Sharma K V 2017 *International Journal of Refrigeration* **80** 1-10.
[37] Ahamed J U, Saidur R, Masjuki H H 2011 *Renewable and Sustainable Energy Reviews* **15** 1593-600.

IOP Conf. Series: Materials Science and Engineering **863** (2020) 012053 doi:10.1088/1757-899X/863/1/012053

Characterization of TiO_2 nanopaint for automotive application

S Z A Sakinah[1], W H Azmi[1,2,3,*] and J Alias[1]

[1] Department of Mechanical Engineering, College of Engineering, Universiti Malaysia Pahang, Lebuhraya Tun Razak, 26300 Gambang, Kuantan, Pahang, Malaysia.
[2] Automotive Engineering Centre, Universiti Malaysia Pahang, 26600 Pekan, Pahang, Malaysia.
[3] Centre of Excellence for Advanced Research in Fluid Flow, Lebuhraya Tun Razak, 26300 Kuantan, Pahang, Malaysia.

*E-mail: wanazmi2010@gmail.com

Abstract. Nanopaint is a coating that can modify the properties of a surface or substance according to user-defined parameters. Like ordinary paint, nanopaint is applied as a liquid and then hardens. Nanopaint is a suspension of liquid containing metal or non-metallic nanoparticles of typical size (1-100 nm) dispersed into the base coat paint. Single composite of Titanium oxide nanoparticle is consider an extension of research work for single nanopaint, which can be carry out through dispersed composites in basecoat paint. This paper presents characterization of single composition of TiO_2 nanopaint for automotive application. The nanopaint mixture was prepared at volume concentration of 1.0% using two-step method. The nanoparticles used namely TiO_2 is dispersed in a base fluid of polyester white paint. The investigation on the characterization for the nanopaint in the present study is conduct through adhesion cutter, SEM, Gloss meter and surface roughness meter. The findings from the investigations on the characterization of nanopaint show that the paint each all the required value for the normal paint characterization. Comparison data between 1% weight concentrations for nanopaint to the normal paint also confirm the increase in the surface roughness quality which is 0.10 less surface deviation compare to normal paint and gloss value up to 26% more glossier. It can be concluded that the nanopaint were successfully prepared and achieved good characterization.

1. Introduction

Automotive paint is one of the industries that move forward with the automotive technology to improve the surface performance [1]. Several techniques have been carried out to reduce the problem address by the end-user such as corrosion, mitigation and protection, durability, UV and heat resistance, strength and durability in thinner films or light weight substrate water and dirt repellence, anti-fouling, and enhanced colour [2]. Automotive clear coats are subjected to different kinds of environmental damage for instance UV radiation, acid rain, tree sap, bird droppings, stone chipping and marring. Marring is mainly due to car wash brushes, decreases the gloss of automotive paint and alters the colour [3]. Outdoor durability of an automotive coating especially the clear coat layer is one of the main concerns of automobile manufacturers. This concern arises from the fact that for most customers, retained appearance during service life is the most important criterion. Generally, basecoat is the colour producing layer, whilst the clear coat improves physical and chemical properties of the automotive coating system as well as enhancing its appearance [4]. Currently the clear coat formulation was add with HALS (hindered amine light stabilizer) and organic UV-absorbers. These ingredients considerably enhanced weathering performance, in addition to have high price, they can migrate to other layers and prone to undergo decomposition during service life.

The technology of nanoparticle dispersion was used in various engineering applications [5-9]. Currently, nanopaint is considered in the stage of research-and-development phase. Nowadays industries are targeting to explore more on the nanotechnology even in the painting and coating area. Such industries include the construction, automotive, aerospace and electrical component [10]. One of the most common materials to be used in the general coating application is Titanium Dioxide (TiO_2). In 1995, the concept of nanofluids was first introduced by Choi [11]. The attention receive by the researcher for TiO_2 nanoparticles because of its potential application in a wide range of fields such as photo catalytic and electro chromic devices, gas sensors, dye-sensitized solar cells, and energy-storage devices [12]. TiO_2 micro particle used in paints, mainly used as a white pigment. Only micrometres sized TiO_2 causes such white colouring. Nano scale TiO_2 functions as ultraviolet (UV) filter protecting the paints' binder material. Equally, it degrades organic materials via generation of radicals (photo catalytic activity), which is used for self-cleaning surfaces. The physicochemical properties of TiO_2 nanoparticles can be improve by controlling the particle size and morphology of the material. Some research groups have reported the formation of hierarchical urchin-like TiO_2 (U-TiO_2), composed of numerous Nano rods, by a facial hydrothermal process or precipitation at low temperatures, and its enhanced dielectric and photo catalytic properties [13].

2. Methodology

2.1. Material

The preparation of nanopaint involved a single nanoparticles namely TiO_2 mixed and dispersed in the polyester basecoat white paint as the base fluid. The TiO_2 nanoparticles were procured from US Research Nanomaterials, Inc. The respective nanoparticles size for TiO_2 is 30 nm with purity of 99.99%. The properties of TiO_2 nanoparticles are tabulated in table 1.

Table 1. Properties of TiO_2 nanoparticles [14, 15].

Properties	TiO_2
Molecular mass, g mol^{-1}	79.86
Average particle diameter, nm	50
Density, kg m^{-3}	4230
Thermal conductivity, W m^{-1} K^{-1}	8.4
Specific heat, J kg^{-1} K^{-1}	692

2.2. Preparation of single nanopaint

The two-step method is used for the preparation of single nanopaint. The nanopaint was prepared by mixing single nanoparticle (TiO_2) together; undergo a mixing and sonication process. Figure 1 shows the brief process in the nanopaint preparation. The preparation of the nanopaint was initially started with the calculation of the required weight according to the concentration. In the present study, the nanopaint was prepared at weight concentration of 1%. The normal paint was first prepared as per comparison to the nanopaint.

Sample preparation of single coating is done by dispersing nanoparticles into raw basecoat paint. nanoparticles exist in pigments, instead of measuring in volume, we need to consider the equation to calculate the concentration weight. Different concentration weight 1.0% of nanoparticles are dispersed into 100 ml raw basecoat paint. The equation (1) is used to calculate the concentration weight.

$$\omega = \frac{\omega_{np}}{\omega_{np} + \omega_{paint}} \times 100\% \tag{1}$$

The perfect spray viscosity of the paint are measured using viscosity cup (DIN 4@28c) about 13±1s. The mixing ratio of basecoat to solvent (slow thinner) is usually 1:0.8 according to the original viscosity of the paint (without nanoparticle). The perfect spray viscosity of the paint are measured using viscosity cup (DIN 4@28c) about 13 ± 1 s. To ensure the paint is not separated, it was stirred again by using magnetic stirrer for about 20 minutes (single nanopaint) then subjected to sonication process for another 30 minutes (single) to make sure the nanoparticle is dispersed uniformly inside the paint.

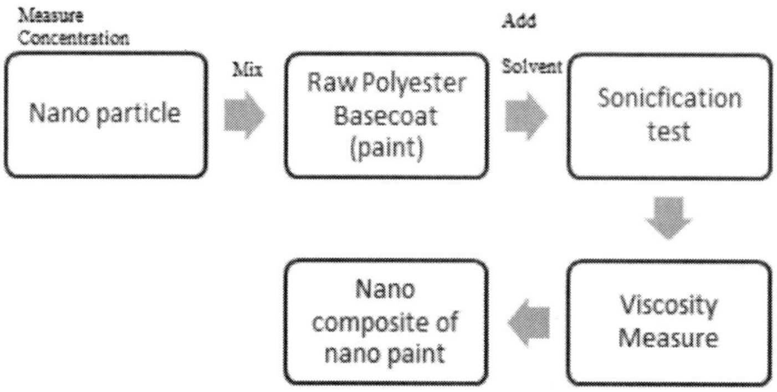

Figure 1. Sample preparation of composite nanopaint.

2.3. SEM
A scanning electron microscope (SEM) was used to observe the surface morphology of nanopaint at an accelerating voltage of 5 kV. A SEM image was taken to view the differences between the surface structures of single nanoparticles basecoat mixture with original basecoat without nanoparticles addition. The best concentration of single nanoparticles basecoat mixture was selected, to compare between the original basecoat. Scanning electron microscopy gave the information of much more value in the study of antifouling paint films than was obtained from the conventional release test.

2.4. Gloss
Gloss is a measurement proportional to the amount of light reflected from a surface. In order to achieve best results the correct measurement geometry should be chosen based on the reflectance of the material. Firstly, measure the gloss unit value at 60°. If the GU is between 50-70, use 60°, if more than >=70 GU, use 20° and if GU<=20, use 85°. These angle 20°, 60° and 85° which where light transmitted to the surface are intended for several measurement purpose. Standardly, for surface matt finish 85° was used as its measurement geometry, while mid gloss surface at 60° and high gloss and metallics at 20°. However, for this study, all three measurements geometry are taken into consideration to simplify the properties characterization.

2.5. Surface roughness
Surface roughness is one of the criteria that need attention when it comes to coating. Surface texture component of a coating often in cooperate with the surface roughness. It was noted that if the deviation is stated bigger, mean the surface not smooth. Smooth surface meaning small deviation. In this present study the comparison of the surface roughness between the concentrations of the nanoparticles for single nanopaint were being measure using the simple surface roughness meter. The unit is in µm. Lt: 5.600 mm, Lc: 0.800*n; n:5. The comparison between normal paint and 1wt% nanopaint are taken into account.

2.6. Adhesion
The coating adhesion experiment was conducted using test method B that is ASTM standard D3002/ D3359 and ISO 2409. For Test Method A, an X-cut was made through the film and the substrate,

pressure-sensitive tape is applied over the cut and then removed. The adhesion was then assessed qualitatively on the 0 to 5 scale. Test Method B, a lattice pattern with either six or eleven cuts in each direction is made in the film and to the substrate, pressure-sensitive tape is applied over the lattice and removed, and then evaluate the adhesion by comparing with the descriptions and illustrations. The apparatus and material needed are; i. Cutting Tool- BYK crosscut hatch, ii. Ruler-tempered steel ruler graduated in 0.5 mm for measuring individual cuts, iii. Tape (Tesa, 4124), iv. Rubber eraser, v. Illumination (aplanatisch) and vi. Magnifying glass. The magnifying glass is an illuminated magnifier to be used while making individual cuts and examining the test area.

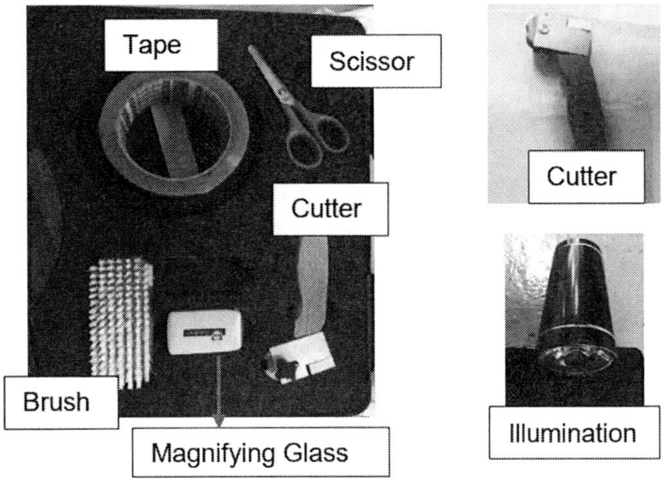

Figure 2. Crosscut hatch apparatus.

3. Results and discussion

3.1. Gloss characterization for TiO_2

The observation of gloss from 1% is shown in figure 3. The gloss of nanopaint is linearly increasing with the increasing of weight concentration. Based on the experiment conducted for gloss measurement for single TiO_2 nanoparticles, the trend in figure 4 shows that with the addition of nanoparticle, there are significant changes happened to the surface appearance, which is the gloss unit value, GU. Roughly, it can be concluded that with the increase of N-TiO_2 weight concentration, the surface is getting glossier. There are many possible reasons for those findings. The nanoparticle concentration is a crucial factor because they can impact rather negatively if there is a surplus of nanoparticles.

Based on figure 4, the comparison between percentage change of gloss between TiO_2 and normal paint were made. TiO_2 nanopaint had an increase of gloss value by approximately 26.45% with 1.0 wt. percentage of nanoparticles measured at an angle of incident 85°, 6.8% at 60°, and 2.3% at 20°. Based on the obtained results, it was evident that the nanopaint with 1.0 wt.% concentration of TiO_2 nanoparticles was the best sample of the nanopaint considered for gloss properties.

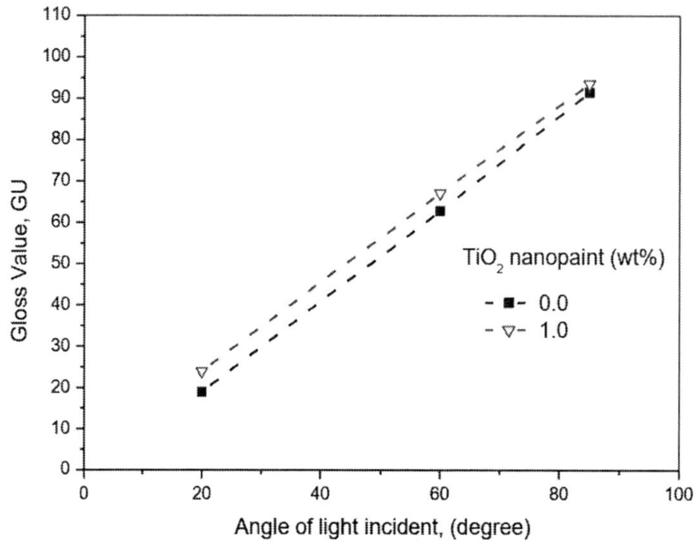

Figure 3. Graph of light incident angle against gloss value for single nanoparticle TiO_2.

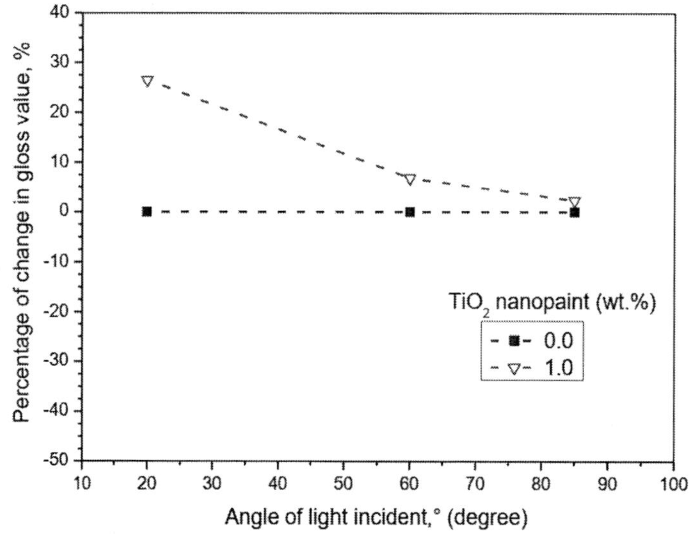

Figure 4. Graph of light incident angle against percentage change of gloss for single nanoparticles TiO_2.

3.2. Surface roughness

Based on figure 5 for single nanoparticles, the surface roughness of the coating can be investigated. The trend in figure 5 shows a decrease in surface roughness with addition of N-TiO_2. This also means that there is less porosity and the gap between the particles are very small resulted in smoother surface. The reason for this is because of TiO_2 does provide glass coating surfaces. Thus, the surface friction is also higher in normal paint surfaces compared with TiO_2 nanopaint.

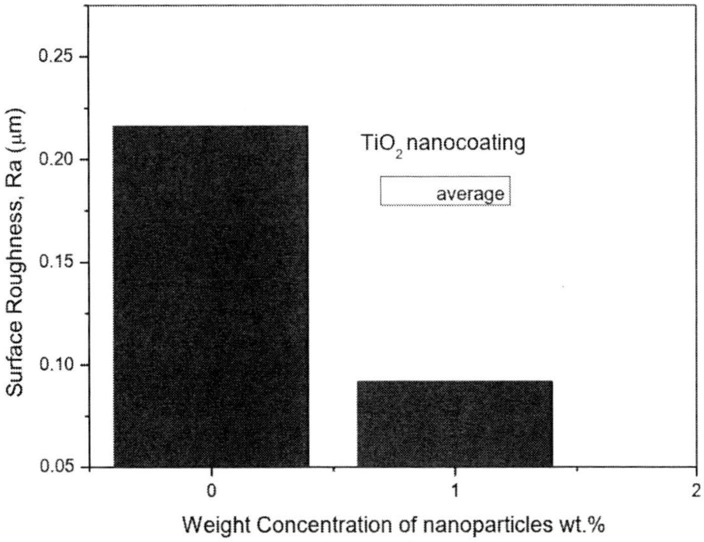

Figure 5. Graph of weight concentration of single nanoparticles against surface roughness TiO_2.

3.3. Adhesion

Adhesion is the most important criteria properties for coating film. Based on the experiment conducted for adhesion testing using crosscut hatch, the result shows a positive outcome for 1.0 wt.% of single TiO_2 nanopaint. The adhesion has then been complied with the classification of ISO2049 based on the adhesion test as shown in figure 6. The result shows none of flakes is detached at the intersection. The edges of the cuts are smooth, no square lattice is detached. This gave the 1.0% weight concentration of TiO_2 adhesion the rating of 5B. Hence, the adhesion evaluation has confirmed the nanopaint is same level with the current basecoat.

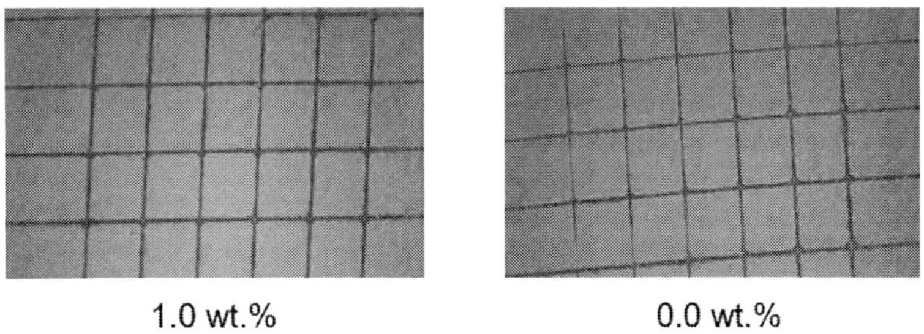

Figure 6. Result of adhesion test for added N-TiO_2 with different weight concentration.

3.4. SEM

Scanning electron microscopy technique (SEM) is used to illustrate the morphology of the nanostructures for the single N-TiO_2 and normal paint. Figure 7 shows that the single nanoparticles; TiO_2 and SiO_2 helps decreases porosity when compared with 0 wt.% nanoparticles. It is observed also in figure 7 that microstructure and pattern of normal paint and single N-TiO_2 are different and can be easily distinguished.

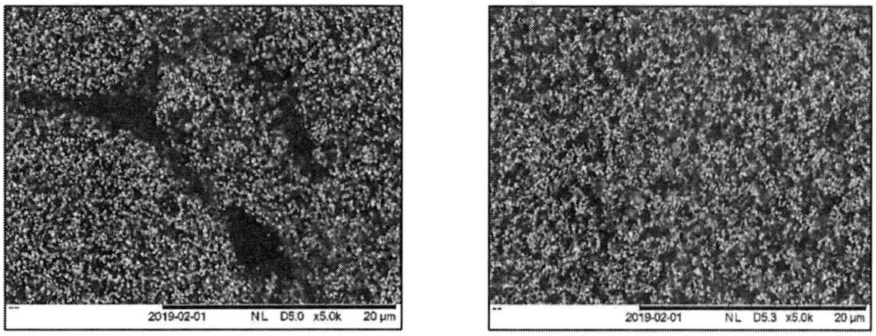

(a) 0.0% (b) 1.0%

Figure 7. SEM images of TiO_2 single nanoparticles.

4. Conclusion

This paper gave a comprehensive and up-to-date result of these new coating formulation to be of value to researchers. In conclusion the result of nanopaint characterization from the experimental investigation were thoroughly summarised. TiO_2 nanoparticles when mixed with basecoat paint lead to better surface appearance and structure such are increased in gloss unit value, reduce the surface roughness and improve adhesion to the substrate-primer surfacer. In view to the fact that the nanoparticles has their significant advantages to the properties mentioned. The final judgement of preliminary testing result based on characteristic selection criteria, 1wt% composition are the best mixing in terms of glossiness, surface roughness, adhesion, and SEM.

Acknowledgements

The authors are grateful to the Universiti Malaysia Pahang for financial supports given under RDU1803169. The authors also thank the research team from Automotive Engineering Centre (EAC) and Advanced Automotive Liquids Laboratory (AALL), who provided insight and expertise that greatly assisted in the research.

References

[1] Verma J, Nigam S, Sinha S, Bhattacharya A 2018 *Vacuum* **153** 24-34.
[2] Mohseni M, Ramezanzadeh B, Yari H, Moazzami M 2012 *New Advances in Vehicular Technology and Automotive Engineering* p.
[3] Bertrand-Lambotte P, Loubet J L, Verpy C, Pavana S 2001.
[4] Kango S, Kalia S, Celli A, Njuguna J, Habibi Y, Kumar R 2013 *Progress in Polymer Science* **38** 1232-61.
[5] Azmi W H, Hamid K A, Usri N A, Mamat R, Mohamad M S 2016 *International Communications in Heat and Mass Transfer* **76** 24-32.
[6] Abdolbaqi M K, Mamat R, Sidik N A C, Azmi W H, Selvakumar P 2017 *International Journal of Heat and Mass Transfer* **108** 1026-35.
[7] Azmi W H, Usri N A, Mamat R, Sharma K V, Noor M M 2017 *Applied Thermal Engineering* **112** 707-19.
[8] Redhwan A A M, Azmi W H, Sharif M Z, Mamat R 2016 *International Communications in Heat and Mass Transfer* **76** 285-93.
[9] Azmi W H, Zainon S N M, Hamid K A, Mamat R 2019 *Journal of Mechanical Engineering and Sciences* **13** 5182-211.
[10] Mathew J, Joy J, George S C 2018 *Journal of King Saud University-Science* In Press Corrected Proof https://doi.org/10.1016/j.jksus.2018.03.015.
[11] Choi S, Eastman J 1995 *IL (United States)*.

[12] Park K-S, Min K-M, Jin Y-H, Seo S-D, Lee G-H, Shim H-W, et al. 2012 *Journal of Materials Chemistry* **22** 15981-6.
[13] Ferrari A M, Pini M, Paolo N, Federica B 2015 *coatings*.
[14] Zawawi N, Azmi W, Redhwan A, Sharif M 2017 *IOP Conference Series: Materials Science and Engineering* **257** 012065.
[15] Hamid K A, Azmi W, Nabil M, Mamat R 2018 *International Journal of Heat and Mass Transfer* **118** 617-27.

Heat transfer characteristics of car radiator using tri-hybrid nanocoolant

A I Ramadhan[1,4], W H Azmi[1,2,3,*] and R Mamat[1,2,3]

[1] Department of Mechanical Engineering, College of Engineering, Universiti Malaysia Pahang, Lebuhraya Tun Razak, 26300 Gambang, Kuantan, Pahang, Malaysia.
[2] Automotive Engineering Centre, Universiti Malaysia Pahang, 26600 Pekan, Pahang, Malaysia.
[3] Centre of Excellence for Advanced Research in Fluid Flow, Lebuhraya Tun Razak, 26300 Kuantan, Pahang, Malaysia.
[4] Department of Mechanical Engineering, Faculty of Engineering, Universitas Muhammadiyah Jakarta, 10510 Jakarta, Indonesia.

*E-mail: wanazmi@ump.edu.my

Abstract. The use of nanoparticle coolant fluid in the car radiator increases the rate of heat transfer and facilitates the reduction of the overall radiator size. In this study, heat transfer characteristics of Al_2O_3-TiO_2-SiO_2 nanofluids-based water/ethylene glycol were analysed experimentally and compared with water/ethylene glycol mixture. Four different nanofluids concentrations were prepared by adding 0.05 to 0.3 vol.% of tri-hybrid nanofluid nanoparticles dispersed a mixture of water/ethylene glycol (60:40). Experiments were carried out by varying the flow rate of coolant between 2 to 12 LPM for working temperature of 70 °C, the velocity of airflow remained at an average of 4 m/s in order to analyse the effect of coolant flow rate on heat transfer. The results showed that the heat transfer coefficient of Al_2O_3-TiO_2-SiO_2 nanofluids or tri-hybrid nanofluids increased with increasing volume concentrations and temperatures. The maximum enhancement of the heat transfer coefficient for coolant side is observed at 39.7% at 0.3% volume concentration. The pressure drop and pumping power have the same pattern which increasing in volume concentrations, the pressure drop and pumping power will increase due to the concentration of the nanofluids. The correlation is applicable for water/ethylene glycol (60:40) mixture and Al_2O_3-TiO_2-SiO_2 nanofluids with volume concentrations of 0.05 to 0.3% at 70 °C working temperature.

1. Introduction

Conventional coolants had been widely hired to deplete heat in the majority of the engineering programs. Ordinary coolants consist of count number in all three states particularly solid, liquid and gas based totally at the requirements of the application and possible mode of heat transfer. However, with the modern technological improvements, an emerging magnificence of new coolants particularly nano-coolants (coolants with dispersed nanoparticles) find their applications in a variety of engineering applications [1-4].

New nanofluid applications use it as a replacement for conventional cooling in a car radiator, an important component of a car engine. Radiator serves as a heat exchanger for conventional car engine cooling using water as an exchange media. The thermal performance of a vehicle engine under the influence of nanofluid has been studied by many researchers, and the main application of nanofluid is as coolant and lubricant on the car radiator in an effort to improve the efficiency of heat removal. Results show that the heat transfer coefficient can be increased by more than 50% compared to conventional cooling but limited by a decrease in fluid pressure. However, experts may conclude that optimal performance can be achieved in low nano-particle volume fraction less than 1% (u <1%) [5]. Heris et al. [6] and Peyghambarzadeh et al. [7] have experimentally investigated the performance of metal oxide with water/Ethylene Glycol as a coolant in vehicle radiator. They found that increased heat transfer turned into observed growth during the overall performance of basic fluids. Similarly to Naraki et al. [8] used CuO/water nanofluid in an automobile radiator. Under laminar flow conditions they investigated nanofluid performance with a concentration of 0.15-0.4%. The overall heat transfer coefficient with nanofluid changes to be determined 6-8% more than water. Leong et al. [9] studying the overall performance of ethylene glycol mainly nanofluid based copper as a coolant in the vehicle radiator. They note that by adding 2% copper particles in EG 3.8% increase in heat transfer can be obtained from ethylene glycol under cooling turbulent flow situation. Ebrahimi et al. [10] conducted an experimental heat transfer study in a car radiator with nanofluid SiO_2 water. They find that the Nusselt range will increase as the temperature of the inlet is cooled, the Reynolds and the volume fraction increase.

In regarding on shape and measurement of the tube models, Park and Pak [11] presented a laminar flow computing study in flat tubes with different shapes and sizes using a mixture of ethylene glycol and water. They apply their calculations in Reynolds range variation from 10 to 200, which includes fluid flow at flow rates in radiator from 18 to 75 l/min for engines with volume transfer of 1800 cc. Vajjha et al. [12] conducted a numerical study to test the thermal performance of nanoparticles of Al_2O_3 and CuO in ethylene glycol and water mixture under laminar flow conditions using a flat tube radiator of the car. The results show a massive increase of the convective heat transfer coefficient beside the flat tube for nanofluid above the base fluid. The heat transfer coefficient of Al_2O_3 nanofluid with volume concentrations is 10% greater than the base fluid with an average of about 91%. In addition, both local and average friction factors and the convective heat transfer coefficient increase with concentration of nanoparticle volume.

Leong et al. [9] investigated analytically the thermal performance of Cu/Ethylene Glycol (EG) nanofluids in an automobile radiator. They utilized records from the literature and empirical correlations to model the thermal properties. The global heat transfer coefficient for the nanofluids was 3.8% for nanofluids higher than to base fluid, for a concentration 2% and Reynolds wide variety of 6000 for the air side and 5000 for the coolant aspect. This enhancement should offer a reduction of 18.7% of the radiator frontal area, with a penalty of 12.1% within the pumping energy.

Namburu et al. [13] the numerically analyzed turbulent flow and heat transfer into several types of nanofluides are copper (CuO), alumina (Al_2O_3) and silicon (SiO_2) in ethylene glycol and water, flowing through circular tubes under constant heat flux. Results found that nanofluid containing tiny lower-density nanosurfer produced higher viscosity and Nusselt numbers. Nusselt numbers are also enlarged on higher particle volume fractions. It was found that at constant heat flux (50 W/cm^2) with a constant Reynolds number (20,000), the heat transfer coefficient of 6% CuO nanofluid has increased 1.35 times the base fluid. At the same fraction of particle volume, CuO nanofluid produces a higher heat transfer coefficient than the different types of nanofluids.

Devireddy et al. [14] conducted a forced conventional heat transfer study in a car radiator with a water-based ethylene glycol and nanofluid TiO_2. Nanofluid was made by taking 40:60 (EG:W) of the mixture as a basic liquid and dispersing the TiO_2 nanoparticles in 0.1%, 0.3% and 0.5% based on volume concentration. They observed a 37% increase in the rate of heat transfer at 0.5% TiO_2 when compared to base fluid. Nambeesan et al. [15] studying characteristics of heat transfer Al_2O_3/water-ethylene glycol cooling in the car radiator. Increased heat transfer of about 37% was obtained with 0.1% Al_2O_3 nanowires. The authors also conducted experiment with a mixture of water/propylene glycol as a basic liquid. The increase of 9% in overall heat conductance was obtained by adding 0.2% alumina nanoparticles into a propylene glycol based coolant liquid. In addition, various experimental works were conducted to investigate the heat transfer performance of nanofluids for various cooling applications [16].

The objective of this study is to investigate of the heat transfer characteristics of the radiator using water/ethylene glycol based tri-hybrid nanofluids with a volume concentration of 0.05-0.3% as coolant was analysed experimentally in automotive radiator setup. Characteristics of heat transfer and behaviour of decreased nanofluid pressure compared to a mixture of base coolant fluid.

2. Methodology

2.1. Material

The preparation of *tri*-hybrid nanofluids involved three different types of single nanofluids namely Al_2O_3, TiO_2 and SiO_2 mixed together and dispersed in the base fluid of water/EG mixture. All the single nanofluids were procured from US Research Nanomaterials, Inc. The respective nanoparticles size for Al_2O_3, TiO_2 and SiO_2 are 13, 50 and 23 nm with a purity of 99.8%, 99% and 99.99%. The properties of each nanoparticle types are given in table 1. The base fluid used in the present study was a mixture of water and EG at a ratio of 60:40 (vol.%). Meanwhile, the properties of ethylene glycol is tabulated in table 2.

Table 1. Properties of Al_2O_3, TiO_2 and SiO_2 nanoparticles [17, 18].

Properties	**Al_2O_3**	**TiO_2**	**SiO_2**
Molecular mass, g mol^{-1}	101.96	79.86	60.08
Average particle diameter, nm	13	50	23
Density, kg m^{-3}	4000	4230	2220
Thermal conductivity, W m^{-1} K^{-1}	40	8.4	1.4
Specific heat, J kg^{-1} K^{-1}	773	692	745

Table 2. Properties of Ethylene Glycol (EG) [19].

Properties	**EG**
Vapour pressure, mmHg at 20 °C	0.08
Boiling point, °C	195–198
Melting point, °C	-13
Density, g ml^{-1} at 25 °C	1.113

2.2. Preparation of tri-hybrid nanofluids

Two-step method is used for the preparation of *tri*-hybrid nanofluids. *Tri*-hybrid nanofluids are prepared by mixing the three nanofluids (Al_2O_3, TiO_2 and SiO_2) together, undergoing mixing and sonication based processes by Ramadhan et al. [20]. Preparation of nanofluids initially begins with the volume calculations required in accordance with concentration. In this study, *tri*-hybrid nanofluides were prepared at volume concentrations of 0.05, 0.1, 0.2, and 0.3%. Nanofluid was first prepared at the highest concentrations of 0.3% and then diluted to a lower concentration.

Single nanofluids Al_2O_3, TiO_2 and SiO_2 are supplied in water suspensions with a weight concentration of 20, 40, 25% for Al_2O_3, TiO_2 and SiO_2 respectively. Equation (1) [21, 22] is used to convert from weight concentration to volume concentration. The dilution from higher volume concentration to lower volume concentration utilized the equation (2) [19] by adding the base fluid (ΔV).

$$\phi = \frac{\omega\rho_w}{\frac{\omega}{100}\rho_w + \left(1 - \frac{\omega}{100}\right)\rho_p} \tag{1}$$

$$\Delta V = (V_2 - V_1) = V_1\left(\frac{\phi_1}{\phi_2} - 1\right) \tag{2}$$

All single nanofluids are mixed together at a ratio of 1/3: 1/3: 1/3 to form *tri*-hybrid nanofluid. Total volume of 100 ml is prepared for each *tri*-hybrid nanofluid concentration. The combined solutions of the three single nanofluids of Al_2O_3, TiO_2 and SiO_2 were mixed together using a magnetic stirrer for 120 minutes. Then, the solution undergoes a sonication process using ultrasonic bath to increase stability.

2.3. *Automotive radiator setup*
The heat transfer performance of nanofluid coolers is measured by experimental use in automotive radiator settings as shown in figure 1. It consists of car radiator, heater storage tank, centrifugal pump, fan blower, flow control valves and thermocouple K-type to measure fluid temperature incoming and outgoing channels. 3.3 kW electric heaters are used to heat the coolant in the reservoir tank. Coolant is distributed using centrifugal pump 1.0 HP. The control valve in the pump is used to vary the flow rate of the coolant fluid into the radiator between 2-12 LPM. Two K-type thermocouples are placed in the inlet and radiator outlet to measure the coolant fluid temperature. Thermocouples are also installed on both sides of the radiator wall surface to measure the air temperature and radiator tubes. The radiator specification as shown in table 3.

2.4. *Automotive radiator procedure*
The experimental heat transfer was performed in radiator experiments preparation of water/EG (60:40) and *tri*-hybrid nanofluids with volume concentration of 0.05 - 0.3%. The coolant in the collection tank with volume 17 litre is heated to the desired temperature and circulated through the radiator using the pump. The temperature of the coolant inlet to the radiator is maintained constant at nominal operating temperature from 60 to 70 °C. The cooling flow rate varies between 2 to 12 LPM. The air flow rate to the radiator is maintained constant at an average of 4 m/s. The temperature of the cooling outlet is recorded using the K-type thermocouple. Furthermore, the K-type thermocouple remains on the radiator wall on both sides to record air temperature and temperature wall tube.

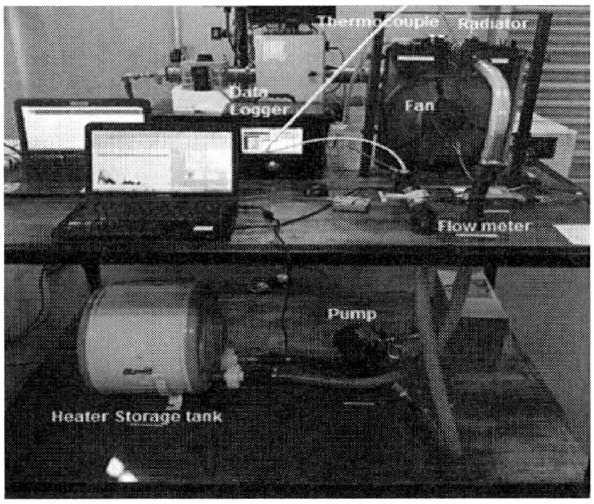

Figure 1. Automotive radiator setup.

Table 3. Components of automotive radiator setup.

Main Section	Description	Function
Radiator	Dimension radiator: 34.0 mm × 1.3 mm × 36.8 mm n: 32 vertical aluminum tubes	Test section for the heat transfer analysis.
Data logger	ADAMView Advantech Data Acquisition	To record the experimental data of temperature and pressure
Fan	Range: 0 to 1250 RPM	For the cooling process of the system.
Heater storage tank	Volume: 17 litre	To store the nanofluid.
Pump	1.0 HP	To circulate working nanofluid to the whole system.
Pressure transducer	Differential type, 5.0 psi	To measure pressure drop of the fluid flow.
Flow rate meter	Range: 2.0 to 14.0 LPM	To measure the fluid flow rate.

2.5. Nanofluid thermo physical properties

The density (ρ_{nf}) and specific heat capacity ($C_{p,nf}$) of *tri*-hybrid nanofluids are obtained by the relation equations (3) and (4) [20].

$$\rho_{nf} = (1-\varphi)\rho_{bf} + \varphi(R\rho)_{Al_2O_3} + \varphi(R\rho)_{TiO_2} + \varphi(R\rho)_{SiO_2} \tag{3}$$

$$C_{p,nf} = \frac{(1-\varphi)\rho_{bf}C_{bf} + \varphi(R\rho C)_{Al_2O_3} + \varphi(R\rho C)_{TiO_2} + \varphi(R\rho C)_{SiO_2}}{\rho_{nf}} \tag{4}$$

Table 4 presented the thermo physical properties of tri-hybrid nanofluids at 70 °C temperature for used in automotive radiator setup.

Table 4. Thermo physical properties of *tri*-hybrid nanofluids at temperature of 70 °C.

ϕ (%)	ρ_{nf} (kg/m^3)	μ_{nf} (Ns/m^2)	C_{nf} (J/kg.K)	k_{nf} (W/m.K)
0.05	1034.59	0.003313	3631.12	0.465
0.1	1035.82	0.003586	3626.25	0.471
0.2	1038.27	0.003475	3616.55	0.473
0.3	1040.71	0.003712	3606.89	0.481

2.6. Experimental data analysis for coolant side

Parameters required for nanofluid side calculations are heat transfer coefficient, nanofluid heat capacity, pressure drop, pumping power. The mathematical formulas used for heat transfer coefficient can be expressed as equation (5).

$$h_{nf} = \frac{Nu_{nf}k_{nf}}{D_{h,nf}} \tag{5}$$

where k_{nf} is obtained from Eastman et al. [23]. The nusselt number for nanofluid is expressed by equation (6) as implemented by Hamid et al. [18].

$$Nu_{nf} = 0.0247\mathrm{Re}^{0.8}\,\mathrm{Pr}^{0.4}\left(1+\frac{\phi}{100}\right)^{6.29} \tag{6}$$

The Reynolds number (Re) expression in equation (6) for nanofluid is given by equation. (7).

$$\mathrm{Re}_{nf} = \frac{G_{nf} D_{h,nf}}{\mu_{nf}} \tag{7}$$

Parameter G_{nf} is define as the Core mass velocity of coolant that can be expressed as equation (8):

$$G_{nf} = \frac{W_{nf}}{A_{fr} \sigma_{nf}} \tag{8}$$

The heat capacity rate, C_{nf} can be expressed as equation (9).

$$C_{nf} = W_{nf} C_{p,nf} \tag{9}$$

Meanwhile, pressure drop, ΔP_{nf} can be expressed as equation (10).

$$\Delta P_{nf} = \frac{G^2{}_{nf} \times f_{nf} \times H}{2 \times \rho_{nf} \times \left(\frac{D_{h,nf}}{4} \right)} \tag{10}$$

where the friction factor correlation of nanofluid is given equation (11) by Hamid et al. [18],

$$f_{nf} = 0.3164 \times (\mathrm{Re}_{nf})^{-0.25} \left(\frac{\rho_{nf}}{\rho_{bf}} \right)^{0.797} \left(\frac{\mu_{nf}}{\mu_{bf}} \right)^{0.108} \tag{11}$$

The pumping power can be expressed as equation (12).

$$P = \left(\frac{W_{nf}}{\rho_{nf}} \right) \Delta P_{nf} \tag{12}$$

3. Results and discussion

3.1. Heat transfer performance

Figure 2 and figure 3 illustrate the heat transfer coefficient and Nusselt number of Al_2O_3-TiO_2-SiO_2 nanofluids at 70 °C. It can be observed that the experimental data for Al_2O_3-TiO_2-SiO_2 nanofluids is in good agreement with the data trend given by Dittus and Boelter [24] and the base fluid. The Reynolds number was influenced by the increasing of heat transfer coefficient and Nusselt number by 39.7% maximum enhancement occurs at 0.3% whereas 45% enhancement occurs at concentration 0.3%. However, all of the concentration is higher than the base fluid in the pattern. The increasing of temperature, volume concentration and Reynolds number were enhanced the heat transfer and Nusselt number. At 70 °C the range of Reynolds number is higher up to 5000. This happened due to the decreasing of viscosity because of the higher temperature. The higher of thermal properties would improve the heat transfer coefficient.

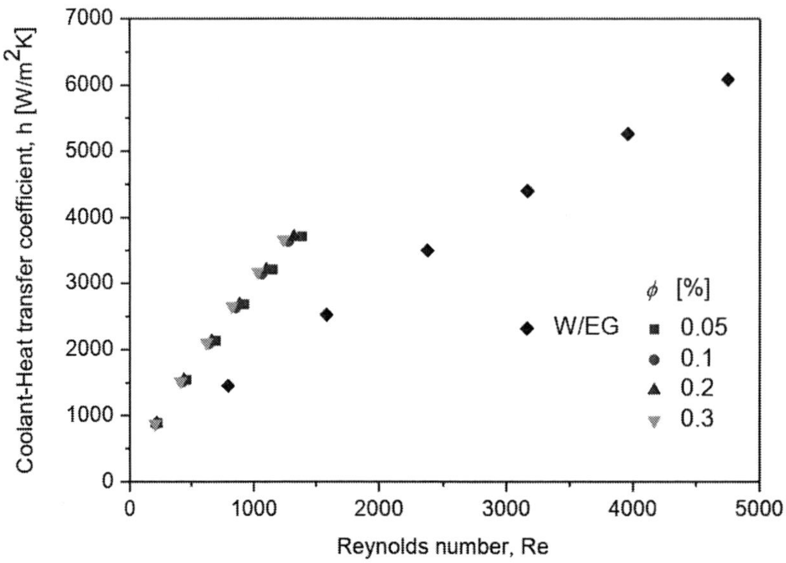

Figure 2. Heat transfer coefficient of Al_2O_3-TiO_2-SiO_2 nanofluids at working temperature 70 °C.

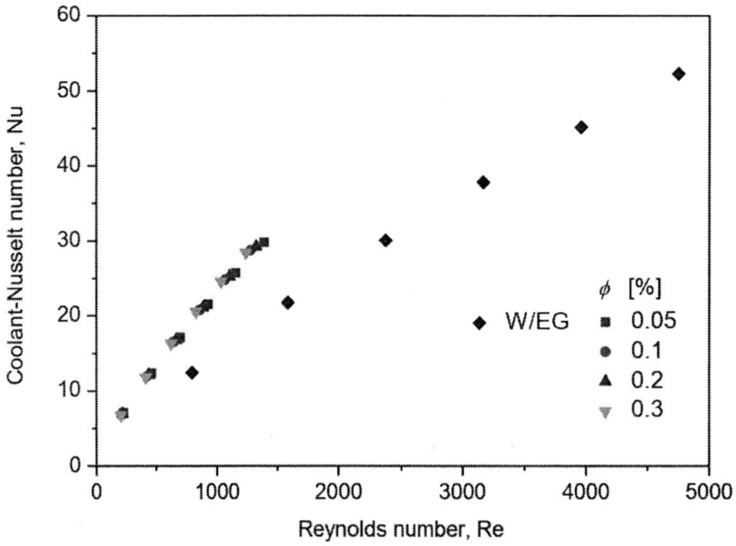

Figure 3. Nusselt number of Al_2O_3-TiO_2-SiO_2 nanofluids at working temperature 70 °C.

3.2. Pressure drop and pumping power

Figure 4 and figure 5 present pressure drops and pumping power from the Al_2O_3-TiO_2-SiO_2 nanofluids at 70 °C working temperature. The pressure drop and pumping power of the *tri*-hybrid nanofluids increased with increasing volume concentration. This is because at higher volume concentration, the viscosity of the nanofluids will also increase. Therefore it will cost pressure and power to increase in the system. The highest pressure drop observed was at 0.3% volume concentration with 180 kPa. Meanwhile for the pumping power, the highest was also observed at 0.3% with 0.035 kW of power. The lowest pressure drop and pumping power could be observed at 0.05% volume concentration with 5 kPa and 0.001 kW. Indicates that all concentrations are above the base fluid which is increased from the base fluid.

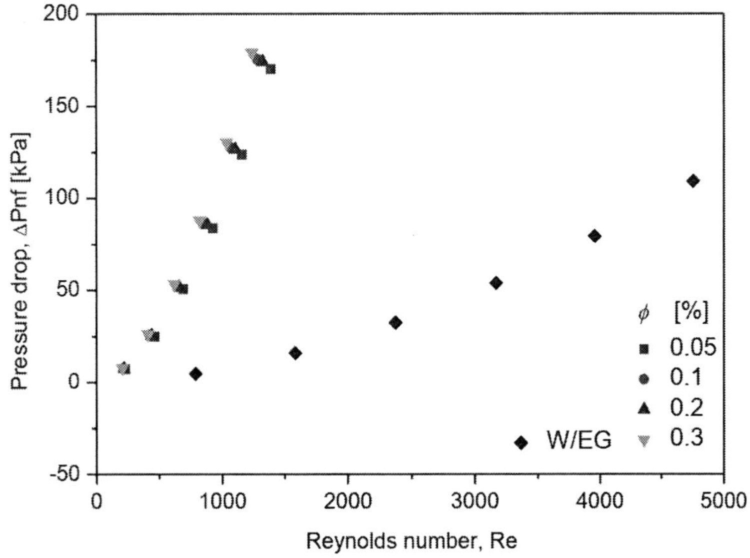

Figure 4. Pressure drop of Al_2O_3-TiO_2-SiO_2 nanofluids at working temperature 70 °C.

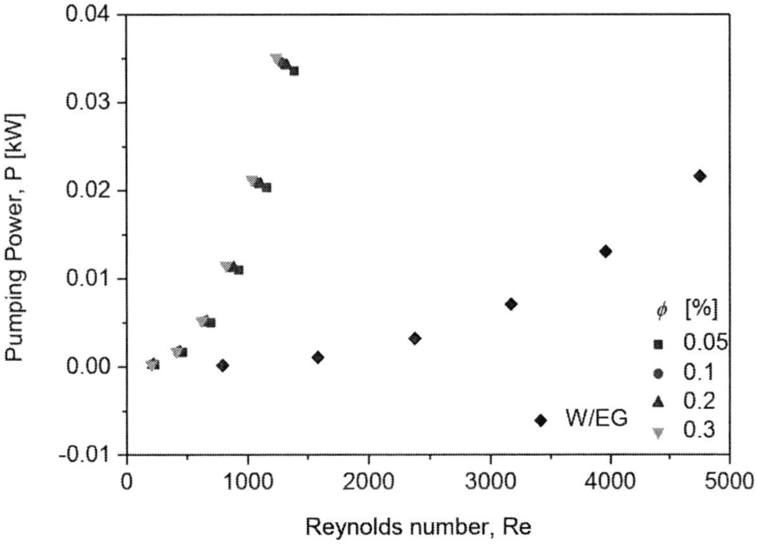

Figure 5. Pumping power of Al_2O_3-TiO_2-SiO_2 nanofluids at working temperature 70 °C.

4. Conclusion

In the present study, thermal performance of Al_2O_3-TiO_2-SiO_2 nanofluids in mixture water/EG (60:40) has been investigated for volume concentrations of up to 0.3% and working temperature of 70 °C. The heat transfer coefficient of Al_2O_3-TiO_2-SiO_2 nanofluids increases with increasing volume concentration and temperature. The maximum enhancement of the heat transfer coefficient for coolant side is observed at 39.7% at 0.3% volume concentration. The pressure drop and pumping power have the same pattern which increases in volume concentrations, the pressure drop and pumping power will increase due to the concentration of the nanofluids. The correlation is applicable for water/EG (60:40) mixture and Al_2O_3-TiO_2-SiO_2 nanofluids with volume concentrations of 0.05 to 0.3% at 70 °C working temperature.

Acknowledgments

The authors are grateful to the Ministry of Education Malaysia and Universiti Malaysia Pahang (www.ump.edu.my) for financial supports given under FRGS/1/2018/TK10/UMP/02/10 and RDU190146. The authors also thank the research team from Automotive Engineering Centre (AEC), Centre of Excellence for Advanced Research in Fluid Flow (CARIFF) and Advanced Automotive Liquids Laboratory (AALL), who provided insight and expertise that greatly assisted in the present research work.

References

[1] Choi S U S, Eastman J A. Enhancing thermal conductivity of fluids with nanoparticles. Argonne National Lab., IL (United States), 1995.
[2] Usri N A, Azmi W H, Mamat R, Hamid K A, Najafi G 2015 *Energy Procedia* **79** 397-402.
[3] Azmi W H, Usri N A, Mamat R, Sharma K V, Noor M M 2017 *Applied Thermal Engineering* **112** 707-19.
[4] Azmi W H, Zainon S N M, Hamid K A, Mamat R 2019 *Journal of Mechanical Engineering and Sciences* **13** 5182-211.
[5] Sidik N A C, Yazid M N A W M, Mamat R 2017 *Renewable and Sustainable Energy Reviews* **75** 137-44.
[6] Heris S Z, Shokrgozar M, Poorpharhang S, Shanbedi M, Noie S 2014 *Journal of dispersion science and technology* **35** 677-84.
[7] Peyghambarzadeh S M, Hashemabadi S H, Hoseini S M, Jamnani M S 2011 *International Communications in Heat and Mass Transfer* **38** 1283-90.
[8] Naraki M, Peyghambarzadeh S M, Hashemabadi S H, Vermahmoudi Y 2013 *International Journal of Thermal Sciences* **66** 82-90.
[9] Leong K Y, Saidur R, Kazi S N, Mamun A H 2010 *Applied Thermal Engineering* **30** 2685-92.
[10] Ebrahimi M, Farhadi M, Sedighi K, Akbarzade S 2014 *IJE Trans B: Appl* **27** 333-40.
[11] Park K W, Pak H Y 2002 *Numerical Heat Transfer: Part A: Applications* **41** 19-40.
[12] Vajjha R S, Das D K, Namburu P K 2010 *International Journal of Heat and fluid flow* **31** 613-21.
[13] Namburu P K, Das D K, Tanguturi K M, Vajjha R S 2009 *International Journal of Thermal Sciences* **48** 290-302.
[14] Devireddy S, Mekala C S R, Veeredhi V R 2016 *International communications in heat and mass transfer* **78** 121-6.
[15] Nambeesan K P, Parthiban R, Kumar K R, Athul U R, Vivek M, Thirumalini S 2015 *International Journal of Automotive & Mechanical Engineering* **12**.
[16] Abdolbaqi M K, Mamat R, Sidik N A C, Azmi W H, Selvakumar P 2017 *International Journal of Heat and Mass Transfer* **108** 1026-35.
[17] Zawawi N N M, Azmi W H, Redhwan A A M, Sharif M Z 2017 *IOP Conference Series: Materials Science and Engineering* **257** 012065.
[18] Hamid K A, Azmi W H, Nabil M F, Mamat R 2018 *International Journal of Heat and Mass Transfer* **118** 617-27.
[19] Lim S K, Azmi W H, Yusoff A R 2016 *International Communications in Heat and Mass Transfer* **78** 182-9.
[20] Ramadhan A I, Azmi W H, Mamat R, Hamid K A, Norsakinah S 2019 *IOP Conference Series: Materials Science and Engineering* **469** 012068.
[21] Azmi W H, Hamid K A, Usri N A, Mamat R, Mohamad M S 2016 *International Communications in Heat and Mass Transfer* **76** 24-32.
[22] Hamid K A, Azmi W H, Mamat R, Sharma K V 2016 *International Communications in Heat and Mass Transfer* **73** 16-24.
[23] Eastman J A, Choi S U S, Li S, Yu W, Thompson L J 2001 *Applied physics letters* **78** 718-20.

[24] Dittus F W, Boelter L M K 1985 *International Communications in Heat and Mass Transfer* **12** 3-22.

TiO_2-SiO_2 nanofluid characterization: Towards efficient with water/ethylene glycol mixture for solar application

M A Fikri[1,5,*], F F Asri[2], W M Faizal[1], H K Adli[1], R Mamat[2], W H Azmi[2], G Najafi[3] and T Yusaf[4]

[1] Faculty of Bioengineering and Technology, Universiti Malaysia Kelantan, 17600 Malaysia.
[2] Department of Mechanical Engineering, College of Engineering, Universiti Malaysia Pahang, Lebuhraya Tun Razak, 26300 Gambang, Kuantan, Pahang, Malaysia.
[3] Tarbiat Modares University, Jalale-E-Aleahmad Highway, Tehran, Iran.
[4] Federation University, Melbourne, Australia.
[5] Engineering Divison, Jabatan Kesihatan Negeri Kelantan, 15590 Kota Bharu, Kelantan, Malaysia.

*E-mail: moamfik.selehor@gmail.com

Abstract. Nowadays renewable energy has been used widely as strong contenders to improve the plight of two billion people, mostly in rural areas, without access to modern forms of energy. In this study, the effect of using water/ethylene glycol-based TiO_2-SiO_2 nanofluid on the characteristic of the solar radiation test rig has been experimentally used to investigate the characteristic of TiO_2-SiO_2 nanofluid suspended in water/ethylene glycol for solar applications. The TiO_2-SiO_2 nanofluids were prepared at a volume concentration of 0.3, 0.5 and 0.7% and dispersed in a base fluid of water/ethylene glycol 60:40 and 70:30 TiO_2-SiO_2 with the one-step method. KD2 Pro measures thermal conductivity set equipment and viscosity by LVDV III Ultra Rheometer. Thermal conductivity results show that the increase with different temperatures but the temperature of viscosity decreases. For solar radiation test rig that the highest point indicates the end of the charging process and the beginning of the discharging process, the maximum concentrations of 0.7% are observed to consistently provide higher temperature output as compared to the other three concentrations of nanofluids based on solar radiation 300, 500 and 700 W/m^2. Therefore, it can be concluded that the higher concentrations give ample time to the test tube to transfer the heat and thus increased its temperature during the charging process.

1. Introduction

Renewable energy sources play a crucial role in the effort of solving environmental problems efficiently and effectively for energy security and sustainability. Renewable energies are supplied (such as from biofuels, solar heat, photovoltaics, wind, hydro, wave, tidal. Geothermal, and ocean-thermal). Future development of renewable energy will enable a further reduction of environmental impacts. Renewable energy sources that meet domestic energy requirements provide energy services with zero emissions of both air pollutants and greenhouse gases [1]. Nowadays, it is about two billion peoples, mostly in rural areas depend on renewable energy sources without utilizing any modern forms of energy [2]. A further benefit of implementing renewable energy technologies in rural areas are provided job opportunities and at the same time, decrease the migration towards urban areas [3]. It shows that renewable energy sources are not only important element in energy sustainability, but also contributes in improving the life of several communities as well.

In recent years, solar energy has become the most important source of energy. The usage of fossil fuels might be reduced in the future due to perceiving a shortage of fossil fuels and for the sake of the environment. Solar energy has become an alternative source of energy to overcome those barriers. Mostly solar energy applications are cost-effective and require only a few kilowatts of power for individual small systems [4, 5]. Utilization of solar energy should be applied in a wide range area since it contributes to energy stability improvement, increase energy sustainability, and increase system efficiency [6-9]. There are many studies conducted in solar, as reported in many kinds of literature [10-12].

Nanofluid is the common fluid which has been suspended with ultra-fine solid particles [13-16]. Commonly used fluids such as water, ethylene glycol, and heat transfer oil have contributed significant benefits to many industrial processes such as power generation, heating or cooling process, chemical processes, and microelectronics. However, the thermal conductivity of these fluids is relatively low and unable to reach high heat exchange rates in thermal engineering devices. In order to overcome this particular obstacle, various investigation of using nanofluid to improve the thermal conductivity of those fluids was carried out. Recently, the effects of nanofluids have been investigated by researchers on the enhancement of heat transfer in thermal engineering devices for experimentally and theoretically. Variety preparation methods, characteristics, and different models used for the calculation of thermo-physical properties of nanofluids has been applied by the researchers [17-24]. The effects of nanofluids on flow and heat transfer in natural and forced convection in different systems also have been summarized in some researches [25-29]. Special properties of nanofluids has given benefits in various applications.

Hybrid nanofluid is the new generation of nanofluid which composed of two variant types of dispersed nanoparticles in a base fluid. Investigation of various applications by using hybrid nanofluids has increased recently [10, 30-32]. Hybridization of nanofluids can increase the thermal conductivity of nanofluids. However, proper hybridization process is needed to get the best performance of hybrid nanofluid in heat transfer [30]. Hybrid nanofluid has served not only better performance in term of thermal conductivity than single-particle nanofluids but also provide high impacts of nanofluid utilization in present technologies [33-35].

Therefore, the objective of this study is to investigate the characteristic of TiO_2-SiO_2 nanofluid suspended in water/ethylene glycol 60:40 and 70:30 TiO_2-SiO_2 with one-step method for solar applications. This study will help researchers to get an idea about the effect of TiO_2-SiO_2 nanofluid based on thermophysical properties to apply to the whole system of solar used.

2. Materials and methods

2.1. Thermo-physical properties measurement

Thermal conductivity of the samples was measured using KD2 Pro thermal property analyzer of Decagon Devices, Inc., USA. The data were collected for a temperature range of 30 to 70 °C after two hours of the sonication process. Various investigators used KD2 pro thermal property analyzer in their measurements of thermal conductivity [36-41]. This instrument applied the transient hot-wire method. The present measurement method allowed the thermal conductivity measurement of nanofluids with minimum free convection effects. The experiment was performed five times for each sample and condition, and data point reported in this study represents an average of five measurements with an estimated error of ±1.2%. The set up for the thermal conductivity measurement is shown in figure 1 consists of a KD2 Pro thermal property analyser, water bath, a single needle sensor and a sample of hybrid nanofluids.

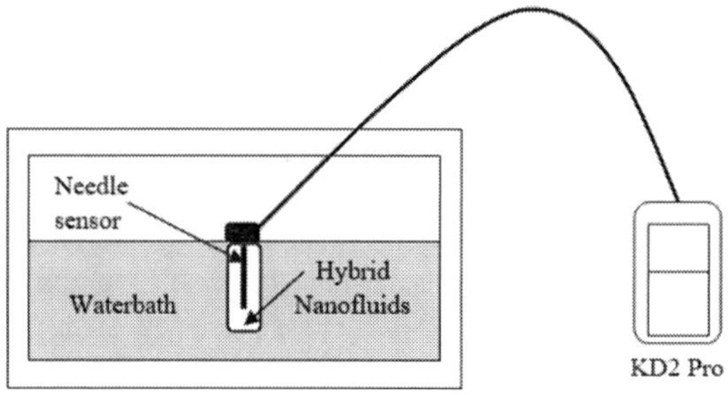

Figure 1. Apparatus and Instrument used in thermal conductivity measurement.

The viscosity of the nanofluids was measured using Brookfield LVDV-III Ultra Rheometer. Several investigators were used the same approach in their work for measurement of viscosity [42-44]. The viscosity was measured in temperatures between 30 °C and 70 °C, and the values were recorded at steady-state conditions, and 30 minutes was allowed to stabilize the temperature. The set up for the viscosity measurement is shown in figure 2.

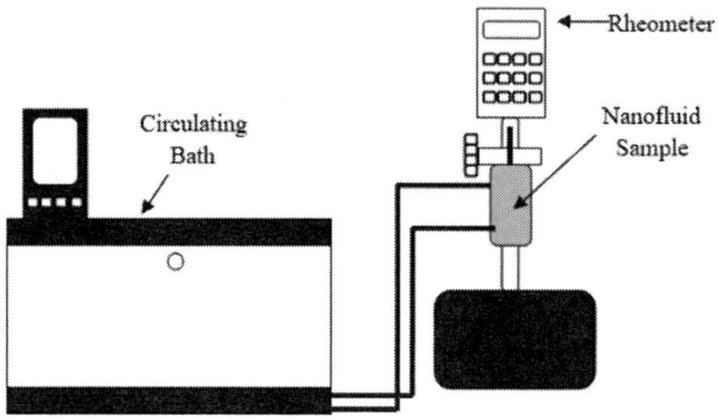

Figure 2. LVDV III Ultra Rheometer used in viscosity measurement.

2.2. The set up for experiments

Figure 3 shows the experimental setup with the solar simulator. Experiments were conducted for nanofluid concentrations ranging from 0.3, 0.5 and 0.7%. The parameters and properties involved in the experiment are given in table 1.

For each concentration, the data in the test tube is collected. The measurement of heat rates on solar simulators was taken using a pyranometer. Each preparation is exposed to short wavelength radiation under the solar simulator with 300, 500 and 700 W/m^2 for 30 minutes, of which 15 minutes is the heating period and the next 15 minutes for a cooling period.

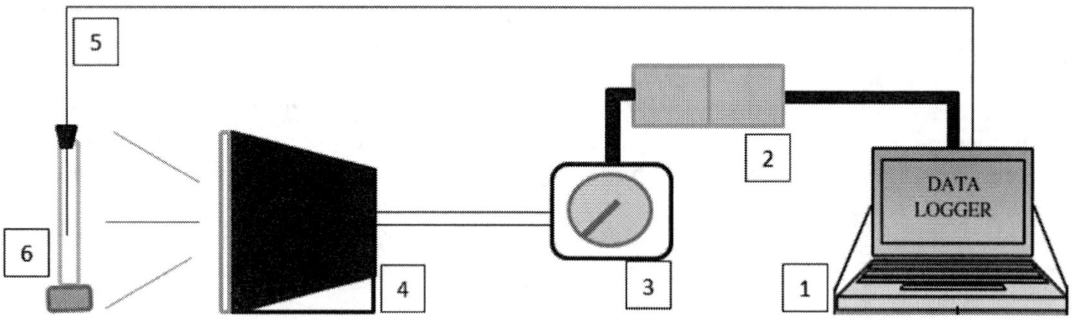

Figure 3. The solar radiation test rig.

Table 1. Detail description of the solar radiation test rig.

Description	Specification	Number
Data logger	ADAMView Advantech Data Acquisition	1
SSO	Switch Socket Outlet (Data Logger & Spotlight Dimmer)	2
Dimmer	Setting up with pyranometer (300, 500 & 700 W/m^2)	3
Spotlight	Halogen 500 W (1 Piece)	4
Thermocouple	K-type	5
Test Tube	100 ml	6

3. Results and discussion

3.1. Viscosity

Figure 4 shows the data analysis for the viscosity measurement under different temperatures for the various concentration of nanofluid. For instrument calibration, viscosity data for ethylene glycol was compared to viscosity in ASHRAE [45] and water/ethylene glycol. The viscosity using water/ethylene glycol was found to be higher for a temperature range of 30–70 °C.

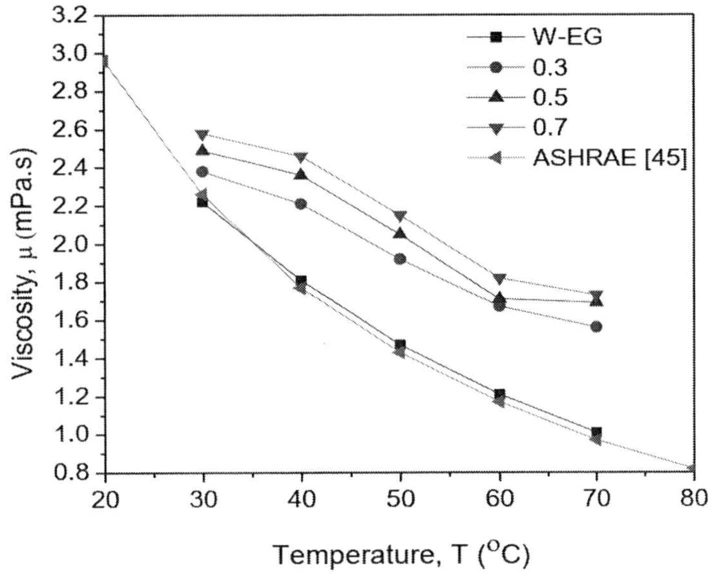

Figure 4. The viscosity of nanofluids with different temperatures.

Viscosity is essential in determining the properties of nanofluids. The viscosity of TiO_2-SiO_2 nanofluid is higher than the base liquid and increases according to the concentration of the liquid. For high volume concentration, it is clearly showing that the increase of viscosity concerning volume concentration and base fluid.

3.2. Thermal conductivity

Figure 5 shows the graph for the thermal conductivity measurement under different temperatures for the various concentration of nanofluid. At the stage of instrument calibration, measured thermal conductivity data for ethylene glycol was compared to thermal conductivity in ASHRAE [45] and water/ethylene glycol. The thermal conductivity using ethylene glycol was found to be higher for a temperature range of 30–70 °C.

The thermal conductivity of TiO_2-SiO_2 nanofluid was a maximum increase higher than the base fluid. The observation of this trend seems to be related to the Brownian movement. At high temperatures, particle collisions occur at higher rates, thereby bringing more kinetic energy.

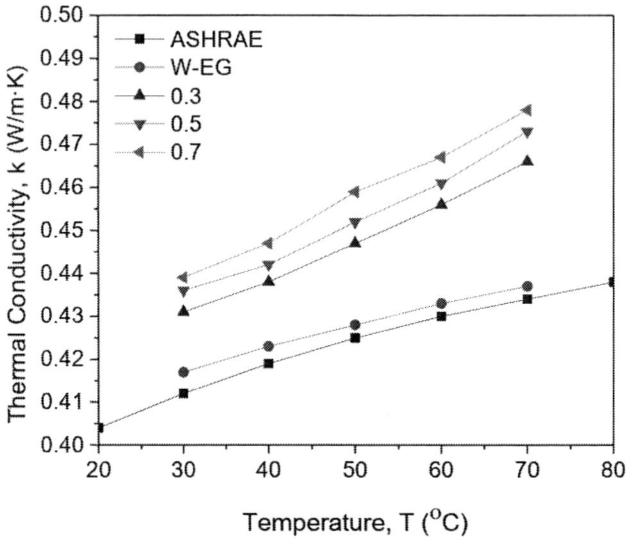

Figure 5. The thermal conductivity of nanofluids at different temperatures and volume concentrations.

3.3. The Effect of TiO_2-SiO_2 nanofluid with water/ ethylene glycol based for absorber solar

The collected data obtained from the data logger, all data related to systems and collected by the k-type thermocouple. The absorber was appraised using the following essential indices, and the states of charge and discharge vary by time in a minute. In our case, to accurately assess the system, the findings were collected during the selected time, which is 15 minutes charging and 15 minutes discharging under different concentrations of TiO_2-SiO_2 nanofluids. All these results based on solar radiation 300, 500, and 700 W/m^2 as shown in figure 6.

For solar radiation of 300 W/m^2 and volume concentration of nanofluids 0.3, 0.5 and 0.7%, the maximum temperature achieved within 15 minutes of charging process is 53.0 °C, 52.5 °C and 52.7 °C respectively. Meanwhile, for solar radiation of 500 W/m^2, the temperature increases until the maximum value within the 15 minutes. However, for solar radiation 700 W/m^2, the volume concentration of nanofluids 0.3, 0.5, and 0.7%, the maximum temperature achieved 66.9, 65.3 and 67.2 °C, respectively. So, the result shows increasing rapidly for solar radiation 700 W/m^2 at the beginning of 15 minutes charging.

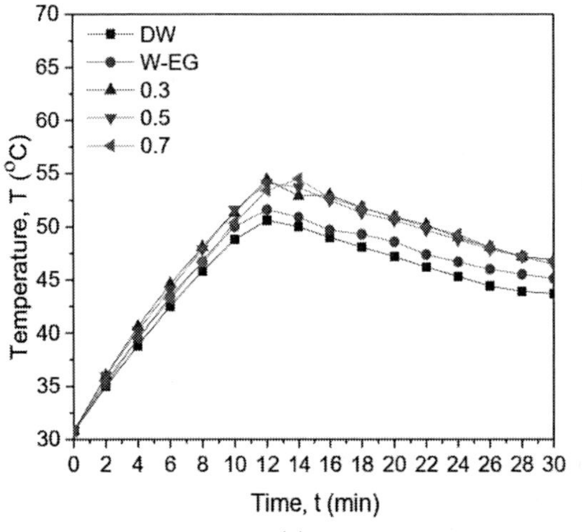

(a)

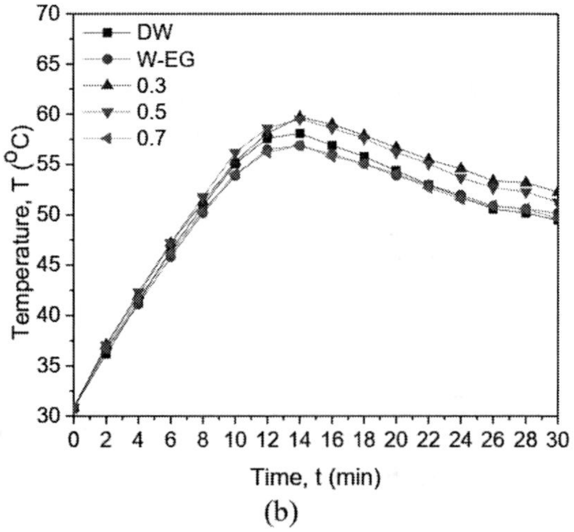

(b)

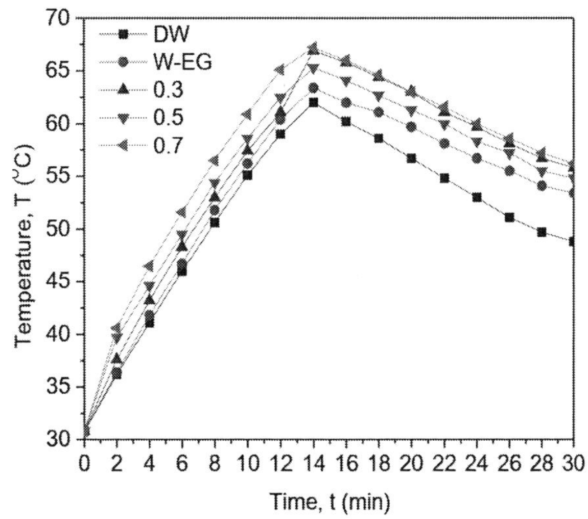

(c)

Figure 6. Temperature output of the absorber based on base fluid and nanofluids. Solar radiation (a) 300 W/m^2, (b) 500 W/m^2, (c) 700 W/m^2.

4. Conclusion

The present study aims to investigate the characteristic of TiO_2-SiO_2 nanofluid suspended in water/ ethylene glycol for solar applications. This study was attentive to the experimental of water/ethylene glycol in mixture ratios of 60:40 based 70:30 TiO_2-SiO_2 nanofluid for a wide range of temperatures from 30 to 70 °C and 300, 500 and 700 W/m^2 for solar radiation. Based on the obtained results, key findings of this investigation can be summarized as follows:

i. The viscosity of water/ethylene glycol-based TiO_2-SiO_2 nanofluid at 30 °C, and 0.3% of volume concentration was 2.72 mPa.s compared to base liquid which was 2.38 mPa.s and increases according to the concentration of the liquid.
ii. The thermal conductivity, *k* of water/ethylene glycol-based TiO_2-SiO_2 nanofluid at 0.3% of volume concentration was 0.431 W/m·K compared to the base liquid which was 0.417 W/m·K and increases according to the temperature difference. This trend seems to be related to the Brownian movement, which is bringing more kinetic energy.
iii. The higher concentrations give ample time to the test tube to transfer the heat and thus increased its temperature during the charging process under the solar radiation test rig. For volume concentration of nanofluids 0.3% and 300 W/m^2 of solar radiation, the maximum temperature achieved was 53.0 °C, and for solar radiation of 700 W/m^2, the maximum temperature achieved was 66.9 °C.

Acknowledgements

The authors would like to thank the Faculty of Bioengineering and Technology, Universiti Malaysia Kelantan and Faculty of Mechanical Engineering, University Malaysia Pahang for providing laboratory facilities.

References

[1] Panwar N, Kaushik S, Kothari S 2011 *Renewable and Sustainable Energy Reviews* **15** 1513-24.
[2] Bamisile O O, Dagbasi M 2015 *International Journal of Engineering Research & Technology (IJERT)* **4** 562-8.
[3] Bergmann A, Colombo S, Hanley N 2008 *Ecological economics* **65** 616-25.

[4] Thirugnanasambandam M, Iniyan S, Goic R 2010 *Renewable and sustainable energy reviews* **14** 312-22.
[5] Sharma A 2011 *Renewable and Sustainable Energy Reviews* **15** 1767-76.
[6] Mekhilef S, Saidur R, Safari A 2011 *Renewable and sustainable energy reviews* **15** 1777-90.
[7] Sowmy D S, Schiavon Ara P J, Prado R T A 2017 *Renewable Energy* **108** 644-51.
[8] Sharafeldin M A, Gróf G 2019 *Renewable Energy* **134** 453-60.
[9] Yadav K, Kumar A, Sastry O S, Wandhare R 2019 *Renewable Energy* **134** 169-77.
[10] Wahab A, Hassan A, Qasim M A, Ali H M, Babar H, Sajid M U 2019 *Journal of Molecular Liquids* 111049.
[11] Ines M, Paolo P, Roberto F, Mohamed S 2019 *Solar Energy* **186** 335-46.
[12] Sahin A Z, Uddin M A, Yilbas B S, Al-Sharafi A 2019 *Renewable Energy*.
[13] Zawawi N N M, Azmi W H, Redhwan A A M, Sharif M Z, Samykano M 2018 *International Journal of Refrigeration* **89** 11-21.
[14] Azmi W H, Abdul Hamid K, Usri N A, Mamat R, Mohamad M S 2016 *International Communications in Heat and Mass Transfer* **76** 24-32.
[15] Azmi W H, Usri N A, Mamat R, Sharma K V, Noor M M 2017 *Applied Thermal Engineering* **112** 707-19
[16] Azmi W H, Zainon S N M, Hamid K A, Mamat R 2019 *Journal of Mechanical Engineering and Sciences* **13** 5182-211..
[17] Li Y, Tung S, Schneider E, Xi S 2009 *Powder technology* **196** 89-101.
[18] Lee J-H, Lee S-H, Choi C, Jang S, Choi S 2011 *International Journal of Micro-Nano Scale Transport*.
[19] Ghadimi A, Saidur R, Metselaar H 2011 *International journal of heat and mass transfer* **54** 4051-68.
[20] Babar H, Ali H M 2019 *Journal of Molecular Liquids* **281** 598-633.
[21] Arshad A, Jabbal M, Yan Y, Reay D 2019 *Journal of Molecular Liquids* **279** 444-84.
[22] Tong Y, Lee H, Kang W, Cho H 2019 *Applied Thermal Engineering* 113959.
[23] Hamid K A, Azmi W H, Nabil M F, Mamat R, Sharma K V 2018 *International Journal of Heat and Mass Transfer* **116** 1143-52.
[24] Usri N A, Azmi W H, Mamat R, Hamid K A, Najafi G 2015 *Energy Procedia* **79** 397-402.
[25] Daungthongsuk W, Wongwises S 2007 *Renewable and sustainable energy reviews* **11** 797-817.
[26] Kakaç S, Pramuanjaroenkij A 2009 *International Journal of Heat and Mass Transfer* **52** 3187-96.
[27] Godson L, Raja B, Lal D M, Wongwises S 2010 *Renewable and sustainable energy reviews* **14** 629-41.
[28] Sarkar J 2011 *Renewable and sustainable energy reviews* **15** 3271-7.
[29] Hamid K A, Azmi W H, Mamat R, Sharma K V 2016 *International Communications in Heat and Mass Transfer* **73** 16-24.
[30] Sarkar J, Ghosh P, Adil A 2015 *Renewable and Sustainable Energy Reviews* **43** 164-77.
[31] Shah T R, Ali H M 2019 *Solar Energy* **183** 173-203.
[32] Fikri M A, Faizal W M, Adli H K, Amran N A 2018 *Proceeding of National Conference on Mechanical Engineer (NCME)* 68-75.
[33] Esfe M H, Amiri M K, Alirezaie A 2018 *Journal of Thermal Analysis and Calorimetry* **134** 1113-22.
[34] Yıldız Ç, Arıcı M, Karabay H 2019 *International Journal of Heat and Mass Transfer* **140** 598-605.
[35] Wei X, Yin Y, Qin B, Wang W, Ding J, Lu J 2019 *Renewable Energy*.
[36] Das S K, Putra N, Roetzel W 2003 *International journal of heat and mass transfer* **46** 851-62.
[37] Ding Y, Alias H, Wen D, Williams R A 2006 *International Journal of Heat and Mass Transfer* **49** 240-50.

[38] Ding Y, Chen H, Wang L, Yang C-Y, He Y, Yang W, et al. 2007 *KONA Powder and Particle Journal* **25** 23-38.
[39] Khdher A M, Sidik N A C, Mamat R, Hamzah W A W 2015 *International Communications in Heat and Mass Transfer* **69** 34-40.
[40] Wen D, Ding Y 2004 *International journal of heat and mass transfer* **47** 5181-8.
[41] Mintsa H A, Roy G, Nguyen C T, Doucet D 2009 *International journal of thermal sciences* **48** 363-71.
[42] Pak B C, Cho Y I 1998 *Experimental Heat Transfer an International Journal* **11** 151-70.
[43] Namburu P, Kulkarni D, Dandekar A, Das D 2007 *Micro & Nano Letters* **2** 67-71.
[44] Koca H D, Doganay S, Turgut A, Tavman I H, Saidur R, Mahbubul I M 2018 *Renewable and Sustainable Energy Reviews* **82** 1664-74.
[45] Handbook A 2009 *American Society of Heating, Refrigerating and Air Conditioning Engineers: Atlanta, GA, USA.*

The effect of Al_2O_3/PAG nanolubricant towards automotive air conditioning (AAC) power consumption

A A M Redhwan[1,*], W H Azmi[2,3], M Z Sharif[2], N N M Zawawi[2], O W Zulkarnain[4] and A R M Aminullah[1]

[1] Faculty of Manufacturing Engineering Technology, TATI University College, 24000 Kemaman Terengganu, Malaysia.
[2] Department of Mechanical Engineering, College of Engineering, Universiti Malaysia Pahang, Lebuhraya Tun Razak, 26300 Gambang, Kuantan, Pahang, Malaysia.
[3] Automotive Engineering Centre, Universiti Malaysia Pahang, 26600 Pekan, Pahang, Malaysia.
[4] Faculty of Electrical and Automation Engineering Technology, TATI University College, 24000 Kemaman Terengganu, Malaysia.

*E-mail: redhwan323@gmail.com

Abstract. The usages of Automotive Air Conditioning (AAC) today is vital because of the hot and humid climate. The AAC is used to provide comfort in the automotive cabin temperature, consequently, increases fuel consumption, and release significant amount of CO_2, and greenhouse gas into the air. Nanotechnologies could be employed into AAC system via lubricant compressor which can decrease power consumption, eventually increase power saving. The main objective of this paper is to experimentally investigate the effect of Al_2O_3/PAG nanolubricant onto the power consumption. The study on the stability and characteristic of the Al_2O_3/PAG nanolubricant has been done prior the experiment. The result shows that the Al_2O_3/PAG nanolubricant is stable event after 28 days with concentration ratio above 70%. The highest power saving percentage of 23.89% is achieved when the AAC system is operated with 0.010% Al_2O_3/PAG nanolubricant. Further, it has been shown that the condenser pressure and evaporator pressure are correlated with the increment of the power consumption with average deviation of 3.129% and 2.919%, respectively. Finally, the power consumption of AAC system reduce with the usage of the Al_2O_3/PAG nanolubricant at low concentration.

1. Introduction

Air conditioning is essential for a vehicle. This is due to the unpredictable weather conditions, especially in countries like Malaysia that have hot and humid weather all year long. In addition, global warming is also a major factor in increasing the use of air conditioning in vehicles. Regular use of air conditioning results in increased load on the engine and thus increases fuel consumption. Therefore, over the last decade, researchers and engineers have been working hard on air-conditioning technology to maximize energy efficiency as well as prioritize the impact on environment. Many efforts have been made by various researchers to improve and optimize the Automotive Air Conditioning (AAC) system through major component improvements as well as in terms of operating systems and electronic control management [1]. However, the authors argue that the working fluid used in the AAC system can still be improved. To date, recent studies have shown that the performance of the vapour compression system can be improved if nanoparticles are added into the refrigerant or into the compressor [2].

Given the current trend, many researchers prefer to use nanoparticles in a compressor because there are not many refrigerants with boiling point higher than room temperature. The lubricating properties used in the AAC compressor change for the better when nanoparticles are mixed with lubricants. According to previous studies, adding nanoparticles in a lubricant compressor promotes the enhancement of the lubricant mixture, having the advantage of thermal properties and improved tribological performance [3-6]. The main idea of using nanolubricant is to reduce friction by improving its rheological properties; thus, reducing the work done by the compressor and improving the efficiency of the system. In addition, nanofluids usually have better thermal conductivity and thus improve heat transfer performance. Therefore, many studies involving cooling systems and lubrication systems have been conducted using nanotechnology to improve system performance.

The first attempt to prove improvement of thermal fluid nanofluid is with the use of Al_2O_3 nanoparticles dispersed in water [7]. With reference to the previous statement, Ahmadi Hojjat tested the effectiveness of nanoparticle nanolubricant use CNT as additives. The results of the experiment, shows the thermal conductivity and the flash point increased by 13.2% and 6.7%, respectively, when the base lubricant was compared with the nanolubricant 0.1% weight percentages [8]. In the previous experiment, the thermal-physical properties of Al_2O_3 nanoparticles dispersed into PAG have been measured. An enhancement of thermal conductivities and the viscosity ratio about 1.04 and 7.58 respectively [9]. In another experiment, the heat transfer performance also enhance up to 129 % when the low concentration R134a/nanolubricant was used instead of R134a/polyolester [10]. In one of the latest finding, it has been shown that a vapour compression system operates at different working pressure gives different result for the compression work thus the power consumption [11]. Basically, to reduce the power consumption of the AAC system, the system is to operate at lower working pressure or lower pressure ratio at a considerable cooling performance. Various experimental studies were conducted to investigate the properties [12, 13] and performance of nanolubricant in refrigeration system.

The objective of this paper to analyse the effect of low concentration Al_2O_3/PAG nanolubricant to reduce the power consumption of the AAC system. The best method to make a stable Al_2O_3/PAG nanolubricant was developed and thoroughly analyse using quantities and qualitative study. The next phase of this work was a systematic experimental investigation of AAC system, which was conducted on a modified automotive air conditioning system for the purpose of this work. The influences of evaporator and condenser pressure were considered to study the effect of nanolubricant as a power consumption reduction and power saving potential in AAC system.

2. Methodology

The methodology is divided into three parts: preparation of Nanolubricant, AAC experimental setup and AAC Power Consumption analysis

2.1. Preparation of nanolubricant

The nanoparticles used were Al_2O_3; with a purity of 99.8%, and 13 nm in size which is obtained from Sigma-Aldrich company. Characterization of nanoparticles such as the size and shape of nanoparticles was confirmed using electron microscopy scanning (FESEM) imaging technique. Figure 1 shows the FESEM image at a magnification of X 300,000. As a result of this scan, it has been determined that Al_2O_3 nanoparticles have a spherical shape with an average size of about 13 nm. The basic properties of Al_2O_3 are shown in more detail in table 1 [14, 15]. The AAC system used in experimental test setup using PAG oil obtained from Denso Corporation. The manufacturer of the AAC system recommends oil type PAG ND8 as it is more compatible with the material used such as aluminum pipe and rubber in the pipe connection. The basic properties of the lubricant PAG 46 at atmospheric pressure are shown in table 2 [16, 17].

Table 1. Properties of Al_2O_3 nanoparticles.

Property	Al_2O_3
Molecular mass, g · mol^{-1}	101.96
Average Particle diameter, nm	13
Density, kg · m^{-3}	4000
Thermal Conductivity, W(m · K)$^{-1}$	36
Specific heat, J(kg · K)$^{-1}$	773

Table 2. Properties of Polyalkylene glycol (PAG 46) lubricant.

Property	PAG 46
Density, g · cm^{-3}@ 20 °C	0.9954
Flash Point, °C	174
Kinematic viscosity, cSt @ 40 °C	41.4-50.6
Pour Point, °C	-51

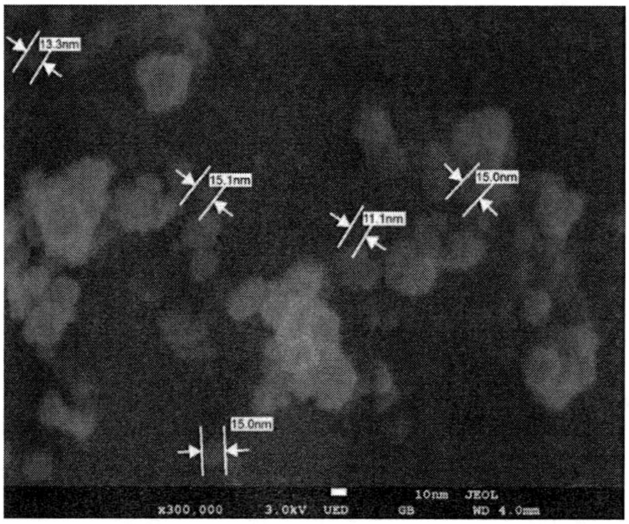

Figure 1. FESEM image of dry Al_2O_3 nanoparticle at X 300,000 magnifications.

Nanolubricant is prepared using the two-step preparation method. Al_2O_3 nanoparticles were mixed into the oil PAG using a magnetic stirrer for 1 hour. Thereafter, the nanolubricant suspension is further stabilized again uses high frequency ultrasonic vibration in the ultrasonic bath (model: FB15051 Fisherbrand) for 1.5 hours; to solve the agglomeration and produce a more uniform dispersion. In another paper, the same method in preparing their PAG-based nanolubricant [9, 18-20]. The stability of nanolubricant was one of the main concerns that affected the properties of nanolubricant and their operation in an application system. The qualitative and quantitative study was used in this paper to assess the stability of the Al_2O_3/PAG suspension. Stabilities for nanolubricant were tested using sedimentation visual observation and UV-Vis Spectrophotometer.

2.2. AAC experimental setup

The original components of the automotive air conditioning system were used to build this experimental test setup. It is adjusted with the original position and condition inside the engine chamber to mimic the situation inside the actual vehicle. The main components of the AAC experimental test system are piston-type compressors, multichannel type evaporator and condenser, and thermostatic expansion valve. The evaporator has been immersed into the cabin calorimetric insulated water containing heated water (60 L capacity) which act as heat load to the evaporator. The cooling capacity and the refrigeration mass flow rate are measured by computing the difference between the water inlet and outlet enthalpy in accordance to the standard of Standard 41.9-2000 (Calorimeter Test Methods for Mass Flow Measurements of Volatile Refrigerants). In this experiment, a 2.2 kW, 3-phase electric motor was used to rotate the pulley connected to the compressor through a belting system. Frequency inverters have been used to control and change the speed of motor rotation. Temperature and pressure were measured using a T-type thermocouple as well as a pressure sensor to read the amount of enthalpy at different points of the cooling cycle, with their accuracy being ± 0.1 ° C and ± 1 kPa. A 3-phase power analyser (Model: Lutron DW-6092) was installed on the motor induction to monitor and measure the power consumption of the AAC system with a precision of ± 0.001 kW. The calibration method was used for all measurement devices in this study and the device has been proven to work perfectly. A schematic diagram for the AAC test setup and power analyser is shown in figure 2 and figure 3.

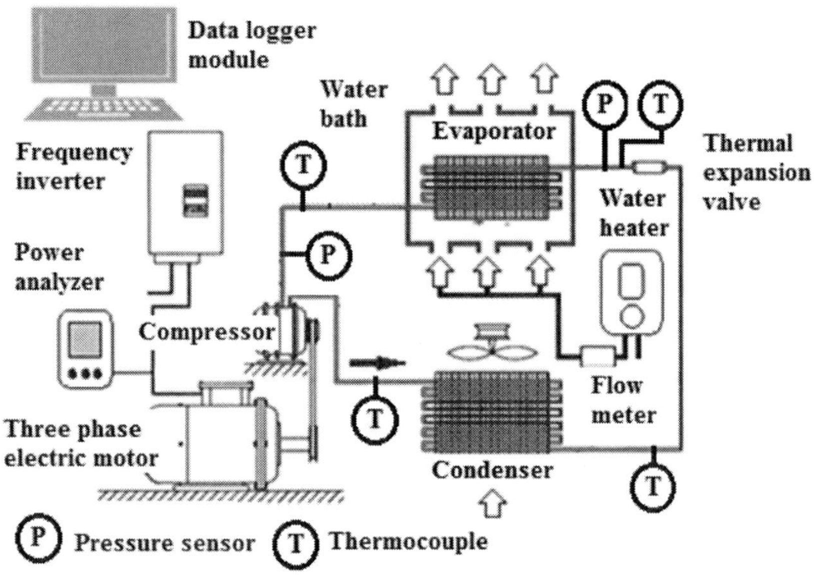

Figure 2. Schematic diagram of AAC system.

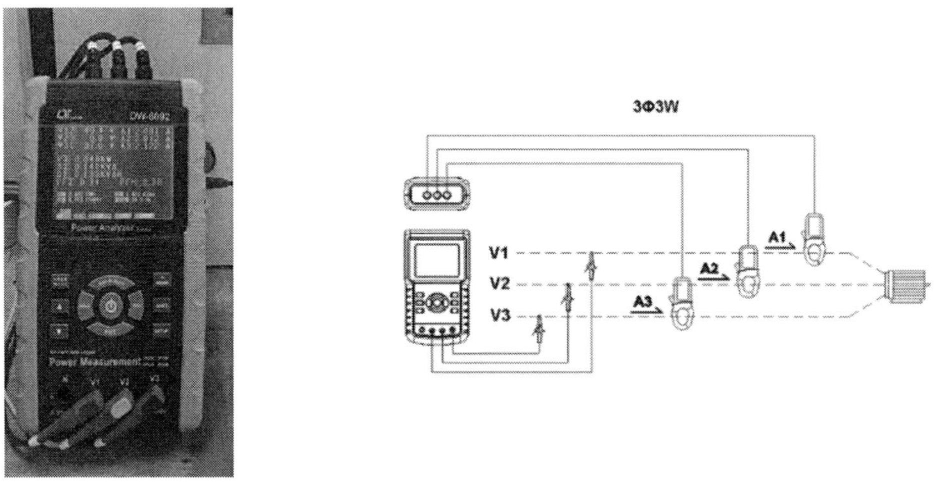

(a) Front view (b) Connection measurement diagram

Figure 3. DW 6092 Lutron power analyser.

2.3. *AAC power consumption analysis*

For ease of calculation, it is assumed that no pressure drops in the condenser, evaporator, and all refrigerant lines in the AAC system. The condensation and evaporation pressure is estimated to be equal to the pressure exerted and the suction measured by the pressure gauge. The data for temperature and pressure are monitored and collected in the data acquisition software, ADAMView.

The test setup continued for 20 minutes for the system to reach a steady-state, so the test data was retrieved, and the logs were taken for 10 minutes after that. Data for pressure and temperature for each given point were used to calculate the enthalpy values using the refrigerant properties table. The experimental procedure was then replicated for various compressor speeds; it varies between 900 and 2100 rpm with 300 rpm intervals. After completing one refrigerant charge, the process is repeated for different refrigerant charges; varies between 90 and 170 grams. After the experiment with all refrigerant charges is completed, the procedure is repeated for different concentrations of Al_2O_3/PAG nanolubricant. In this experimental work, the reliability and consistency of the data are considered. Thus, each set of experiments was repeated three times.

Power saving(E_{saving}) is determined by evaluating the relative power consumption of the AAC system using Al_2O_3/PAG nanolubricant ($\dot{W}_{NL}$) against the AAC system that employed the PAG lubricant ($\dot{W}_{PAG}$) as shown in equation (1).

$$E_{saving} = \frac{\dot{W}_{NL}-\dot{W}_{PAG}}{\dot{W}_{PAG}} \times 100\% \quad (1)$$

Effective power consumption($\dot{W}_{eff}$) for pure PAG ($\dot{W}_{PAG}$) and Al_2O_3/PAG nanolubricant($\dot{W}_{NL}$) is determined by equation (2).

$$\dot{W}_{eff} = \dot{w}_{\text{with load}} - \dot{w}_{\text{w/o load}} \quad (2)$$

where $\dot{W}_{eff}$ is the effective power consumption by the compressor at constant speed with exception of mechanical losses and magnetic losses of motor [21, 22]. Further, $\dot{w}_{\text{with load}}$ is the power consumption measured when the magnetic clutch of the compressor is engaged. Meanwhile, $\dot{w}_{\text{w/o load}}$ is the power consumption measured when the compressor magnetic clutch is disengaged, and the losses is associated in the measurement. DW 6092 Lutron power analyzer as shown in figure 2 is used to measure the electrical power consumption of the 3-phase induction motor which drives the AAC compressor. The connection measurement diagram of the power analyzer is shown in figure 2(b).

3. Results and discussion

The result will discuss on the stability of nanolubricant, power saving and relation between the power consumption and condenser/evaporator pressure.

3.1. *Stability of nanolubricant*

The Al_2O_3/PAG nanolubricant with concentrations between 0.006% and 0.2% were prepared with 1.5 hours sonication time. The concentration ratio against the sedimentation time was plotted for 28 days and shown in figure 4. From the graph, apart from the 0.006% concentration, all other Al_2O_3/PAG nanolubricant concentration ratios ϕ_r were observed to be above 85%. Even though the Al_2O_3/PAG nanolubricant with 0.006% concentration shows the lowest stability in terms of the concentration ratios ϕ_r, but the concentration ratio ϕ_r is kept above 70% even after 28 days. A comparable stability condition was also suggested by [23] and [24]. Hence it can be concluded that the Al_2O_3/PAG nanolubricant are stable for all concentrations.

Two sets of Al_2O_3/PAG nanolubricant which were categorized as low and high-volume concentrations were put aside and kept stationary in test tubes. The test tubes were kept without interruption for a period of a month with the help of a black background studio for clear image capture. The observation of the samples was done daily. The photograph image of both samples was captured on the first day of preparation and after a month for qualitative evaluation. Figure 5(a) shows the sedimentation photograph of Al_2O_3/PAG nanolubricant for the first day of preparation and figure 5(b) shows the sedimentation photograph of Al_2O_3/PAG after a month of preparation for low and high concentrations. Both photograph images were compared and from the image observed, no sedimentation occurred for both low and high concentrations of Al_2O_3/PAG nanolubricant. The samples were stable as signs of sedimentation or colour separation had not occurred. Hence, the Al_2O_3/PAG nanolubricant can be concluded as stable throughout the concentrations.

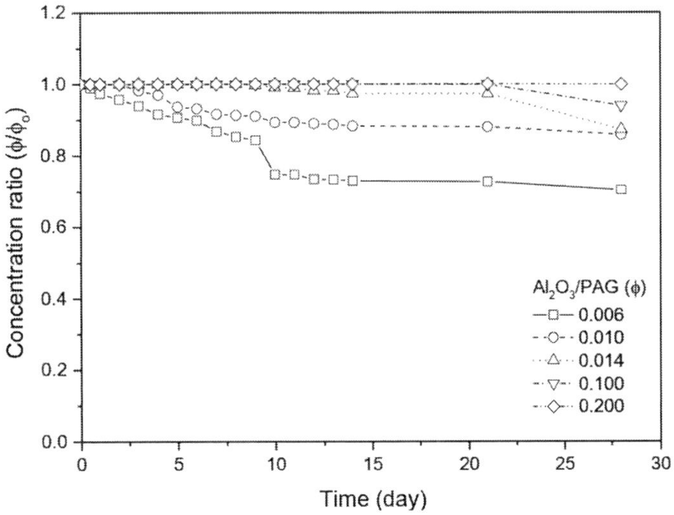

Figure 4. Concentration ratio against the sedimentation time for various concentration of Al_2O_3/PAG nanolubricant.

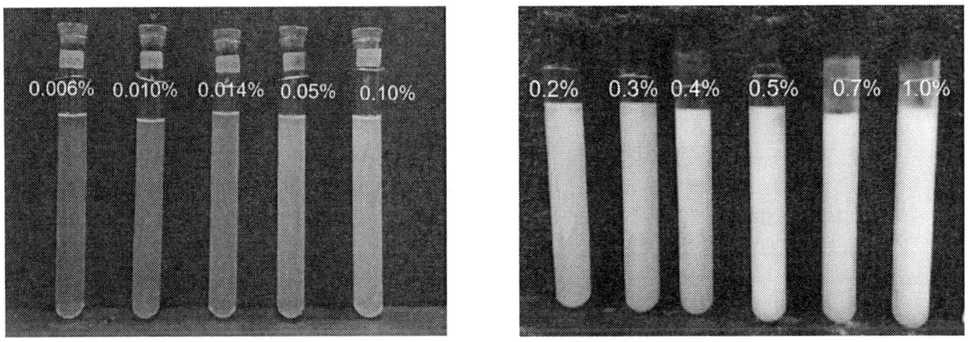

(a) Visual sedimentation for the first day of preparation

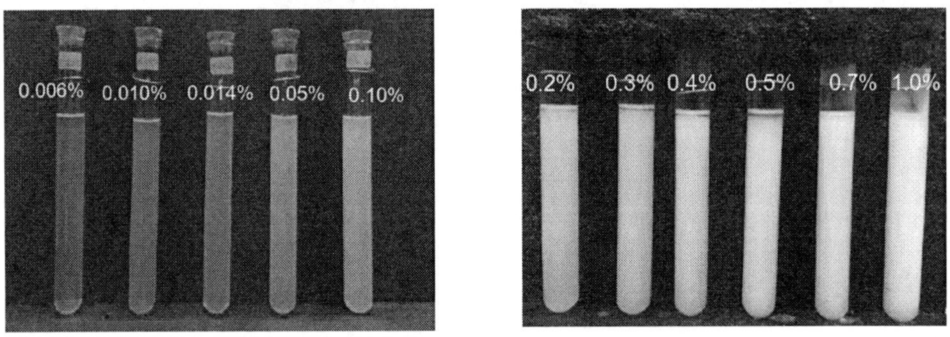

(b) Visual sedimentation after one month of preparation

Figure 5. Al_2O_3/PAG nanolubricant sedimentation photograph.

3.2. Power saving

Figure 6 shows power consumption as a function of refrigerant charge of 0.010% Al_2O_3/PAG nanolubricant as compared to pure PAG lubricant for three variants of compressor speed.

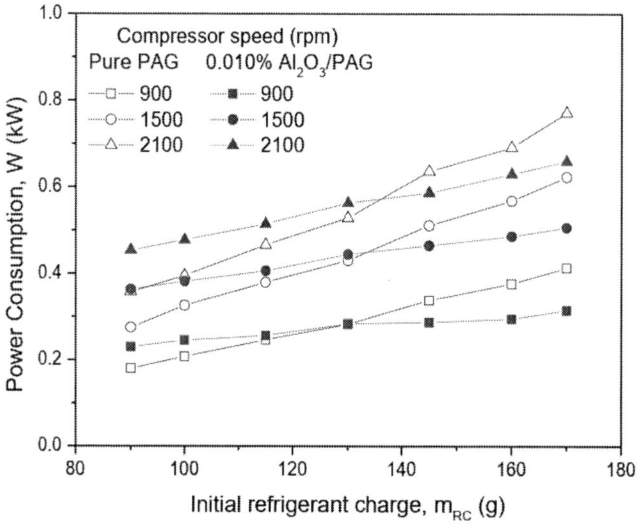

Figure 6. Power consumption as a function of initial refrigerant charge.

At the lower initial refrigerant charge, the power consumption of 0.010% Al_2O_3/PAG nanolubricant is higher compares to pure PAG lubricant. As the refrigerant charge is added, the differences of the power consumption get lesser till the initial refrigerant charge is 130 g. From 130 g initial refrigerant charge onward, the power consumption differences between Al_2O_3/PAG nanolubricant and pure PAG lubricant is increased. At higher refrigerant charge conditions, the power consumption of 0.010% Al_2O_3/PAG nanolubricant is lower compared to pure PAG lubricant. Therefore, 130 g initial refrigerant charge is the interchange point lines of the power consumption between these two kinds of lubricants. At lower initial refrigerant charge, the power consumption of 0.010% Al_2O_3/PAG nanolubricant is higher compared to pure PAG lubricant due to the compressor requiring harder work to pump the refrigerant at the suction side and to overcome resistance from the viscosity of the Al_2O_3/PAG nanolubricant. The higher the refrigerant charge, the less work needs to be done by the compressor to pump the refrigerant at the suction side. At the same time, the rolling friction also takes place since refrigerant (hence Al_2O_3 nanoparticle) are now able to occupy and cover everything in the AAC systems especially inside the compressor. The rolling friction will reduce the COF [25], thus eventually reducing the compressor work. In the absence of Al_2O_3 nanoparticles in the pure PAG lubricant, the rolling effect does not happen in the compressor. Hence at higher refrigerant charges, the power consumption of the Al_2O_3/PAG nanolubricant is lower when compared to pure PAG lubricant.

Table 3 shows the percentage of power saving of Al_2O_3/PAG nanolubricant compared to pure PAG lubricant for 170 g initial refrigerant charge. From the table, it shows that the highest percentage of power saving recorded is 23.89% when the AAC system operates with 0.010% Al_2O_3/PAG nanolubricant and coupled with 900 rpm compressor speed. The average power saving recorded is 11.38%.

Figure 7 shows the power saving as the function of volume concentration for five variants of compressor speeds. The graph shows that the maximum power saving is achieved when the 0.010% volume concentration of Al_2O_3/PAG nanolubricant is used in the AAC system with 900 rpm compressor speed. The graph also shows 0.010% volume concentration of Al_2O_3/PAG nanolubricant is the optimum concentration which gave the highest reading of power saving for all compressor speeds. Hence, it is proven that Al_2O_3/PAG nanolubricant with 0.010% volume concentration give the best performances to the power saving for the AAC system.

Table 3. Percentage of power saving for Al_2O_3/PAG nanolubricant.

Speed, Rpm	Volume Concentration, ϕ (%)				
	0.006	**0.008**	**0.010**	**0.012**	**0.014**
	Percentage enhancement (%)				
900	5.24	16.16	23.89	11.29	9.86
1200	5.79	16.75	21.76	11.85	11.07
1500	3.63	12.41	18.79	10.68	7.87
1800	4.00	13.07	16.00	10.02	6.97
2100	2.80	13.61	14.53	9.14	7.26

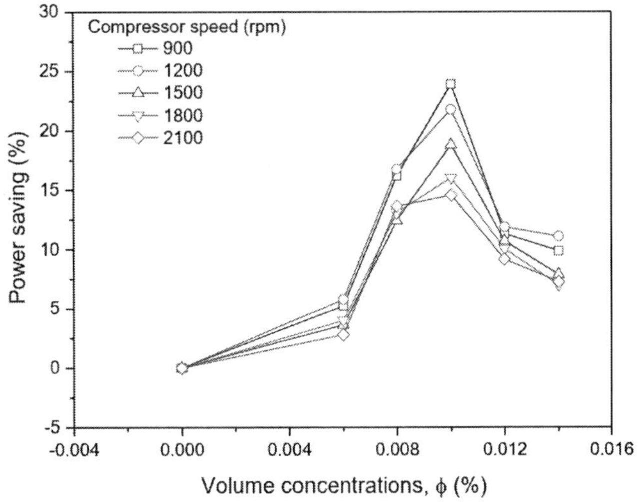

Figure 7. Power saving of Al_2O_3/PAG nanolubricant as a function of volume concentrations.

3.3. Relationship between power consumption and pressure

Figure 8 depicts the evaporator pressure as a function of refrigerant charge while figure 9 shows the condenser pressure as a function of refrigerant charge. From the observation, the pattern of these graphs mimic figure 6 which depicts the power consumption as a function of refrigerant charge. Figure 9 shows that the condenser pressure of Al_2O_3/PAG nanolubricant at 90 g refrigerant charge is higher than pure PAG lubricant and lower at 170 g refrigerant charge. Further investigation on figure 9 reveals that the condenser pressure of Al_2O_3/PAG nanolubricant and pure PAG lubricant lines intersected at 130 g refrigerant charge. A similar intersected pattern is observed for the power consumption graph as depicted in figure 6. Observation of the evaporator pressure as depicted in figure 8 revealed that the evaporator pressure of Al_2O_3/PAG nanolubricant and pure PAG lubricant lines intersected at 140–145 g refrigerant charge. Hence it can be concluded that the condenser pressure graph closely mimics the graph of power consumption. Nevertheless, the evaporator pressure graph also follows a similar pattern with the power consumption graph. Hence both condenser and evaporator pressures can be used to estimate the power consumption utilized by the AAC system.

Accordingly, equation (3) and equation (4) which are developed from the experimental data for the estimation of power consumption related to condenser and evaporator pressures. Other parameters, namely compressor speed, refrigerant charge and Al_2O_3/PAG nanolubricant volume concentration are also considered. Equation (3) is valid for the condenser pressure between 500 kPa to 1123 kPa while equation (4) is valid for the evaporator pressure in the range of 60 to 296 kPa. Both equations shared the same range of other parameters. The compressor speed is in the range of 900 to 2100 rpm, the refrigerant charge is between 90 to 170 g and Al_2O_3/PAG nanolubricant volume concentrations up to 0.014%.

$$\dot{W}_r = \frac{\dot{W}_{NL}}{\dot{W}_{PAG}} = 0.1432\left(\frac{P_{cond}}{1223}\right)^{0.7205}\left(\frac{N_{comp}}{2100}\right)^{-0.1442}\left(\frac{m_{RC}}{170}\right)^{-0.8857}\left(0.01+\frac{\phi}{100}\right)^{-0.4318} \quad (3)$$

$$\dot{W}_r = \frac{\dot{W}_{NL}}{\dot{W}_{PAG}} = 0.2644\left(\frac{P_{evap}}{296}\right)^{0.5105}\left(\frac{N_{comp}}{2100}\right)^{0.3142}\left(\frac{m_{RC}}{170}\right)^{-0.9438}\left(0.1+\frac{\phi}{100}\right)^{-0.6567} \quad (4)$$

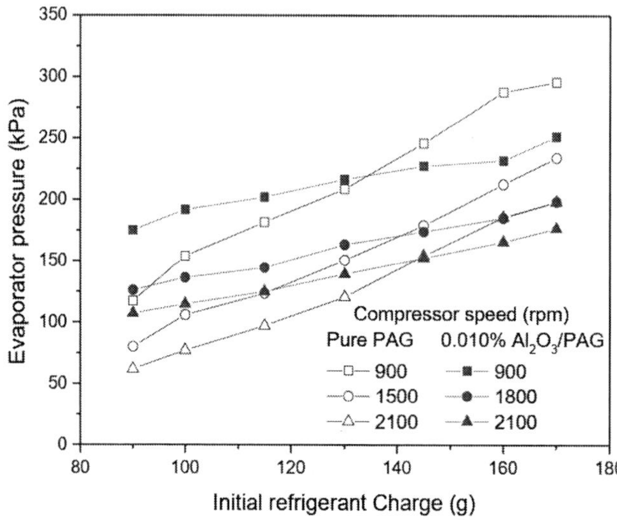

Figure 8. Evaporator pressure as a function of initial refrigerant charge.

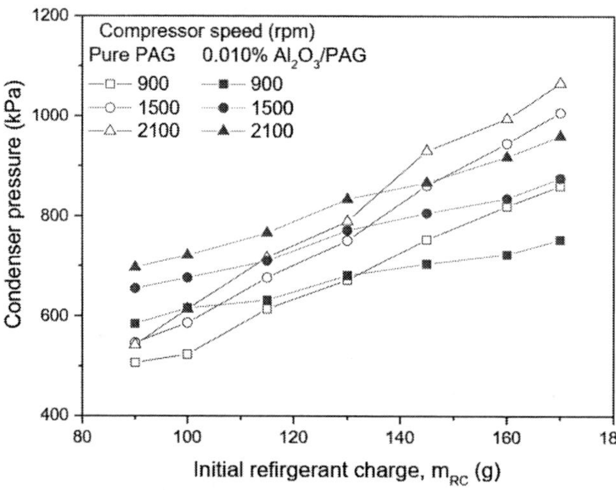

Figure 9. Condenser pressure as a function of initial refrigerant charge.

The average deviation, standard deviation and maximum deviation for power consumption related to condenser pressure were 3.129%, 4.399% and 3.99% respectively while the estimation of power consumption related to evaporator pressure attained at 2.919%, 3.911% and 3.984% for average deviations, standard deviations and maximum deviations, respectively. The distribution of power consumption related to condenser pressure experimental data compared to regression equations are presented in figure 10. Figure 11 shows the distribution of experimental data compared to regression equations of power consumption related to evaporator pressure. From the graphs, the experimental data is in good agreement with estimated equations.

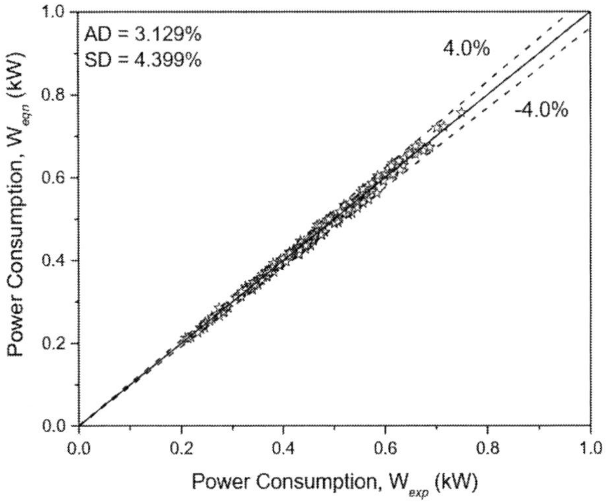

Figure 10. Comparison of power consumption from experimental data and regression estimation related to condenser pressure.

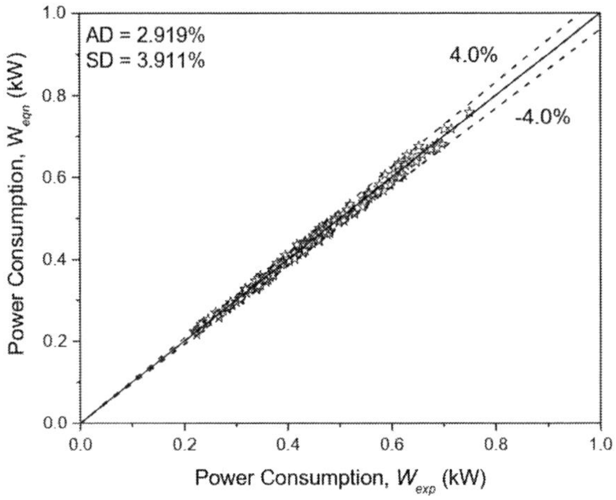

Figure 11. Comparison of power consumption from experimental data and regression estimations related to evaporator pressure.

4. Conclusion

The effect of the Al_2O_3/PAG nanolubricant onto the power consumption of AAC has been investigated. Beforehand, stable Al_2O_3/PAG nanolubricant is established with 1.5 hours sonication. It is found that Al_2O_3/PAG nanolubricant with 0.010% volume concentration give maximum power saving up to 23.89% with average of 18.99%. The relationship between power consumption and AAC condenser/evaporator pressure in mathematical/regression modelling also has been established. It can be concluded that the experimental data is in good agreement with the estimated equations.

Acknowledgments

The authors are grateful to TATI University College, Universiti Malaysia Pahang (UMP) and the Automotive Engineering Centre (AEC) for financial supports given under RDU151411 (RAGS/1/2015/TK0/UMP/03/2) and RDU1603110.

References

[1] Sukri M F, Musa M N, Senawi M Y, Nasution H 2015 *Energy Efficiency* **8** 1201-29.
[2] Zawawi N N M, Azmi W H, Redhwan A A M, Sharif M Z, Samykano M 2018 *International Journal of Refrigeration* **89** 11-21.

[3] Redhwan A A M, Azmi W H, Sharif M Z, Mamat R 2016 *International Communications in Heat and Mass Transfer* **76** 285-93.
[4] Azmi W H, Sharif M Z, Yusof T M, Mamat R, Redhwan A A M 2017 *Renewable and Sustainable Energy Reviews* **69** 415-28.
[5] Sharif M Z, Azmi W H, Redhwan A A M, Mamat R, Yusof T M 2017 *International Journal of Refrigeration* **75** 204-16.
[6] Azmi W H, Zainon S N M, Hamid K A, Mamat R 2019 *Journal of Mechanical Engineering and Sciences* **13** 5182-211.
[7] Choi U S 1995 *Developments and Applications of Non-Newtonian Flows* (New York: American Society of Mechanical Engineers (ASME)) p 99–105.
[8] Ahmadi H, Rashidi A, Nouralishahi A, Mohtasebi S S 2013 *International Communications in Heat and Mass Transfer* **46** 142-7.
[9] Sharif M Z, Azmi W H, Redhwan A A M, Mamat R 2016 *International Journal of Refrigeration* **70** 93-102.
[10] Kedzierski M A 2012 *Journal of Heat Transfer* **134** 051001.
[11] Cabello R, Sánchez D, Llopis R, Arauzo I, Torrella E 2015 *International Journal of Refrigeration* **60** 92-105.
[12] Nabil M F, Azmi W H, Hamid K A, Zawawi N N M, Priyandoko G, Mamat R 2017 *International Communications in Heat and Mass Transfer* **83** 30-9.
[13] Redhwan A A M, Azmi W H, Sharif M Z, Mamat R, Zawawi N N M 2017 *Applied Thermal Engineering* **116** 823-32.
[14] Aldrich S 2013 *Aluminium Oxide*.
[15] Zakaria I, Azmi W H, Mohamed W A N W, Mamat R, Najafi G 2015 *International Communications in Heat and Mass Transfer* **61** 61-8.
[16] Brown W L 1993 *CRC Handbook of Lubrication and Tribology* **3** 253-67.
[17] Dow 2013 *Ucon Refrigerant Lubricant 213*.
[18] Sharif M Z, Azmi W H, Redhwan A A M, Zawawi N N M, Mamat R 2017 *Journal of Mechanical Engineering* **SI 4** 233-47.
[19] Redhwan A A M, Azmi W H, Sharif M Z, Mamat R, Samykano M, Najafi G 2019 *Journal of Thermal Analysis and Calorimetry* **135** 1299-310.
[20] Sharif M Z, Azmi W H, Redhwan A A M, Mamat R, Najafi G 2019 *Journal of Thermal Analysis and Calorimetry* **135** 1285-97.
[21] Shigley J E 2011 Tata McGraw-Hill Education).
[22] Karmakar S, Chattopadhyay S, Mitra M, Sengupta S 2016 Springer).
[23] Habibzadeh S, Kazemi-Beydokhti A, Khodadadi A A, Mortazavi Y, Omanovic S, Shariat-Niassar M 2010 *Chemical Engineering Journal* **156** 471-8.
[24] Lee, Hwang Y, Cheong S, Kwon L, Kim S, Lee J 2009 *Current Applied Physics* **9** e128-e31.
[25] Luo T, Wei X, Huang X, Huang L, Yang F 2014 *Ceramics International* **40** 7143-9.

Influence of detergents types and clean method on PV cells performance: An experimental study

N F Hussein, A A Ismaeel* and A K Ibrahim

Department of Electromechanical Engineering, University of Technology, Baghdad, Iraq.

*E-mail: 50106@uotechnology.edu.iq

Abstract. Accumulation the dust and dirt soiling on photovoltaic (PV) panels' surfaces represents a major problem which inevitably undermines their operational performance. Thereby, removing the soiling and cleaning the panel surfaces is necessary to upgrade the PV cells performance. In this study the influence of detergent types as a cleaning material on PV panels' outcomes has been investigated experimentally. Three detergents types that widely materials using in the home cleaning include: Tap water, mixture of dishwashing liquid with water and glass cleaner were tested. Furthermore, manual cleaning by using a piece of cloth or wiper were utilized as a cleaning tools were also evaluated. The experimental results appeared that enhancement the PV panel performance after using home detergents materials, where the performance increased by 25.35% and 19% when used a glasses cleaner and mixture of water and dishwashing liquid respectively, compared with 7.34% when used tap water only. In addition, using the manual wiper as a cleaning technique was achieved a better PV panel performance than piece of cloth for all detergents materials.

1. Introduction

A photovoltaic (PV) panel is generally considered an essential and indispensable component of any solar energy system which converts solar energy into electric energy. However, the performance of PV panels is decreasing when the dust and dirt density increasing [1, 2]. Therefore, eliminating these negative factors were focused by the previous studies to maintain a high performance level of PV cells. Two groups have been examined by previous studies, the first group analysis the effect of dust deposition characterized on the PV outcomes. In this regard, wide ranges of attempts have been reported. El-Shobokshy and Hussein [3] studied the various types of dust were accumulated on output of PV system. Size of particles, thickness and chemical compositions of the dust accumulated investigated by Pulipaka et al. [4] and Gholami et al. [5]. The second groups of researchers introduced various cleaning techniques and tools to remove the dust and soiling accumulated. In line with this, Al Shehri et al. [6] and Mondal et al. [7] reported an electromechanical water jet systems to clean PV glass surface. Intelligent dry cleaning system with different brushes materials evaluated by AI Shehri [8]. Used Nano coating for self-cleaning PV surface introduced by Zhong et al. [9]. The results of the previous studies were clearly show that the dust accumulated on the PV surface cases blocks of the solar radiation capture by the panel surface and create a partial shading effect. Reduction of solar energy converted into electricity was measured in the study. Hence, cleaning the PV panel surface is very important to steady its operating performance. However, most of the cleaning technology depended on the automated system which is consider complex, costly and non-suitable when installation the PV panel with in small site or home applications. Thereby, using the manual cleaning is considering a simple and economic solution. The main objective of the present study is investigated the influence of home detergent types and manual cleaning tools as a cleaning method on PV cells performance.

2. Experimental model description and procedure

In order to study the effect of detergent types that is utilized in cleaning process of PV panel on overall productive efficiency, three types of cleaning materials were selected, namely tap water, mixture of water and dishwashing liquid (each 5 litters of water were mixed with 10 millilitres of dishwashing liquid) and glass-cleaner liquid. These detergents are cheap and easily obtainable in the local market. Moreover, two manual tools (piece of cloth, rubber wiper) were used to wipe and remove the water and detergent after spraying it on the PV surface also was tested. Four PV panels which are the same specifications as listed in table 1, used as experimental mediums.

Table 1. Characteristics of experimental mediums PV panels.

Description	Value
Module type	JP-22-18 /Bb
Rated Power (Pmax)	22 W
Power tolerance %	±5
Maximum power voltage (Vmp)	6 V
Maximum power current (Imp)	3.67 A
Open circuit voltage (Voc)	7.2 V
Short circuit current (ISC)	3.92 A

One panel was kept as benchmark without cleaning surface, while the others panels were cleaning with the test detergent mediums. The experimental PV panels are shown in figure 1.

Figure 1. Experimental mediums PV panels.

These panels were directed southwards in order to receive the almost of solar radiation during the daytime and placed at tilt angle of 30° based to reference [10]. The measurements were repeated for 20 days (10 days for each manual cleaning tools) to improve the reliability of the result by taking the average reading. The data was collected every one hour starting at 9 am and ending at 5 pm under Baghdad-Iraq environment conditions. The tests were conducted in the sunny days under the same weather conditions, thus, avoiding cloudy or rainy days during the data acquisition. The PV panels were cleaned with cleaning materials about fifteen minutes prior to the beginning of the data measurement.

Average measured data of the solar radiation, current and voltage when glass wiper and piece of cloth as a cleaning tools and for all cleaning materials are listed in table 2 and table 3.

Table 2. Average measured data for different cleaning materials with wiper tool.

Time (hr)	Solar Intensity (W/m²)	Outside air Temp. °C	Without Cleaning		Water only		Glass cleaner		Water + Dish wash liquid	
			I amp	V volt	I amp	V volt	I amp	V volt	I amp	V volt
9:00 am	637.2	20.7	1.52	7.6	1.82	7.6	2.16	7.1	2.1	7
10:00 am	926.1	24.4	2.15	7.5	2.45	7.5	2.91	7.1	2.89	7
11:00 am	1160.4	24.6	2.65	7.5	2.88	7.5	3.46	7	3.43	6.9
12:00 pm	1260.8	25.0	2.86	7.4	3.07	7.4	3.79	7	3.68	6.9
1:00 pm	1243.9	25.7	2.83	7.4	3.01	7.4	3.62	6.9	3.61	6.8
2:00 pm	1100.2	28.4	2.11	7.4	2.68	7.3	3.25	6.8	3.26	6.8
3:00 pm	850.3	25.7	1.2	7.3	2.01	7.3	2.97	6.9	2.95	6.9
4:00 pm	540.1	24.9	0.93	7.4	1.17	7.4	2.14	7	2.1	7
5:00 pm	290.0	24.1	0.27	7.4	0.98	7.4	1.37	7	1.35	7

Table 3. Average measured data for different cleaning materials with cloth tool.

Time (hr)	Solar intensity (W/m²)	Outside air Temp. °C	Without Cleaning		Water only		Glass cleaner		Water + Dish wash liquid	
			I amp	V volt	I amp	V volt	I amp	V volt	I amp	V volt
9:00 am	639.5	25.	1.54	7.5	1.83	7.5	2.16	7	2.14	6.9
10:00 am	912.4	27.7	2.14	7.4	2.37	7.4	2.8	6.9	2.77	6.8
11:00 am	1157.8	31.1	2.61	7.3	2.87	7.3	3.37	6.8	3.3	6.7
12:00 pm	1274.2	33.3	2.77	7.3	2.95	7.3	3.42	6.8	3.4	6.7
1:00 pm	1245.1	33.7	2.72	7.3	2.83	7.2	3.3	6.7	3.23	6.7
2:00 pm	1145.4	34.7	2.25	7.2	2.43	7.2	2.96	6.7	2.92	6.6
3:00 pm	1010.3	30.6	1.42	7.2	2.11	7.2	2.31	6.8	2.3	6.7
4:00 pm	445.2	29.1	0.83	7.2	1.64	7.2	2.11	6.9	2.01	6.8
5:00 pm	289.4	28.6	0.24	7.2	1.16	7.2	1.41	6.9	1.38	6.8

3. Results and discussion

The electricity production from the PV panels for the different manual cleaning tools (wiper and cloth) and various detergent mediums with respect of the test period are described in figures 2 and 3, respectively. From the figures, it is evident that the highest electricity production was observed after cleaning the PV surface due to remove the dust layers and allowed the surface to absorber more sunlight and as a result increase the energy converts into electricity. Also from the figures, it showed that the commercial detergents causes enhanced the power production compared with used the tap water only for PV surface cleaner. This implies that, although the tap water is sterile water, but it still has variable impurities or substances such as lime, chlorine and so on, therefor it cannot treatment the difficult stains or soiling constructed on the PV surface, well.

Figure 4 gives the plot of average PV panel efficiency for different cleaner materials and tools. It is observed that from the figure, that the cleaning tool plays important role on PV performance. The highest efficiency was obtained when used the rubber wiper to wipe the water and detergent mediums soap from the PV glass surface. Rinsed the soap smoothly from the all of PV glass area achieved by wipe tool compare with used a cloth, which is kept some soap or water stains on the surface due to wet and soap stored in the cloth during the cleaning process. Thus, decline the solar radiation absorbing by glass surface and decrease the PV performance.

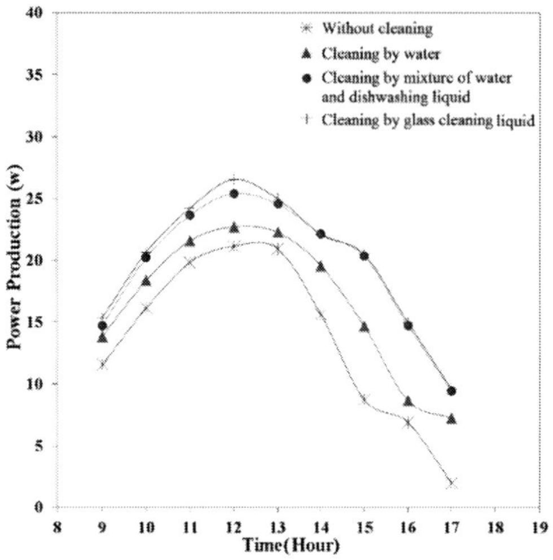

Figure 2. PV-Power Production vs Time for various detergent mediums and with wiper tool

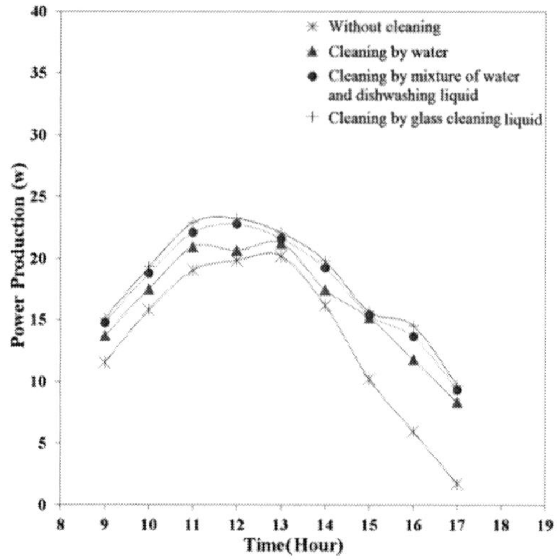

Figure 3. PV-Power Production vs Time for various detergent mediums and with cloth tool.

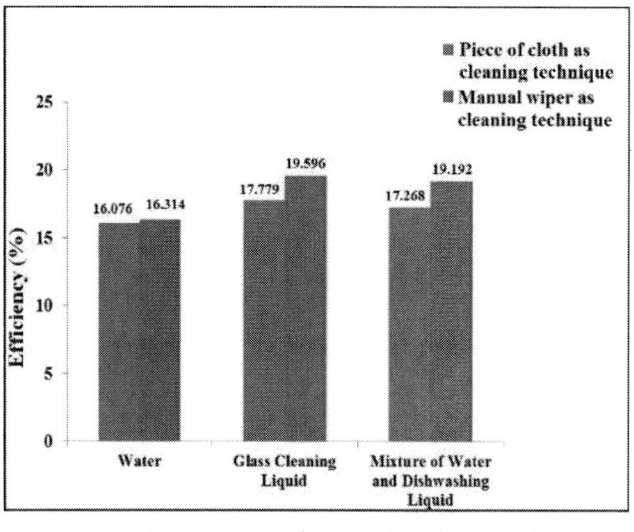

Figure 4. Comparison between cleaning techniques.

Enhancement the PV outcome at the maximum solar radiation (12 pm) is listed in table 4. The table clearly shows that the maximum power production achieved when used glass cleaner with wiper which the enhancement reaches to 25% compared with various detergent mediums.

Table 4. PV electric power enhancement at 12 pm for different cleaning materials and tools.

Cleaning technique	Water only	Glass-cleaning liquid	Mixture of water and dishwashing liquid
Cloth	6.50%	15.01%	12.66%
Wiper	7.34%	25.35%	19.98%

4. Conclusion

In the present study, influence various detergents and manual tools that were used for cleaning on the PV cells in order to maintain the PV performance have been examined experimentally. Depending on the methods of present work, the following points can demonstrate as below:

i) Detergents are most suitable materials for PV surface cleaning compare with using tap water.
ii) Used glass cleaner liquid was obtained a higher improvement in PV power production than dishwashing liquid.
iii) Used ripper wiper as a manual cleaning tools was achieved a better PV performance than piece of cloth.

Acknowledgement
The authors would like to acknowledge University of Technology – Iraq for assistance this study.

References

[1] M. Gürtürk, H. Benli, and N. K. Ertürk, "Effects of different parameters on energy–Exergy and power conversion efficiency of PV modules," Renew.Sustain Energy Rev., vol. 92, pp. 426-439, 2018.
[2] B. Guo, W. Javed, Y. S. Khoo, and B. Figgis, "Solar PV soiling mitigation by electrodynamic dust shield in field conditions," Sol. Energy, vol. 188, pp. 271-277, 2019.
[3] M. S. El-Shobokshy and F. M. Hussein, "Effect of dust with different physical properties on the performance of photovoltaic cells," Sol. Energy, vol. 51, pp. 505-511, 1993.
[4] S. Pulipaka and R. Kumar, "Analysis of irradiance losses on a soiled photovoltaic panel using contours," Energy Conver. Manage., vol. 115, pp. 327-336, 2016.
[5] A. Gholami, I. Khazaee, S. Eslami, M. Zandi, and E. Akrami, "Experimental investigation of dust deposition effects on photo-voltaic output performance," Sol. Energy, vol. 159, pp. 346-352, 2018.
[6] A. Al Shehri, B. Parrott, P. Carrasco, H. Al Saiari, and I. Taie, "Accelerated testbed for studying the wear, optical and electrical characteristics of dry cleaned PV solar panels," Sol. Energy, vol. 146, pp. 8-19, 2017.
[7] A. K. Mondal and K. Bansal, "A brief history and future aspects in automatic cleaning systems for solar photovoltaic panels," Advan. Robotics, vol. 29, pp. 515-524, 2015.
[8] A. Al Shehri, B. Parrott, P. Carrasco, H. Al Saiari, and I. Taie, "Impact of dust deposition and brush-based dry cleaning on glass transmittance for PV modules applications," Sol. Energy,vol. 135, pp. 317-324, 2016.
[9] H. Zhong, Y. Hu, Y. Wang, and H. Yang, "TiO2/silane coupling agent composed of two layers structure: A super-hydrophilic self-cleaning coating applied in PV panels," Appl. Energy, vol. 204, pp. 932-938, 2017.
[10] O. K. Ahmed and Z. A. Mohammed, "Dust effect on the performance of the hybrid PV/Thermal collector," Therm.Scien. Engineering Prog., vol. 3, pp. 114-122, 2017.

Synthesis and properties evaluation of sodium grease formulated from used transformer oil as base oil

N S A Japar[1], M A A Aziz[1,*] and N Hamid[2]

[1] Faculty of Chemical and Process Engineering Technology, University Malaysia Pahang, 26600 Kuantan, Pagang, Malaysia.
[2] Foresight and Strategic Unit, Vice Chancellor Office, University College of Yayasan, 25200 Kuantan, Pahang, Malaysia.

*E-mail: maizudin@ump.edu.my

Abstract. This study aimed to prepare sodium grease using used transformer oil as a base oil and investigate the prepared grease properties. Oil, whether it is new or used, is a significant contributor to water pollution due to its toxicity towards the environment. In this study, used transformer oil (UTO) is utilized as grease's base oil. The properties of the grease were evaluated in terms of their consistency, drop point, oil separation, and bled oil in accordance with ASTM D217, D2265, D1742, and IP 121, respectively. The grease also characterized by using the FTIR-ATR. The results show that the grease can be formulated using UTO with significant grease properties of >180 °C dropping point, −15% to +15% of oil bleeding at both 25 °C and 70 °C, no separated oil during storage, and slightly tarnish on copper strip. The best UTO-based sodium grease formulation was 67.5:32.5 of oil-to-thickener ratio and it was classified as NLGI 2 grease. Moreover, the FTIR spectrum shows no contaminants or unknown chemical changes or additive in UTO and the prepared grease. The addition of MoS_2 in grease shows no significant improvement in sodium grease properties. Thus, UTO was able to work as grease base oil and exhibited overall good properties but, UTO-based grease analysis on its tribology is necessary to study the prepared grease performances during operation.

1. Introduction

Transformer oil is commonly used for insulation and cooling purpose. Petroleum oil, synthetic ester, and silicon oils are traditionally used as transformer oil. Transformer oil is usually a highly-refined mineral oil that is stable at high temperatures and has excellent electrical insulating properties. In the operation of an electrical transformer, the transformer oil is subjected to mechanical and electrical resistance, which results in oil deterioration and degradation. The discarded transformer oil is known as used transformer oil (UTO). The UTO generated in Malaysia is currently does not being utilized into new products or undergo an oil treatment process to reclaim the oil yet. Moreover, the quantity of UTO generated is increasing day by day due to the installation of new transformers and from scrap of old transformer [1]. Malaysia's power industry transformer unit has been introduced with only one type of transformer oil product. This transformer oil is a PCB-free uninhibited mineral oil made from a severely hydro-treated wax-free naphthenic oil. Therefore, the use of UTO generated by this industry could help to ease up lubricating grease composition control since there is no unknown additive associated with this UTO.

Current researches related to UTO are more focused on reclaiming it to produce diesel-like fuel (DLF) [2–4]. There are not many initiatives in using UTO as lubricating grease alternative base oil. The majority types of waste oils being used in lubricating grease formulation are either waste engine oil (WEO) or waste cooking oil (WCO) [5–7]. For example, Ebisike et al. [6] successfully produced sodium grease from spent engine oil, and the grease produced exhibit good properties of dropping point up to 200 °C and classified as NLGI 2 – 3 grease. Razali et al. [8] also formulated grease from waste oil with two types of thickeners (red gypsum and fumed silica), and the grease had a drop point of more than 240 °C. Formulation of grease using WCO was conducted by Abdulbari and Zuhan [7] where they found the formulated grease had no drop point and slightly corrosive to copper.

However, several studies were successfully conducted where transformer oil was utilized as grease base oil [9–11]. For example, transformer oil was utilized by Hassan et al. [12] and blending the oil with lube oil to produce grease with excellent insulating properties. Studies conducted by Cao et al. and Ge et al. [10,11] also found that transformer oil was able to provide insulating properties to the grease and improved with the addition of selected additive into the grease formulation. Therefore, it was expected that UTO could be utilized as lubricating grease base oil in this study.

Therefore, the present study aims to prepare sodium grease using used transformer oil as grease base oil and to investigate the prepared grease characteristics. The greases were prepared by stirring and homogenizing variation of grease component proportions of base oil and thickener, and a fixed amount of additive. The greases properties were evaluated and compared between greases containing fresh and used transformer oil, and greases with and without the presence of the additive.

2. Experimental work

2.1. Materials

Used transformer oil (UTO) collected from an electrical power station located in Selangor, a fresh uninhibited industrial-grade transformer oil (FTO), sodium stearate, and molybdenum disulfide (MoS_2) with 98.5% purity.

2.2. Pre-treatment of used transformer oil (UTO)

Pre-treatment of UTO was carried out through settling, vacuum filtration, and heating to remove unwanted contaminants such as water and suspended dirt in the oil. Water was separated from UTO through gravity-based separation by allowing the water to settle at the bottom of the storage container within 2 weeks. The water was then removed from the container, leaving only UTO. Filtration of UTO via vacuum filtration using a vacuum pump and a glass microfiber filter with 1.2 µm pore size was then carried out to remove any solid particles suspended in the oil. Finally, filtered UTO was heated at 120°C for at least 2 hours with continuous stirring to remove moisture and any volatile compound presented in the oil through the evaporation process. Heated UTO was then cooled down to room temperature before it can be stored in a new cleaned storage container prior to use.

2.3. Preparation of sodium grease

Grease formulation using both fresh transformer oil (FTO) and used transformer oil (UTO) as base oil was formulated. There were two formulations involved, which were formulated grease with and without the addition of additive (MoS_2), respectively. The preparation of sodium greases was started by heating the base oil at 120 °C for at least 1 hour with continuous stirring to remove traces of moisture. The oil temperature was then increased to 180 °C for 15 minutes and stirred continuously. Sodium stearate was added gradually into the oil, and the mixture was stirred for at least 3 hours until the smooth paste was obtained. The homogenization of the grease sample was conducted for at least 30 minutes without the presence of heat to allow uniform dispersion of mixture. As for the additive, MoS_2 was added into the mixture after the smooth paste was formed, and MoS_2 was stirred into the mixture for another 1 hour to allow even dispersion of additive. After homogenization, grease was stored in an enclosed container for 2 days to allow the grease to cool down. Grease preparation steps were as in figure 1, and the formulation of FTO-based grease (FG/FGM) and UTO-based grease (SG_i/SGM_i) was as shown in table 1.

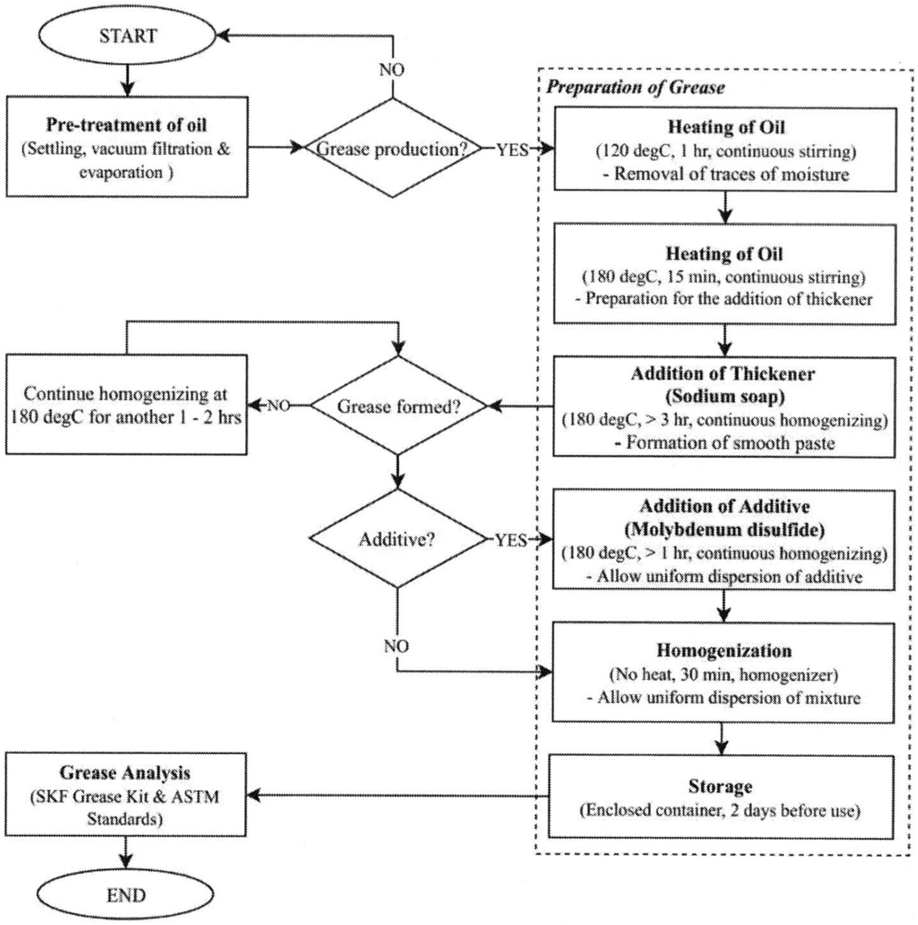

Figure 1. Flowchart of the study procedure.

Table 1. Formulation of the grease.

Sample		Grease composition (wt%):			
		FTO	UTO	Sodium soap	MoS_2
FG	(No additive)	70	–	30	–
FGM	(With additive)	68.6	–	29.4	2
SG_1	(No additive)	–	90	10	–
SG_2		–	90	10	–
SG_3		–	82.5	17.5	–
SG_4		–	75	25	–
SG_5		–	67.5	32.5	–
SG_6		–	60	40	–
SG_7		–	60	40	–
SGM_1	(With additive)	–	88.2	9.8	2
SGM_3		–	80.85	17.15	2
SGM_4		–	73.5	24.5	2
SGM_5		–	66.15	31.85	2
SGM_6		–	58.8	39.2	2
SGM_7		–	58.8	39.2	2

2.4. Grease characterization

2.4.1. Consistency. This test was carried out using the SKF Grease Test Kit TGKT 1. The consistency of the grease was measured using a calibrated measuring scale of NLGI grade. This test method was in accordance with ISO 2137, for which only a small amount of grease sample is required. The consistency level of the greases was classified by NLGI grade (table 2) [13].

Table 2. NLGI classification system based on consistency.

NLGI Grade	Worked Penetration, mm/10	Consistency
000	445-475	Very soft
00	400-430	
0	355-385	Soft
1	310-340	Soft
2	265-295	Creamy texture (buttery)
3	220-250	Semi-solid
4	175-205	Stiff
5	130-160	Stiff
6	85-115	Hard solid

2.4.2. Dropping point. The dropping point test was evaluated according to ASTM D2265 for a wide temperature range [14]. In this test, the sample in the grease test cup was placed in a test tube and placed inside the aluminum block oven. A thermometer was placed in the test tube and positioned that it measures the temperature in the sample cup without coming in contact with the grease. As the temperature increases, at some point, a drop of material fell from the cup, and the reading of thermometer (ODP) and block oven temperature (BT) were recorded to the nearest degree range [14]. The dropping point (DP) of the grease sample was calculated using equation (1).

$$DP\ (°C) = ODP + [(BT - ODP)/3] \quad (1)$$

2.4.3. Oil bleeding. This analysis was conducted using the SKF Grease Test Kit where an only small volume of samples was required [15]. In this analysis, grease sample was placed on a blotter paper and heated on a hotplate at 60 °C for 2 hours [16]. The grease's bleed area percentage differences were measured and calculated by using equation (2) and equation (3) between fresh and aged grease. Where, S_i is the bled area from fresh and aged samples, D_{avi} is the average diameter of the bled area, and *%Diff* represents the bled area difference between fresh and used samples. Aged greases are greases that have been aged for 10 days at both 25 °C and 70 °C [17].

$$S_i = 0.785 \times (D_{avi}^2 - 100) \quad (2)$$

$$\%Diff = 100 \times \frac{(S_{Used} - S_{Fresh})}{S_{Fresh}} \quad (3)$$

2.4.4. Oil separation. The test was carried out on grease that has undergone a 1-month storage period. The amount of oil separated on top of the grease sample during storage was collected and measured in terms of their weight percentage.

2.4.5. Corrosion test. Corrosion tests on both base oil and greases were carried out according to ASTM D130 and ASTM D4048, respectively [18, 19]. This test was performed by preparing a copper strip and totally immersing the strip in a sample of oil or grease. The sample was heated in an oven or liquid bath at a specified temperature for a period of time. The condition for the test was set at 100 °C for 3 hours in a water bath for base oil and at 100 °C for 24 hours in an oven for grease sample. At the end of the test period, the strips were removed from the sample, washed and compared to the ASTM Copper Strip Corrosion Standards (figure 2) [19].

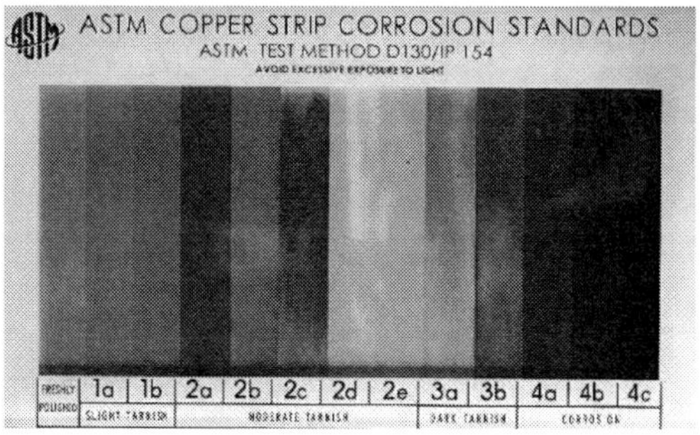

Figure 2. ASTM copper strip corrosion standards.

2.4.6. FTIR characterization. Fourier transforms infrared (FTIR) spectroscopy were used to identify the chemical compound that existed in the grease. FTIR spectrum can provide information on any contamination, additives, and changes presented in the grease. In this experiment, FTIR spectrums were measured using Thermo Scientific Nicolet iS5 FTIR Spectrometer via the FTIR-ATR approach at wavenumber of 500 – 4000 cm^{-1}.

3. Results and discussion

3.1. Physicochemical properties of oil

Table 3 shows the physicochemical properties of FTO and UTO. Grease mainly consisted of more than 80% of base oil where it provides lubrication when applied in an application. Base oil viscosity is the most important properties to be considered when formulating grease. High viscosity oil flows slowly compared to low viscosity oil. In low-speed applications, high viscosity oil is more favourable, and low viscosity oil is recommended for high-speed applications. The choice of base oil viscosity for grease formulation is depended on the grease intended application. Based on table 3, FTO and UTO were a low viscosity oil of 9.84 cSt and 9.93 cSt at 40 °C, respectively when compared to conventional mineral oil hence, these oils can be used in grease formulation for high speed and low load application [20]. However, the oil viscosity can be improved by the addition of proper additive into the oils. The moisture content of UTO after the pre-treatment process was reduced to 0.05%, which was within the range of maximum water content (0.02 to 0.06 %) that can be held by most industrial oil [21].

Table 3. Physicochemical properties of FTO and UTO.

Properties	Fresh transformer oil	Used transformer oil
Appearances	Clear & bright	Bright Yellow
Kinematic viscosity		
at 40 °C, cSt	9.84	9.93
at 100 °C, cSt	2.55	2.55
Viscosity index (VI)	80	77
Moisture content, %	0.002	0.05

3.2. Consistency of greases

Based on table 4, FG, FGM, SG_4, SG_5, SGM_4, and SGM_5 were classified as NLGI 1 – 3 greases, where these consistencies were most common for commercialized grease [22]. NLGI 2 greases were suitable for moderately loaded with medium speed applications. Grease of this class of consistency is formulated to give a good balance of properties required for easy pumping through dispensing systems [23]. Some of the formulated greases that classified with consistency lower than NLGI 1 are considered as soft greases, and this might be due to the insufficient amount of thickener in the formulation to hold the oil in its matrix. However, these soft grease consistency can be increased by decreasing the oil-to-thickener ratio [24]. Conversely, greases with softer consistency do not mean it cannot be used in any application. Greases of such consistencies can be used at low operating temperatures and high-speed applications. Therefore, the grease consistency may be adjusted in the formulations according to the application's operating condition. Grease classified with NLGI > 3 was considered as hard grease for which this grease usually utilized at an application where leakage is a concern.

The addition of MoS_2 in sodium grease shows significant changes in the grease's consistency. For example, FG and FGM were formulated with the same oil and thickener composition of 70:30, but FGM contained 2 percent MoS_2, and FGM's consistency is one NLGI grade stiffer than FG. These results were not inlined with Mohammed [25], where he found that the addition of MoS_2 in different types of greases clearly reduces the grease consistency.

Table 4. Formulated grease properties.

Grease sample	Consistency (NLGI Grade)	Dropping point (°C)	Oil bleeding (%)		Oil separation (%)
			@ 25°C	@ 70°C	
FG	2	194.7	- 7.54	-36.81	0
FGM	3	193.9	- 13.72	-58.79	0
SG_1	000	138	Too soft	Too soft	Too soft
SG_2	000	135			
SG_3	0	158	16.53	-9.54	0.27
SG_4	1	175	9.78	-11.67	0.07
SG_5	2	193	-8.62	-18.04	0
SG_6	4	207	-28.70	-35.57	0
SG_7	4	204	-27.85	-35.84	0
SGM_1	00	143	Too soft	Too soft	Too soft
SGM_2	00	141			
SGM_3	0	170	10.42	3.31	0.33
SGM_4	2	181.2	2.21	-3.67	0.12
SGM_5	3	198.4	-11.16	-16.64	0
SGM_6	5	203.1	-58.02	-60.06	0
SGM_7	5	201.2	-59.08	-61.43	0

3.3. Dropping point of the grease

Table 4 also shows the results of the dropping point test for the formulated greases using both FTO and UTO with and without the additive. According to Totten [26], dropping point is dependent on the type of thickener and the cohesiveness of the oil and thickener of a grease. Based on the results, the grease's dropping point temperature increases with the amount of thickener. This is in line with Buhlak et al. [27], it was found that when the thickener consists of 90:10 percent of sodium-to-lithium soap mixture, the drop point of formulated greases with 20% and 25% of thickener has increased from 119 °C to 128 °C, respectively.

Sodium greases do not have a high dropping point, typically 175 °C. In this study, the dropping point of the prepared greases was increasing with the amount of thickener. The drop point of both FTO-based and UTO-based greases shows high dropping point of more than 175 °C at thickener content of above 17.5%. This result is in agreement with Ebisike et al. [6], where grease formulated using spent engine oil and sodium soap exhibit drop point of 177 °C – 182 °C. Grease produced by Iheme et al. [28] using mineral oil and sodium soap also had comparable high dropping point of 165 °C – 190 °C to this study. Conversely, the addition of MoS_2 into the formulation does not shows an improvement on the dropping point, but it seems to maintain the dropping point within 186 °C ± 15 °C. This result is similar to studies conducted by Mohammed [25], where he stipulated that there was no change in the grease drop point when MoS_2 was added into the formulation of different types of greases.

3.4. Oil bleeding of the grease

Oil bleeding is the tendency of grease to bleed oil. A certain amount of bleeding is desirable in greases to provide continuous lubrication. Oil bleeding from grease depends on the thickener structure, base oil viscosity, and grease firmness. Bots stated that the oil bleeding differences of −15% to +15% are desirable as it indicated that the grease still can be utilized without changing the re-lubrication intervals [29]. The positive value stipulated for more oil bleeding occurs after grease is used where it indicated the possibility of a broken thickener structure during operation hence lead to the need for frequent re-lubrication. A negative value represents the drop of oil bleeding with respect to used grease which indicated grease dry out during operation [29].

Figure 3 shows that only some of the formulated greases achieved the recommended oil bleeding values. The amount of thickener in the formulation was found to affect the grease's oil bleed. Thickener helps to hold the oil in its system but, an excessive amount of thickener was found to hinder the oil from bleeding from the grease. For example, at SG_3, positive value of oil bleeding exceeding +15% shows that more oil was bled due to its soft consistency. As the thickener was increased to 32.5%, the oil bleeding remained within the range. However, at 40% of thickener content, high negative oil bleed values were obtained, which indicates the grease was dried out during the operation as if the oil was unable to bleed out from the thickener or the oil itself has evaporated. This is in line with Dixena et al. [30] where they stated that "an increase in oil bleed tendency was observed with decrease in thickener content and vice versa in both the thickener types PP and mPP over entire temperature range studied."

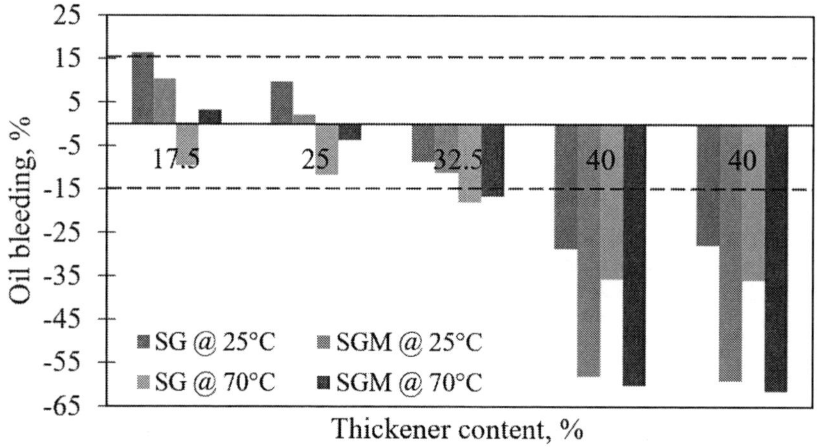

Figure 3. Oil bleeding of grease.

The presence of MoS_2 in grease formulation shows different results where it was found to reduce the tendency of grease to bleed oil. For example, SG_3 and SGM_3 share the same percentage of oil-to-thickener content of 82.5:17.5 percent. SG_3 oil bleed was observed to exceed the recommended values at 25 °C but, with the presence of MoS_2, SGM_3 oil bleed falls within the range from 16.53% to 10.42%. However, oil bleed of greases with 40% of thickener content has exceeded the range. From this, it can be drawn that MoS_2 does not improved or helps in stabilizing the grease oil bleed, but it reduced the tendency of the grease to bleed oil. This finding is similar to Sinitsyn et al. [31] where they found significant changes in grease oil bleed from 31.2% to 23.5% when the amount of MoS_2 was increased from 0% to 3%, respectively. Therefore, MoS_2 can be added to the formulation of grease with high oil bleeding percentage.

3.5. Oil separation of the grease

Oil separation on all grease sample surfaces was observed and tabulated in table 4. According to the ASTM D1742 standard [32], greases of NLGI less than one (1) are not suitable to be tested using the standard test method. However, out of all greases, only SG_{3-4} and SGM_{3-4} show oil separation of top of the sample surface. The oil separated from SG_{1-2} and SGM_{1-2} were not able to be identified as the grease is very soft and in semifluid form. The amount of oil separated from SG_{3-4} and SGM_{3-4} which were 0.27%, 0.07%, 0.33% and 0.12%, respectively were within the desirable value of less than 4% [33]. The oil separation was observed to reduce when the percentage of thickener content was increased.

3.6. Detection of copper corrosion from the grease

According to Verdura et al. [34], "This test does not determine the ability of a grease to inhibit copper corrosion caused by factors other than the grease itself; neither does it determine the stability of grease in the presence of copper. The sole determination is the chemical staining of copper by lubricating grease". Table 5 shows the corrosiveness results of the base oil and the formulated greases towards copper. FTO and UTO were found only slightly tarnish towards copper, thus concluded that the base oil appeared to be not corrosive towards copper strips. According to Mohammed [25], sodium grease was classified as moderately corrosive (2a) towards copper when no additive added, and the addition of 3% MoS_2 into the formulation was found to improve the grease corrosiveness.

In this study, all grease corrosiveness was classified as class 1 on copper strips. According to Mohammed [25] and Sinitsyn et al. [31], MoS_2 acted as a corrosion inhibitor in grease formulation. However, referring to table 5, it was found that all UTO-based greases without MoS_2 (SG_i) demonstrated a slight tarnish to copper strips. Therefore, blending 2% of MoS_2 into the prepared greases (SGM_i) shows no significant effect on grease corrosiveness towards copper strips as both SG_i and SGM_i were classified as class 1 grease corrosiveness. It was concluded that FTO, UTO, sodium stearate and MoS_2 were not corrosive to copper strips when utilized in grease formulations.

Table 5. Results of oil and grease corrosiveness test.

Sample	Corrosiveness
FTO	1a – slight tarnish
UTO	1a – slight tarnish
FG	1a – slight tarnish
FGM	1b – slight tarnish
SG_1	1a – slight tarnish
SG_2	1a – slight tarnish
SG_3	1b – slight tarnish
SG_4	1b – slight tarnish
SG_5	1b – slight tarnish
SG_6	1b – slight tarnish
SG_7	1b – slight tarnish
SGM_1	1b – slight tarnish
SGM_2	1a – slight tarnish
SGM_3	1b – slight tarnish
SGM_4	1a – slight tarnish
SGM_5	1a – slight tarnish
SGM_6	1b – slight tarnish
SGM_7	1b – slight tarnish

3.7. FTIR characterization of the grease

Figure 4 shows the FTIR spectrums of selected sodium greases that represent all samples. FTIR spectroscopy was used to determine the functional group of each chemical compound presented in greases. The measurement was within 500 – 4000 cm^{-1}. The FTIR spectrums show that all samples have the same strong peaks in the region of 2849 – 2951 cm^{-1}, which indicates the CH_2 group asymmetric and symmetric vibration. Weak peaks were observed at 3371 – 3390 cm^{-1} that attributed to O-H stretching vibration [35]. The functional group of symmetrical stretching vibration of COO group was observed at

1558 cm^{-1}. A peak at 1455 cm^{-1} indicates the asymmetric CH_3 deformations [36]. The functional group of alkane C—H bond stretch of CH_3 bond was observed at about 1376 cm^{-1} [37]. Overlapping of CH_2 rocking vibration and long aliphatic hydrocarbon vibration at about 722 cm^{-1}, which also indicates the presence of alkanes [38, 39].

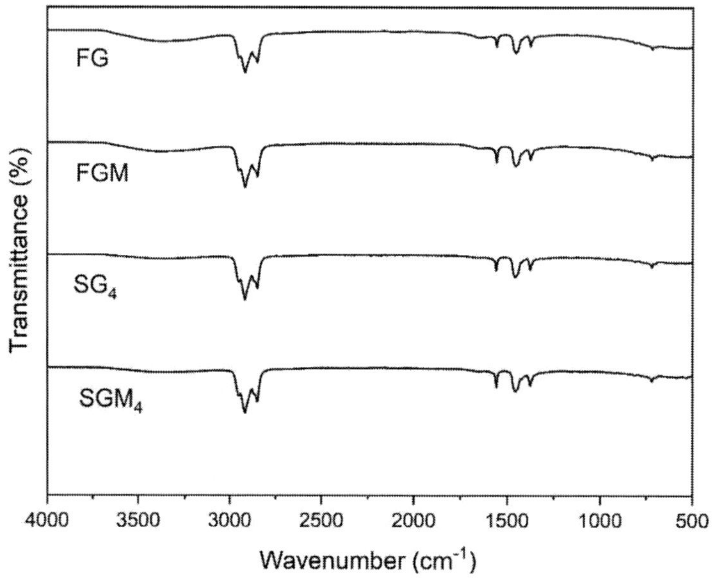

Figure 4. FTIR spectrums of sodium greases.

FTIR spectrums of all types of grease show no significant differences either being formulated with FTO or UTO, with or without MoS_2. There were no clear peaks in the grease, which indicates the presence of contaminants or any by-products such as dissolved gases, oxidation products, and etc. The spectrums show similarity due to the same chemical used in the grease for which the amount varies. These FTIR spectrums proof the absence of any contaminants in the greases.

4. Conclusion

The utilization of used transformer oil (UTO) in grease formulation and the characterization of the grease were evaluated. The best formulation for grease employing UTO is 67.5:32.5 (UTO:Thickener) with 199 °C dropping point, −8.62%, and −18.04% of oil bleeding at both 25 °C and 70 °C. The grease was classified as NLGI 2 grease. Corrosion analysis of the grease indicates that the addition of MoS_2 into the formulation does not affect the grease corrosiveness towards copper strips. FTIR spectrums of all greases show no significant difference. Further grease analysis with regards to its tribology is required to study the grease's performances (wear and friction properties) during operation.

Acknowledgements

The researchers would like to extend their gratitude to Universiti Malaysia Pahang (UMP) for providing the grants for this study under the grant number RDU160382.

References

[1] Mohammed R R, Ibrahim I A R, Taha A H and McKay G 2013 Waste lubricating oil treatment by extraction and adsorption *Chem. Eng. J.* **220** 343–51.

[2] Nabi M N, Akhter M S and Rahman M A 2013 Waste transformer oil as an alternative fuel for diesel engine *Procedia Engineering* vol 56 (Elsevier B.V.) pp 401–6.

[3] Patel T B, Panchal K D and Shah A S 2015 A Technical Review Paper on Use of WTO (Waste Transformer Oil) - Diesel Blends With Additives in a Diesel Engine *Int. J. Sci. Res. Dev.* **3** 665–70.
[4] Mohta V and Chaware K D 2015 Preparation of Alternative Fuel from Waste Transformer Oil and Studying Performance Characteristics On Diesel Engine for Different Blends *Int. J. Sci. Res.* **4** 103–7.
[5] Hegazi S E F 2015 Conversion of used Oil into Lubricating Grease and Characteristics Evaluation *Int. J. Sci. Res.* **4** 1894–8.
[6] Ebisike K, Daniel B, Anakaa M, Kefas H and Olusunle S 2016 Effect of Sodium Hydroxide Thickener on Grease Production *Am. Chem. Sci. J.* **13** 1–8.
[7] Abdulbari H A and Zuhan N 2018 Grease Formulation from Palm Oil Industry Wastes *Waste and Biomass Valorization* 1–11.
[8] Razali M N, Aziz M A A, Hamdan W N A W M, Salehan N A M and Rosli M Y 2017 Synthesis of Grease from Waste Oils And Red Gypsum *Aust. J. Basic Appl. Sci.* **11** 154–9.
[9] Hassan A M, Shahba R M A, Youssif M A, Mazrouaa A M and Youssif M A E 2010 Preparation of Some Dielectric Greases from Different Types of Polymers *J. Appl. Polym. Sci.* **119** 1026–33.
[10] Ge X, Xia Y and Cao Z 2015 Tribological properties and insulation effect of nanometer TiO2 and nanometer SiO2 as additives in grease *Tribol. Int.* **92** 454–61.
[11] Cao Z, Xia Y and Xi X 2017 Nano-montmorillonite-doped lubricating grease exhibiting excellent insulating and tribological properties *Friction* **5** 219–30.
[12] Hassan A M, Mazrouaa A M, Youssif M A, Abou Shahba R M and Youssif M A 2013 Evaluation of Some Insulated Greases Prepared from Rubber and Bitumen Thickeners *Int. J. Org. Chem.* **3** 71–80.
[13] ASTM D217-17 2017 Standard Test Methods for Cone Penetration of Lubricating grease.
[14] ASTM D2265-15e1 2015 Standard Test Method for Dropping Point of Lubricating Grease Over Wide Temperature Range *ASTM Int.*
[15] Lugt P M 2013 Grease Qualification Testing *Grease Lubrication in Rolling Bearings* (West Sussex, UK: John Wiley & Sons, Ltd) pp 346–7.
[16] SKF 2009 SKF Grease Test Kit TKGT 1: Instruction Manual 1–40.
[17] Krusenstierna A Von 2016 *Grease-oil compatibility in tribological applications – Supergrease* (Lund University).
[18] ASTM D4048-16e1 2016 Standard Test Method for Detection of Copper Corrosion from Lubricating Grease.
[19] ASTM D130-18 2018 Standard Test Method for Corrosiveness to Copper from Petroleum Products by Copper Strip Test.
[20] Kumar A 2017 How to Select Right Lubricating Grease For Your Application? *Mach. Lubr. India.*
[21] Noria Corporation 2001 Water In Oil Contamination *Mach. Lubr.* 1.
[22] Rizvi S Q A 2008 *A Comprehensive Review of Lubricant Chemistry, Technology, Selection, and Design* (ASTM International).
[23] Neale M J 1995 *The Tribology Handbook* (Oxford: Butterworth-Heinemann).
[24] Doyle D 2015 General Grease Overview and Bearing Lubrication *ALS Tribol.*
[25] Mohammed M A R 2013 Effect of additives on the properties of different types of greases *Iraqi J. Chem. Pet. Eng.* **14** 11–21.
[26] Totten G E 2006 *Handbook of Lubrication and Tribology* (CRC Press).
[27] Buhlak S, Ibrahim B and Alhamoui M 2014 Manufacturing (Lithium-Sodium) Lubricant Grease Based On Syrian Base Oil And studying Its Physical, Chemical, And Rheological Properties *Int. J. ChemTech Res.* **6** 2247–54.
[28] Iheme C, Offurum J C and Chukwuma F O 2014 Production and Blending of Sodium Based Water-Resistant Lubricating Greases from Petroleum and Petrochemical By-Products *Am. J. Comput. Sci. Eng. Surv.* **2** 70–8.

[29] Bots S 2014 Best Methods for Analyzing Grease *Mach. Lubr.*

[30] Dixena R, Sayanna E and Badoni R 2014 Recycled and Virgin HDPEs as Bleed Inhibitors and Their Rheological Influences on Lubricating Greases Thickened with PP and mPP *Lubricants* **2** 237–48.

[31] Sinitsyn V V, Viktorova Y S and Vakurov P S 1967 Effect of molybdenum disulfide on properties of lithium greases *Chem. Technol. Fuels Oils* **3** 667–71.

[32] ASTM D1742-06 2013 Standard Test Method for Oil Separation from Lubricating Grease During Storage.

[33] Lugt P M 2013 *Grease Lubrication in Rolling Bearings* (Chichester: John Wiley & Sons, Inc.).

[34] Verdura T M, Brunette G and Shah R 2003 Lubricating Greases *Fuels and lubricants handbook: technology, properties, performance, and testing* (ASTM International) p 225.

[35] Kholijah A M S, Yeung S L C, Sazwani S and Yunus R M 2012 Production of High Temperature Grease from Waste Lubricant Sludge and Silicone Oil *J. Appl. Sci.* **12** 1171–5.

[36] Meena Devi R, Subadevi R, Paul Raj S and Sivakumar M 2015 Comparative Studies on Biodiesel from Rubber Seed Oil Using Homogeneous and Heterogeneous Catalysts *Int. J. Green Energy* **12** 1215–21.

[37] Kalathiripi H and Karmakar S 2017 Analysis of transformer oil degradation due to thermal stress using optical spectroscopic techniques *Int. Trans. Electr. Energy Syst.* **27** 1–11.

[38] Patel T B, Panchal K D and Shah A S 2016 Characterization of WTO (Waste Transformer Oil) & Diesel by Fourier Transform Infrared Spectroscopy & Effect of Additives on Properties of WTO-Diesel Blends *Int. J. Sci. Res. Dev.* **3** 747–51.

[39] Pan J, Cheng Y and Yang J 2016 Structural degradation of a lithium lubricating grease after thermal ageing *J. Chem. Eng. Japan* **49** 579–87.

IOP Conf. Series: Materials Science and Engineering **863** (2020) 012059 doi:10.1088/1757-899X/863/1/012059

Comparative CFD power extraction analysis of novel nature inspired vertical axis wind turbines

S N Ashwindran, A A Azizuddin* and A N Oumer

Faculty of Mechanical and Automotive Engineering Technology, Universiti Malaysia Pahang, 26600 Pekan, Pahang, Malaysia.

*E-mail: azizuddin.ump.edu.my

Abstract. The aim of this paper is to investigate the power extraction performance of three biologically inspired vertical axis wind turbine. In this study, the proposed turbines are simulated in 2D in Ansys using URANS approach based on two equation turbulent transport model-SST. The turbines are simulated under the similar solver and numerical configuration. FVM is adapted to analyse the turbines assisted by sliding mesh method (SSM) under non-conformal mesh configuration. The proposed turbine is analysed in terms of moment and power coefficient. Design 1 and Design 2 indicated promising result for a feasible and practical wind turbine. However, Design 3 indicated poor performance in power coefficient at high RPM value.

1. Introduction

Energy harvesting signifies the development of industrial and power growth of economy and socio-economy in a region. Energy harvesting can be classified into two major groups which is renewable and non-renewable energy. Both forms of energy harvesting have different degrees of impact on the sustainability of environment and pollution index. The use of fossil fuels without proper regulation resulted in rise of greenhouse gases (GHG) [1]. In depleting fossil fuels resources and the rise of environment pollution as a result of fossil fuel harvesting [2]. Shamsuddin [3] stated that from 1990 to 2006 Malaysia GHG emission is 7.9% compounded average growth rate (CAGR) placing itself globally among the countries with high GHG emissions. Thus, several countries have taken initiative in the development of renewable energy facilities. As a matter of fact, in 2008 19% the global energy utilization originated from renewable energy resources [4].

Renewable energy can be further classified into several types, the most prominent renewable energy is hydro, solar and wind. The fastest growing renewable energy sector is wind energy harvesting. There are two type of wind power facilities which is onshore and offshore. Wind turbines are classified into two type vertical axis and horizontal axis wind turbines. In terms of lifecycle GHG emission of electricity generation facility, wind power is considered as clean energy with the one of the lowest GHG emission index. Research shows that offshore wind harvesting has the lowest GHG index in comparison to other renewable energy resources [5, 6]. Furthermore, research shows that GHG emissions of offshore wind power facilities are 48% less than onshore facilities [7].

Meanwhile in Malaysia research and development in wind power begin in early 1990's. Inadequate research methodology, poor location selection, inadequate instrumentation and inconsistent data presented failed to convince the general public on the credibility of wind power harvesting in Malaysia [8]. Since most of the research conducted in Malaysia are centered to costal and onshore wind. Hence there is insufficient data on offshore wind power facilities in Malaysia. Pursuant to The Ministry of Energy, Green Technology and Water (KeTTHA) industry in Malaysia adapted the Five Fuel Diversification Strategy energy mix which is which comprise of nature gas, coal, oil, hydropower and renewable energy. The aim of the policy is to control the use of fossil fuels and to promote the use of renewable energy. Under Renewable Energy Act 2011 or Act 725, Sustainable Energy Development Authority (SEDA) is founded to administrate the funds assigned in Feed-in tariff (Fit) for the development of renewable energy resources in Malaysia. On the other hand, 90% of power generation in Malaysia are dependent on fossil fuels which is governed by subsidies. Since the subsidies are politically motivated, it incurs an obstacle for the development of renewable energy [8]. Although currently FiT are not available for wind power harvesting, with the geographical properties of Malaysia makes it suitable for wind energy harvesting. As a matter of fact, Tenaga National Berhad (TNB) was the pioneer in wind power harvesting. TNB installed a 200 kW capacity wind turbine in Perhentian island, in which TNB claims that the power generated by the wind turbine can supply electricity for 200 households [9]. Ministry of Science, Technology and Innovation (MOSTI) supervised the project conducted by Sirim, iWind Energy (M) Sdn Bhd and Sri Waja Resources Sdn Bhd on hybrid photovoltaic solar and wind turbine. Universiti Malaysia Terengganu (UMT) installed 3.3 kW wind turbine in Setiu, Terengganu [10]. More information on the feasibility of wind power farms in Malaysia can be found in [11–13].

Due to the geographical coordinates of Malaysia at 1°22'N latitude and 103°55'E longitude of Southeast Asia. Malaysia experiences strong wind from the Indian Ocean in the Northeast and Southwest. Furthermore, Malaysia wind are mainly influenced by Northeast and Southwest monsoons. Where Northeast monsoon begins from November to March and Southwest monsoon occurs between June to September. Studies shows that, wind speed can reach up to 15 m/s in the east coast region during the monsoon seasons [14]. Furthermore, the occurrence of typhoon in Philippines, influences the wind speed to reach up to 10 m/s in the offshore regions of Sarawak and Sabah [15]. The average offshore wind speed potential is higher than onshore thus able to produce 50% higher electricity than onshore wind farm [16, 17]. In terms of onshore wind farm, the regions of Peninsular Malaysia have a low wind speed potential with an average monthly wind speed of 1.5 to 2 m/s. However, several authors stated that the assessment of wind speed data presented by researchers are inadequate due to lack of substantial data in wind speed assessment methodology and parameters. Methodology and parameters such as the apparatus and instrumentation used for wind data assessments, the height of study, proper air density of the selected location, capacity factor of turbine generator, numerical methodology in wind speed probability distribution analysis and etc.

Onshore wind turbines require vast land site for wind farm construction which is limited. However offshore facilities are not limited to construction space due to the vast spatial area in ocean. Wind behavior varies in onshore and offshore where, offshore wind is stronger, higher energy and less turbulent effect than onshore wind. Furthermore, offshore load centers are more effective in interstate power transmission than onshore sites. On the contrary, there are disadvantages to offshore wind facilities in comparison to onshore. Offshore wind power construction cost is approximately three times higher than onshore wind turbines. This this because site of offshore wind turbines has to constructed with stronger structures that extends to the ocean floor. The turbines were design to withstand strong wind and storm which often happens in the regions of oceans. Furthermore, extensive cost incurred on the power transmission lines which extends from offshore to the onshore power grids.

This aim of this study is to present a comparative study of three novel wind turbine for offshore regions of Malaysia based on power extraction analysis. Based upon the inconsistent wind speed in offshore region due to the seasonal weather influence. Hence a novel blade design is being proposed to accommodate the offshore wind speed potential of Malaysia. Based on the literature review, the overall optimal offshore wind speed potential in Malaysia is 8 m/s. Thus, the theoretical CFD numerical investigation of the proposed turbines were conducted relative to freestream velocity of 8 m/s in *2D*. Since the proposed wind turbine at preliminary stage and to save numerical analysis time, the study was conducted in terms of moment coefficient and power coefficient based on theoretical assumption.

2. Literature survey

2.1. Bio-inspired wind turbines

Herrera et al. [18] presented a horizontal axis wind turbine blade design inspired from Triplaris Americana tree seed. The author analyzed the curvature and aerofoil characteristic of the tree seed for the design of the proposed turbine. Result shows that, a minimum of 2.5 m/s is required to produce energy. Cognet et al. [19] presented a novel elastic bio-inspired wind turbine. The design is inspired by insect flight and plant reconfiguration. The author stated that the elastic bio-inspired blade had improve the efficiency by 35% in comparison to rigid blade design. Fish et al. [20] investigated tubercles of humpback whale's flippers for the integration of wing model. The author stated that by the integration of tubercles on the wing model, the lift value had increased and consequently decreased the drag properties. It is found that the tubercles contributed to the improvement of the motion of flipper by passive flow. As manifested in literature survey, it shows that with proper design methodology and optimization could improvise the performance of the wind turbine.

3. Computational configuration

3.1. Parametric design adaptation

The parametric proposed models were prepared using equation (1) to equation (3) which is relative to vertical axis configuration. CATIA V5 feature such as surface morphing and wireframe modelling were used in the preparation of the models. The models are a result of hybrid fusion of elements from nature. A novel approach has been developed to design and develop a physical design feature of wind turbine by the combination of aerodynamic and non-aerodynamic elements. CFD numerical analysis were conducted to test the credibility of the algorithm in generating feasible and practical wind turbine design for offshore regions. Since the proposed methodology in preliminary stage, the simulations are conducted in *2D* in order to save computational load. In this study, two types of aerodynamic configuration of wind turbine is being modelled which is drag and lift type. The presented designs are model under the similar design configuration.

As expressed in equation (3), the swept area of the turbine, S_A is the product of rotor diameter, D_r and blade height, *H*. Since the simulation approach is in *2D*, the blade height is set to 1 m due to the default setting in ANSYS [21, 22]. *A.R* represents the aspect ratio of the turbine. Since the aim of the study is to analyse the aerodynamic performance of blade, no specific attention was given to the design of rotor hub. For simplicity the proposed models are labelled as Design 1; Design 2; Design 3 as shown in figure 3; figure 4; figure 5 respectively. Design 1 is modelled in lift aerodynamic configuration where else Design 2 and Design 3 are drag driven wind turbines. Design 1 is design based on National renewable energy laboratory (NREL) S819 aerofoil. NREL administrate the classification of aerofoil type according to aerofoil type and families of aerofoils. On the other hand, Design 2 is designed relative to Savonius wind turbine configuration. In order to simplify the analysis, design parameter such as blade height, *H*, overlap ratio, *O.R*, and rotor diameter, *Dr*, of a conventional Savonius wind turbine, were adapted for the modelling of Design 2. Figure 1 shows the proposed parametric blade configuration of Design 2 in comparison to traditional Savonius blade. Meanwhile, Design 3 is a curved non-uniform drag driven blade. The blade is composed of novel shaped cavity bucket vane as shown in figure 2(a). The blade is a result of cavity vane at three different sizes by surface multi-section labelled as A, B and C as illustrated in figure 2(b). Detailed information on Design 3 parameters and simulation configuration can be found on research presented by Ashwindran [23].

$$A.R = \frac{H}{D_r} \tag{1}$$

$$S_{A-Actual} = D_r \times H \tag{2}$$

$$S_{A-simulation} = D_r \times H \tag{3}$$

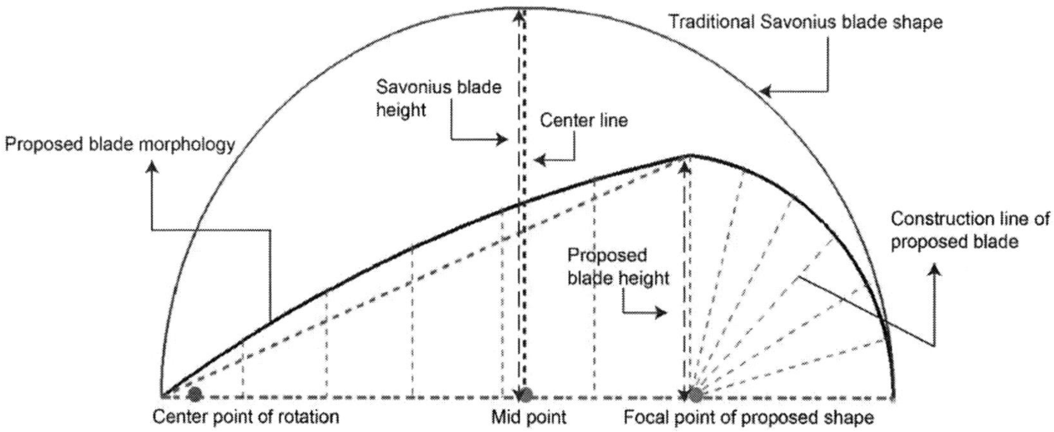

Figure 1. Parametric proposed shape of Design 2 and Savonius configuration.

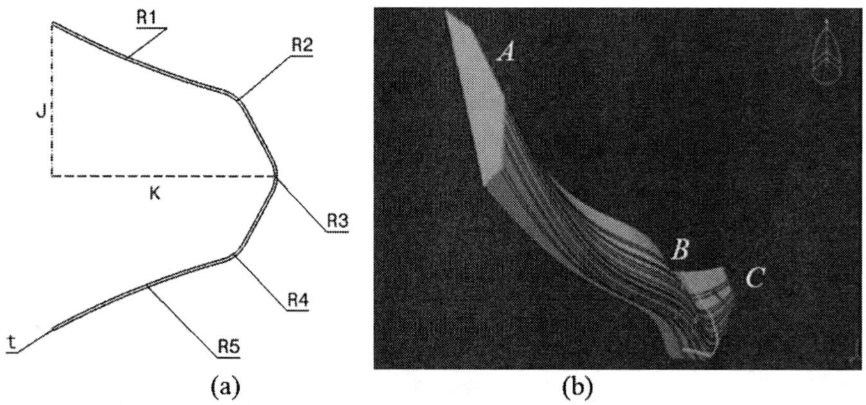

Figure 2. (a) Cavity vane profile of Design 3 and (b) Cavity vane A, B, C with different size.

3.2. Design configuration

Figure 3 shows the *3D* model of Design 1 inspired by epilobium hirsutum and maple seed. The design is separated into two segments as shown in figure 3(a). The top section of the turbine is adapted from epilobium hirsutum and the base section is adapted from maple seed. The blade is composed of NREL S819 aerofoil twisted at 4° angle of attack (AoA). The turbine consists of four blade spaces at 90° to each another. The turbine has a vertical height, H, of 1.33 m; rotor diameter, D_r, of 1.6 m, swept area, S_A, of 2.93 m^2, and aspect ratio, $A.R$ of 0.83. Meanwhile, the solidity of the airfoil, σ, is 0.25, the aerofoil thickness is 56 mm and chord length is 310 mm. As shown in figure 3(b), the planar section labelled as C at the height of 0.8 from datum were selected for the simulation procedure. For more information on Design 1 can be found in research article presented by Ashwindran [24].

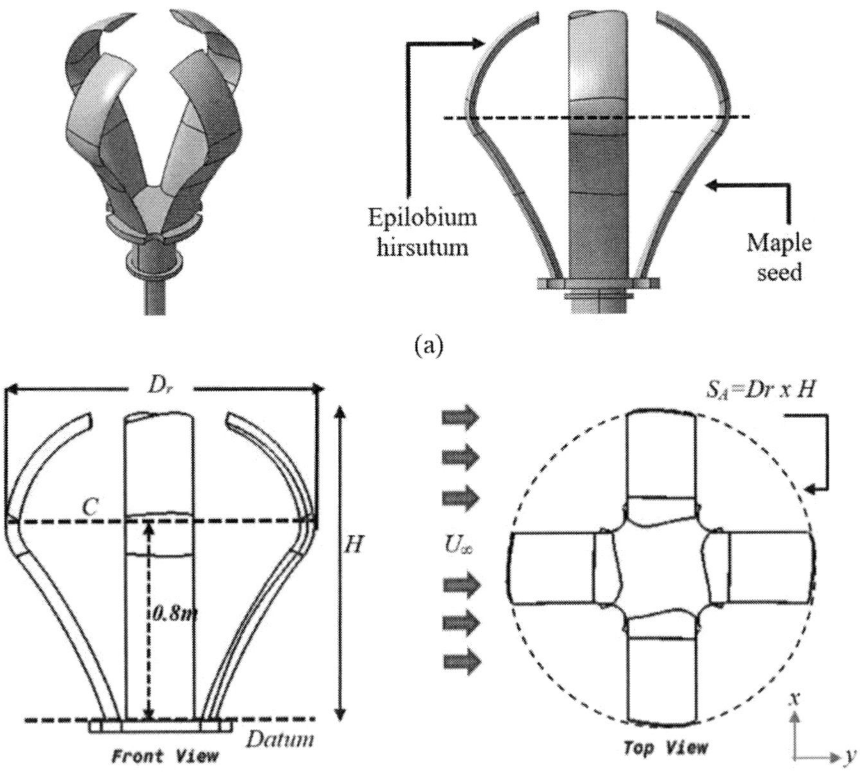

(b)

Figure 3. Proposed wind turbine configuration: (a) adapted morphology, (b) geometric and parametric configuration.

As shown in figure 4, the drag driven wind turbine blade is inspired by SPO algorithm and cycloid curve. In terms of complex geometry analysis, the most abundant design pattern in nature is the spiral geometry. Spiral geometry pattern can be found in flowers, plants, storm pattern, and the very shape of the Milky Way galaxy. This phenomenon even inspired mathematician such as Euler and Fibonacci to study its principals in geometry. Hence in this study, a modified SPO algorithm and cycloid curve were used to generate the desired blade shape in Savonius wind turbine configuration as shown in figure 4(a). More information on spiral configuration and algorithm can be found in [25]. In this study, the design feature of Savonius wind turbine from Sandia Laboratories [26] were used as benchmark to analyse the performance of the proposed wind turbine. Figure 4(b) shows the design configuration of Savonius wind turbine. Both the turbines are designed under the similar design parameters and configuration. The height of the blade, H, is 1 m, rotor diameter, D_r, of 0.9 m and overlap ratio, O.R of 0.1 m.

As for Design 3, the model is a hybrid fusion of three nature elements which is Albatross wing, tulip flower and pitcher plant. As aforementioned the developed algorithm merges aerodynamic and non-aerodynamic elements. The aerodynamic elements are the Albatross wing shape and tulip flower. Due to its large aspect ratio and unique wind shape the Albatross bird is regarded as one of the most aerodynamic bird [27]. Meanwhile, tulip petal curvature shape inspired the design of the proposed cavity vane. The pitcher plant vane configuration was adapted in order to generate curved polynomial blade as shown in figure 4(a) and 4(b) with regards to Design 3, the mid-section was selected for the simulation procedure.

IOP Conf. Series: Materials Science and Engineering **863** (2020) 012059 doi:10.1088/1757-899X/863/1/012059

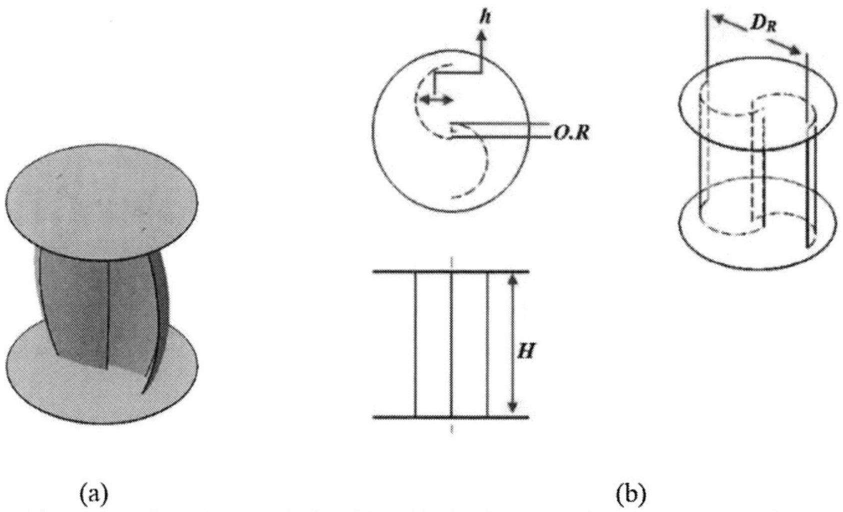

Figure 4. (a) Proposed Design 2 wind turbine blade, (b) Savonius geometric and parametric configuration.

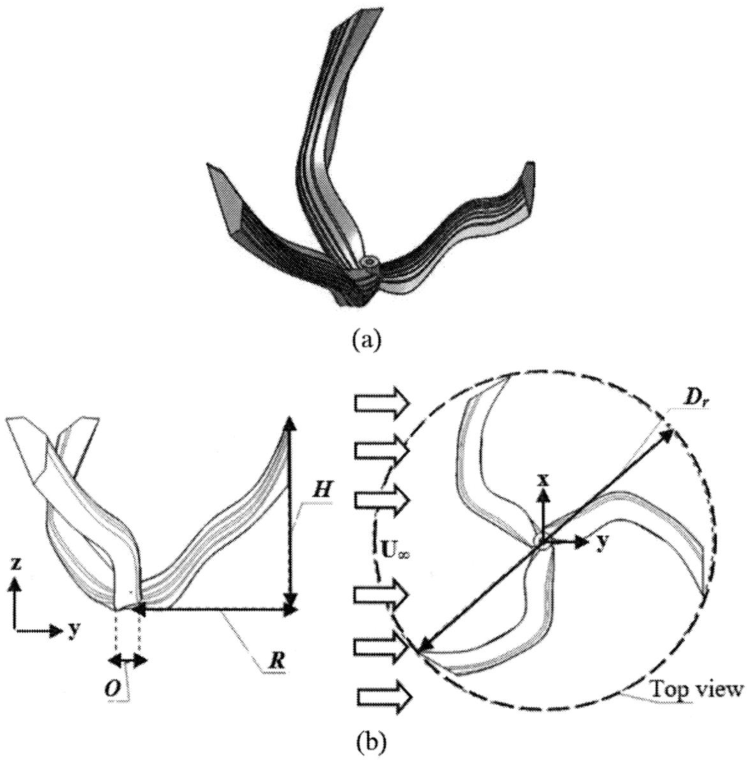

Figure 5. Wind turbine configuration: (a) Proposed drag driven wind turbine design. (b) Rotation configuration and geometric parameter.

3.3. Grid topology and domain configuration

FVM approach were adapted to analyse the proposed wind turbines. FVM is widely used by the wind turbine research community. FVM is a robust model in analyzing rotating machinery by means of analyse the model in a finite volume. FVM works by converting partial differential equation (PDE) into algebraic equation. It analysed finite volume by surface integral using divergence terms.

Due to the complexity presented by each of the geometry, separate domain at different dimension was modelled. For this study, sliding mesh method (SMM) were used to study the turbine. Relative to FVM methodology the domain is modelled as a virtual wind tunnel. The virtual wind tunnel is constructed relative to the rotor diameter of the respective wind turbine. Figure 6(a) shows the general structure of the virtual wind tunnel. The virtual wind tunnel consists of two domains which is main domain and sub-domain. The main domain consists of static mesh configuration. The sub-domain is considered as moving mesh. In order to prevent node connectivity between static and moving mesh domains, non-conformal mesh configuration was utilized. The main and sub-domain is separated by an interface as shown in figure 6(a). The virtual wind tunnel consists of inlet, outlet and two symmetry walls. The virtual tunnel is discretized with unstructured grid. The domain is defined with freestream velocity of 8 m/s from left to right of the virtual tunnel. Figure 6(b) shows the sub-domain of the proposed wind turbines. All the turbines are simulated under the similar simulation configuration. The difference is in the dimension of the virtual wind tunnel and mesh metric. The domain is discretized under fine and medium grid in order to analyse the grid sensitivity.

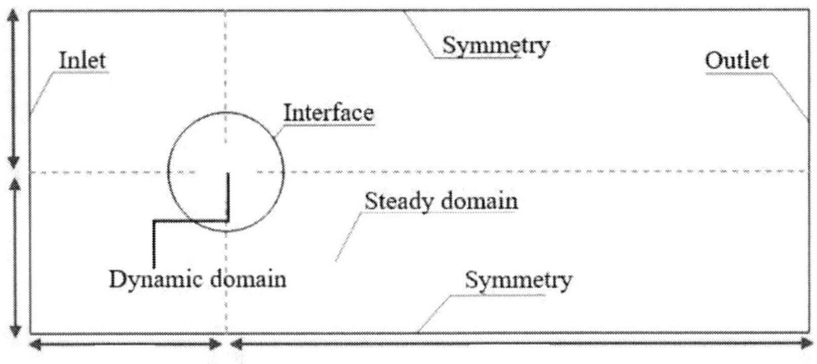

(a)

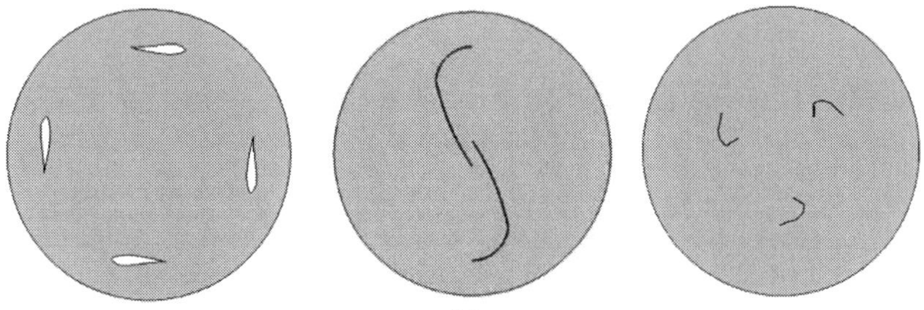

(b)

Figure 6. (a) Virtual wind tunnel configuration and (b) sub domains of the proposed turbines.

3.4 Boundary condition and solver configuration

FLUENT 16.2 were utilized to analyse the flow behaviour of the turbines. In order to simplify the study, the simulations are conducted in a specific direction. Relative to boundary condition and solver configuration, the proposed wind turbine is simulated under the influence of 8 m/s at the inlet from left to right of the virtual wind tunnel. The exit region is set as outlet pressure with 0 pascal. As for the blade regions of the respective wind turbine, were defined as no slip condition. In order to save computational load, the side walls are defined as symmetry type with free slip condition. This boundary condition would avoid solid blockage effect [28, 29]. Turbulent parameters such as turbulent intensity, I, and turbulent viscosity, μ_t/μ, are defined based on empirical observation of residual oscillation behavior. It is found that, Design 1 and Design 2 indicated stable residual oscillation with I=5% and μ_t/μ=10%. Conversely, Design 3 exhibited unstable residual oscillation behavior under I=5% and μ_t/μ=10% turbulent values. Therefore, the turbulent parameters were numerated based on equation (4) to equation (8). Hence the turbulent intensity, *I*, is 2.8 % and turbulent viscosity, μ_t/μ, is 1.71×10^3 for Design 3.

In terms of solver configuration, the turbines are simulated under unsteady formulation. All the turbine is simulated using pressure-based solver based on transient formulation formulation. The proposed turbine is studied under the influence of three turbulent transport formulation such as k-ω; realizable with scalable wall function k-ε; and SST for turbulent sensitivity study. It is found that, SST indicated more stable numerical oscillation in comparison to the counterparts. Hence SST model were chosen for the rest of the simulation procedure. Several authors suggested *SST* turbulent transport model are suitable for CFD analysis pertaining to rotating device because of blending functions of k-ω and k-ε [30–32]. In order to save computational load, URANS approached were adapted to numerate NSE. Since URANS model is the summation of fluctuating and mean components of flow properties, therefore it would be a suitable mathematical model for analyzing the model at preliminary stage. COUPLE scheme was used for all the turbines for pressure velocity coupling. For Design 1 and 2 the Courant number is set to 10, where else for Design 3, the Courant number is set to 20. The explicit relaxation (ERF) factor and under relaxation factor (URF) is set accordingly based on empirical observation of the gather result and numerical behavior. The expression for reference area of the turbine is defined based on the rotor diameter of the respective dimension of the turbine. The turbine is studied under wide range of RPM and tip speed ratios in order to study the moment coefficient and power coefficient performance. In this paper only numerical values from one tip speed ratios will be presented.

Turbulent intensity, I

$$I=\frac{u'}{\overline{u}}=0.16(Re)^{-\frac{1}{8}} \tag{4}$$

Turbulent kinetic energy, k

$$k=\frac{3}{2}(u_{\infty}I)^2 \tag{5}$$

Turbulent viscosity, $\left(\frac{\mu_t}{\mu}\right)$

$$\frac{\mu_t}{\mu}=(C\mu)\rho\frac{k^2}{\mu\varepsilon} \tag{6}$$

3.5. *Numerical parameters*

Since the turbines is studied based on computational numerical analysis and theoretical assumption. Betz limit theoretical value were used as benchmark the power coefficient result of the proposed turbines. Pursuant to Betz theory, the maximum theoretical power coefficient C_p limit of any kind of wind turbine is C_p=0.593 regardless to its design and configuration. Which means the maximum amount of kinetic energy that can be extracted from the flowing wind by the wind turbine is 59.3%.

The proposed turbine is analysed in terms of moment coefficient, power coefficient and tip speed ratio. The numerical parameters are numerated based on equation (7) to equation (9). The equation is expressed relative to turbine properties such as rotor diameter, D_r, rotor radius, R_t, and swept area, S_A.
Moment coefficient

$$C_m=\frac{M}{\frac{1}{2}\rho S_A U_{\infty}^2 R_t} \tag{7}$$

Tip speed ratio

$$TSR(\lambda) = \frac{\omega \times \left(\frac{D_r}{2}\right)}{U_\infty} \quad (8)$$

Power coefficient

$$C_p = C_m \times TSR \quad (9)$$

4. CFD numerical results

As aforementioned the turbine is discretized under fine and medium mesh configuration. Fine and medium mesh indicated trivial dissimilarities. Hence rest of the simulation is carried out under medium mesh configuration. Furthermore, Y^+ study indicated adequate result relative to the turbulent transport model and grid configuration. Y^+ presented an average facet result along the blade region under surface integral option, where the score is within region of $Y^+ > 30$ falls in log-law region layer. This is adequate for two equation turbulent transport model-SST and for the discretized grid at preliminary stage. The result indicated good convergence and stable numerical oscillation.

Figure 7 present the moment coefficient result of Design 1; Design 2; Design 3 at 100 RPM for three revolutions. Result shows that Design 1 and Design 2 indicated stable numerical oscillation. On the contrary, Design 3 exhibited unstable numerical behaviour. Under the influence of constant 8 m/s Design 3 indicated poor performance at high RPM. This is due to the geometry of proposed blade, where the returning incurs high adverse pressure which consequently effected the advancing blade. The presented blade geometry of Design 3 consists of sharp edges which limits the blade for moment generation. Furthermore, the non-uniform geometry is responsible for the rise of form drag along the blade region which resulted in negative moment and power coefficients. The pressure on advancing cavity vane is significantly less than the returning blade. Unstable pressure distribution can be observed along the blade regions. Moreover, the large swept area worsens the credibility of the turbine in moment and torque delivery. Vorticity magnitude indicated strong and high turbulent internal vortices on the advancing blade, which makes it difficult for the advancing blade to initiate rotation. However, at low RPM the turbine indicated positive moment and power coefficient results due to the drop-in form drag and static pressure on retuning blade of the turbine. Design 3 at λ=0.2 generated an average moment coefficient result of C_m = 0.143304 and at λ=0.3 indicated C_m=0.0847. Meanwhile at λ=0.9 the C_m = -0.104.

On the country, Design 1 and 2 performed well at high RPM under the similar freestream velocity. Moment and power coefficient result of Design 2 is compared against a traditional Savonius wind turbine, which is simulated under the similar configuration. At λ=0.59 (100RPM) Design 2 generated a moment coefficient of C_m= 0.401881, where Savonius indicated a moment coefficient result of C_m= 0.375023. Concave and convex blade indicated stable pressure distribution. In terms of flow field properties, strong stagnation point was observed on the convex side of the proposed blade in comparison to Savonius wind turbine. Furthermore, the outer side of the concave blade of the proposed design indicated slightly higher pressure than the Savonius turbine. This explain the small percentage of improvement. The modification of blade and reduction in blade height of the proposed turbine resulted in higher moment coefficient by 4%.

Design 1 outperformed Design 2 and 3 by means of power extraction and moment coefficient. At λ= 0.85 (100RPM) the proposed turbine indicated mediocre result. As the RPM value increase the performance of the turbine in energy capturing increases till the turbine reaches its peak performance. Figure 8 shows the power coefficient result of Design 1; 2; 3 at different tip speed ratios. As shown in figure 8, the peak performance of Design 1 is at λ= 1.7 (200 RPM), with average moment coefficient result of C_m=0.153991. The performance start to decline as the RPM increases after the peak value. Similarly, the performance of Design 2 decline after the peak λ=0.59. As for Design 1 at low RPM, the returning blade exhibited a strong wake property at the leading edge of the aerofoil which consequently effected the advancing blade. Since Design 1 operates in fixed AoA, the performance of the turbine can be further improved by altering the AoA in order to increase the lift generation.

IOP Conf. Series: Materials Science and Engineering **863** (2020) 012059 doi:10.1088/1757-899X/863/1/012059

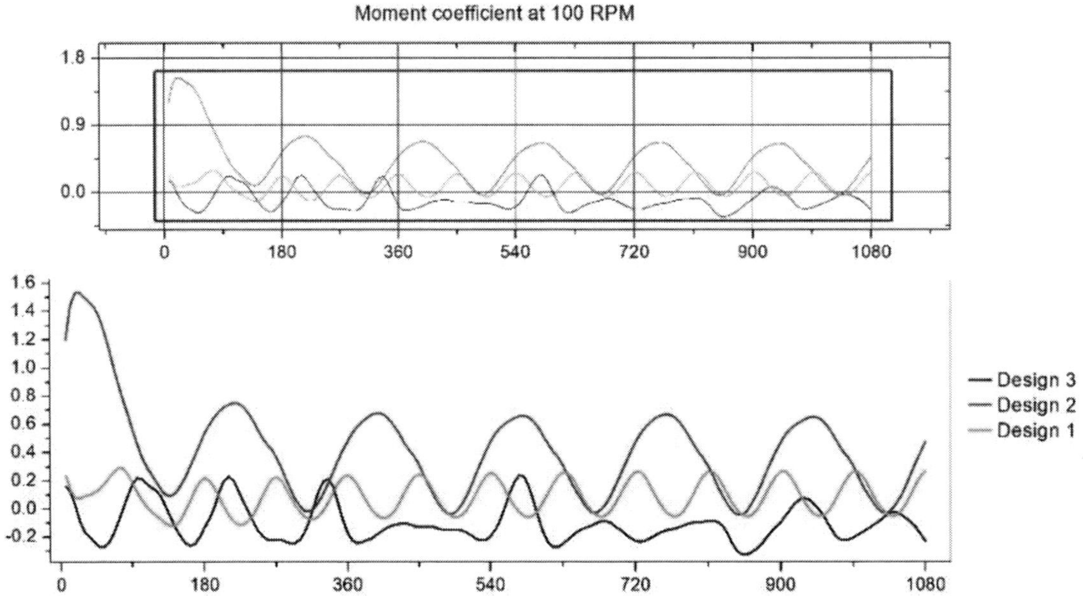

Figure 7. Moment coefficient result at of Design 1; Design 2; Design 3 at 100RPM.

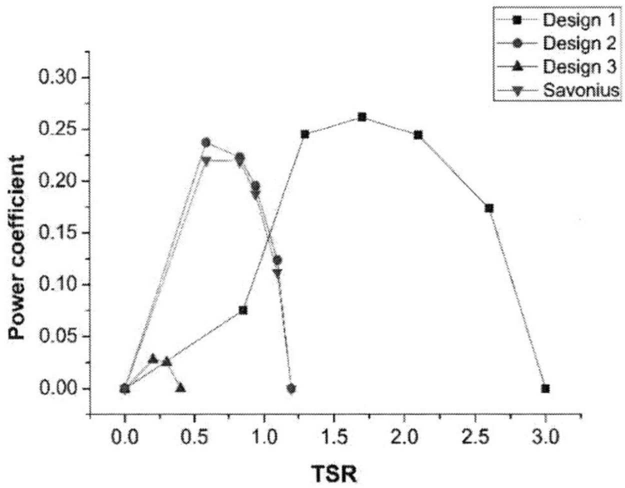

Figure 8. Power coefficient vs TSR of Design 1; Design 2; Design 3.

5. Conclusion

This paper investigated the numerical performance of three biologically inspired wind turbine based on power extraction and moment coefficient. Design 1 and Design 2 indicated satisfying result for a feasible power extraction wind turbine. As for Design 3 further design modification is required in order to reduce the high adverse pressure on the returning blade region. Design 2 indicated higher efficiency in power extraction in comparison to conventional Savonius wind turbine. The proposed blade of Design 2 outperformance Savonius wind turbine in moment coefficient by 4% at λ=0.59. Meanwhile lift driven Design 1 turbine indicated stable power extraction at high RPM. However, in order to improve the performance of Design 1, the AoA of blade needed to be altered to ensure more lift is generated by the rate of change of momentum of flowing across the aerofoil as it faces the freestream wind. Since the simulation is conducted in specific direction, experimental result will be validated against the simulation result in the future for validation and verification.

Acknowledgment
The research is financially supported by Malaysian Ministry of Education under FRGS/1/2017/TK07/UMP/03/1.

References

[1] Kweku D, Bismark O, Maxwell A, Desmond K, Danso K, Oti-Mensah E, Quachie A and Adormaa B 2018 Greenhouse Effect: Greenhouse Gases and Their Impact on Global Warming *J. Sci. Res. Reports* **17** 1–9
[2] Hoel M and Kverndokk S 1996 Depletion of fossil fuels and the impacts of global warming *Resour. Energy Econ.* **18** 115–36
[3] Shamsuddin A H 2012 Development of renewable energy in Malaysia strategic initiatives for carbon reduction in the power generation sector *Procedia Eng.* **49** 384–91
[4] Akorede M F, Mohd Rashid M I, Sulaiman M H, Mohamed N B and Ab Ghani S B 2013 Appraising the viability of wind energy conversion system in the Peninsular Malaysia *Energy Convers. Manag.* **76** 801–10
[5] Nomura N, Inaba A, Tonooka Y and Akai M 2001 Life-cycle emission of oxidic gases from power-generation systems *Appl. Energy* **68** 215–27
[6] Wang Y and Sun T 2012 Life cycle assessment of CO 2 emissions from wind power plants: Methodology and case studies *Renew. Energy* **43** 30–6
[7] Noori M, Kucukvar M and Tatari O 2015 Economic input-output based sustainability analysis of onshore and offshore wind energy systems *Int. J. Green Energy* **12** 939–48
[8] Ho L W 2016 Wind energy in Malaysia: Past, present and future *Renew. Sustain. Energy Rev.* **53** 279–95
[9] Ibrahim H N 2011 Towards Renewable Energy *Tenaga Link Malaysia* **1** 1–20
[10] Albani A and Ibrahim M Z 2017 Wind energy potential and power law indexes assessment for selected near-coastal sites in Malaysia *Energies* **10**
[11] Nor K M, Shaaban M and Abdul Rahman H 2014 Feasibility assessment of wind energy resources in Malaysia based on NWP models *Renew. Energy* **62** 147–54
[12] Irwanto M, Gomesh N, Mamat M R and Yusoff Y M 2014 Assessment of wind power generation potential in Perlis, Malaysia *Renew. Sustain. Energy Rev.* **38** 296–308
[13] Izadyar N, Ong H C, Chong W T, Mojumder J C and Leong K Y 2016 Investigation of potential hybrid renewable energy at various rural areas in Malaysia *J. Clean. Prod.* **139** 61–73
[14] Siti Khadijah N, Azami Z, Ahmad Mahir R, Mohd Said Z, Kamarulzaman I and Kamaruzzaman S 2009 Analyzing the East Coast Malaysia Wind Speed Data *Int. J. Energy Environ.* **3**
[15] Ahmad Zaman A A, Hashim F E and Yaakob O 2019 Satellite-Based Offshore Wind Energy Resource Mapping in Malaysia *J. Mar. Sci. Appl.* **18** 114–21
[16] Junfeng, Li. Shi, Pengfei. Hu G 2010 Chinese Wind Power Outlook 2010 *Chinese Renew. Energy Ind. Assoc.*
[17] IRENA 2012 Renewable Energy Technologies: Cost Anaylsis of Wind Power *Cost Anal. Ser.* **1**
[18] Herrera C, Correa M, Villada V, Vanegas J D, García J G, Nieto-Londoño C and Sierra-Pérez J 2019 Structural design and manufacturing process of a low scale bio-inspired wind turbine blades *Compos. Struct.* **208** 1–12
[19] Cognet V, Courrech Du Pont S, Dobrev I, Massouh F and Thiria B 2017 Bioinspired turbine blades offer new perspectives for wind energy *Proc. R. Soc. A Math. Phys. Eng. Sci.* **473** 20160726
[20] Fish F E, Weber P W, Murray M M and Howle L E 2011 The tubercles on humpback whales' flippers: Application of bio-inspired technology *Integrative and Comparative Biology* vol 51 pp 203–13
[21] Tian W, Song B, VanZwieten J and Pyakurel P 2015 Computational Fluid Dynamics Prediction of a Modified Savonius Wind Turbine with Novel Blade Shapes *Energies* **8** 7915–29
[22] Wenlong T, Baowei S and Zhaoyong M 2015 Conceptual design and numerical simulations of a

vertical axis water turbine used for underwater mooring platforms *Int. J. Nav. Archit. Ocean Eng.* **5** 625–34
[23] Ashwindran S, Azizuddin A A and Oumer A N 2019 Computational fluid dynamic (CFD) of vertical-axis wind turbine: mesh and time-step sensitivity study *J. Mech. Eng. Sci.* **13**
[24] Ashwindran S N, Aziz A and Oumer A N 2019 Unsteady Computational Study of Novel Biologically Inspired Offshore Vertical Axis Wind Turbine at Different Tip Speed Ratios: A Two-Dimensional Study *Int. J. Automot. Mech. Eng.* **16**
[25] Mirjalili S 2015 Moth-flame optimization algorithm: A novel nature-inspired heuristic paradigm *Knowledge-Based Syst.* **89** 228–49
[26] Sheldahl R E, Feltz L V and Blackwell B F 2008 Wind tunnel performance data for two- and three-bucket Savonius rotors *J. Energy* **2** 160–4
[27] Denny M 2009 Dynamic soaring: Aerodynamics for albatrosses *Eur. J. Phys.* **30** 75–84
[28] Nobile R, Vahdati M, Barlow J F and Mewburn-Crook A 2014 Unsteady flow simulation of a vertical axis augmented wind turbine: A two-dimensional study *J. Wind Eng. Ind. Aerodyn.* **125** 168–79
[29] Singh N K 2019 Simulation of Flow over a Rotationally Oscillating Square Cylinder at Low Reynolds Numbers *Int. J. Automot. Mech. Eng.* **16**
[30] Ntinas G K, Shen X, Wang Y and Zhang G 2018 Evaluation of CFD turbulence models for simulating external airflow around varied building roof with wind tunnel experiment *Build. Simul.* **11** 115–23
[31] Ramponi R and Blocken B 2012 CFD simulation of cross-ventilation flow for different isolated building configurations: Validation with wind tunnel measurements and analysis of physical and numerical diffusion effects *J. Wind Eng. Ind. Aerodyn.* **104–106** 408–18
[32] Halim M A A, Mohd N A R N, Nasir M N M and Dahalan M N 2018 The Evaluation of k-ε and k-ω Turbulence Models in Modelling Flows and Performance of S-shaped Diffuser *Int. J. Automot. Mech. Eng.* **15**

IOP Conf. Series: Materials Science and Engineering **863** (2020) 012060 doi:10.1088/1757-899X/863/1/012060

Experimental studies of single cylinder engine run on diesel-biodiesel-butanol blends

I Izzudin[1,*], A F Yusop[1], S Sapee[1], M A Hamidi[1], I M Yusri[1], R Mamat[1] and Erdiwansyah[1,2]

[1] Faculty of Mechanical and Automotive Engineering Technology, Universiti Malaysia Pahang, 26600 Pekan, Pahang, Malaysia.
[2] Fakultas Teknik, Universitas Serambi Mekkah, 23245 Aceh, Indonesia.

*E-mail: izuan.ump@gmail.com

Abstract. Biodiesel and butanol are excellent additive fuels, especially for diesel fuel. Many studies in literature reported that the biodiesel-butanol with various ratios is applied to diesel engines. In this experiment, diesel engines operated using biodiesel-butanol blend in low proportions 5-5%, 5-10% 10-5%, 10-10%, 15-5% and 15-10% mixed with pure diesel, and the blend is characterized. This blend of fuels can be represented as B5Bu5, B5Bu10, B10B5, B10Bu10, B15Bu5 and B15B10 with a numeric number in the fuel blends. This fuel blend is used as test fuel which is operated on a single cylinder diesel engine, four steps, direct injection (DI) at a constant speed of 1200 rpm and engine load of 25% and 50%. The combustion characteristics, performance and engine emissions are analyzed and evaluated by comparing each load and the speed of the engine being operated. Furthermore, fuel additives with pure diesel are needed to check emissions from the engine when the engine is run with a blend of diesel-biodiesel-butanol fuel. Among the six fuel blends samples examined in this experiment, better performance was shown in the B5B10 blend and produced fewer emissions. The results of the whole experiment are presented in full in this paper.

1. Introduction

Alternative fuels from biofuels used for internal combustion engines (IC) have gained momentum in the past few decades to reduce greenhouse gas (GHC) emissions, overcome dependence on fossil fuels and be environmentally friendly. Biofuel is a better choice in the future to meet global demand as transportation fuel as reported by the International Energy Agency (IEA). The low toxicity from the production of vegetable ingredients has released sulphur in biofuels. At present, biodiesel and alcohol are two biofuels that have the potential to be focused as alternative fuels to replace fossil oil [1]. In 2016 global biofuel production increased 2.6% compared to 2015 with only 0.4% of products. Liquid biofuel is very suitable for transportation because it is a very important strategy in reducing air populations, increasing fuel security and supporting rural development. The first generation of biofuels produced from sugar cane corn, etc. is a conventional biofuel commonly referred to as ethanol. Globally, the number of biofuel production reaches more than 100 billion litters each year [2, 3]. Global supply of biofuels for about 70% is produced from Brazil and America in 2015, with each consisting of sugar cane and corn [4, 5]. The use of conventional biofuels has attracted the interest of several researchers in focusing on making biofuels the next generation.

Biodiesel fuel is soluble, sulphur free and bio-degradable into any diesel fuel with any ratio and has an oxygen content of up to 10-15%. Biodiesel has a heating value and an energy content of about 12%

lower than diesel fuel. So it is very suitable for use as a blend of diesel fuel. A biodiesel blend of 20% compared to 80% of diesel fuel (B20) produces comparable engine performance and results in lower emissions of conventional diesel [6, 7]. The combustion process in a diesel engine can be changed in the presence of the properties of fuels such as viscosity, cetane number, density and flame temperature. The time of combustion and fuel injection can be caused by increasing density in biodiesel. Increased NOx emissions are generated from high unsaturated molecules and reduce fuel consumption [8,9]. The problem with regulating injection time is the reason for increasing NOx emissions. Ignition and spray patterns in the combustion chamber can be affected by changes in fuel and rapid NOx formation [10]. Meanwhile, shorter ignition delays and longer burning duration, most of which are indicated by the use of biodiesel. The percentage of fuel blend that increases can reduce NOx emissions. Meanwhile, biodiesel fuel can reduce particulate matter (PM) compared to diesel fuel when operated in diesel engines. The main concern in using biodiesel fuel for diesel engines is to control NOx emissions. Control of NOx emissions on the engine triggered by biodiesel fuel has been carried out with various methods by making several modifications to the engine [11–14]. Blends fuels emulsified with biodiesel and added nanoparticles with higher latent heat evaporation, use of alcohol-biodiesel blends, etc. An example of fuel modification is adopted in controlling NOx emissions from biodiesel fuel for diesel engines [15]. Some of them are steps for exhaust gas recirculation (EGR), water injection, chemical absorption and running dual fuel modes in an effort to reduce NOx emissions from diesel engines using biodiesel fuel from modified engines.

Biodiesel fuel produced from vegetable oil raw materials is not recommended by the Indian government, because it stabilizes food security. Therefore, the focus is on the use of oil from raw materials only in recent years. Some vegetable oils for producing biodiesel such as Karanja oil, castor oil, Neem oil, Mahua oil, cottonseed oil, castor oil and rubber seed oil are currently very important to use in India [16–20]. Among all the vegetable raw materials, Karanja oil is one of the potential raw materials for producing biodiesel fuel because of its adequate availability of oil content to produce biodiesel [21]. Many studies have been done by several researchers relating to KME as an alternative fuel that is applied to IC machines. Examination of the effects of Karanja's biodiesel fuel blend to measure performance, combustion characteristics and emissions of single-cylinder diesel engines with different engine loads for constant speed has been investigated by [22]. The results of their investigation reported that the thermal efficiency of the brakes increased from the consumption of KOME mixed fuel compared to when the engine was operated using pure diesel fuel. However, the engine performance is better when using a 20% KOME blend and reduces emissions compared to diesel fuel. While different studies can be reported that NOx emissions are recorded to increase when the engine is operated using Karnja biodiesel fuel rather than diesel [23].

The effects of biodiesel-n-butanol mixed fuels with performance parameters and emissions from diesel engines have been investigated [24]. Experiments carried out to investigate the performance and emission of the engine are mixed-biodiesel-butanol fuel with varying ratio (B95Bu5, B90Bu10 and B80Bu20). Based on the results of testing they did increase NOx emissions rather than diesel. However, CO emissions were found to decrease for low concentrations of butanol fuel (5% and 10%). While NOx emissions were found to decrease when using butanol fuel with a concentration of 20%. However, it is inversely proportional to the blend of 20% butanol showing an increase in CO and UHC emissions from other blends. The increasing mix of biodiesel-butanol and biodiesel fuels can increase fuel consumption. In different studies as conducted by [24], the concentration of n-butanol blend of 5%, 10% and 20% with Jatropha biodiesel jatropha based on different volumes such as (95JME 5B, 90JME 10B, and 80JME 20B) operated in parallel single cylinder DI diesel engine. Where JME stands for Jatropha methyl ester and butanol abbreviated B. Engine performance can be reported to increase with decreasing viscosity and more oxygen contained in fuel. Meanwhile, engine emissions significantly decreased, but did not occur at UHC emissions when the engine was operated using a blend of n-butanol and JME rather than neat JME. This experiment shows that UHC emissions increase from the results of diesel operations. However, NOx, CO and smoke emissions were recorded to decline. The biodiesel-butanol fuel blend can reduce fuel consumption compared to pure diesel.

In this study experiment, 5%, 10% and 15% biodiesel-butanol fuels were mixed with pure diesel fuel with volumes specified as B5Bu5, B5Bu10, B10Bu5, B10Bu10, B15Bu5 and B15Bu10. The study uses a one-cylinder, four-step engine, constant-speed direct injection (DI) tested with different volume fuel blends to measure combustion parameters, performance, and emissions evaluated, analysed and compared to the operation of both engine load tested. CO and NOx engine emissions as measured parameters, compared to diesel fuel operations and presented in this work.

2. Experiment setup

This experiment was operated using a single cylinder diesel engine, four steps, direct injection (DI) at a constant speed of 1200 rpm. The machine diagram specifications and schematic are shown in figure 1 and table 1. The machines are combined into eddy current dynamometers to provide brake loads. Fuel injection into the engine uses an injection system. Measurement of fuel consumption through the help of a controlled burette to connect to the engine fuel pump. At certain loads and speeds, specific fuel consumption is calculated. The optical axis encoder position serves to give a signal at the centre of the upper die (TDC). TDC position signals and pressure signals are recorded from an analogue-to-digital converter (A/D) that has been connected to the computer. In general, conventional diesel root materials are operated on CI machines. Performance parameters such as Cylinder pressure (CP), Hate Release Read (HRR) and Pressure Rise Rate (PRR) on each load are stored in an Excel sheet. Carbon monoxide (CO) emissions, nitric oxide (NOx) emissions at each load were measured using the A KANE exhaust gas analyser. Each is measured to obtain combustion parameters, cylinder pressure and crank angle and then stored into DAQ. Sample collection was taken as many as 50 cycles continuously. The average value of crank angle pressure, peak pressure, heat release rate and maximum pressure increase rate are calculated using software.

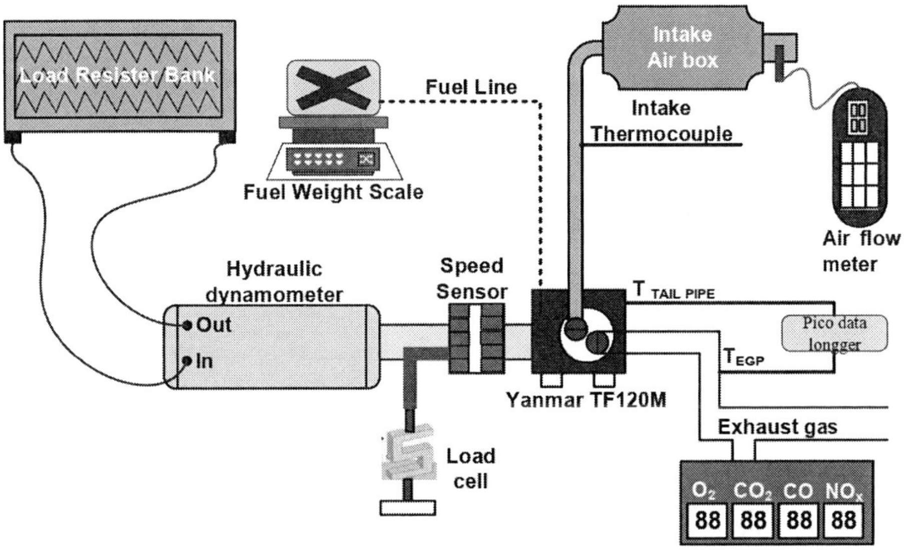

Figure 1. Schematic diagram of single cylinder diesel engine.

The first law for such a system can be written as shown by equation (1).

$$dQ_{hr} = dU + dW + dQ_{ht} \tag{1}$$

where, dQ_{hr} is the instantaneous heat release modeled as heat transfer to the working fluid, dU is the change in internal energy, dW is the work done, dQ_{ht} is the heat transmitted away from the working fluid to the combustion chamber walls. The change of internal energy can be written as equation (2).

$$dU = \left(\frac{C_v}{R}\right) * (PdV + Vdp) \tag{2}$$

Table 1. Detail of engine specifications.

Description	Specification
Engine model	YANMAR TF120M
Engine type	Horizontal single-cylinder 4-stroke diesel engine
Fuel injection type	Direct injection
Bore × Stroke (mm)	92 × 96
Displacement (L)	0.638
Injection timing	17° BTDC
Injection pressure (kg/cm^2)	200
Compression ratio	17:7
Continuous output (HP)	7.82 kW at 2400 rpm
Rated output (HP)	8.94 kW at 2400 rpm
Cooling system	Water cooled (radiator type)

Parameter R in equation (2) represent the gas constant, C_v is the specific heat at constant volume, dW is work done by the working fluid which equals to PdV. Parameters T, P and V represent temperature, pressure and volume, respectively. The heat transfer rate to the wall can be written equation (3).

$$\frac{dQ_{ht}}{dt} = hA_s(T_g - T_w) \tag{3}$$

The contents in the cylinder can be assumed by modeling as an ideal gas and the cylinder contents have a uniform temperature during the combustion process [25]. The above equation is the first thermodynamic law which can be written as equation (4).

$$dQ_{ht} = \frac{y}{y-1} P\frac{dv}{dq} + V\frac{1}{y-1}\frac{dp}{d\theta}\ hA_s(T_g - T_w)\frac{dt}{d\theta} \tag{4}$$

where θ the crank angle in degrees, h is the coefficient of heat transfer, the specific heat ratio of fuel and air is assumed to be y, wall temperature as T_w, while the gas temperature is T_g, where the area in m^3 is through the displacement of gas heat to the combustion chamber wall. The value for pressure is obtained from the cylinder pressure at the appropriate crank angle as shown by equation (5).

$$dQ_{ht} = \frac{y}{y-1} P\frac{dv}{d\theta} + V\frac{1}{y-1}\frac{dp}{d\theta} \tag{5}$$

The heat release rate can be calculated using this relationship. All quantities on the right side are known or can be easily derived after the history of the time the pressure is recorded.

3. Results and discussion

3.1. Cylinder pressure

Figure 2 and figure 3 illustrate the variation in peak cylinder pressure with engine load of 25% and 50% for all fuel samples tested in this study. The air-fuel blend preparation, the delay period and the amount of heat released during combustion largely determine the pressure of the peak cylinder on the CI engine. From the figure it can be observed that, when the engine load increases, the peak pressure also increases for all fuel samples tested. Higher cylinder temperatures and increased amounts of fuel have caused this increasing trend to occur at higher loads. The highest cylinder pressure of 65.5 bar was found for pure diesel fuel at a load of 50%. While the overall fuel blend tested shows cylinder peak pressure reported to be lower than pure diesel fuel at a load of 50%, due to the low heat released. The peak cylinder pressure increases when butanol is added up to 10% compared to biodiesel for overall engine operation. Increased ignition delay and more available oxygen in mixed fuels have resulted in higher amounts of heat in the combustion phase. Furthermore, the percentage of butanol increased in the biodiesel blend with peak cylinder pressure can be reduced. This is because in the combustion phase more amount of heat is released from mixing the fuel. At a load of 50%, the lowest peak cylinder pressure of 56.4 bar is recorded for the B10Bu5 blend. This trend can be caused by combustion that occurs far from TDC due to ignition delays in the fuel blend. The peak cylinder pressure value shows a very small difference. Similar results for n-butanol fuel mixtures in different studies have also been observed by [26].

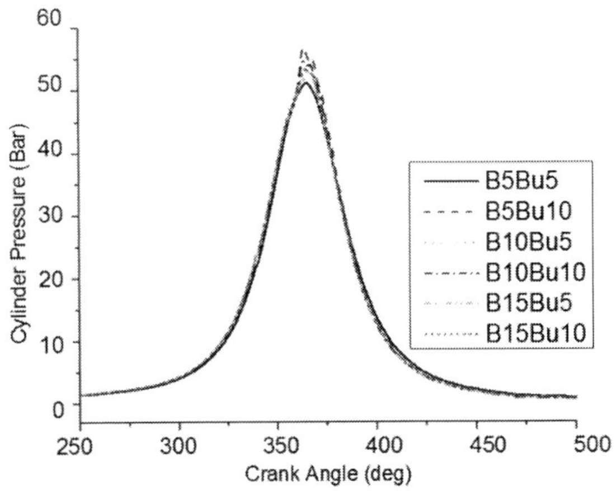

Figure 2. Cylinder pressure for different fuel at engine load 25%.

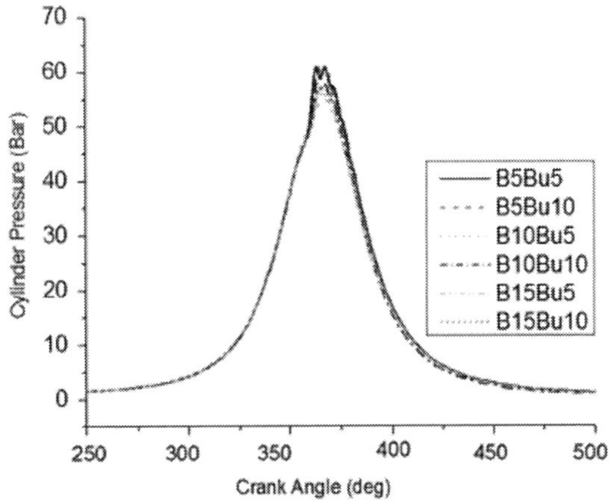

Figure 3. Cylinder pressure for different fuel at engine load 50%.

3.2. Heat release rate

The variation of heat release rate with crank angle using biodiesel-butanol and diesel mixed fuel at a load of 25% and 50% for each sample tested is illustrated in figure 4 and figure 5. The maximum heat release rate is highly dependent on the delay period with the amount of fuel taken from the combustion section of the previous fuel blend and the air-fuel blend formation [27]. The maximum speed of the biodiesel-butanol blend is lower than pure diesel. This is due to a relatively shorter delay period, lower heating values and increased viscosity of the fuel blend into diesel. The rate of heat release, energy content and latent heat of evaporation are more defined by the blend of biodiesel-butanol-diesel fuel than pure diesel. The maximum value of the heat release rate using mixed fuels and diesel at a load of 50% is 54.1 J/°CA and 56.5 J/°CA, respectively. Whereas the lowest heat release rate was found in the B10Bu5 blend among the entire fuel blend tested, with increasing latent heat of evaporation causing a decrease in this blend.

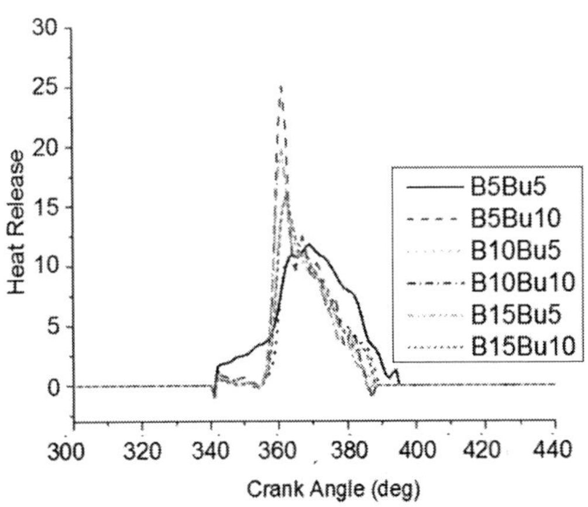

Figure 4. HRR for different fuel at engine load 25%.

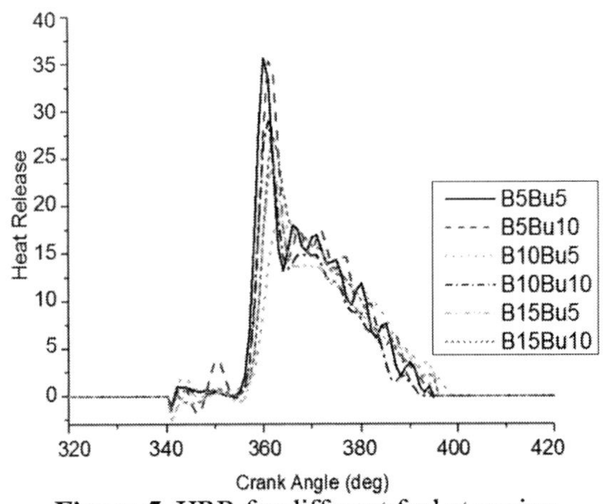

Figure 5. HRR for different fuel at engine load 50%.

IOP Conf. Series: Materials Science and Engineering **863** (2020) 012060 doi:10.1088/1757-899X/863/1/012060

3.3. NOx emission

The variations in NO emissions can be illustrated in figures 6 and figure 7 of both 25% and 50% loads using a blend of B5Bu5, B5Bu10, B10Bu5, B10Bu10, B15Bu5 and B15Bu10 fuels. It can be observed that NO emissions are reduced for a load of 25% compared to a load of 50% for all samples tested. The operation of pure diesel fuel shows NOx emissions are higher than the fuel blend tested in this study, the results show that the heat released in the combustion phase is higher.

NOx emissions are lower when the load is 25% for all fuel blends tested compared to when the engine load reaches 50%. However, this result is lower than the engine operated with pure diesel fuel. Adding small amounts of butanol can reduce ignition delays, causing less amount of injected fuel [28]. Moreover, increasing latent heat from evaporation of butanol-biodiesel has reduced the temperature inside the cylinder and reduced NOx emissions when using the overall blend of fuel tested. For B10Bu5 NO emissions are recorded lower among all fuels tested mainly for diesel.

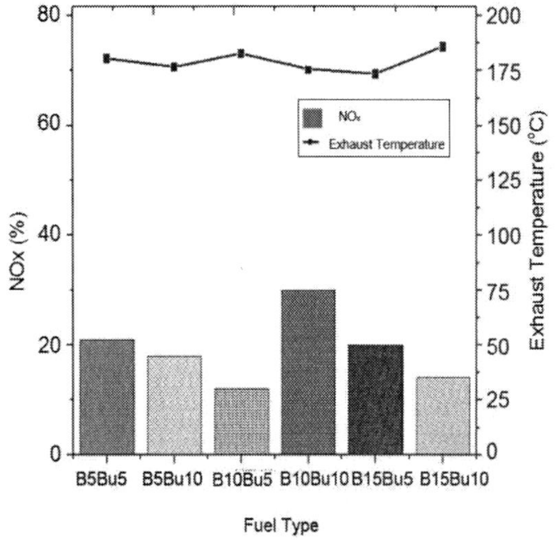

Figure 6. NOx emission compare exhaust temperature at load 25% for different fuel.

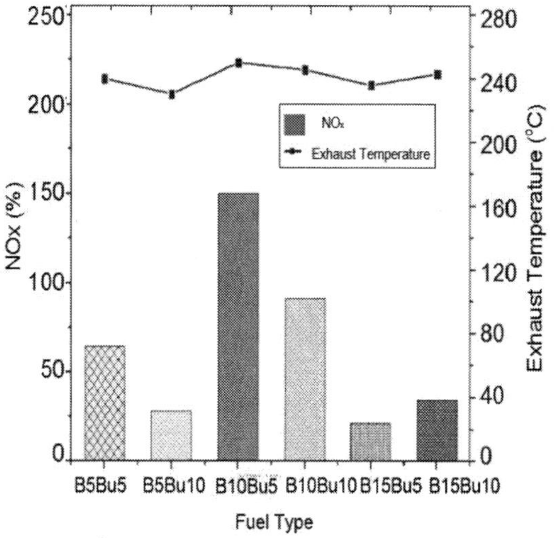

Figure 7. NOx emission compare exhaust temperature at load 50% for different fuel.

3.4. Carbon monoxide (CO) emissions

Figure 8 and figure 9 illustrate trends in CO emissions against combustion temperatures with different engine loads. In general, CO emissions from diesel engines are lower at 25% engine load compared to 50%. Oxygen content, temperature in the cylinder, fuel spray, oxidation rate and formation of ignition centers change because of the effects of oxygenated additives and alcohol [29]. Supply of oxygen can produce inadequate CO emissions in the engine combustion process. This is also due to the formation of the duration of combustion and air-fuel from the fuel blend. The low combustion duration and high oxygen molecules from the B5Bu5 blend can reduce CO emissions compared to pure diesel fuel. As a result, the overall engine operation tested for CO emissions was found to be reduced. However, CO emissions from both engine loads triggered by biodiesel-butanol and diesel fuel blends have a marginal difference.

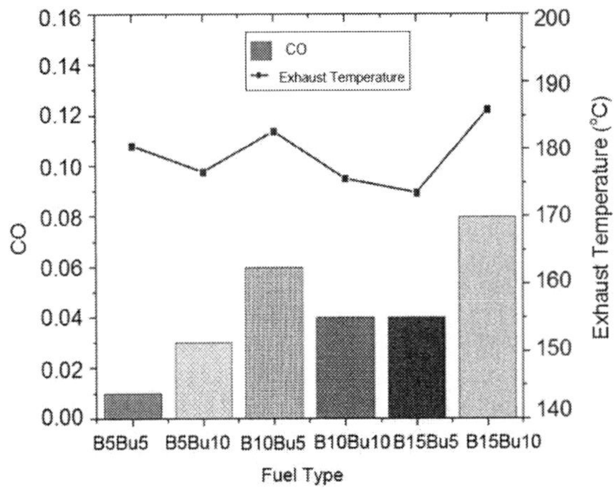

Figure 8. CO emission compare exhaust temperature at load 25% for different fuel.

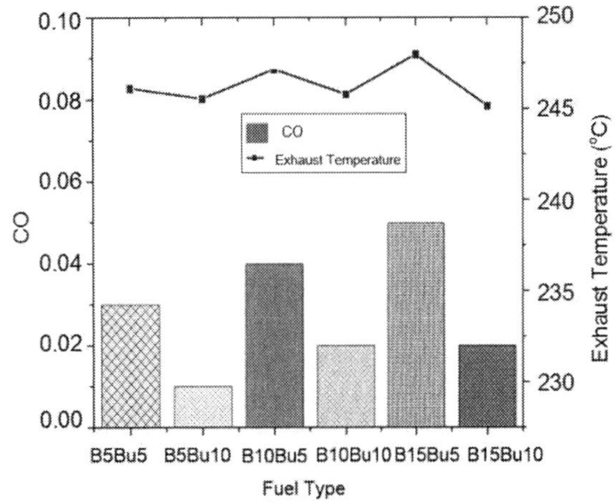

Figure 9. CO emission compare exhaust temperature at load 50% for different fuel.

4. Conclusion

This research, combustion characteristics, engine performance and emission were investigated when the engine was operated using a blend of biodiesel-butanol with pure diesel fuel. This experiment was carried out on a single-cylinder diesel engine to measure performance parameters, combustion and emission from the engine. These blends fuels are as many as six samples banned B5Bu5, B5Bu10, B10Bu5, B10Bu10, B15Bu5 and B15Bu10 tested in the engine at constant speed with engine loads of 25% and 50%. Based on the results of the entire sample tested, the blend of biodiesel-butanol fuel with NOx and CO diesel emissions decreases at low engine loads. Furthermore, some findings that are important from the investigation are explained as follows:

i) The ignition delay of the B10Bu5 blend for a 50% load was found to be shorter around 2 °CA, compared to diesel fuel, while the addition of higher butanol ignition delay periods increased up to 4 °CA when the engine load was 50%.
ii) The increasing the load of the test engine, can increase the duration of combustion. The duration of combustion decreases due to an increase in the percentage of butanol in mixed fuels.
iii) The biodiesel-butanol fuel blends with biodiesel generally reduces NOx emissions in the engine. However, an increase in the B10Bu5 blend occurred compared to pure diesel. This increase is due to the high oxygen content, higher evaporation heat and decreased heat content in the engine.
iv) The butanol fuel blend which increases with a high oxygen supply the average value of CO emissions is found to decrease. However, when the butanol blend decreases with increasing biodiesel it has increased CO emissions.

Acknowledgements

The authors would like to express their deepest gratitude to the University Malaysia Pahang (RDU190355 and RDU172204) for their financial assistance for this project.

References

[1] Ragauskas AJ, Williams CK, Davison BH, Britovsek G, Cairney J, Eckert CA, et al. The path forward for biofuels and biomaterials. Science (80-) 2006;311:484–9.
[2] Owusu PA, Asumadu-Sarkodie S. A review of renewable energy sources, sustainability issues and climate change mitigation. Cogent Eng 2016;3:1167990.
[3] Council WE. World energy resources 2016. World Energy Counc London, UK 2016.

[4] Araújo K, Mahajan D, Kerr R, Silva M da. Global biofuels at the crossroads: an overview of technical, policy, and investment complexities in the sustainability of biofuel development. Agriculture 2017;7:32.
[5] Paramjeet S, Manasa P, Korrapati N. Biofuels: Production of fungal-mediated ligninolytic enzymes and the modes of bioprocesses utilizing agro-based residues. Biocatal Agric Biotechnol 2018;14:57–71.
[6] Meher LC, Sagar DV, Naik SN. Technical aspects of biodiesel production by transesterification—a review. Renew Sustain Energy Rev 2006;10:248–68.
[7] Naik SN, Goud V V, Rout PK, Dalai AK. Production of first and second generation biofuels: a comprehensive review. Renew Sustain Energy Rev 2010;14:578–97.
[8] Mangus M, Kiani F, Mattson J, Depcik C, Peltier E, Stagg-Williams S. Comparison of neat biodiesels and ULSD in an optimized single-cylinder diesel engine with electronically-controlled fuel injection. Energy & Fuels 2014;28:3849–62.
[9] Mattson J, Depcik C. Availability Analysis of Alternative Fuels for Compression Ignition Engine Combustion. Int. Congr. Automot. Transp. Eng., Springer; 2018, p. 542–9.
[10] Mangus M, Kiani F, Mattson J, Tabakh D, Petka J, Depcik C, et al. Investigating the compression ignition combustion of multiple biodiesel/ULSD (ultra-low sulfur diesel) blends via common-rail injection. Energy 2015;89:932–45.
[11] Cecrle E, Depcik C, Duncan A, Guo J, Mangus M, Peltier E, et al. Investigation of the effects of biodiesel feedstock on the performance and emissions of a single-cylinder diesel engine. Energy & Fuels 2012;26:2331–41.
[12] Fattah IMR, Masjuki HH, Kalam MA, Wakil MA, Rashedul HK, Abedin MJ. Performance and emission characteristics of a CI engine fueled with Cocos nucifera and Jatropha curcas B20 blends accompanying antioxidants. Ind Crops Prod 2014;57:132–40.
[13] Churkunti P, Mattson JMS, Depcik C. Influence of Fuel Injection Pressure and Biodiesel upon NO x Emissions. SAE Technical Paper; 2016.
[14] Ganguly A, Ghosh S, Mukhopadhyay T, Muniyappan SK. A Review on Performance, Emission and Combustion Characteristics of Biodiesel Blends in a Diesel Engine With Varying Injection Pressures. Int J Renew Energy Res 2017;7:2130–8.
[15] Jin C, Yao M, Liu H, Chia-fon FL, Ji J. Progress in the production and application of n-butanol as a biofuel. Renew Sustain Energy Rev 2011;15:4080–106.
[16] Arockiasamy P, Anand RB. Performance, combustion and emission characteristics of a DI diesel engine fuelled with nanoparticle blended jatropha biodiesel. Period Polytech Mech Eng 2015;59:88–93.
[17] Tiwari AK, Kumar A, Raheman H. Biodiesel production from jatropha oil (Jatropha curcas) with high free fatty acids: an optimized process. Biomass and Bioenergy 2007;31:569–75.
[18] Leung DYC, Wu X, Leung MKH. A review on biodiesel production using catalyzed transesterification. Appl Energy 2010;87:1083–95.
[19] Achten WMJ, Verchot L, Franken YJ, Mathijs E, Singh VP, Aerts R, et al. Jatropha bio-diesel production and use. Biomass and Bioenergy 2008;32:1063–84.
[20] da Silva César A, Batalha MO. Biodiesel production from castor oil in Brazil: A difficult reality. Energy Policy 2010;38:4031–9.
[21] Gupta AK. Biodiesel production from Karanja oil 2004.
[22] Agarwal AK, Dhar A, Gupta JG, Kim W Il, Choi K, Lee CS, et al. Effect of fuel injection pressure and injection timing of Karanja biodiesel blends on fuel spray, engine performance, emissions and combustion characteristics. Energy Convers Manag 2015;91:302–14.
[23] Maheshwari N, Balaji C and, Ramesh A. A nonlinear regression based multi-objective optimizn of parameters based on experimental data from an IC engine fueled with biodiesel blends. Biomass and Bioenergy 2011;35:2171–83.
[24] Kumar N, Pali HS. Effects of n-Butanol Blending with Jatropha Methyl Esters on Compression Ignition Engine. Arab J Sci Eng 2016;41:4327–36.

[25] Heywood JB. Internal Combustion Engine Fundamentals. vol. 21. 1988. doi:10987654.
[26] Chen G, Yu W, Li Q, Huang Z. Effects of n-butanol addition on the performance and emissions of a turbocharged common-rail diesel engine. SAE Technical Paper; 2012.
[27] Ganapathy T, Gakkhar RP, Murugesan K. Influence of injection timing on performance, combustion and emission characteristics of Jatropha biodiesel engine. Appl Energy 2011;88:4376–86.
[28] Kaimal VK, Vijayabalan P. An investigation on the effects of using DEE additive in a DI diesel engine fuelled with waste plastic oil. Fuel 2016;180:90–6.
[29] Rakopoulos DC. Combustion and emissions of cottonseed oil and its bio-diesel in blends with either n-butanol or diethyl ether in HSDI diesel engine. Fuel 2013;105:603–13.

Stability and density analysis of mango bark and mango leaf nanofluids

J L T Chen*, A N Oumer and A A Azizuddin

Faculty of Mechanical and Automotive Engineering Technology, Universiti Malaysia Pahang, 26600 Pekan, Pahang, Malaysia.

*E-mail: jameslautzechen@gmail.com

Abstract. Thermal conductivity is one of the primary properties for nanofluids application investigated by many researchers. However, there are other factors such as stability and density that need to be considered to reduce sedimentation, pressure drop as well as sustain the enhanced thermal physical properties of nanofluid. The purpose of this study is to investigate and analyse the stability and density of mango bark (MB) and mango leaf (ML) nanofluids as well as hybrid nanofluids between mango bark and leaf with SiO_2 and TiO_2. The stability of nanofluids was measured using sedimentation method. The nanofluids arranged in order of pure ML nanofluid with increasing volume concentration of 0.25%, 0.50% and 1.00% followed by MB nanofluid, ML/SiO_2 nanofluid, MB/SiO_2 nanofluid, ML/TiO_2 nanofluid and MB/TiO_2 nanofluid. The density of nanofluids was measured using KEM DA-640 density meter and calculated using mixture rule equation. The results show that the stability of the nanofluids started to show sedimentation after 1 day of storing. Mango bark nanofluids were more stable than mango leaf nanofluids. At 1% concentration, MB nanofluid has the highest density (0.9981) followed by MB/TiO_2 (0.9970), ML/TiO_2 (0.9964), ML (0.9963), MB/SiO_2 (0.9962) and ML/SiO_2 (0.9962). Density of nanofluids have small increment with volume concentration. Meanwhile, the density of nanofluids have small decrement with volume concentration using analytical calculation. The results also shows small average error of 0.17% between experiment and analytical calculation.

1. Introduction

In past, addition of micron-size particles to fluids was found to enhance the heat transfer rate of the fluids. Then, nano-size particles was found to improve the poor stability and sedimentation of the micron-size particles, erosion, and clogging of the channels, also the decreased pressure drop in heat transfer systems hampered the application of these microfluids. This shows that dispersion stability of nanofluid is an important factor to sustain the enhanced thermal physical properties of nanofluids [1]. Besides that, some researches start to investigate the application of bio particles as drag reduction agent. Waskito et al. investigate the effect of coconut fibre suspensions on drag reduction in circular pipe. The result shows that 7.6% of drag reduction under condition of 25000 Reynolds number with circular pipe ID 38 [2]. Yanuar et al. also shows that biopolymer guar gum solution have effect of drag reduction either on a spiral pipe or circular pipe. The result shows the highest drag reduction can up to 30%. However, it will causes rapid mechanical degradation [3]. Kapal et al. shows the drag reduction of 4% to 6% using bamboo and abaca fiber suspensions in circular pipe [4].

Application of nanofluids in industrial field is strongly depends on the stability of nanofluids. Many researchers have defined stability as the 'validity of nanofluids' [5]. There are many methods for stability analysis such as zeta potential, sedimentation, centrifugation, spectral analysis, electron microscopy and light scattering [6]. However, it is very difficult and big challenge in nanofluid technology to the maintain of colloidal stability [7]. Thus, some techniques have been developed by researchers to enhance the stability of nanofluid [8]. Most common methods used are addition of surfactant, surface modification, pH control and ultrasonication [9,10].

Density can be defined as the heaviness of a material. Thus, density is very dependent on the type of nanoparticle material suspended. For instance, denser material used as nanoparticle suspended in base fluids will caused higher density of nanofluid. However, some parameter like nanoparticle shape, size, zeta potential and additives affect the density of nanofluids insignificantly [11, 12]. Density is one of the important physical properties of nanofluid that affect pressure drop and pumping power in heat transfer systems significantly [13]. Although density of a nanofluid is one of the most important physical properties of the nanofluid, it has not been investigated deeply compared with other physical properties [14]. The objective in this study is to investigate and analyse the stability and density of mango bark and mango leaf nanofluids.

2. Methodology

The technique used for nanofluids preparation was two step method. Bio nano-materials were first synthesized by physical method. The mango barks and leaves were dry grinded, sieved and wet grinded into nano-size. Then, the bio nanoparticles mixed with non-bio nanoparticles and added into distilled water. After that, the nanofluids were stirred homogeneously and ultrasonic bath with constant amplitude for a period of time for nanofluids stability.

2.1. Mango bark and mango leaf nanoparticles preparation

Mango tree barks and leaves have been collected and dried under hot sun. After 8 days of drying process, the mango tree barks and leaves were crushed into smaller pieces using blender with dry mill unit (Panasonic Mill PSN-MXGM1011). Then, sieving process was conducted to filter out the bigger pieces to obtain the smallest pieces. Then, the powder will be used to conduct wet grinding using FRTSCH Planetary Micro Mill PULVERISETTE 7 premium line to gain nano-size nanoparticles. The specification of FRTSCH Planetary Micro Mill PULVERISETTE 7 premium line as shown in table 1.

Table 1. Specification of FRTSCH Planetary Micro Mill PULVERISETTE 7 premium line.

Specifications	Planetary Micro Mill machine
Working principle	Impact force
Optimal for material type	Hard, medium-hard, brittle
Number of grinding station	2
Grinding tools	Grinding bowls and grinding balls
Material of grinding tools	Sintered corundum, zirconium oxide
Grinding bowl sizes	45 ml
Grinding ball diameter	5 mm
Max. feed size (depends on material)	5 mm
Final fineness (depending on material)	< 0.1 micro meter
Rotational speed of main disk	100-1100 rpm
Grinding process	Dry/wet
Grinding in inert gas	Yes
Typical grinding time down to analytical fineness	3 minutes
Transmission ratio planetary disk / grinding bowl	$i_{relative} = 1 : -2$
Effective diameter of main disk	140 mm
Centrifugal acceleration ($g = 9.81\ m/s^2$)	95 g
Electrical details	100-120 V / 200-240 V/1~, 50-60 Hz,
Weight	44 g
Dimension (W x D x H)	Bench top instrument 40 x 58 x 36 cm

IOP Conf. Series: Materials Science and Engineering **863** (2020) 012061 doi:10.1088/1757-899X/863/1/012061

2.2. *Nanofluids preparation*

The material type and base fluid used for nanofluids were mango bark, mango leaf, silicon dioxide (SiO_2), titanium dioxide (TiO_2) and distilled water. Mango bark and leaves nanoparticles were synthesized using top-down method. SiO_2 and TiO_2 nanoparticles were purchased online from NovaScientific manufacturer. Bio and non-bio nanoparticles were weighed using weighing scale and put into containers mixing with distilled water. After that, all the containers were labelled and proceed with stirring and ultrasonic bath. The proportion of each nanoparticles and base fluid in nanofluid at different concentration were as shown in table 2.

Table 2. Proportion of each nanoparticles and base fluid in nanofluid at different concentration.

Nanofluid, %	Bio nanoparticles, %	Non-bio nanoparticles, %	Base fluid, %
0.25	0.125	0.125	99.75
0.50	0.250	0.250	99.50
1.00	0.500	0.500	99.00

2.3. *Stability measurement*

The nanofluids stability measurement technique used is visual inspection of the sedimentation due to the fact that this method is simple and reliable. However, it required long period of visual inspection to determine the stability of nanofluid. The nanofluids were placed inside test tube and left for two weeks. Daily visual inspection of nanofluids was conducted using high quality camera.

2.4. *Density measurement*

Density is very important in identifying unknown materials by analysing the physical, chemical and biological properties. For the nanofluids, volume are determined using graduated containers. With the weight and the volume of the nanofluid, densities can be obtained through equation (1) [15]. Besides, the density of nanofluids were also measured using KEM DA-640 density meter with specification of the density meter is shown in table 3. Both calculation and experiment results will be compared.

$$\rho_{nf} = \phi_w \rho_w + \phi_m \rho_m + \phi_{bio} \rho_{bio} \tag{1}$$

Table 3. Specification of KEM DA-640 density meter.

Specifications	Planetary Micro Mill Machine
Measurement method	Resonant frequency oscillation
Measurement range	0 – 3 g/cm^3
Temperature range	0 – 96 °C
Accuracy (Density)	+/- 1x10^{-4} g/cm^3
Accuracy (Temperature)	+/- 0.5 °C
Repeatability	SD 5x10^{-5} g/cm^3
Resolution (Density)	0.0001 g/cm^3
Resolution (Temperature)	0.01 °C
Viscosity correction	Yes
Sampling	Manual by syringe Auto by sampling pump
Minimum sample required	Approx. 1 ml (Syringe) Approx. 2 ml (Pump)
Method	Saves up to 100 different methods in built-in memory
Measurement time	1 to 4 minutes (Auto) 2 to 10 minutes (Manual)
Dimensions (W x D x H)	320 x 365 x 250 mm
Weight	18 g

3. Results and discussion

3.1. Stability of nanofluids

The arrangement of the nanofluids from left to right is pure mango leaf (ML) nanofluid with increasing volume concentration of 0.25%, 0.50% and 1.00% followed by pure mango bark (MB) nanofluid, ML/SiO_2 nanofluid, MB/SiO_2 nanofluid, ML/TiO_2 nanofluid and MB/TiO_2 nanofluid shown in figure 1. From the results, it shows that all nanofluids have similar results of decreasing stability over time. It shows that the mango bark nanofluids have better stability compared with mango leaf nanofluids. For

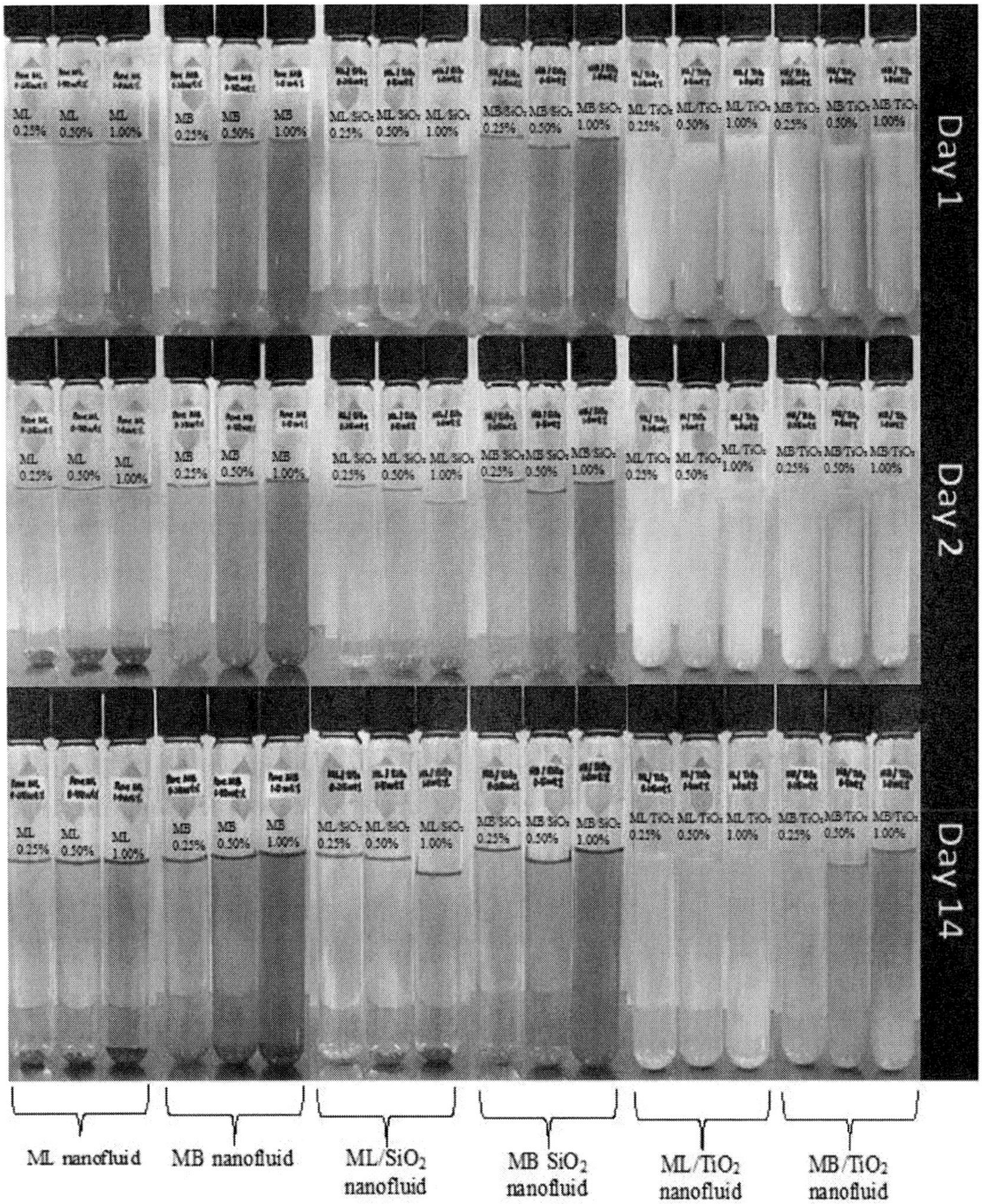

Figure 1. Visual inspection of nanofluids.

hybrid nanofluids, nanofluids with MB/SiO_2 particle type have better stability compared to ML/SiO_2, ML/TiO_2 and MB/TiO_2. Meanwhile, hybrid nanofluids have better stability compared with pure ML and MB nanofluids. For all particle type of nanofluids, the results shows that the higher volume concentration of nanofluids have lower stability than lower volume concentration of nanofluids. However, the longer the time taken for a nanofluid to have sign of sedimentation, the more stable the nanofluid is. For instance, Farbod et al. shown that CNT/water nanofluid have excellent stability with no sign of sedimentation after 80 days [16]. Thus, it proven that the nanofluids in this research have relatively low stability as there have sedimentation after 1 day. After 14 days, it shows small increase of sediment level which means small decrement of stability. Therefore, addition of surfactant needed to further improve the stability of nanofluids.

3.2. Density of nanofluids

The density results of nanofluids obtained from experiment and analytical calculation method at 30 °C are as shown in table 4. From the results, it shows that the error between experimental and analytical calculation are very small (below 1%). Since the error is very small, equation (1) is valid to calculate the density of nanofluids. The density of nanofluids calculated through equation shows that the density decrease with increasing volume concentration. However, the density of nanofluids obtained from experiment shows different trends. For some nanofluids, the density decrease with concentration for most of the nanofluids density increases with concentration. The density of the nanofluids increased with volume concentration except ML nanofluids as shown in figure 2. At 1% of volume concentration, MB nanofluid has the highest density (0.9981) followed by MB/TiO_2, ML/TiO_2, ML, MB/SiO_2 and ML/SiO_2 as shown in figure 3. From the results, nanofluids with mango bark particles tend to have higher density than nanofluids with mango leaf particles.

Table 4. Density of nanofluids using experiment and analytical calculation method at 30 °C.

Material type	Volume concentration, %	Density (Experiment), g/cm^3	Density (Calculation), g/cm^3	Error, %
ML	0.25	0.9976	0.9980	0.04
ML	0.50	0.9961	0.9961	0.00
ML	1.00	0.9963	0.9921	0.42
MB	0.25	0.9960	0.9981	0.21
MB	0.50	0.9960	0.9963	0.03
MB	1.00	0.9981	0.9925	0.56
ML/SiO_2	0.25	0.9960	0.9978	0.18
ML/SiO_2	0.50	0.9961	0.9956	0.05
ML/SiO_2	1.00	0.9962	0.9925	0.37
MB/SiO_2	0.25	0.9962	0.9979	0.17
MB/SiO_2	0.50	0.9961	0.9957	0.04
MB/SiO_2	1.00	0.9962	0.9929	0.33
ML/TiO_2	0.25	0.9961	0.9982	0.21
ML/TiO_2	0.50	0.9961	0.9963	0.02
ML/TiO_2	1.00	0.9964	0.9952	0.12
MB/TiO_2	0.25	0.9960	0.9982	0.22
MB/TiO_2	0.50	0.9965	0.9964	0.01
MB/TiO_2	1.00	0.9970	0.9956	0.14

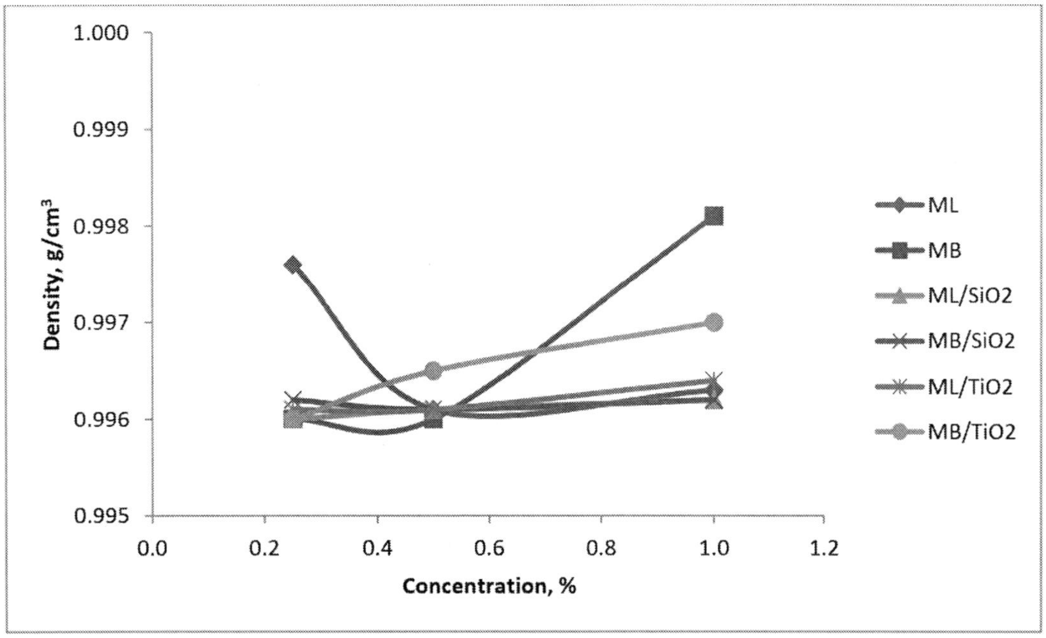

Figure 2. Effect of volume concentration.

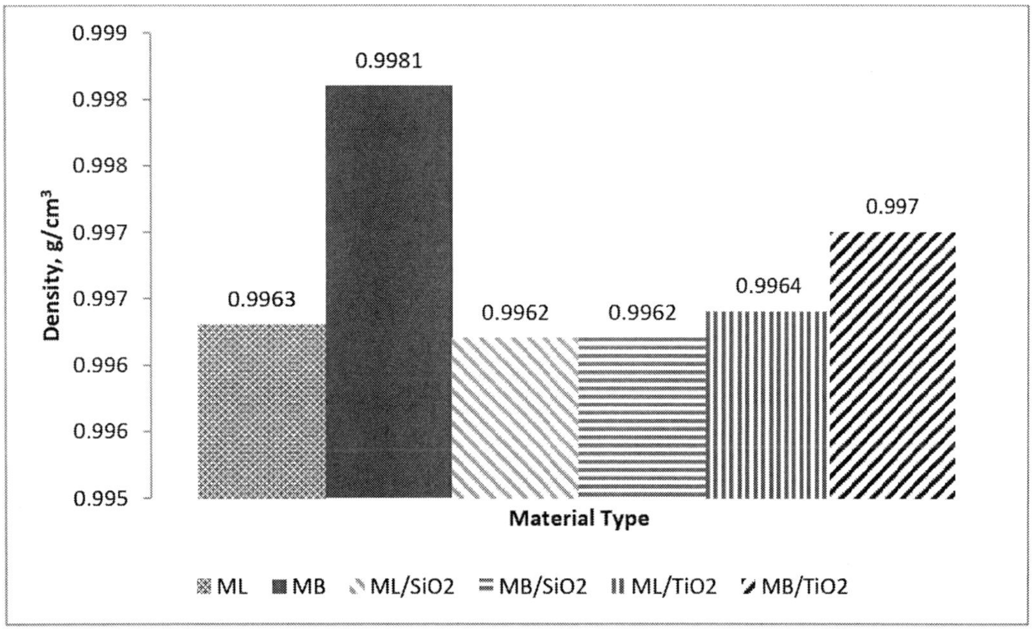

Figure 3. Effect of particle type at 1% volume concentration.

4. Conclusion

The stability of the nanofluids are relatively poor. The nanofluids with high volume concentration have lower stability than nanofluids with low volume concentration. The nanofluids with ML particle type have lowest stability followed by MB, MB/TiO_2, ML/SiO_2, ML/TiO_2 and MB/SiO_2. Therefore, addition of surfactant to nanofluids might improve the stability of the nanofluids. The density of the nanofluids increased with volume concentration except ML nanofluids. At 1% of volume concentration, MB nanofluid has the highest density (0.9981) followed by MB/TiO_2, ML/TiO_2, ML, MB/SiO_2 and ML/SiO_2.

The density of nanofluids obtained from analytical calculation have very small error compared with experimental (below 1%). Therefore, equation (1) is valid to calculate the density of nanofluids.

Acknowledgments
Authors wishing to acknowledge Universiti Malaysia Pahang for providing the assistance and financial support for this paper under the research no. RDU170385 and RDU190319.

References

[1] Sezer N, Atieh MA, Koç M 2019 *Powder Technology*. **344** 404-31.
[2] Waskito KT, Yanuar, Talahatu MA, Wulandari W, Mau S. 2018. Effect of coconut fiber suspensions on drag reduction in circular pipe. *IOP Conf Ser Earth Environ Sci.*
[3] Yanuar, Gunawan, Baqi M. 2012 *J Teknol Sciences Eng.* **58** 95–9.
[4] Kapal G, Baqi M, Fathernas S, Yanuar 2013. *Appl Mech Mater.* **388** 34–9.
[5] Ilyas SU, Pendyala R, Marneni N. 2017 Stability of Nanofluids. **179** 275–321.
[6] Che Sidik NA, Mahmud Jamil M, Aziz Japar WMA, Muhammad Adamu I. 2017 *Renewable and Sustainable Energy Reviews.* **80** 1112–22.
[7] Suganthi KS, Rajan KS. 2017, *Renewable and Sustainable Energy Reviews.* **76** 226–55.
[8] Guardia P, Batlle-Brugal B, Roca AG, Iglesias O, Morales MP, Serna CJ, et al. 2007 *J Magn Magn Mater.*
[9] Fuskele V, Sarviya RM. 2017 *Materials Today: Proceedings* 4049–60.
[10] Sajid MU, Ali HM. 2018 *International Journal of Heat and Mass Transfer.* **126** 211–34.
[11] Mahbubul IM. 2019 *Preparation, Characterization, Properties and Application of Nanofluid.* **c** 113–196.
[12] Żyła G, Vallejo JP, Lugo L. 2018 Isobaric heat capacity and density of ethylene glycol based nanofluids containing various nitride nanoparticle types: An experimental study.
[13] Nabati Shoghl S, Jamali J, Keshavarz Moraveji M. 2016 *Exp Therm Fluid Sci* **74** 339–46.
[14] Sharifpur M, Yousefi S, Meyer JP. 2016 *Int Commun Heat Mass Transf* **78** 168–74.
[15] Jeanmonod DJ, Rebecca, et al. Suzuki K. 2018 *Intech open.* **2** 64.
[16] Farbod M, Ahangarpour A, Etemad SG. 2015. *Particuology.*

Effect of bath temperature on sodium diethyldithiocarbamate trihydrate loading of halloysite nanotube

S S Kumar, S Kakooei*, M C Ismail and M Haris

Centre for Corrosion Research, Department of Mechanical Engineering, Universiti Teknologi PETRONAS, 32610 Bandar Seri Iskandar, Perak, Malaysia.

*E-mail: saeid.kakooei@utp.edu.my

Abstract. Smart coatings have become one of the most sought after solutions for corrosion control in industries. A need to encapsulate the corrosion inhibitor by means of a nano/micro container has come into the picture as the conventional methods are not favourable and has demerits. Halloysite nanotube based smart coatings are effective given that it is environment-friendly unlike most of its competitors. The current work focuses on the effect of bath temperature in a vaporizer on the loading ability of Halloysite nanotube is studied with Sodium diethyldithiocarbamate trihydrate as a corrosion inhibitor. The loading was carried out at three different bath temperatures which are 40, 60 and 80 °C and the resultant samples were investigated by characterization using Scanning Electron Microscope (SEM), Fourier Transform Infrared Spectroscopy (FTIR) and UV-Vis Spectroscopy. It was found that an increase in bath temperature to 80 °C, it will increase the loading capacity of halloysite nanotubes.

1. Introduction

The presence of serious metal corrosion problems has increased the need for using inhibitors in coatings. Direct addition of inhibitor to the coatings is not favourable as it has demerits like pore formation in coating due to removal of inhibitors. Hence, a need to encapsulate the corrosion inhibitors by means of a container in micro/nano scale came into picture. Different types of materials like carbon nanotubes, polymer, and silica capsules can be chosen to be used as carriers. However, halloysite nanotubes had an edge over these due to its exceptional properties. Halloysite nanotubes are inorganic aluminosilicate clay mineral found commonly in countries like China, Brazil, Mexico, United States, New Zealand, and Australia. Halloysite nanotubes consist of two basal surfaces. The outer Si-O-Si layer is tetrahedral, hydrophilic and can be etched with an alkaline, while the inner Al-OH layer is octahedral, hydrophobic and can be etched with acid. These significant features enable HNTs to load and release compounds with different natures. Moreover, HNTs are cheap, biocompatible and easy to work with. The release of inhibitors can be controlled by layer by layer deposition of alternatively charged polymer layers on the outer surface of the HNT or by the formation of end stoppers at the ends of the nanotubes by metal ion-inhibitors compound [1-3]. A number of ways are employed to load halloysite nanotubes (HNT) with corrosion inhibitors. Soaking of the container in a saturated solution of corrosion inhibitors, stirring the mixture of container and inhibitor solution, vacuum negative pressure method, impregnation method, hydrothermal method, layer by layer assembly method are some of the methods used for loading. The factors affecting the loading of HNTs include molar mass or molar weight of inhibitor, the morphology of HNT, pH of inhibitor, its solvent and the solution and nature of inhibitor. If the inhibitor is hydrophilic in nature, the vacuum negative pressure method is employed using a vaporizer as shown in figure 1.

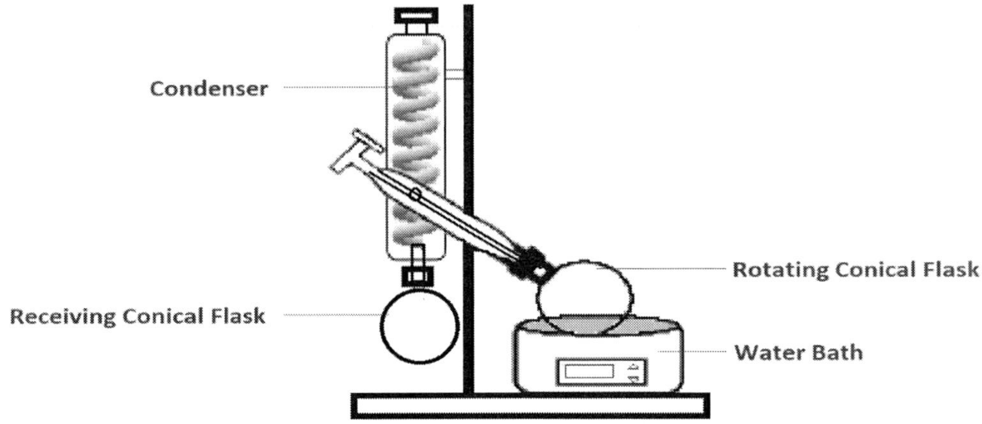

Figure 1. Schematic of a vacuum negative pressure apparatus.

For inhibitors with high solubility, HNT is loaded by blending the inhibitor with appropriate polymer, whereas low melting inhibitors are loaded in liquid form [3-5]. Sodium diethyldithiocarbamate trihydrate ($(C_2H_5)_2NCSSNa.3H_2O$) with IUPAC name sodium diethylaminomethanedithioate trihydrate is a corrosion inhibitor with a molar mass of 225.31 g/mol, readily soluble in water and has a melting point of 93 °C. It is a catenulate organosulfur compound containing nitrogen and sulfur function group and a potent metal-chelating agent. Moreover, it is non-toxic in nature [6, 7].

The loading of the corrosion inhibitor into the halloysite nanotubes is a crucial stage and the many applications of halloysite nanotubes depends on the loading capacity. The loading capacity can be increased by pre-treatment like inner lumen or outer surface etching. The inner lumen etching has been effective and has shown 4 times more loading capacity than the conventional method [8]. However, to the best of our knowledge, much concern has not been given to other factors affecting the loading capacity like the bath temperature, solvents used, pressure, and so on.

This current work focuses on the effect of bath temperature on the loading capacity of halloysite nanotubes using sodium diethyldithiocarbamate as corrosion inhibitor, by loading at different bath temperatures. Characterization using Scanning Electron Microscope (SEM), Fourier Transform Infrared Spectroscopy (FTIR), UV-Vis Spectroscopy were carried out to investigate which of the three samples synthesized at different temperatures had undergone maximum loading.

2. Methodology

2.1. Materials and instrumentation

Halloysite nanoclay, Sodium diethyldithiocarbamate trihydrate were obtained from Sigma Aldrich. Ethanol was obtained from HmbG chemicals. All reagents were of analytical grade and used without further purification. Scanning Electron Microscopy (SEM, Phenom Pro X), Fourier Transform Infrared Spectroscopy (FTIR, Spectrum One/BX, Perkin Elmer Inc.) and UV-Vis Spectroscopy (Cary 100 instrument) were used to characterize the samples.

2.2. Experiment

The method used was based on the literature [9-11]. HNT and sodium diethyldithiocarbamate trihydrate were taken in 5:6 ratios. First, 2 gram sodium diethyldithiocarbamate trihydrate was dissolved in 250 ml ethanol in a magnetic stirrer at 350 rpm for an hour (room temperature). A weight of 2.4 grams HNT was added to the solution and the mixture was stirred at 350 rpm for 24 hours (room temperature). The precipitate formed was transferred to a conical flask and partially submerged in a water bath maintained at 80 °C. Pressure of 175 mbar was applied and the conical flask was set to rotate at 60 rpm. Slight buzzing indicated the evacuation of air from the lumen of HNT. Ethanol evaporated from the solution leaving behind HNT and corrosion inhibitor. The mixture was brought back to atmospheric conditions. The process was repeated twice to ensure an increase in the loading capacity of HNT. The entire experiment was repeated for different bath temperatures of 40, 60 and 80 °C. The obtained loaded HNTs were labelled as HNT40, HNT60, and HNT80 based on the bath temperature.

3. Results and discussion

The bath temperature had a significant effect on the evaporation rate of ethanol. As the bath temperature increased, ethanol evaporated in a short period of time. When the bath temperature was set to be 60 °C and 80 °C, 250 ml of ethanol evaporated from the solution in approximately 8 minutes and 23 minutes respectively. Whereas, it took approximately 530 minutes to evaporate 10 ml ethanol from the solution when the bath temperature was set to 40 °C. The experiment was not continued as it was not practical.

The morphology of the inhibitor loaded halloysite nanotubes can be studied from the scanning electron microscope image shown in figure 2. Clusters of thin, long structured halloysite nanotubes loaded with inhibitor are visible along with some impurities. The impurities are assumed to be sodium diethyldithiocarbamate particles which have failed to load into the lumen of the halloysite nanotubes. The presence of impurities outside the lumen can be due to various intrinsic and extrinsic factors like ratio of carrier to the inhibitor, solvent, time and method of loading, time gap between loading and characterization [11].

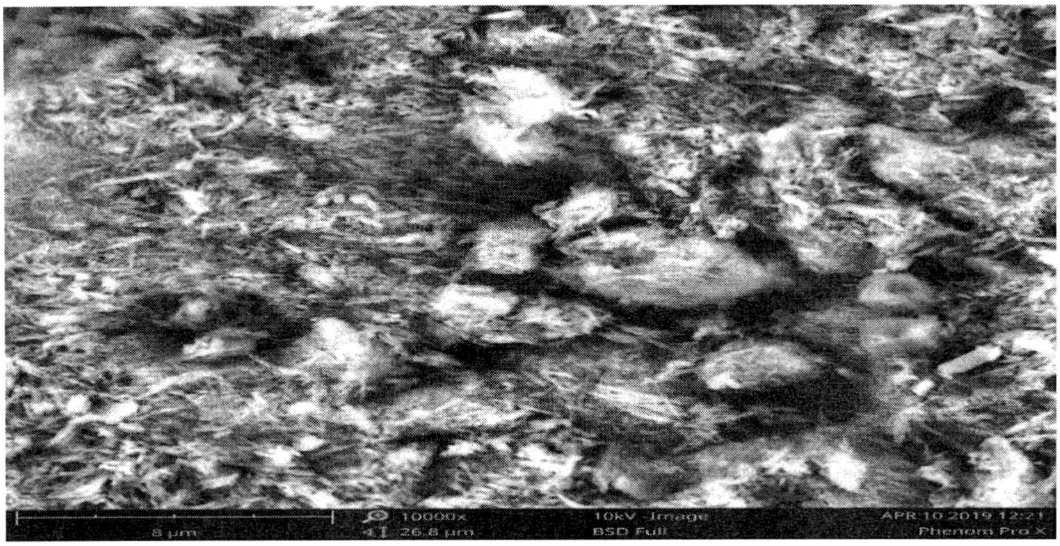

Figure 2. SEM image of halloysite nanotubes loaded with sodium diethyldithiocarbamate taken at 10000X magnification.

The UV-Vis characterization was done using a Cary 100 instrument. Pure ethanol was used as a baseline. The suspensions were made by stirring 2 mg of HNT40, HNT60, and HNT80 in 1-liter ethanol each. The wavelength range was from 200 nm to 400 nm. The scan rate was set at 600 nm/min. The absorbance (%) versus wavelength (nm) plot as shown in figure 3 illustrates that the absorbancy of HNT80 at 208 nm is 0.5, absorbancy of HNT60 at 208 nm is 0.316 and that of HNT40 at 276 nm is 0.083. This clearly shows that the loading was better when the bath temperature was set higher. The loading efficiency of inhibitor can be ranked as HNT40 $<$ HNT60 $<$ HNT80. The results of UV-Vis spectroscopy lies in agreement with the points concluded from the observation of SEM images.

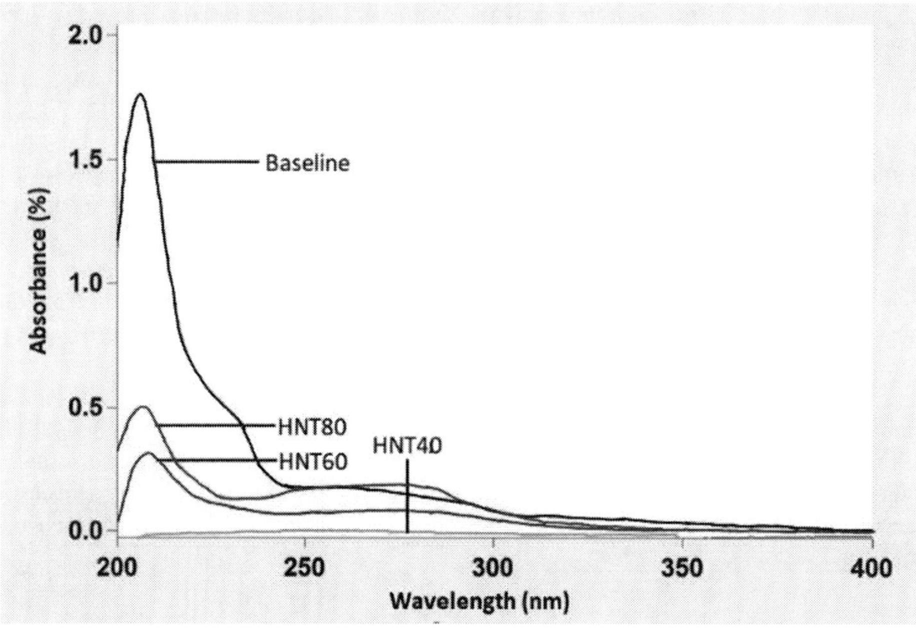

Figure 3. UV-Vis spectra for halloysite nanotubes loaded with the corrosion inhibitor at HNT40, HNT60, and HNT80.

Figure 4 shows the FTIR spectra of HNT40, HNT60, and HNT80. The spectrum of the three are similar and exhibits peaks at the same wavenumber as listed in Table 1. Prominent peaks can be seen in the range of 400 to 3000 1/cm for all three samples of loaded halloysite nanotubes. The peaks at 3696 and 3623 1/cm shows the presence of inner OH stretching vibrations seen in the alumina layers of the halloysite nanotube. The peak at 2978 1/cm maybe due to the interaction of the inhibitor to the inner lumen. The Si-O-Si vibration and interlayer water molecules vibration are indicated by the 1032 and 1655 1/cm peaks respectively. The inner surface hydroxyl groups are the reason for the peaks at 911 1/cm, followed by typical O-H translation, Al-O-Si and Si-O-Si vibration peaks at 754, 536, and 468 1/cm respectively. The transmittance (%) plays a vital role in determining the loading capacity of the halloysite nanotubes. Based on the fact that transmittance decreases with increase in the density of the sample, it can be understood that the samples muc tightly packed or loaded with corrosion inhibitor will allow less infrared light to transmit through it. Thus, it is clear from the reasoning and the spectra in the figure 4 that the loading capacity is higher in halloysite nanotubes loaded with the inhibitor at 80 °C as the transmittance (%) of HNT40 > HNT60 > HNT80. The FTIR results is in agreement to the results of UV-Vis and images of SEM.

Table 1. Significant peaks in FTIR spectra of halloysite nanotubes with corrosion inhibitor of HNT40, HNT60, and HNT80.

Peaks (1/cm)	Assignment
3696, 3623	O-H stretching
1655	Interlayer water
1032	Si-O-Si vibration
911	Inner surface hydroxyl group
754	O-H translation
536	Al-O-Si vibration
468	Si-O-Si vibration

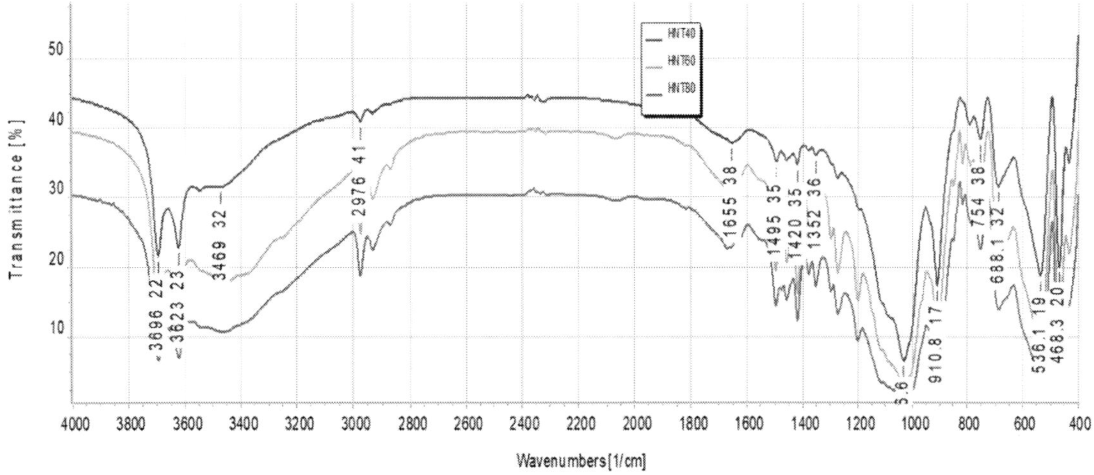

Figure 4. Fourier Transform Infrared spectra of halloysite nanotubes loaded with the corrosion inhibitor at HNT40, HNT60, and HNT80.

4. Conclusion

This study set out to determine the effect of bath temperature on the inhibitor loading of Halloysite nanotubes is done successfully. The investigations on the loaded halloysite nanotubes synthesized at various temperatures like 40, 60 and 80 °C by characterization conclude that the bath temperature has a major role in determining the loading capacity of the halloysite nanotubes. It also confirmed that the loading capacity is enhanced as the bath temperature increases. Overall, this study lays the groundwork for future research by strengthening the idea that loading of Halloysite nanotube at high temperature such as 80 °C will yield better results. However, it is crucial to take notice of the flash point of the solvent and care must be taken to ensure that bath temperature does not exceed the same.

References

[1] J. Antill, S., *Halloysite: A Low-Cost Alternative*. Vol. 56. 2003.

[2] Joussein, E., et al., *Halloysite clay minerals–a review*. 2005, De Gruyter.

[3] Abdullayev, E. and Y. Lvov, *Halloysite for controllable loading and release*, in *Developments in Clay Science*. 2016, Elsevier. p. 554-605.

[4] Zahidah, K.A., et al., *Halloysite nanotubes as nanocontainer for smart coating application: A review*. Progress in Organic Coatings, 2017. **111**: p. 175-185.

[5] Xing, X., et al., *Preparation, release and anticorrosion behavior of a multi-corrosion inhibitors-halloysite nanocomposite*. Chemical Physics Letters, 2019. **718**: p. 69-73.

[6] Li, L., et al., *Sodium diethyldithiocarbamate as a corrosion inhibitor of cold rolled steel in 0.5M hydrochloric acid solution*. Corrosion Science, 2012. **59**: p. 249-257.

[7] Guang Ren, X. and J. Yang, *Sodium Diethyldithiocarbamate Inhibitor Inhibition of 316L Steel*. Vol. 347-353. 2011. 1722-1726.

[8] Abdullayev, E., et al., *Enlargement of halloysite clay nanotube lumen by selective etching of aluminum oxide*. ACS nano, 2012. **6**(8): p. 7216-7226.

[9] Abdullayev, E. and Y. Lvov, *Clay nanotubes for corrosion inhibitor encapsulation: release control with end stoppers*. Journal of Materials Chemistry, 2010. **20**(32): p. 6681-6687.

[10] Abdullayev, E., et al., *Halloysite tubes as nanocontainers for anticorrosion coating with benzotriazole*. ACS applied materials & interfaces, 2009. **1**(7): p. 1437-1443.

[11] Zahidah, K.A., et al., *Benzimidazole-loaded Halloysite Nanotube as a Smart Coating Application*. International Journal of Engineering and Technology, 2017. **7**(4): p. 243-254.

IOP Conf. Series: Materials Science and Engineering **863** (2020) 012063 doi:10.1088/1757-899X/863/1/012063

Rudimentary analysis for solar chimney enhancement

I E Jamil*, C Y Sing and H H Al-Kayiem

Mechanical Engineering Department Universiti Teknologi PETRONAS, 32610 Bandar Seri Iskandar, Perak, Malaysia.

*E-mail: iylia_0008404@utp.edu.my

Abstract. This paper presents fundamental results for solar chimney enhancement. Solar radiation proves to be an important factor in influencing the efficiency of the solar chimney power plant, but its effects on the ambient conditions outside the system has not been considered. Using the Discrete Ordinates radiation model on the semi-transparent collector component of the two-dimensional system, the work presented in this paper simulates the velocity and temperature inside the solar chimney as the ambient temperature of air outside is varied from 307 K to 314 K. At the highest ambient temperature, results show an increase in air velocity of 0.55 m/s at ground level and 1.85 m/s at a height of 1 m along the chimney. To further investigate of this study, a simulation model to include the ambient domain is proposed as a conclusion.

1. Introduction

The solar chimney power plant (SCPP) is an alternative method for electrical power generation. Due to limited resource of conventional energy, the research and development of renewable energy technology has expanded over the decades. An SCPP is a solar thermal system, which converts solar energy into electricity. It consists of typically three main components: a solar collector, a chimney tower and a wind turbine. The working principle of the SCPP is simple (figure 1); solar insolation passes through the semi-transparent collector which heats up the air within. Additionally, it creates a greenhouse effect between the ground and the collector, which further heats the air underneath the collector. The increased air temperature causes difference in buoyancy, which triggers the stack effect. Consequently, continuous updraft is created as hot air leaves through the chimney and ambient air enters from the collector inlet. This flow provides the kinetic energy that runs through the turbine installed at the base of the chimney.

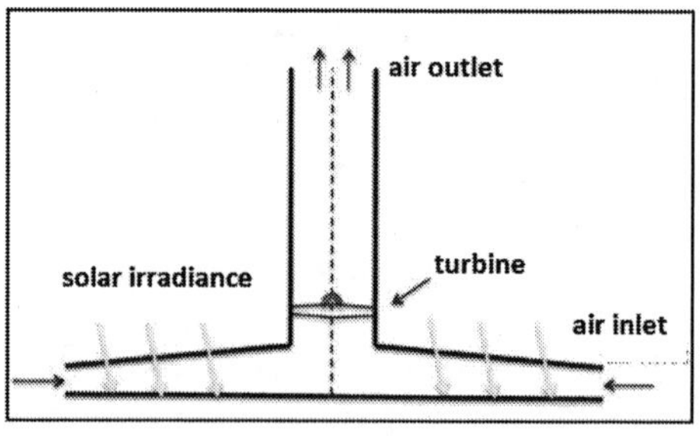

Figure 1. Schematic of the solar chimney [1].

The first pilot solar chimney system was built in Manzenares, Spain in 1982. With a collector radius of 122 m, chimney height and diameter of 194 m and 10 m respectively, the plant was designed to produce a 50-kW electrical power output. From this point, many studies revolve around the Manzenares plant to investigate the defining parameters of the SCPP efficiency. One investigation on analytical and numerical models that consider the influence of ambient conditions and structural parameters on the power output was performed by Bernardes et al. [2]. The results show that the power output increases by increasing chimney height, collector area and transmittance.

To widen the research, development of small-scale prototype of the SCPP has increased popularity up to this day. In Kota, India, a prototype of 8 m height and 12 m collector diameter has been built [3]. The experiment shows that the solar intensity has an effect on air velocity, and ambient air temperature. Kasaeian [4] presented the effects of ambient condition to the performance of SCPP through a scaled model and dimensional analysis. Roozbeh [5] also proved that increase in solar radiation will increase the power output of the system. In another research, Jiakuan [6] showed that power output of the system increases with global solar radiation intensity, collector area and chimney height. It also concluded that a taller chimney would create a higher air velocity through the system.

In University Teknologi PETRONAS, a small-scale pilot plant have been built with 6 m collector diameter and 6.3 m chimney height. From this model, a series of investigations findings have been published. Al-Azawiey et al [7, 8] presented the influence of the collector geometry on SCPP performance. A larger collector area shows a significant increase in the air velocity inside the chimney. In terms of height, the lower height of the collector will provide a higher overall efficiency. An optimum collector height-to-diameter ratio also suggested for future design. To target the hours with low sun radiation, Azeemuddin et al [9], Al-Azawiey et al [10] and Auryby et al [11] have studied various types of energy storage mediums that can sustain the SCPP performance throughout the day.

To this date, no work has been published to investigate the effect of incoming air temperature into the system. Ambient conditions are investigated mainly to see the effect of solar radiation intensity on system performance. The objective of the work presented in this paper is to investigate the effect of inlet air temperature on the flow inside the system. The proposed method to investigate this is by extending the thermal absorption material on the ground beyond the collector area. In this paper, the investigation was conducted by increasing the preset air temperature at the inlet to be higher than the ambient air temperature.

2. Fundamentals of the design

2.1. Governing equations

The fluid motion inside the model is solved through the conservation equation of mass, momentum and energy in *ANSYS Fluent* software. Simulation is carried out assuming steady-state, turbulent, 2-D flow. According to the Fluent User Guide [12], the basic equations can be expressed in tensor form for continuity and momentum equation as shown by equation (1) and equation (2).

$$\frac{\delta}{\delta t} + \nabla.(\rho.\vec{v}) = 0 \tag{1}$$

$$\frac{\delta}{\delta t} + (\rho.\vec{v}) + \nabla.(\rho.\vec{v}.\vec{v}) = -\nabla p + \nabla.\left(\mu[\nabla\vec{v} + \nabla\vec{v}T] - \frac{2}{3}\nabla.\vec{v}I\right) + \rho\vec{g} \tag{2}$$

The incompressible flow assumption is justified for the model since the maximum Mach number is much smaller than 0.1. The buoyancy effect was considered by applying the Boussinesq model in equation (3) to allow a variation of density as a function of temperature. This model treats density as a constant value in all solved equations, except for the buoyancy term in the momentum equation. This model accounts for the full bouncy effect, where the density varies with temperature, and the flow is motivated by force of gravity, which influences the change in density.

$$(\rho - \rho_o)g \approx -\rho_o\beta(T - T_o)g, (\rho) = \rho_o(1 - \beta\Delta T) \tag{3}$$

Equation (3) is valid when $(1 - \beta\Delta T) \ll 1$.

Meanwhile, energy equation as shown by equation (4) has been used in this investigation.

$$\frac{\delta}{\delta t}(\rho . E) + \nabla.\left(\vec{v}(\rho E + p)\right) = -\nabla.\left(k_{eff}\nabla T\right) - h\vec{J} + \left(\mu\left[(\nabla\vec{v} + \nabla\vec{v}T) - \frac{2}{3}\nabla.\vec{v}I\right].\vec{v}\right) + S_h \qquad (4)$$

In equation (4), S_h represents the heat source of the solar radiation. The radiation transfer equation was solved by Discrete Ordinates (DO) model. The DO model solves surface to surface radiation problems and radiation through semi-transparent materials. The radiation equation solved by the software can be written as:

$$\nabla.\left(I_\lambda(\vec{r},\vec{s})\vec{s}\right) + \nabla.\left((a_\lambda + \sigma_s)I_\lambda(\vec{r},\vec{s})\right) = a_\lambda n^2 I_{b\lambda} + \frac{\sigma_s}{4\pi}\int_0^{4\pi} I_\lambda(\vec{r},\vec{s})\varphi(\vec{s},\vec{s}')d\Omega' \qquad (5)$$

The solar intensity in each direction $\vec{s}$ and position $\vec{r}$ can be written as follows:

$$I_\lambda(\vec{r},\vec{s}) = \sum_k I_\lambda(\vec{r},\vec{s})\Delta I_k \qquad (6)$$

In equation (6), the summation represents all wavelength bands. In this study, the collector material is specified Perspex and assumed all the Perspex had the same behavior for all wavelength bands.

2.2. Geometry and boundary conditions

An initial model is built based on geometry from experiment [7] to validate the simulation model. Equations presented in the previous section are solved numerically on *ANSYS Fluent*. The geometries of the model are defined in table 1.

Table 1. Dimensions of the validation model.

Parameter	Dimension
Collector radius (m)	3.00
Collector height at inlet (m)	0.05
Collector angle of inclination from ground (degree)	15.00
Chimney height (m)	6.30
Chimney radius (m)	0.08

Boundary conditions of the simulation model are taken from [7] and described in table 2. Properties of materials used in the simulation are described in table 3. Velocity and temperature values are obtained from post-processing for comparison with the experimental results. It can be observed by simulation that the maximum velocity occurs at the collector inlet, and at the base of the chimney. Increase in solar intensity raises the temperature of the ground and overall air temperature in the system.

Simulation values for maximum temperature at the ground and velocity observed at the chimney base are compared with experimental values for solar intensity of 200 and 800 W/m^2 and shown in table 4. With reasonable percentage difference between the values, this model is acceptable to be used in the proceeding simulation.

Table 2. Boundary conditions for simulation.

Boundary	Type	Condition	Temperature (K)
Ground	Wall	Temperature, T_{ground}	330
Collector	Wall, semi-transparent	Mixed	Not applicable
Chimney	Wall	Adiabatic	Not applicable
Collector inlet	Pressure inlet	Zero gauge pressure	307
Chimney outlet	Pressure outlet	Zero gauge pressure	307

Table 3. Material properties for simulation.

Physical property	Collector	Ground	Chimney
Material	Perspex	Black pebble	PVC
Absorption coefficient	0.06	0.9	0.04
Transmission coefficient	0.92	0	0
Density (kg/m^3)	2700	2640	833
Specific heat (J/kg·K)	840	820	1170
Thermal conductivity (W/m·K)	0.78	1.73	0.19
Emissivity	0.9	0.9	0.91

Table 4. Comparison of simulation and experimental results.

Case		Experiment [7]	Validation Model	Difference %
Velocity at chimney base (m/s)	200 W/m^2	0.71	0.82	0.15
	800 W/m^2	1.85	1.96	0.06
Max T at ground (K)	200 W/m^2	300.0	302.0	0.007
	800 W/m^2	321.0	332.0	0.034
Delta of air (K)	200 W/m^2	1.2	1.3	0.08
	800 W/m^2	14.6	16.0	0.09

The same model is used for other cases presented here with new geometry described in table 5. Figure 2 shows the preliminary model geometry and figure 3 presents the mesh elements at the collector region.

Table 5. Dimensions of the preliminary model geometry.

Parameter	Dimension
Collector radius (m)	4.00
Collector height at inlet (m)	0.07
Collector angle of inclination from ground (degree)	20.00
Chimney height (m)	14.00
Chimney radius (m)	0.08

Quadrilateral elements are used for the mesh and table 6 summarizes the grid independence test. Based on the simulation, mesh model 3 is used for all simulations with the preliminary model.

Table 6. Grid independence test for the preliminary model.

Mesh model	Parameters		
	Number of elements	Velocity at chimney base (m/s)	Error (%)
Mesh 1	27548	1.63	9.2
Mesh 2	61464	1.88	3.7
Mesh 3	111655	1.95	0.5
Mesh 4	145668	1.94	-

The *SIMPLE* algorithm available in *ANSYS Fluent* is employed to solve the system of equations. The convective terms are discretized with a second-order-accurate upwind scheme.

Figure 2. The modelled 2-D geometry.

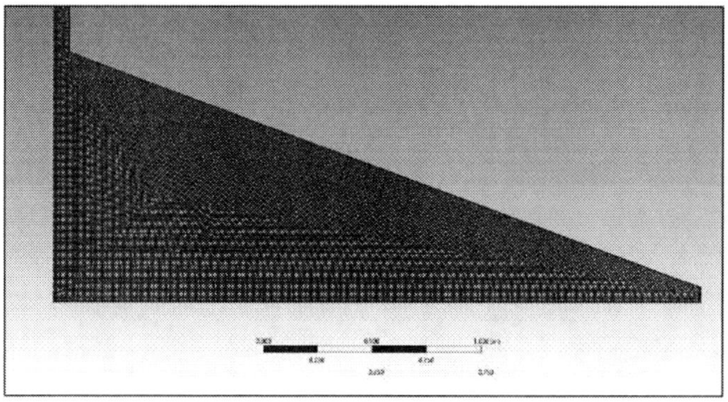

Figure 3. The meshed model with quadrilateral elements.

3. Results and discussion

The simulation is carried out assuming ambient temperature at the collector inlet for the preliminary model. Then, the inlet temperature is increased by a range of values to see the effects on the flow inside the system. Velocity for all results is taken at the base of chimney, unless otherwise stated. For all simulations, the solar radiation is set at 800 W/m^2. The justification of this is when solar radiation is at lower levels, the increase in temperature inside and outside the system is minimal. [3]

From figure 4, it can be observed that air velocity increases with increase in air inlet temperature. As the air temperature at inlet increase by 7 K from ambient temperature, the air velocity doubles from 1.95 m/s to 3.93 m/s.

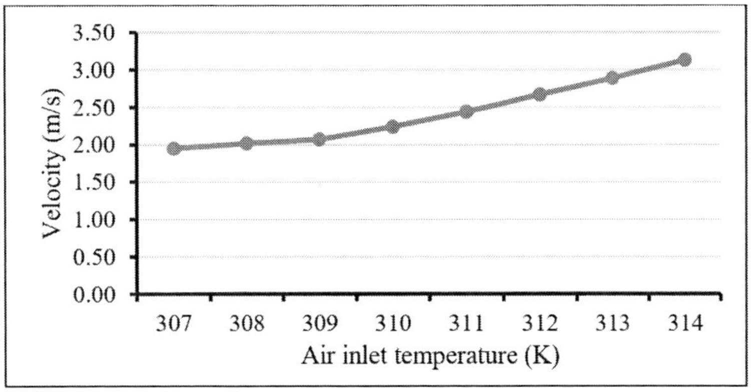

Figure 4. Velocity at the chimney base against air inlet temperature.

At the chimney base, temperature can be observed to increase as the air inlet velocity increases. While at ambient inlet temperature, the air inside is 315.21 K. As the ambient temperature goes up to 314 K, the air temperature inside the chimney increases to 320.52 K (figure 5).

Ground temperature is also observed in the simulation. Figure 6 shows the maximum values of ground temperature as the inlet air temperature is increased from 307 K to 314 K. At the ambient temperature of 307 K, the difference between ground and ambient temperature is 35 K. At 314 K, the difference is 34 K. This shows that the ground temperature is still increasing as the inlet air temperature increases even though the solar radiation level is kept constant.

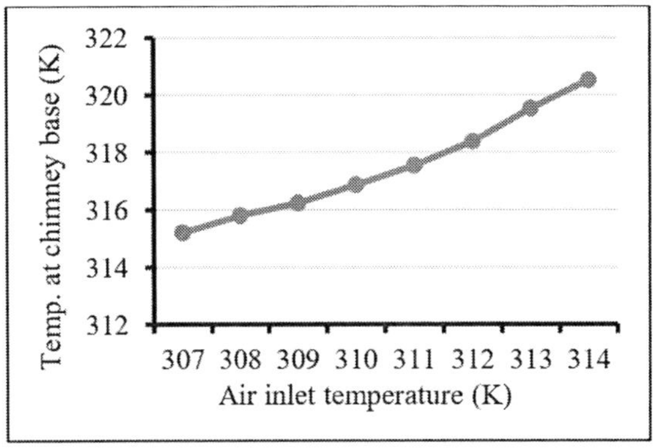

Figure 5. Air temperature at chimney base against air inlet temperature.

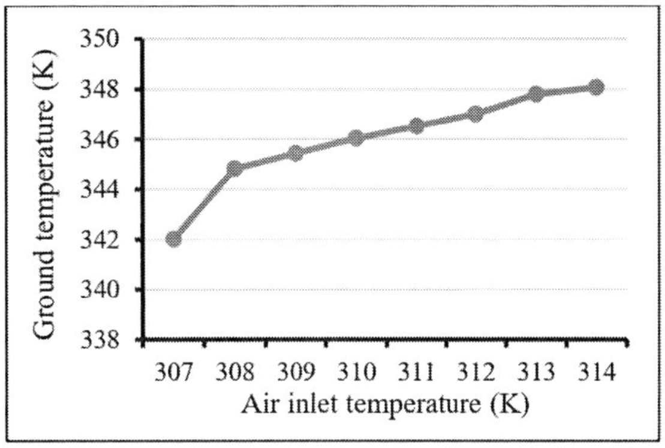

Figure 6. Ground temperature against air inlet temperature.

The velocity values along the chimney height is compared for all values of inlet air temperature. Based on figure 7, it can be observed t3hat at the base of the geometry (y = 0), the difference in air velocity is 0.55 m/s between 307 K and 314 K inlet air temperature. Moving up the chimney height, it is evident that the difference between the lowest and highest velocity increases. At height 1 m from the ground, the difference is 1.85 m/s. It can be concluded that the increase in air velocity along the chimney height is greater as the inlet air temperature increases. As these data are taken at 800 W/m^2 solar intensity and an assumed range of air inlet temperatrues, further investigation is required to see the combiend effect of the solar intensity and changes on the air inlet temperature.

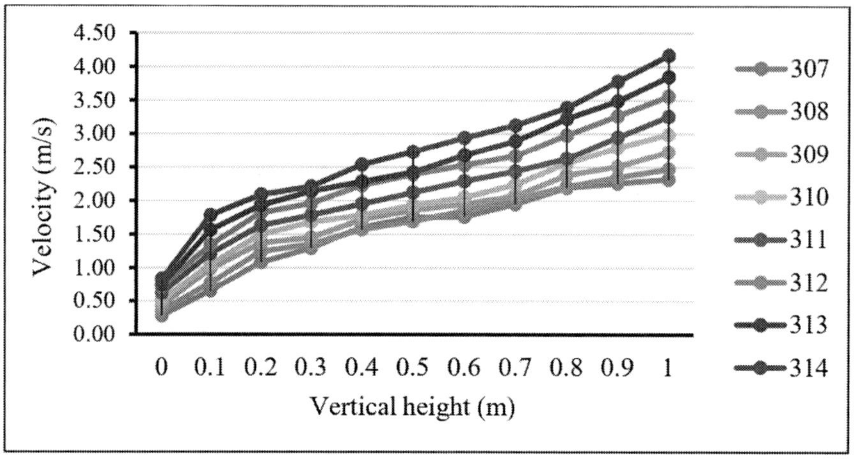

Figure 7. Air velocity along the vertical height of chimney at inlet temperature 307 K to 314 K.

In figure 8, velocity vectors for six air inlet temperatures are compared: from 307 K to 312 K. For all cases, velocity is high at the collector inlet, averaging at 7 m/s. A recirculating flow can be observed along one third of the collector radius upon entrance. However, it can be seen that for a higher inlet air temperature, the recirculation velocity increases and a stronger swirl pattern can be witnessed. For all cases, the air flow inside the collector is driven towards the chimney base. At this location, there is a surge or flow and air velocity increases as it enters the chimney.

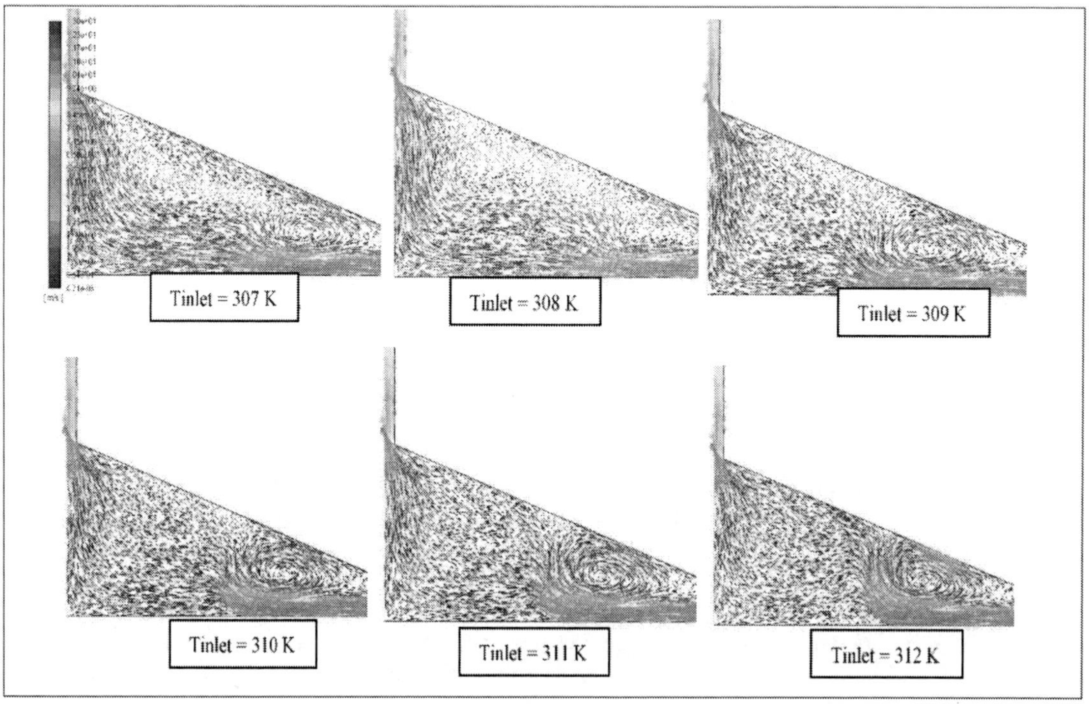

Figure 8. Velocity vector for all inlet air temperature 317 K to 312 K.

Temperature contour of air under the collector are for all cases can be seen in figure 9. For air inlet temperature 307 K, the maximum temperature observed in this region is 315.2 K for, while for 312 K, the maximum temperature observed is 320.6 K. As observed from the contours, the higher temperature regions start to occur at a distance from the collector inlet. This corresponds with the region of the

IOP Conf. Series: Materials Science and Engineering **863** (2020) 012063 doi:10.1088/1757-899X/863/1/012063

swirl formation from figure 8. As air velocity reduces after the swirl, the flow is more settled and convection from the ground increases the air temperature.

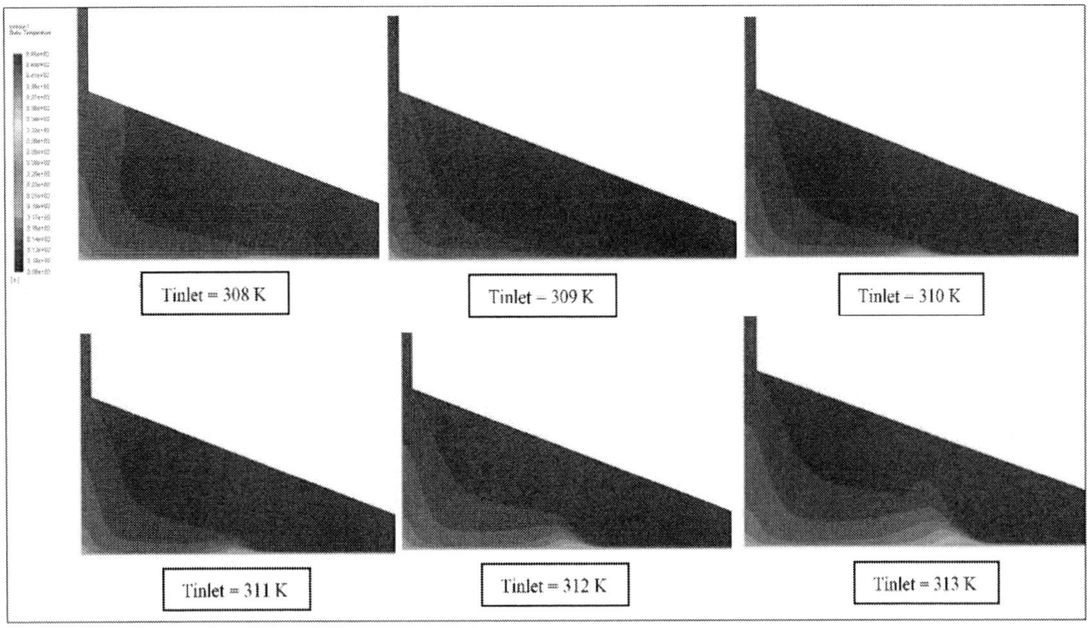

Figure 9. Temperature contour for all inlet air temperature 308 K to 313 K.

4. Conclusion

Solar radiation is one of the most important factors in SCPP performance apart from the design geometry. It influences the ambient condition outside the system by increasing the ambient air temperature. From the results presented, it is apparent that higher temperature of the ambient air result in higher flow velocity in all regions inside the solar chimney. Results show an increase in air velocity of 0.55 m/s at the chimney base and 1.85 m/s at a height of 1 meter from the ground as ambient temperature increases. To further investigate this, a study must be conducted to simulate the SCPP combined with the surrounding ambient domain. Methods to increase the temperature of ambient air are to be included to further enhance the overall SCPP performance. To further investigate this, a simulation model to include the ambient domain is proposed as a conclusion.

Acknowledgements

The authors acknowledge Universiti Teknologi PETRONAS for the financial and technical support to conduct this work under STIRF research grant; CS: 015LA0-005.

References

[1] Fasel H, Shams E and Gross A 2013 CFD analysis for solar chimney power plants *Sol. Energy* **98** 12–22.
[2] Bernardes A S and Weinrebe V A 2003 Thermal and technical analyses of solar chimneys *Sol. Energy* **75** 511–4.
[3] Lal S, Kaushik S C and Hans R 2016 Experimental investigation and CFD simulation studies of a laboratory scale solar chimney for power generation *Sustain. Energy Technol. Assess.* **13** 13–22.
[4] Kasaeian A, Amirifard M, Ahmadi M H and Kasaeian F 2017 Investigation of the effects of ambient temperature and dimensional parameters on the performance of solar chimney power plants *Int. J. Low-Carbon Technol.* **12** 335–48.
[5] Sangi R, Amidpour M and Hosseinizadeh B 2011 Modeling and numerical simulation of solar

chimney power plants *Sol. Energy.*
[6] Zhou X, Yang J, Xiao B and Hou G 2007 Simulation of a pilot solar chimney thermal power generating equipment *Renew. Energy.*
[7] Al-Azawiey S S, Al-Kayiem H H and Hassan S B 2017 On the Influence of Collector Size on the Solar Chimneys Performance *MATEC Web of Conferences.*
[8] Al-Azawiey S S, Al-Kayiem H H and Hassan S B 2016 Investigation on the influence of collector height on the performance of solar chimney power plant pp 12197–201.
[9] Azeemuddin I, Al-Kayiem H H and Gilani S I 2014 Simulation of a collector using waste heat energy in a solar chimney power plant system *WIT Transactions on Ecology and the Environment* p 179.
[10] Al-Azawiey S S and Hassan S B 2016 Heat Absorption Properties of Ground Material for Solar Chimney Power Plants *Int. J. Energy Prod. Manag.*
[11] Aurybi M A, Al-Kayiem H H, Gilani S I U and Ismaeel A A 2017 Numerical assessment of solar updraft power plant integrated with external heat sources *WIT Transactions on Ecology and the Environment.*
[12] Inc. A 2016 ANSYS Fluent Theory Guide v17.1 *ANSYS 17.1 Doc.*

IOP Conf. Series: Materials Science and Engineering **863** (2020) 012064 doi:10.1088/1757-899X/863/1/012064

A study on the performance analysis of bioethanol produced from sugarcane molasses in SI engine

M I N Maárof[1,*], G T Chala[2,*], D Gunness[1] and I Saad[3]

[1] INTI International University, Persiaran Perdana BBN, 71800 Nilai, Negeri Sembilan, Nilai, Malaysia.
[2] International College of Engineering and Management, P.O. Box 2511, C.P.O Seeb, P.C. 111, Muscat, Oman.
[3] Universiti Teknologi MARA (UiTM), 40450 Shah Alam, Selangor Darul Ehsan, Malaysia.

*E-mail: muhammadIzzat.maarof@newinti.edu.my; girma_tade@yahoo.com

Abstract. Recently with ethanol being a renewable, bio-based and ecofriendly, it is projected to be the best replacement for petrol as a fuel in SI engines. Ethanol is derived mainly from sugarcane, cassava, corn and other waste biomass materials. These materials are rich in calories and could be utilized either as complements or as alternatives for conventional fuel sources. This paper aims to experimentally determine the properties of ethanol/gasoline blends and tests its performances in an SI engine. The E10, E20 and E30 ethanol-gasoline blends were studied and compared to gasoline fuel. It was found that the performance of the engine using bioethanol blends was almost comparable to that of pure gasoline, especially at lower ethanol content blends. Yet, increasing the ethanol content only gives adverse effects to the overall engine performance, resulting in 8.86% power loss at 2000 rpm for E20 and 8.18% for E30 compared to E0. Suggestively, with more research and experiments, bioethanol-fuel blends can replace pure gasoline fuels easily and with many added benefits on the same or slightly modified engines.

1. Introduction

Nowadays, most of the fuels used in transportation are in a liquid form due to its ease of storage, and the use of gaseous and solid fuels has gradually declined [1, 2]. There exist two different types of fuels: fuels from fossil resources and biofuels produced from renewable energy sources [3-5]. There is a high demand for alternatives or renewable form of fuels today due to the increase in consumption of motor fuels and the reduction of global oil reserves [6, 7]. Furthermore, global warming has also contributed to the sharp rise in the demand for renewable sources of energy [8, 9]. Bioethanol is currently regarded as the most suitable fuel to substitute petroleum-based fuels in automobiles.

It was reported that the use of biofuels depends on the availability and sustainability of biomass [10]. Another determinant in biofuel use is the biofuel policy, as it can greatly contribute to the successful penetration of the biofuel market. The largest amount of biofuel is produced from plant feedstock sources which must be planted, harvested and processed into biofuel. Other biofuel sources are wastes and animal excrements. From the point of view of recent researches, it is greatly accounted that bioethanol is one of the most promising type of biofuels [11]. Currently, however, most of the ethanol produced globally is used for purposes other than transport-related applications. Nonetheless, as the oil prices continue to rise and more countries adopt policies which place significant favour towards to the use of biofuels, ethanol based fuel is being widely regarded as a very viable alternative [12, 13].

In Mauritius, sugar industry is considered as one of the dominant industries. For decades, it was the main industry in the particular nation, producing sugars to be supplied in various European countries. In recent years, due to increasing competition and diversification of the economy by the government, the sugar industry has been in decline. However, most parts of the country still consist of sugarcane plantations. Although the sugar industry has been using bagasse, a by-product of sugarcane, to produce electricity; and molasses, another by-product, to produce ethanol, there has been little to no attempt to use the ethanol in the transport sector as fuel – especially since Mauritius imports all its oil supply. Mauritius generates 43.5% of its electricity from bagasse, so far, the production of ethanol as a fuel has not been tried [14]. The objective of this study is, therefore, to investigate the performance of bioethanol blends in the SI engine. This would help examine the suitability of bioethanol blends as an alternative, renewable source of fuel for the transport sector of Mauritius.

2. Methodology

2.1. Sample preparation

The pure ethanol (>99% concentration) sample was procured from RCI Labscan Limited. The physical and chemical properties of the ethanol are listed in table 1. The Petron Blaze 95 was chosen as the conventional fuel used in the mixing process. It is a regular gasoline which has been formulated using latest additive technology to provide optimum engine performance. It contains additives which keep the engine's fuel injectors and intake valves clean and efficient in the long run. It also reduces engine friction and enhances power and effectiveness. Additional benefits given by the particular fuel includes optimum power and acceleration, improved fuel economy and better engine protection. The reason for choosing this particular petrol type is because it can be used for both conventional and high-performance gasoline engine [16]. The mixing process consisted simply of pouring and mixing the gasoline and the ethanol in specific proportions in a container. Since both solutions are volatile and flammable, the mixing process was carried out in the chemistry laboratory of INTI International University, Nilai.

Table 1. Properties of ethanol Sample from RCI Labscan Limited [15].

Appearance: Form Colour	**Liquid Colourless**
Odour	Alcohol-like
pH	7.0 at 20 °C
Melting Point	-114.5 °C
Boiling Point	78.3 °C at 1013 hPa
Vapor Pressure	59 hPa at 20 °C
Density	0.790 g/ml at 20 °C
Water Solubility	Soluble at 20 °C
Viscosity	1.2 mPa.s at 20 °C

Four samples were prepared, namely one sample of 7.5 liters of pure gasoline, one sample of 7.5 liters consisting of 750 ml of ethanol and 6750 ml of gasoline, one sample of 7.5 liters consisting of 1500 ml of ethanol and 6000 ml of gasoline, and another sample of 7.5 liters consisting of 2250 ml of ethanol and 5250 ml of gasoline. The pure gasoline sample is the reference or control fuel for the tests. During the mixing process, no distinct phase separation was observed in the mixtures. Figure 1 shows the sample of mixing gasoline and ethanol.

IOP Conf. Series: Materials Science and Engineering **863** (2020) 012064 doi:10.1088/1757-899X/863/1/012064

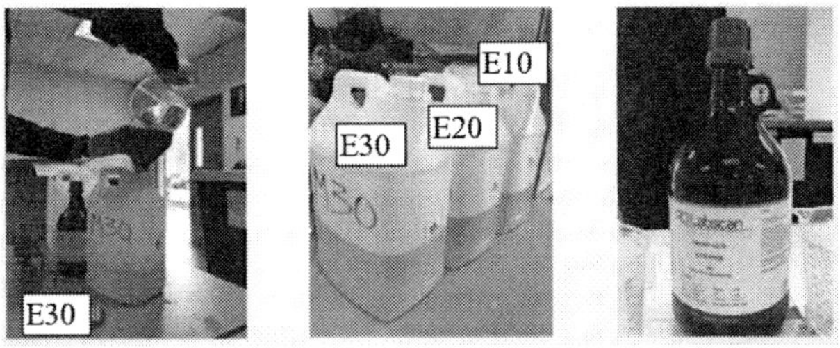

(a) (b) (c)

Figure 1. Mixing sample: (a) mixing ethanol with gasoline, (b) different gasoline-ethanol blends and (c) pure ethanol sample.

Each of these samples were further divided into four different containers, namely four containers containing 6.5 liters of each sample, to be used for SI engine testing; 4 containers containing 200ml of gasoline, E10, E20 and E30 respectively for Properties Tests, and 12 containers with different samples for observation tests. Table 2 shows fuels samples and their abbreviations.

Table 2. Fuel samples and their abbreviations.

No.	Fuel	Abbreviation
1	100% Gasoline (Reference fuel)	Gasoline/ E0
2	90% Gasoline + 10% ethanol	E10
3	80% Gasoline + 20% ethanol	E20
4	70% Gasoline + 30% ethanol	E30

2.2. Properties test

One batch of 4 containers containing 200 ml of Gasoline, E10, E20 and E30 respectively, were sent to the chemical engineering laboratory of the Faculty of Chemical Engineering, Universiti Teknologi MARA (UiTM) to carry out properties testing. The kinematic viscosity test was carried out using a Cannon-Fenske opaque viscometer depicted in figure 2. The calorific value test was carried out using a Bomb Calorimeter IKA C5000 shown in figure 3.

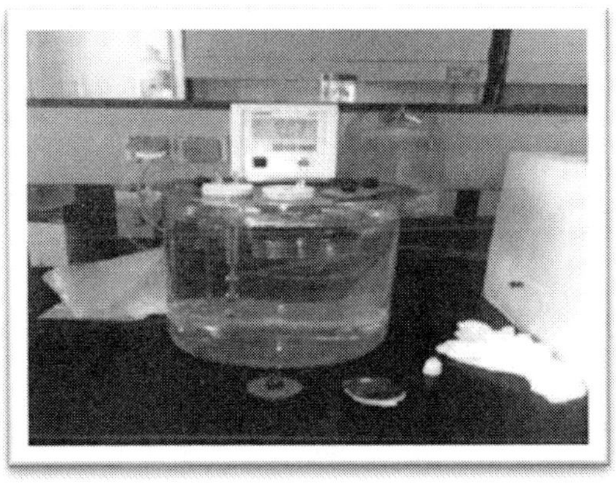

Figure 2. Cannon-Fenske opaque viscometer.

Figure 3. Bomb Calorimeter IKA C5000.

The density of each sample was determined using the formula: Density = mass/volume. The volume of each sample was 200 ml. The mass of an empty beaker was weighed on an electronic scale, then one sample, for example E10, was poured into the same beaker and weighed. The mass of 200 ml E10 sample was found to be [(mass of beaker + E10 sample) – (mass of empty beaker)]. This process was repeated using the gasoline, E20 and E30 samples respectively and their densities were calculated. The room temperature was kept constant at 30 °C.

2.3. SI engine testing

The engine used for this experiment was located at the Mechanical Engineering Laboratory, Faculty of Engineering, Universiti Putra Malaysia (UPM). Figure 4 shows the schematic diagram of the engine. Its specification is depicted in table 3. The engine was connected to a water brake dynamometer and an exhaust gas analyser which generally shows CO and Hydrocarbon emissions. The aim of this test was to use the different blends of ethanol and gasoline in the same engine and carry out a performance analysis. A test run was carried out for 20 minutes to warm up the engine. The petrol tank was then completely drained. 100% gasoline sample (E0) was fed into the engine and the Power and Torque at different speeds (1200-3200 rpm) are noted. This result was the reference for the study. E10 sample was put into the tank and the engine was run to record the output. Similar steps were repeated for E20 and E30 samples, respectively. The results obtained for the E10, E20 and E30 samples were compared to the reference sample (E0).

After setting up the engine, the following measurements were taken. Using a rotating turbine flow meter, flow rate of fuel was measured, which was then converted to mass flow rate. Air flow can also be measured in the same way. After that, the force on the dynamometer was measured using a scale. The exhaust gases emitted (NOx, CO, CO_2 and O_2) were measured using a gas analyser.

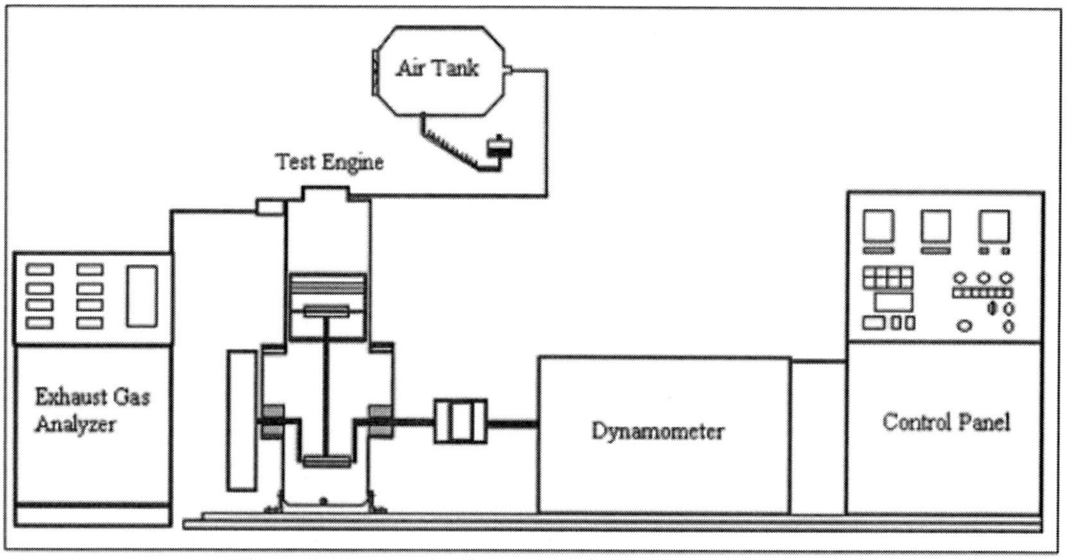

Figure 4. Systematic diagram of the engine test rig [17].

Table 3. Specifiation of the engine.

Model	Subaru EX 21
Type	Air-cooled, 4-cycle, slant single cylinder, OHC,
Bore x Stroke	67 x 60
Piston displacement	211
Continuous Output	3.2(4.4)/3000
Maximum Output	5.1(7.0)/4000
Maximum Torque	13.9/2500
Direction of Rotation	Counter clockwise as viewed from PTO shaft

3. Results and discussion

3.1. Properties test

Viscosity is the measure of flow resistance in a liquid. Viscosity is an important property for fuels as fuel is injected into combustion chambers. If viscosity is too low, the fuel flows very easily and as a result, the lubricating film between static and moving parts in the engine carburetor. On the other hand, if viscosity is too high, fuel atomization will not produce small enough fuel droplet to achieve good combustion in the engine. The values for kinematic viscosity of the samples at 30 °C are shown in figure 5. They were found to be 10.56%, 23.32% and 35.75% more viscous than the petrol sample (0.4876 mm^2/s) for the samples of E10, E20 and E30 respectively. The viscosity values of the blends show a continuous and linear increase, as shown in figure 5.

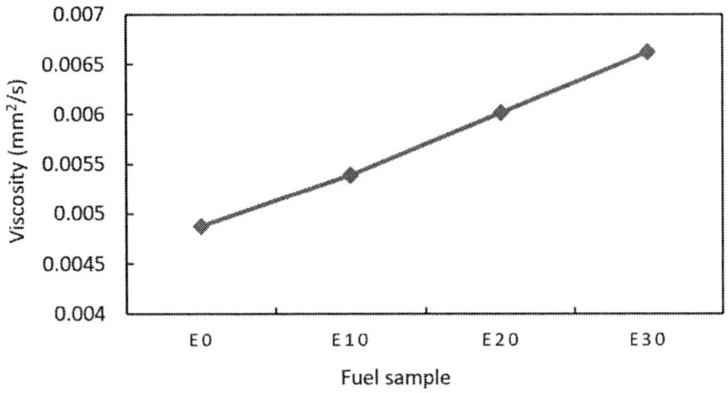

Figure 5. Variation of viscosity values.

The Calorific Value is the energy contained in a sample, and it is determined by the complete combustion of a known quantity of the sample. The calorific values of the different blends along with the gasoline sample are shown in figure 6. The calorific values of the samples show a decline in calorific value with each increase in ethanol content. This decrease in calorific value in the blends is due to ethanol having a lower calorific value of 23625 J/g [18].

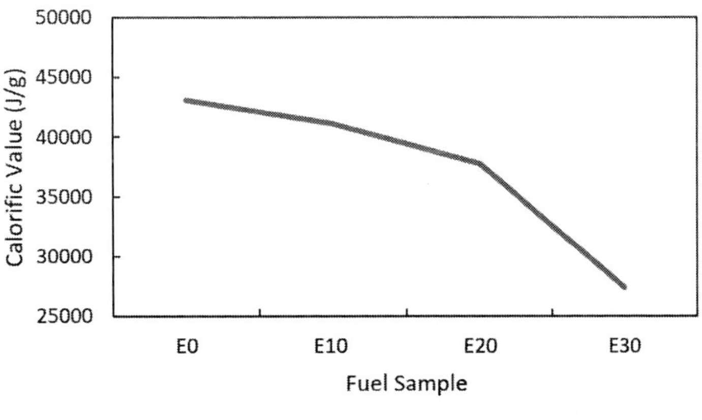

Figure 6. Calorific value trend.

The density of each sample at 30 °C is shown in figure 7. The results show that the density of E10 was slightly lower than that of gasoline (0.7400 kg/L), but the other blends showed a continuous and linear increase in density value. Of the different blends tested, so far all of the blends have provided satisfying results. Further experiments, such as SI engine testing of performance and emission, will determine which blend is the most promising and practical.

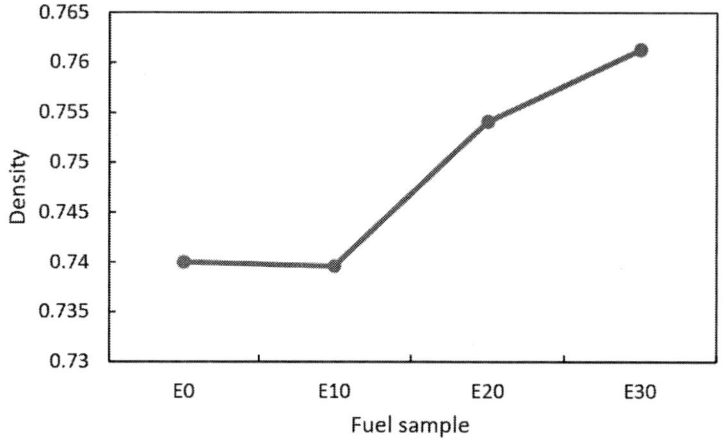

Figure 7. Density of different samples.

3.2. SI engine test

Figures 8 and 9 show the variation of Power and Torque in terms of speed when the engine is running on pure gasoline (E0). The results show that the engine runs smoothly on E0 and there was a little deflection in the results obtained. The graphs were used as reference to compare with the results of the blends.

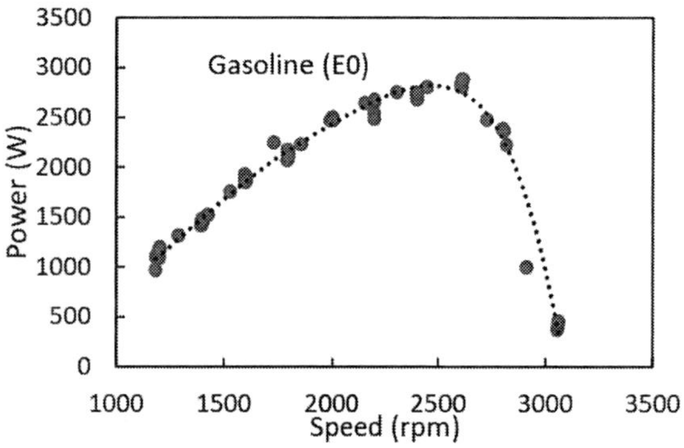

Figure 8. Power (E0).

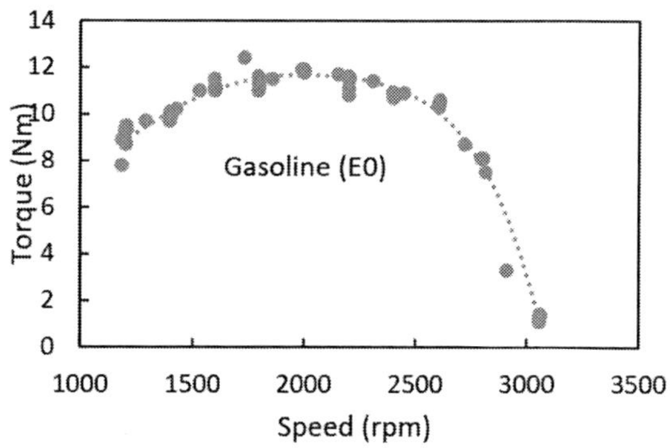

Figure 9. Torque (E0).

Figures 10 and 11 show the variation of Power and Torque in terms of speed when the engine is running on E10. From the graphs, it can be seen that the values are scattered and irregular in comparison to the values obtained from E0. From the trendline curves obtained, it can be deduced that the highest Power of 2800 W was obtained at 2350 rpm while the highest Torque of 12.5 N.m occurs at 1900 rpm.

Figures 12 and 13 show the variation of Power and Torque in terms of speed when the engine was running on E20. From the graphs, similarly, the data obtained are scattered and irregular in comparison to the values obtained from E0. From the trendline curves obtained, it can be deduced that the highest Power of 2600 W was obtained at 2400 rpm while the highest Torque of 11.8 N.m occurs at 1950 rpm.

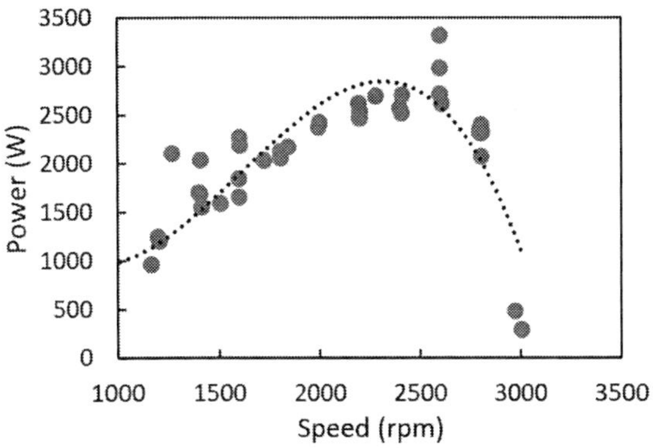

Figure 10. Power (E10).

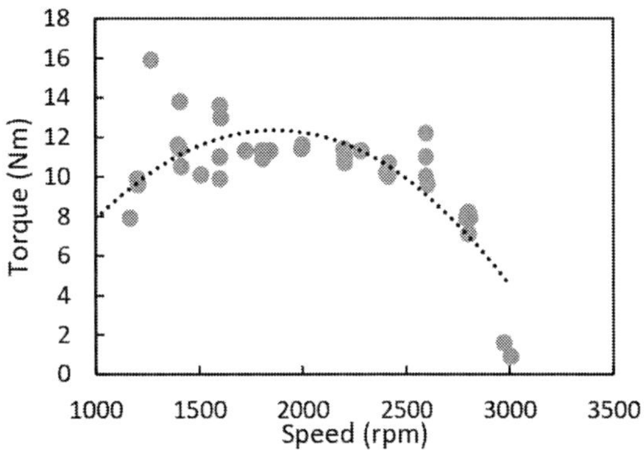

Figure 11. Torque (E10).

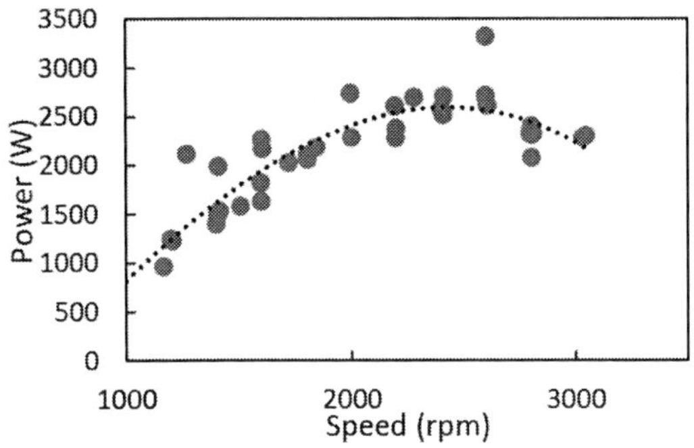

Figure 12. Power (E20).

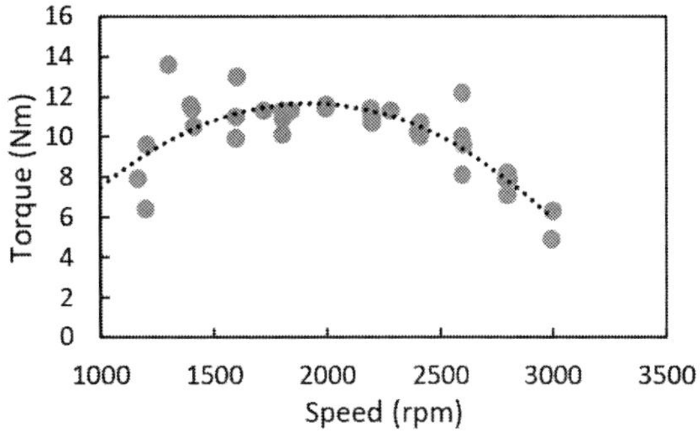

Figure 13. Torque (E20).

Figures 14 and 15 show the variation of Power and Torque in terms of speed when the engine was running on E30. From the graphs also, the values are more scattered and irregular to E0, but more exaggerated than the data obtained from E10 and E20. This was because the engine was not running smoothly as there were constant misfiring. Therefore, the results obtained were inconsistent. All the results of power and torque for the engine running on E0, E10, E20 and E30 have been summarized in one graph plot as shown in figure 16.

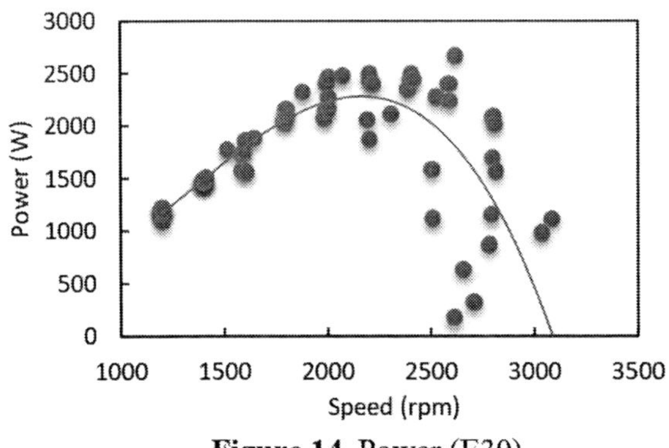

Figure 14. Power (E30).

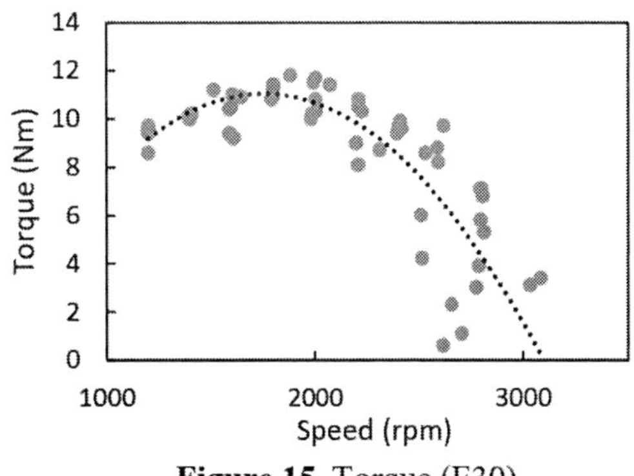

Figure 15. Torque (E30).

IOP Conf. Series: Materials Science and Engineering **863** (2020) 012064 doi:10.1088/1757-899X/863/1/012064

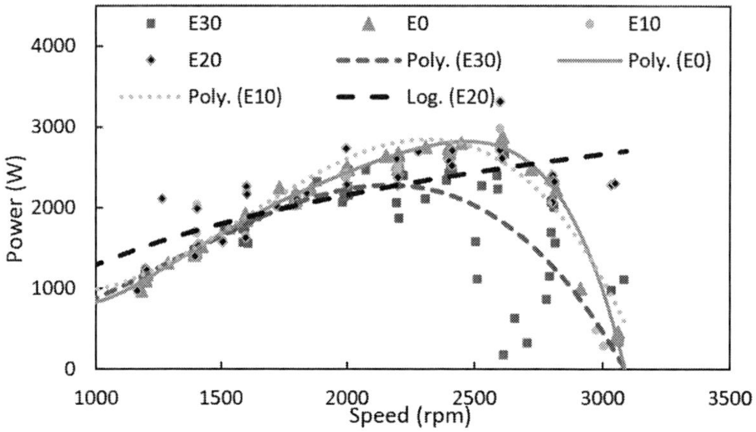

Figure 16. Power versus speed (E0, E10, E20, E30).

Due to the fluctuating values obtained from the experiment, it decided to compare the results obtained for the different samples at a particular speed, in this case, at 2000 rpm. In order to obtain the values to be plotted, an average of all the values obtained for each sample at 2000 rpm was calculated and the graph was plotted as shown in figure 17.

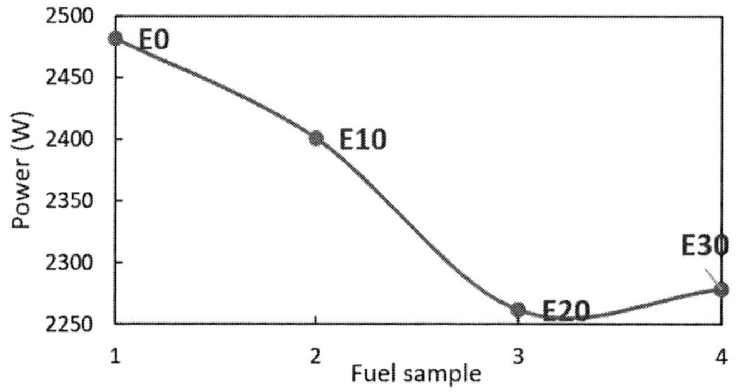

Figure 17. Power at 2000 m/s.

The graph obtained in figure 17 compares the Power output obtained from each sample at 2000 rpm. E0 has the highest value of Power at 2481 W, while at 2261 W, E20 shows the lowest value of Power at that particular engine speed.

4. Conclusion

The continuous studies and development of technologies used in the processing and usage of bioethanol is pushing it to the forefront of renewable fuels. Bioethanol presents many advantages which are suitable for the environment, for energy demand as well as providing a boost in agriculture and trade. As shown by the experiments conducted, the following conclusions can be drawn:

i) The performance of the engine using bioethanol blends is almost comparable to that using pure gasoline, especially at lower ethanol content blends.
ii) Increasing the ethanol content in the mixtures make the conventional SI engine more unstable, resulting in more frequent misfiring due to speed fluctuations and higher variation in the readings obtained. With an increase in the ethanol content there was 8.86% power loss at 2000 rpm for E20, and 8.18% for E30 compared to E0.
iii) With more research and experiments, Bioethanol-fuel blends can replace pure gasoline fuels easily and with many added benefits on the same or slightly modified engines.

iv) For a country like Mauritius, with no oil natural reserves and relying mainly on oil and gas imports as their source of energy, bioethanol is a welcoming source of renewable form of energy which could be produced locally, hence, driving the country on the path to being self-sufficient in energy for the future.

References

[1] Folkson R 2014 *Alternative fuels and advanced vehicle technologies for improved environmental performance: towards zero carbon transportation*: Elsevier).
[2] Mączyńska J, Krzywonos M, Kupczyk A, Tucki K, Sikora M, Pińkowska H, Bączyk A and Wielewska I 2019 Production and use of biofuels for transport in Poland and Brazil–The case of bioethanol *Fuel* **241** 989-96.
[3] Amigun B, Musango J K and Stafford W 2011 Biofuels and sustainability in Africa *Renewable and sustainable energy reviews* **15** 1360-72.
[4] Acharya B, Roy P and Dutta A 2014 Review of syngas fermentation processes for bioethanol *Biofuels* **5** 551-64.
[5] El-Faroug M, Yan F, Luo M and Fiifi Turkson R 2016 Spark ignition engine combustion, performance and emission products from hydrous ethanol and its blends with gasoline *Energies* **9** 984.
[6] Chala G, Abd Aziz A and Hagos F 2018 Natural Gas Engine Technologies: Challenges and Energy Sustainability Issue *Energies* **11.**
[7] Tadesse G and Aziz A R A 2009 Effect of Boost Pressure on Engine Performance and Exhaust Emissions in Direct-Injection Compressed Natural Gas (CNG-DI) Spark Ignition Engine. SAE Technical Paper).
[8] Ma'arof M, Chala G T and Ravichanthiran S 2018 A study on microbial fuel cell (MFC) with graphite electrode to power underwater monitoring devices *International Journal of Mechanical and Technology* **9** 98-105.
[9] Chala G T, Ma'Arof M and Sharma R 2019 Trends in an increased dependence towards hydropower energy utilization-A short review *Cogent Engineering* 1631541.
[10] Raja A S, Arasu A V and Sornakumar T 2015 Effect Of Gasoline-Ethanol Blends On Performance And Emission Characteristics Of A Single Cylinder Air Cooled Motor Bike Si Engine *Journal of Engineering Science and Technology* **10** 3.
[11] Robak K and Balcerek M 2018 Review of second generation bioethanol production from residual biomass *Food technology and biotechnology* **56** 174.
[12] Tutak W, Lukács K, Szwaja S and Bereczky Á 2015 Alcohol–diesel fuel combustion in the compression ignition engine *Fuel* **154** 196-206.
[13] Ershov M, Khabibullin I, Grigor'eva E, Lazarev V and Lazarev E 2017 Results of E30 Bioethanol Fuel Tests in a Full-Scale Injection Engine Vis-A-Vis Standard Gasolines *Chemistry and Technology of Fuels and Oils* **53** 297-307.
[14] Mićić V and Jotanović M 2015 Bioethanol as fuel for internal combustion engines *Zaštita materijala* **56** 403-8.
[15] Rcilabscan.com. RCI Labscan - RCI Labscan. http://www.rcilabscan.com/brand/rci-labscan [Accessed 1 May 2019].
[16] Malaysia. P Products. https://www.petron.com.my/web/site/products [Accessed 5 Mar. 2019].
[17] Joshi S M, Krishna R, Patil B A and Gangal A C 2015 Performance Evaluation of Petrol-Ethanol Blends and Testing on Multicylinder 4-Stroke SI Engine *Journal of Alternate Energy Sources and Technologies* 41-5.
[18] Rcilabscan.com. RCI Labscan - RCI Labscan. http://www.rcilabscan.com/brand/rci-labscan [Accessed 1 June 2019].

IOP Conf. Series: Materials Science and Engineering **863** (2020) 012065 doi:10.1088/1757-899X/863/1/012065

Fundamental study on the effect of Reynolds number on pipe frictional loss reduction by microbubbles

E A Alias[1,*], M N F Saniman[2] and F Dzaharudin[1]

[1] Faculty of Mechanical and Automotive Engineering Technology, Universiti Malaysia Pahang, 26600 Pekan, Pahang, Malaysia.
[2] Faculty of Mechanical, Universiti Kuala Lumpur, 43650 Bandar Baru Bangi, Selangor, Malaysia.

*E-mail: erny@ump.edu.my

Abstract. Microbubbles (MB) injection method is undeniably a promising drag reduction device in water transport. However, the mechanism of drag reduction by MB is complicated and poorly known. This paper presents the effect of Reynolds number on the frictional reduction by MB in a piping system. A fluid friction apparatus and a hydraulic bench were used to execute the experiment. To verify the ability of MB in reducing frictional loss, experiments were carried out in two conditions, without MB and with MB. In order to investigate the effect of turbulent flows, experiments were performed in three different flow rates ranging from 9×10^{-5} to 0.008 m^3/s in two types of pipe diameter (20 mm and 33 mm) to yield different Reynolds number. As the results, MB able to reduce frictional loss in pipe up to 20%. However, regarding the Reynolds number effect, no apparent tendency of the frictional reduction efficiency can be observed between Reynolds number 4000-9500. Overall, MB holds potential in reducing pipe frictional loss in turbulent flow. However, further study in wider range of Reynolds number is required to comprehend its tendency.

1. Introduction

Exhaust emission from ships has been a major concern as it is regarded as a dominant contributor to global air pollution. Based on previous study, shipping industry emits about 5-7 Tg/year of nitrogen oxides, 4.7-6.5 Tg/year of sulphur oxides, and 1.2-1.6 Tg/year of particulate matter (PM) [1]. As the consequence, this industry has been associated with the increment of mortality rate due to cardiopulmonary and lung cancer near the coastal area [2]. While stringent laws are being imposed on land vehicles to improve local air quality, the exhaust emission from marine industry inclining steadily. This matter has triggered many studies on methods to maximise total energy efficiency which will lead to reducing exhaust emissions. Many methods have been proposed and applied in recent years including operational measures (such as moving at low speed and route optimization), incremental measures (improving hull and propeller design), the usage of alternative fuel, and many more. However, implementation of the aforementioned methods requires significant physical modifications which will consume time, money and human labours. Therefore, in order to alleviate these problems, air lubrication is regarded as the most convenient method to cut down the emission by reducing the frictional drag of a ship. By using air, as the drag reduction agent, the frictional drag on a ship's surface can be reduced significantly, and thus enhancing the energy consumption efficiency which will lead to reduction in the emission.

The microbubbles (MB) size is highly controversial and depends on the application. In fluids terminology, bubbles with diameter of few hundred microns or less are regarded as MB. However, in physiological activity study, MB are defined as bubbles with 10-40 μm diameter [3]. Meanwhile, Li et al. interpreted that MB are bubbles with relatively small diameter (size range is from a few tens of microns) and having distinct properties as compared to ordinary bubbles (mili bubbles) [4]. MB are characterized by having unique properties such as low rising speed, self-pressurization, having ion charges and many more. As drag reduction agent, these characteristics play a vital role as they are able to suspend in the water for a long time. As shown in figure 1, unlike mili-sized bubbles, these MB will slowly rise up to the surface of water, shrinking and finally disappear by dissolving all its content to the water.

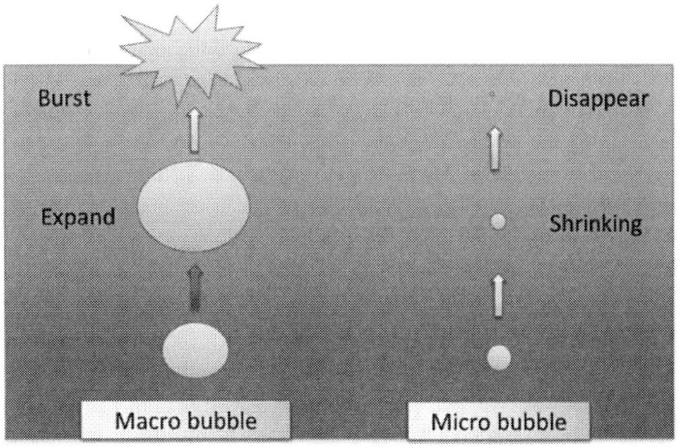

Figure 1. MB characteristics MB as air lubrication method.

Air lubrication methods can be divided into three types, namely, air cavity, air films, and small bubbles injection method. Among these methods, small bubbles injection or also known as microbubbles (MB) is the most prominent method as it is able to save up to 10-15% of total energy consumption [5]. Moreover, unlike air cavity and air films method, MB injection method can be implemented in the existing ships as none modifications on the ship's hull is necessary. The pioneer study of drag reduction by MB was carried out by McCormick and Bhattacharyya [6]. They reported up to 65% of frictional drag reduction was obtained using MB generated via electrolysis. Copper wire was wrapped around floated bodies and MB were generated when electric current passed through it. They stated that that model velocity is proportional to the rate of hydrogen bubbles and changes in viscosity in the near wall is the main key to the frictional reduction mechanism by MB. Thenceforth, research and applications of MB as a drag reduction agent has attracted many attentions including in piping system. However, up until now, the mechanism of drag reduction by MB remains inconclusive. Based on previous studies, this mechanism are due to complex phenomenon which involved viscosity and density [7, 8, 9], positive modification in the turbulent boundary layer [10, 11, 12] and many more. Lacking fundamental knowledge in the mechanism leading to difficulties in optimizing its efficiency. Thus, in this paper, fundamental investigation regarding frictional drag reduction using MB in pipes with different Reynolds Number will be discussed. This study will give insights on the fundamental understanding on parameters affecting MB efficiency as a drag reduction agent in piping system.

2. Methodology

In figure 2 and figure 3, fluid frictional apparatus and hydraulic bench were used to measure pipe frictional loss. These losses are represented in terms of head loss calculated from the manometer reading. This manometer can be attached to the desired section to get the frictional loss value. This study focuses on two sections, straight pipes with 20 mm and 33 mm diameters. MB were generated in sump tank in the hydraulic bench located at the lower part of the fluid frictional apparatus. Pump is used to supply water to the pipelines and the flowrate is adjusted by using a valve. Three types of turbulent flow rates (4100<Re<9100) were carried out for each pipe using water with temperature between 23-29 °C. Both data of head loss with and without the microbubbles were collected for comparison purposes. Data were collected after 10 minutes of switching on the MB generator in order to ensure MB is uniformly distributes in a tank. MB is generated by using a modified 60 watt submersible water pump. An air inlet is attached to the water suction part which cause the air is drawn into the pump and MB are generated due to shear stress by the rotating propeller inside the pump. The range size distribution of generated MB is about 40-130 µm.

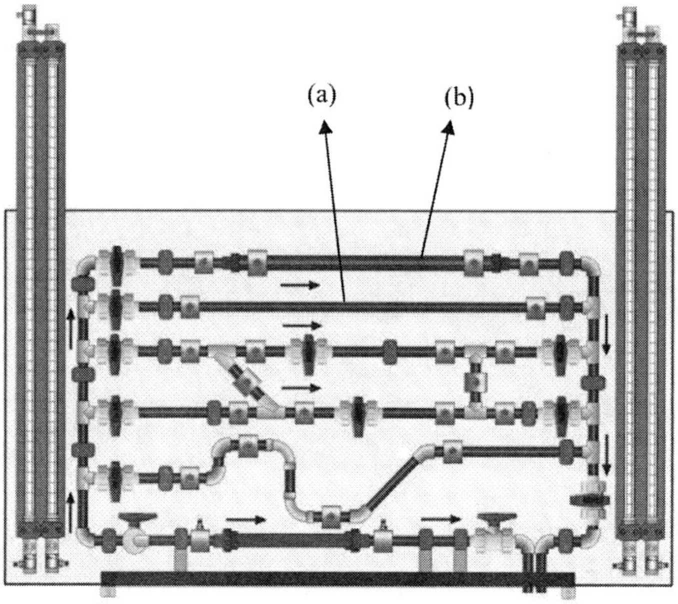

Figure 2. Fluid friction apparatus, (a) pipe with 20 mm diameter (b) pipe with 33 mm diameter.

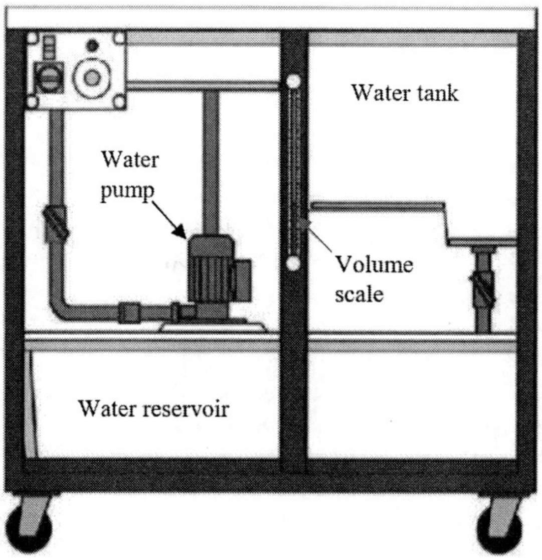

Figure 3. Hydraulic bench.

3. Results

In the former subsection, results of comparison between without MB and with MB will be discussed to analyse the capability of MB in reducing frictional loss in a pipe. In the latter section, the efficiency of MB in reducing pipe frictional loss in regard to Reynolds number will be analysed.

3.1. Straight pipe with 20 mm diameter

Figure 4 shows the frictional head loss in pipe with small diameter correspond to flow rate for both conditions, with MB and without MB. The horizontal and vertical axis represents volume flow rate and frictional head loss, respectively. In general, the frictional head loss increases steadily along with the increment of flow rate. As for comparison between values with bubbles and without bubbles, the head loss for water with bubbles is slightly lower than those without bubbles. However, no apparent tendency can be observed in terms of percentage of frictional reduction in small diameter pipeline. Besides the mass flow rate at 0.0058 m^3/s, the drag reduction for another two points are almost the same. This result suggested that the flow rate does not play a significant role in reducing frictional loss in smaller diameter pipes.

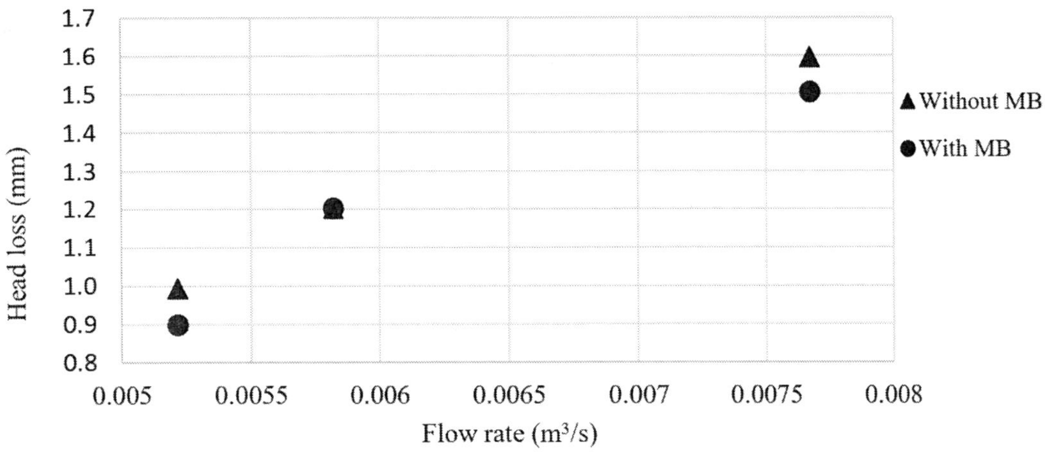

Figure 4. Head loss in regard to flow rate for 20 mm diameter pipe for both conditions, without MB and with MB.

3.2. Straight pipe with 33 mm diameter

Results for the correlation between frictional head loss in pipe with a bigger diameter of pipe and volume flow rate for both with and without MB are indicated in figure 5. Overall, similar to previous sub section, the pipe frictional loss is proportional to the volume flow rate. The head loss for water with MB is noticeably lower in comparison to that for the water without MB. Up to 20% of frictional reduction was recorded by using a bigger diameter of pipelines at the highest mass flow rate. A tendency can be observed as the percentage of frictional reduction by using MB increases as the flow rate increased. The small flow rates obtained minimum value of frictional reductions, while, the highest is recorded at the largest flow rate. However, similar to previous results, no apparent tendency of frictional reduction can be observed.

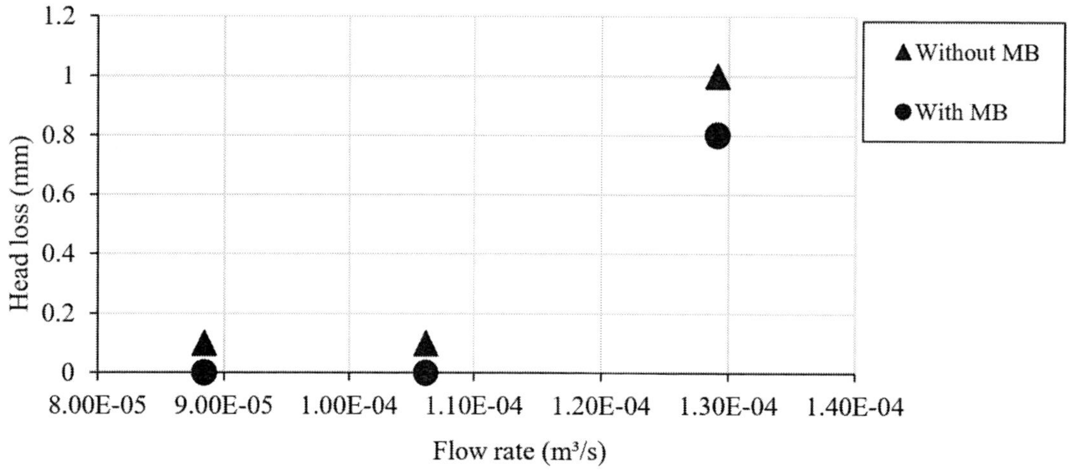

Figure 5. Head loss in regard to flow rate for 33 mm diameter pipe for both conditions, without MB and with MB.

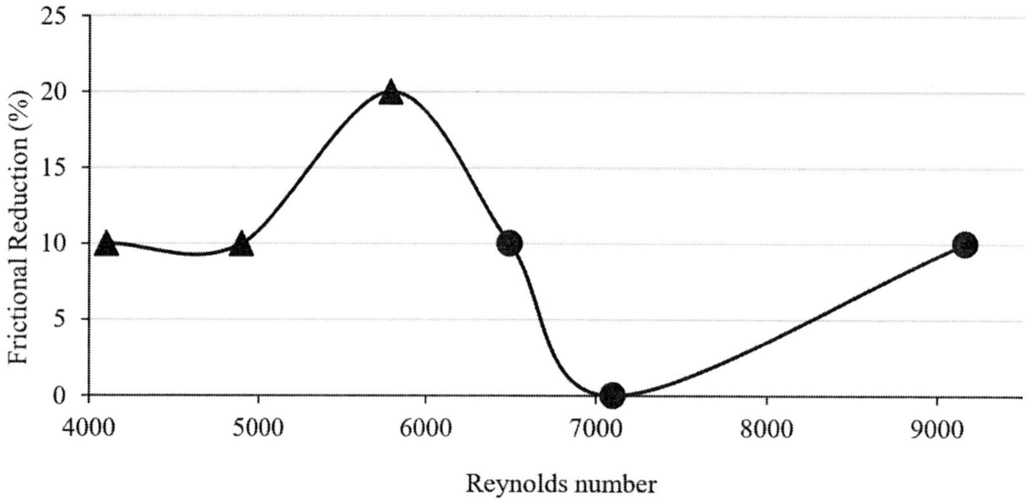

Figure 6. Head loss in regard to flow rate for 33mm diameter pipe for both conditions, without MB and with MB.

3.3. Drag reduction efficiency in regard to Reynolds number

From figure 6, in order to investigate the effect of frictional reduction by MB efficiency, results from subsection 3.1 and 3.2 were analysed in regard to Reynolds number. The triangle dots represent data from smaller pipe, while the circle dots are from larger pipe. In general, data at almost all Reynolds number show a significant frictional reduction by using MB. However, the fluctuated results depict no obvious tendency of MB efficiency can be observed when Reynolds number is between 4000-9500.

4. Conclusion

In this study, the capability of MB in reducing frictional loss in pipelines was investigated by using a frictional loss apparatus. Experiments were performed at different flow rates and pipe diameters to examine the effect of Reynolds number on the MB efficiency as a drag reduction agent. The following conclusions were achieved;

i) By mixing water with microbubbles, a significant reduction in frictional loss in pipe can be achieved.
ii) The efficiency of MB in reducing frictional loss in pipe appears to be independent to Reynolds number between 4000-9500. Larger range of Reynolds Number might yield an obvious tendency of frictional reduction.

Acknowledgments
This research was supported by Universiti Malaysia Pahang, Malaysia (RDU1703155 and RDU1803132) and Ministry of Education, Malaysia (FRGS/1/2017/TK03/UMP/03/1).

References
[1] R. M. Healy, I. P. O'Connor, S. Hellebust, A. Allanic, J. R. Sodeau, and J. C. Wenger, "Characterisation of single particles from in-port ship emissions," *Atmos. Environ.*, vol. 43, no. 40, pp. 6408–6414, Dec. 2009.
[2] J. J. Corbett, J. J. Winebrake, E. H. Green, P. Kasibhatla, V. Eyring, and A. Lauer, "Mortality from ship emissions: A global assessment," *Environ. Sci. Technol.*, vol. 41, no. 24, pp. 8512–8518, 2007.
[3] H. Tsuge, *Micro- and nanobubbles - fundamentals and applications*. 2014.
[4] P. Li, H. Tsuge, and K. Itoh, "Oxidation of Dimethyl Sulfoxide in Aqueous Solution Using Microbubbles," *Ind. Eng. Chem. Res.*, vol. 48, no. 17, pp. 8048–8053, Sep. 2009.
[5] Y. Kodama *et al.*, "A full-scale air lubrication experiment using a large cement carrier for energy saving-result and analysis," *Jpn. Soc. Nav. Archit. Ocean Eng.*, pp. 163–166, Jan. 2008.
[6] M. E. McCORMICK and R. BHATTACHARYYA, "DRAG REDUCTION OF A SUBMERSIBLE HULL BY ELECTROLYSIS," *Nav. Eng. J.*, vol. 85, no. 2, pp. 11–16, 1973.
[7] H. H. Legner, "A simple model for gas bubble drag reduction," *Phys. Fluids*, vol. 27, no. 12, pp. 2788–2790, Dec. 1984.
[8] V. L'vov, A. Pomyalov, I. Procaccia, and V. Tiberkevich, "Drag Reduction by Microbubbles in Turbulent Flows: The Limit of Minute Bubbles," *Phys. Rev. Lett.*, vol. 94, p. 174502, Jun. 2005.
[9] J.-L. Marié, "A simple analytical formulation for microbubble drag reduction," *PCH. Physicochem. Hydrodyn.*, vol. 8, pp. 213–220, Jan. 1987.
[10] A. Kitagawa, K. Hishida, and Y. Kodama, "Two-phase Turbulence Structure in a Microbubble Channel Flow," *Proc. 5th Symp. Smart Control Turbul.*, pp. 1–10, 2004.
[11] H. Kato, K. Miura, H. Yamaguchi, and M. Miyanaga, "Experimental study on microbubble ejection method for frictional drag reduction," *J. Mar. Sci. Technol.*, vol. 3, no. 3, pp. 122–129, 1998.
[12] J. C. S. Meng and J. S. Uhlman, "Microbubble Formulation and Splitting in a Turbulent Boundary Layer for Turbulenee Reduction BT - Advances in Fluid Dynamics," 1989, pp. 168–217.

The development of an aftermarket intercooler spray for turbocharged vehicles using ethylene glycol and hyaluronic acid

M I N Ma'arof[1,*], G T Chala[2,*], Sathurshan Suresh[1] and Shaheerthana Suresh[1]

[1] INTI International University, Persiaran Perdana BBN, Putra Nilai, 71800 Nilai, Negeri Sembilan, Malaysia.
[2] International College of Engineering and Management, P.O. Box 2511, C.P.O Seeb 111, Muscat, Oman.

*E-mail: muhammad.izzat87@gmail.com; girma_tade@yahoo.com

Abstract. Intercooler is used in motor vehicles to cool the air compressed by a turbocharger. Engine heat and weather condition commonly influence the efficiency of the intercooler, and this in turn affects the overall engine's power output. The objective of this study was to design and develop an autonomous cooling system for top-mounted intercooler. The system was powered by Arduino Uno microcontroller that gathers data from the temperature sensor and triggers the spraying mechanism automatically when required. Besides, the system could also be operated manually via a switch by the user. The experimental results showed that the efficiency of the normal intercooler without the spray cooling unit was improved from 56.52% to 90.72% and the minimum possible efficiency with the spray cooling unit was 32.38% more efficient. Upon using ethylene glycol as the spray medium, the system achieved 90.70% efficiency. Even so, hyaluronic acid had the most effective heat absorbing capacity as its efficiency was 90.72%. This proved that the cooling process through spray cooling unit had a vast improvement on the efficiency of the intercooling system. Generally, for all three independent variables, temperature drop increases with the increasing amount of pressure created by the additional pumps, hence, efficiency of the system increases. However, for hyaluronic acid, the efficiency is also dependent on its texture and viscosity. This newly designed portable plug-and-play intercooler spray unit would be beneficial to increase the performance of the intercooler by improving the heat exchange capacity of the unit. For future studies, other types of spray cooling medium or combinations of several cooling mediums could be tested.

1. Introduction

In the automotive industry, turbochargers are used in internal combustion engine for forced induction [1, 2]. Turbochargers consists of two impellers attached to the same shaft spinning around together. One of these impellers called turbine sits on the exhaust steam from the cylinder, whilst, the second impeller – known as compressor is located inside the car's intake, forcing air into the engine's combustion chamber. In the 1970s many experimental applications and research were done, one of which was when an air to air intercooler was attached to a 164 horsepower (hp) tractor and it was observed that the engine power output increased with no side effects [3, 4]. In 1987, Thompson et al suggested that more than 20,000 hp and 30% reduction in fuel consumption were attained by regenerator and intercooler usage in marine engines [5].

In the late 1980s, majority of the studies focused on diesel engine, whilst, in the 1990s there were high portion of both theoretical and experimental studies on force induction with the use of intercooling on diesel and spark ignition engines [6]. The studies mainly suggested that forced induction increases the power output while maintaining low fuel consumption. In the late 1990s, there were various experimental studies on intercooling effects on engine behaviour. Through studies done by Uzun, it was reported that the fuel consumption of a diesel engine decreased by 3% -13% with the use of intercooler [7]. Bilen developed a software predicting performance of an intercooler to its dimension and vice versa by the use of effectiveness transfer unit method [8]. The capacity of gases via exhaust was improved by 15.78%, the fuel consumption reduced by 9.5% and the power output increased by 3%. Exhaust emission proportion through the use of intercooler was studied by Ozulku with European Steady State Test cycle MOD 13 and it showed a similar result for fuel consumption and power output. However, the particle emission dropped from 20 to 50%, NO emission dropped from 4 to 24% and hydrocarbon emission also dropped from 20 to 24% [9]. Finally, from various case studies done recently on homogeneous charge compression ignition engine, the fuel consumption decreased by 50% and had a power output of 146 kW through forced induction via intercooling which passed Euro 6 emission regulations [10].

However, the strength of force induction system has unfortunately its own demise as well. As noted by the ideal gas law, air pressure is directly proportional to its temperature [11]. Henceforth, force induction would result in the increment of air temperature. It should be noted that air at higher temperature is in fact less dense, hence, does not provide the best form of combustion. On top of this the temperature in the engine bay and also the temperature outdoor affect the temperature of the air that enters the air intake. Therefore, the hot air coming from the compressor is passed through a heat exchanger known as an intercooler [12]. With respect the intercooler, at best, this heat (from various sources) is transferred to the surrounding by convection with the help of intercooler fins, so that the air leaving the intercooler is colder and denser, making more oxygen molecules per unit volume for thorough combustion [13]. One of the methods in overcoming the efficiency of the intercooler with respect to air temperature is by utilizing a spray cooling.

Spray cooling is the process of spraying small droplets on heated surfaces in order to increase the effective heat transfer to surrounding when cooling media changes phase [14]. Spray cooling is used in areas such as cooling of hot surfaces in hot strip mill, fire protection, cooling hot gases, cooling high performance electronic devices and dermatological operations in the medical field. Sprinkler system in a fire protection system works by the general law of spray cooling. The discharged water through the nozzle gets atomized into the ambient where tiny droplets are in charge to control or stop the fire. The small droplets spread over a large surface area over the unsaturated ambient air promotes evaporation. As high latent heat is absorbed by the water droplets in the evaporation process, the heat released at the fire can be controlled and prevented from spreading. Moreover, spray cooling is vastly used in metal production and processing industry for high temperatures steel strip casting where temperatures reach over 1800K and the optimization of the microstructure after the hot rolling process. In these processes a gas jet carrying water is sprayed onto the hot surface which needs to be cooled. It can be seen Spray cooling has one of the highest heat flux and heat transfer coefficient. On top of this, Spray cooling is vastly used in various different fields as it uses few components and less complex thus, making it less bulky, more reliable and cheaper to run. Furthermore, cryogenic spray cooling in the medical field is used for precooling of human skin during laser treatment for hair and birthmark removal. This is done to avoid permanent thermal damage on the skin from large temperature differences between the targeted vessels and epidermis. A cryogenic spray is used to spray on the skin surface for a short time interval before the laser treatment is proceeded. In the electric field, spray cooling has the greatest attention because of its ability to remove high heat flux (1200 W/cm^2) while keeping the device temperature below 100 °C with spatial and temporal variations below 10 °C [15]. Spray cooling is being used in high performance computers such as CRAY X-1 and promising results are being shown for the management of heat through high power insulated gate bipolar transistors [16], microwave source components [17] and laser diode laser arrays [18]. The present study is aimed to overcome this issue by promoting heat exchange through spray cooling with an automatic spraying mechanism spraying a certain liquid at a given concentration and pressure to promote evaporative cooling.

2. Methodology

2.1. Identifying critical parameters involved

In identifying the list of critical parameters i.e. controls and variables involved in designing portable plug-and-play intercooler spray unit for automobile; intensive literature review was conducted. The literatures were crucial in providing the necessary information in designing the portable plug-and-play intercooler spray unit for automobile. Through finding out the parameters involved in the whole system, the scope of the project and possible future studies were thought through. The parameters which were set as controlled, independent, and dependent variables for this project are shown in table 1.

Table 1. Controlled, independent and dependent variables in the experiment.

Controlled Variables	Independent Variables	Dependent Variable
Inlet Temperature	Type of Spray Medium	Outlet Temperature
Ambient Temperature	Concentration	
Ambient Humidity	Pump Pressure	
Speed of Air		
Nozzle Type		
Nozzle Diameter		
Pipe Diameter		
Time of Spray		

The outlet temperatures were measured at the intercooler after the use of spraying mechanism with the change of each independent variable. The outlet temperature was chosen as it is the most important parameter in the intercooler/heat exchanger and it is directly linked to the efficiency of the whole unit. The dependent variable, outlet temperature, was measured by a feedback system spraying a fluid automatically for the set spray time just when the temperature reaches an assigned value. Upon the completion of spray cooling, the lowest temperature at the intercooler outlet was measured and temperature drop before and after spraying at the intercooler outlet was calculated. The lowest temperature was taken over the temperature after a certain period of time as this project mainly focused on the maximum effects of the spraying mechanism and not the rate at which the cooling process happened. By using outlet temperature of the intercooler as the dependent variable to calculate the temperature drop, the efficiencies of the Intercooler before and after the spray cooling process were found.

The dependent variable, outlet temperature was measured by a feedback system which sprays a fluid automatically for the set spray time just when the temperature reaches an assigned value. After spraying the lowest temperature at the intercooler outlet was measured and temperature drop before and after spraying at the intercooler outlet was calculated. The lowest temperature was taken over the temperature after a certain period of time as this project focused mainly on the maximum effect of the spraying mechanism and not the rate at which the cooling process happened. By using outlet temperature of the intercooler as the dependent variable to calculate the temperature drop, the efficiency of the Intercooler before and after the spray cooling process were found.

In the experiment, two types of liquid (independent variables) were used: ethylene glycol and hyaluronic acid. For the hyaluronic acid and the glycol solution, different concentration of the medium (mixed with water) were tested. The goal was to determine spray medium that gives the highest decrement of outlet temperature i.e. the best cooling effects and the overall highest efficiency of the system developed.

2.2. Layout of the system

The system works by sensing the temperature at the intercooler exit using the temperature sensor. When the temperature reaches above a certain value, the relay turns on to start the pump and spray water onto the intercooler lowering the temperature. The temperature was analysed by the sensing element throughout, and if the temperature does not drop, the loop continues and sprays until the temperature reaches below the designated value. The independent variables were varied, and the dependent variable was measured to study the effects of these parameters on spray cooling. Schematic diagram of the system is depicted in figure 1.

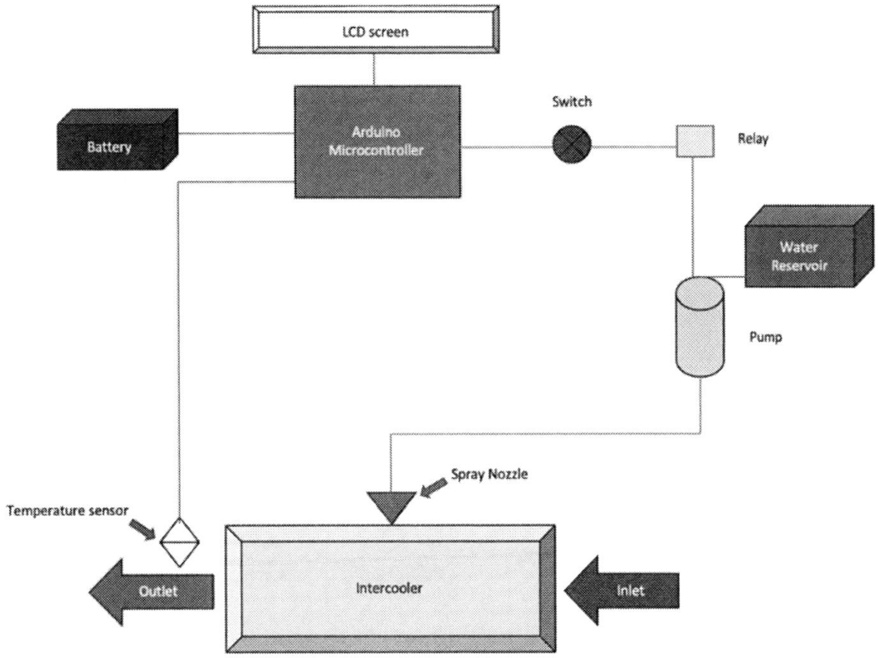

Figure 1. Schematic diagram of the system.

In the feedback system, as the exit temperature reaches above a certain value, the micro controller detects the increase in temperature by the data sent by AM2302 sensor. The microcontroller sends a small current to the relay, turning on the pump and spraying fluid stored in the tank. Through having this feedback system, during testing there was less human error as the Controller controlled the spray mechanism when the set temperature was reached, and this was preferred than waiting for a temperature to reach and spray manually. However, a switch was used for manual use so that the spray can be used manually in case of any malfunction or during fluid change. The whole system was powered by a 12V DC supply, which is similar to that in motor vehicles. Peak inlet and outlet temperatures were measured to find the efficiency of the intercooler without the spray cooling mechanism for comparison with the efficiency of the spray cooling unit. On top of this, the outlet temperature was set as the benchmark temperature to observe the improvement on the system as opposed to standalone unit. Table 2 shows the measured controlled variables.

Table 2. Measured controlled variables.

Ambient Temp.	Ambient Humidity	Peak Inlet Temp.	Peak Outlet Temp. without spraying (Benchmark Temp.)
24.5°C	24.70%	151° C	79.5 ° C

IOP Conf. Series: Materials Science and Engineering **863** (2020) 012066 doi:10.1088/1757-899X/863/1/012066

The collected data for the dependent variable, outlet temperature and calculation was done to obtain the efficiency for comparison with the efficiency of the intercooler without the intercooler spray unit.

$$\text{Temperature Drop} = \text{Inlet Temperature} - \text{Outlet Temperature} \tag{1}$$

$$\text{Efficiency} = \frac{(\text{Inlet Temperature} - \text{Outlet Temperature})}{(\text{Inlet Temperature} - \text{Ambient Temperature})} \times 100$$

$$= \frac{\text{Temperature Drop}}{(\text{Inlet Temperature} - \text{Ambient Temperature})} \times 100 \tag{2}$$

Using equation (2) the efficiency of the intercooler without the spay unit was calculated to be 56.52%.

3. Results and discussion

Table 3 shows experimental results for Ethylene Glycol. Based on the results obtained, it was observed that that all the three (3) independent variables follow a trend, which altered the dependent variable (Outlet temperature) in different ways. The amount of variations of the outlet temperature affected the temperature drop, and this influenced the variation of efficiencies of the system. The overall data collected could have uncertainties as temperature readings were recorded using both sensor and mercury thermometer. There could have been instrumental errors from the readings obtained through the Mercury thermometer; however, the error would be very small, approximately of 0.1 °C.

Table 3. Experimental results for ethylene glycol.

Pressure	Concentration	Inlet Temperature	Outlet Temperature	Temperature Drop	Average Temperature Drop	Efficiency
		151	38.5	112.5		
	20%	151	38.7	112.3	112.60	89.01
		151	38	113		
		151	39.6	111.4		
1	40%	151	40.1	110.9	110.97	87.72
		151	40.4	110.6		
		151	40.8	110.2		
	60%	151	41.3	109.7	109.77	86.77
		151	41.6	109.4		
		151	37.2	113.8		
	20%	151	37.5	113.5	113.47	89.70
		151	37.9	113.1		
		151	39.5	111.5		
2	40%	151	39.5	111.5	111.53	88.17
		151	39.4	111.6		
		151	40.1	110.9		
	60%	151	40.5	110.5	110.60	87.43
		151	40.6	110.4		
		151	36.5	114.5		
	20%	151	36.2	114.8	114.73	90.70
		151	36.1	114.9		
		151	38.6	112.4		
3	40%	151	38.5	112.5	112.60	89.01
		151	38.1	112.9		
		151	39.8	111.2		
	60%	151	39.7	111.3	111.17	87.88
		151	40	111		

Ethyl Glycol solutions followed a general trend where the temperature drop decreased as the concentration increased. However, as the pressure increased, the temperature drop started to increase. In the case of Ethyl Glycol peak efficiency of 90.70% was achieved where the pressure generated was the highest for the lowest concentration of 20% of the spray medium. With the same pressure for the 40% concentrated spray medium the efficiency was 89.01%, followed by 87.88% for the 60% diluted solution. Furthermore, the lowest efficiency with this spray medium was 86.77% with a 60% diluted solution with the lowest pressure set up. It was observed that the concentration is inversely proportional to the temperature drop and the temperature drop increases slightly as the pump pressure increases.

This trend was studied in the literature review, where it was stated that with increasing amount of Glycol in a solution the thermal absorbing capacity of the fluid. As the concentration of glycol increased, the specific heat capacity of the glycol and water mixture reduces, therefore the least optimum amount of glycol was recommended. However, as the pressure was increased, the nozzle sprayed the solution in form of tiny droplets over a wide area which efficiently increased the heat absorbing capacity of the diluted solutions.

Table 4. Experimental results for Hyaluronic acid.

Number of Pumps	Concentration	Inlet Temperature	Outlet Temperature	Temperature Drop	Average Temperature Drop	Efficiency
		151	39.7	111.3		
	20%	151	39.4	111.6	111.57	88.19
		151	39.2	111.8		
		151	40.2	110.8		
1	40%	151	40.1	110.9	110.67	87.48
		151	40.7	110.3		
		151	41.6	109.4		
	60%	151	41.8	109.2	109.40	86.48
		151	41.4	109.6		
		151	38.1	112.9		
	20%	151	37.9	113.1	113.27	89.54
		151	37.2	113.8		
		151	39.7	111.3		
2	40%	151	39.2	111.8	111.73	88.33
		151	38.9	112.1		
		151	40.5	110.5		
	60%	151	40.7	110.3	110.47	87.33
		151	40.4	110.6		
		151	36	115		
	20%	151	36.5	114.5	114.77	90.72
		151	36.2	114.8		
		151	38.1	112.9		
3	40%	151	37.8	113.2	113.17	89.46
		151	37.6	113.4		
		151	40.1	110.9		
	60%	151	39.8	111.2	111.20	87.91
		151	39.5	111.5		

Additionally, hyaluronic acid had the most effective heat absorbing capacity as its efficiency was 90.72%, as depicted in table 4. The hyaluronic acid followed a similar pattern to glycol. The reason behind this was because of high texture and high viscosity of the hyaluronic acid. As the concentration of hyaluronic acid increased, the mixture became thicker and viscous which made the temperature drop fall. As the hyaluronic acid's proportion increased and the pressure is reduced, the thick coarser liquid droplets started to form a gel coating before evaporating. Thus, this hindered the cooling process through evaporation as molecules are trapped from evaporating. The lowest efficiency recorded was 86.48%, which was also among the lowest efficiencies resulted from the thick layer formed with high percentages of hyaluronic acid. However, the use of higher pressure and less percentage of hyaluronic acid, the less viscous spray medium was sprayed over a wide area forming of blanket, covering a large surface area of small droplets that aided the cooling process through absorbing extra moisture from the surrounding and going through evaporative cooling. In the evaporative cooling, the molecules been evaporated after absorbing energy from the hot surface.

4. Conclusion

The present study designed, tested, and optimized aftermarket intercooler cooling spray unit. The independent variables, type and concentration of spray medium, and the pressure that should be generated to have an optimized prototype with good efficiency were studied and utilized for usage. It was found that the most efficient spray medium for spray cooling process was 20% Hyaluronic acid with water, followed by 20% Ethylene glycol solution. The efficiency of the normal intercooler without the spray cooling unit was improved from 56.52% to 90.72% and also the minimum possible efficiency with the spray cooling unit was 32.38% more efficient, indicating that the cooling process through spray cooling unit had a vast improvement on the efficiency of the intercooling system. It was decided that any of the chosen spray medium with the lowest concentration could be used in the optimized unit as they all needed at most 23 psi using 3 pumps, which was replaced by a new single pump in the optimized unit that was readily available in the market. However, it all comes down to the basis of cost and efficiency as the chemical compounds need to be bought but medium such as tap water is readily available.

References

[1] Tadesse G and Aziz A R A 2009 Effect of Boost Pressure on Engine Performance and Exhaust Emissions in Direct-Injection Compressed Natural Gas (CNG-DI) Spark Ignition Engine. SAE Technical Paper.

[2] Muhammad I. N. Ma'arof S S, Shaheerthana Suresh, Yuli P. Asmara, Girma T. Chala 2019 The Development of an Aftermarket Intercooler Spray for Turbocharged Vehicles using Water - *INTI Journal.*

[3] McKeon C and Turney R 1979 Ford Tw-30 Tractor With Air-To-Air Intercooled Engine. SAE Technical Paper.

[4] Andersson J, Bengtsson A and Eriksson S 1984 The Turbocharged and Intercooled 2.3 Liter Engine for the Volvo 760. SAE Technical Paper.

[5] Thomson G, Pratley D and Owen D 1987 Intercooling and regenerating the modern marine gas turbine propulsion system *SAE transactions* 169-75.

[6] Canli E, Darici S and Ozgoren M 2010 Intercooler effect on conventional supercharging systems. In: *International Scientific Conference,* pp 19-20.

[7] Uzun A. 1998 Effects of Intercooling on Engine Performance in a Diesel Engine With Supercharging. Sakarya University.

[8] Bilen K 1998 Intercooler Design for a Diesel Engine With Turbocharger *Kırıkkale University, Kırıkkale TURKEY (in Turkish).*

[9] Akıcı S 1999 Intercooler Effects on Diesel Engine Performance. (Turkey: Marmara University Natural and Applied Sciences Institute).

[10] Watson H C and Mehrani P 2010 The Performance and Emissions of the Turbocharged Always Lean Burn Spark Ignition (TC-ALSI) Engine. SAE Technical Paper.

[11] Tadesse G 2009 Effect of Boost Pressure on Performance, Combustion and Exhaust Emissions in Direct Injection Compressed Natural Gas (CNG) Engine. UNIVERSITI TEKNOLOGI PETRONAS.

[12] Grönman A, Sallinen P, Honkatukia J, Backman J and Uusitalo A 2016 Design and experiments of two-stage intercooled electrically assisted turbocharger *Energy Conversion and Management* **111** 115-24.

[13] Mifdal M, Nuraida M, Norzalina O and Shamil A 2015 Turbo Intercooler Cooling System *International Journal of Engineering Science Invention ISSN (Online)* **2319** 49-56.

[14] Alkhedhair A, Gurgenci H, Jahn I, Guan Z and He S 2013 Numerical simulation of water spray for pre-cooling of inlet air in natural draft dry cooling towers *Applied Thermal Engineering* **61** 416-24.

[15] Bar-Cohen A, Arik M and Ohadi M 2006 Direct liquid cooling of high flux micro and nano electronic components *Proceedings of the IEEE* **94** 1549-70.

[16] Mertens R G, Chow L, Sundaram K B, Cregger R B, Rini D P, Turek L and Saarloos B A 2007 Spray cooling of IGBT devices *Journal of Electronic Packaging* **129** 316-23.

[17] Lin L, Ponnappan R, Yerkes K and Hager B 2004 Large area spray cooling. In: *42nd AIAA Aerospace Sciences Meeting and Exhibit,* p 1340.

[18] Huddle J J, Chow L C, Lei S, Marcos A, Rini D, Lindauer S, Bass M and Delfyett P 2000 Thermal management of diode laser arrays. In: *Sixteenth Annual IEEE Semiconductor Thermal Measurement and Management Symposium (Cat. No. 00CH37068)*: IEEE) pp 154-60.

AUTHOR INDEX